This book describes the progress that has been made towards the development of a comprehensive understanding of the formation of complex, disorderly patterns under far-from-equilibrium conditions.

The application of fractal geometry and scaling concepts to the quantitative description and understanding of structure formed under non-equilibrium conditions is described. Self-similar fractals, self-affine fractals, multifractals and scaling methods are discussed, with examples, to facilitate applications in the physical sciences. Computer simulations and experimental studies are emphasised, but the author also includes a discussion of theoretical advances in the subject. Much of the book deals with diffusion-limited growth processes and the evolution of rough surfaces, although a broad range of other applications is also included. The book concludes with an extensive reference list and guide to additional sources of information.

This book will be of interest to graduate students and researchers in physics, chemistry, materials science, engineering and the earth sciences, and especially those interested in applying the ideas of fractals and scaling to their work or those who have an interest in non-equilibrium phenomena.

Cambridge Nonlinear Science Series 5

Fractals, scaling and growth far from equilibrium

TITLES IN PRINT IN THIS SERIES

G.M. Zaslavsky, R.Z. Sagdeev, D.A. Usikov and A.A. Chernikov

1. Weak chaos and quasi-regular patterns

K. Nakamura

2. Quantum chaos: a new paradigm for nonlinear dynamics

J.L. McCauley

3. Chaos, dynamics and fractals: an algorithmic approach to deterministic chaos

C. Beck and F. Schlögl

4. Thermodynamics of chaotic systems: an introduction

P. Meakin

5. Fractals, scaling and growth far from equilibrium

Fractals, scaling and growth far from equilibrium

Paul Meakin
Department of Physics, University of Oslo

CAMBRIDGE UNIVERSITY PRESS
Cambridge, New York, Melbourne, Madrid, Cape Town,
Singapore, São Paulo, Delhi, Tokyo, Mexico City

Cambridge University Press
The Edinburgh Building, Cambridge CB2 8RU, UK

Published in the United States of America by Cambridge University Press, New York

www.cambridge.org
Information on this title: www.cambridge.org/9780521452533

First published 1998

A catalogue record for this publication is available from the British Library

Library of Congress Cataloguing in Publication data

Meakin, Paul, 1944–
Fractals, scaling and growth far from equilibrium / Paul Meakin.
p. cm. – (Cambridge nonlinear science series: 5)
Includes bibliographical references and index.
ISBN 0 521 45253 8 (hardcover)
1. Fractals. 2. Scaling laws (Statistical physiscs).
3. Mathematical physics. I. Title. II. Series.
QC20.7.G44M43 1997
530.1′3–dc20 96-44263 CIP

ISBN 978-0-521-45253-3 Hardback

Additional resources for this publication at www.cambridge.org/9780521452533

Contents

*This plate section is available for download in color from www.cambridge.org/9780521452533

Preface

The development of a full understanding of the universe around us, in terms of the basic properties of fundamental "particles" and their interactions, has long been a dream of the physicist. Mindful of the difficulties encountered when this approach is used to calculate the behavior of very simple systems, such as molecules containing just a few atoms, the problem of understanding the nature of much more complex systems, such as snowflakes, soot aggregates and rough surfaces produced by processes such as vapor deposition or erosion, might seem to be a daunting prospect. However, during the past one or two decades, substantial progress has been made, based on statistical physics concepts such as scaling and the independent development of fractal geometry, based on late 19th century and early 20th century mathematics, by Benoit Mandelbrot. To a large extent, this progress has been made by giving up the idea that an understanding of complex systems can be based on an ever more detailed knowledge of their microscopic components and focusing instead on the "universal" properties that all materials possess in common, irrespective of their atomic and molecular structure, and the manner in which properties on one length scale relate to those on other length scales. The connection between microscopic and macroscopic behavior is still important, and the theoretical justification for much of the work described in this book is based on models that contain microscopic components and interactions, at least on an abstract level. However, one of the objectives of this book is to illustrate that scaling symmetries can be used, in much the same way as other symmetries, to study a wide variety of systems and phenomena, without taking into account the underlying microscopic physics on a detailed level.

One of the main objectives of this book is to show how a surprisingly wide range of complex, disorderly systems can be quantitatively understood using simple statistical physics concepts and simple mathematical tools. This book contains many equations, but it should be accessible to anyone with a good undergraduate education in the physical sciences. My original idea was to write a single volume on the basics of fractals and scaling and applications in various areas of science and technology. It soon became apparent that I wanted to say more than could reasonably be contained in one book. Consequently this book concentrates on some of the more fundamental aspects of pattern formation, fractals and scaling. I am in the process of writing a second book, focusing on colloidal fractals and aggregation kinetics, and a third monograph on the applications of fractals and scaling in selected areas of science and technology.

My own interest in this area was first stimulated by the work of Thomas Witten and Leonard Sander, more than ten years ago, on diffusion-limited aggregation. The diffusion-limited aggregation model has become one of the most important paradigms for disorderly growth, far from equilibrium, and plays a central role in this book.

Much of the work on this book was carried out during a one year visit to the Center for Advanced Studies at the Norwegian Academy of Science and Letters. The remainder of the work was carried out in the Physics Department at the University of Oslo. I would like to thank Jens Feder and especially Torstein Jøssang for hospitality at the University of Oslo, and Torstein Jøssang for making my stay at the Center for Advanced Studies possible. I have also benefited considerably from stimulating interactions with a quite large number of graduate students and post-doctoral associates at the University of Oslo.

I would like to thank Fereydoon Family, Joachim Krug, Leonard Sander, Lorraine Siperko, Tamás Vicsek and Stephanie Wunder for making valuable comments on a draft of this book and for making suggestions that have led to substantial improvements. I am also grateful to many colleagues and collaborators who have contributed figures; they are acknowledged in the figure captions. Many of the figures have been provided by graduate students in the Cooperative Phenomena Group at the University of Oslo and illustrate various aspects of their own research.

Paul Meakin

Oslo, Norway

Chapter 1

Pattern Formation Far From Equilibrium

The diversity of the natural shapes that surround us has a profound impact on the quality of our lives. For this reason alone, it is not surprising that the origins of these shapes have been the subject of serious study since antiquity.[1] It has long been believed that a quantitative characterization of natural forms is an important step towards understanding their origins and behavior. Unfortunately, there have, until recently, been relatively few general approaches towards the quantitative description of the complex, disorderly patterns that are characteristic of most natural phenomena. The essentially non-equilibrium nature of most pattern-formation processes has also contributed to the comparatively slow development of this field. The systematic and generally well understood techniques of equilibrium statistical mechanics cannot be applied to the majority of pattern-formation processes.

In the past one to two decades, the outlook has improved substantially. The pioneering, interdisciplinary work of B. Mandelbrot [2] has demonstrated that mathematical concepts, once believed to be of no possible relevance to the real world, can provide us with new ways of describing and thinking about an amazingly broad range of structures and phenomena. In addition, the scaling concepts that were originally applied to a relatively narrow range of problems such as critical phenomena [3] and the structure of macromolecules [4] have been successfully applied to a very much broader range of problems. In many cases, the fractal geometry approach developed by Man-

1. For example, references to the six-fold form of snowflakes, that go back many centuries, can be found in the commentary by Mason in Hardie's translation of Kepler's work on snowflakes [1].

delbrot can be used to provide a more intuitive, geometric interpretation of scaling behavior. This has brought a measure of intellectual democracy to previously arcane areas of physics and has enabled physicists to contribute to a wide range of important problems outside of the traditional confines of physics.

Many isolated early applications of the fractal approach to physical phenomena can be found. For example, in 1926 Richardson [5] asked the question "*Does the wind possess a velocity?*" and suggested that the distance x traveled by an air "particle" in time t may have to be described "*by something rather like Weierstrass's function*

$$x = kt + \sum_n (1/2)^n \cos(5^n \pi t).\text{"} \tag{1.1}$$

The Weierstrass–Mandelbrot function, a generalization of equation 1.1, described in chapter 2, is now widely recognized as an example of a self-affine fractal function. Richardson went on to describe studies of the dispersion of tracers in the atmosphere and suggested that this process could be described in terms of a length-dependent diffusion coefficient $\mathscr{D}(\ell) \sim \ell^{4/3}$. Richardson argued that the projection of the density of the dispersing tracers onto a straight line $\rho(\ell, t)$ could then be described by the non-Fickian diffusion equation

$$\partial\rho(\ell,t)/\partial t = \partial[\mathscr{D}(\ell)\partial\rho(\ell,t)/\partial\ell]/\partial\ell. \tag{1.2}$$

Richardson showed that the solution to this equation, starting with a delta function distribution at time $t = 0$, has the form

$$\rho(\ell,t) = t^{-3/2} f(\ell^{2/3}/t), \tag{1.3}$$

where $f(x)$ is an exponentially decaying function. It follows, from equation 1.3, that $< \ell^2(t) >^{1/2} \sim t^{\nu} \sim t^{3/2}$. For a particle moving with a constant velocity, the exponent $1/\nu$ can be interpreted as a fractal dimensionality. However, for the problem studied by Richardson, space and time are "mixed" so that the exponent relating the distance traveled to the elapsed time cannot be given a simple interpretation in terms of a fractal particle trajectory [6].

Similarly, the development of scaling ideas [3, 4, 7, 8], during the last three decades, has provided new ways of quantitatively describing and better understanding the growth kinetics of both fractal and non-fractal objects. During the same period, a new understanding of non-linear phenomena was developed. The work on non-linear systems demonstrated that apparently complex processes could have simple origins and provided paradigms that helped to motivate the work described in this book.

Fractal geometry has been shown to provide a basis for describing objects as small as polymer molecules and as large as the coastlines of continents.

Mandelbrot [2, 9] has also discussed the application of fractals to the distribution of visible matter in the universe. For some time, it has been accepted that the distribution of visible matter is inhomogeneous and can be described in terms of a fractal dimensionality of $D \approx 1.25$ on "short" length scales, less than the "galaxy correlation length" of about 5 Mpc.[2] The ideas that there may be no galaxy correlation length and that the distribution might be fractal out to much longer length scales have been the subject of heated controversy. Some aspects of this problem are discussed in chapter 2.

During the past 15 years, the "big bang" theory for the creation of the universe has been seriously challenged by an inflationary model [10, 11, 12] with an inherently fractal nature. This scenario suggests a fractal universe of almost unlimited size.

Most of the applications discussed in this book are much more "down to Earth". However, it is interesting that theoretical and modeling approaches, similar to those used to simulate non-equilibrium growth and aggregation processes (chapters 3 and 4) and surface growth (chapter 5), have been proposed for the evolution of galaxies [13, 14, 15] and galaxy distributions [16, 17].

While the major applications of fractal geometry have been in the physical and life sciences, many interesting examples have been found that are related to human activities. Examples include the form of urban centers [18, 19], the distribution of weather monitoring stations [20, 21, 22], gravity stations [23, 24] and railroad networks. These fractal distributions did not come about as a result of a design process, but, in some cases, a fractal distribution may be advantageous [25].

It has also been suggested that the distributions of rapidity,[3] characterizing the particles generated by high energy collisions, including hadron–hadron, hadron–nucleus and nucleus–nucleus collisions, and cosmic shower events may have a multifractal character [26, 27, 28, 29, 30, 31, 32]. This has been interpreted in terms of a random cascade model [33].

Fractal geometry has also been used to describe the distribution of events in time. In the case of discrete events, the fractal dimensionality D of the set of times $\{t_i\}$ can be measured. Examples of cases in which this approach appears to be of value include the distribution of reversals in the Earth's magnetic field ($D \approx 0.89$ [34]) and the temporal distribution of earthquakes in a region of limited size ($0.12 < D < 0.26$ [35]).

2. 5 Mpc is 5 megaparsecs, where 1 parsec is the distance at which the mean radius of the Earth's orbit subtends an angle of 1 second, 1 pc $\approx$ 3.2 light years $\approx 3 \times 10^{16}$ m.

3. The rapidity y is defined as $y = (1/2)\ln[(E + p_L)/(E - p_L)]$, where E is the energy of an emitted particle and p_L is the component of its momentum along the collision axis. In practice, the pseudo-rapidity, $\eta = -\ln\tan(\theta/2)$, is usually measured, where θ is the emission angle.

1.1 Power Laws and Scaling

Power law relationships play a central role in the study of fractals and scaling. The power law function $y(x)$, given by

$$y(x) = cx^a, \tag{1.4}$$

has an important symmetry that can be expressed as

$$y(\lambda x) = c(\lambda x)^a = c\lambda^a x^a = const. \cdot y(x). \tag{1.5}$$

This describes the scale invariance of $y(x)$ (the power law function $y(x)$ in equation 1.4 has the same shape on all scales). A trivial, but important, consequence of this scale invariant symmetry is that the exponent a does not depend on the units in which x or y are measured. Functions that satisfy the relationship $y(\lambda x) = \lambda^a y(x)$ are said to be homogeneous. The function

$$y(x) = c_1 x^{a_1} + c_2 x^{a_2} \tag{1.6}$$

does not satisfy equation 1.5 and is an example of an inhomogeneous power law. In practice, the analysis of data from simulations or experiments, in terms of power law exponents, is based on the logarithmic version of equation 1.4

$$\log y(x) = \log c + a \log x, \tag{1.7}$$

so that the exponent a and the amplitude c can be obtained by plotting $\log y(x)$ against $\log x$. The observation of a linear relationship between the logarithms of two quantities over a sufficiently large range of scales is often considered to provide *prima facie* evidence for a power law relationship between these quantities.

In practice, this simple procedure is fraught with hazards. There is no consensus on the standards required for establishing power law relationships from experimental or numerical data. In some areas of physics, it has been possible to observe linear behavior, on a log–log plot, covering more than four orders of magnitude (powers of ten), in both the related quantities. However, data of this quality are quite rare. To observe power law behavior over four decades, from analysis of a recorded image of a physical object, it would be necessary to have a digitized representation with a resolution of better than one part in 10^5 (10^{10} pixels in a 2-dimensional image!). Such images are not routinely available. It would, of course, be possible to extend the range of observation by using images of parts of the structure recorded under different magnifications. If this approach is used, care must be taken to avoid bias towards the selection of "interesting" or even "typical" parts of the pattern.

An important example of a power law relationship is that between the mass M and the characteristic size (average diameter for example) L for a self-similar fractal aggregate, composed of particles with a diameter ϵ. In this case, the mass of the aggregate is given by

$$M(L, \epsilon) \approx C_{\circ} m(L/\epsilon)^D, \tag{1.8}$$

where m is the mass of a single particle ($m \approx \rho\epsilon^d$, where d is the Euclidean dimensionality of the particle and ρ is the particle density) and $C_{\circ}$ is a geometrical constant of order 1. The exponent D, in equation 1.8, is the fractal dimensionality. Very often, the dependence of the mass on L, when all other quantities are held fixed, is the main focus of interest. In this case, a "shorthand" version of equation 1.8

$$M \sim L^D \tag{1.9}$$

is used. However, it should always be remembered that this equation "stands in" for the dimensionally balanced, or dimensionally homogeneous, equation 1.8. In equation 1.9, and others like it, the symbol "$\sim$" should be interpreted as meaning "scales as".

This book focuses attention on the power law part of equations like 1.8. In practice, the "amplitude" (c in equation 1.4) is important and embodies the "real physics" behind power law relationships. In many phenomena, the exponents are universal (invariant to small changes in the physical process or model). Under these conditions, the amplitudes provide the only means to control physical properties and behavior. However, important insights can be obtained from the scaling relationships, described by equations such as 1.9. The amplitudes are omitted (some would say perversely) from almost all of the equations in this book, even when they are well known and/or easily calculated. In most cases, it is much more difficult to calculate the amplitude than the exponent. In many practical situations, the scaling relationship $y(x) \sim x^a$ implied by equations such as 1.9 is all that is required. By ignoring the amplitudes, a much broader range of phenomena can be discussed, so that the power and simplicity of the scaling approach is emphasized. The situation here is similar to that encountered in applications of the quantum theory of angular momentum. In this case, the matrix elements needed to calculate properties of physical interest can be divided into two parts, according to the Wigner–Eckart theorem [36, 37]: a part called the reduced matrix element, analogous to the amplitude c in equation 1.4, that depends on the physical details and that is, in most cases, difficult to calculate, and a part that depends only on the angular momentum quantum numbers, which can easily be calculated using group theory.

Phenomena that require the description of the properties of a large ensemble of similar structures, rather than a single sample, are frequently encountered. In this case, equation 1.8 should be replaced by

$$< M(L, \epsilon) > \approx C_{\circ} m(L/\epsilon)^D, \tag{1.10}$$

where $< M(L,\epsilon) >$ implies averaging the masses of a large number of samples with sizes in a narrow range $L \pm \delta L$, centered on L. This equation, and equations like it, will be replaced by equation 1.9 or $M \sim \epsilon^{-D}$, for clusters with the same overall size composed of monodisperse particles of different sizes ϵ. The notation $< \ldots >$, to indicate averaging, will appear only if emphasis is required or to avoid ambiguity. The omission of amplitudes and averaging symbols considerably reduces the complexity of equations, and their presence is usually clear from the physical context in which the equations are being used.

Empirical power laws can be expected to arise in two quite different ways. They may be the consequence of a single process that has no inherent length scales, apart from inner and outer cut-off lengths, that define the range over which the power law applies. In this case, robust, homogeneous scaling and universality can be expected. Power law behavior can also arise as a result of many processes, each with its own characteristic length(s). Under these circumstances, relatively weak, non-universal scaling may be expected, and the scaling properties may be inhomogeneous (vary from place to place or time to time, within the same system). An effective power law may also be found as a result of a crossover (slow transition) between two regimes that themselves are characterized by different amplitudes and/or exponents. In practice, it can be difficult to distinguish between these alternatives. However, when a single power law is found over a wide range of scales it is reasonable to take seriously the possibility that a single process, with no inherent length scales, is dominant.

A good example of simple, empirical power law behavior over a large range is provided by studies of impact cratering. In many applications, the cratering efficiency Π_1, defined as the volume of material removed from the crater divided by the volume of the impacting body, is of central interest. Figure 1.1 shows the dependence of the cratering efficiency on the gravity-scaled yield Π_2, for transient craters formed by dropping water drops onto water at velocities of $1 - 20\,\mathrm{m\,s^{-1}}$ and for hypervelocity impacts onto water [38]. The gravity-scaled yield is defined as

$$\Pi_2 = (2g/V^2)(m/\rho)^{1/3} = (2rg/V^2)(4\pi/3)^{1/3}, \tag{1.11}$$

where m is the mass of the impacting body, r is its radius, V is the impact velocity, ρ is the density of the impacting body and g is the acceleration due to gravity. Within a factor of order unity, the dimensionless group Π_2 is equal to $1/\mathscr{F}_1$, where $\mathscr{F}_1$, the Froude number, is the ratio between the initial dynamic pressure $V^2\rho$ and the "lithostatic pressure" (hydrostatic pressure in this case) $\rho_2 gh$, where ρ_2 is the "target" density, at a characteristic depth h, equal to the impactor diameter. If the densities ρ and ρ_2 are equal, $\mathscr{F}_1 = V^2/(gh)$.

The data shown in figure 1.1 span almost four decades of velocity, over eight decades in impactor radius (24 decades of mass) and eight decades in gravity. This implies that the dependence of Π_1 on Π_2 can be represented by the power law

$$\Pi_1 \sim \Pi_2^{-a}, \tag{1.12}$$

with $a \approx 0.648$, for a wide range of impact conditions. The cratering efficiency for impacts onto water is determined by the potential energy needed to remove water from the crater. Under these conditions, simple theoretical arguments [45] indicate that, if the cratering efficiency depends only on the kinetic energy of the impactor, then the exponent a should have a value of $3/4$, while if the cratering efficiency depends only on the momentum, then a value of $3/7$ $(0.4285\ldots)$ is predicted. The observed value of about 0.648 can be interpreted in terms of energy coupling or momentum coupling parameters that depend on Π_2. The momentum coupling parameter is larger than 1, because of the momentum of the material ejected from the crater. Consequently, a crossover from an exponent of $a = 3/4$, at large Froude numbers, to $a = 3/7$, at small Froude numbers, might be expected.

In general, the combination of data from different sources to cover a wide range of scales is a hazardous procedure, since the amplitude c in equation 1.4 may be different for the different sources, even if the exponents are the same. Figure 1.1 combines results from two quite different experiments, using different projectile materials (glass at high velocities and water at low velocities). In this case, the reliability of the results and the quality of the evidence for a single power law rest on the "Froude scaling" used in the data analysis. The evidence for a simple power law, covering the full range of Froude numbers, is compromised by the large gap between the data sets from the hypervelocity impact and low velocity, water drop experiments. The effective exponent of $a \approx 0.648$ could be the result of a slow crossover (chapter 2) from the small Froude number limit to the large Froude number limit. Schmidt and Housen [46] have measured a value of about 0.51 for the exponent a in equation 1.12, for impact cratering in dry, unconsolidated granular materials (sand and iron particles).

A scaling approach, similar to that used for impact cratering, can be used to describe and model explosion cratering. Experiments at normal and elevated gravity [46] indicate that the crater radius r, depth δ and volume $\mathscr{V}$ grow algebraically with increasing time ($r \sim t^b$, $\delta \sim t^{b'}$ and $\mathscr{V} \sim t^{3b''}$, with $b \approx b' \approx b'' \approx 0.36$), at least during the early stages of crater growth, before the effects of gravity or material strength become important. Data from explosion-generated craters in water [46] indicated that $\mathscr{V} \sim t^{3b''}$, with $3b'' \approx 1.07$, over five decades in $\mathscr{V}/\mathscr{V}^*$ and t/t^*, where $\mathscr{V}^*$ and t^* are the final crater volumes and characteristic time, respectively. This value for b'' corresponds to a value of

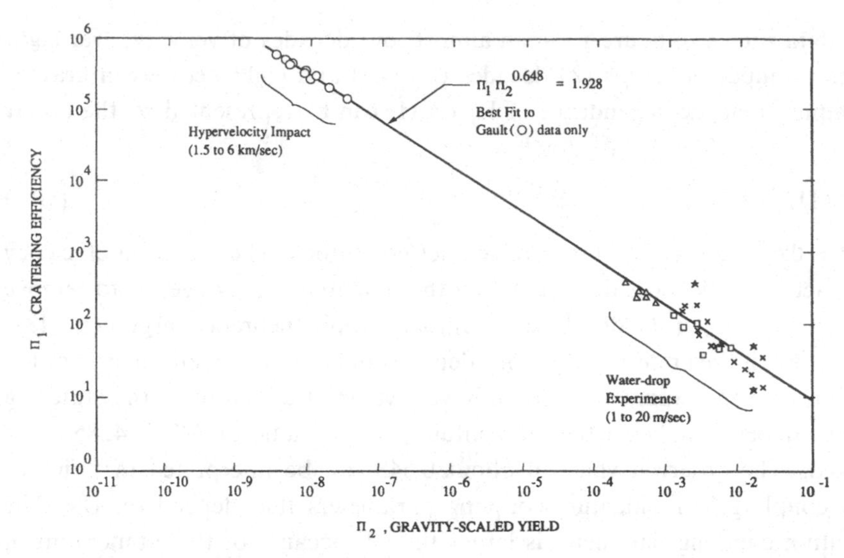

Figure 1.1 Dependence of the cratering efficiency Π_1 on the gravity-scaled yield or inverse Froud number Π_2, for cratering in water. This figure was provided by K. A. Holsapple. See references [39, 40, 41, 42, 43, 44] for original data.

≈ 0.65 for the exponent a in equation 1.12 ($a = 3\mu/(2+\mu)$ and $b'' = 3\mu/(1+\mu)$ [38, 45, 46], where μ is the exponent that relates the far field effects of the impact to the velocity).[4] For $\mu = 0.55$, $a = 1.65/2.55 = 0.65$ and $b'' = 1.65/1.55 \approx 1.06$.

In the case of impacts onto water or unconsolidated granular material, the size of the crater is limited by the potential energy needed to remove material from the crater. If the target material is a cohesive solid, the mechanical properties of the target will be more important than gravitational effects for small craters. Under these conditions, there will be a crossover from a strength dominated regime for small values of the Froude number to a gravity dominated regime at large Froude numbers [47]. The Froude number at which the crossover takes place depends on the Cauchy number, $C = \rho V^2/Y$, which is the ratio between the dynamic pressure and the material strength Y. In the strength dominated regime, a variety of simulations [47, 48] and theoretical models indicate that the cratering efficiency is related to the projectile velocity by $\Pi_1 \sim V^c$ with $c \approx 1.74$. Again, the measured exponent has a value that is intermediate between the value of $c = 2$ expected if the cratering is dominated by the transfer of kinetic energy and the value of $c = 1$ expected if the cratering is dominated by the momentum of the impactor. At very high velocities, phenomena such as melting of solids and vapor-

4. The magnitude of the effects of the impact can be characterized by the quantity $\mathcal{M} = \rho r^3 V^{3\mu}$, where $3\mu = 2$ if the effects of the impact are determined by the kinetic energy of the impactor alone, and $3\mu = 1$ if the effects of the impact are determined by the impactor momentum alone. These values for μ (2/3 and 1/3, respectively) are limiting values; in practice, effective values lying in the range $1/3 < \mu < 2/3$ have been found to give the best representation of experimental data and simulation results.

ization of liquids and solids become important. The simple empirical power law relationships found at lower velocities cannot be expected to extend to conditions under which these effects become important. A discussion of the understanding of the scaling of impact processes at the time of writing can be found in the review of Holsapple [38] and in a numerical study by O'Keefe and Ahrens [47].

Dimensionless groups, also called dimensionless ratios or dimensionless numbers, play an important role in scaling theory [49, 50, 51]. They can be used to compare systems on very much different time and length scales, and reduce to a minimum the number of variables needed to describe a physical system. If all the relevant dimensionless numbers have the same values, for two different systems, then they will behave in the same way. Such systems are often said to be "similar". If a system is described by n physical quantities or a model is described by n parameters, or independent variables, then, a property of the system, or a dependent variable, x_0 can be expressed as

$$x_0 = f(\{x_i^{(n)}\}) = f(x_1, x_2, \ldots, x_n), \tag{1.13}$$

or

$$F(x_0, \{x_i^{(n)}\}) = F(x_0, x_1, x_2, \ldots, x_n) = 0, \tag{1.14}$$

where the set of variables $\{x_i^{(n)}\}$ includes dimensional physical constants such as the acceleration due to gravity g, the mechanical equivalent of heat J and the gas constant R. The physical quantities x_0 and $\{x_i^n\}$ can be expressed in terms of m "fundamental units" $u_1, u_2, \ldots, u_m$. In the case of a mechanical system, mass, length and time are usually used as the set of fundamental units and $m = 3$. The dimensions $[x_i]$ of the physical quantities x_i have the form

$$[x_i] = u_1^{a_{i,1}} u_2^{a_{i,2}} \cdots u_j^{a_{i,j}} \cdots u_m^{a_{i,m}}, \tag{1.15}$$

and the dependent variable has the dimensional form

$$[x_0] = u_1^{a_{0,1}} u_2^{a_{0,2}} \cdots u_j^{a_{0,j}} \cdots u_m^{a_{0,m}}. \tag{1.16}$$

Equation 1.13 can be expressed in terms of dimensionless quantities and written as [49]

$$\Pi_0 = F(\{\Pi^{(l)}\}) \tag{1.17}$$

or

$$\Phi(\Pi_0, \{\Pi^{(l)}\}) = \Phi(\Pi^{(l+1)}) = 0, \tag{1.18}$$

where $\{\Pi^{(l)}\} = \{\Pi_1, \Pi_2, \ldots, \Pi_l\}$ is a complete set of dimensionless groups, constructed from the independent variables $\{x^{(n)}\}$, and F is a characteristic function that describes the behavior of the system. In equations 1.17 and 1.18,

Π_0 is a dimensionless property of the system and is the only dimensionless group containing x_0. The dimensionless groups have the form

$$\Pi_k^{(l)} = x_1^{b_{k,1}} x_2^{b_{k,2}} \cdots x_n^{b_{k,n}} \tag{1.19}$$

and

$$\Pi_{(0)} = x_0 x_1^{b_{0,1}} x_2^{b_{0,2}} \cdots x_n^{b_{0,n}}, \tag{1.20}$$

so that equation 1.17 can be written as

$$x_0 = (x_1^{b_{0,1}} x_2^{b_{0,2}} \cdots x_n^{b_{0,n}})^{-1} F(\{\Pi^{(l)}\}), \tag{1.21}$$

where x_0 and $(x_1^{b_{0,1}} x_2^{b_{0,2}} \cdots x_n^{b_{0,n}})^{-1}$ have the same units, and $F(\{\Pi^{(l)}\})$ is dimensionless.

The dimensionless groups are independent of each other in the sense that one dimensionless group cannot be expressed in terms of a product of powers of the others. In most cases, the number of dimensionless groups, $L = l + 1$, in an equation describing the relationship between physical properties (an equation like equation 1.17) is given by $L = l + 1 = n + 1 - m$ or $l = n - m$.[5]

In many problems, there is no natural distinction between dependent and independent variables, and equation 1.18 can be written as

$$\Phi(\{\Pi^{(L)}\}) = \Phi(\Pi_1^{(L)}, \Pi_2^{(L)} \cdots \Pi_k^{(L)} \cdots \Pi_L^{(L)}) = 0. \tag{1.22}$$

In this case, it is common to construct the set of dimensionless groups $\{\Pi^{(L)}\}$ so that $\Pi_1^{(L)}$ is the only member of the set that contains the variable that is selected as the dependent variable.

The choice of fundamental units is arbitrary [49, 50, 51]. Both the number and nature of the fundamental units can be changed. For example, mass, length and time can be replaced by force, length and time, or the number of fundamental units can be increased by using mass, length, force and time. The number of fundamental units can also be increased by adding units such as temperature or electromagnetic quantities. This does not change the number of dimensionless groups l in $\{\Pi^{(l)}\}$, since addition of a new fundamental unit also adds a new member to the set $\{x_i^{(n)}\}$ of physical quantities (n increases by one). For example, if units of thermal energy are added to the set of units $\{u_j^{(m)}\}$ ($m \to m + 1$) then the mechanical equivalent of heat J must be added to the set of physical quantities $\{x_i^n\}$, so that $n \to n + 1$ and $l = n - m$ remains unchanged. If units of temperature are also added to $\{u_j^{(m)}\}$ then the Boltzmann constant k_B must be added to the set $\{x^{(n)}\}$, and the number of independent dimensionless groups is again unchanged. Similarly, if the number of fundamental units is decreased, the number of physical parameters is decreased correspondingly. In general, the

5. In general, the number of dimensionless groups is given by $l = n - r$, where r is the rank (the maximum number of independent rows or columns) of the matrix $\mathbf{A}$ of the coefficients a ($A_{i,j} = a_{i,j}$) in equation 1.15, with n rows and m columns [50].

units of any subset of $\{x_i^{(n)}\}$ can be used as a set of fundamental units. The number of fundamental units can also be decreased by using quantities such as the gravitational constant or the velocity of light to relate mass, length and time.

In equation 1.17, the function F depends on the boundary conditions, including the geometry. In some cases, it is convenient to supplement the set of physical quantities $\{x^{(n)}\}$ with the dimensionless ratios, such as the length/width ratio, that describe the geometry of a system. This allows results from systems with different sizes *and* shapes to be represented by a single dimensionless function.

The use of dimensional analysis in the study of complex phenomena can be illustrated by its application to impact cratering. In impact cratering, the impacting body may be described by its density ρ, mass m and velocity V; the target may be described by its density ρ_2 and strength Y; and the acceleration due to gravity g is a sixth independent parameter. These six independent parameters can be used, together with the crater volume $\mathscr{V}$, to construct four independent dimensionless groups $\{\Pi^{(L=4)}\} = \{\Pi_1, \Pi_2, \Pi_3, \Pi_4\}$ [45] if mass, length and time are used as the set of fundamental units. The dependence of the crater volume $\mathscr{V}$ on the six parameters m, r, V, ρ_2, Y and g can be described by an equation of the form $\Phi(\Pi_1, \Pi_2, \Pi_3, \Pi_4) = 0$ or $\Pi_1 = F(\Pi_2, \Pi_3, \Pi_4)$, where Π_1 is the only dimensionless product of m, r, V, ρ_2, Y, g and $\mathscr{V}$ that contains the crater volume $\mathscr{V}$. The set of four independent, dimensionless parameters can be constructed in many ways [45]. One choice is $\Pi_1 = \mathscr{V}\rho_2/m$, $\Pi_2 = (g/V^2)(m/\rho)^{1/3}$, $\Pi_3 = Y/(\rho V^2)$ and $\Pi_4 = \rho_2/\rho$. For impact cratering, the relationship between dimensionless groups Π_1, Π_2, Π_3 and Π_4 can be represented quite well by

$$\Pi_1 = c\Pi_2^{\alpha}\Pi_3^{\beta}\Pi_4^{\gamma}, \tag{1.23}$$

over a surprisingly wide range of values of Π_2, Π_3 and Π_4. In general, there is no reason why the relationship between dimensionless groups should have such a simple form. For impact on water, powders or other very weak materials, $\Pi_3 \to 0$, and $\Pi_4 \approx 1$ for many impact processes. Under these conditions,

$$\Phi(\Pi_1, \Pi_2, \Pi_3, \Pi_4) \approx \Phi(\Pi_1, \Pi_2, 0, 1) = \Phi'(\Pi_1, \Pi_2), \tag{1.24}$$

or $\Pi_1 = F'(\Pi_2)$, corresponding to the Froude scaling limit in which Π_1 depends only on the Froude number Π_2. If equation 1.23 is valid, then equation 1.12 is recovered, with $a = \alpha$. On the other hand, if phenomena such as melting are important, new dimensionless groups $(\Pi_5, \ldots)$ containing quantities such as the latent heat of melting must be included in the arguments of Φ. For example, surface tension effects play a significant role in some of the low velocity water drop experiments used to construct figure 1.1 [45]. This results in a breakdown of simple Froude scaling and requires the introduction of a new dimensionless group containing the surface tension. Surface tension effects are the most important for those points closest to the lower right-hand side of figure 1.1.

In the case of impact cratering, the choice of the crater volume as the dependent variable and the set $\{m, r, V, \rho_2, Y, g\}$ of independent variables is quite natural. In other cases the distinction between dependent and independent variables is less clear, and the members of the set $\{\Pi^{(L)}\}$ can be treated on an equal basis.

Equation 1.17 provides the basis for the study of large scale physical phenomena in the laboratory and the use of laboratory models to study phenomena that take place on time scales beyond human experience. Much of our understanding of the world around us is based on equation 1.17. For example, a laboratory-scale experiment using a clay model with a characteristic size of a few inches and duration in the minutes to hours range, can be used to model the behavior of the Earth's crust over a period of millions of years [52]. It is not practical or possible to construct models that are truly similar to complex systems, such as the Earth's crust. In these cases, the design of the model and the interpretation of the model results pose serious challenges. In the cases of models for the Earth's crust, the important dimensionless groups include the Deborah number $\mathscr{D}_e$, defined as $\mathscr{D}_e = \dot{\gamma} t_r$, where $\dot{\gamma}$ is the strain rate and t_r is the relaxation time of the "material", and the ratios between viscous forces, pressure gradient forces and gravity forces (inertial forces can usually be neglected). In comparing the viscous forces with other forces, the non-Newtonian rheology must be taken into account. The materials must be rheologically similar [53], which requires that the exponents that relate strain rates to stresses must be the same for the Earth's crust and the model materials. Such models were used to study complex geological problems long before the application of formal dimensional analysis [54]. Another important example is the use of laboratory-scale models, called flumes or stream tables, for the study of rivers [55, 56]. In this case, the most important dimensionless groups include the Froude number (the ratio between the mean flow velocity V and the gravity wave velocity, given by $\mathscr{F}_2 = V/(gh)^{1/2}$, where h is the mean flow depth)[6] and the Reynolds number, given by $Re = Vh/\nu$, where ν is the kinematic viscosity ($\nu = \eta/\rho$, where η is the viscosity and ρ is the fluid density). Other examples of dimensionless groups include the Bond number (chapter 2, section 2.3) the capillary number [57, 58] (chapter 4, section 4.1.2), the Cauchy number (chapter 1, this section), the Peclet number (chapters 1, sections 1.3.1 and 1.4.7, and 3, section 3.7.2), the Marangoni number (chapter 1, section 1.4) and the Rayleigh number (chapter 1, section 1.3.5).

Empirical power laws have been found in almost all areas of science and technology. A variety of other interesting examples, including the distribution

6. The Froude numbers $\mathscr{F}_1$, described above, and $\mathscr{F}_2$ are defined in different, but closely related, ways in impact cratering and fluid flow studies.

of incomes, insurance claims, community populations, travel distances and the distribution of word frequencies in the use of language, can be found in a book by Zipf [59]. The distribution of funding among universities, an example of the "Matthew effect"[7] that can be described by a power law [60], is a subject of perennial interest. Another example is the distribution of cosmic ray energies [61] that can be described by the power law $J(E) \sim E^{-a}$, where $J(E)\delta E$ is the cosmic ray flux for energies in the range E to $E+\delta E$ and the exponent a has a value slightly larger than 3. In this case, there is significant deviation from the power law at about $10^{18.5}$ eV, which indicates that the higher energy cosmic rays may have a different origin than the lower energy particles. A similar change in exponent is seen in the distribution of earthquake magnitudes, and appears to be related to a crossover from 3-dimensional behavior, for small earthquakes, associated with faults within the brittle crust, to 2-dimensional behavior, for larger earthquakes on faults that span the brittle crust [57].

The concepts of scaling and universality are closely related, and both contribute to the development of a better understanding of complex materials and phenomena. Systems that share common behavior, despite differences on a microscopic level, are said to belong to the same universality class. One of the best known, and most important, examples of universality is the law of corresponding states, which can be written in the form

$$f(P/P_c, T/T_c, \mathcal{V}/\mathcal{V}_c) = 0, \tag{1.25}$$

which is similar to equation 1.18. In equation 1.25, P, T and $\mathcal{V}$ are the pressure, temperature and molar volume of a fluid. The symbols P_c, T_c and $\mathcal{V}_c$ represent the pressure, temperature and molar volume of the same fluid at the critical point. According to the law of corresponding states, the function $f(P^*, T^*, \mathcal{V}^*)$ (where $P^* = P/P_c$, $T^* = T/T_c$ and $\mathcal{V}^* = \mathcal{V}/\mathcal{V}_c$) is a universal function (the function is the same for all "normal" materials). Equation 1.25 implies that, if any two of the three quantities P^*, T^* and $\mathcal{V}^*$ are known, then the other is also known through equation 1.25. Within the van der Waals approximation [62],

$$(P^* + (3/\mathcal{V}^{*2}))(3\mathcal{V}^* - 1) - 8T^* = 0. \tag{1.26}$$

Another example is provided by the success of lattice gas models in the simulation of fluid dynamics. In this case, all "normal" fluids can be said to belong to the Navier–Stokes universality class. This universality class encompasses completely unrealistic models, such as the lattice gas models, provided that the microscopic interactions or collisions conserve energy and momentum.

Many of the most exciting applications of fractals and scaling are in areas such as the earth sciences and biological sciences. In these cases, it is often difficult to obtain large quantities of data, covering wide ranges of time and

7. Matthew, chapter xxv, verse 29.

length scales. In addition, the processes are more complex, and inhomogeneous power laws, such as equation 1.6, or more complex relationships are generally more realistic. In principle, all of the amplitudes (c_1, $c_2 \ldots$) and exponents (a_1, $a_2 \ldots$) can be obtained from the dependence of $\log y$ on $\log x$, but, in practice, an even wider range of scales and more accurate data are needed to obtain reliable results. Another difficulty is that, by their very nature, fractal structures have highly correlated geometries. Consequently, the uncertainties in the parameters obtained by fitting the data sets ($\{(x_n, y_n)\}$ or $\{(\log x_n, \log y_n)\}$) by power laws or straight lines cannot be assessed using the familiar methods developed for uncorrelated Gaussian data. In general, it is better to analyze each data set individually and to use the variance of the results obtained from different data sets to estimate the statistical uncertainties, instead of combining results from different data sets (different realizations of a growth process, for example) and using the total data set to estimate an exponent. Even more important sources of uncertainty arise from "corrections to scaling" (chapter 2) and finite-size effects (chapter 2). For these reasons, the uncertainties in the exponents, given in the literature, should always be regarded with suspicion. They are often mentioned at the insistence of pernicious referees and are commonly understated by an order of magnitude or more.

In many complex structures and phenomena, power law behavior is observed over only about one decade. The value of such data is uncertain and depends on the intended application, as well as the existence of an underlying theoretical interpretation. This does not mean that attempts to apply fractal geometry and scaling to geology, geophysics and biology should be abandoned. On the contrary, it is in these areas that such ideas may eventually be of the greatest value. Because of the difficulty of formulating geological and biological phenomena in terms of continuum equations and the difficulties of obtaining solutions to these equations, if they exist, approaches based on fractal geometry, scaling and algorithmic modeling should be of great value. More emphasis should be placed on including crossovers between different scaling regimes, finite-size effects and resolution effects in the data analysis. There is a strong temptation to apply more complex multifractal models (chapter 2, section 2.8, and appendix B) to data that do not appear to yield to simple fractal analysis. This is a powerful approach that is appropriate in some cases. However, multifractal scaling has not yet lived up to its early promise, and it should be used cautiously, after more simple approaches have been exhausted.

There are many examples of important systems that appear to exhibit some sort of scale symmetry. However, in only relatively few cases have they been analyzed in a serious and critical manner. It is also apparent that many complex phenomena may require complex models and that the simple fractal scaling models used today may not be adequate in all cases. There are strong indications that even structures generated by very simple mechanisms may have a quite complex scaling structure.[8]

The important role played by the development of an ever more quantitative and complete characterization of physical objects and processes, in the advancement of scientific understanding, has been described many times. In the words of Thompson ([63], page 1029),

> If no chain hangs in a perfect catenary and no raindrop is a perfect sphere, this is for the reason that forces and resistance other than the main one are inevitably at work. The same is true of organic form, but it is for the mathematician to unravel the conflicting forces which are at work together. And this process of investigation may lead us on step by step to new phenomena, as it has done in physics, where sometimes a knowledge of form leads us to the interpretation of forces, and at other times a knowledge of the forces at work guides us towards a better insight of form. After the fundamental advance had been made which taught us that the world was round, Newton shewed that the forces at work upon it must lead to its being imperfectly spherical, and in course of time its oblate spheroid shape was actually verified. But now, in turn, it has been shewn that its form is still more complicated, and the next step is to seek for the forces that have deformed the oblate spheroid.

In general, simple Euclidean shapes are formed under close-to-equilibrium conditions. Familiar examples include the simple shapes of soap films, the faceted shapes of slowly grown mineral crystals and the flat surface of water in an undisturbed, open container. However, it is only in the most limited scenarios that the ideas of equilibrium thermodynamics and statistical mechanics provide us with an adequate understanding. Figure 1.2 shows a pattern formed by the condensation of water on a cold polyethylene surface. While the spherical cap shape of the individual droplets can be understood in terms of simple equilibrium concepts, the more complex pattern formed by the droplets is a consequence of non-equilibrium kinetic processes [64, 65].

The formation of a fractal structure, on small length scales, requires the formation of a large surface that separates the fractal structure from the surrounding space. In general, a large energy will be associated with this surface, and fractals cannot be expected under equilibrium conditions. However, there are two important exceptions to this generalization. The first exception is systems near a critical point. As the critical point is approached, the surface tension or interfacial energy approaches zero and a very large "surface" may develop. At the critical point, the idea of a surface separating two "components" or "phases" loses all meaning. The second important exception consists of a broad class of constrained systems. Some examples include polymers, membranes, froths etc. In these cases, the system is in equilibrium, subject to a constraint

8. See, for example, the discussion in chapter 3, section 3.4.5, concerning the 2-dimensional DLA model, which now appears to have a considerably more complex structure than was envisaged a decade ago.

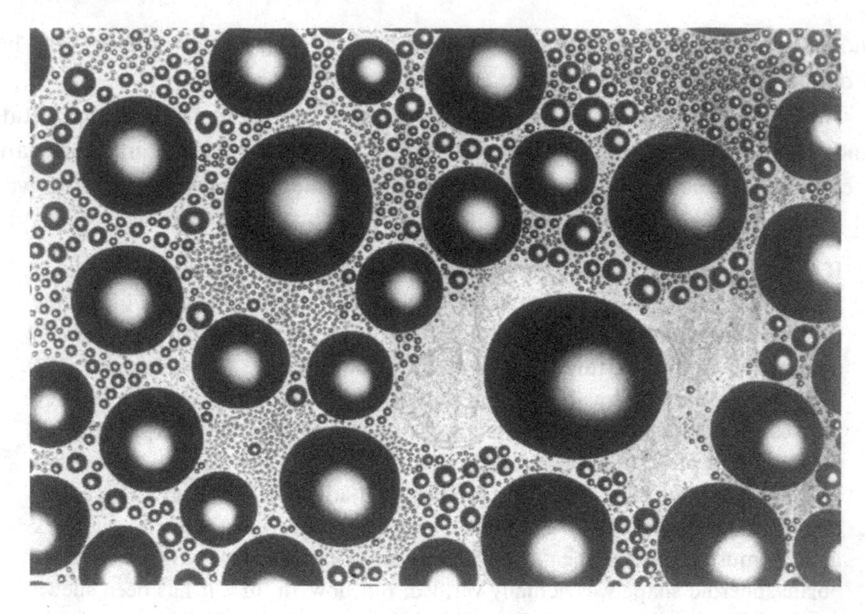

Figure 1.2 A pattern formed by the condensation of water droplets on a cold polyethylene surface. The image is about 1 cm wide. This figure was provided by K. Galvin.

such as the way the monomers are joined, in the case of a polymer, for example. This allows large "surfaces" to exist without either collapse or "dissolution". Without the constraint, a very different, non-fractal, structure would be formed.

1.2 The Logistic Map

As a system is driven further away from equilibrium, there is a tendency for shapes to become more complicated. In many cases, the structure progresses from a uniform or structureless form via a series of regular or periodic patterns to a disorderly chaotic form, as it is forced further from equilibrium, by increasing an external "driving force". For example, the transition from regular Euclidean shapes generated by crystallization near to equilibrium to irregular dendritic shapes under far-from-equilibrium, rapid growth conditions is well known [66], as is the transition from laminar to turbulent flow in fluids and the oscillatory and chaotic behavior of some strongly driven chemical reaction processes.

A simple mathematical model that illustrates this idea is provided by the logistic map $x \to x'$, with

$$x' = f(x) = ax(1 - x), \tag{1.27}$$

where x lies in the range $0 < x < 1$ and the "driving parameter" a lies in the range $0 < a < 4$. This equation provides a paradigm for the way in which a system can progress from simple to complex behavior, as it is forced further from equilibrium, by increasing the "driving parameter" a. This model also illustrates how complex behavior can have simple origins and how "chaotic" behavior can arise in deterministic systems. The logistic map is now one of many paradigms for "routes to chaos", but it retains a special place as the first and most simple.

Equation 1.27 can also be written as

$$x_{n+1} = f(x_n) = ax_n(1 - x_n). \tag{1.28}$$

Starting with an initial value (x_0), the variable x is repeatedly "mapped" onto a new value ($x_0 \to x_1 \to x_2 \to \ldots$) using equation 1.28. If the parameter a is sufficiently small, this procedure quickly converges to a "fixed point" or stable attractor ($x_n \to x_\infty$ for $n \to \infty$). Figure 1.3 shows the "attractor" for the map in equations 1.27 and 1.28. In this figure, the values for $x_{500} - x_{1000}$ are shown for 500 values of a in the range $0 \leq a \leq 4.0$. Each point (a, x_n) is indicated by a short horizontal line centered on (a, x_n) with a length of 0.01. If the parameter a is increased from 0, the fixed point at first remains at $x = 0$. When the parameter a exceeds a critical value, $a_{c1} = 1.0$, the fixed point $x_\infty = 0$ becomes unstable and the mapping converges to a non-zero limit $x_\infty > 0$, given by

$$x_\infty = ax_\infty(1 - x_\infty) \tag{1.29}$$

or

$$x_\infty = (a - 1)/a. \tag{1.30}$$

This is illustrated in figure 1.3. As the parameter a is increased through a second critical value ($a_{c2} = 3.0$), the fixed point, given by equation 1.30, also becomes unstable. The "attractor" is no longer a single point. Instead, the locus of the asymptotic fixed points bifurcate, so that there are two distinct fixed points, x_{2n} and x_{2n+1}, in the limit $n \to \infty$. These fixed points are given by

$$x_\infty = f^{(2)}(x_\infty), \tag{1.31}$$

where $f^{(n)}(x)$ stands for application of the mapping $f(x)$, given in equations 1.27 and 1.28, n times ($f^{(2)}(x) = f(f(x)), f^{(3)}(x) = f(f(f(x))), \ldots$). The fixed points $x = 0$ and $x = (a - 1)/a$ remain as unstable fixed points.

Figure 1.4 shows the values of x_n generated by the logistic map, starting with $x_0 = 0.095$ and $a = 3.3$. This figure shows the map $f^{(1)}(x) = f(x)$ and the lines connecting the coordinates (x_n, x_n) to (x_n, x_{n+1}) and (x_n, x_{n+1}) to (x_{n+1}, x_{n+1}). In addition, the line from $(x_0, 0)$ to $(x_0, x_1) = (x_0, f(x_0))$ is shown. The first 200 values of x were used to construct this figure. However, iterations of the map soon converge onto two asymptotic values of x (x_{2n} and x_{2n+1}, with $n \to \infty$), and only a few stages can be resolved in the figure. The behavior can be seen more clearly in terms of the map $f^{(2)}(x)$. Figure 1.5(a) shows the results of

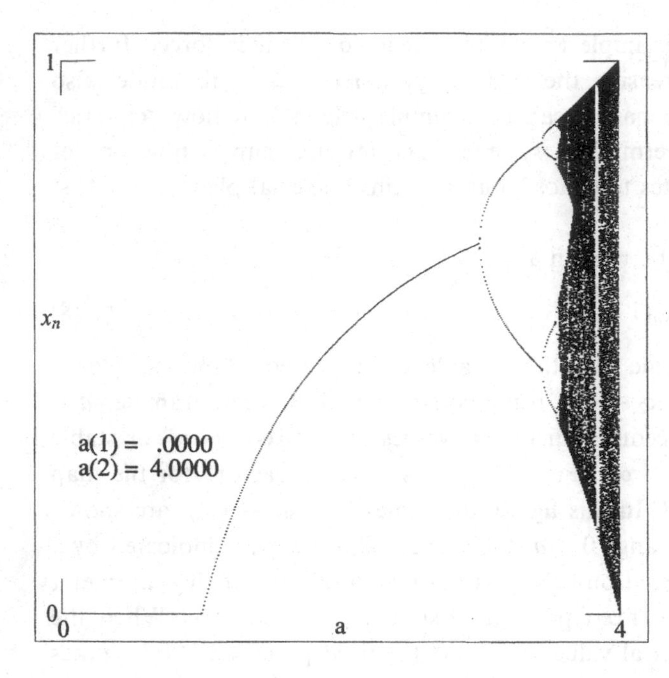

Figure 1.3 The attractor for the logistic map $x' = ax(1-x)$, starting with an initial value of $x_0 = 0.3$. The map was iterated 500 times to approach the "attractor". The values of x_n from the next 500 iterations are shown in this figure. This procedure was repeated for 500 values of a ($a = 0.008n$, $n = 1 - 500$) in the range $0 < a \leq 4$. Equivalent results will be obtained for almost all x_0 in the range $0 < x_0 < 1$.

iterating the map $f^{(2)}(x)$ with $a = 3.3$ and $x_0 = 0.095$. The numbers x_n soon converge onto a fixed point x_∞. Figure 1.5(b) shows the results of iterating the same map with $x_0 = f^{(1)}(0.095)$. The function $f^{(2)}(x)$ has four fixed points, including the one at $x = 0$. However, only two of these are stable and they form the attractor for the map. The sequential mapping converges onto the stable, attracting fixed points for almost all initial values x_0.

As a is increased further, another bifurcation takes place, the fixed points, given by equation 1.31 become unstable, and there are four distinct values for x associated with the attractor ($x_{4n}, x_{4n+1}, x_{4n+2}$ and x_{4n+3}, as $n \to \infty$); the trajectory of x visits them in turn. These points are solutions of the equation $x = f^{(4)}(x)$. As a is increased, more and more bifurcations are encountered, at smaller and smaller intervals of a. After m bifurcations the attractor consists of the 2^m stable fixed points of $f^{2^m}(x)$, separated by the unstable fixed points corresponding to the solutions of $x = f^{2^{m-1}}(x), x = f^{2^{m-2}}(x), \ldots, x = f^1(x)$.

As a approaches the limiting value $a_{c3} = 3.5699456\ldots$ [67], the period of the bifurcations (2^m, where m is an integer) approaches infinity. If Λ_m is the value of a at which the mth bifurcation appears, then, as m increases, the ratio r_m, defined as

$$r_m = [\Lambda_{m+1} - \Lambda_m]/[\Lambda_{m+2} - \Lambda_{m+1}], \tag{1.32}$$

approaches a limiting value of $r_\infty = 4.6692016\ldots$ [68] and the interval $\Lambda_\infty - \Lambda_m$ decreases geometrically

$$\lim_{m\to\infty} (\Lambda_\infty - \Lambda_m) \propto r_\infty^{-m}. \tag{1.33}$$

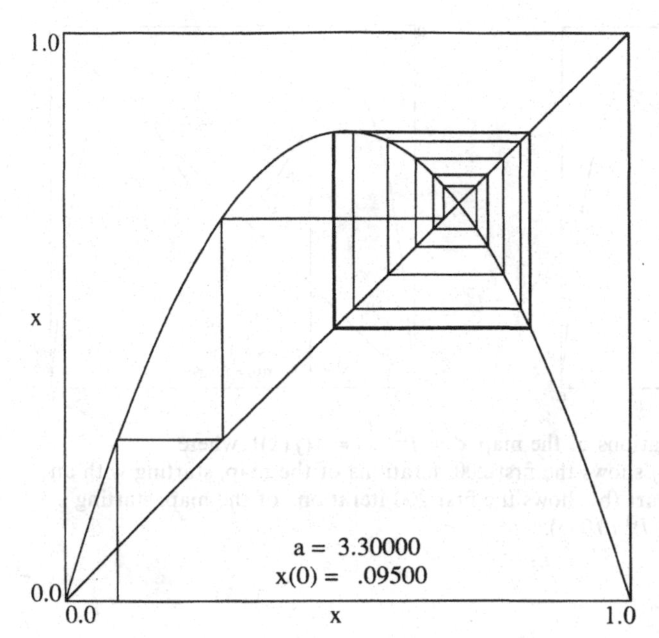

Figure 1.4 The first 200 iterations of the map $x' = 3.3x(1-x)$, starting with an initial value of $x_0 = 0.095$.

The ratio r_∞ is a universal number that is characteristic of the "period doubling route to chaos" and does not depend on the precise form of the map $f(x)$ that generates the period doubling sequence [68]. This universal ratio has the same value everywhere within the map, as well as for a broad class of other maps that have the same (parabolic) shape near their maxima.

As a is increased beyond the value $a = a_{c3} = \Lambda_\infty$, a qualitatively new type of behavior emerges and the attractor becomes chaotic. In the chaotic regime, the trajectory $\{x_n\}$ visits all points in the region $x_{min} \leq \{x_n\} \leq x_{max}$, but not with equal probability. This is by no means the end to the fascinating behavior exhibited by the simple logistic map. As a is increased further, islands of periodicity appear in the "sea" of chaos. The most prominent of these is the region of three-fold periodicity that can be seen clearly in figures 1.3 and 1.6. Odd-period regions first appear at $a = 3.6786\ldots$ [67], and the region between $a = 3.6786\ldots$ and $a = 4.0$ contains periodic regions of all odd orders. Figure 1.6 shows an enlargement of the attractor for the logistic map for driving parameters in the range $3.5 < a < 4.0$. This figure shows that, as a is increased, in the largest regime of three-fold periodicity, a bifurcation to six-fold periodicity takes place. This is followed by additional bifurcations to 12-fold, 24-fold, $\ldots, 3 \times 2^n$-fold,$\ldots$ periodicity. The ratio of the intervals in a between successive bifurcations is given by equation 1.33 with the universal value of $r_\infty = 4.6692016\ldots$. The structure towards the large a side of this "three-fold window" consists of three miniature, but somewhat distorted, replicas of the entire logistic map. Within these miniature maps, windows of all odd orders, with their own miniature

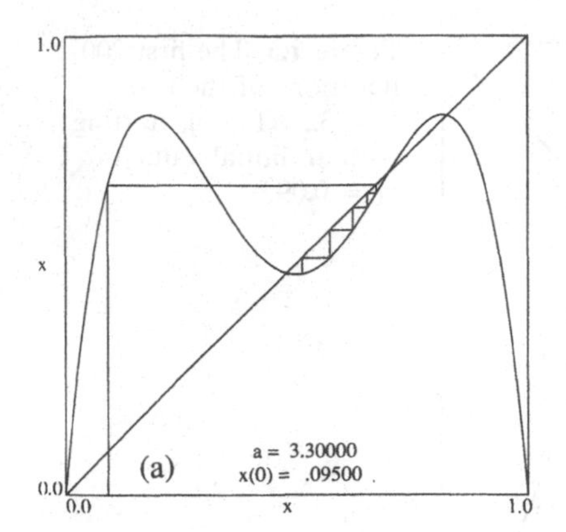

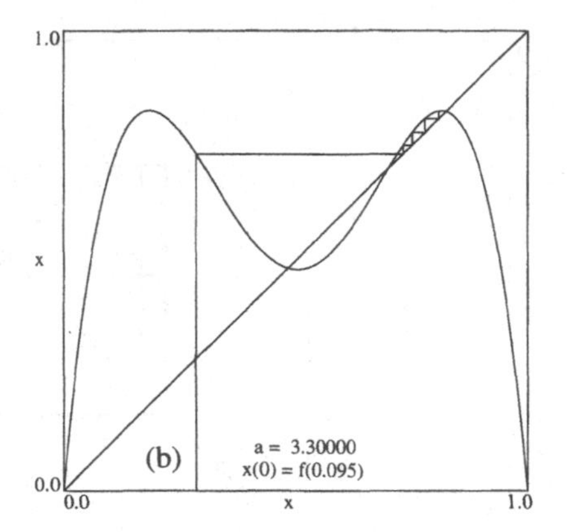

Figure 1.5 The first 200 iterations of the map $x' = f^{(2)}(x) = f(f(x))$, where $f^{(1)}(x) = 3.3x(1-x)$. Part (a) shows the first 200 iterations of the map, starting with an initial value of $x_0 = 0.095$. Part (b) shows the first 200 iterations of the map, starting with an initial value of $x_0 = f^{(1)}(0.095)$.

maps, can be found. Although these miniature maps are distorted versions of the logistic map, the ratio r_∞ retains its universal value of 4.6692016... for the bifurcations within the miniature maps. In this sense, the attractor for the logistic map is infinitely nested within itself. Figure 1.7 shows a "blow up" of that part of the logistic map, near to the large three-fold window, which contains a replica of the entire logistic map.

Interesting behavior is also found in the chaotic regime, just to the low a side of an odd-period window. In this region, the "trajectory" $\{x_n\}$ is almost periodic for long sequences of successive n values. This is illustrated in figure 1.8(a). Here, the first 200 values of x_n are shown with $a = 3.82842$ and $x(0) = 0.095$. It can be seen that the distribution of x_n is concentrated into three narrow ranges of x corresponding to almost periodic behavior, most of the time. The nature of this behavior can be seen better in figure 1.8(b), which shows the first 200 iterations of the map $f^{(3)}(x)$, with the same value of a and the same initial value for x ($x(0) = 0.095$). It can be seen that the almost periodic behavior occurs when the line $y = x$ is almost a tangent to $f^{(3)}(x)$ and the trajectory $x_n, x_{n+1}, x_{n+2} \cdots$ becomes "trapped" between $f^{(3)}(x)$ and $y = x$. These "periods" of almost regular behavior are followed by chaotic behavior. For obvious reasons, this behavior is often referred to as "intermittency". Figure 1.9 illustrates that if the parameter a is increased slightly, the function $f^{(3)}(x)$ intersects the line $y = x$ eight times corresponding to three attractive fixed points and five repelling fixed points.

The logistic map also exhibits a rich scaling structure. For example, at the bifurcation limit, $a = a_{c3}$, the attractor of the logistic map is a multifractal

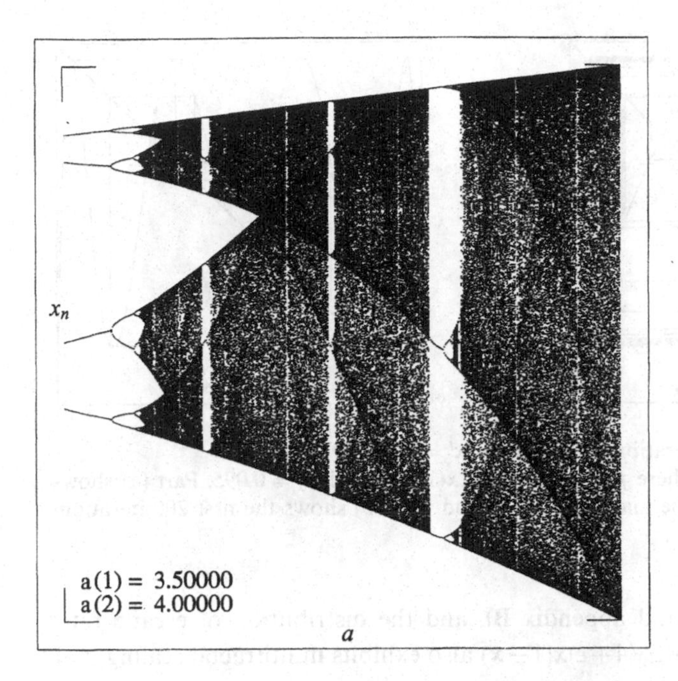

Figure 1.6 A "blow up" of the logistic map shown in figure 1.3, for 500 values of a in the range $3.5 < a < 4.0$. The ordinate x_n lies in the range $0 \leq x_n \leq 1$.

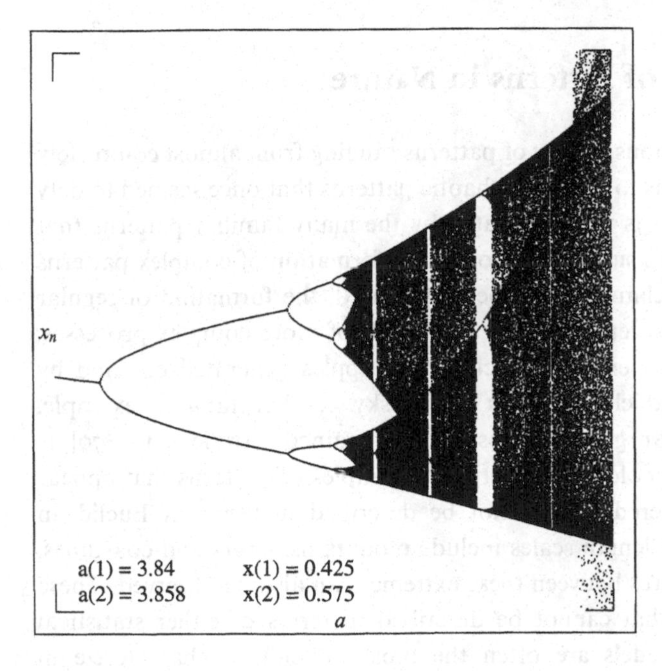

Figure 1.7 A "blow up" of the logistic map shown in figure 1.6. This blow up contains the miniature logistic map in the middle of the largest three-fold window. The numbers x_n for $1501 \leq n \leq 3000$ that fall in the range $0.425 \leq x \leq 0.575$ are shown for 500 values of a in the range $3.84 \leq a \leq 3.858$.

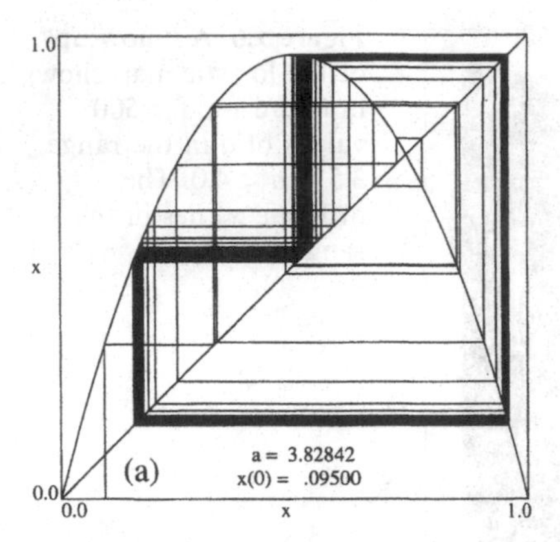

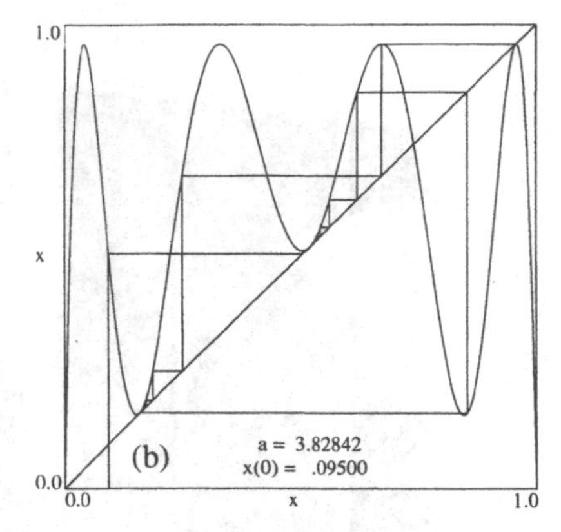

Figure 1.8 The first 200 iterations of the maps $x' = f(x) = f^1(x)$ and $x' = f^{(3)}(x) = f(f(f(x)))$, where $f^{(1)}(x) = 3.82842x(1-x)$ and $x_0 = 0.095$. Part (a) shows the first 200 iterations of the map $x' = f^{(1)}(x)$ and part (b) shows the first 200 iterations of the map $x' = f^{(3)}(x)$.

(chapter 2, section 2.8, and appendix B), and the distribution of escape rates from the map $f(x) = \lim_{\epsilon \to 0}(4+\epsilon)x(1-x)$ also exhibits multifractal scaling [69].

1.3 The Variety of Patterns in Nature

Nature exhibits an enormous variety of patterns ranging from almost completely regular Euclidean patterns to complex, chaotic patterns that once seemed to defy description. This richness is well illustrated by the many familiar patterns that can be seen on the Earth's surface. Although the formation of complex patterns as a result of simple mechanisms will be emphasized, the formation of regular patterns with well defined length scales, as a result of more complex processes, is also common. Familiar examples include the ripples generated on sand by flowing water or air and cloud streets in the sky. A less familiar example, the formation of circular gravel ridges on fine-grained, Arctic soils [70] is shown in figure 1.10 (see color plate section). Examples of patterns that appear to be completely disordered and cannot be described in terms of Euclidean geometry or well defined length scales include mountains, rivers and coastlines. A wide variety of structures between these extremes can also be observed. These intermediate structures that cannot be described in terms of either statistical models or Euclidean models are often the most difficult to characterize in quantitative terms.

The beautiful patterns formed by the growth of ice crystals in the form of snowflakes have inspired a broad range of research on pattern-formation [1, 71]. The formation of symmetric but intricate patterns points to a deterministic growth mechanism that is exquisitely sensitive to the growth conditions. The

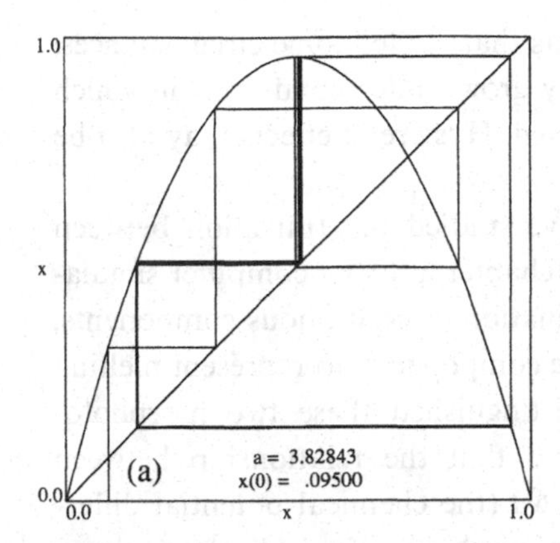

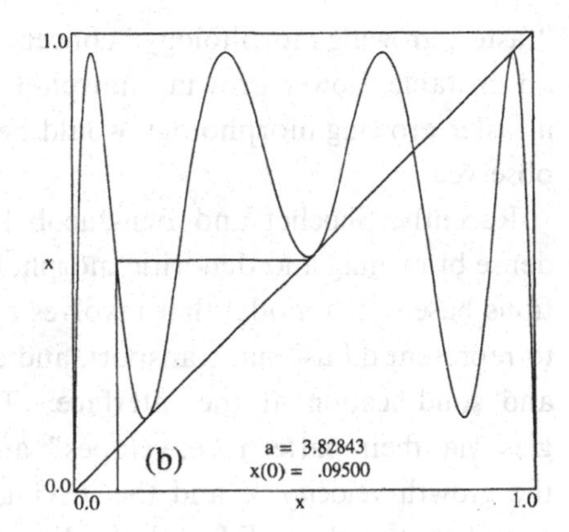

Figure 1.9 The first 200 iterations of the maps $x' = f(x) = f^1(x)$ and $x' = f^{(3)}(x) = f(f(f(x)))$, where $f^{(1)}(x) = f(x) = 3.82843x(1-x)$ and $x_0 = 0.095$. Part (a) shows the first 200 iterations of the map $x' = f^{(1)}(x)$ and part (b) shows the first 200 iterations of the map $x' = f^{(3)}(x)$. This figure should be compared with figure 1.8.

growing arms of a snowflake are near enough to each other to experience the same growth conditions, such as temperature, humidity and impurities, but too far apart for the growth of one arm to completely control that of its neighbors. The overall six-fold symmetry of most snowflakes is related to the underlying six-fold symmetry of the internal crystalline structure. However, the development of a quantitative understanding of the overall morphology of snowflakes presents a challenge that has been at the center of pattern-formation research for several decades [71].

In many processes, such as electrochemical deposition discussed in chapter 4, sections 4.1.1 and 4.2.1, several quite different patterns can be observed if the growth conditions are changed. In these cases, a "morphological phase diagram" or "morphology diagram" can be constructed in the control parameter space. These morphological phase diagrams are reminiscent of ordinary equilibrium phase diagrams, but, in most cases, the nature of the "transitions" between different "morphological phases" is not understood and there is, as yet, no general systematic way of classifying morphological phase transitions. Ben-Jacob *et al.* [72, 73] have proposed a dynamical morphology-selection principle based on the idea that, if more than one morphology is possible, the most rapidly growing one is "selected". This selection principle might appear to be tautological, but an important consequence of this idea is that morphological transitions of this type would be expected to lead to an abrupt change in morphology as the growth conditions are changed. In practice, most morphological phase transitions appear to be relatively gradual, but definitive experiments are difficult and, in many cases, the appropriate way to characterize the morphologies, so that the transition can be studied quantitatively, is not clear. The "fastest growing morphology" concept suggests that, under some circumstances, an unstable, slower growing morphology may grow under conditions in which a faster growing morphology would be expected. Hysteretic effects may also be observed.

Recently, Shochet and Ben-Jacob [74] have studied the transition between dense branching and dendritic morphologies (chapter 4) using computer simulations based on a model that involves a combination of continuous components, to represent diffusional transport, and discrete components, to represent melting and solidification at the interface. They distinguished these two morphologies via their pattern "envelopes" and found that the relationship between the growth velocity V and the driving force $\Delta\mu$ (the chemical potential difference between the solid and the liquid, at infinity) was $V \sim (\Delta\mu)^3$ for dense branching morphologies with convex envelopes, and $V \sim (\Delta\mu)^{3/2}$ for concave envelope, dendritic patterns. This model generates dendritic patterns at small $\Delta\mu$ and dense radial patterns at large $\Delta\mu$. However, by carrying out the simulations in channels oriented in the direction of one of the axes of a square lattice, the dense radial morphology could be grown with small driving forces at which the dendritic morphology usually grows. Under these circumstances, the velocity at which the dense radial morphology grows is smaller than that at which the dendritic morphology grows, and it corresponds to an extrapolation of the dependence of V on $\Delta\mu$ found for the dense radial morphology, at large driving forces. In some simulations, usually in wide channels, spontaneous transitions between the dendritic and tip-splitting patterns were observed. This lends support to the idea that microscopic solvability criteria[9] [75, 76] may not provide a completely reliable way of predicting the growth morphology.

There is experimental evidence that, in some cases [77], a discontinuous change in the growth velocity may be associated with a morphological phase transition. Using an analogy with equilibrium phase transitions, such a transition can be called a "first order" morphological transition, to distinguish it from transitions at which only the slope of the velocity $\partial V/\partial\Delta\mu$ changes.

1.3.1 Euclidean Patterns

There are many classes of patterns that are more or less regular and can be described quite adequately in terms of simple Euclidean shapes. Very often, the most simple patterns are formed near to equilibrium and they become more and more complex as the system is driven further from equilibrium. This is similar to the behavior found in the logistic map, described earlier in this chapter,

9. The solvability criterion is based on the idea that, in the presence of surface tension, the tip of a growing dendrite should have a smooth, cuspless shape. The surface tension introduces the microscopic capillary length and, for this reason, the solvability criterion is often referred to as "the microscopic solvability criterion".

as the "forcing parameter" a is increased. While it is not possible to make generalizations, a series of "transitions" between different classes of patterns is often found. Hysteresis is common, and several patterns can often be formed under the same conditions and boundary conditions, depending on the history of the events leading up to the pattern formation. The shapes of the patterns and the sequence of events, as the system is driven further from equilibrium, is often similar for a number of phenomena that appear to be quite different. This suggests an underlying universality. However, it is dangerous to assume that, because the patterns are "the same", their origins can be explained in the same way. The almost spherical shapes of pebbles in a stream bed, pearls and water droplets have quite different origins.[10] In particular, the similarity between the patterns obtained using oversimplified models and processes observed in the laboratory or field must not be taken too seriously.

Lamellar, columnar or striped structures constitute a common class of patterns. Prominent examples of columnar structures include anodic films on aluminum [79, 80, 81] and basaltic column arrays such as the Giant's Causeway, County Antrim, Northern Ireland. More or less regular lamellar structures can be formed in many ways. The formation of stripped patterns in arctic and alpine [82] ground is an interesting example related to the pattern shown in figure 1.10 (see color plate section). The patterns formed during directional eutectic growth [83, 84] constitute one of the most important classes of lamellar patterns, from a technological point of view. A typical example is shown in figure 1.11. In this case, the patterns consist of alternating layers of material that are rich in one, or the other, of the two components. These lamellar structures can be described in terms of the characteristic spacing, repeat distance or lamellar thickness ξ_m. In other structures, there may be a broad distribution of spacings. In eutectic solidification, a lamellar morphology is favored if the volume fractions x_1 and x_2 of both phases lie approximately in the range $0.3 \leq x_1, x_2 \leq 0.7$. Outside this range, the minority phase forms columns embedded in the continuous majority phase.

During eutectic solidification, the two components must diffuse in the liquid phase, over a distance of the order of ξ_m, to form the lamellar structure. From this, it follows that $\xi_m \sim V^{-1/2}$, where V is the velocity of the solidification front. It can usually be assumed that the solid phases and the liquid are at local equilibrium, so that the local geometry at the lines where all three phases are in contact are determined by the three interfacial energies. As a result, the exposed edges of the lamellae are curved with a curvature that is proportional to $1/\xi_m$. This means that the undercooling ΔT at the growing front of the solid lamellar structure is proportional to ξ_m^{-1}, so that $\Delta T \sim V^{1/2}$. These relationships between ξ_m, ΔT and V have been known for some time and were placed on a sound theoretical foundation by Jackson and Hunt [83].

10. In practice, the shapes of pebbles abraded by transport in a stream bed can be represented much better by ellipsoids than spheres [78].

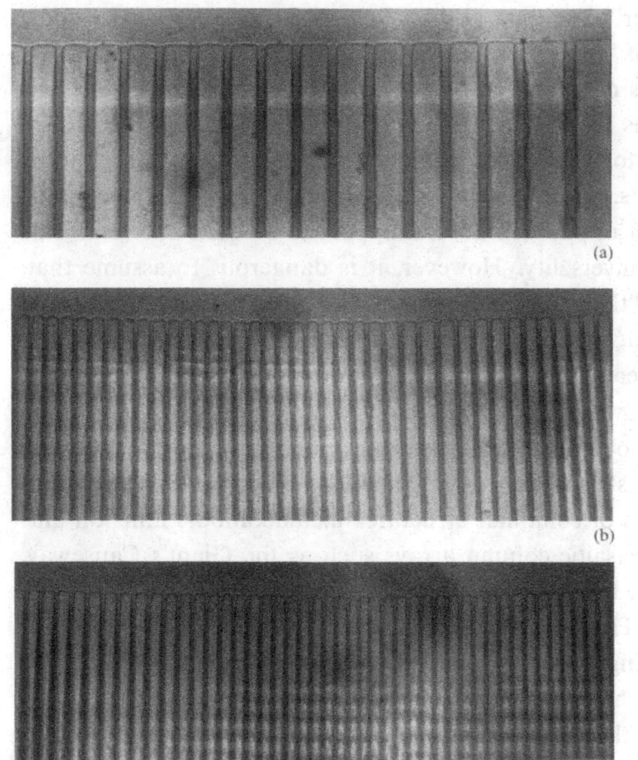

Figure 1.11 Lamellar patterns generated by the directional, eutectic solidification of 8.4 weight % CBr_4/C_2Cl_6 alloys. Parts (a), (b), and (c) show the effects of the growth velocity (0.2 μm s^{-1}, 0.5 μm s^{-1} and 1.0 μm s^{-1}, respectively) on lamellar spacing. This figure was provided by R. Trivedi.

Kassner and Misbah [85] have recently shown that, for a theoretical eutectic growth model, the selected length scale is given by

$$\xi_m = V^{-1/2} f(\xi_{\mathscr{D}}/\xi_T), \tag{1.34}$$

in the small Peclet number limit, $Pe = \xi_m/\xi_{\mathscr{D}} \ll 1$, where $\xi_{\mathscr{D}}$ is the diffusion length and ξ_T is the thermal length. The diffusion length is given by $\xi_{\mathscr{D}} = \mathscr{D}/V$, where $\mathscr{D}$ is the diffusion constant. The thermal length is given by $\xi_T = (c_1 - c_2)|m/g_T|$, where c_1 and c_2 are the concentrations in the two solid phases ($c_1 > c_2$), g_T is the imposed thermal gradient and m is the slope of the liquidus line in the phase diagram. This theoretical approach shows that $\xi_m \sim V^{-1/2}$ only at relatively high velocities, where the scale of the morphology ξ_m is controlled by diffusion and the function $f(x)$ in equation 1.34 has a constant value. At smaller velocities, the effective exponent is smaller than $1/2$ ($\xi_m \sim V^{-a}$, with $a < 1/2$) and the morphology is controlled by the thermal gradient. The form of equation 1.34 arises because the directional solidification process, or at least the model for the process, is invariant under the transformation $\xi_m \rightarrow b\xi_m$, $V \rightarrow b^{-2}V$, $g_T \rightarrow b^{-2}g_T$. Equation 1.34 implies that $\xi_m \sim V^{-1/2}$, even at low velocities, providing that $\xi_{\mathscr{D}}/\xi_T$ ($\xi_{\mathscr{D}}/\xi_T \propto g_T/V$) is held constant.

Similar considerations apply to the other common eutectic solidification morphology consisting of rods of one composition embedded in the other. These embedded rods often form a quite regular array. As the growth velocity is increased, defects begin to appear [83]. Similar morphologies are formed during the directional solidification of impure "single component" systems and systems that are far from the eutectic composition. In this case, the morphology at small growth velocities consists of layers or cells of the majority component separated by walls of the second component or material into which impurities are concentrated [86, 87, 88]. If the front velocity V is changed, the characteristic length ξ_m, in equation 1.34, changes, and a variety of interesting "defect" patterns can be generated as the system adjusts to the new growth conditions [83, 89].

1.3.2 Cellular Patterns

Cellular patterns are quite common. Familiar examples include foams, biological tissues, [1, 90, 91] honey-combs [1], mud cracking patterns and the grain structure of many metals, ceramics and minerals [92]. Publicity has recently been given to the idea that the visible matter in the universe is organized into empty bubbles separated by relatively dense walls [93]. This is not inconsistent with a fractal geometry. Other examples include a packing of peas swollen in a confined space [94], phospholipid monolayers in the liquid-condensed/liquid-expanded regime [95], ecological territories [96], the stress distribution in 2-dimensional packings, the surfaces of shrunken gels [97, 98] and the Voronoi–Dirichlet [99, 100] polygons ($d = 2$) or polyhedra ($d = 3$) used to characterize random sets of points, granular materials, glasses and other disordered systems. A variety of cellular structures has been found in porous polymer membranes. In general, polymer membranes are formed by quite complex phase separation processes [101]. In some cases, the resulting cellular structures are highly ordered [102]. Cellular patterns have also been studied in flame fronts [103, 104]. The hexagonal patterns formed by the growth of anisotropic materials such as NH_4Cl and $(NH_4Cl)_x(CuSO_4)_{1-x}$ from thin supersaturated layers [105] provide other interesting examples.

Figure 1.12 shows a shaddowgraph picture of a particularly regular hexagonal pattern formed by convection of CO_2, a non-Boussinesq fluid.[11] Bodenschatz *et al.* [106] obtained this pattern in a horizontal cell with an aspect ratio (radius/depth) of 86 (8.941 ± 0.001 cm diameter and 0.0520 ± 0.0005 cm depth) with a silver bottom and a sapphire top that was heated uniformly from

11. In non-Boussinesq fluids the temperature dependencies of properties, other than the density, play a significant role.

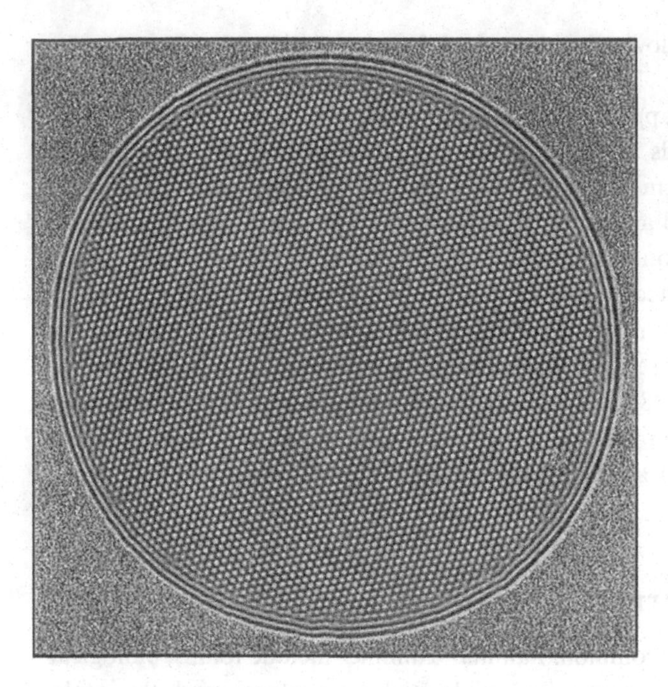

Figure 1.12 A hexagonal pattern of convection cells in a high aspect ratio, circular cell. In this experiment, $\epsilon_R = (R_a/R_c) - 1 = 0.06$, where R_a is the Rayleigh number and R_c is its critical value. This figure was provided by E. Bodenschatz.

the bottom. This figure and figure 1.15 illustrate the degree of order that can be obtained in experiments using very accurately constructed cells and well-controlled experimental conditions. The columnar patterns formed during three-dimensional eutectic solidification can also be regarded as being cellular, particularly if a column and the surrounding material closest to the column are considered to be a "cell".

Cellular patterns are commonly formed by fracture processes. A familiar example is the cracking of drying mud, and similar patterns are formed in a wide variety of related fragmentation processes covering a very wide range of length scales. Figure 1.13 shows a cellular pattern formed as the result of a different type of fracture process.

In the Voronoi–Dirichlet (or Wigner–Seitz [107]) construction, space containing a set of points is divided into cells surrounding each particle, so that the cell associated with a particular point contains all the positions in the embedding space that are closest to that point. Determination of the distributions of the sizes and shapes of the Voronoi–Dirichlet cells is an important approach to the quantitative characterization of many disorderly structures.

Soap froths or foams, such as those shown in figure 1.14, have, for a long time, been used as models for other cellular structures such as granular metallic alloys and ceramics [108]. The essential similarity between cellular patterns of quite different origin arises because they all fill space randomly with, more or less, polygonal or polyhedral shapes, and evolve via the slow transfer of material

Figure 1.13 A digitized image of a 36 cm × 47 cm part of a fracture pattern generated by the accidental damage of a highly tempered 86 cm × 100 cm glass window, during sand blasting. This fracture network is topologically equivalent to a 2-dimensional network of polygons. The original photograph was provided by P. D'Onfro and the digitized image was prepared by T. Walmann, using a high resolution CCD camera.

from small cells to larger cells. However, a foam is in mechanical equilibrium at all times, except during the topologic adjustments that are almost instantaneous on the time scale of most experiments. Metallurgical systems can support non-equilibrium stress/strain distributions, that decay only slowly. Attention has been focused on the topological properties of cellular networks [109, 110, 111], including correlations between one cell and its neighbors [92, 110, 112]. In particular, much attention has been focused on the distribution of the number of vertices (N_0), edges (N_1), faces (N_2) etc. These numbers are related by the Euler equation [113]

$$\sum_{m=0,d} (-1)^m N_m = N_0 - N_1 + N_2 - N_3 \ldots = \Xi(d), \qquad (1.35)$$

where $\Xi(d)$ is a small integer called the Euler characteristic. For cellular structures embedded in 3-dimensional space, $\Xi(3) = 0$ (including the infinite cell surrounding a finite network) and $\Xi(2) = 2$. Equation 1.35 is of fundamental importance in the study of polyhedral networks. In equation 1.35, N_m is the number of m-dimensional components of a simply connected network (a network in a region without holes). In practice, most theoretical work has been concerned with "topologically stable" structures in which the number of edges converging at each vertex in a d-dimensional network is $d + 1$. Such cellular structures have been called froths [110]. The "stability" assumption implies that the angles between edges at a vertex, in a 2-dimensional froth, are all 120°, and, similarly, that the angles between contacting faces in a 3-dimensional froth are also 120°. However, experiments with 2-dimensional soap bubble networks indicate that significant deviations from this ideal geometry may occur [114]. The Euler equation can be used to show that, for an infinite 2-dimensional network, the average number of sides per polygon is six if three edges meet at each vertex (for 2-dimensional froths). No such simple relationship exists for $d = 3$ and higher dimensionality cellular systems, but equation 1.35 places important constraints on the composition $\{n_m\}$, where $n_{m,i}$ is the number of m-dimensional elements associated with the ith cell. For example, in the case of a 3-dimensional froth,

$$< n_f > = 12/(6- < n_e >), \tag{1.36}$$

where $< n_f >$ is the average number of faces per cell and $< n_e >$ is the average number of edges per face.

Maximum entropy ideas [109, 115] have been used in an attempt to develop a better understanding of the geometry of cellular systems. This approach is based on the idea that the entropy S of a system in which P_n is the probability of finding a cell with n sides is given by

$$S = -\sum_n P_n \ln P_n. \tag{1.37}$$

The problem of calculating the distribution P_n of the number of sides then reduces to the maximization of the "entropy" in equation 1.37, subject to the known constraints. These constraints arise from the requirements that the cellular pattern must be space filling, must satisfy the Euler equation (equation 1.35) and that the probabilities $\{P_n\}$ must be normalized ($\sum_n P_n = 1$). This can be accomplished using the "Lagrange multiplier" approach, which is important in statistical mechanics [116]. Despite the appeal of the maximum entropy idea, it remains controversial. Similar approaches, based on optimization concepts, have been used widely in the struggle to develop a coherent understanding of pattern-formation phenomena. In some cases, this approach appears to be well motivated, but it is not a general principle for pattern formation [117].

Although the structure of foams can be idealized in terms of polygons or polyhedra, the curvature of the interfaces between adjacent cells is also an important aspect of their geometry. The curvature is associated with a pressure differences and net diffusive transport across the interface. This is the main mechanism for the evolution of the foam geometry. The curvatures of the interfaces between adjacent cell must be taken into account, if the structure and dynamics are to be fully understood [118].

An interesting example of cellular structures, which illustrates the diversity of the systems in which this type of morphology can be found, is "magnetic froth" [119]. In magnetic froths, the majority of the material has a magnetization in one direction, corresponding to the cell interiors, while the rest of the material, corresponding to the cell walls, has the opposite orientation. Such structures have been observed in garnet films [120]. The cellular domains can form a regular hexagonal array, but a variety of disordered structures, including patterns that resemble those found in soap froths, can also be formed. Like soap froths, magnetic froths can coarsen or ripen. In this case, coarsening can be brought about by increasing the external magnetic field. The fraction of the material in the majority orientation increases, while the fraction in the cell walls decreases. While the mechanism of coarsening is different in the magnetic froths from that in the soap froths, the patterns evolve in a remarkably similar manner.

1.3.3 Spiral and Helix Patterns

Spiral and helix patterns are another example of a relatively simple class of structures that can arise in many simple and complex systems. Some important examples are spiral galaxies, screw dislocations [121] and a variety of biologically important macromolecules, including DNA. Spirals are found in many sea shells, and the spiral form is a recurring theme in the morphology of plants [122]. The formation of spiral patterns in "excitable media" that evolve by a combination of chemical reactions and diffusion is a common phenomenon. One of the most studied examples of spiral pattern-formation in reaction-diffusion systems is the Belousov–Zhabotinsky reaction [123, 124, 125, 126]. Many studies of the formation of spiral patterns and related forms can be found in the proceedings of a NATO workshop [127]. Quite striking spiral patterns have also been found in smectic membranes [128] and in convection experiments. A fine example is shown in figure 1.15. This figure shows a double spiral pattern obtained by Bodenschatz *et al.* [106], using CO_2 in a high aspect ratio circular convection cell. The pattern rotates clockwise with a period of $\approx 2400\ t_v$, where t_v is the vertical thermal diffusion time, defined as $t_v = b^2/\mathcal{D}_T$, where b is the gap between the top and bottom plates and $\mathcal{D}_T$ is the thermal diffusion coefficient. In general, the patterns consisted of an n-arm spiral surrounded by concentric rolls. The n spirals terminated in n dislocations and sometimes extended to the cell boundary. Bodenschatz *et al.* observed patterns with $n = 0$ (concentric rolls) to $n = 13$ arms.

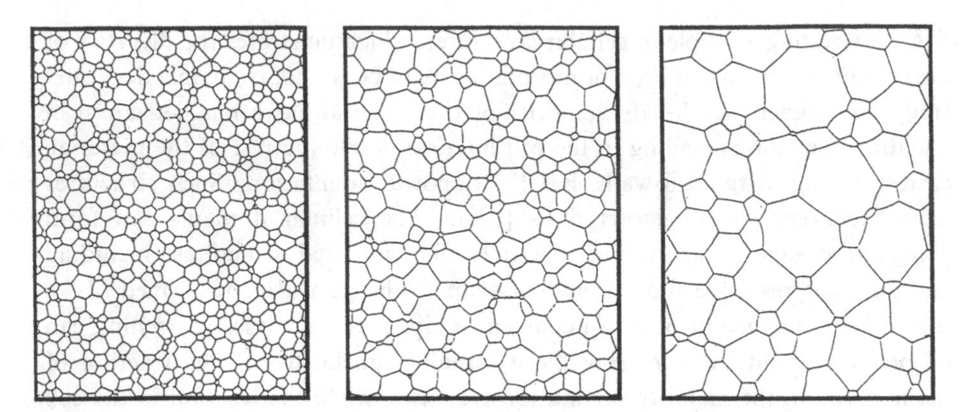

Figure 1.14 Three stages in the evolution of a 2-dimensional soap froth. The froth was confined to the narrow gap between two parallel sheets of glass and was allowed to drain to maintain the film properties during the experiment. This figure was provided by J. Stavans.

1.3.4 Labyrinthine Patterns

Many systems have a lamellar or almost lamellar structure on short length scales, but form a complicated maze-like or labyrinthine pattern on longer length scales. Such patterns are common; they are frequently seen in nematic liquid crystals and magnetic domain arrays [129]. Labyrinthine patterns are also generated by reaction-diffusion processes and by lipid monolayer domains in the 2-dimensional liquid-condensed/liquid-expanded coexistence range [130]. Figure 1.16 shows a labyrinthine pattern formed in a 6 μm thick garnet film. Labyrinthine patterns are also frequently found in polymer systems, such as block copolymers [131]. In this case, the system is close to a constrained equilibrium, and the characteristic distance between adjacent channels in the labyrinth is controlled directly through the molecular architecture. As is the case for many other pattern-forming systems, ordered cylindrical and spherical domains, lamellar morphologies and other patterns can also be observed under different conditions.

Labyrinthine patterns are found in Rayleigh–Bénard [132, 133] convection experiments, in which a thin layer of fluid is heated from below. Complex labyrinthine patterns are also formed when a magnetic field is applied to a thin layer of magnetic fluid, confined together with a non-magnetic fluid, in a thin gap between two vertical glass plates [134] (in a vertical Hele-Shaw cell). In a typical experiment, the dense magnetic fluid, consisting of magnetite particles in kerosene, for example, fills the bottom part of the vertical cell and the remainder is filled by an immiscible, non-magnetic fluid. The labyrinthine pattern grows at the fluid–fluid interface. The characteristic length scale of these labyrinthine patterns appears to be determined thermodynamically.

Figure 1.15 A shaddowgraph picture of a two-armed, spiral convection pattern. In this experiment, $\epsilon_R = R_a/R_c - 1 = 0.15$. This figure was obtained using the same cell as that used to generate the cellular pattern shown in figure 1.12. This figure was provided by E. Bodenschatz.

Figure 1.16 A labyrinthine pattern formed in a 6 μm thick garnet film. The film was mounted in a magnetic coil and observed, using crossed polarizers, via the Faraday magneto-optical effect. A digitized image from an area of approximately 1100 μm $\times$ 700 μm is shown. This figure was provided by A. T. Skjeltorp.

Other pattern-formation processes have been studied using magnetic fluids under different conditions. If a drop of magnetic fluid, surrounded by non-magnetic fluid in a Hele-Shaw cell, is subjected to a rotating magnetic field, the droplet deforms and eventually forms a branched pattern, as the magnetic field is increased [135]. Recently, Bacri *et al.* [136] studied the deformation of buoyant magnetic droplets in rotating magnetic fields. A variety of patterns was observed. In the absence of a magnetic field, the droplets were spherical. At high magnetic fields, a large number of arms that rotated with the magnetic field emerged from the droplet. The number of arms grew as $H_\circ^2$, where $H_\circ$

is the magnitude of the magnetic field, and appeared to be independent of the frequency.

1.3.5 Fluid Convection Patterns

In Rayleigh–Bénard convection experiments, a fluid is held between two horizontal, parallel plates separated by a distance b and held at different temperatures. Usually, the upper plate is held at a temperature of T and the lower plate at a temperature of $T + \Delta T$. As the temperature difference ΔT is increased, a series of increasingly complex convection patterns is formed. In practice, the exact form of these patterns depends on the boundary conditions and initial conditions. Typically, the most simple pattern in a square cell, which has the form of parallel rolls, appears when the Rayleigh number R_a exceeds a critical value R_c. The Rayleigh number is defined as

$$R_a = \frac{\alpha g \nabla T b^4 \rho}{\mathcal{D}_T \eta} = \frac{\alpha g \Delta T b^3 \rho}{\mathcal{D}_T \eta}, \tag{1.38}$$

where α is the thermal expansion coefficient, ∇T is the temperature gradient, g is the acceleration due to gravity, ρ is the fluid density, $\mathcal{D}_T$ is the thermal diffusion coefficient and η is the viscosity. When $R_a \ll 1$, the buoyancy forces are insufficient to make a hot bubble rise, or a cold bubble sink. Thermal diffusion cools (or heats) the bubble before it can rise (or fall) and the bubble does not move. Under these conditions, thermal diffusion is the main heat transfer process between the top and bottom plates. If $R_a \gg 1$, buoyancy forces move hot or cold bubbles before they can equilibrate with their surroundings. If the Rayleigh number is large enough ($R_a > R_c$), convective motion takes place and becomes the main heat transport process. The critical Rayleigh number is $R_c \approx 1707.76$, and the corresponding critical wave vector $q_\circ$ has a value of $q_\circ = 3.117$, which is very close to π, so that the roll width ($\lambda/2$, where λ is the wavelength) is close to the cell depth. At very high Rayleigh numbers, the fluid flow becomes completely chaotic. For a particular configuration, the system can be characterized by the reduced Rayleigh number $\epsilon_R = (R_a - R_c)/R_c$. The roll patterns formed at small reduced Rayleigh numbers can be described quite well, using Euclidean geometry. As ΔT or ∇T is increased, defects appear, and the number of defects increases with increasing ϵ_R. Similar defects are formed in a wide range of other systems, including dislocations in single crystals, ripple patterns formed by the flow of water or air over granular materials and sand dunes. Eventually, as ϵ_R is increased, the defects become too numerous and the cells too distorted for the concept of a simple roll-cell pattern with defects to be useful. At this stage, the patterns have a complex "labyrinthine" appearance. In some respects, the behavior observed in Rayleigh–Bénard experiments is similar to that found in the logistic map and the reduced Rayleigh number plays a role similar to the control parameter a in equation 1.27.

Rayleigh–Bénard convection experiments can be described in terms of the 3-dimensional fluid and heat transport equations. The flow of an incompressible fluid is described by the Navier–Stokes equation [137]

$$\rho(T)d\mathbf{V}/dt = -\nabla(P + \rho(T)gh) + \eta\nabla^2\mathbf{V} = -\nabla(\Phi) + \eta\nabla^2\mathbf{V}, \tag{1.39}$$

with the continuity equation

$$\nabla \cdot \mathbf{V} = 0, \tag{1.40}$$

where $\mathbf{V}$ is the velocity, g is the acceleration due to gravity, h is the height (in the direction of the gravitational field), P is the pressure, η is the viscosity and Φ is the hydraulic potential. It is usually assumed that $\rho(T)$, the density at temperature T, is given by $\rho(T) = \rho_\circ - \alpha(T - T_\circ)$. The Navier–Stokes equation must be coupled with the heat transport equation

$$\partial T/\partial t = \mathscr{D}_T\nabla^2 T - \mathbf{V} \cdot \nabla T. \tag{1.41}$$

In principle, equations 1.39, 1.40 and 1.41 should be solved, with the appropriate boundary conditions, to reproduce the formation of convection patterns.

The qualitative features observed in Rayleigh–Bénard experiments are reproduced by the 2-dimensional Swift–Hohenberg equation [138]

$$d\Psi/dt = \overline{a}\Psi - (\nabla^2 + q_\circ^2)^2\Psi - \Psi^3, \tag{1.42}$$

where $q_\circ$ is a characteristic wave vector and Ψ is a scalar field ($\Psi(\mathbf{x}, t)$). This equation is unstable with respect to the growth of perturbations with a wave vector $q_\circ$, and instabilities of the form $e^{i\mathbf{q}_\circ\mathbf{x}}$ will grow and eventually saturate at an amplitude of about $\overline{a}^{1/2}$. In this equation, the parameter $\overline{a}$ plays a role similar to that of the reduced Rayleigh number $\epsilon_R = (R_a - R_c)/R_c$. This equation is much easier to solve, both analytically and numerically, than the full Navier–Stokes/heat transport equation. In most cases, $\Psi(\mathbf{x}, t)$ is a real field, but complex fields and a variety of modified Swift–Hohenberg equations have been investigated. In some cases a noise term $\eta(\mathbf{x}, t)$ with the form $<\eta(\mathbf{x}, t)\eta(\mathbf{x}', t')>= 2\mathscr{D}\delta(\mathbf{x} - \mathbf{x}')\delta(t - t')$, like that used in the surface growth equations in chapter 5, is added to the right-hand side to represent the effects of thermal fluctuations and other noise sources. The Kuramoto–Sivashinsky equation [139, 140] (chapter 5, section 5.1.5) has also been used as a simplified model for a variety of hydrodynamic pattern-formation processes.

In the original Rayleigh–Bénard experiments [132], a layer of fluid with a free upper surface was heated from below. In this case, fluctuations in the surface tension Γ, resulting from fluctuations in the surface temperature, play an important role [141]. Gradients in the surface tension $\Gamma(T(\mathbf{x}))$ drive the fluid at the upper surface, and, in most cases, this "Marangoni" effect is dominant.

Under these conditions, the first pattern to be formed, as the temperature gradient is increased, is an array of "hexagonal" convection cells in which the fluid rises in the centers of the cells and falls near their margins. However, buoyancy-driven convection with a free surface also leads to hexagonal cells, so hexagonal cells are not a reliable indicator of Marangoni convection.

1.4 Moving-Boundary Processes

A wide range of pattern-formation processes can be described in terms of the evolution of a moving boundary that is controlled by transport processes on one or both sides of the boundary and the energetics of the boundary itself. In the most simple cases, the transport processes can be described in terms of the diffusion or Laplace equations. However, in many important processes, such as those involving the motion of liquid/solid phase boundaries, convective transport may play an important, if not dominant, role. In many cases, the growth of the interface is not controlled by the material that eventually forms the pattern, but by impurities that may accumulate near the advancing interface and modify the local equilibrium. The kinetics of growth at the interface also plays an important role in many pattern-formation processes. In these kinetically controlled processes, small quantities of impurities that become adsorbed in the growth front may completely change both the growth kinetics and the resulting morphology.

Convection can be suppressed by carrying out experiments in microgravity or by using gels and porous materials. The use of gels and porous media adds a variety of new effects, so that microgravity experiments are important in bringing together experimental studies and theoretical work on tractable models [142, 143]. However, convective transport may be important, even under reduced gravity (space shuttle) conditions [144, 145]. If free surfaces are present, convection can be driven by surface tension gradients, generated by concentration and/or temperature gradients. The onset of convection, forced by a temperature gradient induced surface tension gradient, depends on the Marangoni number

$$M = \frac{\partial \Gamma}{\partial T}(\nabla T L^2/\nu\rho\mathscr{D}_T) = \frac{\partial \Gamma}{\partial T}(\Delta T L/\nu\rho\mathscr{D}_T), \tag{1.43}$$

where ρ is the fluid density, $\nu = \eta/\rho$ is the kinematic viscosity, Γ is the interfacial energy per unit area, or surface tension, and L is a characteristic size for the liquid container. On the other hand, buoyancy-driven convection depends on the Rayleigh number, which scales as gL^4, for a fixed temperature gradient ∇T, where g is the acceleration due to gravity. Consequently, buoyancy-driven convection dominates at large g and large fluid volumes, while Marangoni convection dominates at small g and/or small container sizes. As Hurle [145] has pointed out, a reduction in g by a factor of 10^4 has only the same effect on the onset of buoyancy-driven convection as a ten-fold reduction in L, at the same temperature gradient.

1.4.1 Solidification

Considerable theoretical and experimental effort has been expended in an attempt to develop a better understanding of growth processes that are controlled primarily by diffusional transport and the energetics of the curved phase boundary. The most simple example of this class of moving-boundary problems is the growth of a solid shape, starting from a small "seed" that is large enough to grow in an "infinite" melt of the pure material that is initially undercooled to a temperature T_∞, everywhere. In many materials, the kinetics of molecular attachment to the growing surface is fast, and the advance of the interface into the undercooled liquid is controlled by the rate at which the latent heat of solidification can be transported from the surface to the surroundings. In such materials, the surface is microscopically rough, but this roughness does not play a direct role in the larger scale aspects of pattern-formation. Consequently, the growth of the liquid/solid phase boundary can be represented by a continuum equation of motion for the coarse-grained surface.

To keep things simple, it will be assumed that the molecular volumes, or partial molecular volumes in a mixture, are the same in both phases. This is an implicit assumption in most lattice models. In practice, unequal molecular volumes can lead to a variety of complications during the growth of phase boundaries.

Heat transport within the solid should be taken into account in detailed comparisons between experimental results and theory or computer simulations, but the essential features of solidification are captured by simplified models, called "one-sided models", that take into account only heat transport in the liquid phase, or "symmetric models" [146], that assume that the thermal diffusivity and heat capacity are the same in both phases. In general, the motion of the interface is given by

$$\mathbf{V}_n(\mathbf{x}_s) = -Q_T \mathscr{D}_T \nabla_n T(\mathbf{x}_s) + Q'_T \mathscr{D}'_T \nabla_n T'(\mathbf{x}_s) \tag{1.44}$$

or

$$\mathbf{V}_n(\mathbf{x}_s) = -Q_T [\mathscr{D}_T \nabla_n T(\mathbf{x}_s) - A\mathscr{D}'_T \nabla_n T'(\mathbf{x}_s)], \tag{1.45}$$

where $\mathbf{V}_n(\mathbf{x}_s)$ is the velocity of the interface at position $\mathbf{x}_s$ on the surface in a direction perpendicular to the surface, $\mathscr{D}_T$ and $\mathscr{D}'_T$ are the thermal diffusion coefficients in the liquid and solid phases, respectively, $T(\mathbf{x})$ is the temperature field in the liquid phase and $T'(\mathbf{x})$ is the temperature field in the solid phase. In equation 1.45, the symbol $\nabla_n \Phi(\mathbf{x}_s)$ stands for the gradient of the scalar field $\Phi(\mathbf{x})$ normal to the surface at position $\mathbf{x}_s$ on the surface. The gradient is measured in a direction pointing away from the surface, in the liquid phase, so that $\nabla_n \Phi(\mathbf{x}_s)$ is positive if $\Phi(\mathbf{x}_s)$ increases with increasing distance from the surface, in the liquid phase, and $\nabla_n \Phi(\mathbf{x}_s)$ is negative if $\Phi(\mathbf{x}_s)$ increases with increasing distance from the surface, in the solid phase. Calculation of the temperature distribution in the solid phase would add little to the difficulty of obtaining numerical solutions to the equations describing the interface motion, but would add considerably to the difficulty of an already difficult theoretical problem, without contributing much to basic understanding. Consequently, simplified models are often used in theoretical work. In the one-sided model, the growth of the interface is given by

$$\mathbf{V}_n(\mathbf{x}_s) = -Q_T \mathscr{D}_T \nabla_n T(\mathbf{x}_s), \tag{1.46}$$

and in the symmetric model by

$$\mathbf{V}_n(\mathbf{x}_s) = -Q_T \mathscr{D}_T [\nabla_n T(\mathbf{x}_s) - \nabla_n T'(\mathbf{x}_s)]. \tag{1.47}$$

The constant Q_T, in equations 1.45, 1.46 and 1.47, represents the dependence of the amount of solid phase formed on the amount of heat transported away from the interface, and is given by $Q_T = c_p/H_L$, where c_p is the heat capacity of the liquid and H_L is the latent heat of solidification per unit volume of solid.

In order to obtain a material-independent equation of motion and to facilitate comparison with other growth phenomena, it is convenient to describe the temperature distribution in the liquid in terms of the dimensionless scalar field

$$U_T(\mathbf{x}) = (T(\mathbf{x}) - T_\infty)Q_T = (T(\mathbf{x}) - T_\infty)c_p/H_L. \tag{1.48}$$

Equation 1.46 can then be written in the form

$$V_n = -\mathscr{D}_U \nabla U(\mathbf{x}_s), \tag{1.49}$$

and the diffusion of the dimensionless thermal field $U = U_T$ can be described by the diffusion equation

$$\partial U/\partial t = \mathscr{D}_U \nabla^2 U, \tag{1.50}$$

where $\mathscr{D}_U = \mathscr{D}_T$ is the thermal diffusion coefficient, in this case.

The model is completed by the boundary conditions for the scalar field $U(\mathbf{x})$. Far from the growing pattern, $T = T_\infty$ and $U_\infty = 0$. The temperature field on the liquid/solid boundary is usually obtained by assuming a local equilibrium between the liquid and solid phases,[12] so that $\mu_L(\mathbf{x}_s) = \mu_S(\mathbf{x}_s)$, where $\mu_L(\mathbf{x}_s)$ and $\mu_S(\mathbf{x}_s)$ are the chemical potentials in the liquid and solid phases at position $\mathbf{x}_s$

12. Although the local equilibrium assumption can be justified under slow growth conditions, the overall pattern-formation process remains a far-from-equilibrium process on longer length scales. As the growth velocity increases, the local equilibrium assumption may become inadequate and new phenomena may occur.

on the interface. The chemical potential μ_i, for the ith component or phase, is given by

$$\mu_i = \partial G/\partial n_i, \tag{1.51}$$

where G is the Gibbs free energy and n_i is the number of molecules of component i. For a curved interface, the chemical potential is increased as a consequence of the increased surface energy, and the equilibrium temperature at the curved interface T_s is given by the Gibbs–Thompson equation

$$T_s = T_m - (\Gamma/\Delta S)\mathcal{K} = T_m[1 - (\Gamma\mathcal{K}/H_L)], \tag{1.52}$$

where $\Delta S = H_L/T_m$, T_m is the equilibrium melting point for a flat liquid/solid interface, Γ is the interfacial energy and $\mathcal{K}$ is the curvature. The curvature is considered to be positive if the solid bulges into the liquid. In three dimensions $\mathcal{K} = (1/r_1) + (1/r_2)$, where r_1 and r_2 are the principal radii of curvature. If the temperature is expressed in terms of the dimensionless thermal field U_T, defined in equation 1.48, equation 1.52 can be expressed as

$$U(\mathbf{x}_s) = \Delta - (T_m c_p \Gamma \mathcal{K}(\mathbf{x}_s)/H_L^2), \tag{1.53}$$

where $\Delta = \Delta_T = (T_m - T_\infty)c_p/H_L$ is the dimensionless undercooling. Equations 1.49, 1.50 and 1.53 completely define the simplified solidification model. Although these equations are quite simple, and easily derived, the solution of these equations has provided a major theoretical challenge for several decades. Until quite recently, it was even impractical to obtain accurate numerical solutions for times that were long enough to explore the asymptotic, long growth time, morphology.

1.4.2 Growth from Solution

Similar equations of motion can be written for a wide variety of other growth processes. One of the most important classes is that of growth processes that are controlled by the diffusion of material rather than heat. In most practical situations of interest in material science, the diffusion of *both* heat and materials is important. Because thermal diffusion coefficients $\mathcal{D}_T$ are usually much larger than chemical diffusion coefficients $\mathcal{D}_C$, relatively small amounts of "impurity" can have a large impact on pattern-formation processes. Extremely pure materials are needed to study solidification of melts if unwanted effects due to impurities are to be avoided [147, 148]. A relatively simple example of a process that is controlled by material diffusion is the growth of a solid from a supersaturated solution with a uniform concentration of C_∞ at time $t = 0$. Experiments are often carried out in thin cells in which an essentially constant temperature can be maintained. Under these conditions, the growth of the interface is given by

$$\mathbf{V}_n(\mathbf{x}_s) = Q_C C(\mathbf{x}_s)\mathcal{M}_C \nabla_n \mu(\mathbf{x}_s) = Q_C \mathcal{D}_C \nabla_n C(\mathbf{x}_s), \quad (1.54)$$

where $\mathcal{M}_C$ is the chemical mobility and $\mu(\mathbf{x})$ is the chemical potential of the solute that forms the solid phase, at position $\mathbf{x}$ in the solution. If the concentration of the solute, in equilibrium with the growing solid, C_{eq} is small compared with the concentration C_s in the solid ($C_{eq} \ll C_s$), the material-dependent quantity Q_C is the molecular volume v_m or molar volume $\mathcal{V}_m$, depending on the units used for the concentration. More generally, $Q_C = 1/\Delta C$, where $\Delta C = C_s - C_{eq}$, and equation 1.50 can be used to describe the diffusion of solute, where the dimensionless scalar field U is given by $U(\mathbf{x}) = (C_\infty - C(\mathbf{x}))/\Delta C$ and equation 1.46 can be replaced by

$$V_n = -\mathcal{D}_C \nabla U(\mathbf{x}_s). \quad (1.55)$$

In general, the velocity calculated from the heat flow should be the same as that calculated from the chemical diffusion [149], so that

$$V_n = Q_C \mathcal{D}_C \nabla_n C(\mathbf{x}_s) = -Q_T \mathcal{D}_T \nabla_n T(\mathbf{x}_s), \quad (1.56)$$

or

$$V_n = -\mathcal{D}_C \nabla_n U_C(\mathbf{x}_s) = -\mathcal{D}_T \nabla_n U_T(\mathbf{x}_s), \quad (1.57)$$

in the one-sided approximation.

The chemical potential at the solid interface is given by $\mu_s = \mu_{eq} + \Gamma \mathcal{K} v_m$, where μ_{eq} is the chemical potential of a flat interface, under the same conditions. It follows, from the definition of the chemical potential given in equation 1.51, that the solute concentration at the interface $C(\mathbf{x}_s)$ is given by

$$C(\mathbf{x}_s) = C_{eq} + C_{eq}(\Gamma v_m / k_B T)\mathcal{K}(\mathbf{x}_s) = C_{eq} + C_{eq}(\Gamma \mathcal{V}_m / RT)\mathcal{K}(\mathbf{x}_s), \quad (1.58)$$

where C_{eq} is the concentration at equilibrium with a flat interface, v_m is the molecular volume, $\mathcal{V}_m$ is the molar volume, k_B is the Boltzmann constant and R is the gas constant. This equation can be written in terms of the dimensionless concentration field U as

$$U(\mathbf{x}_s) = \Delta - C_{eq}\Gamma \mathcal{K}(\mathbf{x}_s) v_m / \Delta C k_B T, \quad (1.59)$$

where $\Delta = \Delta_C = (C_\infty - C_{eq})/\Delta C$ is the dimensionless supersaturation.

Patterns generated by these and a wide variety of similar processes can be partially described in terms of characteristic lengths such as the thickness of branches, the distance between adjacent branches and the radii of curvature of branch tips. In many simple pattern-formation processes, all of these characteristic lengths are (approximately) equal [150] and may be replaced by a single "morphology length" ξ_m. The characteristic lengths that appear in pattern-formation are related to the characteristic lengths associated with the transport processes and interfacial phenomena that control the interfacial dynamics.

The identification of these characteristic lengths, and the way in which they depend on growth conditions, using experimental, theoretical and computer simulation approaches, have played an important role in the emerging understanding of the origins of some of the major pattern classes. For example, equations 1.53 and 1.59 can both be written in the form

$$U(\mathbf{x}_s) = \Delta - \xi_c \mathscr{K}(\mathbf{x}_s), \tag{1.60}$$

where the quantity ξ_c serves as a length scale, called the capillary length, which characterizes the properties of the interface under the growth conditions. A comparison of equations 1.53 and 1.60 shows that the thermal capillary length is given by

$$\xi_{cT} = (T_m c_p \Gamma / H_L^2). \tag{1.61}$$

The capillary length ξ_{cT} has a magnitude of the order of nanometers for typical materials. It is the length that corresponds to the curvature in equilibrium with the melt, at one unit of dimensionless undercooling. The chemical capillary length ξ_{cC}, obtained from equation 1.59, is given by

$$\xi_{cC} = (C_{eq} \Gamma v_m / \Delta C k_B T). \tag{1.62}$$

This corresponds to the equilibrium radius of curvature at one unit of supersaturation.

In cases where the kinetics of attachment to the surface is important, equation 1.60 can be replaced by

$$U(\mathbf{x}_s) = \Delta_T(\mathbf{x}_s) - \xi_c \mathscr{K}(\mathbf{x}_s) - k^{-1} V_n(\mathbf{x}_s), \tag{1.63}$$

where k^{-1} is a kinetic parameter.

A second type of characteristic length $w_{\mathscr{D}}$ describes the width of the hot boundary layer in the thermal transport limited processes, or the width of the chemically depleted boundary layer next to the interface in the mass transport limited case. These lengths are determined by the undercooling, or supersaturation, and the transport coefficients. For a planar interface moving at constant velocity V into a supersaturated solution or undercooled liquid, the steady-state diffusion equation can be written as

$$\mathscr{D}[\nabla^2 U - \nabla U / \xi_{\mathscr{D}}] = 0, \tag{1.64}$$

where $\xi_{\mathscr{D}}$ is the diffusion length. Since the velocity V of the interface is given by $V = -\mathscr{D}\nabla_n U$, and the gradient of the scalar field U in the boundary layer is given by $\nabla_n U \approx -\Delta / w_{\mathscr{D}}$, then $V \approx \mathscr{D}\Delta / w_{\mathscr{D}}$, or

$$w_{\mathscr{D}} \approx \mathscr{D}\Delta / V. \tag{1.65}$$

In the thermal case,

$$w_{\mathscr{D}_T} \approx (T_m - T_\infty)\mathscr{D}c_p/H_L V = \mathscr{D}_T \Delta_T / V, \tag{1.66}$$

and in the chemical diffusion case,

$$w_{\mathscr{D}_C} \approx (C_\infty - C_{eq})\mathscr{D}/\Delta C V = \mathscr{D}_C \Delta_C / V. \tag{1.67}$$

The diffusion length $\xi_{\mathscr{D}}$ is the width of the boundary layer at $\Delta = 1$ and is given by

$$\xi_{\mathscr{D}} \approx \mathscr{D}/V \approx w_{\mathscr{D}}/\Delta. \tag{1.68}$$

Very often, the diffusion length is defined as $\xi_{\mathscr{D}} = 2\mathscr{D}/V$.

1.4.3 Solidification of Impure Materials

Another important pattern-formation process is the isothermal growth of a solid from an impure liquid [146]. The growth of solid from an impure liquid in a thin cell can often be approximated, quite well, by a 2-dimensional, isothermal model. This process is very similar to the solidification of a pure material in an undercooled melt. In the growth of a solid from an impure liquid, the growth process is limited by the diffusion of impurity that is rejected by the growing solid, away from the interface, in much the same way that growth is limited by the diffusive transport of latent heat, away from the interface, in the solidification of a pure melt. The velocity of the interface is given, in the one-sided approximation, by [146]

$$\mathbf{V}_n(\mathbf{x}_s) = -QC(\mathbf{x}_s)\mathscr{M}_C(\nabla_n \mu(\mathbf{x})_s) = -Q\mathscr{D}_C(\nabla_n C(\mathbf{x})_s), \tag{1.69}$$

where $C(\mathbf{x})$ is the impurity concentration in the liquid phase, $\mu(\mathbf{x})$ is the impurity chemical potential, $\mathscr{M}$ is the chemical mobility of the impurity and $Q = 1/\Delta C$. In this process, ΔC plays the same role as the latent heat in the solidification of a pure, undercooled melt. Here, ΔC is the miscibility gap ($\Delta C = C_L - C_S$, where C_L is the impurity concentration in the liquid and C_S is the impurity concentration in the solid, in equilibrium with each other at the temperature of the interface). Equation 1.69 can be written as

$$\mathbf{V}_n(\mathbf{x}_s) = (\mathscr{D}_C \Delta C)\nabla_n \mu(\mathbf{x})\partial C(\mathbf{x})/\partial\mu = \mathscr{D}_C \nabla U, \tag{1.70}$$

where the dimensionless field U is given by

$$U(\mathbf{x}) = (\mu(\mathbf{x}) - \mu_{eq})/[\Delta C(\partial\mu/\partial C(\mathbf{x}))]. \tag{1.71}$$

The concentration of impurity required to melt the impure solid is smaller near to a convex solid surface than it is near to a flat surface, and the chemical potential of the impurity, in equilibrium with a curved interface, is given by

$$\mu = \mu_{eq} - (\Gamma\mathscr{K}/\Delta C), \tag{1.72}$$

so that

$$U(\mathbf{x}_s) = -\mathscr{K}\Gamma(1/\Delta C)^2/(\partial\mu/\partial C(\mathbf{x})); \tag{1.73}$$

the capillary length is given by

$$\xi_{cC} = \Gamma((1/\Delta C)^2/(\partial\mu/\partial C(\mathbf{x})); \tag{1.74}$$

and the diffusion length is given by

$$\xi_{\mathscr{D}C} = \mathscr{D}_C/V. \tag{1.75}$$

In general, the transport of both heat and material is important in the solidification of impure materials. In a typical experiment to study the basic aspects of this process, a thin cell is pulled at a constant velocity V from a hot region to a cold region. A cellular morphology, consisting of an array of parallel "fingers", is generated under a quite wide range of conditions. In this case, the melting point at the surface is reduced by both curvature and impurity concentration effects

$$T_s = T_m - (T_m\Gamma\mathscr{K}/H_L) + f(C_i), \tag{1.76}$$

where C_i is the impurity concentration in the liquid. The function $f(C_i)$ in equation 1.76 describes the shape of the liquidus line (the line that divides the liquid-phase region from the liquid-plus-solid region). For small impurity concentrations, the function $f(C_i)$, in equation 1.76, can be approximated by $f(C_i) = mC$, and

$$T_s = T_m - (T_m\Gamma\mathscr{K}/H_L) + mC, \tag{1.77}$$

where m is the slope of the liquidus line.

The impurity effects increase as either the pulling velocity V or the impurity concentration is increased and the sizes and shapes of the fingers change. Since the diffusion of both heat and impurity may be important in this class of pattern-formation processes, the theoretical difficulties are exacerbated. In most cases, it has been assumed that the temperature varies linearly between hot and cold regions. This is equivalent to assuming that the thermal diffusivities in the liquid and solid phases are large and equal, or that the latent heat is small and the thermal diffusivities are equal. The temperature is given by $T(h) = T_\circ + gh$, where h is the distance measured in the direction of the thermal gradient, in a coordinate system moving at the pulling velocity, and g is the magnitude of the gradient. Equilibrium at the interface implies that

$$T_s = T_m - (T_m\Gamma\mathscr{K}/H_L) + mC = T_\circ + gh. \tag{1.78}$$

1.4.4 Viscous Fingering

Another phenomenon that has been extensively studied and has contributed to the growing understanding of pattern-formation processes is the immiscible displacement of an incompressible viscous fluid by a very much less viscous, incompressible fluid in a Hele-Shaw cell in which the fluids are confined to a narrow gap between two transparent, rigid walls. The basic transport equation, in a 2-dimensional approximation, for this fluid–fluid displacement process is

$$\mathbf{V}(\mathbf{x}) = -(b^2/12\mu)\nabla P(\mathbf{x}) = -\mathcal{M}\nabla P(\mathbf{x}), \qquad (1.79)$$

where μ is the fluid viscosity, b is the width of the gap between the cell walls, $P(\mathbf{x})$ is the pressure field and $\mathbf{V}(\mathbf{x})$ is the fluid velocity at position $\mathbf{x}$ in the cell. In writing equation 1.79, it is assumed that the cell is horizontal and that $\mathbf{x}$ represents the position in the plane of the cell. In many experiments, the viscosity ratio ($m = \mu_2/\mu_1$, where μ_2 is the viscosity of the more viscous fluid) is very large, and it is usual to ignore the pressure drop in the less viscous displacing fluid.

The pressure difference δP across the fluid–fluid interface is given by $\delta P = -\mathcal{K}\Gamma$. The pressure in the viscous fluid at the interface is then given by

$$P_s = \Delta P[1 - (\mathcal{K}\Gamma/\Delta P)], \qquad (1.80)$$

where ΔP is the pressure difference across the viscous fluid. In some cases, it may be necessary to include a correction to equation 1.80 that accounts for the motion of the interface. Equation 1.80 then becomes

$$P_s = \Delta P - \mathcal{K}\Gamma + f(V). \qquad (1.81)$$

The function $f(V)$ includes a term of the form $k^{-1}V_n$, like that in equation 1.63, but the major term in $f(V)$ is proportional to $V_n^{2/3}$ and represents the effects of a thin wetting film on the plates of the Hele-Shaw cell [151].

Since the pressure field is described by the Laplace equation, there is no characteristic length associated with this field. However, near to the tip of an unconfined, advancing finger, the pressure difference $\delta P'$ between the tip of the finger and the side of the finger occurs over a distance that is approximately equal to $1/\mathcal{K}$, where $\mathcal{K}$ is the curvature of the interface. The pressure difference between the tip of a finger and the side of the same finger is given by $\delta P' \approx \Gamma\mathcal{K}$. Consequently, the pressure gradient near the tip of the finger can be estimated as $\nabla P \approx -\Gamma\mathcal{K}^2$, and the velocity with which the tip penetrates into the viscous fluid is given by

$$V = -\nabla P\mathcal{M} \approx \Gamma\mathcal{K}^2(b^2/12\mu), \qquad (1.82)$$

so that the characteristic length associated with the tip of the finger (the radius of curvature or finger width) is given by

$$r_\circ = \xi_m = \mathcal{K}^{-1} \approx \left[\frac{12\mu V}{\Gamma b^2}\right]^{-1/2} = (\Gamma/12\mu V)^{1/2} b, \tag{1.83}$$

where V is the velocity of the finger tip, with respect to that of the fluid, far from the tip.

If the finger is confined to a narrow cell, with a width w ($w \gg b$), then the width of the cell serves as a second length scale. If $w \gg \xi_m$, the finger is essentially unconfined and the morphological length is given by ξ_m. If, on the other hand, $w \ll \xi_m$, then it can be anticipated, in agreement with experiments, that the finger width will be cw, where c is a constant of order 1.

1.4.5 Pattern Selection

The growth of a solid from an undercooled melt was studied by Ivantsov [152, 153], assuming that the temperature inside and on the surface of the advancing pattern is the equilibrium melting temperature T_m for a flat interface. The Ivantsov equation is invariant to the rescaling transformation

$$\ell \to \lambda\ell \tag{1.84}$$

and

$$t \to \lambda^2 t, \tag{1.85}$$

where ℓ represents the spatial coordinates and t is time.[13] This means that if the function $f(\mathbf{x}, t)$ is used to describe the shape of the moving interface, $f(\mathbf{x}, t) = f(\lambda\mathbf{x}, \lambda^2 t)$. The rescaled velocity is $\lambda^{-1}V$, where V is the velocity before rescaling. Consequently, the Ivantsov equation leads to an infinite family of patterns that are also related by equations 1.84 and 1.85. These solutions to the Ivantsov model are a continuous series of parabolas, or elliptical paraboloids in three dimensions [154], satisfying the condition

$$r_{tip} V_{tip} = c(\Delta) \tag{1.86}$$

13. Ivantsov assumed that the motion of all of the constant temperature surfaces, including the solidification front, could be described by a Stefan equation of the form $V_n \sim -\nabla T$. This implies that the evolution of the temperature field can be described by an equation of the form $\partial T/\partial t = -f(T)|\nabla T|^2$. Because of this assumption, this approach determines only those solutions for which each isotherm moves with a normal velocity V_n, which is a function of the temperature and the normal temperature gradient only. If both sides of this equation are multiplied by $f(T)$, an equation of the form $\partial F(\mathbf{x}, t)/\partial t = -|\nabla F(\mathbf{x}, t)|^2$ is obtained, where $|F(\mathbf{x}, t)| = f(T)|\nabla T|$. This equation is invariant to the transformation given in equations 1.84 and 1.85. Although Ivantsov did not use the diffusion equation, this equation is also invariant to the same transformation. Consequently, the solutions to the full problem, with isothermal boundary conditions, can also be anticipated to be invariant to the same rescaling.

or

$$V_{tip} = (\mathscr{D}_T/r_{tip})c'(\Delta). \tag{1.87}$$

In equations 1.86 and 1.87, r_{tip} is the radius of curvature of the parabolic tip, V_{tip} is the velocity at which the tip advances into the liquid and $c(\Delta)$ is a constant that depends on the dimensionless undercooling Δ. Both the existence of an infinite family of solutions for each value of Δ, and the instability of these solutions to perturbations of any length scale, can be regarded as being consequences of the lack of a characteristic length scale, other than the diffusion length $\mathscr{D}/V$. The existence of a continuous family of solutions, containing thin parabolas at high velocities and fat ones at low velocities, for each value of Δ, raised the issue of selection among the infinite number of possible solutions [154]. Experiments, such as those of Glicksman *et al.* [147, 148, 155] are in sharp contrast to the scenario of a continuous family of solutions. They indicate a unique tip radius, velocity and side-branch spacing for each value of the undercooling Δ [147], irrespective of the initial shape of the "seed".

These and similar observations suggest the need for a "selection principle" to differentiate the infinite number of possible solutions and/or a different approach to this class of pattern-formation processes. The issue of selection principles has dominated theoretical work in this area for several decades. The realization that inclusion of surface energy effects provides a second length scale (the capillary length) was a key step towards the development of a better understanding of a wide range of pattern growth phenomena. This led to the development of the idea of microscopic solvability, based on the microscopic capillary length. Another major difference between theory and experiments is the existence of a much richer range of morphologies in experiments than is generated by models based on diffusional transport and isotropic surface energies. The gap between theory and experiments was substantially narrowed, and the level of understanding was enhanced by the introduction of anisotropy into the models.

1.4.6 Anisotropy and Growth Velocity

The growth of single-crystal dendrites has been the focus of much of the basic research on pattern-formation. Despite the manifest importance of anisotropy in the growth of six-armed snowflakes and other crystalline dendrites [1], it is only within the past decade that the crucial role played by anisotropy in the stabilization of the parabolic shape of dendrite tips and the "selection" of their radii of curvature has been fully appreciated. The anisotropic properties of the solid dendrites cannot be ignored, even in the most basic models for the shapes of dendrite tips. In particular, the anisotropy of the surface energy density (per unit area) $\Gamma(\Omega)$, or the anisotropy of the capillary length, is an important factor in determining the dendritic morphology. The effects of anisotropy appear in dramatic and sometimes unexpected ways in some of the simple growth models discussed in chapter 3. In the 2-dimensional case, the form of this anisotropy can be approximated by

$$\Gamma(\theta) = \Gamma_\circ[1 - \varepsilon_m \cos(m\theta)], \tag{1.88}$$

where ε_m is the strength of the anisotropy and θ is the angle between the local normal to the surface and a symmetry axis of the crystalline material. The kinetic parameter k^{-1} in equation 1.63 will have a similar, direction dependent form. If the surface energy density has the form given in equation 1.88, then the surface tension $\Gamma'(\theta)$ is given by

$$\Gamma'(\theta) = \Gamma(\theta) + (d^2\Gamma(\theta)/d\theta^2) = \Gamma_\circ[1 + (m^2 - 1)\varepsilon_m \cos(m\theta)]. \tag{1.89}$$

The relationship between the growth velocity V and the undercooling Δ is one of the central issues in the kinetics of pattern growth. The scaling form, given in equation 1.87, can be written in the form

$$V = (\mathscr{D}/\xi_m)f(\Delta), \tag{1.90}$$

where ξ_m is a length scale that is characteristic of the morphology. While the shape of the scaling function $f(\Delta)$ depends on the details of the growth process, the scaling form, given in equation 1.90 is quite universal.

Theoretical models and experimental studies indicate that the scaling function $f(\Delta)$ has a power law form for small values of Δ, so that

$$V \sim \Delta^n, \tag{1.91}$$

in the small Peclet number, large diffusion length, quasi-stationary limit. Equation 1.91 appears to work quite well for processes such as solidification in systems with microscopically rough surfaces, for which there are no barriers to molecular attachment to the surface, and impurities do not play a significant role. Glicksman *et al.* [147] measured a value of about 2.6 for the exponent n from 3-dimensional experiments on the free dendritic growth of high purity succinonitrile. The experimental data covered about a decade in the undercooling ($0.04 \leq \Delta_T \leq 0.3$, where Δ_T is the dimensionless undercooling) and about two to two-and-a-half orders of magnitude in the growth velocity. As the growth rate increases, the role of molecular attachment kinetics becomes more important, and the scaling relationships that characterize relatively slow growth break down.

Figure 1.17 shows the shapes of the dendrite tips for succinonitrile growing at six different undercoolings, or velocities [148]. This figure shows that the parabolic shape of the tip is independent of the undercooling, despite the wide range of growth tip velocities. However, the shape of the entire region

Figure 1.17 Succinonitrile dendrites growing into a pure melt at six different undercoolings. The magnifications have been selected so that the tip radii appear to be the same. This figure was provided by M. E. Glicksman.

is not completely independent of the tip velocity. At larger undercoolings, side-branching occurs nearer to the tip.

For the full, non-local, 2-dimensional dendritic growth problem [156, 157, 158], near the $\Delta \to 0$ limit, the velocity scales as

$$V \sim \mathscr{D}\Delta^4 \varepsilon^{7/4}/\xi_c \tag{1.92}$$

and the tip radius r_{tip} scales as

$$r_{tip} \sim \xi_c \Delta^{-2} \varepsilon^{-7/4}, \tag{1.93}$$

in the $\Delta \to 0$ limit where ε is the strength of the anisotropy defined in equation 1.88.

1.4.7 Laplacian Growth

An important quantity that characterizes the growth of dendritic patterns is the Peclet number Pe defined as

$$Pe = r_{\circ}V/2\mathscr{D} = \xi_m/\xi_{\mathscr{D}}, \tag{1.94}$$

where $r_{\circ} = \xi_m$ is the radius of curvature of the advancing tip. The Peclet number provides a measure of the deviation of the growth process from "stationarity". In the limit $Pe \ll 1$, the field $U(\mathbf{x})$ adiabatically follows the changing shape of the interface and the diffusion equation can be replaced by the Laplace equation. For large Peclet numbers, the field $U(\mathbf{x})$ becomes essentially constant outside of a boundary layer adjacent to the interface and changes rapidly within the boundary layer, near to the interface.

Growth models that require the solution of the Laplace equation on a moving boundary provide some of the most simple paradigms for pattern-formation. Such moving-boundary problems are frequently called "Stefan" problems. Despite their relative simplicity, there are few exact analytical results. The most simple Stefan problem is defined by the equations

$$\nabla^2\Phi = 0, \tag{1.95}$$

$$\Phi(\mathbf{x}_s) = 0, \tag{1.96}$$

$$\Phi(\infty) = 1 \tag{1.97}$$

and

$$V(\mathbf{x}_s) = \mathbf{n}(\mathbf{x}_s) \cdot \nabla\Phi(\mathbf{x}_s) = \nabla_n\Phi(\mathbf{x}_s), \tag{1.98}$$

where Φ is the scalar field that obeys the Laplace equation, $\mathbf{x}_s$ is a position on the moving boundary and $\mathbf{n}(\mathbf{x}_s)$ is the unit vector perpendicular to the moving boundary at $\mathbf{x}_s$. In 2-dimensional systems, it is often convenient to use the arc length s, measured along the interface, to distinguish points on the interface.

1.4.8 Instabilities

While the simple parabolic shape observed near the tips of growing dendrites has provided the focus for a large body of theoretical work, the manner in which these parabolic tips become unstable and evolve into dendritic patterns is of more interest, from a pattern-formation point of view. A first step towards understanding this process was provided by a linear analysis of the stability of the regular "Ivantsov" shapes with respect to small perturbations. The work of Mullins and Sekerka [149, 159] and Saffman and Taylor [160] on the effects of perturbations on the "regular" solutions to simplified solidification and fluid–fluid displacement models was an important step towards an understanding of the origins of more complex patterns. This indicates how a characteristic morphological length scale ξ_m can arise as a result of the competing effects of stabilization by interfacial energetics and destabilization by the proces(es) that drive the pattern-formation. These simple theoretical models have now been superseded. However, the general approach that they represent is still of value and often constitutes the first step towards a theoretical understanding of a very wide range of pattern-formation processes. An outline of this pioneering work is given in appendix A.

1.4.9 Characteristic Lengths

Linear analysis leads to the idea that the conditions under which instability will take place and the characteristic wavelength, or length scale, of the instability can be understood qualitatively in terms of more fundamental characteristic lengths such as the capillary and diffusion lengths, which provide natural length scales for the growth process. This provides a rough indication of the characteristic morphological length(s) in the patterns that grow as a result of these instabilities. Linear stability theory leads to stability functions $\mathscr{S}(\omega, \{\xi\})$ or $\mathscr{S}'(\xi_m, \{\xi\})$ (like that derived by Mullins and Sekerka [149] for perturbations of the form $\delta \sin(\omega x)$ on a flat growing interface, where x is the distance measured in the direction along the growing interface; see appendix A), where $\{\xi\}$ indicates the set of all relevant length scales. The stability functions are usually defined so that the interface is stable if $\mathscr{S}'(\xi_m, \{\xi\}) < 0$ and unstable if $\mathscr{S}'(\xi_m, \{\xi\}) > 0$. In general, these stability functions are quite complex. However, they often depend on only a few characteristic lengths near to morphology transitions, and the location of these transitions can often be expressed in terms of simple stability parameters, like those given in equations A.6 and A.8, in appendix A. In a similar manner, the characteristic wavelength (the wavelength of the most unstable mode) can be expressed in similar terms. For example, in dendritic solidification the characteristic length scale ξ_m of the morphology (which is represented by the tip radius of curvature, the width of the stem and side-branches, the separation between side-branches ...) is given by

$$\xi_m \approx (\xi_{\mathscr{D}} \xi_c)^{1/2}. \tag{1.99}$$

Under many circumstances, the length ξ_m that characterizes the morphology is related to characteristic lengths such as ξ_c, $\xi_{\mathscr{D}}$ and ξ_T by

$$\xi_m \approx \xi_{\mathscr{D}}^a \xi_c^b \xi_T^c, \tag{1.100}$$

with $a + b + c = 1$. When the morphological length ξ_m is determined by just two characteristic lengths, ξ_1 and ξ_2, associated with the pattern-formation mechanism, the morphological length ξ_m is often the geometric mean $\xi_m \approx \xi_1^{1/2} \xi_2^{1/2}$.

1.4.10 Beyond Linear-Stability Analysis

Phenomena such as the Mullins–Sekerka and Saffman–Taylor instabilities are often referred to "linear" instabilities since their onset can be demonstrated by a simple linear analysis of the growth or decay of shape perturbations. To develop a better understanding of pattern-formation processes, it is necessary to go beyond the linear theory typified by the Mullins–Sekerka and Saffman–Taylor analyses. This can be accomplished by a more detailed non-linear analysis or by resorting to computer simulations. In general, the process can be characterized by a control parameter a that describes the strength of the perturbation or driving force that generates the instability. Often, the linear analysis indicates that there is a critical value a_c for a, beyond which the instability will grow. However, the linear analysis does not indicate how the resulting pattern will evolve or if it will grow to a non-zero amplitude. In other systems, the linear analysis is completely inadequate. Sometimes, non-linearity may lead to the development of instabilities at control parameters smaller than the value of a_c obtained from a linear stability analysis for perturbations with a non-zero amplitude. In other cases, a linear stability analysis may indicate that the system is stable with respect to pattern-formation at all values of a, but the system may become unstable if the amplitude A of the shape perturbation exceeds a critical value $A_c(a)$. Guo *et al.* [161] have shown that $A_c(a) \to 0$ as $a \to \infty$ for a Saffman–Taylor-like process in which the surface tension Γ has the form $\Gamma = \Gamma_0 + c\mathcal{K}$, where $\mathcal{K}$ is the local curvature of the fluid–fluid interface and c is a constant that serves as the control parameter a.[14] Guo *et al.* [161] studied the process in which the displacing fluid is more viscous than the displaced fluid. Without the curvature dependent surface tension, this would be a stable process. If c is negative, the surface tension is reduced when the interface bulges into the displaced fluid, and the stability of the viscous finger is reduced. The curvature dependence of the surface tension generates a term of the form $-c\mathcal{K}^2$ ($\delta P = -[(\Gamma_\circ + c\mathcal{K}]\mathcal{K}$) in the pressure difference δP across the interface that has no effect on the perturbative linear stability analysis. The threshold $A_c(c)$ is always non-zero for this process and $A_c \to 0$ as $c \to -\infty$. The instability that would not be revealed by a linear stability analysis would, in fact, take place in an experimental realization if $A_c(c)$ ($A_c(a)$) became small relative to the fluctuations and disorder that are present in all real systems.

14. Guo *et al.* suggested that the surface tension might depend on the curvature, in this manner, for a fluid–fluid interface in the presence of a polymeric surfactant.

1.5 Solution of Interface Equations of Motion

The deceptively simple and easily derived equations of motion for phase boundaries, described above, have provided a major theoretical challenge for several decades. Difficulties arise because of the non-local and non-stationary nature of these moving-boundary problems. Exact analytical solutions are very difficult to obtain. An alternative approach is to develop growth algorithms, based on a more microscopic understanding of the physics of non-equilibrium growth. In some cases, the algorithms may be considered to be atomistic representations of the growth process. The diffusional part of these discrete algorithms is represented by particles that follow random walk paths. The randomly walking particles in the algorithm may represent one or more particles in a real growth process or may be considered to simulate the distribution of the scalar fields, as is the case in most physical realizations of the diffusion-limited aggregation model, described in chapter 4.

More recently, hybrid algorithms [74, 162, 163] have been developed in which a discretized version of the diffusion equation is solved on the lattice, without using random walkers. In these algorithms, the dynamics of the interface due to processes such as freezing and melting are represented by a Monte Carlo algorithm with local rules, which are dictated by the physical principles such as conservation of mass and detailed balance, to represent local equilibrium at the interface, which controls the evolution of the interface.

1.5.1 Numerical Solution of the Non-Local Equations

The numerical solution of non-local moving-boundary problems with retardation (slow relaxation of the field $U(\mathbf{x})$) requires the full capabilities of modern computers and numerical methods. It was only since the late 1980s, and as a result of considerable effort, that reliable numerical solutions to 2-dimensional pattern growth equations appeared [164]. It is still difficult to carry out simulations covering a sufficiently large range of time and length scales to allow questions associated with the asymptotic structure and kinetics to be addressed. The demands on computer resources are reduced, but are still quite substantial, in the low Peclet number limit, in which the diffusion equation can be replaced by the Laplace equation and may be solved numerically by using the Laplacian Green's function to calculate the Laplacian field gradient at the interface [165, 166, 167, 168]. There are still very few results from 3-dimensional calculations.

A good example of the results of this endeavor is provided by the work of Jasnow and Viñals [167] on the numerical solution of equations 1.50, 1.55 and 1.60, for the evolution of an initially straight 1-dimensional interface in a 2-dimensional system, with a constant flux boundary condition far from

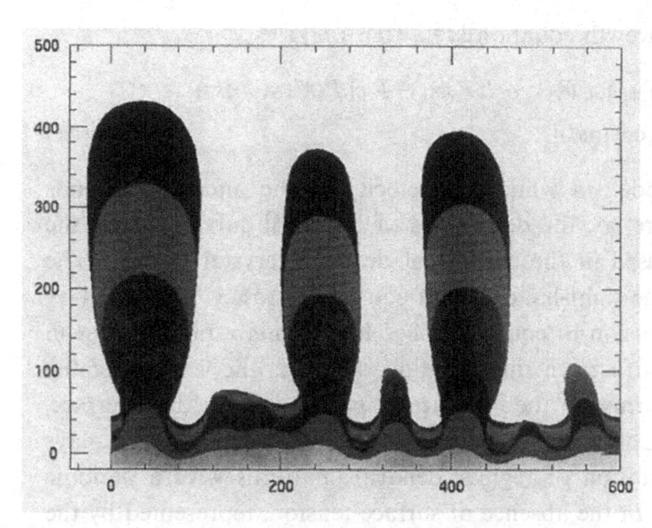

Figure 1.18 Development of the interface during a simulation of surface growth, using a one-sided model. This model includes surface tension but no anisotropy. The patterns are shown at times of 0, 1000, 2000, 4000, 6000 and 8000. This figure was provided by J. Viñals.

the advancing interface. In this numerical study, the diffusion equation was replaced by the Laplace equation. This quasi-stationary approximation neglects the effects of retardation due to finite diffusion constants. Figure 1.18 shows the results of a typical simulation, carried out using isotropic surface tension boundary conditions, characterized by the capillary length $\xi_c = 1.0$ at the growing interface and a constant flux (constant growth velocity) at the distant boundary, corresponding to a diffusion length of $\xi_{\mathscr{D}} = 40$. A linear combination of the unstable modes ($q = 2\pi n/L$), with $L = 600$, was used as the initial condition for the interface, where L is the width of the system. In most simulations, $n = 1 - 11$. The simulations [167, 168] indicated that there is a single morphological length scale, which grows linearly with increasing time, so that the patterns at times t and bt are related (statistically) to each other, by a change of length scale by a factor of b. This pattern-formation process can also be studied using Monte Carlo simulations [169]. Both approaches give similar results. However, it is difficult to reduce the noise [170, 171] in Monte Carlo simulations to realistic levels, and a numerical solution of the integral equation for the normal velocities [167] is a more practical approach to this class of problems.

1.5.2 Local Models

Although the non-local aspects of transport-limited pattern growth phenomena must be taken into account to describe fully most growth processes, local models have played an important role in the development of a better understanding of pattern-formation. These relatively simple models can be solved numerically and have been studied theoretically, leading to the development of a new shape selection principle.

For example, the local growth equation [172, 173, 174]

$$V_n(\mathbf{x}_s) = [\mathcal{K}(\mathbf{x}_s) + a\mathcal{K}(\mathbf{x}_s)^2 - b\mathcal{K}(\mathbf{x}_s)^3 + c(d^2\mathcal{K}(\mathbf{x}_s)/ds^2)]$$
$$[1 + \varepsilon_m \cos(m\theta)], \tag{1.101}$$

called the geometrical model, in which the velocity of the interface depends only on the local curvature $\mathcal{K}$, the derivatives of the local curvature and the anisotropy ε_m, has been used in the context of dendritic crystal growth. The "$d^2\mathcal{K}(\mathbf{x}_s)/ds^2$" term, on the right-hand side of equation 1.101, is motivated by the form of the surface tension in equation 1.89. It prevents rapid changes in the curvature along the surface. In this equation, $V_n(\mathbf{x}_s)$ is the velocity of the interface, in a direction normal to the interface at position $\mathbf{x}_s$ on the interface, and $s = s(\mathbf{x}_s)$ is the arc length at $\mathbf{x}_s$. In this model, the anisotropy ε_m plays a crucial role in the generation of regular dendritic patterns with a periodic emission of side-branches. In the absence of surface tension, represented by the $d^2\mathcal{K}(\mathbf{x})/ds^2$ term in equation 1.101, a continuous family of solutions, analogous to the continuous family of solutions for the Ivantsov problem, was found. If $c \neq 0$ in equation 1.101, a discrete family of solutions, each with its own velocity, was obtained. A finite anisotropy ε_m^* is required to stabilize the regular dendritic solution. This critical anisotropy is a function of the model parameters a, b, c and m. If the anisotropy parameter ε_m is smaller than ε_m^*, the pattern evolves via a tip-splitting process.

Similar conclusions were reached using the more realistic boundary layer model [175, 176]. In the boundary layer model, it is assumed that the characteristic decay length of the diffusion field is much shorter than the radius of curvature. This model includes diffusive transport along the moving boundary and is, in this sense, a non-local model. However, it does not fully represent the long range memory effects described by the full diffusion equation. In the boundary layer model, it is assumed that the dimensionless temperature $U_T(\mathbf{x}_s)$ at the solid surface is given by equation 1.63 with a direction dependent kinetic parameter and capillary number. The growth velocity V_n is given by

$$V_n = (\nabla_n U)_s \mathcal{D}_T \approx \mathcal{D}_T U_s/w, \tag{1.102}$$

where w is the width of the boundary layer. The transport of heat in the boundary layer is expressed in terms of the heat content per unit length of interface

$$h = U_s w, \tag{1.103}$$

which is given by the heat balance equation

$$dh/dt = V_n(1 - U_s) + \mathcal{D}_T \frac{\partial}{\partial s} w \frac{\partial U_s}{\partial s} - \mathcal{K} V_n h, \tag{1.104}$$

where s is the arc length coordinate of a point on the surface. The first term on the right-hand side of equation 1.104 represents the latent heat released by solidification, the second term describes diffusion along the boundary layer and the third term represents the effects of stretching the boundary layer due to growth. Equations 1.102 to 1.104 fully describe the boundary layer model. In addition, the rate of evolution of the curvature and arc length coordinates are given by

$$d\mathscr{K}/dt = -(\mathscr{K}^2 + \partial^2/\partial s^2)V_n \tag{1.105}$$

and

$$ds/dt = \int_0^s \mathscr{K} V_n ds'. \tag{1.106}$$

In this model, the boundary layer thickness or heat content is large at concave parts of the surface and small near the convex tips. Consequently, the boundary layer approximation should work best near to the most exposed tips and will fail completely if the boundary layer thickness approaches the separation between side-branches, in the interior.

Simulations carried out using the geometrical and boundary layer models indicated that surface tension alone is not sufficient to select a parabolic, needle-like solution of the surface growth equations. Instead, a complex pattern arises via tip-splitting dynamics. The lowest temperature on the surface of a growing dendrite would be found at the tip, since the curvature is highest at the tip and a lower temperature is required to bring the solid and liquid phases to a local equilibrium. Because of the variation of the temperature along the interface, heat will flow from the regions near the interface, behind the tip, towards the tip, reducing the rate of growth at the tip. This process leads to a fattening and eventually splitting of the dendrite tips. It is represented in both the geometrical and boundary layer models, but appears more directly and realistically in the boundary layer model. This process will take place, however small the magnitude of the surface tension. The surface tension is said to be a singular perturbation, because a small surface tension completely changes the solutions to the equation of motion for the interface. Anisotropic surface tension, which reduces the surface tension at the tip of the growing dendrite relative to parts of the interface that are behind the tip, moves the lowest temperature on the interface away from the tip. This stabilizes the tip and induces side-branching.

The boundary layer and geometrical models led to the idea that the selection of the velocity and width of needle shaped dendrite tips are determined by a solvability criterion [177]. The solutions of the equations of motion are required to have reflection symmetry about the tip. Consequently, the first derivatives of the curvature, the normal velocity and other quantities in the models with respect to the arc length coordinate are required to be zero at the dendrite tip.

These derivatives are also zero far from the tip. The requirement that the tip of the dendrite must be smooth, even in the presence of anisotropy, provides a basis for selection of the solutions for the interface growth equations. These ideas have more recently been extended to the full problem of 2-dimensional dendritic growth, with long range diffusive transport, and to 3-dimensional dendritic growth. The "microscopic solvability" criterion appears to provide a good understanding of the behavior of a single dendrite tip. However, much remains to be done in order to obtain a similar level of understanding of the growth of more complex patterns that involve the simultaneous growth of many competing branches.

In many non-equilibrium growth processes, there is a physical boundary layer, and the boundary layer model can be expected to work well when the width of this boundary layer is much smaller than the radius of curvature of the interface. For systems that satisfy this condition, the boundary layer model provides a realistic description of the growth of dendritic patterns, well into the non-linear regime. However, the failure of this model to take the non-local spatial aspects of dendritic growth fully into account leads to disastrous long time behavior. At the early stages in a simulation, quite realistic patterns develop, and one or more generations of branching may take place. The problems arise because two points on the surface, which may be far apart in the arc length coordinate system in which distances are measured along the interface in the models, may be separated by only a small distance in the 2-dimensional embedding space. Consequently, this model allows more than one branch to occupy the same space. Similar problems are associated with the geometrical model, which, in addition, neglects all memory (retardation) effects associated with the diffusion process. A similar problem can be anticipated with 3-dimensional versions of these models.

In solidification processes, a thin boundary layer is formed when the dimensionless undercooling Δ_T is large. If $\Delta_T = 1$, the latent heat released by the advance of the solidification front is just sufficient to maintain the boundary layer. Under these conditions, the interface dynamics will be described by the boundary layer model, except when two parts of the surface that are separated by a large arc length ($\gg \xi_m$) become separated by a distance of $\approx \xi_m$ in the embedding space. The growth of these parts of the interface will then become substantially slower, and this process may lead to a densely branched morphology. In densely branched morphologies (chapter 4, section 4.2) the pattern is composed of closely spaced dendrites that fill a fraction ϕ_b of the space inside a more or less circular ($d = 2$) or spherical ($d = 3$) envelope that advances with a constant velocity V_e. On length scales larger than the widths of individual branches, the internal structure is uniform. If $\Delta_T < 1$, the liquid inside the envelope cannot absorb all of the latent heat that would be released by solidification of the liquid inside the envelope, but the latent heat can be absorbed by cooling the liquid inside the envelope if $\phi_b = \Delta_T$. A decrease in Δ_T would increase the width of the boundary layer and decrease the envelope velocity V_e. If $\Delta_T > 1$, the undercooled liquid can absorb all of the latent heat locally. The solidification process will lead to the growth of a compact solid with constant interface velocity that is limited only by local interface kinetics. Similarly, in material transport-limited growth of densely branched structures from supersaturated solutions, $\phi_b = \Delta_C$.

Figure 1.19 shows a morphological phase diagram for processes that can be described by equations 1.46, 1.50 and 1.60 with

$$\xi_c(\theta) = \xi_{c_\circ}(1 - \varepsilon_4 \cos(4\theta)), \qquad (1.107)$$

where the surface tension anisotropy has been expressed in terms of an anisotropic capillary length. This morphological phase diagram was constructed by Brener *et al.* [157], using results from a number of experiments and simulations. They identified four distinct morphologies, which they called "compact dendritic", "compact seaweed", "fractal dendritic" and "fractal seaweed" patterns. Fractal scaling can only extend over an unlimited range of length scales, in the $\Delta \to 0$ limit. Figure 1.20 shows an example of a "fractal seaweed" pattern from simulations carried out by Ihle and Müller-Krumbhaar [164]. A fractal dimensionality (chapter 2) of $D \approx 1.7$ was measured for the "fractal seaweed" patterns, using a box counting analysis and from the dependence of the perimeter length on the overall radius of the pattern. This value for D is similar to the value of $D \approx 1.715$ found for the diffusion-limited aggregation (DLA) model, described in chapter 3.

1.6 Complex and Disorderly Patterns

In nature, complex and/or disorderly patterns are the most common. Familiar examples include breaking waves on a beach, the internal structure of the human body and natural landscapes. A few decades ago, the task of describing such patterns in quantitative terms, let alone developing an understanding of how they were formed, appeared to be overwhelming. Today, these problems seem to be less formidable, though the road still appears to be very long and many interesting problems will remain unsolved for the foreseeable future. The development of high speed computers and high resolution graphics has played an important role in this increased optimism. Mathematicians have appreciated, for a long time, that simple equations, like 1.27, can have complicated solutions. However, it is only within the last one to two decades that this has been appreciated by a broad range of scientists. This simple idea calls for a re-examination of what is meant when a pattern is described as "complex". It becomes natural to ask questions such as "Is a

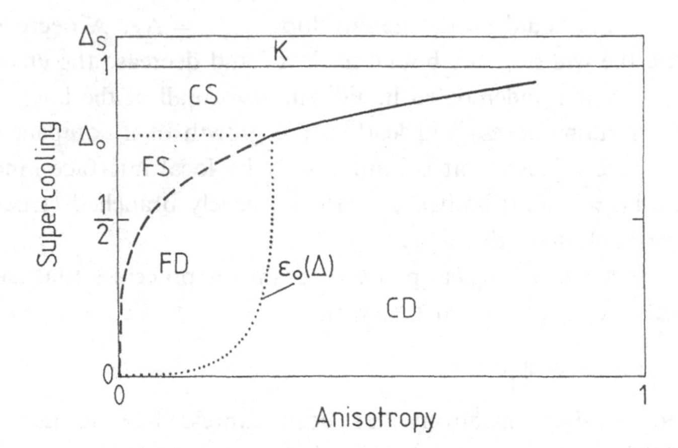

Figure 1.19 A schematic morphology diagram for crystal growth into an undercooled melt. The conditions under which compact dendritic, compact seaweed, fractal dendritic, fractal seaweed and dense compact patterns are generated are indicated by the symbols CD, CS, FD, FS and K, respectively. The transition between the FD and FS morphologies, in the region surrounded by the dotted lines, is sensitive to noise. The anisotropy is specified by the parameter ϵ_4 in equation 1.107. This figure was provided by H. Müller-Krumbhaar.

pattern that can be generated by a simple equation or algorithm complex?" and "Does it make a difference whether or not the equation or algorithm is known?".

The stylized paintings of breaking waves by Hokusai Katsushika (1760–1849) have often been used as a qualitative illustration of complex fractal forms. More recently, Cartmill and Su [178] measured the size distribution of micro-bubbles, with diameters in the range 34 – 1200 μm, formed by breaking waves in large laboratory tanks. They found that the distribution of bubble diameters ℓ could be represented by a power law probability density of the form

$$P(\ell) \sim \ell^{-\tau_\ell}, \qquad (1.108)$$

where $P(\ell)\delta\ell$ is the probability that a randomly selected bubble will have a diameter in the range $\ell \rightarrow \ell + \delta\ell$. A value of about 3 was found for the exponent τ_ℓ over a range of just over one decade in ℓ. This corresponds to a power law probability density for the size s (volume) with an exponent τ_s of about 5/3, over more than three decades in the bubble volumes. It is not known if this power law distribution of bubble sizes can be related to the structure of the breaking wave. This might be an interesting direction for further research.

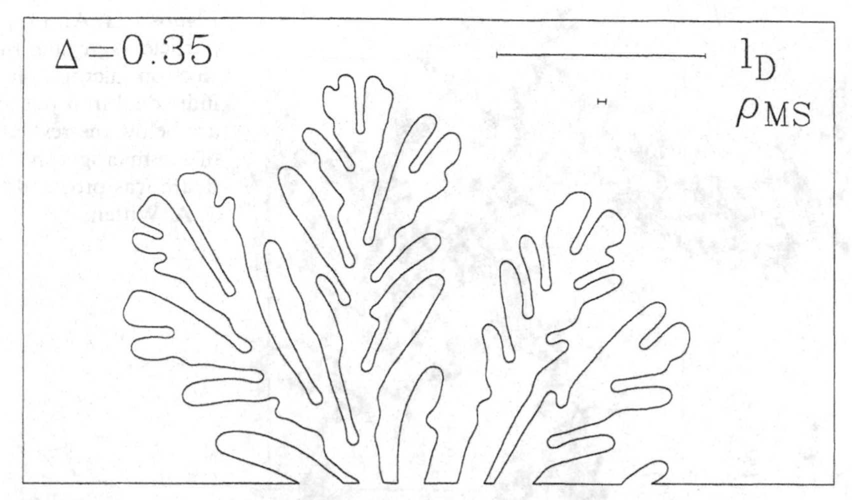

Figure 1.20 Fractal seaweed patterns obtained from numerical solution of equations 1.46, 1.50 and 1.60. These patterns were obtained in the $\varepsilon_4 = 0$ (zero anisotropy) limit. The scale labeled l_D indicates the diffusion length; the scale labeled ρ_{MS} is the Mullins–Sekerka length (appendix A: an estimate of the morphology length ξ_m obtained from a linear stability analysis); and Δ is the supercooling. This figure was provided by H. Müller-Krumbhaar.

1.6.1 Aggregates

The process of aggregation (the self-assembly of small particles to form larger structures) has played an important role in our growing understanding and appreciation of fractal geometry. The low density structure of large aggregates of small particles found in systems such as smoke and colloids has been described for many years, using terms such as "tenuous", "branched" "filamentous" and "fluffy" [179].

Figure 1.21 shows an electron micrograph of an iron particle aggregate that was studied by Forrest and Witten [180]. Because this aggregate is so tenuous, its projection does not cover a plane. This means that essentially all of the structure can be seen in a projection and its fractal scaling properties can be completely characterized, using a digitized image of the projection seen in the electron micrograph. In this manner, Forrest and Witten were able to demonstrate that the aggregates, formed under these conditions, could be described in terms of fractal geometry. The aggregates are self-similar fractals with a fractal dimensionality of 1.7 to 1.9. This was one of the first applications of fractal geometry to an experiment in condensed matter physics and materials science, other than problems in polymer physics, which were generally discussed in terms of a different but closely related language [4]. This work was important because it stimulated the later development of the diffusion-limited aggregation, or DLA, model by Witten and Sander [181], and it was the first demonstration of the fractal nature of a real aggregate.

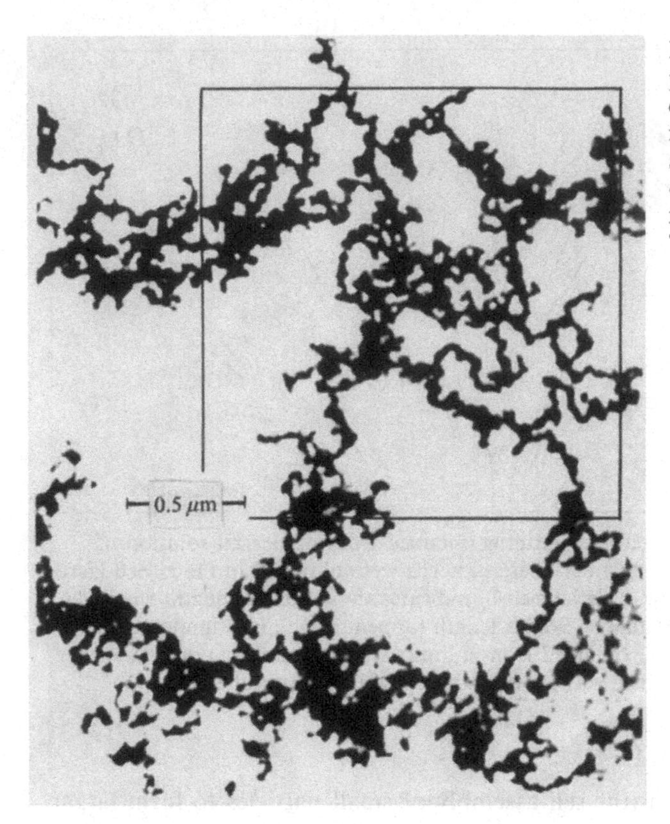

Figure 1.21 An iron particle aggregate. In this electron micrograph, the individual iron particles are below the resolution of the micrograph. This figure was provided by T. A. Witten.

It has turned out that structures such as that shown in figure 1.21 are better explained in terms of cluster–cluster aggregation models [58, 182, 183] than particle–cluster aggregation models such as the DLA model (chapter 3). Because of the hierarchical nature of cluster–cluster aggregation phenomena, it is less surprising that they lead to fractal structures. Cluster–cluster aggregation is now one of the most firmly established and extensively studied of all processes leading to the formation of fractals.

1.6.2 Polymers

Scaling ideas have been used in the study of polymers for many years, and polymer molecules provide some of the most important examples of real fractal structures. This is a well developed area with an extensive literature. The simplest models for linear polymers – the random walk model for polymer melts and the self-avoiding walk model for polymers in good solvents – are among the most extensively studied and best understood examples of statistically self-similar fractal scaling (chapter 2). In addition, branched polymers and gels may have a hierarchical topology, which can also be expressed in terms of fractal scaling exponents such as the fracton or spectral dimensionality D_s. The spectral dimensionality is an example of an "intrinsic" property that characterizes the connectivity and does *not* depend on the way in which the polymer is arranged in space. Polymers with a hierarchy of flexible branches or loops that can be described by a fracton dimensionality D_s are call "polymeric fractals" [184]. The hierarchical topology of polymeric fractals has a profound influence on both their structure and dynamics [184]. Some of the models discussed in chapter 3, such as the lattice animal model and the percolation model, have been used with considerable success to obtain a better understanding of branched polymers and gels [185].

The techniques and concepts developed in a polymer science context have been used with considerable success to investigate other fractals, and the topics discussed in this book owe much to earlier and contemporary developments in polymer physics. Similarly, fractal and scaling concepts developed outside of polymer science have contributed much to its development. This book is primarily concerned with applications in other areas. Many excellent books and reviews have been published on the application of fractals and scaling to polymer systems. These include [4, 185, 186].

1.7 Scaling Symmetry

Much of this book is concerned with scale symmetry (the manner in which systems change when their length scales or magnifications are changed). As is the case with more familiar symmetries, self-similarity, and other scaling symmetries, can be used to obtain answers to "difficult" problems without the need to develop a deep understanding of the underlying mechanisms. Results obtained in this manner are generally more reliable than those obtained by employing more detailed physical arguments, and can often be obtained with relatively little effort, using a minimum of mathematics. Results obtained in this manner can provide a valuable contribution to the development of a more complete understanding. An important aspect of scaling symmetries is that the transformations must obey the relationships that are characteristic of other groups. For example, if an object is dilated by a factor of λ_1 and then dilated again by a factor of λ_2, the result must be the same as a dilation by $\lambda = \lambda_1\lambda_2$. It is also apparent that the order of the dilations does not change the effect. These properties can be expressed as

$$\mathcal{T}\lambda = \mathcal{T}\lambda_1\mathcal{T}\lambda_2 = \mathcal{T}\lambda_2\mathcal{T}\lambda_1, \quad (1.109)$$

where $\mathcal{T}\lambda$ indicates magnification of all distances by λ. Self-affine transformations, in which the magnification is different in different directions, obey similar group properties.

1.8 Notation

There is no standard notation for fractal geometry and its applications. Following Mandelbrot [2], the symbol D, with appropriate subscripts and superscripts, will be reserved for scaling exponents that can be considered to be fractal dimensionalities. Consequently, the symbol $\mathcal{D}$ will be used for diffusion coefficients instead of the more usual D. Similarly, d will be used to denote ordinary, Euclidean dimensionalities (usually the dimensionality of the space in which the fractal is embedded), and τ will be used exclusively for the exponent that describes power law distributions of sizes, lengths, etc. Similarly, τ_c will be used for the corresponding cumulative distributions. For example, if the number of objects with sizes S in the range $s \rightarrow s + \delta s$ is given by $N_s \delta s$ (in the limit $\delta s \rightarrow 0$) and N_s has the power law form $N_s \sim s^{-\tau}$, then the corresponding cumulative distribution $N(S > s)$ is given by $N(S > s) \sim s^{-\tau_c}$, where $N(S > s)$ is the number of objects with sizes greater than s and $\tau_c = \tau - 1$. The symbols ϵ, L and ξ are reserved for characteristic lengths or correlation lengths and ℓ is used for other lengths. Apart from these "reserved" symbols, an attempt has been made to follow established conventions such as t for time and T for temperature. An effort has also been made to use a consistent notation within each of the individual chapters but, in the interest of simplicity, no attempt has been made to employ a consistent notation throughout the entire book. The word "size" is useful but ambiguous; it can be used to indicate the overall scale of an object (its diameter, radius, etc.), its mass, or the number of components that it contains. In most cases, it will be clear from the context what is meant by the word size. In order to reduce possible confusion, the symbols s or S will be used when size refers to the number of components, such as pixels, lattice sites or particles, the symbols m or M will be used when size should be interpreted as a mass and symbols such as ℓ, L, ξ and ϵ will be used when size means length.

1.9 Monte Carlo Methods

The term "Monte Carlo simulation" is used for any calculation that relies on the use of random numbers. The fast generation of long period pseudo-random numbers R, uniformly distributed over the range $0 < R < 1$, without correlations that would influence the outcome of a simulation [187], is a crucial part of any Monte Carlo calculation. There are many pitfalls for the unwary, since all random number generators are deterministic algorithms that may contain hidden idiosyncrasies that are too subtle to be easily detected, but are sufficient to cause unacceptable errors in simulations. Improvements continue to be made, but a discussion of random number generation is beyond the scope of this book.

The Monte Carlo simulation method was developed to study the properties of equilibrium systems. This approach to equilibrium problems is based on a configuration of interacting particles and an algorithm that allows the configuration to be changed. At each stage in the simulation, the configuration Ω is changed from Ω_n to Ω_{n+1}, and the energies E_n $(E(\Omega_n))$ and E_{n+1} $(E(\Omega_{n+1}))$ are calculated before and after the configuration change. To simulate the fluctuations of an equilibrium system, the rules used in the simulation must ensure that detailed balance is obeyed by the dynamics of the simulation and that all parts of the configuration space can be reached. There are two basic approaches that are still in common use, but many other algorithms have been devised which may be much more efficient for specific applications. In the Metropolis algorithm [188], the new configuration Ω_{n+1} is retained if $\delta E = E_{n+1} - E_n$ is negative. If δE is positive, then the new configuration is accepted if $R_n < e^{-\delta E/k_B T}$, and if $R_n \geq e^{-\delta E/k_B T}$, the system is returned to the configuration Ω_n. Here R_n is the nth random number (distributed uniformly over the range $0 < R_n < 1$) produced by the random number generator. In the "heat bath" algorithm, the new configuration is accepted if the random number R_n satisfies $R_n < e^{-\delta E/k_B T}/[1 + e^{-\delta E/k_B T}]$ and returned to Ω_n if $R_n \geq e^{-\delta E/k_B T}/[1 + e^{-\delta E/k_B T}]$. Typically, the system is started in a more or less arbitrarily selected configuration Ω_0 with "properties" that are, in general, much different from those characteristic of the equilibrium ensemble. The first part of the simulation consists of following the Markov chain $\{\Omega_i\} = \Omega_0, \Omega_1, \Omega_2 \ldots$ until all "memory" of the initial configuration has been lost and the subsequent configurations have properties that are representative of the equilibrium ensemble. Only after this initial period of N steps is the remaining part of the Markov chain $\{\Omega_i^{eq}\} = \Omega_{N+1}, \Omega_{N+2}, \Omega_{N+3}, \ldots$ used to calculate the equilibrium properties of the system. In most cases, this part of the Markov chain must be very long in order to obtain a properly weighted, representative sample of the configuration space and to reduce statistical uncertainties to a low level.

Because the Monte Carlo method was initially developed to study equilibrium systems, it is still not universally appreciated that this method can also be used to simulate the behavior of systems far from equilibrium. However, if a system is simulated using the standard Metropolis or heat bath algorithms, the dynamics of the simulation is not generally the same as the dynamics of the corresponding physical system.

1.10 Additional Information

Many books have been written on patterns and pattern-formation. Introductory texts include *Patterns in nature* by Stevens [189]. The beautifully written book entitled *On growth and form* by Thompson [63] is relatively descriptive by modern standards. However, it is a real classic and continues to inspire interest in the origins of patterns in nature. On the more theoretical side, the review on "Pattern formation outside of equilibrium", by Cross and Hohenberg [117], is recommended. In their review, Cross and Hohenberg state that

> Indeed, states with power-law correlation and distribution functions are the exception, not the rule in the experimental systems we have considered. Although scaling phenomena do exist in nature, they apparently result from special circumstances which remain to be fully elucidated.

In this respect, their point of view is much different from that taken in this book. This conclusion appears to be more a result of the current limitations of the theoretical approaches on which their review is based, than a scarcity of scaling phenomena or fractal structures in nature.

A collection of key papers entitled *Dynamics of curved fronts*, with a substantial introduction, has been prepared by Pelcé [190]. This book is concerned primarily with dendritic growth, direction solidification, flame fronts and viscous fingering processes. It brings together important contributions from a wide range of sources.

Chaos has become a popular subject for books on all levels. Cvitanović [191] and Bai-Lin [192] have published collections of important papers that have appeared in this area up until the early 1980s. The second edition of Cvitanović's collection [193] includes more recent contributions.

Chapter 2

Fractals and Scaling

The related concepts of fractal geometry and scaling have proven to be extremely valuable assets in the description and understanding of a wide range of disorderly structures and their origins. There are many types of fractals and they may be described qualitatively and quantitatively, in many ways. In this chapter, those classes of fractals that have proven to be of the most value in applications to the physical sciences are described, and the problems of measuring their corresponding fractal dimensionalities are discussed.

2.1 Self-Similar Fractals

One of the first fractals studied from a mathematical point of view was the Cantor set [2], illustrated in figure 2.1. This Cantor set can be constructed by first removing the middle third of a line segment covering the interval $0-1$ (figure 2.1(b)). In the next stage of the hierarchical construction process (figure 2.1(c)), the middle thirds of the two remaining line segments are removed. During each subsequent stage, the middle third is removed from each of the remaining line segments, so that after the nth stage the number of line segments has grown to 2^n and their total length has decreased to $(2/3)^n$. In the limit $n \to \infty$, a fractal set will be generated. This "object", with zero measure (total length of the remaining line segments) and an infinite number of pieces was considered to be a paradox, with no possible relevance to the real world!

The Cantor set possesses several properties in common with many other fractals. It contains holes (gaps) on all length scales and fills a negligible

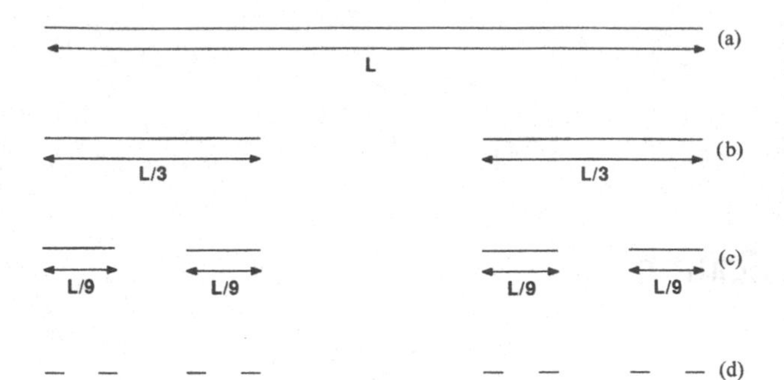

Figure 2.1 Three stages in the construction of a Cantor set. Part (a) shows the original line segment. Parts (b), (c), and (d) show the prefractals after the first three stages in the construction process.

fraction of the space that it occupies. If the Cantor set is dilated by a factor of 3^m, it can be covered by a minimum of 2^m replicas of itself. Structures with this property are said to be scale invariant, self-similar or to have dilation symmetry. In general, if M^m is the minimum number of replicas of the original pattern that are required to cover itself after dilation (change of length scale) by a factor of N^m, its geometric scaling properties can be characterized by a fractal dimensionality (the similarity dimensionality) D given by

$$D = \log(M)/\log(N). \tag{2.1}$$

This equation can be written in the more general form

$$D = \log(R_\mu)/\log(R_\ell), \tag{2.2}$$

where $R_\mu = \mu_2/\mu_1$ is the change in the measure μ (amount of "stuff") and $R_\ell = \ell_2/\ell_1$ is the change in the length scale. In the simplest case, the measure μ corresponds to the Lebesgue measure. The Lebesgue measure of a set $\{S\}$ can be defined as

$$\mathscr{L}^d(\{S\}) = \inf \sum_{i=1}^{i=\infty} C_i^d(V_i^d), \tag{2.3}$$

where $\{S\} \subset \cup_{i=1}^{i=\infty} V_i^d$ and where C_i^d is the length, area, volume ... of the ith line segment, parallelogram, parallelepiped, or other simple d-dimensional Euclidean shape, V_i^d, used to cover the set. This means that the Lebesgue measure of a set $\{S\}$ embedded in a d-dimensional space is the "volume" of the optimal (smallest volume) covering by d-dimensional Euclidean elements V_i^d. It corresponds to common intuitive ideas about the length, area or volume of a set.

For ordinary, Euclidean objects, such as continuous lines, rectangular planes or cubes, equation 2.2 gives the familiar, integer, Euclidean dimensionalities of 1, 2 and 3, respectively. For the Cantor set, illustrated in figure 2.1, the fractal dimensionality is $\log(2)/\log(3)$ or 0.6309.... This result ($0 < D < 1$) is intuitively reasonable. The Cantor set is more than a 0-dimensional point (it contains an infinite number of them) but less than a 1-dimensional line (the total "occupied" length is zero). In this case, the fractal dimensionality can be regarded as a measure of how effectively the Cantor set fills the 1-dimensional space in which it resides (in which it is embedded).

Functions such as the algebraic spiral $r(\phi) = \phi^{-a}$ can also be considered to be self-similar. However, this function is only self-similar about its "center" at $r = 0$. This book is concerned with globally self-similar fractals rather than locally self-similar patterns such as the algebraic spiral.

The structures shown in parts (b), (c) and (d) of figure 2.1 are not strictly fractals; they are uniform empty spaces or filled lines on short length scales. Such structures, generated at intermediate stages in the construction of fractals, are often called "prefractals". In practice, it is often useful to think of these prefactals as having the geometric scaling properties of the corresponding asymptotic $n \to \infty$ fractals over length scales in the range $\epsilon \leq \ell \leq L$, where $\epsilon = (1/3)^n$ is the inner cut-off for the Cantor set illustrated in figure 2.1, and L (1 in this case) is the outer cut-off for the fractal scaling range.

The procedures used to generate Cantor sets, and their analogs in higher-dimensional Euclidean spaces, can be generalized in many ways. For example, the quantities M and N in equation 2.1 can be varied systematically or randomly from generation to generation [194], or the lacunarity, discussed below, can be varied from generation to generation by varying M and N whilst keeping $\log M/\log N$ fixed. In some cases, such as alternating generators, this simply corresponds to using a more complex generator, but in other cases the attractive simplicity of self-similar scaling is lost. In nature, systems that have structure on all scales but do not have the *same* structure on all scales are frequently encountered. In some cases, a description in terms of different generators acting on different length scales that would create fractals with different dimensionalities or different lacunarities on different length scales may be useful. However, use of the oxymoron "scale dependent fractal dimensionality" to describe systems that do not scale appears much too frequently in the literature.

The Sierpinski gasket, shown in figure 2.2 is another simple example of a self-similar fractal. This pattern can be constructed by fitting together three identical, equilateral triangles, with sides of length ϵ, to form a new triangle with sides of length 2ϵ and with a triangular hole in the middle. In the next stage of the hierarchical construction process, three of the triangles with sides of length 2ϵ, with triangular holes in their middles, are fitted together to form a triangle with sides of length size 4ϵ, with one triangular hole of size 2ϵ and three triangular holes of size ϵ. After n generations, a triangular pattern with sides of length $2^n\epsilon$ containing 3^n solid triangles, each with sides of length ϵ, is formed. Each time the length scale is increased by a factor of 2, the area or measure increases by a factor of 3, and the fractal dimensionality is given by $D = \log 3/\log 2$. Very often, only the edges of the triangles of size ϵ are used to form a "wire-frame" Sierpinski gasket that also has a fractal dimensionality of $D = \log 3 / \log 2$.

An alternative procedure would be to start with a triangle with sides of length $L = 1$ and remove a triangle with sides of length $L/2$ from its middle, to generate a pattern of three triangles with sides of length $L/2$. This procedure would be repeated for all of the triangles with sides of length $L/2$ and all of the sub-triangles during each stage in the generation process. After the nth stage, the sides of the sub-triangles would have lengths of $t = L/2^n$.

The process of constructing a Sierpinski gasket prefractal from smaller prefractals resembles a systematic aggregation process. Consequently, this approach is useful for illustrating the concepts of fractal geometry in the context of problems such as aggregation. It is less useful from a mathematical point of view, since in most mathematical definitions of a fractal the limit $\epsilon \to 0$ is taken. However, this is equivalent to building prefractals of size L from pieces of size ϵ and taking the limit $\epsilon/L \to 0$ or rescaling the prefractal to a size of 1. Within any of the triangles of length ϵ used to construct a Sierpinski gasket via the aggregation approach, the 2-dimensional space is completely filled and $D = 2$. On longer length scales, $\ell > \epsilon$, the measure μ or amount of stuff (filled area in this case) grows as $\ell^{(\log 3/\log 2)} = \ell^D$.

In many systems of practical and/or scientific interest, the measure μ corresponds to the mass M and the length scale ℓ corresponds to the overall size (length scale) L. In this case,

$$M = \Lambda_M m(L/\epsilon)^D \sim L^D, \tag{2.4}$$

where Λ_M is an amplitude related to the lacunarity, described later in this chapter, and m is the mass of a particle of size ϵ. This scaling relationship between mass and size is consistent with the corresponding scaling relationships $\mu \sim L^d$ or $M \sim L^d$, for Euclidean objects, where μ is the length, area, mass or volume, ..., for $d = 1, 2, 3, \ldots$. Equation 2.4 indicates that, for such systems, the mean density ($\rho = M/V$, where V is the d-dimensional volume that contains the fractal, in the embedding space) is not constant, but decreases with scale according to

$$\rho \sim L^{D-d} \tag{2.5}$$

or

$$\rho \sim M^{(D-d)/D}. \tag{2.6}$$

This decrease in density with increasing length scale or size is one of the most important characteristics of self-similar fractal objects.

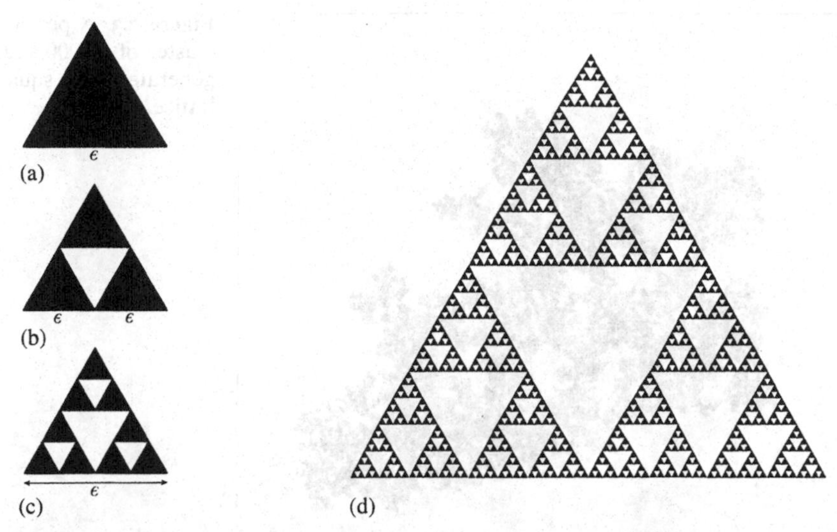

Figure 2.2 Parts (a) to (c) show the first two stages in the construction of a Sierpinski gasket ($D = \log 3/\log 2$) and part (d) shows the eighth-generation prefractal.

2.1.1 Statistical Self-Similarity

Symmetries, such as invariance to rotation, reflection, inversion and translation, have played an important role in Euclidean geometry and its applications in the physical sciences. The structures discussed earlier in this chapter all possess some of these symmetries. In addition, they possess the symmetry of self-similarity (invariance to a uniform change of length scales in all "directions"). Unfortunately, these fascinating structures do not represent, at all well, the structures that we see around us. A much more useful paradigm for most natural structures is provided by the percolation cluster shown in figure 2.3 or the random walk shown in figure 2.4. These structures clearly are not invariant to the symmetry operators associated with Euclidean geometry and they cannot be mapped onto themselves after a change of length scales. However, they do retain an important vestige of both the symmetry of self-similarity and some of the symmetries of Euclidean geometry. These, and many other structures, are said to be statistically self-similar fractals, because all of the statistical quantities that can be used to characterize them are invariant to a change of length scales ℓ (providing, of course, that the length scales remain in the range $\epsilon < \ell < L$). They may also be invariant, in a similar statistical sense, to Euclidean symmetry operators. Despite the relatively weak character of this "statistical symmetry", it can be used, in much the same way as the symmetries associated with Euclidean systems, to reduce substantially the difficulty of solving important problems in the physical sciences. While the applications of fractal geometry are not yet as important as those of group theory in areas such as spectroscopy, nuclear and

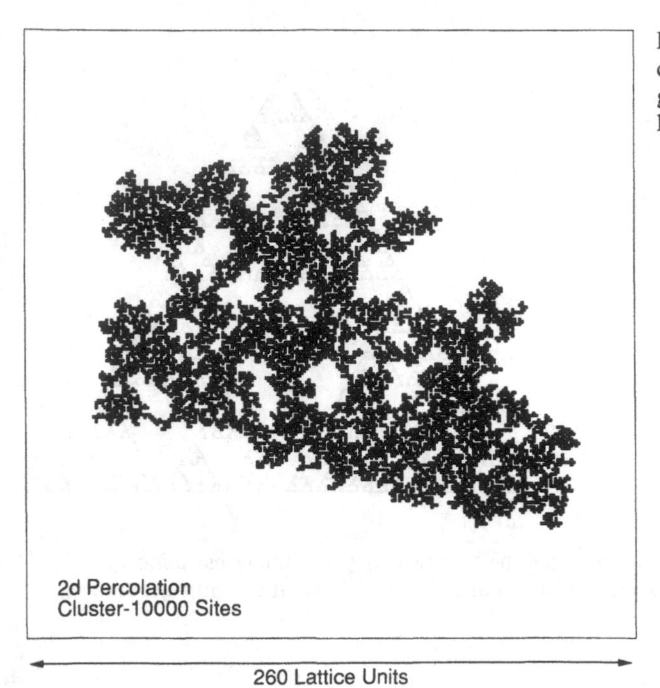

Figure 2.3 A percolation cluster of 10 000 sites, generated on a square lattice.

high energy physics, materials science and engineering, it is apparent that many important applications will be found, if for no other reason than the existence of a very wide variety of systems that appear to exhibit fractal properties. Some of these applications are discussed elsewhere [57].

2.1.2 Lacunarity

Although the fractal dimensionality provides a valuable index that can be used to describe structures such as the Cantor set prefractals shown in figure 2.1, it does not provide a complete characterization of this set. This is illustrated in figure 2.5, which shows the "generators" for five other Cantor sets with fractal dimensionalities of $\log(2)/\log(3)$. The corresponding fractals are obtained by replacing each of the line segments in the nth stage prefractal by scaled down copies of the generator. The fractals obtained using the generators shown in figure 2.5 can be distinguished by their "lacunarity" [2, 9, 195]. Lacunarity can be described in terms of the amplitudes Λ in power law relationships, such as Λ_M in equation 2.4, used to measure the fractal dimensionality. For a simple self-similar fractal, the measure $\mu(\ell)$ within a distance ℓ from a point on a fractal set is given by $\mu(\ell) = \Lambda\mu(\epsilon)(\ell/\epsilon)^D$, where $\mu(\epsilon)$ is the measure of an occupied

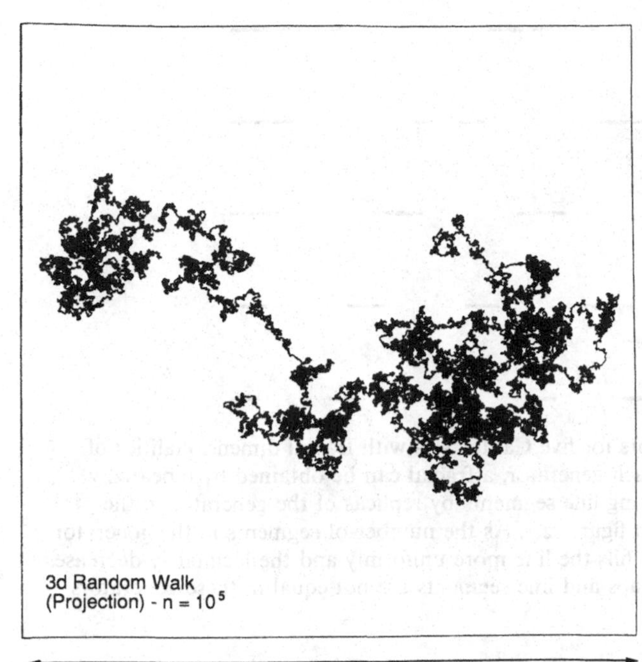

Figure 2.4 An off-lattice random walk with 10^5 steps of equal length. This figure shows a projection of a random walk in three dimensions onto a 2-dimensional plane. The random walk is a random fractal with a fractal dimensionality of exactly 2.0.

region of size $\epsilon(\mu(\epsilon) \approx \epsilon^d)$,[1] and the mass of a fractal of size L is given by equation 2.4. Fractals with large values of the amplitude Λ_M fill space more uniformly, and they are said to be less lacunar. The amplitude Λ_M can be used to characterize quantitatively the lacunarity. Both the lacunarity and fractal dimensionality are needed to describe how a fractal object fills space. Sincc the visual estimation of fractal dimensionalities is based to a large extent on the degree of space filling, these two quantities can easily be confused.

Since the lacunarity describes the uniformity or fluctuations in the way space is filled, it can also be defined and measured in terms of these fluctuations. The fluctuations in density or space filling can be characterized by measuring the momcnts $M_n(\ell)$ of the distribution of "mass" or measure in regions of size ℓ, centered on occupied points on the fractal. If a large number N of such "samples" is taken, the probability distribution (probability density)

$$P(\mu, \ell) = N_\mu(\ell)/N \tag{2.7}$$

1. In many simple cases $\mu(\epsilon)$ is the mass of a particle or the volume of a filled lattice site and ϵ is the diameter of the particle or the lattice constant a, respectively. However, in more complex systems the basic "construction units" may be difficult to identify. The basic construction units may have a distribution of shapes and/or sizes. They may also have a complex internal structure. In some cases, it is sufficient to replace ϵ in equation 2.4 by a characteristic particle size ξ_M associated with the size/shape distribution. In other cases, the amplitude Λ_M in equation 2.4 may not provide a useful characterization of the lacunarity.

Figure 2.5 Generators for five Cantor sets with fractal dimensionalities of log(2)/ log(3). For each generator, a fractal can be obtained by repeatedly replacing the remaining line segments by replicas of the generator in the manner illustrated in figure 2.1. As the number of segments in the generator increases, the fractal fills the line more uniformly and the lacunarity decreases. The lengths of the gaps and line segments are not equal in these generators.

can be measured. Here, $N_\mu(\ell)\delta\mu$ is the number of boxes, or regions of size ℓ that contain a mass or measure in the range $\mu - \delta\mu/2$ to $\mu + \delta\mu/2$. In a lattice model or digitized image, μ is the number of occupied sites or pixels. The moments $M_n(\ell)$ can then be defined as

$$M_n(\ell) = \frac{1}{N} \sum \mu^n N_\mu(\ell), \tag{2.8}$$

or

$$M_n(\ell) = \int \mu^n P(\mu, \ell) d\mu, \tag{2.9}$$

and the lacunarity $\Lambda(\ell)$ can be described in terms of moment ratios [2, 196, 197] such as

$$\Lambda_{2,1}(\ell) = M_2(\ell)/(M_1(\ell))^2. \tag{2.10}$$

For statistically self-similar fractals, the probability distribution $P(\mu, \ell)$ has the scaling form

$$P(\mu, \ell) = \mu^{-1} f(\mu/\ell^D), \tag{2.11}$$

and the moment ratios (equation 2.10) are independent of ℓ for $\epsilon \ll \ell \ll L$ [198]. The scaling function $f(x)$ in equation 2.11 characterizes the lacunarity of the fractal and may be called a lacunarity function.

Fractals can also be characterized by covering the fractal with a randomly positioned grid consisting of elements of size ℓ. If the distribution $N'(\mu, \ell)$ of the measure within the elements of size ℓ (volume ℓ^d) is measured, it can be represented by the scaling form $N'(\mu, \ell) = \mu^{-2} g'(\mu/\ell^D)$, and the corresponding probability distribution has the form

$$P'(\mu, \ell) = \mu^{-1} f'(\mu/\ell^D), \quad (2.12)$$

for a self-similar fractal. Here, $f'(x)$ is another lacunarity function, which can be used to provide a more complete characterization of fractals than the fractal dimensionality alone.

Blumenfeld and Ball [199] have proposed another approach to the characterization of the distribution of mass or measure in fractal structures. For a fractal set of points $\{\mathbf{r}\}$, the mass or measure $\mu_i(r)$ is determined in spherical shells centered on each of the points in the set. Each shell has the same width measured in units of the log of the distance from the ith point in the set. The distribution about the ith point, including the fluctuations, can be characterized via the quantities

$$m_i(r) = \mu_i(r)/ < \mu(r) >, \quad (2.13)$$

where $\mu_i(r)$ is the mass or number of points in the shell with a mean radius of r centered on the ith point or particle and $< \mu(r) >$ is the mean value of the mass in shells of radius r, averaged over all of the points. These relative fluctuations can be described using the correlation matrix $\mathcal{M}(x, x')$ defined as

$$\mathcal{M}(x, x') = < m_i(r) m_i(r') >, \quad (2.14)$$

where $x = \ln r$, $x' = \ln r'$ and the average is over all of the points. The correlation matrix $\mathcal{M}(x, x')$ should be scale invariant for a simple self-similar fractal. It can be represented by the "radial lacunarity function"

$$\mathcal{L}(x - x') = \mathcal{M}(x, x'). \quad (2.15)$$

Blumenfeld and Ball [199] measured the correlation matrix $\mathcal{M}(x, x')$ for fractals generated by the self-avoiding walk, cluster–cluster aggregation and DLA models (chapter 3) and concluded that $\mathcal{M}(x, x')$ depended only on $x - x'$, except for the largest and smallest values of $x - x'$, where cut-off effects are important. They proposed that the radial lacunarity function $\mathcal{L}(x - x')$ could be used to distinguish fractals that had the same or similar fractal dimensionalities. Although this idea has not been fully evaluated, it appears to be a promising approach to a more complete characterization of fractal structures.

The characterization of disorderly structures can be further generalized [199] via the vector correlation matrix $\mathcal{M}(\mathbf{r}, \mathbf{r}')$ defined as

$$\mathcal{M}(\mathbf{r}, \mathbf{r}') = < m_i'(\mathbf{r}) m_i'(\mathbf{r}') >, \quad (2.16)$$

where $m_i'(\mathbf{r})$ is $\mu_i(\mathbf{r})/ < \mu(\mathbf{r}) >$ and $\mu_i(\mathbf{r})$ is the mass or measure in a segment of a shell with a mean radius $r = |\mathbf{r}|$ and mean direction $\mathbf{r}/r$, centered on the ith particle. This vector correlation matrix can be measured by dividing the spherical shells around each of the points or particles into sections with equal angular sizes. A correlation matrix $\mathcal{M}(\mathbf{x},\mathbf{x}')$, like $\mathcal{M}(x,x')$, can then be defined as

$$\mathcal{M}(\mathbf{x},\mathbf{x}') = < m_i'(\mathbf{r}) m_i'(\mathbf{r}') >, \tag{2.17}$$

where $\mathbf{x}$ is a vector parallel to $\mathbf{r}$ with a length of $x = \ln r$. For a simple statistically self-similar fractal, the correlation matrix can be expressed in terms of the radial lacunarity function $\mathcal{L}(\log R, \theta)$ given by

$$\mathcal{L}(\log R, \theta) = \mathcal{M}(\mathbf{r},\mathbf{r}'), \tag{2.18}$$

where θ is the angle between the vectors $\mathbf{r}$ and $\mathbf{r}'$ and R is r'/r, or $\log R = x' - x$. The lacunarity function $\mathcal{L}(x - x')$ given in equation 2.15 is the angular average of $\mathcal{L}(x - x', \theta)$, with the proper solid angle $(d\Omega/d\theta)$ weighting.

2.1.3 Determination of the Fractal Dimensionality

A variety of different practical approaches to the measurement of the fractal dimensionality of "self-similar" structures have been developed and evaluated. These have been used quite extensively to characterize structures obtained from both experiments and computer simulations. While most of the experience with the measurement of fractal dimensionalities has been obtained during studies of specific experimental systems or models, some work [200] has been concerned primarily with the characterization of fractals.

2.1.3.1 Correlation Functions

Correlation functions have been used for a long time in attempts to develop a better theoretical understanding of disorderly systems and to characterize their structures. One of the most important of these is the family of density–density correlation functions $C^n(\mathbf{r}_1, \mathbf{r}_2, \ldots, \mathbf{r}_n)$ or $C^n(\{\mathbf{r}\})$ defined as

$$C^n(\{\mathbf{r}\}) = C^n(\mathbf{r}_1, \mathbf{r}_2, \ldots, \mathbf{r}_n) = < \rho(\mathbf{r}_0)\rho(\mathbf{r}_0 + \mathbf{r}_1) + \cdots + \rho(\mathbf{r}_0 + \mathbf{r}_n) >, \tag{2.19}$$

where $\rho(\mathbf{r})$ is the density at position $\mathbf{r}$ and the product of densities on the right-hand side of equation 2.31 is averaged over all possible origins $\mathbf{r}_0$, with equal weight or probability.[2] In equation 2.19, n is the number of vectors, $\mathbf{r}_1, \mathbf{r}_2, \ldots, \mathbf{r}_n$, describing the relative positions of the $n+1$ points. Correlation functions such as these can be used to describe both fractal and non-fractal structures. For self-similar fractals, these $n+1$-point correlation functions have a homogeneous power law form

$$C^n(\lambda\mathbf{r}_1, \lambda\mathbf{r}_2, \ldots, \lambda\mathbf{r}_n) = \lambda^{-n\alpha} C^n(\mathbf{r}_1, \mathbf{r}_2, \ldots, \mathbf{r}_n). \tag{2.20}$$

2. In many applications to systems with a lower cut-off scale ϵ, a normalized density–density correlation function $C^n_N(\{\mathbf{r}\})$ with $C^n_N(\{\mathbf{0}\}) = 1$ is used. In general, $C^n_N(\{\mathbf{r}\}) = C^n(\{\mathbf{r}\})/C^n(\{\mathbf{0}\}) = C^n(\{\mathbf{r}\})/ < \rho^{n+1} >$. If an on $(\rho = 1)$/off $(\rho = 0)$ pixel image or lattice representation is used, $C^n_N(\{\mathbf{r}\}) = C^n(\{\mathbf{r}\})/ < \rho >$.

The exponent α in equation 2.20 is called the codimensionality and is equal to $d - D$, where d is the Euclidean dimensionality of the embedding space. Very few studies of fractal structures have gone beyond the two-point density–density correlation function $C(r)$ given by

$$C(r) = C^1(r) = \langle\langle \rho(\mathbf{r}_0)\rho(\mathbf{r}_0 + \mathbf{r}) \rangle\rangle_{|\mathbf{r}|=r} . \tag{2.21}$$

In equation 2.21, $\langle\langle \ldots \rangle\rangle$ implies averaging over all origins $\mathbf{r}_0$ and all orientations. The two-point density–density correlation function is important in practice, since it determines the small-angle single-scattering properties of any structure. In some cases, in which $D < 2$, this has been exploited to measure the fractal dimensionality by obtaining diffraction patterns from photographic images. The dependence of the light-scattering intensity on the scattering angle θ for a self-similar fractal with a fractal dimensionality of D is given by[3]

$$S(q) \sim q^{-D}. \tag{2.22}$$

Here, $q = 4\pi \sin(\theta/2)/\lambda$, where λ is the wavelength. Consequently, D can be measured from the shape of $S(q)$. In practice, the measurement of D involves more than just "plotting" the dependence of $\log S(q)$ on $\log q$ and measuring the slope. There is only a limited range of scaling, so the effects of the crossovers on both short length scales and long length scales must be taken into account. In addition, the instrumental contributions to $S(q)$ can also require important corrections. The value of the scattering approach is compromised because scattering from a fractal structure cannot be easily distinguished from scattering from a power law distribution of Euclidean objects of different sizes [201]; see section 2.3. Another way of measuring $C(r)$ is to superimpose two photographic negatives in which the "fractal" is transparent and the surroundings are opaque, and then to measure the transmitted light intensity as a function of the relative displacement $r = |\mathbf{r}|$ of the negatives [202].

Equation 2.20 implies that the two-point density–density correlation function has the algebraic form

$$C(r) = C^1(r) \sim r^{-\alpha} \sim r^{D-d}, \tag{2.23}$$

for simple self-similar fractals. The power law form of equation 2.23, which is invariant to a change of length scales, reflects the scaling symmetry of the underlying structure.

Measurement of the two-point density–density correlation function has proven to be a robust and reliable way to characterize both fractal and non-fractal structures from digitized images and computer-generated structures. In practice,

3. This form for the scattering intensity can be obtained by subdividing the fractal into blobs of size $\xi_q \sim q^{-1}$. The scattering intensity $S_b(q)$ from each blob is proportional to ξ_q^{2D} or q^{-2D}, and the number of blobs $N_b(q)$ is proportional to ξ_q^{-D} or q^D. Consequently, the total scattered intensity scales as $S(q) \sim S_b(q)N_b(q) \sim q^{-D}$.

$C(r)$ may be measured by using a large number of origins in a large structure (a structure covering a large range of length scales), and $C(r)$ is often averaged over an ensemble of "samples" as well. The most frequently used method is based on the relationship

$$C(r) \sim << \rho'(r) >> = \lim_{\delta r \to 0} << \mu(r, \delta r)/V(r, \delta r) >>, \tag{2.24}$$

where $\rho'(r)$ is the mean density measured at a distance r from an *occupied* point on the structure. In equation 2.24, $\mu(r, \delta r)$ is the measure in the range $r - \delta r$ to $r + \delta r$ (with $\delta r/r \ll 1$) from the occupied point and $V(r, \delta r)$ is the corresponding volume in the embedding space. The symbols $<< \ldots >>$ imply averaging $\mu(r, \delta r)/V(r, \delta r)$ over all occupied points and over a sufficiently large ensemble of structures. In practice, $\rho(r_n)$ is measured in a number of annular regions with a mean radius of r_n centered on a sufficiently large number of randomly selected *occupied* points. Very often, data from a lattice model or a pixel representation are used, and the structure is represented by points with equal measure, lying at the centers of each of the occupied pixels. In this case, it is important, for small values of r, to use the number of lattice sites or pixels with centers in the range $r - \delta r$ to $r + \delta r$ for the volume $V(r, \delta r)$, instead of the volume of the d-dimensional embedding space lying between $r - \delta r$ and $r + \delta r$.

Alternatively, $C(r)$ can be obtained by measuring $< \mu(r_n) >$, where $\mu(r_n)$ is the measure contained within a distance r_n from a randomly selected point. The fractal dimensionality can then be obtained from

$$< \mu(r_n) > \sim r_n^D \tag{2.25}$$

or from

$$C(r) \sim \frac{d\mu(r)/dr}{dV(r)/dr} \tag{2.26}$$

and equation 2.23. Sometimes, rectangular boxes of size $\ell_n = 2r_n$, with a volume of ℓ_n^d, are used instead of hyperspheres. If necessary, simple strategies [203] can be used to calculate $C(r)$ using only points that are separated by distances in the range, $r_1 \le r \le r_2$, over which $C(r)$ is to be calculated. In many cases, r_1 is selected to avoid large short-length-scale corrections due to lattice and finite-particle-size effects. Similarly, r_2 is selected to avoid large finite-size effects. This can substantially reduce the amount of computer time required to calculate $C(r)$, for length scales in the scaling range.

The density–density correlation function $C(r)$ can also be obtained from the relationship $C(r) \sim r^{1-d}P(r)$, where $P(r)$ is the distribution of lengths between pairs of points selected at random ($P(r)$ is the probability that a pair of points selected with probabities $\rho(\mathbf{r}_1)$ and $\rho(\mathbf{r}_2)$ will be separated by a distance $|\mathbf{r}_1 - \mathbf{r}_2|$ of r).

In many systems, it is useful to think of the structure in terms of particles joined by bonds. A further simplification can then be obtained if the particles are replaced by point nodes at their centers $\mathbf{r}_i$. The bonds, associated with a particular node at $\mathbf{r}$, can be described in terms of a bond density function [204]

$$\rho(\mathbf{r}, \theta), \tag{2.27}$$

which consists of a delta function at each angle corresponding to the direction of a bond that joins the node at $\mathbf{r}$. The bond angle distribution function $\rho(\mathbf{r}, \theta)$ can be represented by its Fourier components

$$\rho(\mathbf{r}, \theta) = \sum_{m=-\infty}^{m=\infty} \theta_m(\mathbf{r}) e^{-im\theta}. \tag{2.28}$$

The orientational order of the bonds can then be characterized by the correlation functions

$$C_m(\mathbf{r}) = < \phi_m(\mathbf{r}_\circ)\phi_m^*(\mathbf{r}_\circ + \mathbf{r}) >, \tag{2.29}$$

where $< \ldots >$ implies averaging over all nodes in the network. For most disordered systems, these correlation functions decay exponentially. Hexatic phases, for which the correlation function $C_6(\mathbf{r})$ has the algebraically decaying form

$$C_6(\mathbf{r}) = < \phi_6(\mathbf{r}_\circ)\phi_6(\mathbf{r}_\circ + \mathbf{r}) > \sim |\mathbf{r}|^{-\eta_6} \tag{2.30}$$

are important exceptions. In practice, the bond orientation correlations can be characterized by measuring the quantity $< \cos m\theta > (r)$, where θ is the angle between the directions of pairs of bonds and $< \ldots > (r)$ implies averaging over all bonds separated by a small range of distances ($r - \delta r$ to $r + \delta r$) with an average distance of r.

Except in astrophysics, there has been very little effort devoted to the characterization of disorderly structures in terms of three-point and higher order correlation functions. The three-point (two-vector) density–density correlation function $C^2(\{\mathbf{r}\})$ for a system that is statistically invariant under translation and rotation can be written in the form

$$C^2(\{\mathbf{r}\}) = C(|\mathbf{r}_\circ - \mathbf{r}_1|)C(|\mathbf{r}_\circ - \mathbf{r}_2|)f(|\mathbf{r}_\circ - \mathbf{r}_1|/|\mathbf{r}_\circ - \mathbf{r}_2|, \theta), \tag{2.31}$$

where θ is the angle between the vectors $\mathbf{r}_\circ - \mathbf{r}_1$ and $\mathbf{r}_\circ - \mathbf{r}_2$, $C(r)$ is the two-point density–density correlation function and $f(x, y)$ is a scaling function. The correlation matrix $\mathscr{M}(\mathbf{r}, \mathbf{r}')$ defined in equation 2.15 is related to the three-point correlation function by $C^2(\mathbf{r}_\circ - \mathbf{r}_1, \mathbf{r}_\circ - \mathbf{r}_2) = C^2(\mathbf{r}, \mathbf{r}') = C(r)C(r')\mathscr{M}(\mathbf{r}, \mathbf{r}')$, so that the correlation matrix is equal to the scaling function $f(|\mathbf{r}_\circ - \mathbf{r}_1|/|\mathbf{r}_\circ - \mathbf{r}_2|, \theta)$ or $\mathscr{L}(\ln(r'/r), \theta)$. The radial lacunarity function $\mathscr{L}(x - x')$, defined by equation 2.15, can then be obtained as an average of the scaling function $\mathscr{L}(\ln(r'/r), \theta)$ over the angular coordinates θ.

2.1.3.2 The Hausdorff–Besicovitch Dimensionality

The Hausdorff–Besicovitch dimensionality [2] played an important historic role in the development of the mathematical background leading to the more recent flourishing of fractal geometry. The definition of the Hausdorff–Besicovitch dimensionality D_{HB} is based on the Hausdorff measure μ_H, which is determined by the optimal covering of the fractal (set) by "balls" of radius ϵ *or less.* A ball $B(r)$ of radius r contains everything within a radius r of its center. The Δ-dimensional Hausdorff measure of a set $\{S\}$ is then given by

$$\mu_H(\Delta) = \lim_{\epsilon \to 0} \inf \sum_i C_\Delta(B_i(r_i \leq \epsilon)) = \lim_{\epsilon \to 0} \inf \sum_i \gamma(\Delta)(r_i \leq \epsilon)^{(\Delta)}, \qquad (2.32)$$

where $\{S\} \subset \cup_i B_i$ and B_i is the ith ball in the covering, with radius r_i. Here, $\cup_i B_i$ is the union of all the balls. In equation 2.32, "inf" indicates the optimal (minimum) covering, $C_\Delta(B(r \leq \epsilon)) = \gamma(\Delta)r^\Delta$ is the volume of a Δ-dimensional hypersphere of radius $r \leq \epsilon$ and the sum is over all of the covering balls with radii $r \leq \epsilon$. In general, Δ is not an integer and the geometric factor $\gamma(\Delta)$ in equation 2.32 is given by $\gamma(\Delta) = \Gamma(1/2)^\Delta / \Gamma(1 + \Delta/2)$, where Γ is the "gamma" function. The Hausdorff–Besicovitch dimensionality is then the critical dimensionality Δ_c for which the Hausdorff measure $\mu_H(\Delta)$ has the value 1. For $\Delta > (D_{HB} = \Delta_c)$ the Hausdorff measure is zero, and for $\Delta < (D_{HB} = \Delta_c)$ the Hausdorff measure is infinite. This concept of dimensionality does not provide a practical approach to the measurement of the fractal dimensionality of objects generated in experiments or computer simulations. Because the Hausdorff measure is based on an optimum covering with balls of radius ϵ or less rather than an optimum covering with balls of radius of equal size ϵ, it may give different values for D than other methods, even for some quite simple examples.

Another definition of dimensionality [205] is based on the minimum number $N(\ell)$ of d-dimensional hypercubes with sides of length ℓ needed to cover the fractal. The capacity dimension D_k is then defined as

$$D_k = -\lim_{\ell \to 0} \log N(\ell) / \log(\ell). \qquad (2.33)$$

In general, $D_{HB} \leq D_k$, but, in most cases of physical interest, $D_{HB} = D_k = D$, where D is the all-purpose fractal dimensionality for a self-similar fractal.

Because they require an approach to the zero length scale limit, the "mathematical" definitions of dimensionality such as those given in equations 2.32 and 2.33 do not lead to practical ways of characterizing real (prefractal) structures, though they can be used to describe some mathematical models for real objects and phenomena. Additional practical difficulties are encountered for methods that require optimal coverages to be determined.

2.1.3.3 The Minkowski Dimensionality

Many important applications of fractal geometry depend on the unoccupied volume $V(\ell)$ contained within a distance ℓ of a fractal, or the volume of the fractal after "fattening" by a length ℓ minus the volume of the fractal itself. A set, fattened by ℓ, consists of the original set plus all of the points within a distance ℓ from the original set. The fattened set consists of the union of balls with the same radius ℓ, centered on every point on the set. The Minkowski measure can then be defined as

$$\mu_M(\ell) = V(\ell)\ell^{\delta-d}, \tag{2.34}$$

where d is the dimensionality of the Euclidean embedding space. However, $\mu_M(\ell)$ may not have a limiting value as $\ell \to 0$. In this case, distinct upper and lower limits $\mu_M^+(\ell)$ and $\mu_M^-(\ell)$ may be defined.

If $\mu_M^+(\ell) = \mu_M^-(\ell)$ as $\ell \to 0$, the Minkowski dimensionality is then the critical value for the exponent δ in equation 2.34, such that $\mu_M(\ell)$ becomes infinite in the limit $\ell \to 0$ for $\delta < D_M$ and $\mu_M(\ell)$ is zero if $\delta > D_M$. In general, the upper and lower values of the Minkowski measure, $\mu_M^+(\ell)$ and $\mu_M^-(\ell)$, in the $\ell \to 0$ limit may lead to two distinct values for D_M (D_M^+ and D_M^-). However, it appears that $D_M^+ = D_M^- = D_M$ in most cases. Curves, such as the Koch curve described below, that have been fattened are often called "Minkowski sausages".

The volume (Lebesgue measure) $V(\ell)$ within a distance ℓ of the surface of a complex object, such as an electrode or catalyst support, often plays a major role in physico-chemical phenomena. For a fractal surface, this volume can be described in terms of the Minkowski–Bouligand dimensionality D_{MB}

$$D_{MB} = \lim_{\ell \to 0} \log[V(\ell)\ell^{-d}]/\log[1/\ell]. \tag{2.35}$$

In practice, it can be assumed that the upper and lower limits give the same value for D_{MB}, and no distinction is made between these limits in equation 2.35.

Compact objects with fractal surfaces (fractals with a finite volume) are often referred to as "fat fractals" [206]. If the fractal is fattened by an amount ℓ, and $V'(\ell)$ is the volume of the fattened fractal $V(\ell)$ minus that of the unfattened fractal $V_\circ$, then the "exterior capacity dimensionality" D_x is defined by Grebogi *et al.* [206] as

$$D_x = \lim_{\ell \to 0} \log V'(\ell)/\log \ell. \tag{2.36}$$

An analogous interior fractal dimensionality can be defined by fattening the unoccupied part of the embedding space or "thinning" the fat fractal. The volume of a fattened fractal is given by

$$V(\ell) = V_\circ + A\ell^{D_x}. \tag{2.37}$$

For the case of a digitized image, the dimensionalities discussed in this section can be estimated by fattening or thinning the image in stages. This can be accomplished by filling the unoccupied perimeter sites or by removing the occupied perimeter sites at each stage in the fattening or thinning process.

2.1.3.4 Box Counting

One of the simplest methods used to characterize fractals is known as "box counting". In this approach, the fractal lying in a d-dimensional space is covered by a d-dimensional grid with elements of size (length scale) ℓ. The box counting or capacity dimension is then given by

$$D_B = -\lim_{\ell \to 0} \log(N(\ell))/\log(\ell) = \lim_{\ell \to 0} \log(N(\ell))/\log(1/\ell), \qquad (2.38)$$

where $N(\ell)$ is the number of non-empty grid elements. In practice, it is not possible to approach the limit $\ell \to 0$, so D_B is estimated by measuring the dependence of $\log(N(\ell))$ on $\log(\ell)$, for length scales in the range $l_1 < \ell < l_2$, where $l_1 > \epsilon$ (the lower cut-off length), and $l_2 < L$ (the upper cut-off length), and equation 2.38 can be replaced by

$$D'_B = - < d(\log N(\ell)/d(\log \ell) >_{l_1<\ell<l_2} . \qquad (2.39)$$

Similarly, if the two-point density–density correlation function $C(r)$, defined in equation 2.21, is used to estimate the fractal dimensionality of a real structure characterized in an experiment, or a pattern generated in a computer simulation, then the fractal dimensionality is usually obtained by fitting a straight line to the data points $(\log C(r), \log r)$ over that range of length scales in which $\log C(r)$ depends linearly on $\log r$.

2.1.3.5 Some General Comments

In this book, any object that exhibits a power law mass–length scaling relationship, over a useful range of length scales, will be called a "fractal". In practice, most methods used to estimate fractal dimensionalities involve the fitting of straight lines to data sets on log–log scales. The fractal dimensionality is then calculated from the slope of the straight line. A degree of judgement and experience is required to select the range of length scales used in the fitting procedure. If the data are of high enough statistical quality, it is useful to determine local slopes over relatively small length scale ranges that together cover the entire range of length scales for which data are available. These local slopes can be used to calculate "local fractal dimensionalities", and the fractal regime can be identified as the range of length scales over which the local fractal dimensionalities are constant. This procedure is also useful in the identification of crossovers. It is also helpful to plot the data in such a way that the log–log plots are essentially horizontal, so that deviations from a theoretical or estimated fractal dimensionality appear as fluctuations or systematic trends with respect to a horizontal line. The use of these simple procedures would substantially reduce the ambiguity associated with many published fractal dimensionalities, and would demonstrate that many structures reported to be fractal are not fractal at all.

It is also important to examine the raw data and log–log plots directly, before attempting to measure the fractal dimensionality by numerically fitting a straight line to a log–log plot. In some cases, a structure may exhibit quite convincing linear log–log plots of the type usually used to measure fractal dimensionalities but the apparent fractal dimensionality does not capture the most important characteristics. Farin's rabbit [207] and works of art [208] are good examples. In other cases, strong deviations from linearity may suggest better ways to analyze the data, in terms of crossovers, algebraic corrections, etc.

2.1.4 The Devil's Staircase

The construction of a "Devil's staircase" is shown in figure 2.6. The generator, shown in figure 2.6(a) consists of sloping linear segments from the origin $(0, 0)$ to $(1/3, 1/2)$ and from $(2/3, 1/2)$ to $(1, 1)$, joined by a horizontal line from $(1/3, 1/2)$ to $(2/3, 1/2)$. At each step in the construction process, the horizontal segments are unchanged and the sloping segments are replaced by appropriately (affinely) scaled replicas of the generator shown in figure 2.6(a). After n stages, the total length of the horizontal component of the sloping segments has decreased to $(2/3)^n$ and their projection onto the horizontal x axis is the nth stage prefractal Cantor set, illustrated in figure 2.1. Horizontal segments in the Devil's staircase correspond to *gaps* in the corresponding Cantor set. In the asymptotic, $n \to \infty$ limit, the total length of the horizontal segments is 1, as is the total length of the inclined (vertical) segments.

The Devil's staircase is related to the Cantor bar. The Cantor bar is constructed in the same way as the Cantor set except that the measure μ is distributed equally between the remaining segments, so that after n generations the density of the measure ρ_μ is $(3/2)^n$ in each of the segments. This can be thought of in terms of the width or mass per unit length of the remaining segments, so that, after n generations, $\rho_\mu(n)$, the width or density of the measure in the remaining segments, is given by $\rho_\mu(n) = (3/2)^n$, the length is given by $\ell = (1/3)^n$ and the measure contained in each segment is given by $\mu = (1/2)^n$. This means that the relationship between the length ℓ of the segments and the mass of measure is given by

$$\mu \sim \ell^{\alpha_H}, \tag{2.40}$$

where $\alpha_H = \log 2/\log 3$ is known as the Holder, or Lipshitz–Holder, exponent. This exponent has become important in the study of multifractal measures described in section 2.8 and appendix B. The height of the Devil's staircase at a distance x from the origin is then given by

$$h(x) = \int_0^x \rho_\mu(x')dx', \tag{2.41}$$

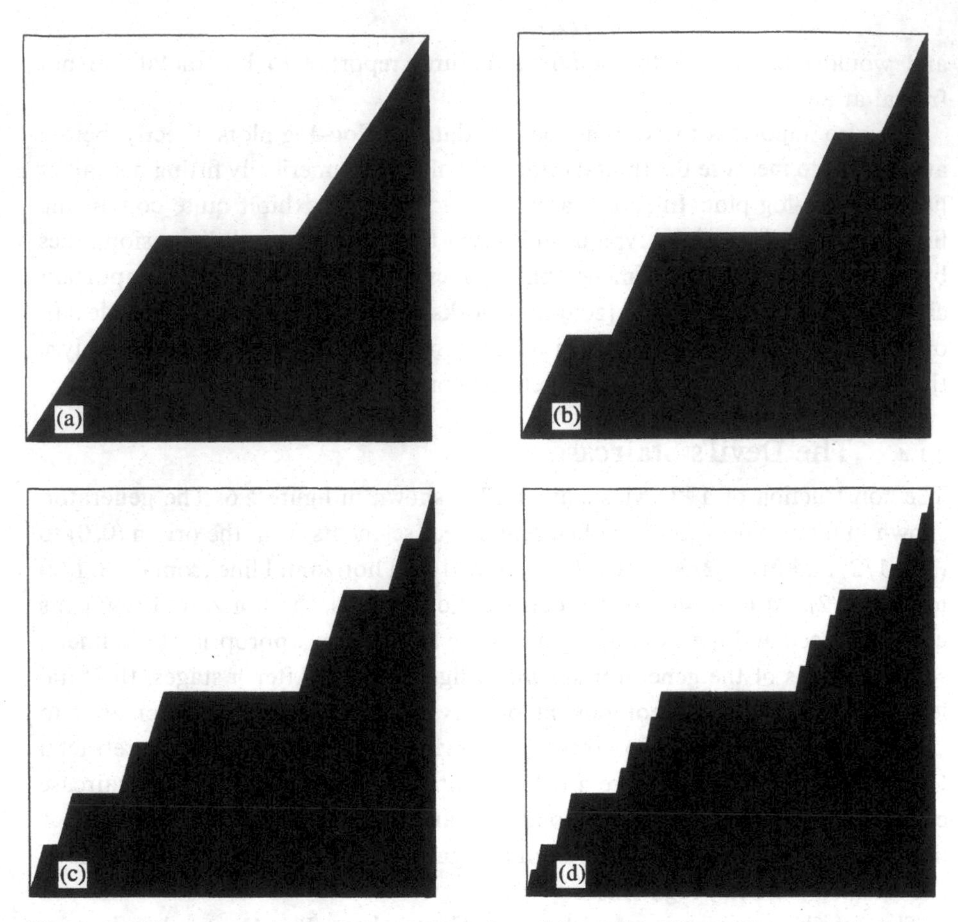

Figure 2.6 Four stages in the construction of a subfractal Devil's staircase based on the Cantor set shown in figure 2.1. Here, the Devil's staircase prefractal is the boundary separating the gray and white areas.

where $\rho_\mu(x')$ is the "linear density" (measure per unit length) in the corresponding Cantor bar at position x' (the measure or amount of "stuff" in an interval of length ϵ at position x is given by $\lim_{\epsilon \to 0} \mu(x, \epsilon) = \epsilon \rho_\mu(x)$).

The process of constructing a fractal by concentrating the measure onto smaller and smaller regions of the embedding space has been called "curdling" [2]. It is analogous to real curdling processes in which material is conserved, but concentrated onto a small fraction of the space. The construction process described above can be generalized in many ways. One of the simplest is to base the construction process on other Cantor sets.

The "Devil's staircase" is neither self-affine nor self-similar, except about "special" points, such as the origin. However, there is a strong motivation to classify it as a fractal and to characterize its "fractal" scaling properties. The distribution of the lengths of the horizontal segments is a power law with an exponent τ given by $\tau = 1 + D_c$, where D_c is the fractal dimensionality of the associated Cantor set. If the "Devil's staircase" shown in figure 2.6 is rotated by 90°, this distribution of segment lengths can be interpreted as the distribution of step heights. Otherwise, it represents the distribution of distances separating steps of equal height. Pfeifer and Obert [209] have suggested that the Devil's staircase, and similar patterns, can be characterized by measuring the volume V_r surrounding the Devil's staircase that is inaccessible to balls of radius r. The dependence of V_r on r is given by

$$V_r \sim r^{d-D_i}, \tag{2.42}$$

where D_i is called the "irregularity dimensionality". For the Devil's staircase, D_i is the dimensionality of the corresponding Cantor set ($D_i = D_c$). For a self-similar fractal, the irregularity dimensionality and the "all purpose" fractal dimensionality are equal.

Patterns like the Devil's staircase that are not themselves fractals but have fractal subsets or other associated fractal patterns are called "subfractals".

2.2 Simple Rules

While much of our understanding of fractals and their properties has been obtained using more or less *ad hoc* approaches, a few simple "rules" or ideas have proven to be extremely valuable. Many of the applications of fractal geometry are concerned with structures with a characteristic size (length scale) L made up of compact subunits (particles) of size ϵ. Familiar and important examples include polymer molecules and colloidal aggregates. For such structures, the total number of particles s is given by

$$s \sim (L/\epsilon)^D, \tag{2.43}$$

and the mean density ρ in the volume of size (length scale) L occupied by this structure is given by

$$\rho \sim (L/\epsilon)^{D-d}. \tag{2.44}$$

If two such objects are placed randomly in a region of space of size L, the average number of particle–particle overlaps n_o will be given by

$$n_o \approx s^2/N \approx s\rho \approx (L/\epsilon)^{2D-d}, \tag{2.45}$$

where N is the number of regions of size ϵ in the volume of size L ($N \approx (L/\epsilon)^d$). In the asymptotic limit ($L/\epsilon \to \infty$), the number of overlaps is zero if $2D < d$ and infinite if $2D > d$.

Similar arguments show that two fractals, with dimensionalities D_1 and D_2, will overlap in a d-dimensional space if $D_1 + D_2 > d$ and will "coexist" without overlap if $D_1 + D_2 < d$. This may have important implications for measurement strategies [21]. If events occur on a fractal subset with a dimensionality of D_e in a d-dimensional space and the measuring network or device has a dimensionality of D_m, then the events will be detected with a high probability only if $D_e+D_m > d$. For true self-similar fractals, this result is valid, whatever the resolution of the detecting system. In practice, crossovers at large and/or small scales will complicate this simple picture.

In general, the fractal dimensionality of the intersection between n fractals in a d-dimensional space is given by

$$D(\cap) = D(S_1 \cap S_2 \cap \ldots \cap S_N) = D_1 + D_2 + \cdots + D_n - (n-1)d, \tag{2.46}$$

where D_i is the fractal dimensionality of the set S_i. In those cases where the fractal dimensionality of the intersection is negative, the index $D(\cap)$ indicates how rare the intersections are, how empty the intersection is, or how rapidly the probability of an intersection decreases as the system size, or range of length scales, L/ϵ, increases [210].

Equation 2.46 is also valuable when one or more of the intersecting sets is Euclidean. In particular, the intersection of a D-dimensional fractal and a d_x-dimensional Euclidean form in a d-dimensional embedding space has a fractal dimensionality of $D(\cap) = D - d + d_x$, and this relationship is used to determine the fractal dimensionality using d_x-dimensional cuts through fractal patterns.

If a fractal with a dimensionality D moves through a d-dimensional space, via a path with dimensionality D_w, then it will sweep out a "volume" with a dimensionality of $min(D_w + D, d)$. Consequently, if two fractals with dimensionalities D_1 and D_2 move simultaneously through a space with a dimensionality of d, and the path of one fractal with respect to the other has a dimensionality D_w, then they will collide if $D_1 + D_2 + D_w > d$ and will "miss" each other if $D_1 + D_2 + D_w < d$.

If a D-dimensional fractal is projected from a d-dimensional space onto a Euclidean "surface" with a dimensionality d_s, then the dimensionality of the projection in the d_s space is given by

$$D_p = \begin{matrix} d_s & \text{if } D \geq d_s \\ D & \text{if } D < d_s. \end{matrix} \tag{2.47}$$

This result implies that the dimensionalities of fractals in 3-dimensional space can be directly measured from 2-dimensional digitized pictures, such as electron or optical micrographs, if $D < 2$. In practice, this procedure is unreliable if D is too close to 2.

All of these rules apply equally well to Euclidean objects; they can be regarded as "special case" fractals. For Euclidean shapes, these results are quite obvious, and this makes the more general fractal rules more intuitively reasonable.

While many fractals, such as the DLA clusters discussed in chapter 3, are essentially loopless structures, many others, such as 2-dimensional percolation clusters, consist of a network of loops. In such cases, the distribution of the areas of the regions enclosed by the loops has a power law form

$$N_s \sim s^{-\tau}, \tag{2.48}$$

where $N_s \delta s$ is the number of regions with areas in the range $s - \delta s/2 \rightarrow s + \delta s/2$. In general, equation 2.48 describes the size distribution of "holes" in any fractal that subdivides the embedding space in which it resides into non-contacting regions (the holes). The exponent τ in equation 2.48 is given by

$$\tau = (d + D)/d. \tag{2.49}$$

This relationship can be used to measure the fractal dimensionality from the size distribution of the holes. For example, Néda and Mocsy [211] have used this approach to measure the fractal dimensionality of fracture networks in glass plates from the fragment size distribution. However, caution should be exercised in using this approach, since a power law distribution of fragments does not necessarily imply that the surrounding space is a fractal.

2.3 Finite-Size Effects and Crossovers

Fractal geometry provides an idealized picture for a wide variety of real structures, in much the same way that Euclidean geometry provides an idealization for many others. In both cases, it is rare for simple fractal or Euclidean models to provide an accurate description of nature over more than a few orders of magnitude in length scale. For example, many surfaces that appear to be "flat" turn out to be rough on sufficiently short length scales, and, as is shown in chapter 5, this roughness can often be described quite well in terms of self-affine fractal geometry. This breakdown of the Euclidean, smooth surface model can have important implications for tribological, chemical, optical and other properties. Similarly, all real "fractals" exhibit geometric scaling over length scales ℓ confined to only a limited range $\xi_1 < \ell < \xi_2$, where ξ_1 and ξ_2 are the "inner" and "outer" cut-off lengths). In many cases, the outer cut-off length, ξ_2, corresponds to the overall size L of a growing structure or to the system size. The inner cut-off length, ξ_1, often corresponds to atomic sizes or the size of the "particles" ϵ from which a structure is assembled. For example, in colloidal aggregates ϵ often corresponds to the particle diameter. In some cases, it is possible to observe different fractal scaling behavior over different length scale ranges. In experiments, it is usually difficult to carry out accurate measurements over a sufficiently wide range of length scales to establish the existence of more than one fractal regime. In any event, crossover from Euclidean to fractal behavior or between different fractal regimes is common, and the quantitative interpretation of both experiments and computer simulations frequently requires an understanding of the nature of these crossovers.

A crossover from fractal to Euclidean behavior can often be observed and studied systematically by starting with a fractal model or experimental system and perturbing it. A simple example that has been studied both experimentally [212] and by computer simulations [212] is the 2-dimensional gradient (gravity) stabilized invasion percolation process. In the simplest, 2-dimensional invasion percolation model for the slow displacement of a wetting fluid by a non-wetting fluid in a porous medium [213, 214, 215], random numbers, t_i, uniformly distributed over the range $0 - 1$ are assigned to each of the sites on a lattice. These random numbers then represent the randomly distributed "thresholds" (capillary pressures) that the fluid–fluid interface must overcome to advance through the porous medium. The invaded region is represented by a cluster of filled sites on a lattice. The fluid–fluid displacement process is then simulated by finding and filling the site with the lowest threshold on the unoccupied external perimeter (chapter 3, section 3.1) of the cluster of filled sites. The growth is restricted to the external perimeter to represent the effects of "trapping" (the enclosure of regions of wetting fluid by the invading, non-wetting fluid, which prevents growth into the enclosed, or trapped, regions). This process is repeated many times, to represent the fluid–fluid displacement process. The external perimeter consists of all the vacant sites that have one or more occupied nearest neighbor, and can be reached from "infinity" (from outside of the region occupied by the pattern) by a path that consists only of steps between nearest-neighbor unoccupied sites. The clusters generated by this model have a fractal dimensionality of about 1.82. Like the fractal dimensionalities associated with many non-equilibrium growth models, there is, at present, no theory capable of providing an exact value for this fractal dimensionality. On the other hand, the fractal dimensionality of the external perimeter is known exactly; its value is 4/3 [216]. This is a non-rigorous result, but it is almost certainly correct.

In the corresponding experiments of Birovljev *et al.* [212, 217, 218], air (the non-wetting fluid) was injected via a channel along one edge of a quasi 2-dimensional cell consisting of a monolayer of glass beads trapped between two glass sheets. To represent this process, a complete row of lattice sites, lying along one edge of the lattice, was filled at the start of each simulation. At the start of the experiments, the model 2-dimensional, porous medium was saturated with the wetting fluid (glycerin/water). In the experiments, the "sides" of the 2-dimensional cell were sealed and the two ends were open to allow fluids to be injected and withdrawn. Figure 2.7(a) shows a displacement pattern obtained with a horizontal cell. In the experiments, a perturbation can be added in a controlled manner by tilting the cell. Parts (b), (c) and (d) in figure 2.7 show displacement patterns obtained from experiments in which the more dense fluid was withdrawn very slowly, via a channel running along the lower edge of an inclined cell. In the corresponding simulations, the perturbation is introduced

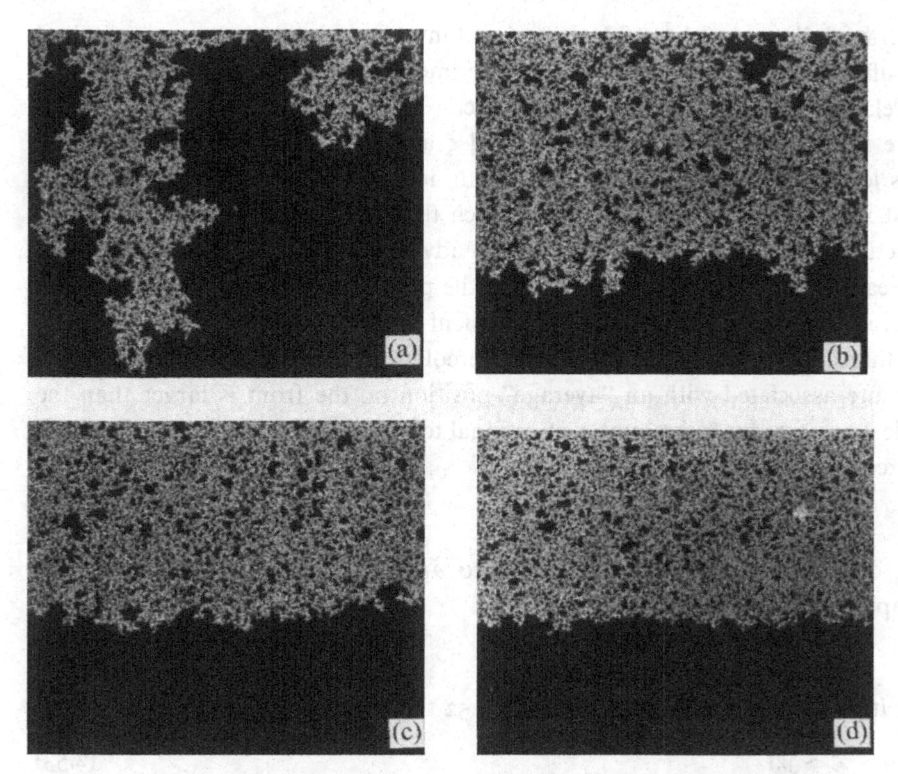

Figure 2.7 Patterns generated by the slow displacement of a wetting fluid (glycerin/water colored with 1 g of nigrosin per kilogram of liquid) by a non-wetting fluid (air). Part (a) shows results obtained for a horizontal cell. Parts (b), (c) and (d) show patterns from cells inclined at 1.5°, 4° and 13°, respectively. In these experiments, the dense fluid (black) was withdrawn from the bottom of the inclined cell, which is located at the bottom of each part of the figure. This figure was provided by A. Birovljev.

by adding a gradient to the distribution of thresholds representing "quenched disorder" on the lattice. In the perturbed model, the thresholds are given by

$$t_i = R_i + gh_i, \tag{2.50}$$

where R_i is a random number selected from a uniform distribution over the range $0 < R_i < 1$ at the ith site, h_i is the distance from the ith site to the edge of the lattice at which "growth" begins and g is the magnitude of the threshold gradient. The simulated patterns are qualitatively and quantitatively similar to those shown in figure 2.7.

In gradient stabilized displacements, the advancing front can be described in terms of a string of self-similar fractal "blobs" of size ξ. On short length scales, the front has a fractal dimensionality of 4/3, and on long length scales it is flat. As the non-wetting fluid advances through the porous medium, it can surround regions of wetting fluid and isolate them from the main body of wetting fluid. This process is called "trapping". Although trapping changes the fractal dimensionality of 2-dimensional invasion percolation clusters, it does not affect the structure of the external perimeter. The relationship between the correlation length ξ and the gradient g can be obtained, if it is assumed that there is only one correlation length [219] ξ associated with the invasion front. This length describes both the front width, in a direction parallel to the mean front velocity, and the distance over which the lateral correlations in the front structure persist. In order for the front to advance through the porous medium, the leading edge of the front must reach the percolation threshold (the pressure difference across the front must be sufficient to overcome a fraction $p \approx p_c$ of the thresholds, where p_c is the critical percolation fraction). Consequently, the pressure associated with an "average" position on the front is larger than the critical pressure by an amount proportional to $g\xi$, and the percolation threshold is exceeded by an amount δp given by

$$\delta p = (p - p_c) \sim g\xi. \tag{2.51}$$

The correlation length is also related to δp by ordinary percolation theory (chapter 3, section 3.5) [220]

$$\xi \sim |p - p_c|^{-\nu}, \tag{2.52}$$

and it follows from equations 2.51 and 2.52 that

$$\xi \sim |g|^{-\nu/(\nu+1)}. \tag{2.53}$$

This relationship between the characteristic length ξ and the gradient g has been confirmed by both experiments and computer simulations [212].

In the experiments, the correlation length scales as $\xi \sim B_\circ^{-\nu/(\nu+1)}$, where $B_\circ$ is the Bond number, which expresses the ratio between the gravitational forces and the capillary forces. The Bond number is defined as $B_\circ = |\delta\rho|g'\epsilon/(\Gamma/\epsilon) = |\delta\rho|g'\epsilon^2/(\Gamma)$, where g' is the acceleration due to gravity, $\delta\rho$ is the density difference between the two fluids, Γ is the interfacial surface tension and ϵ is the characteristic size of the pores in the porous medium. In a series of geometrically similar porous media, such as random packings of particles with the same size distribution and the same composition (the same contact angle), the correlation length $\xi(g', \Gamma, \epsilon)$ will scale as $\xi \sim B_\circ^{-\nu/(\nu+1)}$ if ξ is measured in units of ϵ. In this case, the length ϵ characterizes both the particle size and the pore size. In absolute units, the correlation length ξ scales as $\xi \sim \epsilon B_\circ^{-\nu/(\nu+1)}$.

After the front has passed, a pattern is left behind that can also be characterized by the same correlation length ξ. In this case, the internal structure can be thought of in terms of a dense packing of blobs of size ξ that have a self-similar fractal internal structure, like that of an unperturbed invasion percolation cluster ($g \to 0$). The length ξ also characterizes the maximum hole size in the internal structure. The hole size distribution is given by

$$N_s(g) = s^{-\tau} f(s/s^*(g)), \tag{2.54}$$

where $\tau = (D + d)/d$ (equation 2.49) and $s^*(g) \sim \xi^d$ ($s^*(g) \sim g^{-\nu d/(1+\nu)}$) is the maximum or characteristic hole size. The function $f(x)$ in equation 2.54 is constant for $x \ll 1$ and decays faster than any power of x for $x \gg 1$.

Similar ideas can be used in the case of a destabilizing gradient. In a destabilizing gradient, a single "finger" with a width ξ propagates along the direction of the gradient [221, 222]. In this case, the pressure across the fluid–fluid interface is smaller behind the advancing tip than it is at the tip. This prevents the finger from spreading too far in the lateral direction(s). Equations 2.51 and 2.52 can still be applied, and the blob size is related to the magnitude of the gradient by equation 2.53. Now, the displacement pattern can be described in terms of a self-affine string of blobs directed along the gradient. (Self-affine fractals are discussed later in this chapter.) If the fluid is injected through a surface that is perpendicular to the direction of the gradient instead of at a point, the surface becomes covered by a layer of blobs of size ξ, and a single finger emerges from a random position on the surface. This is illustrated in figure 2.8.

Fractal blob models have been used for many years in polymer physics [4]. A good example is provided by the work of Pincus [223] on stretched polymer chains. In the unstretched state, the mean end-to-end distance Z is given by

$$Z \sim N^\nu, \tag{2.55}$$

where N is the number of monomers in the chain. This power law relationship can be interpreted in terms of a fractal structure with $D = 1/\nu \approx 5/3$, in the 3-dimensional case. If the chain is stretched by a small amount δZ, the restoring force is given by

$$F = 3k_B T \delta Z / R_F^2, \tag{2.56}$$

where k_B is the Boltzmann constant, T is the temperature and R_F is the "Flory radius" ($R_F \sim N^\nu$). It follows from equations 2.55 and 2.56 that for small forces $\delta Z \sim N^{2/D}$. However, at large chain extensions, the chain forms a string of blobs with a characteristic size ξ, aligned in the direction of the extensional force. Within each blob, the structure is like that of an unperturbed self-avoiding random walk ($D \approx 5/3$). The structure inside a blob is not perturbed very much by the force F, and equation 2.56 is approximately correct. The characteristic blob size depends on the force F and can be estimated from equation 2.56 since $\delta Z_b \approx \xi \approx R_F$ for a typical blob, where $\delta Z_b \approx Z_b$ is the end-to-end distance for a polymer blob, so that $\xi \sim F^{-1}$. This implies that the mass (number of monomers) in a blob scales as F^{-D}, so that the length L of the stretched chain is given by the product of the number of blobs (NF^D) and the correlation length $\xi \sim F^{-1}$. Consequently,

$$L = \delta Z = \sum \delta Z_b \approx N\xi F^D \sim NF^{D-1}. \tag{2.57}$$

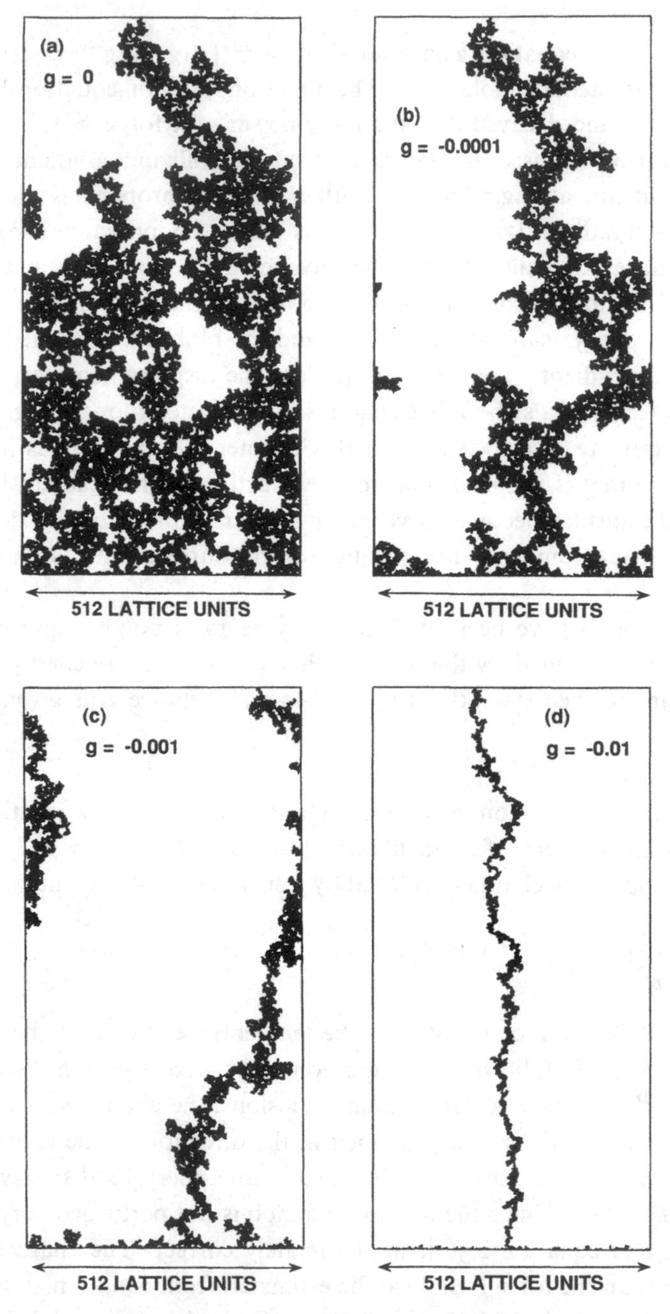

Figure 2.8 Patterns generated by simulations of invasion percolation with a destabilizing gradient. Parts (a), (b), (c) and (d) show patterns generated with destabilizing gradients of 0, -10^{-4}, -10^{-3} and -10^{-2}, respectively. The decrease of the blob size or correlation length ξ with increasing magnitude of the destabilizing gradient can be clearly seen. The length ξ corresponds to both the finger width and the depth of penetration of those parts of the invading fluid that are not part of the dominant finger.

The crossover between the small and large strain scaling regimes can be described by the scaling form

$$L = \delta Z = \sum \delta Z_b \sim F N^{2/D} f(N^{1/D} F), \qquad (2.58)$$

where the scaling function $f(x)$ has the form $f(x) = const.$, for $x \ll 1$. To have the correct dependence of L on the chain length N for large stresses ($x \gg 1$), the scaling function $f(x)$ must have the form $f(x) \sim x^{D-2}$, for $x \gg 1$, which is consistent with equation 2.57.

Similar models can be used to describe polymers in confined geometries. For example, a polymer confined to a layer of good solvent in a narrow gap between two parallel walls with a separation a can be described in terms of a 2-dimensional self-avoiding random walk of blobs of size $\xi \approx a$ that have an internal fractal structure like that of a 3-dimensional self-avoiding random walk [224]. Similarly, a polymer chain confined to a narrow pore or capillary of diameter a can be described in terms of a linear array of blobs of size $\xi \approx a$, with a similar 3-dimensional self-avoiding random walk internal structure [224].

The stretched polymer and gradient stabilized and destabilized invasion percolation models provide good illustrations of the "fractal blob" picture for describing systems with characteristic length scales. Similar fractal blob models will be employed frequently in this book. Under many circumstances, naturally generated systems can be described as self-similar fractals for length scales in the range $\epsilon \leq \ell \leq \xi$ and as Euclidean structures ($D = d$) outside this range, where ξ is the blob size.

Most attempts to analyze the spatial distribution of galaxies have been based on the assumption that the two-point position correlation function $C(r)$ has the form [225]

$$C(r) = <\rho> + f(r), \qquad (2.59)$$

where r is distance and $<\rho>$ is the mean number density of galaxies in the observable universe or the galaxy sample [225]. Observation supports the idea that the function $f(r)$ has the form $f(r) = Ar^{-\alpha}$ with $\alpha \approx 1.75$. This form for $C(r)$ indicates that there is a galaxy correlation length ξ given by $f(\xi) \approx <\rho>$. It is often supposed that the function $f(r)$ in equation 2.59 decreases more rapidly than a power law with increasing r for distances r greater than about 2ξ. This suggests a fractal blob model with blobs of size $\xi \approx 5$ Mpc (megaparsecs) and a short length scale fractal dimensionality of $D = 1 - \alpha = 1.25$. If, instead, the distribution of galaxies is fractal on all length scales larger than an inner cut-off, then the mean number density $<\rho>$ will approach zero, for a sufficiently large sample volume. However, the Earth is located in a region of high density,[4] and

4. In a fractal the density measured from *any* occupied point decays as a power of the distance (with large fluctuations). This power law decay, with the Earth as origin, does *not* imply that the Earth occupies a special position in the universe.

any estimate of $< \rho >$, based on data from a finite volume with the Earth as origin, will lead to a finite value for $< \rho >$ that will depend on the length scale over which ρ is measured. Consequently, the value estimated for ξ will depend on the size (length scale) L of the sample. Only if $L \gg \xi$ can reliable estimates for the mean density and correlation length be obtained. This condition is not satisfied by the available observational data. In fact, the data available at the time of writing are consistent with the idea that the universe has a fractal (or multifractal) structure [9] on the longest length scales. While the details of the scaling structure await refinement, it appears that the two-point correlation function for galaxies, galaxy clusters and quasars may have a power law form ($C(r) \sim r^{-\alpha}$), with $\alpha \approx 1.75$ corresponding to a fractal dimensionality D of ≈ 1.25, on length scales up to at least 100–200 Mpc [13, 226], the largest length scale for which useful data are available.

It has been argued [225] that the homogeneity of maps of the angular positions of galaxies indicates that the galaxy distribution is uniform on long length scales. However, the lacunarity of a fractal [2] has a strong influence on the perception of homogeneity, and a low lacunarity fractal will appear to be much more homogeneous than a high lacunarity fractal with the same fractal dimensionality [2]. The orthogonal projection of a 1.25-dimensional fractal universe, based on a 3-dimensional galaxy map, onto a plane, will result in a 2-dimensional map in which the galaxy distribution also appears to have a fractal dimensionality of 1.25. On the other hand, an angular map is not the same as a projection onto a plane, and it will appear to be more homogeneous than a projection onto a plane [227]. Based on these and other considerations, Pietronero and Coleman [226, 227] have concluded that the 5 Mpc correlation length is an artifact of the data analysis and that the fractal correlations extend to at least 200 Mpc.

It has also been proposed that the distribution of mass in the universe can be described in terms of a multifractal model (multifractals are discussed later in this chapter). Coleman and Pietronero [227] analyzed a sample of 442 galaxies for which the galaxy masses were estimated from their luminosities. Evidence was found for a multifractal distribution of galaxy masses lying on a support with a fractal dimensionality $D_{q=0}$ of 1.5 ± 0.1. This corresponds to the ordinary fractal dimensionality, if the galaxy masses are not taken into account. This value for the fractal dimensionality is larger than the value of $D \approx 1.25$ obtained in most other studies. A value of 1.5 ± 0.1 was estimated for the information dimensionality $D_{q=1}$, and a value of 1.3 ± 0.1 was estimated for $D_{q=2}$. The family of dimensionalities D_q that can be used to characterize multifractals is discussed later in this chapter (section 2.8). This is not a large data set, and the uncertainties in both the interpretation and analysis of the data set are quite large. This may account for the discrepancy between the value of about 1.5 found for D_0 and the somewhat smaller values found for D_0 in other studies.

At present, it appears that the universe has structure on length scales up to at least 100–200 Mpc. It is not yet clear if the statistics of this structure can be adequately represented by a *simple* fractal model. The development of a consensus is thwarted by sampling biases and interpretation difficulties. This topic continues to be an active, controversial research area [228], and the origin of the large scale structure is an important unresolved problem.

Another important type of crossover is "resolution dependent crossover" in which details on short length scales are absent due to the limitations of measurement techniques or subsequent analysis. A familiar example is provided by vascular networks. On short length scales, most of the vascular network consists of a capillary bed in which a blood vessel (capillary) is close to every piece of tissue. In this sense, the vascular network is space filling and $D = d$ [2]. If a vascular network is examined at much lower resolution, only the major blood vessels will be seen. This low resolution network may be described in terms of a tree of linear ($D = 1$) segments, and the whole network may appear to be fractal. However, in many cases, it seems to be more appropriate to describe such structures in terms of a crossover from $D = 1$ on short length scales to $D = d$ on longer length scales with a crossover length scale ξ that depends on the resolution ϵ_S, the diameter of the smallest blood vessel that can be seen. Other examples of resolution dependent crossovers include rivers and fracture networks. These examples will be discussed elsewhere [57]. In some cases, the width or related quantities such as the flow rate or area drained through the segment, can be used to define a measure on an otherwise linear segment, and the pattern with this measure may be a multifractal (see section 2.8 below and appendix B).

A simple model that can be used to illustrate the idea of a resolution dependent crossover [229] is based on the Eden growth model, described in chapter 3, section 3.4.1. In the standard Eden model, unoccupied sites on the perimeter of a growing cluster are selected randomly and filled. In the Eden tree model, a bond is formed between the randomly selected, unoccupied perimeter site that is filled at each stage in the growth process and one of its previously occupied nearest-neighbor sites, which is selected at random.[5] In a completed cluster, each site at position $\mathbf{r}$ connects $n(\mathbf{r})$ other sites to the origin of the cluster via the tree of bonds ($n(\mathbf{r})$ sites would be disconnected from the main part of the cluster by removing the site at $\mathbf{r}$). In figure 2.9(a), each of the sites in the cluster is represented by a circle with a radius proportional to $n(\mathbf{r})^{1/2}$. In terms of river networks, $n(\mathbf{r})$ can be thought of as the area of land surface that drains through the river channel at position $\mathbf{r}$. In this case, the empirical data support

5. In the original Eden model (Eden model B), described in chapter 3 section 3.41, an occupied perimeter site is selected at random and one of its randomly selected nearest neighbors is filled, at each stage in the simulation. In this model, the bonds (between the occupied perimeter site and the newly occupied site) appear in a more natural manner.

a power law relationship between the river width $w(\mathbf{r})$ and the flow $Q(\mathbf{r}) \sim n(\mathbf{r})$ ($w(\mathbf{r}) \sim Q(\mathbf{r})^a$ with $a \approx 1/2$) [57]. Consequently, the "Eden tree" model, with branch widths at $\mathbf{r}$ proportional to $n(\mathbf{r})^{1/2}$, can be thought of as a primitive river model. Parts (b), (c), (d) and (e) in the same figure show only those parts of the cluster in which $n(\mathbf{r})$ exceeds a threshold N_t. In parts (b) and (c) ($N_t = 10$ and $N_t = 100$), it is clear that those sites for which $n(\mathbf{r})$ exceeds the threshold form a network that is uniform on all but quite short length scales. The clusters shown in parts (d) and (e) look more "fractal". However, this is an illusion. If the growth process were continued to a much larger length scale and the sites with $n(\mathbf{r}) > 1000$ or $n(\mathbf{r}) > 10\,000$ were displayed, then the resulting networks would look like those shown in figures 2.9(b) and 2.9(c).

Figure 2.10(a) shows the two-point density–density correlation functions ($C(r)$) obtained from 116 clusters grown to a maximum radius of 500 lattice units. Figure 2.10(b) shows the result of an attempt to scale these correlation functions using the threshold N_t as a scaling parameter with the scaling form

$$C(r) = N_t^{-\nu} f(r/N_t^{\nu}). \tag{2.60}$$

A reasonably good data collapse was obtained with $\nu \approx 0.37$. This scaling collapse and the form of the correlation function indicate that the structure can be described in terms of a crossover from $D \approx 1.15$, on short length scales, for which $\ell < \xi = N_t^{\nu}$, to $D = 2$, on longer length scales, for which $\ell > \xi = N_t^{\nu}$. The effective short length scale fractal dimensionality of 1.15 is consistent with an asymptotic fractal dimensionality of 1.0 for $\epsilon \ll \ell \ll \xi$.

Sandau and Kurz [230] have reported fractal dimensionalities in the range $1.25 \leq D \leq 1.45$ for the blood vessels in the chorioallantoic membrane of incubated chicken eggs and for patterns obtained from a model for the growth of the blood vessels that leads to patterns very similar to those shown in figure 2.9. The fractal dimensionalities were obtained from a box counting analysis of CCD camera images of the chorioallantoic membrane blood vessels and prints generated from the model. Only a very small range of length scales was used in the analysis (7–15 pixels). It appears that the results obtained in this manner should be regarded as effective fractal dimensionalities, resulting from a resolution dependent crossover similar to that described above.

A richer crossover behavior can be found if the widths of the bonds or links in a network are taken into account. If all of the widths w of the elements in a network in d-dimensional space are much smaller than their lengths l_b (ξ_3), then $D = d$ for length scales ℓ given by $\ell \ll w$, $D = 1$ for $w \ll \ell \ll l_b$, and the effective fractal dimensionality is d for $\ell \gg l_b$ if the network is uniform on long length scales. In many cases, the links have a distribution of widths. Figure 2.11 shows two examples. Loopless networks are common in structures such as branched rivers, vascular systems, trees and many other biological structures. Figure 2.11(a) can be considered to be a crude 2-dimensional model

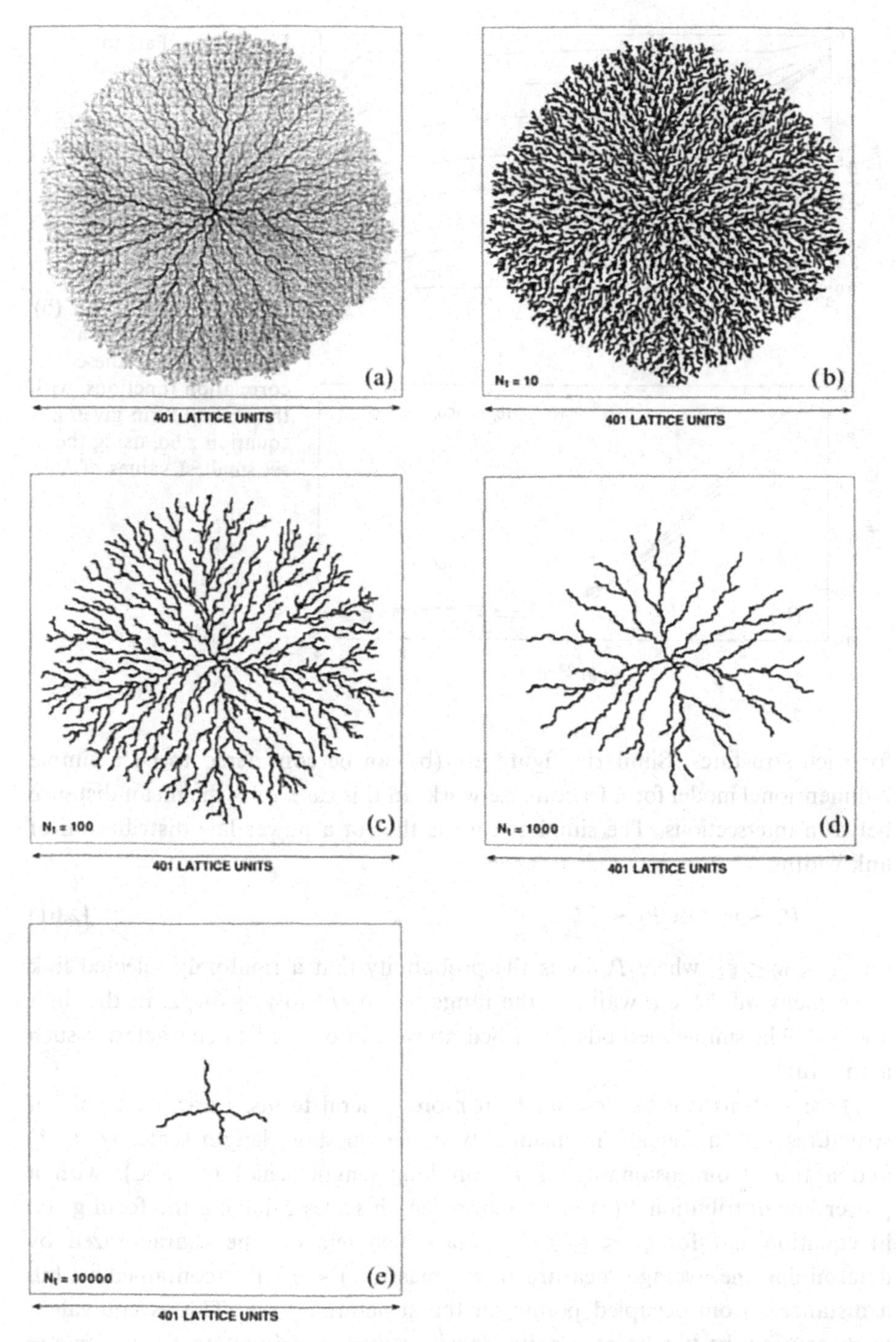

Figure 2.9 Part (a) shows an Eden model cluster in which all of the sites at positions **r** are represented by circles with diameters proportional to $n(\mathbf{r})^{1/2}$, where $n(\mathbf{r})$ sites would be disconnected from the cluster if the site at **r** was removed. Parts (b), (c), (d) and (e) show those parts of the cluster for which $n(\mathbf{r})$ exceeds a threshold N_t of 10, 100, 1000 and 10 000, respectively.

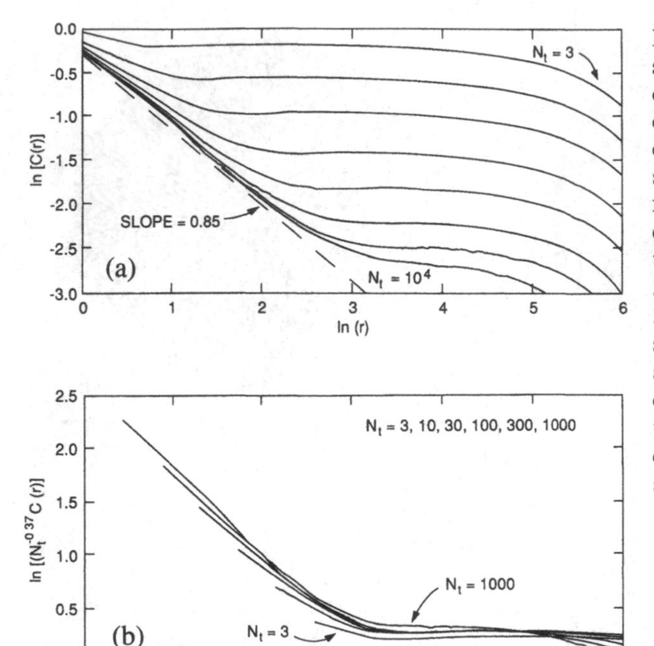

Figure 2.10 Part (a) shows the two-point density–density correlation functions, for clusters similar to those shown in figures 2.9(b)–(e), with eight values for the threshold N_t ($N_t =$ 3, 10, 30, 100, 300, 1000, 3000 and 10 000). Part (b) shows the results of an attempt to scale these correlation functions, with the scaling form given in equation 2.60, using the six smallest values of N_t.

for such structures. Similarly, figure 2.11(b) can be considered to be a simple 2-dimensional model for a fracture network. In this case, ξ_3 is the mean distance between intersections. The simplest case is that of a power law distribution of link widths

$$P_w \sim w^{-\tau} \text{ or } P_\xi \sim \xi^{-\tau} \tag{2.61}$$

for $\xi_1 \leq w \leq \xi_2$, where $P_w \delta w$ is the probability that a randomly selected link or segment will have a width in the range $w - \delta w/2$ to $w + \delta w/2$, in the limit $\delta w \to 0$. The simple methods described above can be used to characterize such a structure.

These systems can be described, in more general terms, as an ensemble of structures with a fractal dimensionality of D_1 on short length scales ($\ell \ll \xi$) and a fractal dimensionality of D_2 on long length scales ($\ell \gg \xi$), with a power law distribution $P(\xi)$ of crossover length scales ξ having the form given in equation 2.61 for $\xi_1 \leq \xi \leq \xi_2$. Such a system can be characterized by determining the average measure (area, mass ...) $< \mu(r) >$ contained within a distance r from occupied points on the structure or set. This is equivalent to measuring the two-point density–density correlation function $C(r)$. For any distance r, in the range $\xi_1 < r < \xi_2$, two cases must be considered. For structures in which $r \ll \xi$, essentially all balls of radius r centered on the structure will contain a measure proportional to r^{D_1}, and the number of possible centers for the balls is proportional to $\xi^{D_1}\xi^{-D_2}$. Consequently, the contribution μ_1 from structures in which $r < \xi$ to $< \mu(r) >$ can be approximated by

$$\mu_1(r) \sim r^{D_1} \int_r^{\xi_2} P(\xi)\xi^{D_1-D_2} d\xi. \tag{2.62}$$

A second contribution comes from those members of the ensemble for which $\xi < r$. If $\xi \ll r$, a ball of radius r will contain a measure that is proportional to $\xi^{D_1}(r/\xi)^{D_2}$, and the number of possible origins for the balls is again proportional to $\xi^{D_1}\xi^{-D_2}$. Consequently, the contribution μ_2 from these structures to $< \mu(r) >$ can be approximated by

$$\mu_2(r) \sim r^{D_2} \int_{\xi_1}^{r} \xi^{D_1-D_2} P(\xi)\xi^{D_1-D_2} d\xi. \tag{2.63}$$

It follows from equations 2.62 and 2.63 that

$$\mu_1(r) \sim r^{2D_1-D_2-\tau+1}, \qquad \text{if } \tau > D_1 - D_2 + 1 \text{ and } r \ll \xi_2, \tag{2.64}$$

$$\mu_1(r) \sim r^{D_1}\xi_2^{D_1-D_2-\tau+1}, \qquad \text{if } \tau < D_1 - D_2 + 1 \text{ and } r \ll \xi_2, \tag{2.65}$$

and

$$\mu_2(r) \sim r^{2D_1-D_2-\tau+1}, \qquad \text{if } \tau < 2D_1 - 2D_2 + 1 \text{ and } r \gg \xi_1, \tag{2.66}$$

$$\mu_2(r) \sim r^{D_2}\xi_1^{2D_1-2D_2-\tau+1}, \qquad \text{if } \tau > 2D_1 - 2D_2 + 1 \text{ and } r \gg \xi_1. \tag{2.67}$$

Consequently, the sum of these two contributions $< \mu(r) > \approx \mu_1(r) + \mu_2(r)$ is given by

$$< \mu(r) > \sim r^{2D_1-D_2-\tau+1}, \tag{2.68}$$

for $\xi_1 \ll r \ll \xi_2$, $D_1 - D_2 + 1 < \tau < 2D_1 - 2D_2 + 1$ and $D_1 > D_2$.

A third contribution from members of the distribution that do not satisfy the inequalities $\xi \ll r$ and $\xi \gg r$ has not yet been considered. In the limit $\xi_2/\xi_1 \to \infty$, the contribution from components with $\xi \approx r$ will be negligible[6] and

6. The total contribution to $\mu(r)$ is given by

$$\mu(r) \sim r^{D_1} \int_{\xi_1}^{\xi_2} P(\xi)\xi^{D_1-D_2} f(\xi/r) d\xi, \tag{2.69}$$

where $f(x) = f(\xi/r)$ is a scaling function (see below) with the form $f(x) \sim x^{(D_1-D_2)}$ for $x \ll 1$ and $f(x) = const.$ for $x \gg 1$. Making the substitution $x = \xi/r$ in this equation, it is apparent that

$$\mu(r) \sim r^{D_1} \int_{x_1}^{x_2} r^{-\tau} x^{-\tau} r^{(D_1-D_2)} x^{(D_1-D_2)} f(x) r dx \tag{2.70}$$

or

$$\mu(r) \sim r^{(2D_1-D_2-\tau+1)} \int_{x_1}^{x_2} x^{(D_1-D_2-\tau)} f(x) dx. \tag{2.71}$$

Since $f(x) \sim x^{(D_1-D_2)}$ for $x \ll 1$ and $f(x) = const.$ for $x \gg 1$, this integral will have a constant, finite value if $D_1 - D_2 + 1 < \tau < 2D_1 - 2D_2 + 1$, $x_1 \ll 1$ and $x_2 \gg 1$. Under these conditions, $\mu(r) \sim r^{(2D_1-D_2-\tau+1)}$.

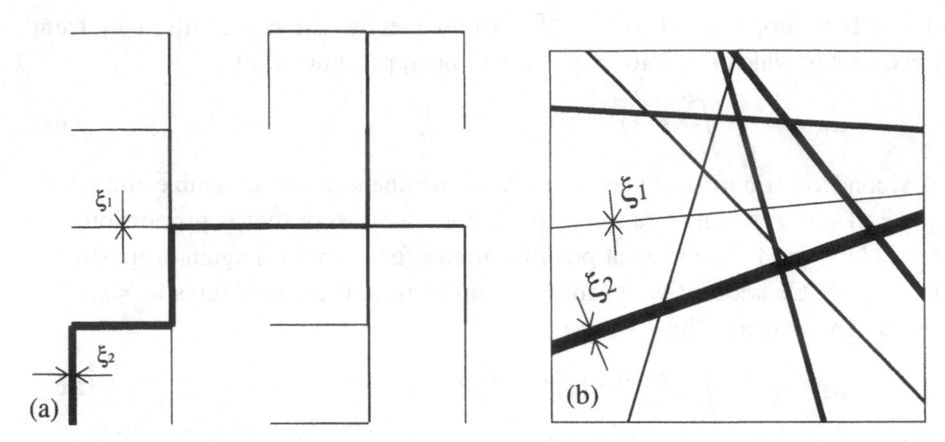

Figure 2.11 Examples of patterns containing linear elements with a distribution of widths. Part (a) shows a "tree", like that associated with branched river networks. Part (b) shows a network of intersecting lines that can be used as a crude model for fracture networks. This figure was provided by T. Sun.

$< \mu(r) > \sim r^{2D_1-D_2-\tau+1}$. If the range of length scales $\xi_1 < \ell < \xi_2$ is large enough, equation 2.68 should be accurate for distances r in the range $\xi_1 \ll r \ll \xi_2$, and an effective fractal dimensionality of $D_x = 2D_1 - D_2 - \tau + 1$ will be measured in the crossover between the short length scale $D = D_1$ and long length scale $D = D_2$ regimes if $D_1 - D_2 + 1 < \tau < 2D_1 - 2D_2 + 1$.

These results apply equally well to ensembles of both Euclidean and fractal structures. A distribution of rods, in 3-dimensional space, or strips in 2-dimensional space ($D_1 = d$ and $D_2 = 1$) are important special cases. For such a distribution of linear elements,

$$< \mu(r) > \sim r^{2d-\tau}, \qquad \text{if } d < \tau < 2d-1 \text{ and } \xi_1 \ll r \ll \xi_2. \tag{2.72}$$

In this case, an effective fractal dimensionality of $D = 2d - \tau$ may be measured from $< \mu(r) >$. Figure 2.12 shows the qualitative form of the two-point density–density correlation function $C(r)$ for a random distribution of strips in 2-dimensional space ($D_1 = d = 2$ and $D_2 = 1$). This effective fractal behavior can be regarded as being a consequence of the superposition of a distribution of crossovers from $D = d$ for $r \ll \xi_i$ to $D = 1$ for $r \gg \xi_i$, where ξ_i is the width of the ith bond or link. Another important case is that of sheets with a distribution of thicknesses in 3-dimensional space ($D_1 = 3, D_2 = 2$). Such a model could be used to represent a fractured rock. In this case, ξ_i would represent the aperture of the ith fault or joint.

Another frequently encountered example is that of a power law distribution of Euclidean or fractal objects, all having a fractal dimensionality of $D_\circ$ ($D_1 = D_\circ$, $D_2 = 0$). In this case, $\mu_1(r) \sim \mu_2(r) \sim r^{2D_\circ-\tau+1}$, if $2D_\circ + 1 > \tau > D_\circ + 1$. This means that the density–density correlation function will have the form $C(r) \sim$

$r^{2D_\circ-\tau+1-d}$ and the scattering intensity $S(q)$ will have the form $S(q) \sim q^{-D'}$, where $D' = 2D_\circ - \tau + 1$.[7] This complicates the problem of interpreting power law scattering intensity structure factors $S(q)$, since a power law structure factor can arise from large fractal scatters or a power law distribution of Euclidean or fractal scatterers [231]. Consequently, a power law scattering structure factor *cannot* be interpreted in terms of scattering from a fractal, in the absence of additional information. Size distributions must also be taken into account in the interpretation of scattering experiments [231]. In the case of scattering from Euclidean objects with a power law distribution of sizes ($P(r) \sim r^{-\tau}$, where r is the radius), $S(q) \sim q^{-(2d-\tau+1)}$ if $2d+1 > \tau > d+1$ [201].

A box counting analysis will give results that are quite different from those obtained from the correlation function analysis described in the previous few paragraphs for systems with a power distribution of crossover lengths. In this case, the number of boxes $N(\ell)$ of size ℓ, in the range $\xi_1 \ll \ell \ll \xi_2$, required to cover a pattern characterized by the short length scale fractal dimensionality D_1 and the long length scale fractal dimensionality D_2, is given by

$$< N(\ell) > = \ell^{-D_1} \int_{\xi_1}^{\xi_2} P(\xi)\xi^{D_1-D_2} f(\xi/\ell) d\xi, \tag{2.73}$$

where the function $f(x) = f(\xi/\ell)$ has the form $f(x) \sim x^{(D_2-D_1)}$ for $x \ll 1$ and $f(x) = const.$ for $x \gg 1$. The number of boxes of size ℓ can also be represented by the scaling form

$$< N(\ell) > = \ell^{-D_2} \int_{\xi_1}^{\xi_2} P(\xi)\xi^{D_1-D_2} g(\xi/\ell) d\xi, \tag{2.74}$$

where $g(x) = const.$ for $x \ll 1$ and $g(x) \sim x^{D_1-D_2}$ for $x \gg 1$. Equation 2.73 can be written in the form

$$< N(\ell) > = \ell^{1-D_2-\tau} \int_{x_1}^{x_2} x^{D_1-D_2-\tau} f(x) dx, \tag{2.75}$$

where $x = \xi/\ell$, $x_1 = \xi_1/\ell$ and $x_2 = \xi_2/\ell$. If the exponent τ is large ($\tau > 1$), there will be very few members of the ensemble with large values of ξ or x. Consequently, the integral in equation 2.73 will be dominated by the lower cut-off and $< N(\ell) > \sim \ell^{-D_2}$. Similarly, if τ is small enough ($\tau < 1 + D_1 - D_2$), the integral will be dominated by the upper cut-off and $< N(\ell) > \sim \ell^{-D_1}$. If $\tau < 1 + D_1 - D_2$ and $\tau > 1$, both cut-offs will be important, and $< N(\ell) >$ will have the form $< N(\ell) > = c_1\ell^{-D_1} + c_2\ell^{-D_2}$, for lengths in the range $\xi_1 \ll \ell \ll \xi_2$. An intermediate scaling regime, with $N(\ell) \sim \ell^{1-D_2-\tau}$, can be found only if $1 > \tau > 1 + D_1 - D_2$. Consequently, an intermediate scaling regime is expected only if $D_2 > D_1$.

7. If the distribution of cluster sizes (number of particles s) in an ensemble of aggregates is given by $N_s \sim s^{\tau_s}$, then $S(q) \sim q^{-D'}$, where $D' = D_\circ(3 - \tau_s)$ if $3 > \tau_s > 2$.

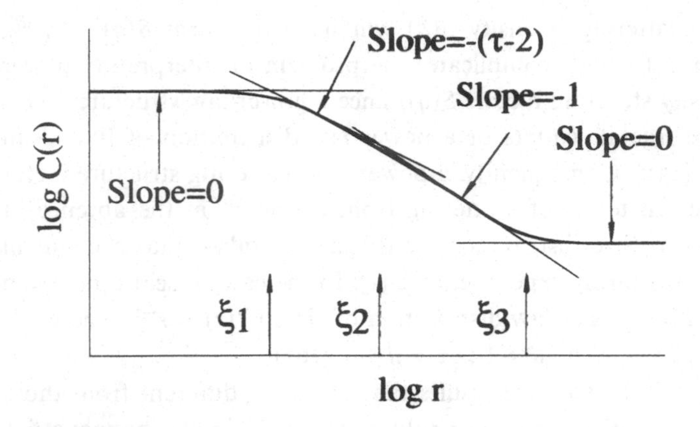

Figure 2.12 A qualitative indication of the form of the density–density correlation function $C(r)$, for patterns like those shown in figure 2.11 ($D_1 = d = 2$, $D_2 = 1$) that cover a broad range of length scales. The difficulty of obtaining reliable estimates of the exponents, for length scales in the range $\xi_1 < \ell < \xi_3$, is illustrated in this figure, which was provided by T. Sun.

Crossovers can also be a consequence of a change in amplitudes rather than a change of exponents. For example, the function $y(x)$, which may be the result of analyzing a pattern, may be described by the power law

$$y(x) = c_s x^a, \tag{2.76}$$

for small values of x ($x \ll \xi$), and by

$$y(x) = c_l x^a, \tag{2.77}$$

for large values of x ($x \gg \xi$), so that

$$y(x) = f(x/\xi)x^a, \tag{2.78}$$

where $f(x) = c_s$ for $x \ll 1$ and $f(x) = c_l$ for $x \gg 1$. It is not unusual for $f(x)$ to "look like" a power law over one or two decades of x in the crossover so that an effective exponent, different from a, may be measured in this regime.

Very often, the curvature associated with crossovers is apparent under careful inspection. All too often this curvature is ignored, and effective exponents are interpreted as real (asymptotic) exponents.

2.4 Power Law Distributions

In many areas of science and technology, the distribution of some measured quantity is an important characteristic. Familiar examples include the height distribution in a human population, the distribution of crater sizes on the moon and the distribution of incomes in a particular profession or country. In many cases, these distributions have a power law form

$$P(x) \sim x^{-\tau}, \tag{2.79}$$

where $P(x)\delta x$ is the probability that, for a randomly selected sample, the quantity X lies in the range x to $x+\delta x$, with $\delta x \rightarrow 0$. The function $P(x)$ is often called the probability density for X. In many cases, it is more convenient to characterize experimental or simulation data in terms of the cumulative distribution

$$P(X > x) \sim x^{-\tau_c}, \tag{2.80}$$

where $P(X > x)$ is the probability that, for a randomly selected sample, the measured quantity X will be larger than x. The exponents τ_c and τ are related by $\tau_c = \tau - 1$. In the same way that the geometrical scaling in physical fractals or prefractals is limited to length scales in the range $\epsilon < \ell < L$, power law distributions, or "Pareto distributions", also extend over a limited range of scales in real systems ($x_{min} < x < x_{max}$).

The measurement of the characteristic exponents τ or τ_c in equations 2.79 and 2.80, is the main objective of a large body of research. In a typical experiment or simulation, the data consist of a set of numbers $\{x\}$, representing the values measured for the quantity X. These numbers can be converted into estimates for the probability $P(x)$ by counting how many of the numbers $x_i \subset \{x\}$ lie in the ranges or bins b_n, where ($x_n^{min} < x_i < x_n^{max}$). If N_n members of $\{x\}$ lie in bin b_n, then the probability density P_n in the bin is proportional to $N_n/(x_n^{max} - x_n^{min})$. The most straightforward way to estimate the exponent τ is to fit a straight line to the data points ($\log < x_n >, \log P_n'$), where $< x_n >$ is the average "position" of the nth bin and $P_n' = N_n/(x_n^{max} - x_n^{min})$. In practice, it is best to use a logarithmic distribution of bin widths ($x_n^{max}/x_n^{min} = const.$), and to allow the bins to overlap.[8] However, when the data set $\{x\}$ consists of integers, there is usually sufficient data, for small integer values of x, to use a bin for each integer up to a value of, say, 20 and then to use logarithmic bins for larger values of x. In the case of integer data, the quantities P_n' should be calculated from $P_n' = N_n/I(x_n^{max}, x_n^{min})$, where $I(x_n^{max}, x_n^{min})$ is the number of integers in the range x_n^{min} to x_n^{max}.

Very often, this simple procedure gives poor results, because of large statistical uncertainties and the effects of cut-offs and crossovers. In many cases, the distribution has the form

$$P(x) \sim x^{-\tau} e^{-[(x/x^*)^a]}, \tag{2.81}$$

8. It may appear to be preferable to use the data points ($< \log x_n >, \log P_n'$), where ($< \log x_n > = \log[(x_n^{max} x_n^{min})^{1/2}]$. However, if the bins have a constant logarithmic width ($x_n^{max}/x_n^{min} = const.$, or $\log x_n^{max} - \log x_n^{min} = const.$), these two procedures will give exactly the same exponent.

where x^* is a characteristic cut-off size for x. If sufficient data are available, a better estimate of the exponent τ can be obtained by fitting the data with the form given in equation 2.81, using a non-linear least squares fitting procedure.

The use of the cumulative distribution $P(X > x)$ has important advantages. It reduces the effects of statistical uncertainties and there is no need for binning, with its attendant uncertainties. The cumulative exponent τ_c can then be obtained by fitting a straight line to the coordinates $(\log x, \log P(X > x))$ or $(\log x, \log N(X > x))$, where $N(X > x)$ is the number of samples with $X > x$. However, the effects of cut-offs in the range of power law scaling are even more serious than they are in the measurement of τ from $\{(\log < x_n >, \log P'_n)\}$. This can be seen from the case where the distribution $\{x\}$ follows a power law up to $x = x^*$, and there are no larger events or samples. The corresponding cumulative distribution $P(X > x)$ is given by $P(X > x) \sim x^{-\tau_c} - (x^*)^{-\tau_c}$, where $\tau_c = \tau - 1$. Consequently, a least squares fit to the data $(\log x, \log P(X > x))$ that includes points with x near to x^* will give a value for τ_c that is too large. Similar effects occur for distributions of the form given in equation 2.81. In general, a least squares fit of a straight line to data on a log–log scale will give poorer values for the exponent τ_c from the cumulative distribution than the effective value obtained for τ from the ordinary probability distribution, for ranges of x that are near (within a few decades) to the cut-off x^*, providing that statistical effects are not too large. Very often, data from computer simulations are wasted by not measuring the cumulative distribution and/or not using appropriate binning to estimate the probability density for the rarer events that usually correspond to large values of x. Hard won experimental data are not as commonly abused in this manner.

In some cases, corrections are absent or minimal, even if very large values of the measured variable x are not present in the distribution. For example, in the growth of clusters from a line using a DLA or ballistic deposition model, there are no large clusters if the deposit is grown to a finite height. However, in this case, there is an excess of clusters with sizes near to $x^*(h)$, where h is the height of the deposit. These excess clusters correspond to clusters that would have grown larger if the growth process had been continued and would have contributed to the power law distribution of clusters with sizes greater than x^*. These excess clusters compensate exactly for the missing clusters, with sizes larger than x^*, in the cumulative distribution.

Equation 2.49 indicates that the exponent τ_c that describes the power law form of the cumulative distribution of the characteristic sizes (lengths) of the holes in a fractal, which subdivides the space in which it is embedded into separated regions, is equal to the fractal dimensionality D. If the exponent $\tau_{c,1}$ is measured by taking 1-dimensional "cuts" through the system, then it follows from equation 2.49 that $\tau_{c,1} = D_1$ will be obtained, where D_1 is the fractal dimensionality of a 1-dimensional cut through the fractal. Similarly, a

2-dimensional cut will give a distribution of 2-dimensional holes with a power law characteristic length (radius, diameter ...) distribution given by $\tau_{c,2} = D_2$. In general, $\tau_{c,n} = D_n$ for an n-dimensional cut and $\tau_{c,n} - \tau_{c,m} = m - n$. For this reason, the exponent τ_c for a power law size (radius, diameter ...) distribution of fragments is often called the "fractal dimensionality" of the distribution. In many simple cases, this fractal dimensionality is the fractal dimensionality of the "gaps" that divide the fragments, if all the gaps have the same, infinitesimal, widths or (in practice) if all the gap widths are smaller than the power law cut-off length ϵ. In general, the exponent τ_c is not related to an identifiable fractal structure and may have a value larger than that of the physical embedding space (d).

Random numbers sampled from a power law probability density $P(x) \sim x^{-\tau}$ are often needed in computer simulations. They can be obtained by first generating a random number R uniformly distributed over the range $R_1 < R < R_2$. The value x associated with the corresponding random sample is then given by

$$x = R^{-1/(\tau-1)}, \tag{2.82}$$

and x lies in the range $R_2^{-1/(\tau-1)} < x < R_1^{-1/(\tau-1)}$. The values of R_1 and R_2 can be selected to generate random power law distributions over any range of values $x_{min} < x < x_{max}$ using the values $x_{max}^{-(\tau-1)}$ and $x_{min}^{-(\tau-1)}$ for R_1 and R_2, respectively.

It is often necessary to express power laws in terms of several different, related quantities $\{x_i\}$. For example, the distribution of (spherical) particles may be expressed in terms of the particle radii r or volumes v. In general, the probability densities $P(x_1)$ and $P(x_2)$ are related by

$$P(x_1)dx_1 = P(x_2)dx_2 \tag{2.83}$$

or

$$P(x_1) = P(x_2)\frac{dx_2}{dx_1}. \tag{2.84}$$

For example, if the probability density for the particle radii $P(r)$ has the form $P(r) \sim r^{-\tau_r}$, and the probability density for the particle volumes $P(v)$ has the form $P(v) \sim v^{-\tau_v}$, then

$$P(v) = P(r)\frac{dr}{dv} \tag{2.85}$$

and

$$P(v) \sim v^{-\tau_v} \sim r^{-\tau_v d} \sim r^{-\tau_r} / \left(\frac{dv}{dr}\right) \sim r^{-\tau_r} r^{-d+1}, \tag{2.86}$$

so that

$$d\tau_v = \tau_r + d - 1 \tag{2.87}$$

or

$$\tau_v = (\tau_r/d) + 1 - (1/d). \tag{2.88}$$

It is evident that the corresponding cumulative distribution function exponents τ_{cv} and τ_{cr} are related by $\tau_{cv} = \tau_{cr}/d$. Since $\tau_{cv} = \tau_v - 1$ and $\tau_{cr} = \tau_r - 1$, it follows that $\tau_v - 1 = (\tau_r - 1)/d$, which is equivalent to equation 2.88.

2.5 Scaling

Both fractal and non-fractal structures can be characterized by the mean density $\rho(r)$ at a distance r measured from the center of mass or some other well defined point, or $C(r)$, the density–density correlation function, which is equivalent to $\rho(r)$ averaged over all positions in the structure as the origin at which $\mathbf{r} = 0$. For the case of a square with sides of length L, the density profile $\rho(r)$ with the center of the square as origin is given by

$$\rho(r) = \begin{array}{ll} 1 & \text{for } r \leq L/2 \\ 1 - 4\cos^{-1}(L/2r)/\pi & \text{for } L/2 \leq r \leq 2^{-1/2}L \\ 0 & \text{for } r > 2^{-1/2}L. \end{array} \tag{2.89}$$

Equation 2.89 can be written as

$$\rho(r) = f(r/L), \tag{2.90}$$

where the function $f(x)$ has the form $f(x) = 1$ for $x \leq 1/2$, $f(x) = 1 - (4/\pi)\cos^{-1}(1/2x)$ for $1/2 \leq x \leq 2^{1/2}$ and $f(x) = 0$ for $x > 2^{1/2}$. Similarly, the density–density correlation function $C(r)$ can be written as

$$C(r) = f'(r/L). \tag{2.91}$$

The function $f'(x)$ has the form $f(x) = 1$ for $x \ll 1$, and $f'(x)$ decays rapidly to zero for $x \gg 1$. The calculation of $f'(x)$ is, in many cases, a straightforward but tedious exercise. Similarly, the mass $M(r)$ within a distance r measured from the center of mass of a square of size L can be expressed in the form

$$M(r) = \pi r^2 f''(r/L), \tag{2.92}$$

where the function $f''(x)$ has the form $f''(x) = 1$ for $x < 1/2$, $f''(x)$ decays from 1 to zero in the range $1/2 > x > 2^{-1/2}$ and $f''(x) = 0$ for $x > 2^{-1/2}$. Equations 2.90, 2.91 and 2.92 are very simple examples of scaling forms, and they express the simple but important idea that objects of different sizes may have the same shape.

For structures of finite size, the two-point density–density correlation function will depend on the overall size of the structure described by a characteristic length R as well as the internal length r. In this case, the correlation function can be written as $C(r, R)$. The function $C(r, R)$ represents the density–density correlation function averaged over a large number of structures of the same size or characteristic length R. If large structures are, on average, related to small structures by a change of length scale, it is natural to assume that $C(r, R)$ is a homogeneous function of its arguments, so that

$$C(\lambda r, \lambda R) = \lambda^{-\alpha} C(r, R). \tag{2.93}$$

Taking a value of $1/R$ for λ, equation 2.93 becomes

$$C(r/R, 1) = R^{\alpha} C(r, R) \tag{2.94}$$

or

$$C(r, R) = R^{-\alpha} f(r/R). \tag{2.95}$$

According to this scaling assumption, the density–density correlation functions for systems of different sizes (R) can be expressed in terms of the scaling form given in equation 2.95. The function $f(x)$ is called the scaling function, and the exponent α is a scaling exponent. Equation 2.95 can be written in the form $C(r, R)R^{\alpha} = f(r/R)$, which can be compared with equation 1.17.

If equation 2.95 provides a valid description of the geometric scaling properties, then plots of $R^{\alpha} C(r, R)$ against r/R for structures of different sizes will fall on a common curve, the scaling function $f(x)$. This is illustrated for clusters generated using a 3-dimensional off-lattice model for diffusion-limited cluster–cluster aggregation [58] in figure 2.13. Figure 2.13(a) shows the two-point density–density correlation functions $C(r)$ for clusters of five different sizes (100, 300, 1000, 3000 and 10 000 particles). In this case, it is more convenient to select clusters of a particular size (number of particles, s) rather than clusters with a particular value, or small range of values, of the characteristic length R. Since s and R are related by

$$< R > \sim s^{1/D} \tag{2.96}$$

for fractal objects, equation 2.95 can be replaced by

$$C(r, s) = s^{-\alpha/D} f(r/s^{1/D}). \tag{2.97}$$

The exponent α in equation 2.97 is the codimensionality of the self-similar fractal in the embedding space, defined as $\alpha = d - D$, and equation 2.97 can be written as

$$C(r) = C(r, s) = s^{(D-d)/D} f(r/s^{1/D}). \tag{2.98}$$

In figure 2.13(b), plots of $\ln[s^{(d-D)/D} C(r)]$ against $\ln(s^{-1/D} r)$ are shown for the five different cluster sizes. The fact that these curves overlap, almost perfectly, confirms the simple scaling assumption expressed in equation 2.98. This shows that both the internal density correlations and the global scaling properties of

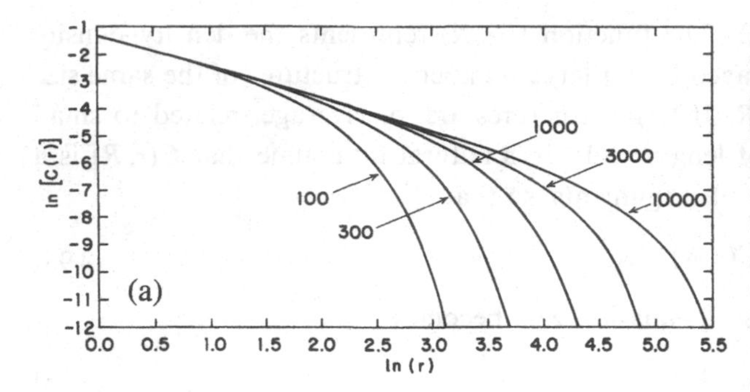

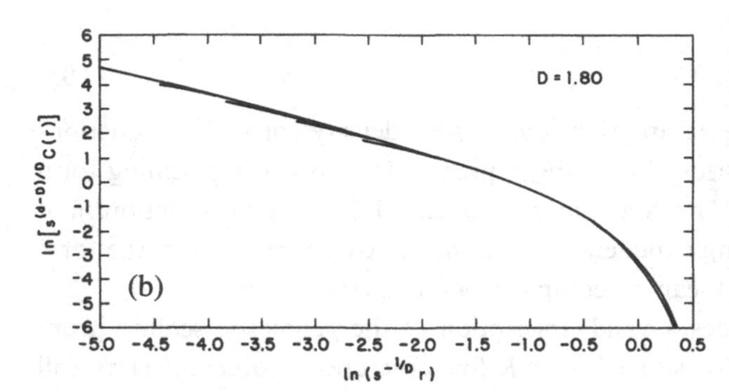

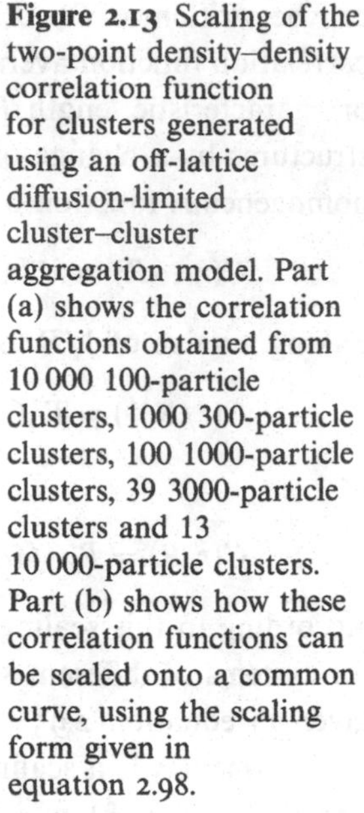
Figure 2.13 Scaling of the two-point density–density correlation function for clusters generated using an off-lattice diffusion-limited cluster–cluster aggregation model. Part (a) shows the correlation functions obtained from 10 000 100-particle clusters, 1000 300-particle clusters, 100 1000-particle clusters, 39 3000-particle clusters and 13 10 000-particle clusters. Part (b) shows how these correlation functions can be scaled onto a common curve, using the scaling form given in equation 2.98.

the cluster, such as the dependence of the radius of gyration on the mass, can be described in terms of the same fractal dimensionality. For structures generated during experiments or computer simulations, the density–density correlation functions have simple power law forms, such as that given in equation 2.23 for the two-point correlation function over only a limited range of length scales, $\epsilon \ll r \ll L$. Here, ϵ is a lower cut-off length for the fractal scaling that might, in practical terms, correspond to a few particle diameters, or lattice units in a lattice model. The upper cut-off length L is often related to the overall size of the structure $L \approx R$. However, in real systems, L may be controlled by processes such as thermal fluctuations, collapse due to external fields such as gravity, or a finite overall density. In practice, the range of length scales over which $C(r)$ or the corresponding scaling function $f(x)$ in equation 2.98 exhibit simple power law behavior can be limited or non-existent (see figure 2.13(a), for example). Consequently, it can be quite difficult to obtain a reliable estimate of the codimensionality α or the corresponding fractal dimensionality $D_\alpha = d - \alpha$, from the dependence of $\log(C(r))$ on $\log(r)$. The data collapse obtained using the scaling form given in equation 2.98 provides a more reliable way of measuring the fractal dimensionality, by varying the value used for D until the "best" data collapse is obtained. Equation 2.98 can also be written as

$$C(r) = r^{(D-d)} g(r/s^{1/D}), \tag{2.99}$$

where $g(x)$ has the form $g(x) = const.$ for $x \ll 1$ and $g(x) \to 0$ ($g(x)$ decreases more rapidly than any power of x) for $x \gg 1$. Equation 2.99 expresses explicitly the r dependence of $C(r)$ for $\epsilon \ll r \ll s^{1/D}$.

In a structure such as a colloidal aggregate, the particle size (radius or diameter) ϵ also provides an important characteristic length scale. Over short length scales, the density–density correlation function can be represented by a simple homogeneous scaling form, similar to that given in equation 2.99

$$C(r,\epsilon) = r^{-\alpha}h(r/\epsilon), \tag{2.100}$$

where $h(x) \sim x^{\alpha}$ for $x \ll 1$ and $h(x) = const.$ for $x \gg 1$. The correlation function can be represented on all length scales by the more general scaling form

$$C(r,\epsilon,R) = r^{-\alpha}F((r/\epsilon),(r/R)), \tag{2.101}$$

which should be compared with equation 1.21 in chapter 1. Over long length scales ($r/\epsilon \to \infty$), equation 2.101 can be written as

$$C(r,\epsilon,R) = r^{-\alpha}F(\infty,(r/R)) = r^{-\alpha}g(r/R), \tag{2.102}$$

and over short length scales ($r/R \to 0$)

$$C(r,\epsilon,R) = r^{-\alpha}F((r/\epsilon),0) = r^{-\alpha}h(r/\epsilon). \tag{2.103}$$

If $r/R \ll 1$ and $r/\epsilon \gg 1$, then $C(r,\epsilon,R) = r^{-\alpha}F(\infty,0) = cr^{-\alpha}$.

In general, the scaling form places no restrictions on the shape of the scaling function in the crossover region. This is illustrated quite well by the clusters generated by simple models for the colloidal aggregation of monodisperse particles. For distances comparable to the particle size, the density–density correlation function can be represented quite well by equation 2.100, and the scaling function $h(x)$ has a simple algebraic form in the $x \ll 1$ and $x \gg 1$ limits. However, near to the crossover at $x \approx 1$, $h(x)$ has a complex form that reflects the local particle configuration and is sensitive to model details [232, 233, 234, 235, 236, 237].

The origins of the applications of scaling theory in modern statistical physics can be found in the study of critical phenomena in finite systems [238, 239, 240, 241]. Near to a critical point, the fluctuations are correlated over distances in the range $\epsilon < \ell < \xi$. If $\xi \ll L$, where L is the system size, no significant finite-size effects are expected. If, on the other hand, $\xi_{\infty} \gg L$, where ξ_{∞} is the correlation length in the limit $L \to \infty$, then the system can be expected to exhibit behavior characteristic of a corresponding 0-dimensional system. There will be crossover from d-dimensional behavior, far from the critical point, to 0-dimensional behavior (or d'-dimensional behavior if the system is restricted along $d - d'$ coordinates) near to the critical point at which $\xi_{\infty} \to \infty$. This crossover can be described in terms of finite-size scaling [238, 239, 241] forms such as

$$\mathscr{P}(\xi) = \xi^G f_{\mathscr{P}}(\xi/L), \tag{2.104}$$

where $\mathscr{P}$ is a property of the system that diverges as $\mathscr{P} \sim \xi^G$ as the critical point is approached and $\xi \to \infty$ in an infinite system. Near to a critical point, the correlation length ξ is related to the temperature T by $\xi \sim |T - T_c|^{-\nu}$, so that

$$\mathscr{P}(T) = |T - T_c|^{-\nu G} f'_{\mathscr{P}}(|T - T_c|^{-\nu}/L) \tag{2.105}$$

or

$$\mathscr{P}(T) = L^G g_{\mathscr{P}}(L^{1/\nu}|T - T_c|). \tag{2.106}$$

This means that, far from the critical temperature, $|T - T_c| \gg L^{-1/\nu}$, the property $\mathscr{P}(T)$ will exhibit behavior characteristic of an infinite system, but that near to the critical temperature ($|T - T_c| \ll L^{-1/\nu}$) the non-analytic critical behavior will break down.

Renormalization group methods [242, 243] provide theoretical support for the finite-size scaling approach [239, 241] in thermodynamic systems. The basic approach in renormalization group analysis, for systems that can be described by a Hamiltonian $\mathscr{H}$ which depends on the interaction constants $\{a_i\}$, is to study the effects of successive coarse-graining on the system [244]. In the case of a system that can be described in terms of interacting "spins" on a lattice, the coarse-graining consists of grouping the spins into regions or "blocks" of size $\lambda\epsilon$, where ϵ is the original lattice spacing and λ is the change in length scales. If the interactions between the spins have a short range, the grouping of the spins will not change the long wavelength fluctuations. Under these circumstances, it is assumed that the long wavelength fluctuations are characteristic of a new Hamiltonian $\mathscr{H}'$ that has the same form as the original Hamiltonian, with new short range interactions $\{a'_i\}$. The partition function for the new, coarse-grained system, with interactions $\{a'_i\}$, will have the same form as that for the old system, and the free energy per unit volume, with the original coupling constants $\phi(\{a_i\})$, can be written as

$$\phi(\{a_i\}) = f(\{a_i\}) + \lambda^{-d}\phi(\{a'_i\}), \tag{2.107}$$

where the first term on the right-hand side represents the contribution from fluctuations with wavelengths in the range ϵ to $\lambda\epsilon$, and the second term represents the contribution from the longer wavelength fluctuations, associated with the unique behavior at the critical point. The original correlation length $\xi(\{a_i\})$ and the correlation length in terms of the block spins $\xi(\{a'_i\})$ are related by

$$\xi(\{a_i\}) = \lambda\xi(\{a'_i\}). \tag{2.108}$$

The process of gathering spins (or coarse-graining) can be repeated to create larger blocks with interaction constants $\{a_i'\}(\lambda_1\lambda_2)$ given by

$$\{a_i'\}(\lambda_1\lambda_2) = \mathscr{T}(\lambda_1\lambda_2)\{a_i\} = \mathscr{T}(\lambda_1)\mathscr{T}(\lambda_2)\{a_i\} = \mathscr{T}(\lambda_2)\mathscr{T}(\lambda_1)\{a_i\}, \quad (2.109)$$

where $\mathscr{T}(\lambda)$ is the transformation that maps the interaction constants for blocks of size ℓ to interaction constants for blocks of size $\lambda\ell$. Equation 2.109 implies that the transformations $\mathscr{T}(\lambda)$ have group properties (chapter 1, section 1.7).[9] If the transformation $\mathscr{T}(\lambda)$ is applied repeatedly, then eventually, after n applications, the correlation length ξ will be exceeded (when $\lambda^n\epsilon > \xi$) and the group properties will be lost. Only at the critical point will the mapping $\mathscr{T}(\lambda)$ reproduce the system exactly. Most points in the parameter space $\{a_i\}$ will evolve towards fixed points that correspond either to a completely ordered "low temperature" state or a completely disordered "high temperature" state under the coarse-graining/rescaling transformation of the renormalization group. In this sense, the critical interaction constants $\{a_i\}_c$ must correspond to an unstable fixed point.[10] At the fixed point, the set of coupling constants satisfy the rescaling relationship

$$\{a_i\}_c = \mathscr{T}(\lambda)\{a_i\}_c. \quad (2.110)$$

Near the critical point, λ can be considered to be continuous, and the transformation $\mathscr{T}(\lambda)$ can be linearized in the $\lambda \to 1$ limit.

A variety of arguments [245] have been advanced to motivate homogeneous scaling near to critical points [7], and this scaling form has been demonstrated for some specific models. It has also been confirmed for a wide range of second order phase transitions, using accurate experimental approaches and computer simulations. However, its universality remains a hypothesis. The scaling approach has also been extended to dynamical phenomena near to a critical point [246, 247, 248]. It is assumed that the correlation functions[11] $C(\mathbf{k},\omega)$ describing the dynamical behavior have the scaling form

$$C(\mathbf{k},\omega) = k^a f(\omega/\omega^*(\mathbf{k})), \quad (2.111)$$

where $\omega^*(\mathbf{k})$ is a characteristic frequency. It is also assumed that $\omega^*(\mathbf{k})$ is a homogeneous function of wave number $k = |\mathbf{k}|$ and the correlation length ξ,

9. The transformations $\mathscr{T}(\lambda)$ do not form a true group, since $\lambda > 1$ and there is no inverse transformation.
10. There are both stable and unstable directions at the fixed point, but most trajectories are eventually repelled from the fixed point under the coarse-graining/rescaling transformation.
11. The correlation function $C(\mathbf{k},\omega)$ is the Fourier transform of the real space-time correlation function $C(\mathbf{r},t) = < P(\mathbf{r},t)P(0,0) >$, where $P(\mathbf{r},t)$ is the value of the dynamical variable corresponding to the property P at time t and position $\mathbf{r}$ [3, 247, 248, 249]. If a quantum mechanical model is required, the correlation function $C(\mathbf{r},t)$ can also be defined as $C(\mathbf{r},t) = \frac{1}{2} < \{\mathscr{P}(\mathbf{r},t)\mathscr{P}(0,0)\} >$, where $\mathscr{P}$ is the operator for the property P and "$\{\ldots\}$" indicates the anticommutator.

and that the dependence of $f(\omega/\omega^*(\mathbf{k}))$ on k and $\xi(T)$ depends only on the product $k\xi$.

The extension of scaling theory from equilibrium to non-equilibrium phenomena is on a much less firm theoretical foundation. It is a valuable tool when it works, but must be approached with caution. For non-equilibrium systems, the structure is not controlled by a Hamiltonian and it is not possible to construct a renormalization group based on interaction constants. The behavior of equilibrium systems near a critical point can be a valuable guide in the analysis of non-equilibrium systems. In particular, the corrections to power law scaling (discussed below) often have similar forms. However, care must be taken not to push these analogies too far.

The trajectories followed by particles and clusters in a fluid can be described in terms of fractal geometry and scaling. The Brownian motion [250] of particles in a dense fluid has been studied for many decades [251]. It is well known that the mean distance $<r>$ traveled by a random walker from its position at time $t_\circ$ is given by

$$<r>\sim(t-t_\circ)^{1/2}. \tag{2.112}$$

Since the distance ℓ, measured along the path, is directly proportional to the elapsed time, it follows that

$$<r>\sim(\ell)^{1/2}, \tag{2.113}$$

so that the Brownian trajectory has a fractal dimensionality D_w of 2.0. Equation 2.112 has been confirmed experimentally, over a very wide range of length scales.

In a low pressure gas, small particles and aggregates can move over distances large compared with their own size before their direction is changed significantly by collisions with the surrounding gas. In this case, their trajectories can be described, to a good approximation, in terms of "ballistic flights" with fractal dimensionalities of 1 ($D_w = 1$). In general, a crossover from a fractal dimensionality of 1 on short length scales to a dimensionality of 2 on long length scales may be expected. In this case, the dependence of the distance traveled $r(t)$ on time can be written as

$$<r(t)>=<v>tf(t/t^*), \tag{2.114}$$

where $<v>$ is the mean velocity and t^* is the mean time required to change the direction of travel. Here, the crossover function $f(x)$ has the form $f(x) = 1$ for $x \ll 1$ and, since $<r(t)>\sim t^{1/2}$ at long times ($t \gg t^*$), $f(x)$ must have the form $f(x) \sim x^{1/2}$ for $x \gg 1$. Depending on the exact shape of the crossover function, the mean distance traveled $<r(t)>$ may appear to have an algebraic time dependence with an exponent other than 1 or 1/2. Such relationships may have empirical value, but give no fundamental insight into the behavior of particles in quiescent fluids. In systems with random velocity fields, such as turbulent flows and flow in porous media with correlated disorder, superdiffusive or subdiffusive dynamics with non-classical exponents is common [252, 253].

Simple scaling models have also been used with considerable success to describe the kinetics of non-equilibrium growth. For example, the distribution of cluster sizes in simple aggregation processes [58] can often be described by the simple scaling form

$$N_s(t) = s^{-2} f(s/t^z). \tag{2.115}$$

Here, $N_s(t)$ is the number of clusters of size s at time t, and the exponent z can have integer or non-integer values. The arguments of the scaling function are always dimensionless ratios. This does not, of course, mean that the cluster size s has units of $(\text{time})^z$. Instead, equation 2.115 stands for

$$N_s(t) = s^{-2} f(s/S(t)), \tag{2.116}$$

where $S(t)$ is the mean cluster size, which grows as t^z. Equation 2.116 is much more general than equation 2.115, since it applies to a broad range of processes in which $S(t)$ does not grow algebraically with increasing time.

2.5.1 Corrections to Scaling

In practice, many of the simple models used to represent fractal structures do not exhibit a pure power law relationship,

$$M(L) \sim L^D, \tag{2.117}$$

between mass M and length L. Instead, it is often found that the mass/length relationship can be represented very well by the non-homogeneous form

$$M(L) = A_1 L^{D_1} + A_2 L^{D_2} + A_3 L^{D_3} + \cdots, \tag{2.118}$$

with $D_1 > D_2 > D_3 > \ldots$. In the asymptotic, $L \to \infty$, limit, the first term on the right-hand side of equation 2.118 becomes completely dominant, and the fractal dimensionality D_1 is measured. Unfortunately, at more accessible length scales, the other terms may become important, and a plot of $\log M(L)$ against $\log(L)$ over the range $L_1 \leq L \leq L_2$ will be a slightly curved line with an average slope that is not equal to D_1.

Equation 2.118 is reminiscent of the behavior of thermodynamic properties near a critical point in an infinite system [254], which can be represented by

$$\mathscr{P}(T) = A_1 t^{\Delta_1} + A_2 t^{\Delta_2} + A_3 t^{\Delta_3} + \cdots, \tag{2.119}$$

Here, $t = (T - T_c)/T_c$, and $\mathscr{P}(T)$ is the value of the thermodynamic property $\mathscr{P}$ at temperature T. A theoretical foundation for the inhomogeneous power law form of equation 2.118 is provided by renormalization group theory [239]. For thermodynamic properties, in finite systems, a finite-size scaling representation of the form

$$\mathscr{P}(T,L) = \lambda^k f(t\lambda^{y_1}, \zeta_2\lambda^{y_2}, \zeta_3\lambda^{y_3}, \ldots) \tag{2.120}$$

is expected. In equation 2.120, $\lambda = L/\xi$, the quantities ζ_n are called "scaling fields" and the exponents y_n describe how these fields evolve under the renormalization group transformation. Fields ζ_n with $y_n < 0$ are said to be irrelevant fields;[12] they give rise to the correction terms in equation 2.119. The renormalization group equations also show how logarithmic corrections can arise [254].

Aharony *et al.* [255] have shown that, at least in some simple examples, the exponents D_n in equation 2.119 can be interpreted as the fractal dimensionalities of well identified substructures associated with the fractal. They found that the dimensionalities D_n can be related directly to the eigenvalues of the transfer matrix [256] that describes how the numbers of the basic structural elements evolve, during each generation in the iterative procedure used to generate the fractal ($D_n = \log(\Lambda_n)/\log(\lambda)$, where Λ_n is the nth largest eigenvalue of the transfer matrix and λ is the change of length scales between generations). Similar approaches can be used to characterize random fractals [255], but the potential value of this approach has not yet been properly explored.

2.5.2 Multiscaling

In the simplest scaling model, the density profile $\rho(r,R)$ for structures, such as invasion percolation clusters, grown from a single seed or growth site can be described in terms of the scaling form

$$\rho(r,R) = r^{-\alpha} f(r/R), \tag{2.121}$$

where $\rho(r,R)$ is the mean density at a distance r from the origin, seed or initial growth site of the cluster and R is a distance characteristic of the overall

12. If the coupling constants $\{a\}$ are written as $\{a\} = \{a_c\} + \{a^*\}$, then $\mathscr{T}(1+\delta)\{a\} = \{a_c\} + \{a^*\} + \mathbf{T}\{a^*\}\delta$, with $\delta \to 0$. The elements of $\mathbf{T}$ describe how the coupling constants evolve under a continuous coarse-graining/rescaling transformation and are given by $\mathbf{T}_{i,j} = \partial a_i'/\partial a_j$. The eigenvectors $\tilde{a}_i^*$ of the linearized transformation matrix $\mathbf{T}$ with positive eigenvalues Λ_i are said to be relevant since the corresponding coupling constants grow away from the fixed point ($\{\tilde{a}_i^*\} = \{0\}$). In order for the trajectory of the coupling constants $\{a\}$ to reach the critical point $\{a_c\}$, the amplitudes of the relevant scaling fields (the eigenvectors of $\mathbf{T}$ with positive eigenvalues) must be zero. This defines a sub-space of the parameter space (the coupling constant space), which is the basin of attraction for the fixed point, or the "critical surface". The eigenvectors of the linearized transformation matrix T with negative eigenvalues are said to be irrelevant and those with eigenvalues of zero are said to be marginal. Equation 2.107 can be written in terms of the coupling constants $\{a^*\}$ as $\phi(\{a^*\}) = f(\{a^*\}) + \lambda^{-d}\phi(\{a^{*\prime}\})$. Since the free energy per unit volume is singular at the critical point and $f(\{a^*\})$ is analytic, this allows the critical exponents to be calculated from the eigenvalues Λ_i of $\mathbf{T}$. Finite-size effects can be included by using L^{-1} as one of the scaling fields, or scaling variables [241].

pattern size. In some systems, a more general scaling form may be required [257] to describe the growth and form of complex structures. For example, equation 2.121 might be replaced by

$$\rho(r,R) = r^{f_2(r/R)} f(r/R), \tag{2.122}$$

where the function $f_2(r/R)$ can be written as

$$f_2(r/R) = D(r/R) - d = -\alpha(r/R). \tag{2.123}$$

For the case where $D(r/R)$ is a constant ($D(r/R) = D$), equation 2.121 is recovered. This "multiscaling" picture has been used to describe the structure factor $S(q,t)$ associated with spinodal decomposition [258] and the structure of DLA model clusters [259, 260] discussed in chapter 3, section 3.4.5.4. Equation 2.122 implies that the *local* fractal dimensionality should be a function of r/R. Although it has been shown for a variety of models that a better data collapse of the density profiles can be obtained using equation 2.122 than when using equation 2.121, with a constant value for α, this idea does not seem to have been directly tested by explicitly measuring the *local* fractal dimensionality as a function of r/R. High quality data, covering a very wide range of length scales, would be required for this purpose.

It must be emphasized that the multiscaling form $\rho(r,R) = r^{D(r/R)-d} f(r/R)$ is very different from the more simple scaling form $\rho(r,R) = r^{D-d} g(r/R)$ with constant D. This is discussed further in chapter 3, in a DLA context. Equation 2.122 obeys the group properties $\mathscr{T}(\lambda_1)\mathscr{T}(\lambda_2) = \mathscr{T}(\lambda_1\lambda_2) = \mathscr{T}(\lambda_2)\mathscr{T}(\lambda_1)$, where $\mathscr{T}(\lambda)$ represents the transformation $r \rightarrow \lambda r$, $R \rightarrow \lambda R$.

A similar generalization of the scaling form describing the free energy of thermodynamic systems near to a phase transition has been presented by Hilfer [261].

2.6 Fractal Trees and Inhomogeneous Fractals

The simplest self-similar fractals fill space in a very non-uniform manner, and, in this sense, they are extremely inhomogeneous. However, they are homogeneous in the sense that the fractal dimensionality can be measured locally, in a region that is occupied by the fractal and is the same everywhere. The term "inhomogeneous fractal" will be used to indicate a fractal for which the local scaling properties vary from place to place *on the fractal*. The lacunarity may also vary with position on the fractal, but this possibility will not be considered. The local fractal dimensionality can be obtained from

$$\lim_{r_n \rightarrow 0} \mu(r_n, \mathbf{x}) \sim r_n^{D(\mathbf{x})} \tag{2.124}$$

or

$$D(\mathbf{x}) = \lim_{r_n \to 0} \log \mu(r_n, \mathbf{x}) / \log r_n, \tag{2.125}$$

where $\mu(r_n, \mathbf{x})$ is the measure within a distance r_n from a point at position $\mathbf{x}$, on the fractal set. Alternatively, the Hausdorff dimensionality or the capacity dimensionality (equation 2.33) could be calculated for different parts of a fractal. Figure 2.14 shows the first three stages in the generation of a simple, self-similar fractal called the Koch curve. The generator for this fractal is obtained from a line on the unit interval by removing a fraction f and inserting two segments of length $(1-f)/2$ to bridge the gap and form a pattern consisting of four lines of equal length $(1-f)/2$, such as that shown in figure 2.14(b). The fractal obtained from this generator can be covered by four replicas of itself after a change of length scale by a factor of $2/(1-f)$. Consequently, the similarity dimensionality is given by $D = \log 4/\log(2/(1-f))$, and D can lie in the range 1 (for $f = 1/2$) to 2 (for $f = 0$). The fractal dimensionality is also given by $D = \log 4/\log(2(1 + \sin\theta))$, where 2θ is the angle between the two bridging segments. For the prefractals shown in figure 2.14, the angle θ is 30° and D is $\log 4/\log 3$. The corresponding fractal is homogeneous.

Figure 2.15 shows an inhomogeneous self-similar fractal, constructed in a very similar manner. At each stage in the construction process, when a line segment of length δ is replaced by four line segments of length $[(1-f)/2]\delta$, the fraction f was made equal to $(1-x)/2$, where x is the position of the center of the line segment projected along the unit interval. Consequently, the *local* fractal dimensionality changes continuously from $D = 1$ ($f = 1/2$) on the left-hand side of figure 2.15 to $D = 2$ ($f = 0$) on the right-hand side of figure 2.15.

It is useful to think about many natural structures in terms of trees (loopless networks). Important applications include rivers, branched fracture patterns, a broad range of biological structures (such as vascular systems, nerve cells and real trees), viscous fingering patterns (chapter 4) and other dendritic growth patterns (chapter 4).

There are many types of trees, and a comprehensive discussion would be beyond the scope of this book. However, Mandelbrot [2] has described a class of symmetrically branched trees that provide a simple illustration of the concept of inhomogeneous fractality. The geometry of these trees is based on a symmetric, "Y"-shaped generator, shown in figure 2.16, consisting of three line segments (a "stem" of length 1 and two "arms"). The two arms are attached to one end of the stem and are separated by an angle θ. The ratio between the lengths of the arms r_a and the length of the stem r_s is $R = r_a/r_s$. To construct the tree, the arms of the "Y", generated in the previous stage of the construction process, are covered by scaled replicas of the "Y"-shaped generator, so that the stem of the scaled replica replaces the arms of the initial structure. In this

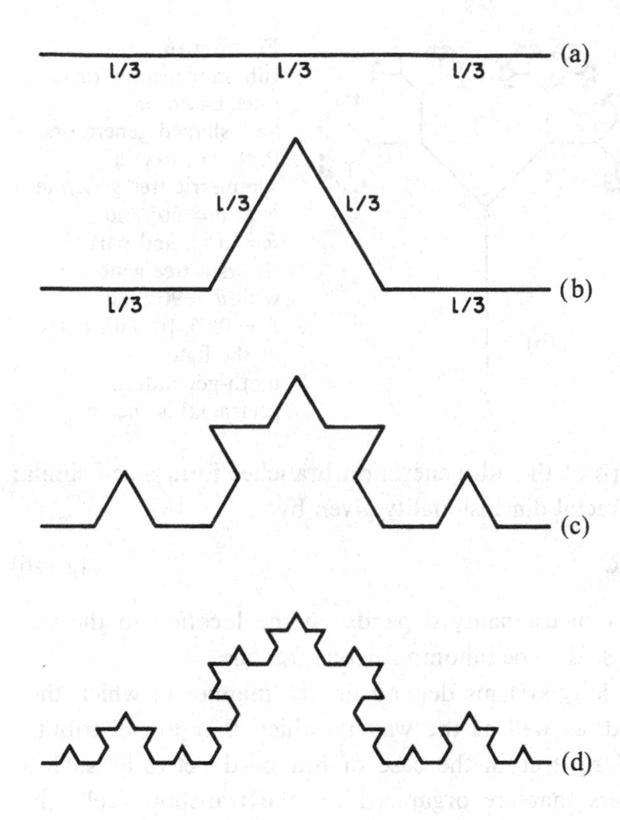

Figure 2.14 The first three stages in the generation of a homogeneous, self-similar Koch curve fractal with a similarity dimensionality of $\log 4/\log 3$. In the first stage, the line of length 1 shown in part (a) is replaced by the generator shown in part (b). In each subsequent stage, every line segment is replaced by an isotropically scaled version of the generator.

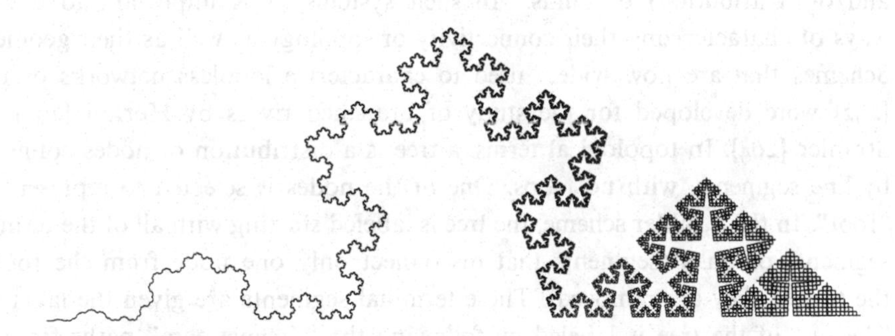

Figure 2.15 The tenth-generation prefractal corresponding to an inhomogeneous, self-similar fractal based on the Koch curve construction shown in figure 2.14. In this case, the angle θ varies continuously from 90° ($D = 1$) on the left-hand side to 0° ($D = 2$) on the right-hand side.

manner, the tree is grown, generation by generation, so that after n generations the tree contains branches of order (defined below) 1 to $n + 1$, corresponding to segments with $n + 1$ different lengths in this case. The new branches generated in the mth generation have lengths of R^m and there are 2^m of them.

It is clear that if the tree is examined in the region near to the original stem, then a single line segment with $D = 1$ will be observed. On the other hand,

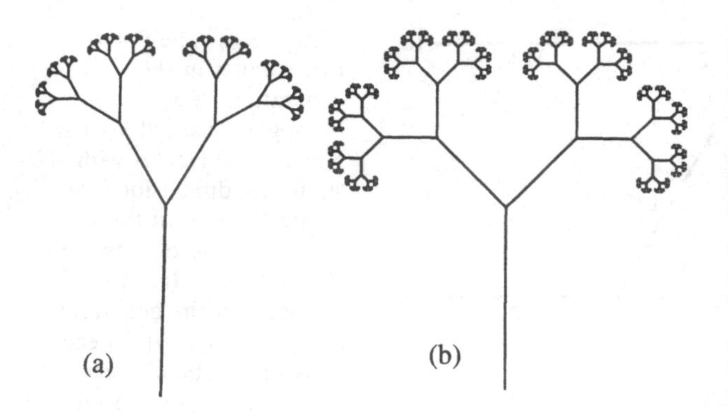

Figure 2.16 Inhomogeneous fractal trees based on "Y"-shaped generators. Part (a) shows a symmetric tree generated with $\theta = 60°$ and $R = 0.50$, and part (b) shows a tree generated with $\theta = 90°$ and $R = 0.55$. In both parts of the figure, the tenth-generation prefractal is shown.

in the limit $n \to \infty$, the tips of the nth generation branches form a self-similar subset of the tree with a fractal dimensionality given by

$$D = -\log 2/\log R. \tag{2.126}$$

Because the local fractal dimensionality depends on the location in the tree, structures of this type are said to be inhomogeneous fractals.

The properties of disorderly systems depend on the manner in which their components are connected, as well as the way in which they are distributed in space. This is particularly true in the case of branched networks such as vascular systems and rivers that are organized for the transport (collection and/or distribution) of fluids. In such systems, it is important to develop ways of characterizing their connectivity or topology as well as their geometry. Schemes that are now widely used to characterize loopless networks or trees [262] were developed for the study of branched rivers by Horton [263] and Strahler [264]. In topological terms, a tree is a distribution of nodes connected by line segments, with no loops. One of the nodes is selected to represent the "root". In the Strahler scheme, the tree is labeled starting with all of the terminal segments or links (segments that disconnect only one node from the root of the tree, if they are removed). These terminal segments are given the label "1". The rest of the tree is labeled by following the "downstream" paths from the terminal segments to the root. If only a single labeled segment contacts a node, the "downstream", unlabeled segment is given the same label as the labeled, upstream segment.[13] If more than one labeled segments meet at a node, then the label given to the previously unlabeled, downstream segment is either $\sup\{l_i\}$, where $\{l_i\}$ is the set of "tributary" labels and $\sup\{l_i\}$ is the largest member of this set, or $\sup\{l_i\} + 1$, if more than one tributary has the same maximum value. A downstream segment can be labeled only after all of the upstream segments have been labeled. The link labels obtained in this manner are called

13. A node connected to only two segments is topologically equivalent to a single segment, and it is usual to replace all such configurations by a single segment.

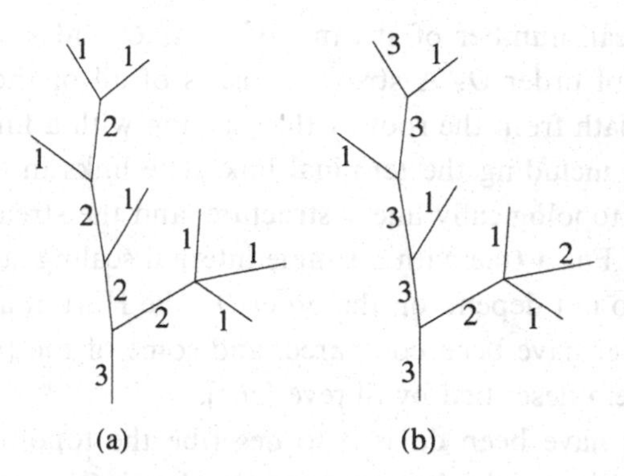

Figure 2.17 A small tree labeled according to the Strahler (a) and the Horton (b) schemes. This figure was provided by T. Sun.

the Strahler orders of the links or segments. A small tree labeled according to the Strahler scheme is shown in figure 2.17(a).

The earlier Horton scheme is more complicated. Whenever two or more tributary links with the same maximum label l_{max} meet, then the downstream link is given the label $l_{max} + 1$, just as in the Strahler scheme. However, once this has been done, all of the segments along one of the upstream paths, from the segment labeled $l_{max} + 1$ up to and including one of the terminal segments, are relabeled $l_{max} + 1$. This relabeling is propagated along an upstream path, always following the tributary link with the highest, previously assigned label. If two or more tributary links have the same highest order or label, the tie can be broken in several ways. In a real system, such as a river, the tributary with the longest length or the tributary draining the largest area could be chosen. One of the highest order tributaries could also be selected at random. After this relabeling has been completed, the labeling process proceeds in a downstream direction, by selecting nodes for which all of the tributaries have been labeled and labeling the downstream link or segment. Each time a node with two or more tributaries having the same highest label is identified, the upstream relabeling procedure, described above, is carried out. When the "root" of the tree is finally reached by all of the links contacting the root, the relabeling is carried out for a final time, if necessary. The labels on each of the links are then their Horton orders. When the labeling process is complete, a single path of links from the "root" to one of the terminal links will carry the same label, corresponding to the highest order. The order of an entire river or a tributary is equal to the highest order of any of its links. The links with this highest order form the "mainstream".

It is common to express the results of a Horton or Strahler analysis in terms of the ratios

$$r_N(O) = < N_O > / < N_{O-1} > \tag{2.127}$$

and

$$r_\ell(O) = < \ell_O > / < \ell_{O-1} >, \tag{2.128}$$

where $< N_O >$ is the mean number of streams of order O and $< \ell_O >$ is the mean length of streams of order O. A stream consists of all of those links of the same order along a path from the root or the junction with a link of higher order to a terminal link, including the terminal link. The links in each stream are connected to form a topologically linear structure, and the stream length is the length of this path.[14] For a tree with a simple internal scaling structure, the ratios $r_N(O)$ and $r_\ell(O)$ do not depend on the order O. The Horton and Strahler definitions of stream order have been compared, and some of the fundamental relationships between them described by Shreve [262].

Several other schemes have been devised to describe the topology of trees. For example, in his studies on the human lung, Weibel [265] simply started with the "trunk", which was given a label of "0" and increased the label by 1 following each branched node encountered along every outwards directed (upstream) path.

The Horton–Strahler approach to the quantitative description of branched networks has been used the most widely in hydrology and biology. There have been few applications in physics, and in most of these applications the more simple Strahler method has been used. An interesting exception is the work of Hinrichsen *et al.* [266, 267, 268] on diffusion-limited aggregation model clusters (chapter 3). They used the Horton scheme for branch orders and found that the ratios r_N and r_ℓ were essentially constant (independent of the order at which they were measured). They estimated a fractal dimensionality of $D \approx 1.6$ using the relationship

$$D_H = -\frac{\ln r_N}{\ln r_\ell}. \tag{2.129}$$

Essentially the same value was obtained for r_ℓ, if the Pythagorean end-to-end distance of the mainstream, measured in the embedding space, or the distance measures along the mainstream path were used. This is consistent with a minimum path dimensionality (chapter 3, section 3.5) of $D_{min} = 1$. Hinrichsen *et al.* also carried out a box counting analysis of the sets formed by the links of different orders. They found that the number of non-empty boxes of size ℓ for the region of order O could be represented by the scaling form

$$N_O(\ell) = s\ell^{-D} f(\ell/\xi_O), \tag{2.130}$$

where s is the number of particles in the cluster. The function $f(x)$ in equation 2.130 has the form $f(x) = const.$ for $x \gg 1$ and $f(x) \sim x^{D-D_{min}}$ for $x \ll 1$. The crossover length ξ_O is the mean length of the tributaries of order O. Equation 2.130 expresses the uniform distribution of tributaries of order O throughout the entire structure, on length scales greater than $\xi_O = \ell_O$, and describes the local mainstream structure, on length scales smaller than ℓ_O.

14. In some cases, the stream is a fractal, and the length measured along the stream must be distinguished from the end-to-end distance, measured in the embedding space.

This type of analysis can be carried out for any growth model, even models that generate compact structures. For example, in the compact Eden growth model, described in chapter 3, section 3.4.1, an Eden tree can be constructed using the procedure described above. Hinrichsen [267] has shown that the ratios $r_\ell(O)$ and $r_N(O)$ approach constant values, at sufficiently large values of O, for a variety of model structures.

The distribution $P(\theta_{O_S}, s)$ of the angles θ_{O_S} between the vectors from the origin to the bases of branches with a Strahler order of O_S and the vectors from the bases to the tips of the branches has recently been measured by Lam *et al.* [269], using 50 2-dimensional DLA clusters, each containing 10^6 particles. The distributions were found to approach the isotropic limit $P(\theta_{O_S}, s) = 1/2\pi$ quite rapidly, with increasing Strahler order O_S.

2.7 Self-Affine Fractals

In general terms, self-affine fractals have different scaling properties in different directions. Figure 2.18 shows the first four stages in the construction of a self-affine fractal. In this case, the iterative process used to generate the fractal is like that used to generate a Cantor set. In the first stage, figure 2.18(a), a "filled" rectangular area of size $l_x \times l_y$ ($l_x = l_y = 1$ in this case) is subdivided by removing those regions with x coordinates lying in the range $l_x/5 < x < 2l_x/5$ or $3l_x/5 < x < 4l_x/5$ or y coordinates in the range $l_y/3 < y < 2l_y/3$. In subsequent stages, the remaining filled areas are decimated in the same manner, so that after n generations the pattern consists of 6^n filled regions each having a size of $(1/5)^n \times (1/3)^n$. If such a self-affine fractal is characterized using the approaches that work well for self-similar fractals, the results obtained will be contradictory. For example, in the small length scale $\ell \to 0$ limit, the structure will appear to be the product of a line and a Cantor set with a fractal dimensionality of $\log 3/\log 5$, so that the self-affine fractal will appear to have a short length scale, local fractal dimensionality of $D = 1 + (\log 3/\log 5) = 1.68260\ldots$. On the other hand, if the construction process is continued towards the $\ell_x \to \infty$, $\ell_y \to \infty$ limit, the ratio l_y/l_x will approach zero and the fractal will "look like" a Cantor set with a fractal dimensionality of $D = \log 3/\log 5 = 0.68260\ldots$. It could also be argued that, since the measure (filled area) changes by a factor of 6 every time the area is changed by a factor of 15, the fractal dimensionality given by equation 2.6 is $(D - d)/D = \log(6/15)/\log(6)$ or $D = 1.32328\ldots$. In addition, it could be argued that the fractal shown in figure 2.18 is the product of two Cantor sets with fractal dimensionalities of $D = \log 3/\log 5 = 0.68260\ldots$ and $D = \log 2/\log 3 = 0.6309\ldots$, so that the fractal dimensionality of the pattern shown in figure 2.18 is $D = 0.68260\ldots + 0.6309\ldots = 1.3135\ldots$.

The failure of the methods that give reliable results for self-similar fractals to give consistent results in this case is a consequence of the failure to identify correctly the self-affine scaling symmetry. The structure shown in figure 2.18 cannot be rescaled by isotropic dilation or contraction. Instead, this fractal can be rescaled by simultaneously changing the length scale in the y direction by $\lambda_y = 3^n$ and the length scale in the x direction by $\lambda_x = 5^n$. For the fractal shown in figure 2.18, it is easy to identify the correct scaling symmetry. In the case of random fractals generated by natural processes, it is more difficult to determine the scaling symmetry. If an isotropic self-similar fractal is deformed, it may look like a self-affine fractal, but it will remain self-similar. In the past, the use of methods developed for self-similar fractals to characterize self-affine fractals has led to considerable confusion, and the results that have been reported in a quite large body of literature must be re-evaluated.

Another example of a self-affine fractal is shown in figure 2.19. In this case, seven discs with a diameter of ϵ are assembled, during the first stage, to form a cross with a length of 5ϵ in the "parallel" direction and a width of 3ϵ in the "perpendicular" direction. In the second stage, seven of the first stage crosses are assembled to form a structure with a length of 25ϵ and a width of 9ϵ. After the nth stage, the prefractal has a length of $5^n\epsilon$ and a width of $3^n\epsilon$ and contains 7^n "particles". The length l and width w are related to the number of discs or particles s by

$$l = s^{\nu_\parallel} \tag{2.131}$$

and

$$w = s^{\nu_\perp}, \tag{2.132}$$

with $\nu_\perp = \log 3/\log 7$ and $\nu_\parallel = \log 5/\log 7$. The asymptotic, $n \to \infty$, structure can be rescaled by changing the length in the parallel direction by a factor of 5^n and simultaneously changing the length scale in the perpendicular direction by a factor of 3^n, where n is a positive or negative integer.

The differences between the transformations that rescale self-similar and self-affine fractals are illustrated in figure 2.20.

Many structures and phenomena can be described in terms of a single-valued function $f(\mathbf{x})$. Important examples include time dependent signals $\mathcal{S}(t)$, surfaces ($h(\mathbf{x}) = h(x, y)$, the height at a lateral position of $\mathbf{x}$) and oil well logs ($\mathcal{P}(z)$ [270], the value of some measured rock property at a depth of z). Very often, these functions have self-affine scaling properties. In many cases, the self-affine structure arises naturally as a result of fundamental differences between the f and $\mathbf{x}$ directions. For example, the horizontal $\mathbf{x}$ and vertical h directions are quite different on the Earth's surface because of the dominant role played by gravity.

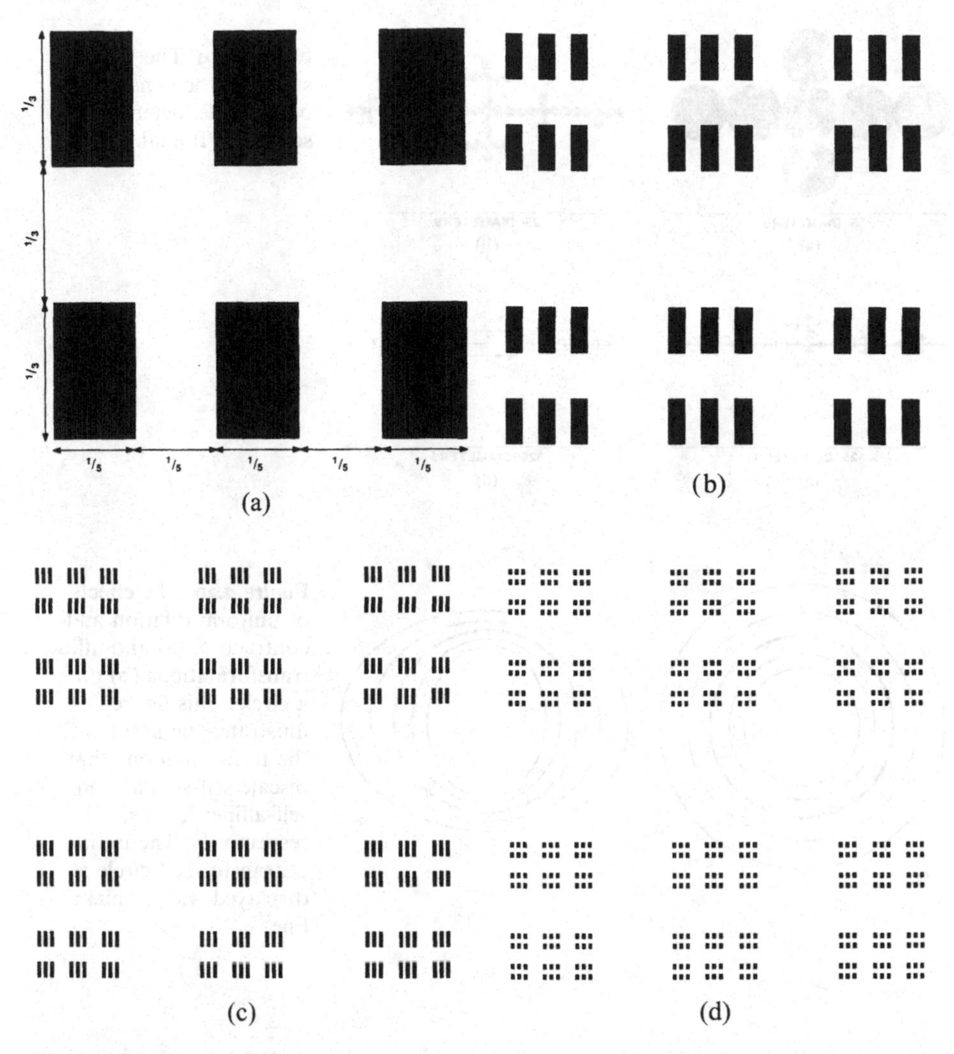

Figure 2.18 The first four stages in the construction of a self-affine fractal. The fourth stage prefractal shown in part (d) is not fully resolved

A simple deterministic fractal of this type is shown in figure 2.21. In this case, the generator consists of four line segments of equal length (the "long" segment in the generator should be thought of as two concatenated "short" segments). In the second stage, shown in figure 2.21(b), each of the four line segments in the generator has been replaced by a replica of the generator. The length in the horizontal direction has been increased by a factor of 4 while the height has been increased by a factor of 2. After n generations, the self-affine prefractal has a length of 4^n and a height of 2^n. The construction process can be extended to smaller length scales by replacing each line segment by an appropriately scaled version of the generator, at each stage in the generation process. In the

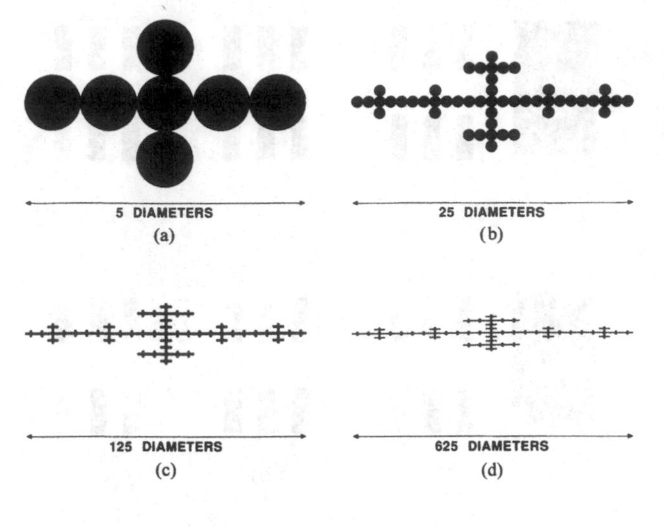

Figure 2.19 The first four stages in the construction of a simple, deterministic, self-affine fractal.

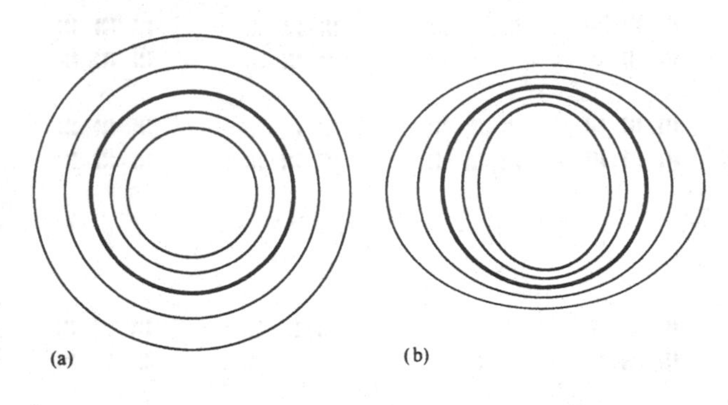

Figure 2.20 The effects of uniform dilation and contraction (a) and affine transformations (b) on a circle. This figure illustrates the nature of the transfomations that rescale self-similar and self-affine fractals, respectively. The initial, untransformed circle is displayed with a thicker line.

asymptotic limit, the fractal curve $f(x)$ can be scaled onto itself by changing the horizontal length scale by a factor of $\lambda_x = 4^n$, while the vertical length scale is changed by a factor of $\lambda_y = 2^n$, so that $\lambda_y = \lambda_x^{1/2}$ and $\lambda^{1/2}f(x) = f(\lambda x)$ or $f(x) = \lambda^{-1/2}f(\lambda x)$.

As is the case for self-similar fractals, random structures that can be statistically rescaled by a self-affine transformation are of much more scientific interest and practical importance than the precisely hierarchical fractals shown in figures 2.18, 2.19 and 2.21. In this case, the statistical rescaling involves changing the horizontal coordinates by a factor λ_x while the vertical coordinates are changed by a factor of $\lambda_y = \lambda_x^H$. The exponent H is called the Hurst exponent. In many important cases, the random self-affine fractal can be thought of in terms of fluctuations about a straight line or flat surface. In this case, the Hurst exponent characterizes the relationship between the height differences between pairs of points on the surface $h(\mathbf{x})$ that lie above or below points

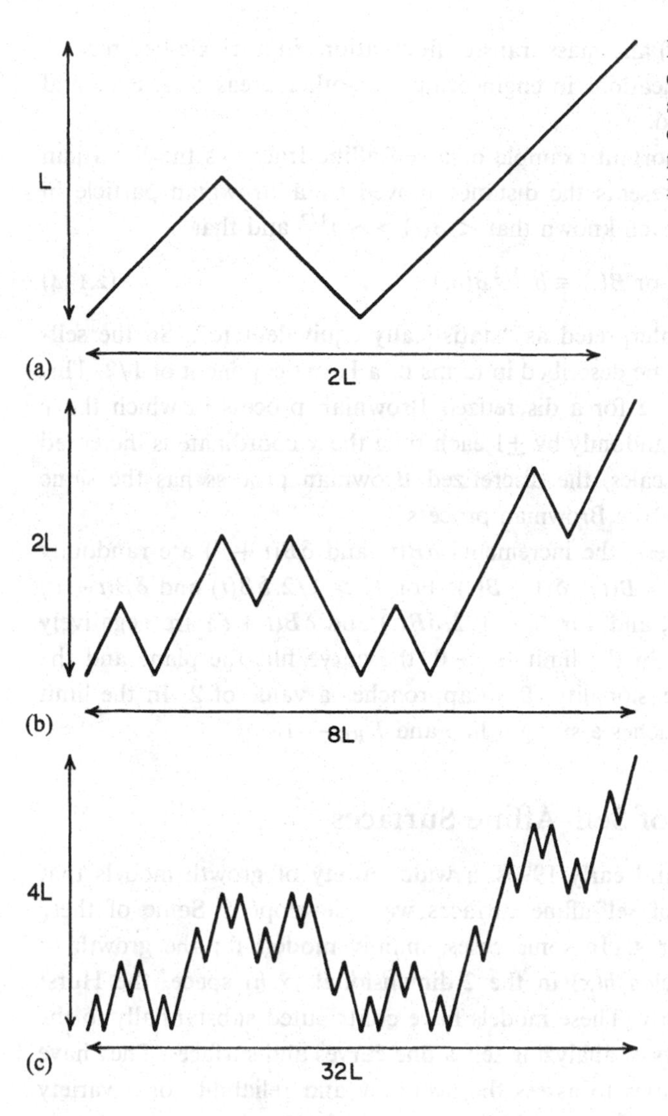

Figure 2.21 The construction of a simple, single-valued, deterministic self-affine fractal curve with a Hurst exponent of 1/2.

separated by a distance $x = |\mathbf{x}|$ on the flat reference plane. For a self-affine surface,

$$< |h(\mathbf{x}_1) - h(\mathbf{x}_2) >_{|\mathbf{x}_1 - \mathbf{x}_2| = x} \sim x^H. \tag{2.133}$$

In general, the scaling behavior can be different in more than two directions, and constructions like those shown in figures 2.18 to 2.21 can be carried out in embedding spaces with $d \geq 3$.

Self-affine scaling is often found in association with time dependent phenomena. Representative chemical engineering applications include pressure fluctuations in bubble columns ($H \approx 0.9$ for "homogeneous" bubbling and $H \approx 0.6$ for turbulent bubbling [271]) and mass transfer fluctuations in a trickle-bed reactor [272]. Many other applications in engineering and other areas of science and technology could be cited.

Perhaps the most important example of a self-affine fractal is the Brownian process $B(t)$, which represents the distance moved by a Brownian particle in time t. In this case, it is well known that $< B(t) > \sim t^{1/2}$ and that

$$b^{1/2}B(t) \equiv B(bt) \text{ or } B(t) \equiv b^{-1/2}B(bt), \tag{2.134}$$

where "$\equiv$" should be interpreted as "statistically equivalent to". So the self-affine scaling of $B(t)$ can be described in terms of a Hurst exponent of $1/2$. This is illustrated in figure 2.22 for a discretized Brownian process in which the y coordinate is increased randomly by ± 1 each time the x coordinate is increased by 1. On long length scales, the discretized Brownian process has the same scaling properties as the true Brownian process.

In the Brownian process, the increments $\delta B(t)$ and $\delta B(t + t')$ are randomly correlated, where $\delta B(t) = B(t + \delta t) - B(t)$. For $H > 1/2, \delta B(t)$ and $\delta B(t + t')$ are positively correlated, and, for $H < 1/2, \delta B(t)$ and $\delta B(t + t')$ are negatively correlated (figure 2.23). In the limit $H \to 0$, the curve fills the plane and the local box counting dimensionality (D_{BL}) approaches a value of 2. In the limit $H \to 1$, the curve approaches a straight line and $D_{BL} \to 1$.

2.7.1 Generation of Self-Affine Surfaces

During the late 1980s and early 1990s, a wide variety of growth models that lead to the formation of self-affine surfaces were developed. Some of these are described in chapter 5. In some cases, mainly models for the growth of 1-dimensional height fields $h(x)$ in the 2-dimensional (x, h) space, the Hurst exponent is known exactly. These models have contributed substantially to the development of new ways of analyzing self-affine curves and surfaces. They have also provided opportunities to assess the accuracy and reliability of a variety of approaches to the measurement of Hurst exponents. However, most of these models do not allow the Hurst exponent to be tuned continuously over all or part of the range $0 < H < 1$. Consequently, the generation of self-affine surfaces with known, asymptotic, Hurst exponents is usually carried out using simple statistical models.

Statistically self-affine fractal surfaces with controlled Hurst exponents are required for many purposes, such as to test procedures for measuring Hurst exponents, to explore the physics of rough surfaces using computer simulation, to generate "forged" landscapes and to set the initial conditions in geophysical modeling. Self-affine fractal curves were first generated by Mandelbrot and Van Ness [273] and used by Mandelbrot and Wallis [274], in the 1960s, as a model for hydrological and geological records. Mandelbrot *et al.* studied several

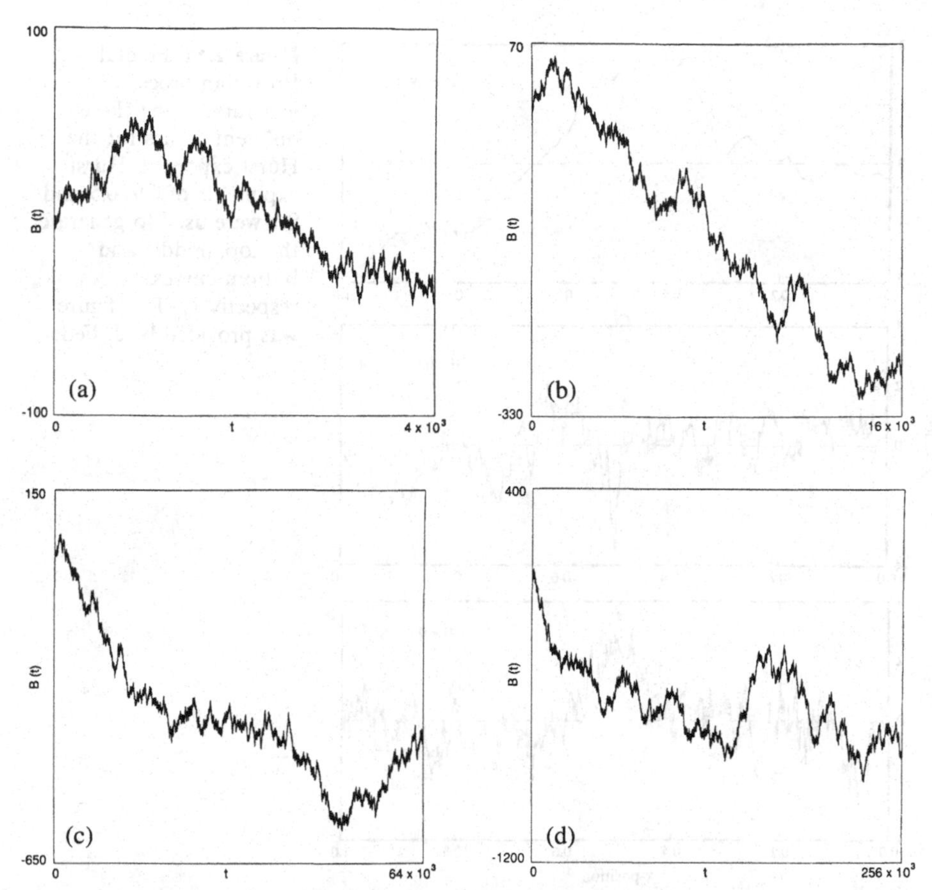

Figure 2.22 Four sections of the same discretized Brownian process. Parts (a), (b), (c) and (d) show the first 4×10^3, 16×10^3, 64×10^3 and 256×10^3 steps. The curves have been scaled to illustrate that $B(t)$ is statistically equivalent to $b^{-1/2}B(bt)$. In each successive stage ((a) → (b), etc.), the horizontal scale is increased by a factor of 4 and the vertical scale is increased by a factor of 2 ($4^{1/2}$).

approximations to fractional Brownian motion $B_H(t)$ defined as [273]

$$B_H(t) - B_H(0) = \frac{1}{\Gamma(H+(1/2))}\left\{\int_{-\infty}^{0} [(t-s)^{H-(1/2)} - (-s)^{H-(1/2)}]dB(s) + \int_0^t (t-s)^{[H-(1/2)]}dB(s)\right\}, \tag{2.135}$$

as models for these processes. This equation can be written in the form

$$B_H(t) - B_H(0) = \frac{1}{\Gamma(H+(1/2))}\left\{\int_{-\infty}^{t} K(t-s)dB(s)\right\}, \tag{2.136}$$

where $K(t-s)$ is the kernel for the integral, which has the form

$$K(t-s) = \begin{array}{ll} (t-s)^{H-(1/2)} & \text{for } 0 \le s \le t \\ (t-s)^{H-(1/2)} - (-s)^{H-(1/2)} & \text{for } s < 0. \end{array} \tag{2.137}$$

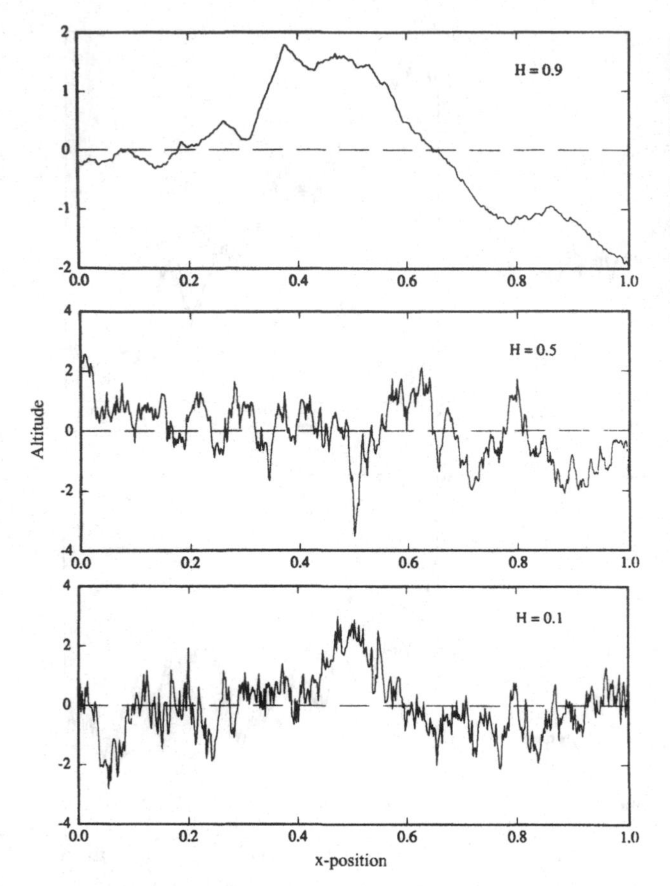

Figure 2.23 Fractal Brownian processes generated using three different values for the Hurst exponent. Hurst exponents of 0.9, 0.5 and 0.1 were used to generate the top, middle and bottom curves, respectively. This figure was provided by J. Feder.

$\Gamma(x)$ is the gamma function. It follows from equation 2.136 that

$$B_H(bt) - B_H(0) = \frac{1}{\Gamma(H + (1/2))} \int_{-\infty}^{t} K(bt - bs)dB(bs). \qquad (2.138)$$

Since $dB(s)$ represents uncorrelated Gaussian increments, $dB(bs) = b^{1/2}dB(s)$, and if the kernel $K(x)$ has the form $K(bx) = b^{[H-(1/2)]}K(x)$ it follows that

$$B_H(bt) - B_H(0) \equiv b^H [B_H(t) - B_H(0)], \qquad (2.139)$$

where "≡" means "statistically equivalent to".

"Fractional Brownian noise" models are discretized approximations to equation 2.138. The simplest of these is

$$X(t) = \frac{1}{\Gamma(H + (1/2))} \left\{ G(t) + \sum_{s=t-N}^{t-1} (t - s)^{[H-(1/2)]} G(s) \right\} \qquad (2.140)$$

or

$$X(t) = \frac{(\delta s)^{1/2}}{\Gamma(H + (1/2))} \left\{ G(t) + \sum_{s=t-N\delta s}^{t-\delta s} (t-s)^{[H-(1/2)]} G(s) \right\}, \tag{2.141}$$

where $G(s)$ and $G(t)$ are uncorrelated random numbers, selected from a Gaussian distribution, with a mean value of zero and a unit variance. The discretization is based on replacing $dB(s)$ in equation 2.135 by $(\delta s)^{1/2} G(s)$ in equation 2.141. The discretized random process $X(t)$ generated in this manner behaves like a fractal Gaussian process over time intervals in the range $\delta t < t < N\delta t$. The quantity N in equations 2.140 and 2.141 is a cut-off in the range of persistent ($H > 1/2$) or antipersistent ($H < 1/2$) correlations. This parameter should be made as large as possible, but is limited by practical considerations. A "fast fractal Gaussian noise generator" has been developed by Mandelbrot [275].

A variety of methods are now available for the generation of self-affine surfaces. Two of the most simple, and most widely used, are "random midpoint displacement" and "Fourier synthesis".

A fractal curve can be generated via random midpoint displacement by first taking a "horizontal" line segment on the unit interval. The midpoint is then displaced randomly in the vertical direction by an amount X_g, where X_g is a number selected at random from a Gaussian distribution, with a mean value of zero and a unit variance. This creates two inclined line segments covering the vertical intervals $0 - 1/2$ and $1/2 - 1$. In the next stage, the second generation, the midpoints of these two segments are displaced vertically by "Gaussian" random numbers with variances of $(1/2)^{2H}$. This creates four inclined line segments with horizontal projections of $1/4$ each. The process of displacing the midpoints of each of the line segments vertically is continued, generation after generation, to create a self-affine fractal. The vertical displacements in the nth generation are selected at random from a Gaussian distribution with a variance of $(1/2)^{2(n-1)H}$. This process is continued for N generations, until the "curve" consists of 2^N segments each having a horizontal projection of $(1/2)^N$. This procedure is not completely satisfactory because the points generated in an early generation are not statistically equivalent to points generated later. This deficiency can be remedied by adding a random Gaussian displacement with a variance of $(1/2)^{2(n-1)H}$ to *all* of the points, those generated at earlier stages as well as the new ones, during the nth stage of the construction process [196]. This approximately doubles the computer time required, but the algorithm is still of order 1 (the time required grows linearly with the number of points in the final approximation to a statistically self-affine curve). To generate a self-affine fractal covering the widest possible range, the process may be started by displacing vertically one of the ends of the horizontal line segment by an amount selected randomly from a Gaussian distribution with a variance of 2^{2H}. This approach can easily be extended to generate self-affine surfaces in higher-dimensionality embedding spaces, and the random displacements can be selected from a non-Gaussian distribution.

To generate a self-affine curve by Fourier transformation, the frequency domain representation is first constructed by filling an array of size L with Gaussian random numbers that have random phases. In most cases, L is a power of 2, but the only requirement is that L can be factored into small prime numbers, so that a 'fast Fourier transformation' algorithm can be used. This array of numbers can then be regarded as the reciprocal space representation $Z(\mathbf{k})$ of a periodic surface $h(\mathbf{x})$. Spatial correlations are then introduced into $h(x)$, by transforming $Z(\mathbf{k})$ into a new random function $Z'(\mathbf{k})$ given by

$$Z'(\mathbf{k}) = k^{-[H+(1/2)]}Z(\mathbf{k}). \tag{2.142}$$

The random field $Z'(\mathbf{k})$ is then Fourier transformed to obtain the height field $h(\mathbf{x})$. This is not as fast as the random midpoint displacement since the computer time required to generate a rough surface with N points grows as $N \log N$ instead of linearly with N. This approach can also be easily extended to higher dimensions. For a $d+1$-dimensional surface, equation 2.142 is replaced by

$$Z'(\mathbf{k}) = |\mathbf{k}|^{-[H+(d/2)]}Z(\mathbf{k}). \tag{2.143}$$

Self-affine fractal curves, with well defined fractal dimensionalities or Hurst exponents, can also be generated using the Weierstrass–Mandelbrot function [2, 276]

$$h(x) = \sum_{n=-\infty}^{n=\infty} [(1 - e^{i\gamma^n x})e^{i\phi_n}]/\gamma^{(2-D)n}, \tag{2.144}$$

with $1 < D < 2$ $(D = 2 - H)$ and $\gamma > 1$, where the ϕ_n are arbitrary phases, which are often selected randomly. The graph of the real or imaginary part of $h(x)$ is a fractal.

The Weierstrass–Mandelbrot cosine series with $\phi_n = 0$,

$$h(x) = \sum_{n=-\infty}^{n=\infty} (1 - \cos \gamma^n x)/\gamma^{(2-D)n}, \tag{2.145}$$

the corresponding truncated series

$$h_c(x) = \sum_{n=n_{min}}^{n=n_{max}} (1 - \cos \gamma^n x)/\gamma^{(2-D)n} \tag{2.146}$$

and the alternating sine series

$$h(x) = \sum_{n=-\infty}^{n=\infty} (-1)^n \sin \gamma^n x/\gamma^{(2-D)n}, \tag{2.147}$$

which does not converge for $D = 1$, are important special cases. The parameter

n_{min} in equation 2.146 determines the lower cut-off frequency and is given by $\gamma^{n_{min}} \approx 2\pi/L$, where L is the length of the sample. The function defined in equation 2.145 satisfies the self-affine scaling relationship

$$h(\gamma x) = \gamma^{(2-D)} h(x) = \gamma^H h(x), \tag{2.148}$$

so that $h(x)$ is self-affine with a Hurst exponent of $H = 2 - D$. This can easily be seen, since one of the terms on the right-hand side of equation 2.145 ($h_n = (1 - \cos \gamma^n x)/(\gamma^{(2-D)n})$) becomes

$$(1 - \cos \gamma^{(n+1)} x)/(\gamma^{(2-D)(n+1)}) = h_{n+1}, \tag{2.149}$$

under the transformation $x \to \gamma x$, $h \to \gamma^{2-D} h$. The power spectrum is given by [276]

$$S(k) \sim k^{-(5-2D)} \sim k^{-(1+2H)}. \tag{2.150}$$

Another way of generating a self-affine fractal curve is to follow a self-similar perimeter with a fractal dimensionality D_s [277]. If the distance r, measured from some arbitrary point, is then plotted as a function of the distance t measured along the path on the perimeter, then the distances r and t at two different points on the perimeter are related by

$$< |r_2 - r_1| > \sim (|t_2 - t_1|)^{1/D_s}, \tag{2.151}$$

and the curve $r(t)$ is self-affine with a Hurst exponent of $1/D_s$.

If the increments of ± 1 in the discretized Brownian process are replaced by increments taken from the Levy power law distributions of the form

$$P(|X| > x) = \begin{matrix} x^{-\alpha} & \text{for } x \geq 1 \\ 1 & \text{for } x < 1, \end{matrix} \tag{2.152}$$

then the resulting random process $B_L^{(\alpha)}$ can be rescaled by the transformation $B_L^{(\alpha)}(bt) = b^{1/\alpha} B_L^{(\alpha)}(bt)$ for $1 < \alpha < 2$ and the Hurst exponent is $H = 1/\alpha$. Self-affine models of this type, with antipersistent correlations, appear to provide an attractive approach to the modeling of stratified sedimentary rocks, such as those found in oil reservoirs [278]. Here, the large magnitude increments from the tails of the distribution represent major changes in lithology. In this case, the self-affine fractal is persistent if $H > 1/\alpha$ and antipersistent if $H < 1/\alpha$. Distributions with $\alpha > 2$ obey the central limit theorem, and the boundary between persistent and antipersistent behavior lies at $H = 1/2$.

Levy flight and Levy walk models are also finding a broad range of applications to the dynamical behavior of complex systems. For example, Shlesinger *et al.* [6] have shown how Levy walk models in which the velocity $V(\delta)$ during a step of length delta is given by $V(\delta) \sim \delta^{\gamma}$ (so that the time required for a step of length δ scales at $t(\delta) \sim \delta^{1-\gamma}$) can be used to simulate superdiffusive transport processes in which the mean square end-to-end distance $< R^2 > (t)$ after a time

t is given by $< R^2 > (t) \sim t^a$, with $a > 1$. The Richardson model [5] for the dispersion of tracers in the atmosphere, discussed at the beginning of chapter 1, is an interesting application of this approach.

2.7.2 The Geometry and Growth of Rough Surfaces

As a result of the importance of rough surfaces in many areas of science and technology, considerable effort has been devoted to the development of ways of classifying them and describing them in quantitative terms. For the most part, this effort has led to complicated empirical equations with a large number of parameters that have little fundamental significance. One of the most important characteristics of a rough surface is the dependence of the average height difference $< \delta h(\delta x) >$ on the "horizontal" separation δx between pairs of points. For a self-affine surface,

$$< \delta h(\delta x) > = A(\delta x)^H, \tag{2.153}$$

and most of the methods used to measure the Hurst exponent are based on equation 2.153 or the equivalent equation 2.133.

One of the most robust and direct approaches towards the characterization of rough surfaces is via the family of height difference correlation functions

$$C_q(\mathbf{x}) = < (\delta h(x))^q >^{1/q} = < |h(\mathbf{x}_\circ + \mathbf{x}) - h(\mathbf{x}_\circ)|^q >^{1/q}, \tag{2.154}$$

which are a generalization of equation 2.153. In equation 2.154, $h(\mathbf{x})$ is the height of the surface at lateral position $\mathbf{x}$ above a reference surface $\mathscr{H}(\mathbf{x})$ that provides a smooth, coarse-grained approximation to the rough surface. In other words, $h(\mathbf{x})$ describes the fluctuations about a "general trend" $\mathscr{H}(\mathbf{x})$, and the surface is described by the function $f(\mathbf{x})$, where $f(\mathbf{x}) = \mathscr{H}(\mathbf{x}) + h(\mathbf{x})$. In practice, the "general trend" can be difficult to identify and separate from the fluctuations. In some cases, the reference surface is determined from physical considerations. For example, in the case of growth from a planar substrate, a plane parallel to the substrate in the coordinate system moving with the rough surface can be used as the reference surface. In other cases, the reference surface can be taken as the plane that minimizes

$$\int_A |h(\mathbf{x})|^q d\mathbf{x}, \tag{2.155}$$

where the integral is over the area A of the reference surface and $h(\mathbf{x})$ is the height measured from the reference surface at $\mathbf{x}$. While this height difference correlation function approach has many advantages, it transfers the problem of describing the geometry of the rough surface to that of describing the shape(s) of the correlation function(s). However, this can have important advantages, since many of the properties of rough surfaces can be more or less directly described in terms of the correlation functions $C_q(x)$.

For many rough surfaces, it has been found that the height difference correlation functions have the form

$$C_q(x) \sim x^{H_q}, \tag{2.156}$$

over a substantial range of length scales. Here, $C_q(x)$ is given by

$$C_q(x) = < C_q(\mathbf{x}) >_{|\mathbf{x}|=x}, \tag{2.157}$$

where $< C_q(\mathbf{x}) >$ implies that $C_q(\mathbf{x})$ has been averaged over all origins $\mathbf{x}_o$ in the smooth reference surface.

In most cases, the exponents H_q in equation 2.156 all have the same value H, at least for positive values of q. However, it has been suggested by Barabási *et al.* [279, 280, 281] that some rough surfaces may exhibit a more complex, multi-affine scaling structure for which the exponents H_q depend on q. The subscript "q" will be omitted from the exponents α_q, β_q and H_q in most equations. However, it is worthwhile to measure the exponents using several values of q in analyzing experiments and simulations. The scaling functions always depend on q. Equation 2.156 implies that the rough surface is a statistically self-affine fractal over a corresponding range of length scales with a characteristic Hurst exponent H, also called the wandering exponent [282, 283]. For such surfaces, the mean height difference between pairs of points separated by a "horizontal" distance δx is given by equation 2.153.

In most studies of rough surfaces generated in experiments, found in nature or obtained from computer simulations, the $q = 2$ correlation function $C_2(x)$ has been used to measure the Hurst exponent. It is not clear that this is the best choice for q. For $q \geq 2$, the largest height differences at a given range x are given a high weight. Since there may be few of these large height differences, the statistical uncertainties may increase with increasing q. In physical systems, the self-affine scaling regime is bounded by upper and lower correlation lengths, ξ^+ and ξ^-, in both the horizontal ($\parallel$) and vertical ($\perp$) directions. Self-affine scaling is found over the range

$$\xi_\parallel^- < \delta x < \xi_\parallel^+ \tag{2.158}$$

$$\xi_\perp^- < \delta h < \xi_\perp^+. \tag{2.159}$$

These correlation lengths are related by

$$(\xi_\perp^+/\xi_\perp^-) = (\xi_\parallel^+/\xi_\parallel^-)^H. \tag{2.160}$$

The height difference correlation functions $C_q(x)$, for simple self-affine surfaces, with cut-offs in the range of scaling, can be represented by the scaling form $C_q(x) = x^H F_q(x/\xi_\parallel^+, x/\xi_\parallel^-)$, and for $x \gg \xi_\parallel^-$

$$C_q(x) = x^H f_q(x/\xi_\parallel), \tag{2.161}$$

where $f_q(y) = F_q(y, \infty)$. The scaling function $f_q(y)$ has the form $f_q(y) = const.$ for $y \ll 1$ and $f_q(y) \sim y^{-H}$ for $y \gg 1$. Here, and in most other equations, the symbols $\xi_\parallel$ and $\xi_\perp$, without superscripts, are used for the upper cut-off lengths $\xi_\parallel^+$ and $\xi_\perp^+$. The height difference correlation functions can also be represented by the scaling form

$$C_q(x) = \xi_\perp g_q(x/\xi_\parallel), \tag{2.162}$$

where $g_q(y) \sim x^H$ for $x \ll 1$ and $g_q(y) = const.$ for $y \gg 1$ ($g(x \gg 1) = \sqrt{2}$ for $q = 2$ if $\xi_\perp$ is defined as the rms surface width, $\xi_\perp^{(2)}$ in equation 2.171). Figure 2.24 illustrates the form of the height difference correlation functions $C_q(x)$, given in equation 2.161, for simple self-affine surfaces with finite correlation lengths $\xi_\perp$ ($\xi_\perp^+$) and $\xi_\parallel$ ($\xi_\parallel^+$).

For the important case of growth from a smooth surface, the correlation lengths $\xi_\parallel$ ($\xi_\parallel^+$) and $\xi_\perp$ ($\xi_\perp^+$) are frequently found to grow algebraically with increasing time, or mean thickness t. Under these conditions, the growth of the surface width can be written as

$$\xi_\perp \sim t^\beta. \tag{2.163}$$

The correlation lengths $\xi_\parallel$ and $\xi_\perp$ are related by

$$\xi_\perp \sim \xi_\parallel^\alpha, \tag{2.164}$$

and if the vertical correlation length $\xi_\perp$ grows according to equation 2.163, then

$$\xi_\parallel \sim t^{\beta/\alpha} \sim t^{1/z}, \tag{2.165}$$

where the exponent α is often called the roughness exponent.[15] In general the roughness exponent α and the Hurst exponent H are not equal. Equations 2.160 and 2.164 are not inconsistent. Equation 2.160 can be written as

$$\xi_\perp = \xi_\perp^+(t) = [\xi_\perp^-(t)/(\xi_\parallel^-(t))^H](\xi_\parallel^+(t))^H = A(t)\xi_\parallel^H, \tag{2.166}$$

and $\alpha = H$ only if the amplitude $A(t) = \xi_\perp^-(t)/(\xi_\parallel^-(t))^H$ is time independent, or is confined to a finite range. In the simplest processes, the Hurst exponent and the roughness exponent do have the same value. The exponents H and α were used interchangeably until the mid 1990s, when surface growth models that require them to be distinguished were discovered. Because it is unusual to find situations in which the roughness exponent α and the Hurst exponent H are different, considerable confusion has surrounded processes in which $\alpha \neq H$. There do not appear to be any known experimental systems in which α and H are

15. Unfortunately, the symbol α is commonly used for both the codimensionality $d - D$ and the roughness exponent. This well established convention will be used here, even though it can cause confusion in some situations, such as in the analysis of the self-similar horizontal cuts through self-affine surfaces. However, the codimensionality of the horizontal cuts and the roughness exponent do have the same numerical values in the majority of cases, where the Hurst exponent and roughness exponent are equal.

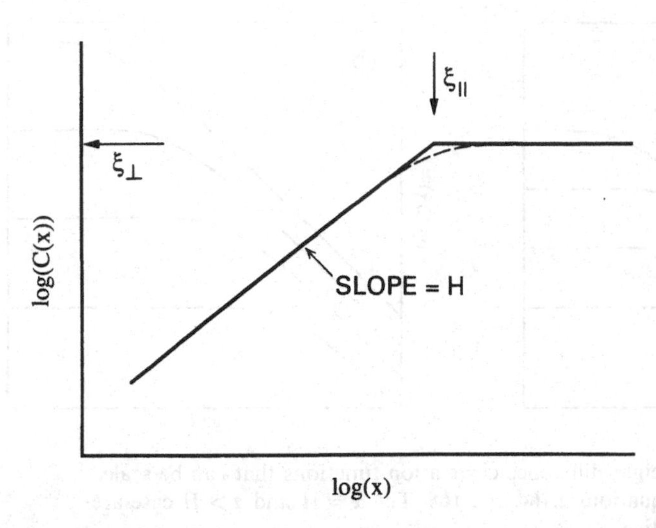

Figure 2.24 Schematic representation of the height difference correlation function $C(x) = C_q(x)$ for a self-affine surface, with long length scale cut-offs in the self-affine scaling at vertical and horizontal distances of $\xi_\perp$ and $\xi_\parallel$, respectively.

different. However, there are very few cases in which both exponents have been measured. In most experimental studies, the Hurst exponent has been measured, whereas the roughness exponent has been the focus of most simulations. In many physical growth processes and simulations, $\xi_\perp^-(t) \approx \xi_\parallel^-(t) \approx \epsilon$, where ϵ is the particle size or the size of a lattice site. In these cases, $A(t)$ in equation 2.166 is constant and $\alpha = H$. The exponent z, which characterizes the growth of the correlation length $\xi_\parallel$ in equation 2.165, is called the "dynamical exponent".

The self-affine rescaling of the height difference correlation function for a growing, self-affine surface can be expressed as

$$C_q(\lambda x, \lambda^z t) = \lambda^\alpha C_q(x,t), \tag{2.167}$$

if $x \gg \xi_\parallel^-$. This is consistent with the scaling form for the height difference correlation function $C_q(x)$ given in equation 2.161, equation 2.164 and equation 2.165. It implies that the evolution of the correlation function for a growing surface can be represented by the scaling form [284]

$$C_q(x,t) = \xi_\parallel(t)^\alpha g_q(x/\xi_\parallel(t)), \tag{2.168}$$

or

$$C_q(x,t) = \xi_\parallel(t)^\zeta x^H f_q(x/\xi_\parallel(t)), \tag{2.169}$$

where $\zeta = \alpha - H$. The scaling function $g_q(y)$ in equation 2.168 has the form $g_q(y) \sim y^H$ for $y \ll 1$, so that $C_q(x,t) \sim x^H$ for $x \ll \xi_\parallel$ and $g_q(y) = const.$ for $y \gg 1$, so that $C_q(x,t) \approx \xi_\perp$ and $C_q(x,t) \sim \xi_\parallel^\alpha$ for $x \gg \xi_\parallel$. Similarly, the scaling function $f_q(y)$ in equation 2.169 has the form $f_q(y) = const.$ for $y \ll 1$ and $f_q(y) = y^{-H}$ for $y \gg 1$. Figure 2.25(a) illustrates correlation functions that can be scaled in this manner with $H = \alpha$ ($\zeta = 0$) and Figure 2.25(b) illustrates correlation functions that can be scaled with $H \neq \alpha$ ($\zeta \neq 0$). The Hurst exponent always lies in the range $0 \leq H \leq 1$ [285]. In some cases, the roughness exponent

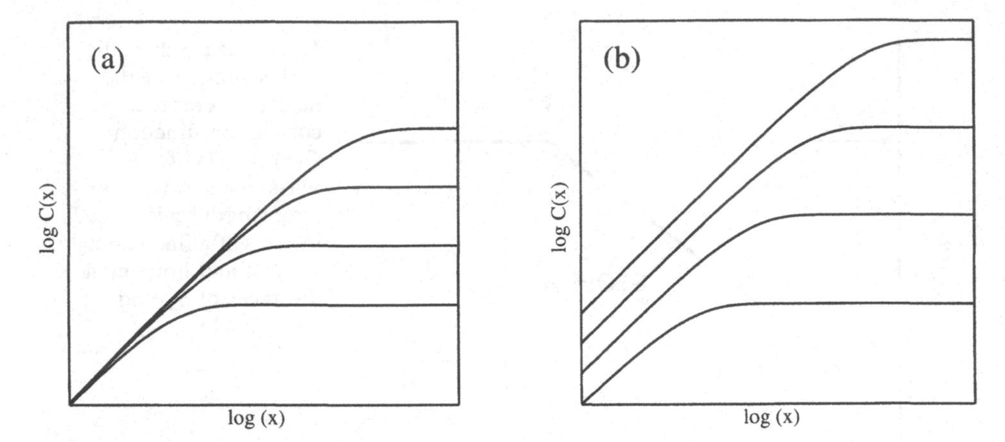

Figure 2.25 Two series of height difference correlation functions that can be scaled using the scaling forms in equations 2.169 or 2.168. The $\alpha = H$ and $\alpha > H$ case are shown in (a) and (b), respectively.

α can be greater than unity. This corresponds to cases where $\alpha \neq H$ and the exponent ζ in equation 2.169 is greater than zero. In some simple growth models, the exponent ζ in equation 2.169 is non-zero as a result of a continuous growth of the mean height differences between parts of the surface that are separated by a distance of one lattice unit in the lateral **x** coordinate system. Such unlimited growth of the step heights between adjacent parts of the surface does not seem to be likely in most physical systems.

The Hurst exponent H does not indicate how rough a surface is. Instead, it indicates how the roughness, or variance in the height, changes as the lateral length scale, over which it is measured, changes. Both the amplitude (A in equation 2.153, for example) and the exponent, as well as the lateral correlation lengths, are needed to describe surface roughness. Similarly, the roughness exponent α indicates how the surface width responds to a change in the lateral correlation length, during processes in which both correlation lengths evolve. In addition, information such as the distribution of the height fluctuations $P(\delta h)$ with respect to the mean height are needed for a full characterization. The skewness and other moment ratios are important parameters that partially describe $P(\delta h)$.

In practice, growth from an initially rough surface is probably more common than growth from a smooth surface, but there is very little known about the effects of initial roughness on surface growth. However, some aspects of this problem have been addressed theoretically [286]. It is reasonable to suppose that the height difference correlation functions will have the form

$$C_q(x) = x^{H_n} f_q(x/\xi_{\|}(t)), \tag{2.170}$$

where the scaling function $f_q(y)$ has the form $f_q(y) = const.$ for $y \ll 1$ and $f_q(y) = y^{H_s - H_n}$ for $y \gg 1$, where H_s is the Hurst exponent, defined in equations 2.154 and 2.156, for the self-affine substrate and H_n is the Hurst exponent for the growth process.

2.7.3 Characterization of Self-Affine Rough Surfaces

One of the most important applications of self-affine scaling is to surfaces growing from an extended d-dimensional substrate in a $d+1$-dimensional space. Such processes or models are said to be "$d+1$ dimensional". This notation is recommended, since it leaves no ambiguity concerning the dimensionality of either the substrate or the space in which the growth process is taking place. It will be assumed that the growing surface can be described by the $d+1$-dimensional, single-valued function $h(\mathbf{x}, t)$, where $h(\mathbf{x}, t)$ is the height of the surface at time t above position $\mathbf{x}$ on the d-dimensional substrate or reference plane. For most self-affine fractal surfaces, this is not a serious assumption, and, in most cases, $h(\mathbf{x}, t)$ can be interpreted as the highest occupied position above position $\mathbf{x}$. Growth from a $d = 2$, planar substrate in 3-dimensional space is the most interesting, from a physical point of view.

Much of the early work on the characterization of self-affine surfaces was carried out using methods such as box counting or "walking yardstick" analysis that were developed for self-similar fractals. While these methods can be used to analyze self-affine surfaces, they have often been applied in an inappropriate manner and have led to ambiguous results. In general, it is better to characterize self-affine fractals by measuring the Hurst exponents H_q. The measurement of these exponents using the height difference correlation functions defined in equation 2.154 and equation 2.157 appears to provide a convenient and reliable means of characterizing the surface roughness. As in other cases, an adequate range of length scales and good statistics are required. Corrections to the asymptotic scaling relationships may be important and must be carefully taken into account, if reliable results are to be obtained. Figure 2.26 illustrates the use of the height difference correlation function approach for two $1+1$-dimensional surface growth models, described in chapter 5, section 5.2. In figure 2.26, $x^{-1/2}C_2(x)$ is plotted as a function of $\ln(x)$. Both of these models are believed to lie in the universality class described by the surface growth equation of Kardar *et al.* [287], the KPZ equation described in chapter 5, section 5.1.1, for which the Hurst exponent or wandering exponent H has a value of exactly $1/2$ for the $1+1$-dimensional case. Consequently, horizontal curves might be expected in figure 2.26.

Figure 2.26(a) shows height difference correlation functions, at several stages, obtained from a simulation carried out using the single-step solid-on-solid deposition model [288] described in chapter 5, section 5.2. The latest stage corresponds to a mean height $\overline{h}$ of about 3600 lattice units. In this case, the correlation function $C_2(x)$ converges quite rapidly to the expected form, with no

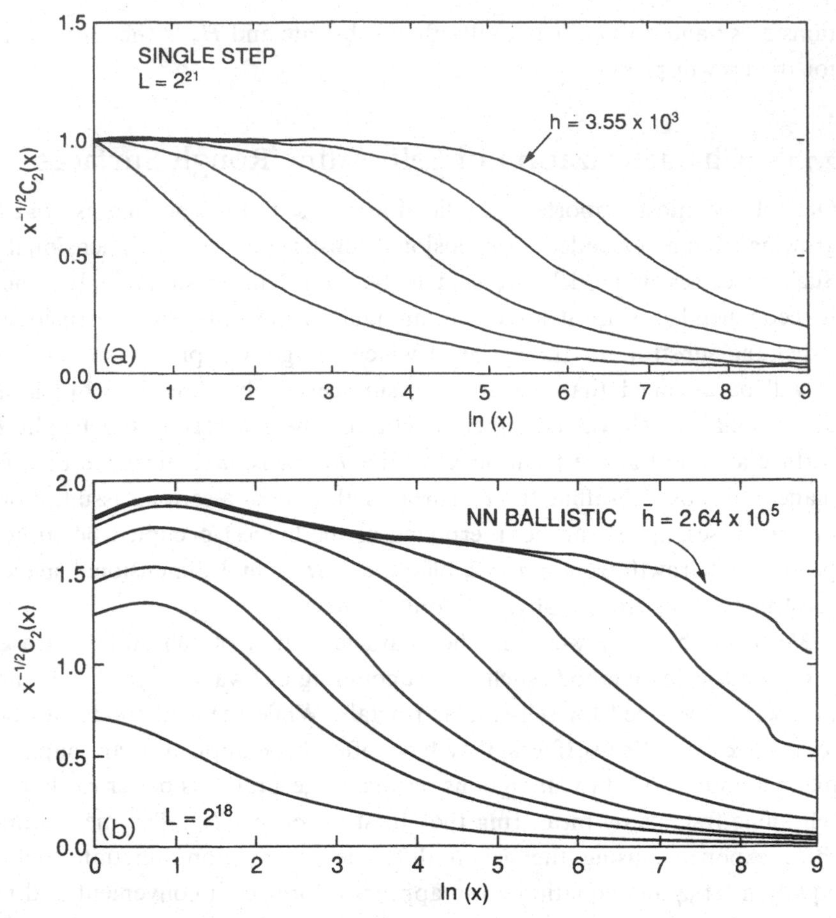

Figure 2.26 Height difference correlation functions calculated at several stages during simulation of surface growth using simple $1 + 1$-dimensional, square lattice models with widths of $L = 2^{18} = 262\,144$ lattice sites. Part (a) shows results from the "single-step" solid-on-solid model, and part (b) shows results from the "ballistic deposition" model. Both models belong to the same (KPZ) universality class with a Hurst exponent of $H = 1/2$. In both parts of the figure, $x^{-1/2}C_2(x)$ has been plotted as a function of $\log x$ to emphasize deviations from the asymptotic KPZ scaling ($C_2(x) \sim x^{1/2}$), which would correspond to a horizontal curve. This figure shows that the asymptotic form is approached much more quickly for the single-step model than the square lattice ballistic deposition model.

short length scale corrections. The rapid decrease of $x^{-1/2}C_2(x)$ with increasing x, on length scales larger than $\xi_\|$, can be seen quite clearly in this figure. At the later stages, the correlation functions can be superimposed by displacing them horizontally on the log–linear scale shown in figure 2.26(a). This demonstrates that the correlation functions have the scaling form $C_2(x,t) = x^{1/2}f(x/\xi_\|(t))$.

Figure 2.26(b) shows a similar set of correlation functions obtained from the standard, $1 + 1$-dimensional, square lattice, ballistic deposition model, also described in chapter 5, section 5.2. In this case, the mean height, measured from the substrate, at the latest stage is much larger, about 260 000 lattice units. However, there are now quite substantial short length scale corrections, which make it much more difficult to measure the exponent H. In practice, the data shown in figure 2.26(b) could be interpreted in terms of an effective exponent of about 0.47. This is not a very good estimate for the Hurst exponent, despite the large scale of the simulations in which a total of about 3.5×10^{10} sites were deposited. The relatively inaccurate values found for key exponents, even from very large scale simulations, illustrates the need for either a quantitative theoretical understanding leading to exact values of the exponents, or at least a good qualitative understanding of the corrections to scaling that can be used to guide the interpretation of simulation results. Most experiments cover a range of length scales similar to those shown in figure 2.26. The coincidence of the correlation functions over larger and larger length scales as the number of sites increases demonstrates that the amplitude of the correlation functions does not change and $\alpha = H$.

Ko and Seno [289] have recently obtained similar results, and they have interpreted the effective exponent of $H \approx 0.45$ for ballistic deposition in terms of a genuine discrepancy with the theoretical value of 1/2, which they attribute to the overhangs in the surface. They also interpreted differences between ballistic deposition and restricted step height solid-on-solid models in higher-dimensional embedding spaces in similar terms. It may prove to be difficult to resolve the different interpretations of these simulations on the basis of larger scale simulations or more careful analysis of existing data. It seems unlikely that the exponentially decreasing distribution of large step heights [288] could change the asymptotic scaling exponents in the standard ballistic deposition model. In addition, data from ballistic deposition models and models that definitely do lie in the KPZ universality class indicate that there is a stronger amplitude universality [290, 291] as well as an equivalence of exponents. However, the measurement of amplitudes is uncertain in the presence of significant corrections to scaling, and more work is needed to develop a better understanding of how the large steps in the surface of deposits generated by ballistic deposition influence the approach to the asymptotic scaling behavior.

In many growth processes, the time dependence of the correlation length $\xi_\perp$ (the surface width) is also a quantity of considerable interest. In the case of a lattice model, the quantities $\xi_\perp^{(q)}$, defined as

$$\xi_\perp^{(q)} = \left[1/N \sum_{i=1}^{N} |h_i - \overline{h}|^q \right]^{1/q} \tag{2.171}$$

are measured. Here, $\overline{h}$ is the mean surface height and $N = L^d$ is the number of sites in the surface. Equation 2.171 can also be used to measure the vertical correlation lengths $\xi_\perp^{(q)}$ for surfaces generated in experiments, if the height h is measured at N equally spaced positions. In most cases, $\xi_\perp^{(q)} \approx \xi_\perp^{(q')} \approx \xi_\perp$ for all $q > 0$, and, for models in the $1+1$-dimensional KPZ universality class, the exponent β in equation 2.163 has a value of $1/3$, so that $\xi_\perp$ is expected to grow as $t^{1/3}$, for all $q > 0$. Figure 2.27 shows results from the two simulations used to obtain the results shown in figure 2.26. For both models, $\ln(t^{-1/3}\xi_\perp^{(2)})$ has been plotted against $\ln(t)$, so that a straight, horizontal line would indicate that $\beta = 1/3$. In both cases, the curves are horizontal, within statistical uncertainties, for times larger than about 50, in units of growth events per lattice column.

For off-lattice simulations, the surface width $\xi_\perp$ can be defined as

$$\xi_\perp(N + M/2) = \frac{1}{2N}\sum_{i=M+1}^{M+N} |h_i - h_{i-1}|, \tag{2.172}$$

where h_i is the height at which the ith particle is deposited. In this case, $N/M \ll 1$, but N is large enough to reduce statistical uncertainties to a reasonable level. In most cases, $N/M = 0.05$ is a reasonable compromise that provides a logarithmic binning of the deposition height fluctuations. For lattice models, with a single-valued height function $f(\mathbf{x})$, the surface width is often defined as the square root of the variance of the surface height, $w = (< h^2 > - < h >^2)^{1/2} = < h_i^2 - < h >^2 >^{1/2}$, and this can also serve as a measure of the correlation length $\xi_\perp$. If this definition is accepted, then $\xi_\perp^{(2)} = w = C_2(r \gg \xi_\parallel)/2^{1/2}$. In some applications, the height–height correlation function $G_2(x)$ defined as

$$G_2(x) = < [(h(\mathbf{x}_\circ) - < h >)(h(\mathbf{x}_\circ + \mathbf{x}) - < h >)] >_{|\mathbf{x}|=x} \tag{2.173}$$

is measured. The correlation functions $G_2(x)$ and $C_2(x)$ are related by

$$G_2(x) = \{[C_2(x' \gg \xi_\parallel)]^2 - [C_2(x)]^2\}/2 \tag{2.174}$$

or

$$G_2(x) = w^2 - [C_2(x)^2/2], \tag{2.175}$$

where w is the surface "width".

Most simulations have been carried out using lattices of width L, with periodic boundary conditions in the lateral direction(s). Under these circumstances, $\xi_\perp$ and $\xi_\parallel$ grow according to equations 2.163 and 2.165, respectively, until $\xi_\parallel \approx L$. At this stage, both $\xi_\perp$ and $\xi_\parallel$ stop growing, and, in the limit $t \gg L^z$, the surface width is given by

$$\xi_\perp \sim L^\alpha, \tag{2.176}$$

in accord with equation 2.164, so that the roughness exponent α can be determined by carrying out simulations in strips of different widths (or columns of different widths for the $2+1$-dimensional case) and measuring the dependence

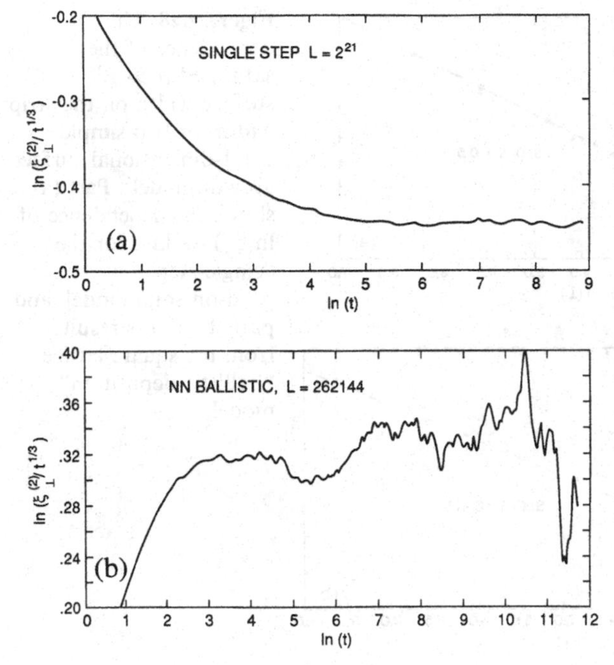

Figure 2.27 The dependence of the surface width $w = \xi_{\perp}^{(2)}(t)$ on the deposition time t or mean surface height for the $1+1$-dimensional "single-step" solid-on-solid model (a) and the $1+1$-dimensional ballistic deposition" model (b). In both parts of the figure, $\ln[\xi_{\perp}^{(2)}(t)/t^{1/3}]$ has been plotted as a function of $\ln t$, to emphasize deviations from the asymptotic $\xi_{\perp}^{(2)}(t) \sim t^{1/3}$ scaling that would correspond to a horizontal curve.

of the stationary (asymptotic) value of $\xi_{\perp}$ on L. This is illustrated in figure 2.28 for the $1+1$-dimensional single-step and ballistic deposition models. As in figure 2.26, the corrections to the asymptotic scaling behavior are much larger for the ballistic deposition model than for the single-step model.

In this growth scenario, the dependence of $\xi_{\perp}$ on t and L can be represented by the scaling form

$$w(t,L) = \xi_{\perp}(t,L) = L^{\alpha} f(t/L^{z}), \tag{2.177}$$

discovered by Family and Vicsek [292] and Jullien and Botet [293]. The exponent z in equations 2.165 and 2.177 is the "dynamical exponent" and is given by $z = \alpha/\beta$. The scaling function $f(y)$ has the form $f(y) = const.$ for $y \gg 1$ (so that $\xi_{\perp} \sim L^{\alpha}$, at late times) and $f(y) \sim y^{\beta}$ for $y \ll 1$ (so that $\xi_{\perp} \sim t^{\beta}$, at early times). Both of the exponents in equation 2.177 can be measured by collapsing $\xi_{\perp}(t,L)$ onto a single curve $(f(y))$ using the scaling form given in equation 2.177. Here, $L^{-\alpha}\xi_{\perp}$ is plotted against t/L^{z} and the exponents α and z are varied systematically to obtain the "best" data collapse. In practice, this is often done by "eye".

Unfortunately, substantial corrections to the asymptotic scaling behavior are present for many models (see figures 2.26 and 2.28, for example). Presumably, similar corrections occur in physical realizations. These corrections may seriously compromise the reliability of exponents obtained from experimental work and

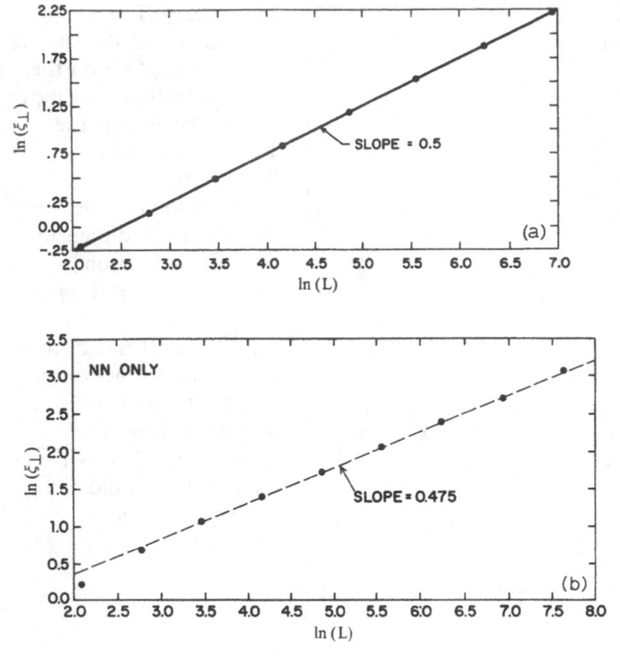

Figure 2.28 The dependence of the saturated ($t \gg L^{1/z}$) surface width on the strip width for two simple 1 + 1-dimensional surface growth models. Part (a) shows the dependence of $\ln(\xi_\perp)$ on $\ln L$ for the "single-step" solid-on-solid model, and part (b) shows results from the square lattice "ballistic deposition" model.

simulations. In most cases, the uncertainties due to corrections to scaling are much larger and more difficult to assess than the statistical uncertainties.

The widespread use of equation 2.176 to measure the roughness exponent α has most probably been influenced by the important role played by equation 2.177 in the establishment of a general scaling model for the growth of rough surfaces. However, it is generally preferable to determine the surface growth exponents α, H, β and z by measuring the height difference correlation functions $C_q(x,t)$ during a single simulation with a large system size L. In experimental work, it is not generally convenient to vary the system size (L), and in many experiments this would be impossible. Under these conditions, the surface growth exponents can be measured from the shape of the correlation function and from a data collapse of the correlation functions based on the scaling form

$$C_q(x,t) = t^\beta f'_q(x/t^{1/z}), \tag{2.178}$$

where $f'_q(y) = y^H$ for $y \ll 1$ and $f'_q(y) = const.$ for $y \gg 1$. This is equivalent to the scaling form given in equation 2.168 if $\xi_\parallel \sim t^{1/z}$. In most cases, the correlation functions $C_2(x,t)$ and $C_1(x,t)$ have been used. This approach is particularly valuable if the dynamical exponent z is large and it is difficult to reach the stationary $t \gg L^z$ regime for large values of L.

In many experiments, a slightly different procedure is used. Instead of measuring the correlation functions, the scale dependence of the mean surface width $w(\ell)$ is determined by taking segments of size ℓ from the surface. For a self-affine, 1 + 1-dimensional interface

$$w(\ell) = \left\langle \left[1/\ell \sum_{j=1}^{j=\ell} (h_{i+j} - \overline{h}_{(i,\ell)})^2 \right]^{1/2} \right\rangle, \qquad (2.179)$$

where $\overline{h}_{(i,\ell)}$ is the average surface height over the interval $i+1$ to $i+\ell$ and $< \ldots >$ implies averaging over i. For lengths ℓ that are smaller than the correlation length $\xi_\parallel$ ($\ell \ll \xi_\parallel$)

$$< w(\ell) > \sim \ell^H. \qquad (2.180)$$

On much longer length scales, where $\ell \gg \xi_\parallel$, the width saturates at the value $< w(\ell \gg \xi_\parallel) > = \xi_\perp^{(2)}$. The growth of the surface width $w(\ell, t) = < w(\ell) > (t)$ can be described using scaling forms like those for the height difference correlation function in equations 2.168 and 2.169 and all of the exponents H, α, z and β can be determined.

One approach that appears to be quite reliable is to measure the fluctuations $\delta h(x)$ over a range ℓ with respect to a baseline connecting the points $(x, h(x))$ and $(x + \ell, h(x + \ell))$. The Hurst exponent can then be estimated from the relationship

$$< R(\ell) > \sim \ell^H, \qquad (2.181)$$

where $< R(\ell) >$ is the average value of $R(x, \ell)$, and $R(x, \ell)$ is the range (difference between the maximum and minimum value of δh) over the interval x to $x + \ell$. This approach is called the "bridge" method (the line from $(x, h(x))$ to $(x+\ell, h(x+\ell))$ is the bridge). Similarly, the Hurst exponent can be measured from the dependence of $w_\delta(\ell)$ on ℓ, where $< w_\delta(\ell) >$ is the width of $\delta h(x)$ measured over ranges of size ℓ. The analysis of fluctuations with respect to bridges (applied consistently at all scales[16] ℓ), to remove the effects of linear trends, is particularly important in the analysis of experimental data. Unfortunately, this approach has not been widely used, in practice. Arneodo *et al.* [294] have pioneered the use of wavelet analysis methods for removing the effects of trends $\mathcal{H}(x)$ that have linear and non-linear components. This approach is based on a family of wavelet basis functions $\Psi^{(n)}$ (chapter 3, section 3.10.4) for which the first n moments are zero (Arneodo *et al.* used the $n+1$st derivatives of the Gaussian as their wavelet functions). At the time of writing this was a new method that had not been extensively tested. It appears to be a promising way to detrend data.

16. It is not sufficient to analyze the fluctuations about a single bridge connecting the first and last data points. This merely changes the linear trend by imposing an artificial constraint on the data.

In many cases, the surface width in the stationary limit ($t \gg L^z$) can be described in terms of contributions from an "intrinsic" part that depends on the local structure and a "scaling" part that depends on L. Under these circumstances, the surface width $\xi_\perp(t, L)$ can be written as

$$[\xi_\perp^{(2)}(t, L)]^2 = w^2(t, L) = L^{2\alpha}(f(t/L^z))^2 + w_i^2, \tag{2.182}$$

where w_i is the intrinsic contribution [295, 296]. A more reliable estimate of the value of the roughness exponent may be obtained by using equation 2.182 to analyze the L and t dependence of $\xi_\perp(t, L)$. In the Eden and ballistic deposition models, the major contribution to w_i comes from large steps in the surface. More accurate values for H and α can often be obtained by modifying the model, so as to restrict the maximum step size. This can be achieved directly by excluding growth events that would generate large steps [288, 297] or indirectly using approaches such as noise reduction [298, 299, 300]. In order to employ these approaches, it is imperative that the modifications introduced to control the intrinsic width do not change the universality class of the model.

At this stage, a quite good understanding of the corrections to scaling associated with a number of the characteristic properties of ballistic deposition and Eden growth models is emerging [301]. Extremely large scale simulations can be carried out using these models, and quite reliable values can be obtained for the exponents α, β, H and z, if corrections to scaling are either small or well enough understood to be taken into account. However, for more complex models, it is not generally possible to carry out simulations on a scale comparable with those used to obtain the results shown in figures 2.26 to 2.28. In addition, less may be known about the corrections to scaling for these models. Similar uncertainties are associated with the analysis of most experimental data.

For a very wide range of surface growth models, the exponents α and z are related by

$$\alpha + z = 2 \tag{2.183}$$

or

$$\alpha + (\alpha/\beta) = 2 \tag{2.184}$$

[288, 302, 303]. Such theoretical scaling relationships, which depend only on fundamental symmetries, provide valuable checks on simulation results. However, experience with surface growth simulations indicates that equation 2.183 is often satisfied quite well when the individual exponents α and β are far from their asymptotic values. In experiments, and even in some quite simple models (see chapter 5), it is not always easy to determine to which universality class the process belongs. If the universality class cannot be established, comparison with equation 2.184, or similar exponent scaling relationships, can be interesting but cannot be used to assess the quality of the measurements.

Additional information about the structure of rough surfaces can be obtained by measuring the distribution of height fluctuations $\mathscr{P}(\delta h)$, where δh is given by

$$\delta h(\mathbf{x},t) = h(\mathbf{x},t) - < h(\mathbf{x}',t) > . \quad (2.185)$$

Plischke and Rácz [304] measured $\mathscr{P}(\delta h)$ for the active zone (external perimeter) of Eden model clusters and found that this distribution had an essentially Gaussian form. In general, non-equilibrium surface growth processes are inherently asymmetric, and the function $\mathscr{P}(\delta h)$ should reflect this asymmetry. With some exceptions, such as the work of Amar and Family [305], this aspect of surface growth has been neglected. The shape of $\mathscr{P}(\delta h)$ will most likely prove to be important in understanding the properties of rough surfaces, and this aspect of surface growth deserves more attention.

Measurement of the scaling exponents, for experimentally generated rough surfaces, is a much more difficult undertaking. In general, a smaller range of length scales is available, and it is often not practical to generate a large number of samples. In addition, the processes leading to the formation of rough surfaces may be poorly understood and/or poorly controlled. In some cases, measurement of the roughness exponent may be complicated by crossover effects and large corrections to scaling. These difficulties have sometimes been exacerbated by a poor selection of the method(s) used to characterize the surface roughness.

2.7.3.1 Slit Island Analysis

"Slit island" methods [306] have been applied quite successfully to a variety of experimentally generated rough surfaces. In one implementation of this approach, the rough surface is coated with a second material and carefully ground and polished parallel to the flat reference surface (see above) to reveal a series of horizontal cuts. As material is removed, "islands" of the surface material will appear in a sea of the coating material. As further material is removed, these islands will grow and merge. Eventually, islands of the coating material will appear in a sea of the surface material.

In a region of size ℓ, the distribution of height fluctuations $P(\delta h)$, where $\delta h(\mathbf{x}) = h(\mathbf{x}) - < h >_\ell$ and $< h >_\ell$ is the average height in the region of size ℓ, will have the form

$$P(\delta h) = w(\ell)^{-1} f(\delta h / w(\ell)), \quad (2.186)$$

where $w(\ell)$ is the width of the distribution. Since $w(\ell) \sim \ell^H$ for a self-affine surface, the density in a cross-section of size ℓ will be given by

$$\rho(\ell) \sim (\ell)^{-H}. \quad (2.187)$$

This suggests that the boundary between the two materials, in these cross-sections parallel to the reference plane, is a self-similar fractal with a fractal dimensionality given by

$$D = d - H, \tag{2.188}$$

where d is the Euclidean dimensionality of the reference surface. This means that the methods that have been developed to analyze self-similar fractals may be used indirectly to measure the roughness exponent H. In practice, cross-sections near to the middle of the distribution $\mathscr{P}(\delta h)$, defined in equation 2.185, corresponding to small values of δh, should be used. In the $d = 1$ case, the intersection with a "horizontal" line is a fractal dust[17] with $D = 1 - H$. The fractal dimensionality can be measured via the distribution of holes or gaps, using equation 2.49 [307, 308]. If the distribution of hole sizes, or the distribution of first return lengths (the minimum value of x for which $h(x_\circ + x) = h(x_\circ)$) is given by $N_x \sim x^{-\tau}$, then $D = \tau - 1$ and

$$H = 2 - \tau. \tag{2.189}$$

This method has been used to study the roughness of fractured surfaces [307, 309].

The slit island method is illustrated in figure 2.29, using surfaces generated by a solid-on-solid ballistic deposition model (chapter 5, section 5.2.3) in which the steps in the surface height between adjacent (nearest neighbor) columns of the lattice are always +1 or −1 lattice units [288]. Figure 2.29 shows all of those columns with heights equal to the mean surface height $< h >$ for three different mean heights. These are the coastlines referred to above. Figure 2.29(d) shows the two-point density–density correlation functions $C(r)$ obtained for similar coastlines at a number of heights $< h >$. In this case, the correlation function is expected to have the form $C(r) \sim r^{-\alpha} \sim r^{D-d}$ for length scales in the range $1 \ll r \ll \xi_\parallel$ (lattice units).[18] It follows, from equation 2.188, that the codimensionality of the horizontal cut α and the Hurst exponent H are equal. Other computer simulations indicated that the roughness exponent, measured via the dependence of $\xi_\perp$ on L in the stationary regime, and the Hurst exponent H, measured from the height difference correlation function, have a value of about 0.36 for the model illustrated in figure 2.29. Figure 2.29(d) indicates a reasonably good agreement between these two methods for measuring the Hurst exponent.

A convenient way of characterizing rough surfaces, via the slit island approach, is to examine the dependence of the island perimeters $\mathscr{P}$ on their areas A. Assuming that the islands are 2-dimensional regions with a fractal boundary,

17. A fractal of disconnected points.
18. Here α is the codimensionality of the horizontal cut, *not* the roughness exponent.

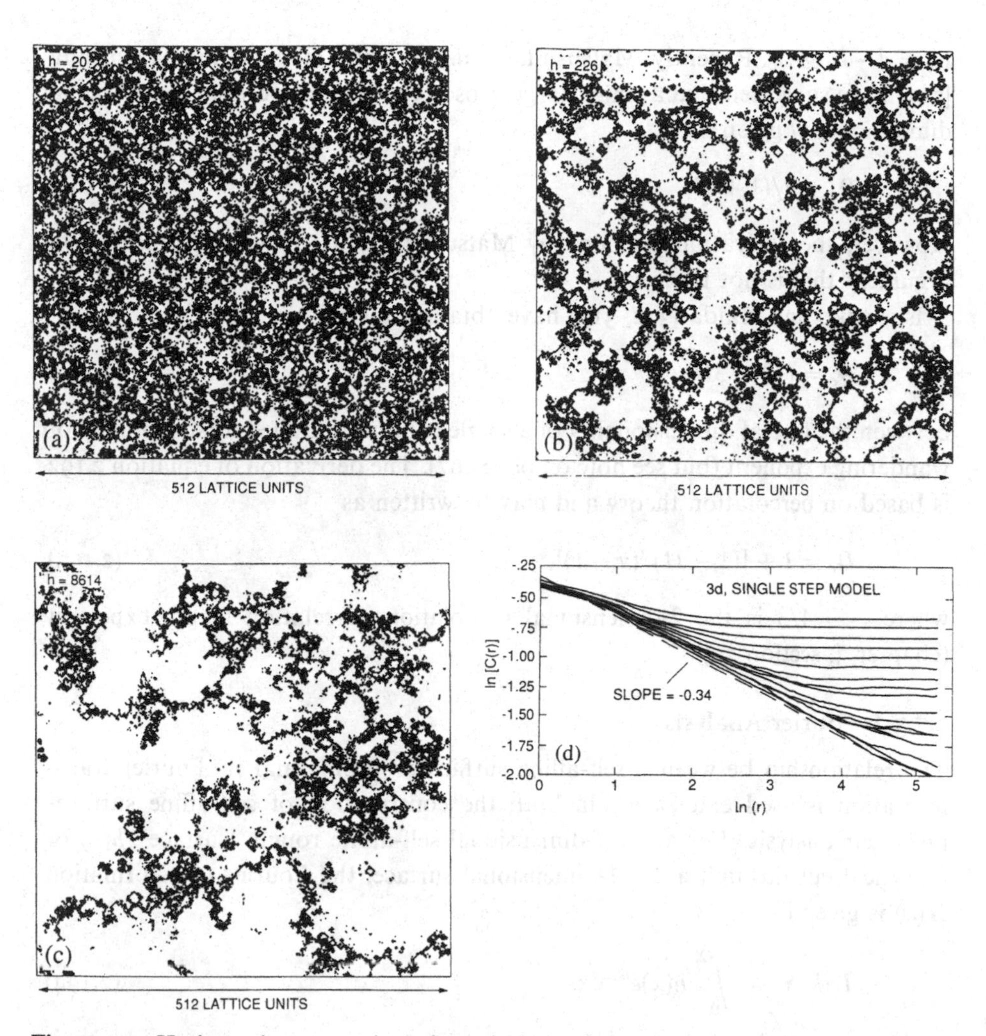

Figure 2.29 Horizontal cross-sections through the surface generated by the $2 + 1$-dimensional (cubic lattice) single-step surface growth model. Parts (a), (b) and (c) show cross-sections at the mean surface height after the mean height $< h >$ has reached $< h > = 20$, 226 and 8614 lattice units, respectively. Part (d) shows log–log plots of the two-point density–density correlation functions obtained from cross-sections at 18 stages in a simulation similar to that used to obtain parts (a)–(c). This clearly shows a crossover from $C(r) \sim r^{-\alpha}$ on short length scales ($r < \xi_{\parallel}(< h >)$) to $C(r) = const.$ on long length scales ($r > \xi_{\parallel}(< h >)$).

then the (asymptotic) relationship between $\mathscr{P}$ and A is given by

$$\mathscr{P} \sim (A^{1/2})^{D_c} \sim A^{D_c/2}, \tag{2.190}$$

where D_c is the fractal dimensionality of the island perimeter or "coastline". In general, the self-similar fractal coastline measured using the slit island method includes the coastlines of many islands, lakes within the island and islands in the lakes, etc. This "total coastline" will have a fractal dimensionality D_t with

$D_t = d - H \geq D_c$. Recently, Matsushita *et al.* [310] and Isogami and Matsushita [311] have suggested that a single, large, closed coastline or contour has a fractal dimensionality given by

$$D_c = 2/(1 + H). \tag{2.191}$$

However, the derivation provided by Matsushita *et al.* relies on a mean field argument that is not rigorous.

Isichenko and Kalda [312, 313] have obtained the relationship

$$D_c = (10 - 3H)/7 \tag{2.192}$$

between the fractal dimensionality of a single contour and the Hurst exponent or wandering exponent (but see note on page 167). The derivation of equation 2.192 is based on percolation theory and may be written as

$$D_c = 1 + [(1 - H)/(v + 1)], \tag{2.193}$$

where $v = 4/3$ is the 2-dimensional percolation correlation length exponent (chapter 3, section 3.5).

2.7.3.2 Fourier Analysis

The relationship between a self-affine surface or profile and its Fourier transformation is used extensively in both the construction of self-affine surfaces and their analysis. For a $1 + 1$-dimensional self-affine rough "surface" $h(x)$ or a vertical cut through a $2 + 1$-dimensional surface, the Fourier transformation $H(k)$ is given by

$$H(k, X) = \int_0^X h(x)e^{ikx}dx, \tag{2.194}$$

where the surface is recorded over the interval 0 to X. If the surface is self-affine, then

$$h(x) \equiv \lambda^{-H}h(\lambda x), \tag{2.195}$$

so that

$$H(k, X) \equiv \int_0^X \lambda^{-H}h(\lambda x)e^{ikx}dx, \tag{2.196}$$

and making the substitution $x' = \lambda x$, equation 2.196 can be written as

$$H(k, X) = \lambda^{-(H+1)} \int_0^{\lambda X} h(x')e^{ikx'/\lambda}dx'. \tag{2.197}$$

A comparison between equations 2.194 and 2.197 shows that

$$H(k, X) = \lambda^{-(H+1)}H(k/\lambda, \lambda X). \tag{2.198}$$

The power spectrum is defined as

$$S(k) = (1/X)|H(k,X)|^2, \tag{2.199}$$

and this definition of $S(k)$, with equation 2.198, requires that

$$S(k) = \lambda^{-(2H+1)} S(k/\lambda) \tag{2.200}$$

or

$$S(k) \sim k^{-\gamma}, \tag{2.201}$$

where $\gamma = 2H + 1$. In general[19]

$$S(k) \sim k^{-(d+2H)}, \tag{2.202}$$

In practice, it is common to find that $\gamma > d + 2$. In this event, the large value for γ indicates that $h(x)$ is smooth at high frequencies and is differentiable; it should be interpreted as $H = 1$.

Hough [314] has discussed some of the difficulties encountered in the application of the power spectrum method. Hough demonstrated that non-stationary but non-fractal processes can lead to power law spectra and argued that a power spectrum of the form $S(k) \sim k^{-\gamma}$ can be interpreted in terms of a Hurst exponent of $H = (\gamma - 1)/2$, only for processes that are known to be stationary and random.

Nakagawa [315] has suggested that the Hurst exponent can be obtained from the power spectral density via the integrals

$$I(\gamma') = \int_{k_\circ}^{k_{max}} S(k) k^{\gamma'} d\mathbf{k}, \tag{2.203}$$

where $k_\circ$ is the lower cut-off wavenumber for the scaling regime. A critical value γ'_c for γ' can then be obtained by fitting straight lines to the dependence of $\log I(\gamma')$ on γ', for large positive and large negative values of γ'. In these limits, $\log I(\gamma')$ has the form $\log I_1(\gamma') = C + (\log k_{max})(\gamma' - 2H) + f(\gamma')$, and $\log I_2(\gamma') = C + (\log k_\circ)(\gamma' - 2H) + f(\gamma')$, respectively. Here, C is a constant, and $f(\gamma')$ is a slowly varying function of γ' ($f(\gamma') = -\log(|\gamma' - 2H|)$ for $d = 1$, $f(\gamma') = -\log[(|(\gamma' - 2H)(\gamma' - 2H - 1)|])$ for $d = 2$, $\cdots$). If the effects of the function $f(\gamma')$ are ignored, then it is apparent that the lines $\log I_1(\gamma')$ and $\log I_2(\gamma')$ intersect when $\gamma' = 2H$, and the intersection can be used to estimate the value of the Hurst exponent H. This method appears to be robust, but it has not yet been adequately evaluated. The integration in equation 2.203 uses all of the data in the power spectrum and appears to be an effective way of

19. This follows from the rescaling of the real space function $h(x)$ by the transformation $x \to \lambda x$ ($k \to \lambda^{-1} k$), $h \to \lambda^H h$. The rescaling increases the amplitude of the corresponding Fourier components by λ^H and their contribution to the power spectrum by a factor of λ^{2H}. An additional factor of λ^d comes about because expansion in real space corresponds to compression in the Fourier space. The same relationship applies to the power spectrum $P(\omega)$ of a self-affine, time dependent signal $x(t)$.

reducing the effects of the fluctuations in $S(k)$. However, for large positive and large negative values of γ' the integrals depend on only the very large and very small wavenumber parts of $S(k)$. Consequently, the statistical uncertainties will grow as the corrections due to $f(\gamma')$ decrease.

Sayles and Thomas [316] have suggested that the power spectra from a large number of rough surfaces can be represented quite well by a single curve (straight line on a log–log scale) covering eight orders of magnitude in the horizontal length scale ($\ell = 2\pi/k$) with the form

$$S(k) \sim k^{-2}. \tag{2.204}$$

However, the data from each source were normalized by a parameter called the "topothesy" $\mathcal{T}$ defined by

$$S(k) = 2\pi\mathcal{T}/k^2. \tag{2.205}$$

The topothesy $\mathcal{T}_i$ for the ith data set was obtained by fitting the data by equation 2.205 so that, after normalization, the "center" of each data set lay on the line $\log S(1/\lambda) = 2\log(\lambda)$, where $\lambda = 2\pi/k$ is the wavelength. Consequently, the impressive looking log–log plot presented by Sayles and Thomas is primarily a consequence of this normalization procedure and does not imply a universal Hurst exponent of $1/2$ for rough surfaces. The Hurst exponents measured for most individual rough surfaces are larger than $1/2$, if the analysis is confined to those length scales over which convincing power law behavior is observed.

Yordanov and Nickolaev [317] have demonstrated that sharp spectral cut-offs on the power law form expected for a self-affine fractal have an important effect on the power spectrum, leading to substantial errors in the determination of the Hurst exponent. They showed that if the power spectrum is fitted by the form expected for random processes, with power law spectra and sharp spectral cut-offs, then accurate values can be obtained for H through the entire range $0 \leq H \leq 1$. In practice, high frequency and low frequency cut-offs are unavoidable. They must be properly taken into account to obtain reliable estimates for the Hurst exponents from a power spectrum analysis. Unfortunately, this is rarely done.

2.7.3.3 Vertical Cuts

Rough surfaces are frequently characterized via 1-dimensional height profiles. These may be obtained using an approach similar to that described above for the slit island method by grinding and polishing in a direction perpendicular to the reference plane of the surface. Frequently, a stylus profilometer with a stylus tip radius on the order of a micrometer and a similar vertical resolution is used. Information on shorter length scales can be obtained using a scanning tunneling microscope (STM). The Hurst exponent H can then be obtained using the height difference correlation functions, but many investigators prefer to obtain this exponent from the power spectrum $S(k)$. The power spectrum is the Fourier transform of the correlation function $G_2(x)$ defined in equation 2.173. For a self-affine $d+1$-dimensional surface, the power spectrum $S(k)$ has the form given in equation 2.202, so that H can be determined from the shape of $S(k)$ [318]. In practice, the power spectrum is often very noisy. This can be ameliorated by smoothing, binning or filtering. Equation 2.202 provides the basis for generating statistically self-affine fractals in real space via Fourier transformation.

2.7.3.4 Variation Method

Another approach to the measurement of the fractal dimensionality or Hurst exponent is known as the variation method [319]. This method is based on the measurement of the maximum variation $v(\mathbf{x}_\circ, r)$ in the height field $h(\mathbf{x}_\circ)$, within a distance r of $\mathbf{x}_\circ$, measured in the d-dimensional lateral coordinate space, so that

$$v(\mathbf{x}_\circ, r) = [\sup(h(\mathbf{x}_1)) - \inf(h(\mathbf{x}_1))]_{|\mathbf{x}_\circ - \mathbf{x}_1| < r} \tag{2.206}$$

or

$$v(\mathbf{x}_\circ, r) = \sup|h(\mathbf{x}_1) - h(\mathbf{x}_2)|, \tag{2.207}$$

where $|\mathbf{x}_1 - \mathbf{x}_\circ| < r$ and $|\mathbf{x}_2 - \mathbf{x}_\circ| < r$. The "variation" $V(r)$ of $h(\mathbf{x})$ is then defined as

$$V(r) = \int \ldots \int v(\mathbf{x}_\circ, r) d\mathbf{x}_\circ, \tag{2.208}$$

where $d\mathbf{x} = dx_1 \ldots dx_d$, and the Hurst exponent H_v is given by

$$H_v = \lim_{r \to 0} [\log V(r) / \log(r)]. \tag{2.209}$$

As with other approaches, it is necessary to work with discretized versions of these equations and measure the dependence of $\log V(\ell)$ on $\log \ell$ over the accessible range of length scales. For example, in the case of a discretized $2+1$-dimensional surface, the maximum height fluctuation above regions of size $\ell \times \ell$ in the d-dimensional reference plane (with $\ell = 2n+1$, where n is an integer and distances are measured in pixel units) might be measured. In the case of $1+1$-dimensional data sets, this method can easily be used to analyze the fluctuations about bridges from $(x_\circ, h(x_\circ))$ to $(x_\circ + r, h(x_\circ + r))$, averaged over all origins x_o, to eliminate the effects of linear trends. Dubuc *et al.* [319] have shown how the variation method can be implemented efficiently for $1+1$-dimensional profiles. Miller and Reifenberger have used this method to analyze scanning tunneling microscopy images of amorphous carbon fracture surfaces [320].

2.7.3.5 Rescaled Range Analysis

Under a wide range of circumstances, data from experiments, simulations or natural phenomena consist of measurements of some quantity of interest x at a large number of times t. In many cases, it is natural or convenient to measure $x(t)$ at equally spaced times $x(t) = x(t_n) = x(n\delta t)$, where δt is the constant time interval. Under these circumstances, the process $x(t)$ can be characterized by the fluctuations about its average value $< x(t) >$, measured over an interval of $N\delta t$. An equivalent description is provided by the deviations of the running sum of $x(t)$ from the average trend over the same period. This quantity is given by

$$X(n, N) = \sum_{m=1}^{m=n} (x(t_m) - < x(t) >_N), \tag{2.210}$$

where $< x(t) >_N = \frac{1}{N}\sum_{m=1}^{m=N} x(t_m)$. It is often appropriate to think of $x(t_n) - < x(t) >_N$ as the increase in X during the nth time interval. The range $R(N)$ of $X(n, N)$ is then defined as the difference between the maximum and minimum values of $X(n, N)$, with $1 \leq n \leq N$. If $X(n, N)$ is a self-affine fractal, then $R(N) \sim N^H$, and the exponent H can be measured from the dependence of $R(N)$ on N. This is essentially what is done if the variation method is applied to real data.

Hurst [321] spent much of his life studying the Nile river. In this case, $x(t_n)$ is the annual discharge and $R(N)$ can be interpreted as the smallest reservoir storage capacity that would be required to maintain the mean discharge $< x(t) >_N$, during the N-year period [322]. Hurst measured the quantity $R(N)/S$, where S is the standard deviation of $x(t_m) - < x(t) >_N$ over the N-year period. He found [321, 323] that

$$R/S = (N/2)^K. \tag{2.211}$$

This indicates that $X(n, N)$ is a self-affine fractal with a Hurst exponent of $H = K$. Using data from 329 sources, including 72 phenomena, Hurst found effective values for the exponent H ranging from 0.46 to 0.96, with a mean of 0.73. From a more extensive study, including a wide range of phenomena such as tree rings, sunspot numbers and the thickness of lake mud deposit layers (varves) Hurst found that $H = 0.73$, with a standard deviation of about 0.09. In most cases, the length of the data set N was less than 1000. Similar results were obtained for phenomena relating to human activity, including the lengths of reigns of kings and popes, and statistics related to sports and economic data. Large effective Hurst exponents ($H > 1/2$) do not usually arise because of large fluctuations in $x(t_n)$. These increments are considered to possess finite moments. Instead, the self-affine scaling has been attributed to long range correlations between these quantities, which decay only algebraically with the interval $(t_n - t_m)$.

Mandelbrot and Wallis [324] emphasized that the estimation of the Hurst exponent should be based on the measurement of $R(N)/S(N)$ for many values of the lag N, with many starting points spread over the entire record. Unfortunately, this recommendation is not always followed. Since the R/S method analyzes fluctuations about a linear trend, there is no need to "detrend" the data.

2.7.3.6 Short Length Scale Fractal Dimensionalities

In the past, considerable confusion existed concerning the relationship between self-affine fractals that can be rescaled by transformations, which require different changes in length scale in different directions, and self-similar fractals that can be rescaled via the same change of length scales in all directions. Unfortunately, papers are still being published in which self-affine fractals are treated as if they were self-similar.

The self-affine scaling properties of a surface $h(\mathbf{x})$ (with $h(\mathbf{0}) = 0$) can be represented as

$$h(\mathbf{x}) \equiv \lambda^{-H} h(\lambda \mathbf{x}), \tag{2.212}$$

where "$\equiv$" implies equivalent statistical properties. Such a surface *can* be characterized using the approaches developed for self-similar fractals, *but* the results must be analyzed with more care. One method is to cover the set (surface) by hypercubic boxes of size (side length) ℓ. If $< h(\mathbf{x}_\circ) - h(\mathbf{x}_\circ + \mathbf{x}) >_{|\mathbf{x}|=\ell} > \ll \ell$, then only one box with sides of length ℓ will be required to cover the fluctuations about a region of size ℓ^d on the d-dimensional reference surface. Consequently, the number of boxes $N(\ell)$ of size (ℓ) required to cover the surface is given by

$$lim_{\ell \to \infty} N(\ell) = \ell^{-d}. \tag{2.213}$$

In the local regime, $< h(\mathbf{x}_\circ) - h(\mathbf{x}_\circ + \mathbf{x}) >_{\mathbf{x}=\ell} \gg \ell$, things are quite different: the number of boxes with sides of length ℓ required to cover the fluctuations about a region of size ℓ^d on the d-dimensional reference surface is proportional to ℓ^{H-1}, so that the number of boxes of size (ℓ) required to cover the entire surface is given by

$$lim_{\ell \to 0} N(\ell) = \ell^{H-(d+1)}. \tag{2.214}$$

For a self-similar fractal,

$$N(\ell) \sim \ell^{-D_B}, \tag{2.215}$$

for all values of ℓ, where D_B is the "box counting" dimensionality. This implies that, when examined using box counting, a self-affine surface appears to be flat for $\ell \to \infty$ (larger length scales) and to have a fractal dimensionality D of

$d+1-H$ on short length scales. The crossover between these two regimes will be found for box sizes ℓ^* given by

$$< |h(\mathbf{x}_\circ + \mathbf{x}) - h(\mathbf{x}_\circ)| >_{|\mathbf{x}|=\ell^*} = \ell^*. \tag{2.216}$$

In many cases, the local $|\mathbf{x}| < \ell^*$ regime is not physically accessible [325] and only the global regime ($D_B = d$) will be seen. However, the surface can always be brought into the local regime by an affine transformation that exaggerates the vertical coordinates. The Earth's surface illustrates the crossover from local to global roughness. On long length scales, the height difference $< \delta h(\delta x) >$ between pairs of points is always smaller than the "horizontal separation" δx (equation 2.153) and the Earth's surface appears to be "smooth" when viewed from the Moon. On short length scales, the height difference may be larger than the "horizontal separation". In most locations this short length scale regime is physically inaccessible, because of a cut-off in the range of fractal scaling.

An approach that has frequently been used to measure the fractal dimensionality of self-similar fractal curves such as coastlines is the "divider" or "walking yardstick" method. The total length of the curve $C(\ell)$ is measured using a constant spacing (ℓ) between the two tips of the divider, which is always rotated in the same direction between points on the curve, without crossing the curve. This method is popular, since it can be applied to maps and photographs, with essentially no equipment (apart from the divider!) and a minimum of effort. The process is repeated for a number of divider spacings (ℓ) and the dependence of $C(\ell) = \ell N(\ell)$ on ℓ is measured. For a self-similar fractal,

$$C(\ell) \sim \ell^{1-D_c}, \tag{2.217}$$

where D_c is the fractal dimensionality of the curve. For a self-affine curve, a quite different result is obtained. On long length scales, the result $D_c = 1$ is obtained for the global dimensionality. It is not generally practical to explore the limit $\ell \to 0$, but in this limit the local fractal dimensionality is given by [326, 327]

$$D_c = 1/H. \tag{2.218}$$

A "walking yardstick" method was used by Richardson [328] to analyze coastlines and international boundaries, and his results were later interpreted by Mandelbrot [329] in terms of self-similar fractal scaling.

2.7.4 Finite-Size Effects and Crossovers

In practical applications, self-affine scaling may be limited by crossovers and finite-size effects, in much the same way as self-similar fractal scaling. A simple model to illustrate this is the "biased, discrete Brownian process". In this model, a particle executes a 1-dimensional walk on a lattice. The probability of moving towards the origin ($X = 0$) is $0.5 + k|X|$ and the probability of moving away from the origin is $0.5 - k|X|$, where $|X|$ is the displacement from the origin. This may be regarded as a model for the motion of a Brownian particle in a harmonic potential. This process generates a curve $X(t)$ that is self-affine on short length scales and "flat" on long length scales. Figure 2.30 shows 2.5×10^5 steps from simulations carried out with $k = 10^{-4}, 10^{-3}$ and 10^{-2}. All three curves are displayed on the same scale. Figure 2.31(a) shows the height difference correlation functions $C^2(\delta t)$ obtained for eight different values of k. This figure shows clearly the crossover from a slope of $1/2$ ($H = 1/2$) on short scales ($X \ll \xi_\perp, \delta t \ll \xi_\|$) to a slope of 0, characteristic of a flat profile, on long scales. The process can be described in terms of a self-affine blob model. The blobs are anisotropic with sizes (length scales) of $\xi_\perp \times \xi_\|$. Within the blobs, the structure is like that of the unperturbed, $H = 1/2$, Brownian process. On longer length scales, the blobs form a flat array ($H = 0$). Figure 2.31 demonstrates that the correlation functions can be represented very well by the scaling form

$$C_2(x) = C_2(\delta t) = k^{-1/2} h(kx), \tag{2.219}$$

where the scaling function has the form $h(y) \sim y^{1/2}$, for $y \ll 1$ and $h(y) = const.$ for $y \gg 1$. This means that the surface profile $X(t)$ is self-affine for $y \ll 1$ ($\delta t \ll k^{-1}$) and flat for $y \gg 1$ ($\delta t \gg k^{-1}$).

In some situations, this picture must be generalized. For example, in the case of the growth of an anisotropic surface there will, in general, be three correlation lengths ($\xi_{\|x_1}$, $\xi_{\|x_2}$ and $\xi_\perp$), two Hurst exponents and two roughness exponents that may be different from the Hurst exponents

$$\xi_\perp \sim [\xi_{\|x_1}]^{\alpha_{x_1}}, \tag{2.220}$$

$$\xi_\perp \sim [\xi_{\|x_2}]^{\alpha_{x_2}} \tag{2.221}$$

and

$$\xi_{\|x_1} \sim [\xi_{\|x_2}]^{\alpha_{x_2}/\alpha_{x_1}}. \tag{2.222}$$

In this case, the surface can be covered by self-affine blobs with sizes of $\xi_{\|x_1} \times \xi_{\|x_2} \times \xi_\perp$.

2.7.5 Status

The Fourier power spectrum, height difference correlation function, R/S analysis and variation methods described above are the most direct and most widely used methods for characterizing self-affine surfaces. Experience with all of these approaches has been obtained by using them to analyze "synthetic" self-affine fractals with known Hurst exponents [272, 319, 330, 331, 332, 333].

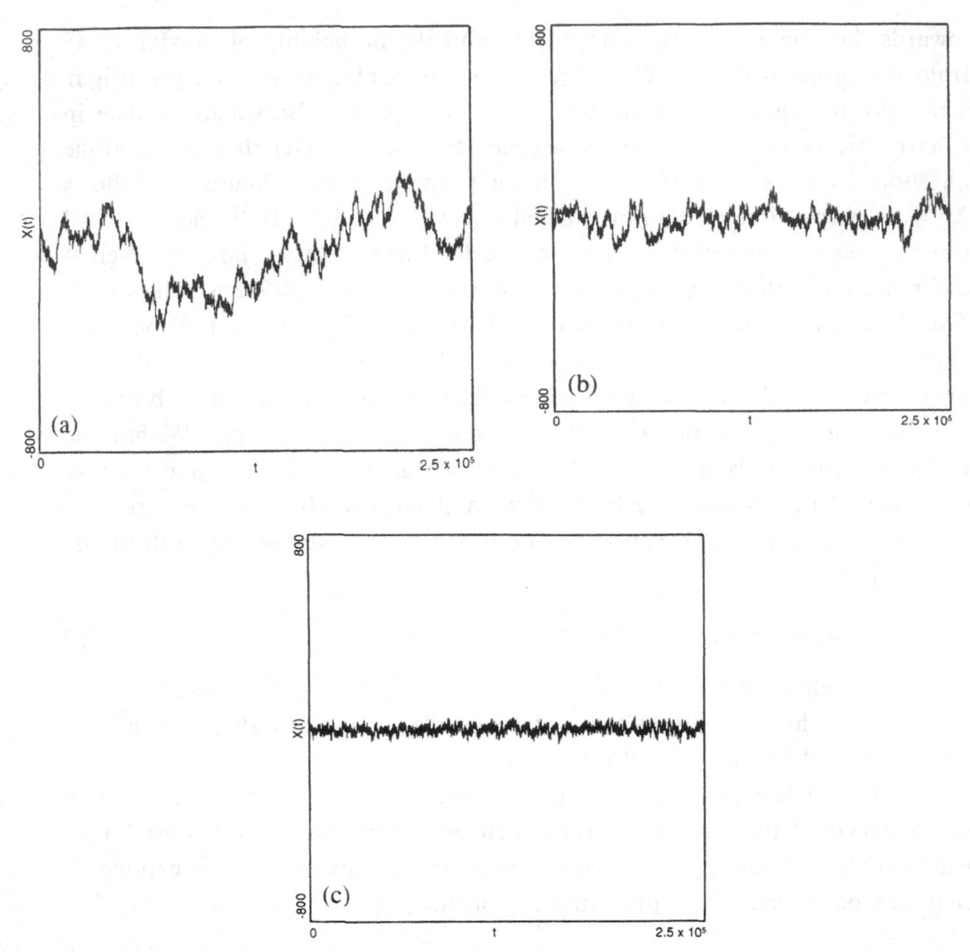

Figure 2.30 Displacement curves $X(t)$ obtained from a simple model for a Brownian particle in a harmonic potential. Parts (a), (b) and (c) show results for "restoring force constants" k of $10^{-4}, 10^{-3}$ and 10^{-2}, respectively.

These studies indicate that very large, high quality data sets are needed to get reasonably reliable results. The uncertainties become larger as $H \to 1$ or $H \to 0$ [330]. At present, sufficient experience with these methods has not been accumulated. Most of the methods that have been used cannot be relied on to give values for H with uncertainties smaller than ± 0.1, over the full range of H values. For values of H near to 0.5 (say $0.25 \leq H \leq 0.75$), these methods appear to become more reliable, and H can be measured with uncertainties of ± 0.05 or smaller with reasonably sized data sets. The height difference correlation function, slit island and variation methods have the advantage that they can be easily modified to cases where data points are missing and/or are insensitive to missing data points. The measurement of the surface width $w(\ell)$ as a function of ℓ, the variation method and R/S analysis have the advantage

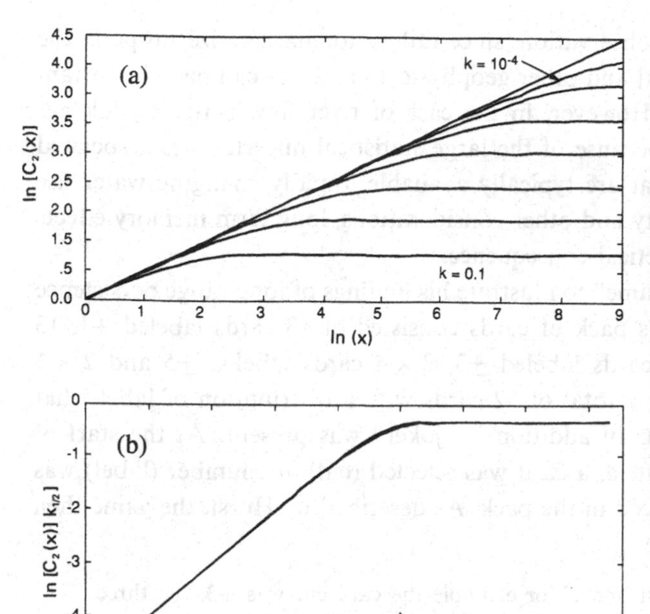

Figure 2.31 Height difference correlation functions $C_2(x) = C_2(\delta t)$ obtained from the discrete model for a Brownian particle in a harmonic potential. Part (a) shows the correlation functions (on a log–log scale) for eight values of the force constant k $(0, 10^{-4}, 3 \times 10^{-4}, 10^{-3}, 3 \times 10^{-3}, 10^{-2}, 3 \times 10^{-2}$ and $10^{-1})$. Part (b) shows how these correlation functions can be collapsed onto a common curve using the scaling form in equation 2.219.

that they can easily be modified to include linear detrending or they explicitly include linear detrending. There is a need for a clearer understanding of the effects of finite record sizes, errors in data measurement and recording, missing data, drifts in recording devices and physical background, etc. Unfortunately, advances towards more reliable ways of measuring Hurst exponents, such as that of Yordanov and Nickolaev [317], are scattered throughout a large body of literature, and they have not been disseminated effectively.

2.7.6 Long Range Persistence

If the results obtained by Hurst are taken at face value, they imply a long range persistence[20] (or, rarely, antipersistence if $H < 1/2$), for a large range of natural phenomena, that has no obvious physical origin. It has been suggested

20. The persistence or antipersistence of a self-affine fractal curve $h(x)$ can be seen by considering the increment in $h(x)$ over two successive intervals, δx_1 and δx_2, of equal length δx [2, 334]. For a self-affine fractal with a Hurst exponent H, $<\Delta h^2> = <(\delta h_1)^2> + <(\delta h_2)^2> + 2 <\delta h_1 \delta h_2>$, where $\Delta h = \delta h_1 + \delta h_2$ is the increment over the period $2\delta x = \delta x_1 + \delta x_2$. Since $<\delta h^2> \sim \delta x^{2H}$ and $<\Delta h^2> \sim (2\delta x)^{2H}$, it follows that $<\delta h_1 \delta h_2> = 2^{2H-1} - 1 <\delta h^2>$, where $<\delta h_1^2> = <\delta h_2^2> = <\delta h^2>$. Consequently, $<\delta h_1 \delta h_2>$ is greater than zero if $H > 1/2$, and $<\delta h_1 \delta h_2>$ is less than zero if $H < 1/2$. Since this relationship is equally valid for successive intervals of any length, the persistence or antipersistence has a long range nature.

that this is an important observation, since failure to characterize properly the statistics of river flow [335] and other geophysical processes can have important economic consequences. However, in the case of river flow statistics, Klemes *et al.* [336] argued that, because of the large statistical uncertainties associated with the short records that are typically available, rapidly changing water use practices, demand elasticity and other considerations, long term memory effects are of relatively little practical consequence.

Hurst devised a "card game" to illustrate his findings of long range persistence [321, 322]. One of Hurst's pack of cards consisted of 13 cards labeled $+1$, 13 cards labeled -1, 2×8 cards labeled ± 3, 2×4 cards labeled ± 5 and 2×1 cards labeled ± 7, making a total of 52 cards with a distribution of labels that approximated a Gaussian. In addition, a "joker" was present. At the start of a game the deck was shuffled, a card was selected (cut), its number (label) was recorded and it was replaced in the pack. As described by Hurst, the game then proceeds as follows:

> Two hands are then dealt and if for example the card cut was $+3$, the three highest positive cards in one hand are transferred to the other, and from this the three highest negative cards are removed. This hand then has a definite bias. A joker is now placed in it and it is shuffled and a card is cut from it. The number on this card is the first of the series. It is replaced and the hand then reshuffled, and another card is cut, recorded and replaced. This cutting and shuffling goes on until the joker is cut. Then the joker is replaced and all the cards are put together and the pack is reshuffled, after which the process is repeated.

Hurst carried out a series of four experiments [322], each consisting of 1000 cuts, using the pack of labeled cards described above to generate the data set $\{x(t)\}$ from which the range R and standard deviation S were calculated. The results from all four games were combined and the Hurst exponent was determined by fitting a straight line to a plot of $\log(R/S)$ against $\log N$. The data could be fit quite well with a slope of 0.71. Later [321], the results from two more experiments were added. In the fifth experiment, the pack consisted of 62 cards labeled ± 1 (12 of each), ± 3 (9 of each), ± 5 (6 of each), ± 7 (3 of each) and ± 9 (1 of each). The pack of 62 cards was dealt into three hands. One of the hands containing 21 cards was biased, in a manner similar to that in games $1 - 4$, by combining the two other hands and exchanging the n highest cards from the deck of 41 cards with the n lowest (most negative) cards from the deck of 21 cards, where n was determined by random selection from the whole deck at the beginning of the experiment. The joker was then added to the biased hand of 21 cards and the rest of the game was played with this hand. The sixth experiment was carried out with yet another variant of Hurst's card game. The results from all six games were combined and the data could be fit quite well with a Hurst exponent of 0.71.

However, it is clear that there is no long range persistence in the Hurst card game, and the data record generated in this game must belong to the same universality class as the simple $H = 1/2$ Brownian process and other processes for which the central limit theorem applies.

Feder [327] has simulated the Hurst card game on the computer and has analyzed much longer records of up to 10^5 "cuts". Figure 2.32(a) shows the dependence of $\log_{10}(R/S)$ on $\log_{10}\tau$, obtained from this simulation, where $\tau = N$ is the number of cuts. Because of the low level of statistical fluctuations, resulting from averaging over the large data sets, systematic deviations from a simple power law are apparent. However, it is clear that for more noisy data, an R/S analysis of this process could easily be interpreted in terms of a Hurst exponent in the range $1/2 \leq H \leq 1$, with the value of H depending on the range of N used to fit the data. In figure 2.32(b), the dependence of $\log_{10} N^{-1/2}(R/S)$ on $\log_{10}\tau$ is shown. This reveals the true nature of the Hurst card game. It is apparent that there is a crossover from a short "time" regime with $H = 1/2$ or $R/S = A_1 N^{1/2}$, corresponding to a horizontal curve in this figure, to a long time regime, also with $H = 1/2$, but with a larger amplitude ($R/S = A_2 N^{1/2}$, with $A_2 > A_1$). If the game is followed over short intervals, the random distribution of the cards in the two hands is dominant and $H = 1/2$. Over larger intervals the effects of the biasing of the hands becomes dominant and $H = 1$. On still longer time scales, the bias is averaged by shuffling the cards and $H = 1/2$. It is quite clear, from the "R/S" analysis results presented by Feder, that the effective Hurst exponent approaches a value that is much closer to 1/2 than 0.7 as the interval N increases and that the data are fully consistent with an asymptotic $N \rightarrow \infty$ value of 1/2. The Hurst card game illustrates how a crossover from two simple limits can masquerade as a non-trivial power law characterized by an effective Hurst exponent in the range $1/2 < H < 1$.

Hurst's card game illustrates how apparent or "effective" exponents, that have no fundamental meaning, can arise in all manner of circumstances. This exercise should be taken as a warning that even quite convincing empirical power laws may be chimeras with no foundation. The evidence for persistence ($H > 1/2$) obtained by Hurst for 690 time series records of 75 geophysical phenomena and similar results obtained by others has become known as the "Hurst effect". Unfortunately, most of the records available to Hurst were quite short.

In the analysis of self-similar fractals, it is often revealing to plot the deviations from a theoretical model on a log–log plot. Similarly, in the analysis of geological records, one of the main issues is whether or not the rescaled range $(R/S)(N)$ deviates from the $N^{1/2}$ behavior expected for phenomena without long range persistence. In this case, $\log[(R/S)(N)/N^{1/2}]$ can be plotted as a function of $\log(N)$. Mesa and Poveda [337] have reanalyzed the data of Hurst, in a similar manner, by plotting $(R/S)(N)/N^{1/2}$ as a function of N and have concluded that either there is no Hurst effect ($H \approx 1/2$), or that the series is not long

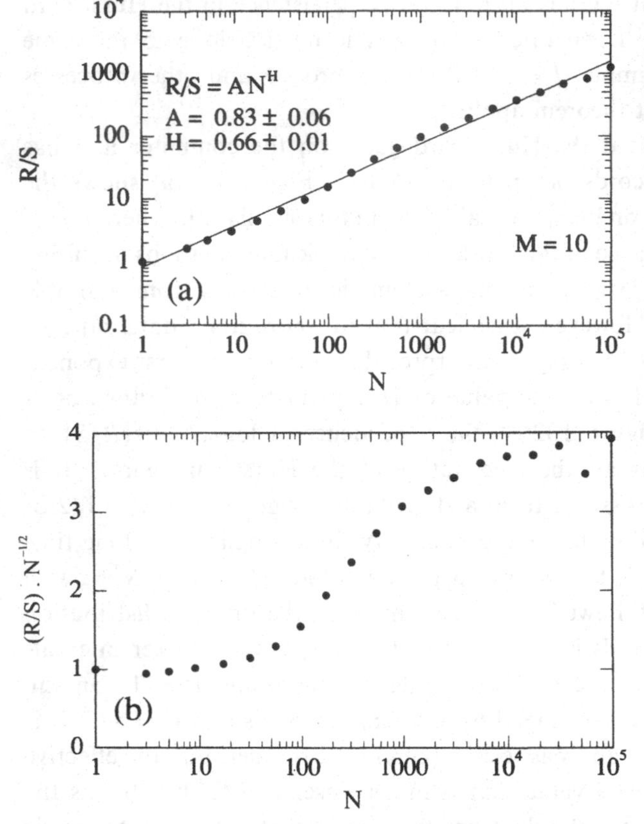

Figure 2.32 Results of an R/S analysis of the Hurst card game with a large number of cuts. In this figure, the interval N or number of cuts over which R/S is measured is indicated by "τ". Part (a) shows a standard R/S plot, which indicates how an effective Hurst exponent of about 0.66 might be obtained. Part (b) shows the dependence of $\log_{10} N^{-1/2}(R/S)$ on $\log_{10} N$. This shows very clearly the crossover from the small N, $H = 1/2$, regime to the large N, $H = 1/2$, limit with no intervening $1/2 < H < 1$ power law behavior. This figure was provided by J. Feder.

enough to provide definitive results. Only for the mud varves studied by Hurst does $(R/S)(N)/N^{1/2}$ appear to increase with increasing interval t for all the intervals, including the largest. Mesa and Poveda also analyzed a series of 18 000 wind velocity measurements recorded at 0.1 s intervals. A least squares fit of a straight line to the coordinates $(\log[(R/S)(N)], \log N)$ gave a Hurst exponent of $H \approx 0.773$. They argued that the dependence of $(R/S)(N)/N^{1/2}$ on N suggests that the data are consistent with a Hurst exponent of 1/2, for large N. However, the sampling fluctuations were large for large N, and it is difficult to distinguish between $H \approx 0.773$ and $H = 0.5$ for large N on the basis of these results. They also measured the variance $V(N)$ of $(R/S)(N)/N^{1/2}$. The variance $V(N)$ increased with increasing interval N but appeared to approach a constant value for large N. This was interpreted as another indication that there is no Hurst effect in this wind velocity data.

Feder [327] has used the "R/S" analysis approach to analyze an extensive wave-height data set obtained from Tromsøflaket in the Norwegian Sea. In this study, wave heights were measured at intervals of three hours for the 1980–1983 period. A clear crossover from a large effective Hurst exponent $H \approx 0.92$ at short times to $H \approx 0.52$ (indistinguishable from $H = 1/2$) at longer times was found, with a crossover time of $10 - 20$ days.

Bhattacharaya *et al.* [338] have shown that if the random variables $x(t_n)$ have the form

$$x(t_n) = y(t_n) + f(t_n), \tag{2.223}$$

where $y(t_n)$ is a stationary sequence of weakly dependent random numbers with a mean value of zero, and a finite variance so that $Y_N = \sum_{n=1}^{n=\infty} y(t_n)$ belongs to the Brownian motion universality class, and $f(t_n)$ has the form

$$f(t_n) = c(t_m + t_n)^{\beta}, \tag{2.224}$$

where t_m is a non-negative parameter and c is not zero, then an asymptotic Hurst exponent of $H = 1 + \beta$ will be obtained from an R/S analysis if $-1/2 < \beta < 0$. For $\beta \leq -1/2$, $H = 1/2$; for $\beta = 0$, $H = 1/2$; and for $\beta > 0$, $H = 1$. Here, the function $f(t_n)$ can be thought of as a "long term monotonic trend". In the asymptotic $N \to \infty$ limit, the range $R(N)$ is dominated by the contribution of $f(t_n)$ if $-1/2 < \beta < 0$. However, the standard deviation is dominated by $y(t_n)$ and approaches the standard deviation of $y(t_n)$ as $N \to \infty$ ($S(N)$ approaches a constant limit). This leads directly to the value obtained for the Hurst exponent ($H = 1 + \beta$) from the R/S analysis. This is a plausible explanation for Hurst exponents in the range $0.5 < H < 1$, since the long term monotonic trend decreases with increasing time and may not be noticed amongst the fluctuations in $y(t_n)$. Nevertheless, a convincing physical explanation for the origin of these long term monotonic trends in a wide range of phenomena is still lacking.

Although compelling arguments have been made in favor of the self-affine model for annual river discharges and other geophysical phenomena [324, 339], this model cannot be thoroughly tested with the relatively short duration records that are presently available. It also appears to be difficult to address this problem using process based or mechanistic models. Many phenomena such as climate, tectonics, landslides, biological processes and, more recently, human activities contribute to the hydrological record. These phenomena are in themselves complex and display statistical characteristics similar to those of the river discharges. In many geological phenomena, it appears that empirical power laws arise because of many interacting processes, each with its own characteristic time and length scale(s). Under these conditions, weak universality, relatively small scaling ranges and *approximate* empirical power laws can be expected. On the other hand, simplified models for geological processes frequently exhibit scaling behavior, and, if the process(es) captured by the model is dominant, a more robust scaling can be anticipated. After more than 40 years of study, there is still no consensus, at the time of writing, concerning the importance or interpretation of the Hurst effect.

2.8 Multifractals

In many physical problems, the distribution of "intensity" or "density" is of fundamental importance [340]. For example, in a turbulent flow the distribution of the regions of different velocity gradients or energy dissipation [341] might be of interest, and, in the case of a fluid flowing through a porous medium, the spatial distribution of regions with different fluid velocities would be of interest. Another example is the distribution in time of the parts of a random signal with widely fluctuating levels. An important example, more central to the subject of this book, is the distribution of growth probabilities on the surface of a diffusion-limited aggregation (DLA) model cluster. In this case, the location of the regions of highest growth probability is of interest in trying to develop a better understanding of how DLA clusters grow. The location of the regions of lowest growth probability is also of interest, for other reasons. Figure 2.33(a) shows an off-lattice DLA cluster containing 50 000 identical particles. The surface of this cluster was then "probed" using 5×10^6 randomly walking particles that follow the same trajectories that would be used to grow the cluster further. However, when a randomly walking particle contacted a particle in the cluster, it was not added to the cluster. Instead, a "counter" associated with the contacted particle was incremented, to record the number of times each of the particles in the cluster was contacted, and the random walker was removed. After 5×10^6 such trajectories, only 8305 of the 50 000 particles in the cluster were contacted, and their locations are shown in figure 2.33(b). Similarly, figures 2.33(c), 2.33(d), 2.33(e) and 2.33(f) show those particles that were contacted ten or more times, 100 or more times, 1000 or more times and 10 000 or more times, respectively. The largest number of contacts for any particle in the cluster was 39 175. It is apparent from figure 2.33 that the particles with the highest growth probabilities are located at the most exposed tips of the branches. The regions of lowest probability are dispersed throughout the cluster [342], but all of those particles near the center of the cluster have very small growth probabilities.

It turns out that many such complex distributions can be described in relatively simple terms. If the regions associated with a narrow range of intensity, density or probability (depending on the particular problem) are examined, then it is often found that each narrow range lies on a subset of the whole system that can be characterized by its own fractal dimensionality. For example, the regions of very high growth probability in a DLA cluster lie at a few points and form a 0-dimensional subset. On the other hand, regions with more typical growth probabilities lie on larger subsets with larger fractal dimensionalities. In fact, DLA is *not* one of the simple cases, and the distribution of the regions of lowest growth probability is still controversial (appendix B, section B.3). This division of a system into an infinite number of (generally intricately interpenetrating) subsets identified by the value of some intensive quantity and each having its own fractal dimensionality ($D(x)$, where x is the value of the intensive property) is the essence of multifractal geometry.[21] This qualitative picture must be made more precise for the quantitative description of natural phenomena.

Spatial distributions can be described in terms of the "measure" or "density" $\mu(\mathbf{x})$, which is positive everywhere and can be defined by

$$\mu_\epsilon(\mathbf{x}) = \int_{V(\epsilon,\mathbf{x})} \mu(\mathbf{y})d\mathbf{y}, \tag{2.225}$$

where $V(\epsilon, \mathbf{x})$ is a volume of size (spatial extent) ϵ in the neighborhood of a point at $\mathbf{x}$. The quantity $\mu_\epsilon(\mathbf{x})$ is the amount of "stuff" in $V(\epsilon, \mathbf{x})$. A normalized measure $\mu'_\epsilon(\mathbf{x})$ can be defined as

$$\mu'_\epsilon(\mathbf{x}) = \mu_\epsilon(\mathbf{x}) / \sum_i \mu_\epsilon(\mathbf{x_i}) \tag{2.226}$$

or

$$\mu'_\epsilon(\mathbf{x}) = \mu_\epsilon(\mathbf{x}) / \int \mu(\mathbf{y})d\mathbf{y}, \tag{2.227}$$

where the integration is over the entire system. The distribution can then be subdivided into regions of size ϵ and described in terms of the partition or distribution of the measure $\mu(\mathbf{x})$ into these regions. If the spatial distribution of the measure is not explicitly taken into account, the partition can be characterized using the measure $p_i(\epsilon)$ in each of the regions of size ϵ. The (normalized) measure $p_i(\epsilon)$, in the neighborhood of $\mathbf{x_i}$, is given by

$$p_i(\epsilon) = \mu'_\epsilon(\mathbf{x_i}). \tag{2.228}$$

One approach towards the characterization of multifractal sets is to use the partition functions [343] $Z_q(\epsilon)$ defined by

$$Z_q(\epsilon) = \sum_i p_i(\epsilon)^q. \tag{2.229}$$

In this way, different parts of the distribution can be selected. If a large positive value is used for q, then $Z_q(\epsilon)$ is dominated by those parts of the measure corresponding to the largest values of $p_i(\epsilon)$, and if a large negative value is used for q, then $Z_q(\epsilon)$ is dominated by those parts of the measure corresponding to the smallest (non-zero) values of $p_i(\epsilon)$. The partition function for a system with a characteristic overall size L can be written in the form of a sum over all of the subsets that make up the measure,

$$Z_q(\epsilon, L) = \sum A_\alpha (L/\epsilon)^{D_\alpha} (L/\epsilon)^{-q\alpha}, \tag{2.230}$$

21. It appears that in many multifractal systems the subsets are not simple self-similar fractals, but they do exhibit the "mass–length" scaling characteristic of self-similar fractals.

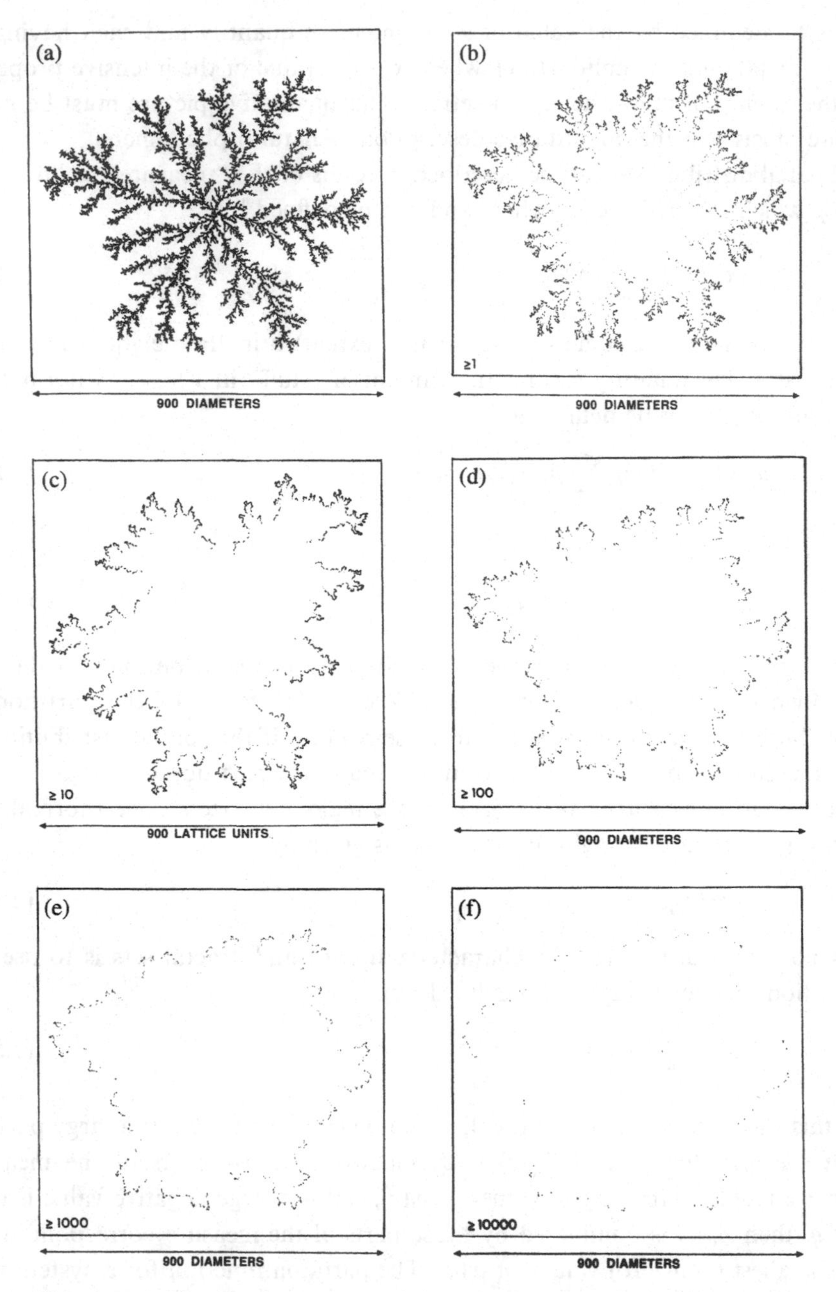

Figure 2.33 The distribution of growth probabilities on the surface of a 50 000-particle, off-lattice DLA cluster. Part (a) shows the cluster. Parts (b)–(f) show those particles that were contacted ≥ 1, ≥ 10, ≥ 100, ≥ 1000 and $\geq 10\,000$ times, respectively, after 5×10^6 randomly walking "probe" particles had contacted the cluster.

where A_α is an amplitude that characterizes the lacunarity of the subset. Equation 2.230 can also be written in the form

$$Z_q(\epsilon, L) = \int \rho(\alpha)(L/\epsilon)^{D_\alpha}(L/\epsilon)^{-q\alpha}d\alpha, \tag{2.231}$$

where the index α, which is a Lipshitz–Holder exponent, identifies each of the subsets in terms of the dependence of the probabilities $p_i(\epsilon)$ on the degree of subdivision ($\lambda = L/\epsilon$). The quantity D_α is the fractal dimensionality of the set(s) with index α and $\rho(\alpha)$ is a bounded density coefficient. It is a common convention to use the symbol $f(\alpha)$ for the fractal dimensionality D_α. Here, the symbol "$D_f(\alpha)$" will be used instead to emphasize that this quantity is a fractal dimensionality and to maintain the convention of using "D" for all quantities that can be considered to be fractal dimensionalities.

The dependence of the partition function $Z_q(\epsilon, L)$ on ϵ and L can be written as

$$Z_q = \sum_i (p_i)^q \sim (L/\epsilon)^{-\tau(q)}. \tag{2.232}$$

In the limit $L/\epsilon \to \infty$, the partition function Z_q will be dominated by the subset with the largest value of $D_f(\alpha) - q\alpha$. It is in this sense that the use of the qth moment of the probabilities p_i selects a subset of the multifractal distribution. This subset is that for which the index $\alpha(q)$ is given by

$$d/d\alpha[D_f(\alpha) - q\alpha]_{\alpha=\alpha(q)} = 0 \tag{2.233}$$

or

$$d[D_f(\alpha(q))]/d\alpha = q. \tag{2.234}$$

This means that the partition function Z_q can be expressed in terms of this dominant subset alone

$$Z_q \sim (L/\epsilon)^{D_f(\alpha(q)) - q\alpha(q))}, \tag{2.235}$$

so that

$$\tau(q) = q\alpha(q) - D_f(\alpha(q)). \tag{2.236}$$

This relationship provides a way of measuring the function $D_f(\alpha)$ that characterizes the multifractal measure. First, the function $\tau(q)$ is determined by partitioning the multifractal with a range of partition ratios $\lambda = L/\epsilon$. Then, $\alpha(q)$ is determined from the relationship

$$\alpha(q) = d\tau(q)/dq, \tag{2.237}$$

which follows directly from equation 2.236. The function $D_f(\alpha)$ can then be determined using equation 2.236.

In theoretical problems, the system size L is generally kept fixed (it can usually be taken to have a value of unity) and the limit $\epsilon \to 0$ is analyzed. This is not practical in experiments or simulations. In a growth process, it is quite natural to keep ϵ fixed at the lower cut-off length scale. This is usually one lattice unit in a computer simulation, and in an experiment it may be the particle size or experimental resolution. The length L is then the size of the growing structure and L/ϵ grows as L grows. In other cases, it may not be possible to follow the evolution of the measure of interest. In this case, L/ϵ can be varied by examining regions with different sizes L and/or with different resolutions ϵ.

The exponent $\tau(q)$ is often written in the form

$$\tau(q) = (q-1)D_q, \tag{2.238}$$

where the D_q are the generalized fractal dimensionalities of Hentschel and Procaccia [344] (see also [345]), defined as

$$D_q = \frac{1}{q-1} \lim_{\epsilon \to 0} \frac{\log \sum_i p_i(\epsilon)^q}{\log(\epsilon)} = \frac{1}{q-1} \lim_{\epsilon \to 0} \frac{\log(Z_q)}{\log(\epsilon)}. \tag{2.239}$$

The information dimension ($D_q \to 1$) is given by

$$D_0 = \lim_{\epsilon \to 0} \frac{\log \sum_i p_i(\epsilon) \log p_i(\epsilon)}{\log(\epsilon)}. \tag{2.240}$$

It follows from equation 2.236 and equation 2.238 that

$$D_q = [q\alpha(q)/(q-1)] - [D_f(\alpha(q))/(q-1)]. \tag{2.241}$$

For the case $q = 0$,

$$D_0 = D_f(\alpha(0)) = D_x. \tag{2.242}$$

Equation 2.234 indicates that $D_f(\alpha(0))$ is the maximum in $D_f(\alpha)$. All of the other subsets have smaller fractal dimensionalities so that $D_x = D_f(\alpha(0))$ is the fractal dimensionality of the "support" of the measure (all of those places where the measure is non-zero). The fractal dimensionality D_x should be interpreted as $D_x = \lim_{q\to 0} \lim_{\lambda \to \infty}[-\tau(q,\epsilon,L)]$, where $\lambda = L/\epsilon$, so that regions of size ϵ in which the measure decreases more rapidly than any power of ϵ with decreasing ϵ will not contribute to D_x. This definition of D_x implies that $D_x \leq D$, where D is the fractal dimensionality of the substrate on which the measure is distributed.[22]

It also follows from equation 2.241 that

$$D_{\pm\infty} = \alpha(\pm\infty). \tag{2.243}$$

Consequently, the quantities D_q that go under the name "generalized fractal dimensionalities" are not related in a simple way to the fractal dimensionalities $D_f(\alpha)$ and in this context are not fractal dimensionalities. They are, however, very useful scaling indices that are used extensively to characterize fractal and multifractal sets.

22. $D = \lim_{\lambda\to\infty} \lim_{q\to 0}[-\tau(q,\epsilon,L)]$.

While equations 2.232, 2.236 and 2.237 together provide a simple prescription for determining the function $D_f(\alpha)$, often called the "multifractal spectrum", this procedure should be used with extreme caution. This method always leads to a smooth convex function. There is little indication of either statistical uncertainties or systematic uncertainties resulting from too small a range of length scales (L/ϵ). In some cases, the dimensionality of the "support" of the measure is known and the maximum value of $D_f(\alpha)$ should be equal to the dimensionality of the support (D_x). Unfortunately, many published $D_f(\alpha)$ curves obtained from experiments or simulations are of little or no value for a variety of reasons, including inadequate data, large corrections to scaling and failure to test the multifractal scaling hypothesis using a data collapse or other means.

The nature of multifractals and some applications to non-equilibrium growth are discussed in appendix B. This appendix also describes the measurement of $D_f(\alpha)$ by scaling the distribution of $p_i(\epsilon, L)$ obtained for different sizes and/or resolutions. This provides a more reliable way of estimating $D_f(\alpha)$ than the more popular "Legendre transformation" procedure described in this chapter.

2.9 Universality

In many systems or models with a well defined scaling structure, the characteristic scaling exponents are often found to be invariant with respect to small changes in the fundamental physical interactions or model parameters. This allows scaling phenomena to be classified in terms of the scaling exponents. Systems with the same scaling exponents are generally said to belong to the same "universality class". However, in some cases, more stringent tests of universality that involve the amplitudes associated with key scaling relationships are applied (chapter 5, section 5.1.3). These ideas were first developed during the study of equilibrium phase transitions and critical phenomena, including percolation. As the parameters in a model are changed, transitions between different universality classes may be encountered, but situations in which the exponents change continuously as a model parameter, or physical parameter in an experiment, is changed are rare. The idea of universality has allowed a broad range of phase transitions to be classified. This classification scheme provides an important framework for describing the essential characteristics of phase transitions in both real materials and models. It reduces the problem of understanding phase transitions to one of manageable proportions and allows attention to be focused more effectively on outstanding problems and new phenomena. While it can be dangerous to apply concepts developed in the study of equilibrium systems to non-equilibrium phenomena, the universality idea has been demonstrated to be a valuable and attractive hypothesis in the study of non-equilibrium pattern formation processes.

Percolation provides a commonly used example of universality. The characteristic exponents, including the fractal dimensionality of the incipient infinite percolation cluster and well defined subsets of the cluster, are found to have the same value for a very broad range of lattice based and continuum percolation models. These exponents depend only on the dimensionality d of the embedding space. Exponents that are independent of d as well are said to be "superuniversal". Although the origins of universality are quite well understood for specific equilibrium models and some non-equilibrium models, there is no general understanding of universality in non-equilibrium growth processes and the universality concept can break down. Breakdown of universality is often found in systems with a broad (power law) distribution of interactions or correlations. For example, in the "Swiss-cheese" percolation model (chapter 3, section 3.5) the random overlapping of voids leads to a broad (power law) distribution of resistances in the equivalent resistor network. For $d = 2$, both the static and conductivity exponents have the same values as 2-dimensional lattice percolation models. For $d = 3$, the fractal dimensionality of the critical percolation cluster is the same as that for the $d = 3$ lattice models, but the conductivity exponent, which describes how the conductivity depends on the void fraction near the critical void fraction, is not the same [346]. The exponents describing the behavior of the fluid permeability and the shear modulus are different for the Swiss-cheese and lattice percolation models for both $d = 2$ and $d = 3$ [346]. In correlated percolation, the exponents are not universal and vary continuously with the correlation exponent, for a range of values of this model parameter.

2.10 Additional Information

Although it was published more than a decade ago, Mandelbrot's instant classic *The fractal geometry of nature* [2] remains the most comprehensive introduction to fractal geometry and can still be profitably mined for valuable ideas and inspiration. However, this book does not seriously address the more practical question of how the concepts of fractal geometry can be applied to characterize and understand the results of laboratory experiments and computer simulations. For those interested in the application of fractal geometry to experimental science and computer simulations, the book by Feder [327] is highly recommended. Mandelbrot [347] has published a list of major publications on fractals including books and special issues of journals. This list also includes popular and science fiction books.

Most recent books on scaling are quite specialized. Cardy [241] has published a collection of important papers on finite-size scaling, and Privman [240] has edited a collection of reviews in the same area. Privman's book starts with a good general introduction to finite-size scaling and goes on to more specialized topics. Finite-size scaling is a very important but relatively narrow application of scaling concepts.

Renormalization group theory is beyond the scope of this book. However, renormalization group methods are based on the concept of self-similarity, and they provide a theoretical foundation and motivation for much of the work on statistically self-similar and statistically self-affine fractals that is described here. A good general introduction to this family of theoretical techniques is provided in *Introduction to renormalization group methods in physics* by Creswick, Farach and Poole [243]. This book discusses a broad range of applications, is relatively unburdened by formalism and provides a good access to more specialized sources of information. Even at this level, renormalization group techniques are quite specialized and demand a strong theoretical background or substantial fortitude and perseverance.

Note added in proof

Kondev and Henley (Jané Kondev and Christopher L. Henley. Geometrical exponents of contour loops on random Gaussian surfaces. *Physical Review Letters*, 74: 4580–4583, 1995.) have used scaling analysis to obtain the relationship $D_c = (3 - H)/2$ between the single contour (island perimeter) fractal dimensionality D_c and the Hurst exponent H for 2+1-dimensional self-affine surfaces. While this result is not rigorous, it appears to supersede equations 2.191 and 2.192.

Chapter 3

Growth Models

The development of extremely simple models has provided much of the motivation for the systematic study of non-equilibrium growth using both theoretical and experimental approaches. In some cases, such as the Eden model [348, 349], these models are now quite well understood from a theoretical point of view. In other cases, such as the diffusion-limited aggregation (DLA) model, we are quite far from a satisfactory understanding, and these models pose a serious theoretical challenge. In the first part of this chapter, some the most important model classes, including Eden growth, ballistic deposition, DLA and percolation are described. Cluster–cluster aggregation models are discussed elsewhere [58]. For each class of models, some of the possible variations are described. No attempt is made to describe all of them. Instead, one or a few of the most simple models from each class are described. However, the approaches used to characterize and understand quantitatively the patterns discussed in this chapter can be applied to the enormous range of models that are not included.

Particular attention is focused on the DLA model. This model can be used as a basis for understanding a wide range of pattern-formation phenomena. It has been studied extensively for the past 15 years and still plays a central role in the study of fractal pattern-formation. Apart from its broad applications, theoretical importance and aesthetic appeal, the DLA model provides an excellent paradigm for the progress that has been made and the challenges that remain in the study of far-from-equilibrium growth processes. The second half of this chapter is concerned primarily with the current state of our knowledge and incomplete understanding of DLA. The sensitivity of the patterns generated by non-equilibrium growth models to perturbations and variations in the rules of the model, particularly the sensitivity of the geometric scaling properties to model details, is an important theoretical and practical issue. Sensitivity to perturbations is also important in experimental work. This issue is discussed, but in the absence of a complete theory for DLA it is not possible to provide a definitive picture.

The DLA and percolation models are probably the two most important and extensively studied models for random fractal patterns. The percolation model is one of the better understood models of this type, and it has been the main subject of many reviews and books [220, 350, 351, 352, 353, 354]. However, there is still much to be learned about percolation processes, particularly for the physically most important $d = 3$ case. The DLA model provides a good example of the approaches that are being taken to define the scaling characteristics of disorderly fractal patterns and to develop a better theoretical understanding of the origins of these properties. At present, there is no general theoretical approach that allows a detailed, quantitative prediction to be made concerning the geometry of the clusters generated by simple growth models from the growth rules alone without making often unjustified assumptions about the nature of the scaling properties of the clusters and/or other approximations.

3.1 Cluster Growth and Cluster Surfaces

In most of the models for non-equilibrium growth and aggregation processes, the growing structures are represented by clusters of filled sites or occupied bonds on a lattice, or clusters of simple shapes, such as spheres. In general terms, a cluster is a set of connected or associated objects. Most lattice based growth models can be described equally well in terms of "broken" and "unbroken" bonds connecting nodes or filled and empty lattice sites. In most cases, a description in terms of filled and empty sites is more natural and this convention is the most commonly used. Many connectivity rules have been used to determine if filled sites belong to the same cluster, but in most cases only occupied nearest-neighbor (*nn*) sites are considered to be connected. Consequently, two occupied sites belong to the same cluster if they are connected by a path consisting of only steps between *nn* occupied sites. Unless an indication to the contrary is given, it can be assumed that this convention is used for all of the lattice models discussed in this chapter. Nearest-neighbor sites on a d-dimensional lattice are defined as sites that share a $d - 1$-dimensional interface. In the case of off-lattice models, where the connected objects are represented by d-dimensional hyperspheres or "particles", two particles belong to the same cluster if they they are connected by a path between contacting particles.

In a typical cluster growth model, a simulation is started by filling a site in the middle of a large lattice. The growth process is then represented by a sequence of steps in which an unoccupied perimeter site is selected and filled. The models differ in the manner in which the unoccupied perimeter sites are selected. In most cases, the unoccupied perimeter sites are selected randomly, but the probabilities with which each site is selected may depend on the local geometry or the geometry of the entire cluster of filled sites. Such models are frequently used to represent macroscopic growth processes. A disadvantage in this context is the essentially uncontrolled noise associated with the growth algorithm. An important strategy that is commonly used to control the role played by this "growth noise", and (hopefully) to accelerate the approach to the asymptotic, large cluster size limit scaling behavior, is called "noise reduction". In noise reduced growth models [355, 356, 357], the selection of unoccupied perimeter sites is unchanged. However, a site is not filled, and the new unoccupied perimeter found, until a perimeter site has been selected m times. Each time a new unoccupied perimeter site is generated, by filling one of its nearest neighbors, it is given a "score" of zero. Each time it is selected, its score is increased by one, until the score reaches m and the site is filled. The number of selections m required for growth is called the noise reduction parameter.

Despite the importance of 3-dimensional growth processes, most computer modeling studies have been restricted to two dimensions. In general, 2-dimensional models can be used to explore a larger range of length scales and the results can be more easily visualized and explained.

In many cases, the "surfaces" of clusters are important in both the cluster growth processes and in simulations that are carried out to explore the physical and chemical properties of the structures that they represent. It is possible to define the surface of a cluster in many ways. In some cases, particularly in 2-dimensional models, these surfaces may have very different structures including different fractal dimensionalities. In most cases, the "physics" of the problem of interest suggests a natural definition of the "surface" of a cluster, but it is important to avoid ambiguity. Unfortunately, there does not appear to be a well established convention.

In general, the surfaces of clusters can be described in terms of filled sites, empty sites or the interfaces between them. In this book, the term "occupied perimeter" will be used to indicate the set $\{\mathcal{S}_p^o\}$ of all occupied sites with one or more unoccupied nearest neighbors (*nn*). Similarly, the term "unoccupied perimeter" will be used to indicate the set $\{\mathcal{S}_p^u\}$ of all unoccupied sites with one or more occupied *nn*. All of those unoccupied sites $\{\mathcal{S}_h^u\}$ that are nearest neighbors or next-nearest neighbors (*nnn*) of an occupied site, and can be reached from outside of the region occupied by the cluster via a path consisting of steps between unoccupied *nn* sites or unoccupied *nnn* sites, will be called the "unoccupied hull", and the occupied sites that can be reached by taking one more step (the occupied sites that are in *nn* or *nnn* positions with respect to one

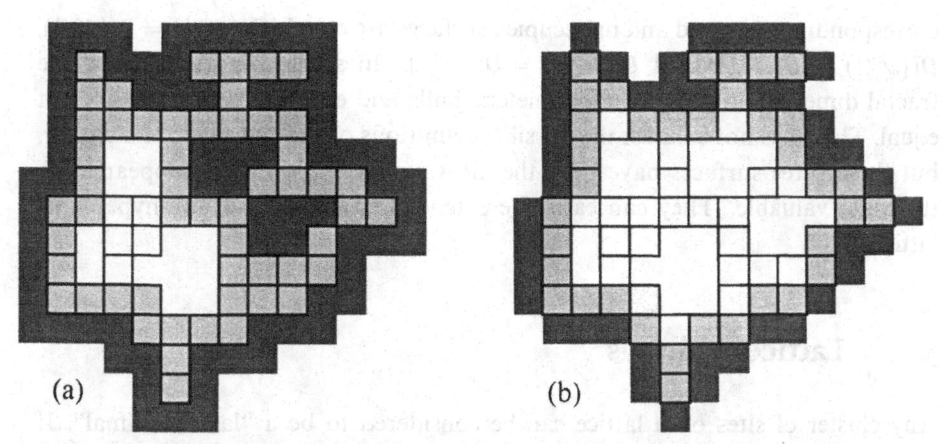

Figure 3.1 Several important surfaces of a small square lattice cluster. The cluster consists of all the gray and completely bordered white sites. Part (a) shows the occupied hull ($\{\mathscr{S}_h^o\}$ gray sites) and the unoccupied hull ($\{\mathscr{S}_h^u\}$, black sites). Part (b) shows the occupied external perimeter ($\{\mathscr{S}_p^o\}$, gray sites) and the unoccupied external perimeter ($\{\mathscr{S}_p^u\}$, black sites).

or more sites in the unoccupied hull) will be called the "occupied hull" $\{\mathscr{S}_h^o\}$. Figure 3.1(a) shows the unoccupied hull and occupied hull of a small square lattice cluster. In some applications, it is more convenient, or more realistic, to include only the subset $\{\mathscr{S}_h'^u\}$ of $\{\mathscr{S}_h^u\}$ that consists of sites that are nearest neighbors of occupied sites (the intersection of $\{\mathscr{S}_h^u\}$ and the external perimeter) in the "unoccupied hull". In general, $\{\mathscr{S}_h^u\}$ and $\{\mathscr{S}_h'^u\}$ will have the same fractal dimensionality.

The occupied hull can also be identified using a "hull walk" approach. The hull walk starts with an occupied site with the largest or smallest x or y coordinate, which is always on the hull. The hull walk then follows a path between occupied *nn* sites on the cluster. At each step, the path turns to the right, if possible. If the site to the right is unoccupied, then the forward direction is selected. If the site in the forward direction is also unoccupied, then the site to the left direction is selected and, if that site is also blocked (empty), the path reverses direction. The direction of the path is initially set in the direction away from the cluster, and the path continues until the first occupied site in the path is reached for a second time. The occupied hull consists of all of the sites visited by the path.

All of those unoccupied perimeter sites that can be reached from outside of the region occupied by the cluster, via a path consisting of steps between unoccupied *nn* sites, will be called the "unoccupied external perimeter" $\{\mathscr{S}_e^u\}$, and the filled sites that are nearest neighbors to the "unoccupied external perimeter" will be called the "occupied external perimeter" $\{\mathscr{S}_e^o\}$. Figure 3.1(b) shows the unoccupied external perimeter and occupied external perimeter for a small square lattice cluster. In general, the fractal dimensionalities of the corresponding occupied and unoccupied surfaces are equal: $D(\{\mathscr{S}^o_p\}) = D(\{\mathscr{S}^u_p\})$, $D(\{\mathscr{S}^o_e\}) = D(\{\mathscr{S}^u_e\})$ and $D(\{\mathscr{S}^o_h\}) = D(\{\mathscr{S}^u_h\})$. In some important cases, the fractal dimensionalities of the perimeters, hulls and external perimeters are not equal. This does not exhaust the possible definitions of the "surface" of a cluster, but these three surfaces have been the most widely explored and appear to be the most valuable. They can easily be extended to $d > 2$ and non-hypercubic lattices.

3.2 Lattice Animals

Any cluster of sites on a lattice can be considered to be a "lattice animal". If all configurations containing n sites are generated with equal probability, then the mean radius of gyration $< R_g(n) >$, or some other length L characterizing the mean cluster extent, is related (in the limit $n \to \infty$) to the cluster size n by

$$< R_g(n) > \sim n^{1/D_a}, \tag{3.1}$$

where D_a is the lattice animal fractal dimensionality. All configurations of n sites, including compact clusters with $D = d$ and linear clusters with $D = 1$, are included in the average indicated in equation 3.1. However, for large n, such "atypical" configurations become extremely rare, and almost all of the clusters are "typical" clusters that can be described individually as self-similar fractals with a fractal dimensionality equal to D_a. The Flory mean field approach gives the result [358]

$$D_a = 2(d+2)/5 \tag{3.2}$$

($D_a = 8/5$, $D_a = 2$ and $D_a = 12/5$, for $d = 2$, $d = 3$ and $d = 4$, respectively). The values of $D_a = 2$ and $D_a = 12/5$ for $d = 3$ and $d = 4$, respectively, appear to be exact [359, 360]. In practice, large lattice animals that look like topologically linear chains, DLA clusters or (compact) Eden model clusters are never generated. For small cluster sizes, lattice animals can be studied by generating *all* possible configurations. The cluster configuration statistics for the random occupation of sites on a lattice, in the limit $p \to 0$, where p is the fraction of sites occupied, is the same as that of lattice animals [361]. One approach, based on this idea, is to generate clusters using the growth algorithm of Leath and Alexandrowicz [362, 363] described in section 3.5.1 with a small growth probability p or large blocking probability $1-p$. More generally, percolation clusters generated with $p < p_c$, where p_c is the critical value for p, with a size $L \gg \xi_p$, where ξ_p is the percolation correlation length, have a lattice animal geometry. However, the probability of growing large clusters becomes very small in the $p \to 0$ or $L \gg \xi(p)$ lattice animal limit. Despite the conceptual simplicity of the lattice animal model, it is not easy to design a simple algorithm for generating large lattice animals (random configurations with a large number of sites) without bias [361, 364, 365, 366].

3.3 Random Walks

Random walks are among the most simple of all random fractals. Knowledge of the properties of random walks provides a foundation for understanding a very wide range of phenomena in physics, chemistry and biology. In particular, they play an important role in polymer physics and in many of the growth models described later in this chapter. Random walks can be generated on simple lattices or in continuous spaces. The most familiar example is the nearest-neighbor walk on a square lattice. In this case, the random walk starts on a site that can be placed at the origin $(0, 0)$. The random walk can be regarded as the path followed by a "random walker". At each step in the walk, the random walker steps from its current position (i, j) to one of its nearest neighbors at $(i-1, j), (i+1, j), (i, j-1)$, or $(i, j+1)$, which is selected at random, with equal probability. It can easily be shown that the radius of gyration, measured from the origin, $(0, 0)$, increases with the number of steps n according to the power law

$$< R_g >\sim n^{1/2}, \tag{3.3}$$

with no corrections! Equation 3.3 is valid for random walks on regular lattices, in spaces of any dimensionality d. This means that the random walk can be regarded as a random fractal with a fractal dimensionality of $D = 2$. This idea becomes more physical if the random walk is used to represent the growth of a polymer molecule, with each step in the walk representing a monomer or a small number of monomers corresponding to the persistence length [367, 368] in the polymer chain.

One of the most important basic properties is the number of distinct sites S_n that the random walker has visited after n steps. In the case of a linear lattice ($d = 1$), the random walker explores the lattice compactly, and the asymptotic dependence of S_n on n is given by

$$S_n \sim n^{1/2}. \tag{3.4}$$

For $d > 2$, the random walk path cannot fill the embedding space and

$$S_n \sim n. \tag{3.5}$$

The $d = 2$ case lies at the boundary between space filling and non-compact paths. In this sense, $d = 2$ is a "critical" dimensionality, and, in the limit $n \to \infty$, the number of distinct sites visited is given by

$$S_n \sim n/\log n. \tag{3.6}$$

For some simple lattices, the amplitudes associated with the power laws in equations 3.4, 3.5 and 3.6 are known exactly [369, 370].

Another important property of random walks on lattices is the probability F_n of return to the origin after n steps [371], or the probability F_t of return to a region of size ϵ at the origin, after a time t, in the case of an off-lattice random walk. For Euclidean lattices, this probability is given by

$$F_n \sim n^{-d/2}, \tag{3.7}$$

since the random walker is distributed over a volume of order $n^{d/2}$. In the case of an off-lattice walker following a Brownian path,

$$F_t \sim t^{-d/2}. \tag{3.8}$$

In Levy flight [2] or Levy walk models [372], the constant step size in the random walk is replaced by a power law distribution of step lengths. The step length distribution satisfies the conditions

$$P(\Delta \geq \delta) = \begin{matrix} \delta^{-f} & \text{if } \delta \geq 1 \\ 1 & \text{if } \delta < 1, \end{matrix} \tag{3.9}$$

where $P(\Delta \geq \delta)$ is the probability that the step length Δ will be greater than δ. In a Levy walk, the "particle" can be considered actually to follow linear segments of length Δ. In a Levy flight, only the ends of the linear segments are visited and the points visited by the flight lie on a fractal with a dimensionality of $D = f$. This means that if $1 < f < 2$, then the flight will compactly cover a 1-dimensional lattice, so that the number of sites (or regions of size ϵ) visited after n steps is given by

$$S_n \sim n^{1/f}. \tag{3.10}$$

Similarly, if $f < 1$, the flight does not cover the 1-dimensional space and

$$S_n \sim n. \tag{3.11}$$

The cases $f = 1$ and $f = 2$ are "critical", and [369]

$$S_n = \begin{matrix} n/\ln n & \text{for } f = 1 \\ [n \ln n]^{1/2} & \text{for } f = 2. \end{matrix} \tag{3.12}$$

3.3.1 Self-Avoiding Random Walks

It is apparent that the simple random walk model cannot be used to represent a physical object such as a polymer molecule, since a random walk may visit a particular site on the lattice many times. A more realistic model (for a polymer molecule in a "good" solvent [4, 186], for example) is the self-avoiding random walk. In principle, self-avoiding walks of length n can be obtained by generating all walks of length n and selecting those that do not self-intersect. Self-avoiding random walks can be generated more efficiently by growing ordinary random walks and "throwing away" each walk that enters a previously occupied site. A new walk is started after a self-intersecting walk has been rejected or after a self-avoiding walk has grown to the required size.[1] The efficiency of the algorithm can be improved by excluding steps that would reverse the direction of the random walk, returning it to the previously occupied site, but very few walks will survive to a long length for $d < 4$.[2] The probability that a random walk will "survive" without intersecting itself decreases exponentially as the length of the walk increases. This procedure favors walks that have a more extended configuration than the typical random walk. This "bias" is so strong that the relationship between the radius of gyration R_g and n is changed to

$$< R_g > \sim n^{\beta} \sim n^{1/D_{\beta}}, \tag{3.13}$$

with $D_\beta < 2$.[3] A simple mean field theory [375] gives the result

$$D_\beta = (d+2)/3. \tag{3.14}$$

Equation 3.14 gives the correct value for D, in the cases $d = 1, 2$ and 4. The critical dimensionality is $d_c = 4$. Equation 3.14 indicates that $D_{RW} - D_{SAW}$ decreases with increasing d, where $D_{RW} = 2$ is the random walk fractal dimensionality and D_{SAW} is the self-avoiding random walk fractal dimensionality. For $d \geq 4$, the fractal dimensionality is the same as that of an ordinary random walk ($D = 2$).

For $d > 4$, two random walks can occupy the same region of space, or two parts of the same walk can occupy the same region, without intersecting ($D+D < d$, chapter 2, section 2.2). Consequently, the self-avoiding "interactions" are of no consequence, as far as the asymptotic scaling is concerned. Self-intersection of parts of the walk that are well separated, in terms of their

1. A "true self-avoiding random walk" [373] can be generated by modifying the random walk algorithm described above. In the strong avoidance or low temperature limit, the random walk moves from its current position (i, j) to one of its *unoccupied* nearest neighbors at $(i-1, j), (i+1, j), (i, j-1)$, or $(i, j+1)$, which is selected at random, with equal probability. If all of the nearest-neighbor sites at $(i-1, j), (i+1, j), (i, j-1)$, or $(i, j+1)$ are occupied, a random selection is made among those nearest-neighbor sites that have been occupied the least number of times. For this model the upper critical embedding space dimensionality is $d_c = 2$ and $< R_g > \sim (n/\log n)^{1/2}$ for $d = 2$. Family and Daoud [374] have suggested that the true self-avoiding random walk model may be relevant to the structure of very polydisperse polymer solutions.
2. Even for $d > 4$, *local* attrition (rejection of random walk configurations in which a pair of sites separated by only a small number of steps overlap each other) is important, but the probability that the walk will enter a site that was visited at a much earlier stage is small.
3. Unfortunately the symbol β is commonly used both for the exponent that relates R_g to s and for the exponent that relates the width ($\xi_\perp$ or w) of a growing surface to the time that the surface has been growing or the distance advanced by the growing surface. This convention will be used here, even though it could lead to some confusion.

positions on the walk, rejects only a relatively small fraction of the configurations and the fractal dimensionality is not changed.

For $d = 3$, the mean field theory does not give the exact value for D; the value from equation 3.14 ($D = 5/3$) is too low by a small percentage. An exact value of the fractal dimensionality has not yet been obtained for this simple model. A renormalization group calculation [376] gives $D = 1/(0.588 \pm 0.001) = 1.701 \pm 0.003$, which is in good agreement with computer simulation results [377]. Despite the conceptual simplicity of the self-avoiding walk model, it is difficult to generate large walks for the $d = 2$ and $d = 3$ cases [378]. It appears that some of the algorithms that have been devised to generate self-avoiding random walks in an efficient manner may generate configurations that are not completely independent of each other or else are biased selections from the ensemble of all possible self-avoiding walks. For this reason, and because of corrections to the asymptotic scaling behavior, it has only recently been possible to claim, with a reasonable degree of confidence, that the estimates for D obtained from Monte Carlo simulations are consistent with the renormalization group calculations, for $d = 3$ [379]. Substantial progress has been made towards the development of efficient self-avoiding random walk algorithms for $d = 2$ and $d = 3$ [380, 381]. The "pivot" Monte Carlo algorithm [382] appears to be the most efficient so far. In this algorithm one end of the chain remains fixed at the origin and a new configuration is obtained by randomly selecting one of the points p_k visited by the walk after n steps and using this point as a "pivot" for the part of the chain further from the origin ($\{p_{k+1}, p_{k+2}, \ldots, p_n\}$), which is transformed to a new configuration using a randomly selected element of the symmetry group of the lattice, or a subset of the symmetry group, selected to ensure that all allowed configurations would be "visited" with equal probability after an infinite number of steps in the algorithm. If the new configuration is self-intersecting it is rejected, otherwise it becomes the next configuration in a Markov chain and provides the basis for additional configuration changes. This algorithm owes its efficiency to the surprisingly high acceptance of large configuration changes. While this relatively crude algorithm is remarkably efficient, the design of fast, bias free algorithms remains a challenge.

3.3.2 Indefinitely Growing Walks

Because of the difficulty of generating long, self-avoiding random walks on 2-dimensional and 3-dimensional lattices, other models have been devised for generating random paths that do not self-intersect. However, these random paths do not, in general, have the same fractal scaling properties as self-avoiding walks. One of these is the indefinitely growing self-avoiding walk [383], or IGSAW, illustrated in figure 3.2. This algorithm is usually implemented on a square lattice, though other 2-dimensional lattices have been used. For the square lattice model, trapping is avoided by examining the three sites that could be entered in the next step of the IGSAW and the two next nearest neighbors that would be nearest-neighbours of the site occupied by the randon walk if one more step were taken in the direction of the last step in the walk. The sites that have been occupied by the walk are labeled with a "winding number" that increases each time the walk takes a step to the right and decreases each time a step is taken to the left (figure 3.2). Figures 3.2(a) and (b) show two potential trapping configurations. These configurations can be avoided by using the winding numbers associated with the sites near the tip of the growing end of the walk [383]. Computer simulation results indicate that this model generates a fractal "chain" with a fractal dimensionality of $D = 1.75 \pm 0.03$, distinctly different from the value $D = 4/3$ associated with ordinary self-avoiding walks.

Weinrib and Trugman [384] studied a version of this model on the bonds of a honeycomb lattice. They showed that the walk generated by this model separates occupied and unoccupied sites on the external perimeter of a percolation cluster on the dual triangular lattice.[4] The IGSAW model does not generate closed loops. However, if an allowed path to the origin of the walk is always maintained, and a loop is allowed to be formed if the origin is reached, this model can be used to generate the external perimeter of the corresponding critical percolation cluster with a fractal dimensionality of $D = 7/4$ exactly. For this model the hull and the external perimeter are equivalent, and both have a fractal dimensionality of 7/4. It is not possible to generate 3-dimensional IGSAW clusters using models based on examination of the local environment of the growing end of the walk and a generalization of the winding number concept since blocked and unblocked channels may have the same local configuration of winding numbers. For example, "tubes" with open and closed ends cannot be distinguished via local winding numbers.

3.3.3 The Diffusion-Limited Growth Walk

Another way of avoiding trapping is to grow the walk, which will be called "the polymer" to avoid confusion with the ordinary random walks used in

4. The term "triangular lattice" will be used to describe configurations in which the centers of "particles" lie on the nodes of a regular tiling of 2-dimensional space by equilateral triangles. The Voronoi polyhedra surrounding the centers of each of the particles are hexagons and each particle has six nearest neighbors. In many growth algorithms it is useful to think of the growth process in terms of filling the hexagonal cells in the dual honeycomb lattice. In some cases the "particles" are placed on the nodes of a honeycomb lattice. In this case the particles have three nearest neighbors and the growth can be thought of in terms of filling the triangles in the corresponding Voronoi tesselation (the dual triangular lattice). Confusion can be avoided by referring to the coordination number (the number of nearest neighbors) of the sites or cells that are filled in the simulation.

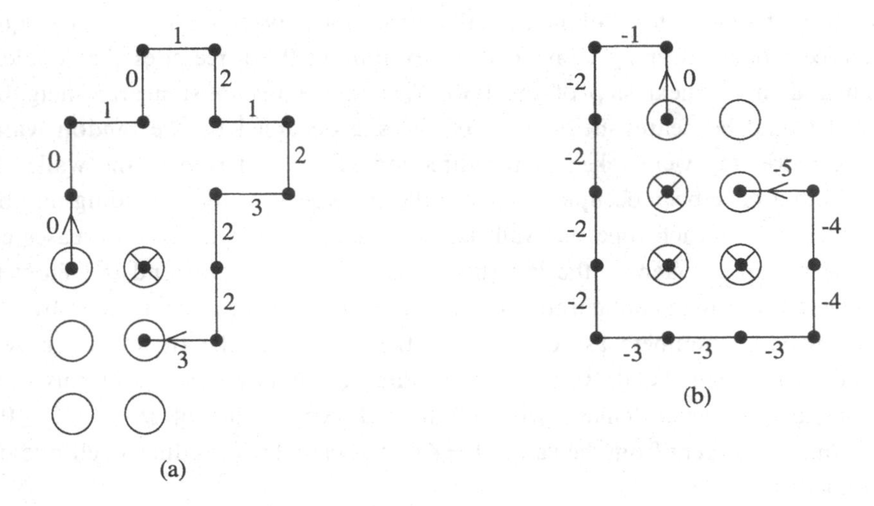

Figure 3.2 Potential trapping configurations in an indefinitely growing self-avoiding random walk (IGSAW). The bond winding numbers (the number of turns to the right minus the number of turns to the left) associated with each step in the walk are shown and the sites are labeled with the winding number of the bond that leaves them. If the walk enters the sites labeled "⊗" it will be trapped. Only the sites labeled "○" can be entered in the next step. Trapping can be avoided by using the winding number information from the sites labeled "⊙" to determine the steps that can be taken without inevitably trapping the walk. Without the "global" winding number information, the local configuration at the growing end of the IGSAW is the same in parts (a) and (b). However, the walk must turn in different directions to avoid trapping. All of the six sites labeled "○", "⊙" and "⊗" near the growing end of the IGSAW must be examined to avoid trapping. In this example, only two of the six "end sites" carry winding number information, and these two sites are occupied by the ends of the IGSAW. In general, trapping can occur whenever the growing end of the IGSAW approaches previously occupied sites, whether they are at the origin of the IGSAW or at some other location in its path.

the simulation, like a DLA cluster, described below, starting with a single occupied site on a lattice [385, 386]. The polymer is then grown by launching a random walker from outside of the region occupied by the growing polymer and allowing it to follow its random walk path until it reaches a site that is adjacent to the occupied site. This vacant perimeter site is then occupied and becomes the active "tip" of the growing chain or polymer. The process is repeated with a new random walker, which is not allowed to step onto sites occupied by the polymer. This time, growth takes place when the random walker finds the "living" tip of the polymer. Since a random walker can always find the path to the growing tip that was found by the previous random walker, growth will always continue. Simulations can be carried out with reflecting boundary conditions on the "inactive" polymer sites or the random walkers can be adsorbed if they enter sites occupied by the polymer. If the random walker is adsorbed, then a new random walker must be started, outside of the region occupied by the growing polymer. The methods, described later in this chapter, that have been developed to grow large DLA clusters can be used to make this a practical algorithm. Equivalent and more efficient algorithms in which the random walkers are launched from the living tip of the growing polymer could also be used. In such a model, the first unoccupied nearest-neighbor of the living tip that was occupied by the random walker is filled to represent the growth process, if the random walker reaches a sufficiently large distance from the growing polymer without encountering an absorbing site or returning to the living tip.

In a related model, the random walker is "killed" when it first enters the last filled site (the "active site") at the growing tip of the polymer. The site that the random walker stepped from, to enter the filled tip site, is then filled to represent the growth process and becomes the new active site. The relationship between these two models is like that between the DLA and dielectric breakdown models, described later in this chapter (section 3.4.3). Polymers can also be generated by numerically solving the Laplace equation with the appropriate boundary conditions, instead of using random walkers [387]. The "trade-off" between these two approaches is similar to that between the standard DLA and dielectric breakdown models. 2-Dimensional simulations with "dielectric breakdown model" boundary conditions (growth in the unoccupied perimeter site from which the random walker entered the active growing tip site) give fractal dimensionalities of $D \approx 1.23$ and $D \approx 1.48$ for absorbing and reflecting "inactive" sites, respectively [387, 388, 389]. More recently, it has been shown that the dielectric breakdown version of this model with absorbing boundary conditions is equivalent to the "loop-erased self-avoiding walk" [390, 391], for which the fractal dimensionality is exactly 5/4 [392]. The fractal dimensionality estimated from computer simulations is in remarkably good agreement with this result.

Surprisingly, simulations with "DLA" boundary conditions (growth in the first unoccupied perimeter site entered by the random walker) do not generate polymers with the same fractal dimensionality as those generated using dielectric breakdown model boundary conditions [389, 393]. With reflecting "inactive" sites and DLA boundary conditions at the active tip, a fractal dimensionality of ≈ 1.30 was measured [385, 386, 389]. Similarly, the fractal dimensionality with absorbing "inactive" sites and DLA boundary conditions at the active tip is smaller than 1.23 and appears to be sensitive to other model details [389]. These results strongly suggest that the fractal dimensionality is non-universal for these models. Simulations have also been carried out to explore "polymer" growth on cubic lattices [386, 389] and growth from 1-dimensional and 2-dimensional

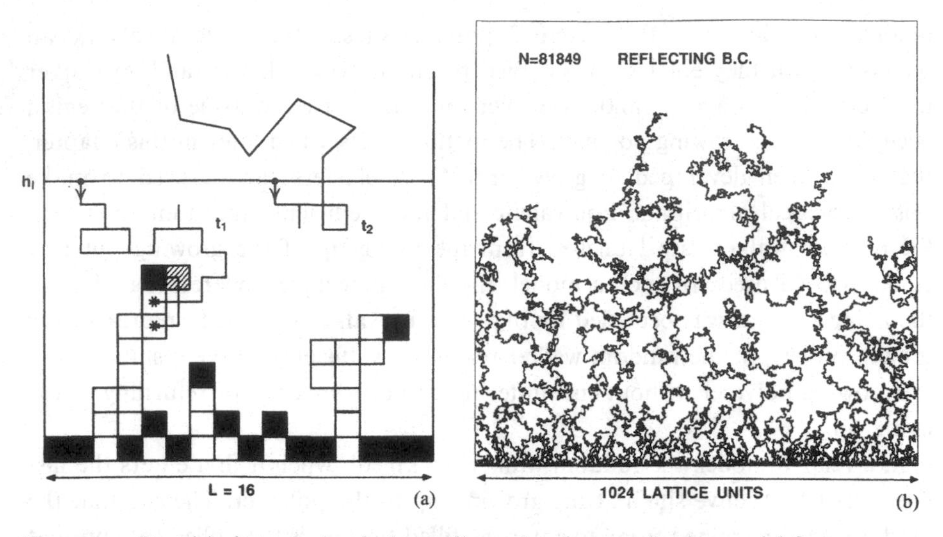

Figure 3.3 Part (a) shows a simple 1 + 1-dimensional square lattice model for diffusion-limited polymerization with dielectric breakdown model growth rules and reflecting boundary conditions at the inactive filled sites. The filled active zone sites at the tip of each growing polymer are black and the other, innactive, filled sites are bordered. The random walkers are launched from randomly selected positions on the "horizontal" launching surface at a height h_1, above the interface. Trajectory t_1 eventually moves the random walker onto a filled active zone site. The shaded unoccupied site, from which it entered the active site, is then filled to represent the growth process. This trajectory also illustrates the reflecting boundary conditions at the inactive filled sites; an asterisk indicates the inactive filled sites that are entered by the random walkers and return them to the unoccupied sites from which they entered. Trajectory t_2 moves the random walker away from the interface. If the random walker reaches the "killing surface", far from the interface, it will be terminated and a new random walker will be started from a randomly selected position on the launching surface. Trajectory t_2 illustrates the long, off-lattice steps, far from the interface. Part (b) shows a pattern generated by this model.

surfaces [389]. Figure 3.3 illustrates the $1+1$-dimensional surface growth model and the patterns that it generates.

Miyazima *et al.* [394] have investigated a related model in which particles in the growing cluster remain "sticky" to the random walkers, for a "time" τ. Here, τ is not the natural time scale for the process; instead the simulation time is incremented by one unit after each particle is added. In the limit $\tau \to \infty$, the standard DLA model, described later in this chapter, is recovered. For $\tau = 1$, only the tip remains sticky and the model becomes a diffusion-limited polymerization model or diffusion-limited growth walk, with reflecting boundary conditions on the "inactive" tail and DLA boundary conditions at the "active" head. In this limit, Miyazima *et al.* obtained a fractal dimensionality of $D \approx 1.04$ that is quite different from the values of $D \approx 1.30$ obtained by Bradley and Kung [386], Debierre and Turban [385] and Meakin [389]. The reason for this discrepancy is not clear. The clusters grown by Bradley and Kung and Debierre and Turban were quite small (32 sites and 45 sites, respectively). However, the clusters grown by Meakin were substantially larger (800 clusters each containing 1000 sites were generated). These clusters are somewhat larger than those used by Miyazima *et al.*, for small values of τ. Consequently, the discrepancy cannot be accounted for by the small cluster sizes used by Bradley and Kung or Debierre and Turban.

For $1 < \tau < \infty$, Miyazima *et al.* found that the growth of the mean square end-to-end distance $< R^2 >$ could be described by the scaling form

$$< R^2 > = t^{2\beta} f(t/\tau^{\phi}), \tag{3.15}$$

where β is the exponent that describes the growth of the radius of gyration in DLA clusters ($\beta \approx 0.584$) and $\phi \approx 1.85$. To fit the asymptotic behavior in the limits $\tau \to 1$ and $\tau \to \infty$, the crossover function $f(x)$ in equation 3.15 must have the form $f(x) = const.$ for $x \ll 1$ and $f(x) \approx x^{2(\beta'-\beta)}$ for $x \gg 1$, where $\beta' = 1/D_p$ and D_p is the fractal dimensionality for the corresponding diffusion-limited polymerization model.

3.3.4 Random Walks on Random Substrates

There is considerable interest in random walks on random substrates, including fractal substrates, as models for transport phenomena in disordered systems. In particular, the properties of random walks on percolation clusters have been studied extensively. In many cases, an exact enumeration approach is valuable [395, 396]. The substrate is represented by a cluster of occupied, unblocked sites surrounded by unoccupied, blocked sites, and a randomly selected site on the substrate is assigned a value of 1, to indicate that the initial probability of finding the random walker at this origin is 1. If there are m occupied (unblocked) nearest-neighbor sites, then this probability is divided equally between each of them, so that each of the m occupied nearest-neighbor sites has a value of $1/m$ and the origin has a value of 0. At each stage in the simulation, each site distributes its probability equally to all of its occupied (unblocked) nearest neighbors. This represents the steps in all possible random walks, and, in this manner, a probability map, which indicates the probability $P_i(n)$ that after n steps the random walker will have reached the ith site, evolves. Since the number of possible random walks grows exponentially with increasing n and the number of operations needed to generate the probability map grows as only a small power of n, this is an efficient way of generating the probability distribution for all walks of length n.

The procedure can then be repeated, many times, with different origins to provide a more representative sampling of the substrate. The procedure described above corresponds to a "myopic" random walker which steps with equal probability onto unblocked nearest-neighbor sites, but is not influenced by the configuration of blocked and unblocked sites on longer length scales. To obtain the site-occupation probability distribution for a "blind" walker which attempts to step onto any nearest-neighbor site with equal probability, but is returned if it attempts to step onto a blocked site, a fraction m/k of the probability in each of the sites is distributed equally to its unblocked nearest neighbors and a fraction $(k - m)/k$ remains on that site. Here, k is the coordination number of the lattice (the number of directions in which a random walker can move, if it is not blocked) and $k - m$ is the number of blocked nearest-neighbor sites that the random walker cannot enter. This process is repeated n times to calculate the site-occupation probability distribution for walks of n steps. Depending on the nature of the substrate and the type of information required, this approach may or may not have advantages over the simulation of many individual random walks from many origins. This method gives complete information with respect to a number of selected origins, but in some cases it may be more important to average over the structure of the substrate. In most cases, it appears to offer very substantial advantages [395].

Random walks have also been studied extensively on Euclidean substrates with quenched disorder. If a lattice model with quenched disorder is used, the effects of the quenched disorder can be represented in terms of site occupation probabilities P_i or transition probabilities W_{ij}. If the distribution of occupation probabilities is small enough ($P_{min}/P_{max} > 0$), then the asymptotic scaling properties of the random walk will not be changed by the quenched disorder. If P_{max}/P_{min} diverges, then the fractal dimensionality of the random walk may be changed. For example, P_{max}/P_{min} diverges for random walks on multifractal lattices [397] and the fractal dimensionality of the random walk depends on the multifractal scaling exponents.

3.3.5 Active Random Walk Models

In active random walk models, the random walk(s) take place on a substrate that completely or partially controls the random walk path(s). At the same time, the random walker(s) change the substrate on which the random walk(s) take place. In many cases, the substrate can be thought of in terms of a surface represented by the single-valued function $h(\mathbf{x})$. For example, in a simple erosion/deposition model for the evolution of the Earth's surface [330, 398], the random walkers might be biased to follow a "downhill" path on the surface, to represent the transport of water, and might transport material from regions with high slopes, where the erosion is rapid, to regions with small slopes where there is net deposition. In such a model, the random walkers can be allowed to "store" material so that the balance between erosion and deposition depends on the amount of stored material as well as the local slope.

In a Monte Carlo simulation, the surface $h(\mathbf{x})$ can often be thought of in terms of an energy surface $E(\mathbf{x})$, and the random walker (or walkers) may be moved randomly from site i to site j with probabilities $P_{i\to j}$ given by

$$P_{i\to j} = \begin{cases} 1 & \text{for } E_j \le E_i \\ e^{(E_j - E_i)/T} & \text{for } E_j > E_i, \end{cases} \tag{3.16}$$

where E_i and E_j are the energies (heights) associated with the neighboring sites i and j. Equation 3.16 corresponds to the Metropolis Monte Carlo algorithm. Other algorithms, such as the heat bath algorithm (chapter 1, section 1.9), could also be used. Another common approach is to use an activated hopping model in which the random walker moves with a probability given by $P = e^{E_a/T}$ to a randomly selected nearest neighbor, at each step in the simulation. Here, E_a is an activation energy that depends on the local configuration.

Schweitzer and Schimansky-Geier [399] have investigated a model in which a large number of random walkers of type A move on a surface and emit walkers of type B at a constant rate. The "B" walkers decay exponentially and move randomly without being influenced by the surface. On the other hand, each B atom reduces the height in the sites that it occupies at a constant rate. The "A" walkers do not move completely randomly on the surface. Instead, their paths are biased in a "downhill" direction. Starting with an initially flat substrate $h(\mathbf{x})$ and a random distribution of A walkers, the distribution of A and B walkers becomes very inhomogeneous, and eventually all of the A and B walkers become concentrated into a single "cluster". Models such as this may prove to be helpful in understanding processes such as the chemo-tactically controlled aggregation of organisms such as slime molds.

There is an endless variety of active random walk models. The results are likely to be the most rewarding if the models are kept as simple as possible to illustrate theoretical concepts and/or if the model is motivated by a real physical application [400].

3.4 Cluster Growth Models

A wide range of cluster growth models has now been developed. These models have been motivated by and contribute to our understanding of problems in almost all areas of science and technology. In this section, some of the most fundamental and important of these models, including the Eden, ballistic aggregation and DLA/dielectric breakdown models, are described. Some of the earliest and most important models were motivated by applications in biology, colloid science and materials science. The development of new models has also been motivated by theoretical questions and by the hope that important applications for simple models will found in the future. In some cases this optimistic hope has proven to be well founded.

3.4.1 The Eden Model

The Eden model [348, 349] was initially developed to investigate the growth of biological cell colonies. In general, the simulation can be carried out on almost any lattice or tessellation. At the start of the simulation, one site or cell is selected and "filled". In the original Eden model, often called Eden model B, an occupied site on the perimeter of the cluster of filled sites is selected randomly, with equal probabilities, and one of its nearest-neighbor unoccupied perimeter sites is then selected randomly and filled to represent the growth process. This process can be repeated many times to generate a sufficiently large cluster. In a simplified version of this model, Eden model A, which is the most frequently used by physicists, perimeter sites are simply selected at random, with equal probabilities, and filled. Figure 3.4 shows a cluster of 2.5×10^6 sites grown on a square lattice using this algorithm. In a third simple variant of the Eden model [293, 401], Eden model C, unoccupied perimeter sites are selected randomly, with probabilities proportional to their number of occupied nearest neighbors, and filled.

Plischke and Rácz [402, 403] measured the width of the active zone consisting of all those sites in which growth is possible. For the Eden model, the width of the active zone is a measure of the width of the surface. They found that, for growing Eden clusters, the surface width $\xi_\perp$ grew algebraically with increasing cluster size s, so that

$$\xi_\perp \sim s^{\overline{\nu}}, \tag{3.17}$$

where s is the number of occupied sites and $\overline{\nu} = 0.18 \pm 0.03$. It is not surprising that the surface width $\xi_\perp$ grew more slowly than the mean radius of the active zone (unoccupied perimeter, in this case) $< r_s >$ or the radius of gyration R_g, which grew as

$$R_g(s) \sim < r_s > \sim s^{1/2}. \tag{3.18}$$

However, the power law growth of $\xi_\perp$ indicated that interesting new scaling behavior might be associated with the evolution of the rough surfaces of growing, compact forms. This subject blossomed during the 1985–1995 decade and is discussed in more detail in chapter 5.

A variety of modifications of the Eden model have been developed. The first of these was a model for skin cancer introduced by Williams and Bjerknes [404]. In this model, "filled" sites represent cancerous cells and empty sites represent normal cells. The simulations of Williams and Bjerknes were carried out using a triangular lattice. At the start of a simulation, one of the six coordinate sites (nodes in the triangular lattice) in the middle of the lattice is labeled to represent a single cancerous cell in an otherwise normal array of cells. In this model, a lattice site (site a) on either the occupied or unoccupied perimeter is selected

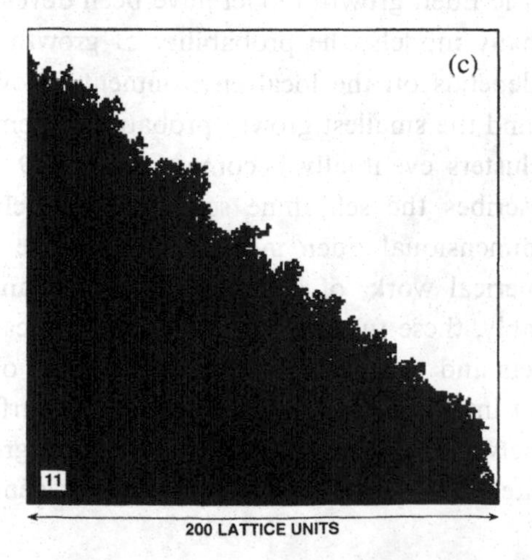

Figure 3.4 A cluster of 2.5×10^6 sites grown on a square lattice using Eden model A. Part (a) shows the entire cluster, while parts (b) and (c) show parts of the surface growing in the (01) and (11) directions.

at random and one of its nearest neighbors (site b) is selected randomly. If sites a and and b have different labels, indicating whether or not the sites are "cancerous", then the label associated with site b is changed to that of site a. The initial, a site, selection is biased in favor of occupied perimeter sites, so that the probability of selecting a particular occupied site is k times as large as the probability of selecting a particular unoccupied site. The parameter k is called the "carcinogenic advantage" and may be interpreted as the rate of division of cancerous cells with respect to the rate of division of normal cells. A value greater than $1/2$ provides a bias in favor of growth and in the $k \to \infty$ limit the Willaims Bjerknes model becomes equivalent to version B of the Eden model. The simulations are started with a single occupied site, so that, even for values of k substantially larger than 1, the cluster may disappear. The probability of this happening is $1/k$. Because the algorithm includes both growth and decay, the cluster of occupied sites can become fragmented. In the large size limit, the pattern consists of a single large cluster with a rough surface accompanied by a large number of small islands. The rough surface and the islands occupy an active zone that decreases in width, relative to the diameter of the cluster, as it grows. Williams and Bjerknes found that the number of sites in the occupied perimeter N_p was related to the total number of occupied sites N by

$$N_p \sim N^{0.55}, \tag{3.19}$$

for all values of k. Similar results were obtained using square, six-coordinate, triangular and three-coordinate, honeycomb lattices. Williams and Bjerknes were aware of the work of Mandelbrot [2] and Hausdorff [405], a decade before fractals became popular, and interpreted this observation in terms of a (fractal) dimensionality of $D = 1.1$. However, more recent computer simulations [406] and theoretical considerations [407] do not support this idea. It appears that the surfaces of Williams–Bjerknes clusters are self-affine, on sufficiently long length scales, and the exponent in equation 3.19 has a value of exactly $1/2$.

Many other variations of the basic Eden growth model have been developed [401, 408] and investigated. In many models, the probability of growth in a particular unoccupied perimeter depends on the local environment. Provided that the ratio between the largest and the smallest growth probabilities remains finite, as the clusters grow, the clusters eventually become compact ($D = d$) and the Hurst exponent that describes the self-affine scaling of the cluster surface assumes the universal d-dimensional Eden model value. There have been many claims, based on numerical work, of self-similar surfaces and/or internal structures. Almost invariably, these turn out to be short time scale or short length scale "transient" effects and the long length scale structure of the Eden model, consisting of compact internal structures with self-affine surfaces, is preserved [408]. If the ratio between the largest and the smallest growth probabilities does not remain finite, then Eden growth models may generate fractal shapes. For example, computer simulations indicate that Eden growth on multifractal substrates [409, 410], with a compact support (appendix B, section B.1), generates compact $D = d = 2$ clusters with self-similar surfaces. It appears that long range screening effects or "interactions" are required to generate the diverging growth probabilities needed to change the basic scaling properties.

Off-lattice Eden models have also been investigated by Botet [411] and by Wang *et al.* [412, 413]. In the model of Botet, the new particles are placed with equal probability at any position in which they can contact the growing cluster, without overlap. In the model of Wang *et al.*, a particle in the cluster is selected randomly and a new particle is placed with equal probability at any position in which it can contact the selected particle, without overlapping the cluster. If it is not possible to find a contacting and non-overlapping position for the new particle, then the selected particle is removed from the list of "active" particles and is not selected at later stages in the simulation. There are important differences between these two models. In Botet's model, small parts of the "active zone" may remain buried for a long time, deep in the growing cluster. In the model of Wang *et al.*, small, isolated segments of the active zone have a relatively large probability of being eliminated. In this model, the width of the active zone was found to grow as

$$\xi_\perp \sim < r >^a, \tag{3.20}$$

where the exponent a has a value of ≈ 0.396. Taking statistical and systematic uncertainties such as corrections to scaling into account, this value for the exponent a is consistent with the value of $1/3$ expected for a growing rough surface in the KPZ universality class [287] (chapter 5, section 5.1.1). A value of ≈ 0.5 was measured for a by Botet. In this case, the simulations were carried out on too small a scale to obtain reliable results, but the high effective value for a may be a consequence of the small fragments of active zone that are left behind as the main part of the active zone advances, and the asymptotic value of the exponent a may be larger than $1/3$. Figure 3.5 shows small, 10 000-particle, 2-dimensional and 3-dimensional clusters generated using the Botet version of off-lattice Eden growth.

3.4.2 Ballistic Aggregation

In ballistic aggregation models, particles are added, one at a time, to a growing cluster or aggregate of particles via linear (ballistic) paths that are selected randomly, without bias, from all possible paths that could result in a "collision" between the particle and the cluster. The particles are added irreversibly to the growing cluster in the positions in which they first make contact with the cluster. 3-Dimensional, off-lattice simulations of this type were first carried out by Vold

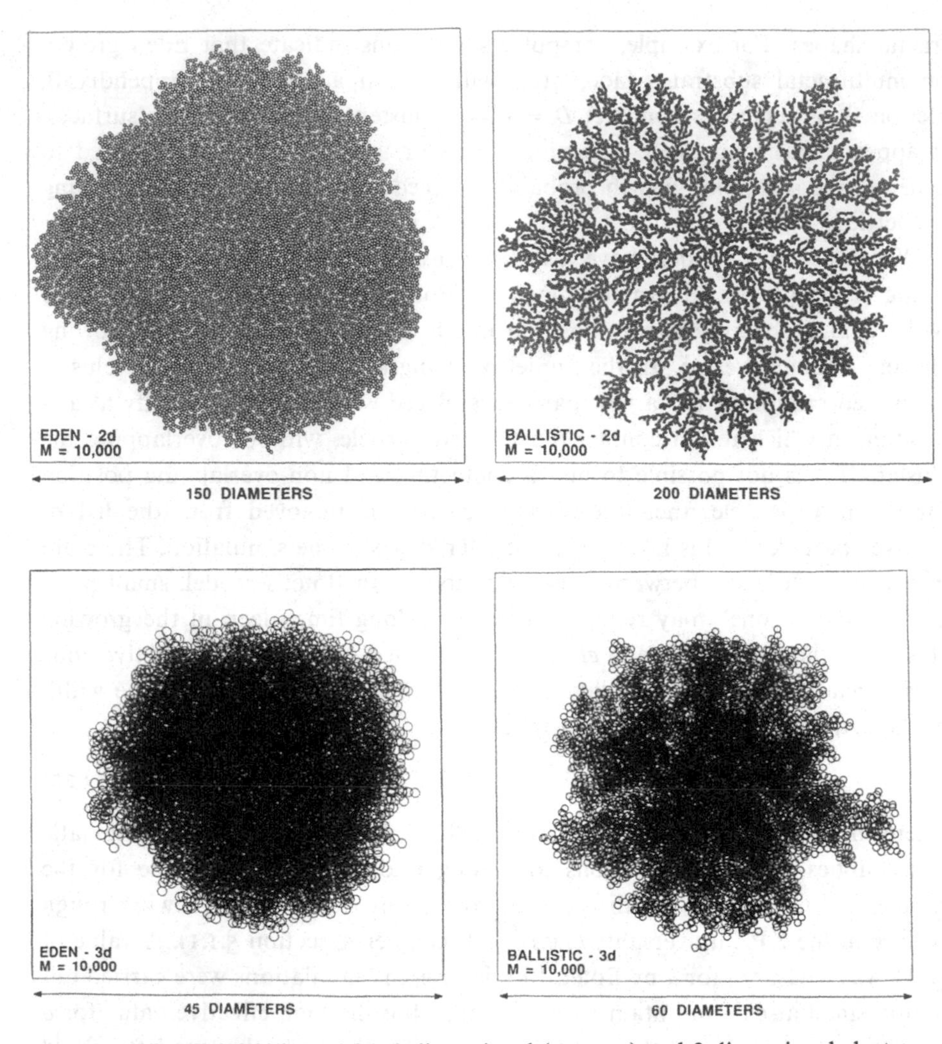

Figure 3.5 Small (10 000-particle), 2-dimensional (top row) and 3-dimensional clusters generated using off-lattice Eden (left side) and ballistic aggregation models.

[414]. Vold found that the mean density $\overline{\rho}$ for the core region containing s particles is related to the core size by

$$\overline{\rho} \sim s^{-a}, \tag{3.21}$$

with $a \approx 0.29$. For a self-similar fractal cluster, the exponent a could be interpreted as $a = (d/D) - 1$ or $D = d/(1+a)$, corresponding to an effective fractal dimensionality of $D \approx 2.3$. Similarly, the density inside a thin shell that just encloses the innermost s particles scales as $s^{a'}$, with $a' \approx 0.29$, which can also be interpreted in terms of the same effective fractal dimensionality of ≈ 2.3. In addition, measurement of the dependence of the number of particles $s(r)$ within a distance r, measured from the center of mass, indicated that $s(r) \sim r^{1/0.429}$, corresponding to a fractal dimensionality of $1/0.429 \approx 2.3$. However, the clusters used in this study were very small and the corrections to scaling for this model are large. In addition, Sutherland and Hutchison [415, 416] showed that a biased distribution of trajectories was generated in Vold's simulations. The simulations of Vold were repeated using unbiased trajectories, and it was found [416, 415] that $s(r) \approx r^{\approx 1/0.36}$, corresponding to a fractal dimensionality of $D \approx 1/0.36 \approx 2.78$. Sutherland [415] concluded that

> It seems highly probable that as the floc size increases the core reaches a constant porosity of 0.83,

which implies that $D = d = 3$. Here, the term "floc" is used to indicate a cluster or aggregate of spherical particles. Much larger scale simulations and simple but convincing theoretical arguments, outlined below, indicate that $d = D$ for these models. The clusters have a quite small density (figure 3.5 and chapter 5, section 5.2.2), particularly for large d [417, 418], but the structure is uniform on all but short length scales.

3.4.3 The Diffusion-Limited Aggregation Model

The diffusion-limited aggregation (DLA) model [181, 419] is one of the most striking examples of the generation of a complex disorderly pattern by a simple model or algorithm. It was developed at a time when interest in fractals was growing rapidly and soon became the principal paradigm for pattern generation far-from-equilibrium. After almost 15 years, interest in this model is still at a high level. This interest is maintained by a wide range of applications in both the physical and biological sciences as well as applications in other areas [420]. The DLA model also continues to present an important theoretical challenge that has not yet been fully met. Many aspects of the structure of DLA clusters can be described in terms of a statistically self-similar fractal scaling model. However, it is apparent that the structure of DLA may be more complex than that of a homogeneous, statistically self-similar fractal, such as a percolation cluster. At present, there is no convincing scaling model for the structure of DLA. Unfortunately, most theoretical approaches either directly or implicitly assume a scaling model for DLA. It appears that very large scale computer simulations are required to discern its asymptotic scaling properties. More direct theoretical approaches that assume no knowledge of the asymptotic scaling properties have not yet been very successful. There is, at present, no systematic theoretical framework for the class of non-equilibrium, non-local growth problems that DLA represents.

Despite the lack of a fundamental understanding of DLA in more basic terms, the DLA model itself provides a basis for understanding a wide range of natural phenomena. It may have to be accepted, for a long time, that many far-from-equilibrium processes may be understood in terms of simple algorithms but that understanding in terms of analytical solutions to equations describing their growth may not be available. The idea that understanding natural processes in terms of simple algorithms provides an alternative to understanding in terms of the analytical solution of equations is gaining acceptance. In some cases, the "quality" of understanding may be greater for the "algorithmic" approach, and simple algorithmic models frequently lead to a more intuitive understanding.

In the original diffusion-limited aggregation model of Witten and Sander [181], "particles", represented by lattice sites, are added, one at a time, to a growing cluster or aggregate of particles via random walk paths starting outside the region occupied by the cluster. In the most simple version of the model [181], a simulation is started by occupying a site in the center of a square or triangular lattice to represent the "seed", "growth site" or "nucleation site". A site far from the cluster is then selected, and a random walk is started from the selected site. If the random walker moves too far from the growing cluster, it is terminated and a new random walk is started. If the random walker eventually reaches a site that is the nearest neighbor to a previously occupied site, the random walk is stopped and the unoccupied perimeter site (the last site occupied by the random walker) is filled to represent the growth process. The process of launching random walkers from outside the region occupied by the growing cluster and terminating them when they wander too far from the cluster or "stick" to the growing cluster by reaching an unoccupied perimeter site is repeated many times to simulate the cluster growth process.

Because many random walks must be generated, this algorithm is very time consuming, and the largest clusters that can be grown, using reasonable amounts of computer time, contain only a few thousand lattice sites. The algorithm can be improved by starting the random walk trajectories at a randomly selected point on a circle, called the launching circle, that just contains the cluster, and transferring the random walker to the nearest lattice site. This is justified, since a random walker started at a random position far outside of a circle will first cross the circle at random, with equal probability at all positions on the circle. For small clusters, it is satisfactory to terminate random walkers if they reach a "killing circle" centered on the original growth site (the cluster origin) and having a radius two to three times the radius of the launching circle. This model is illustrated in figure 3.6. Clusters containing a few tens of thousands of sites can be generated using this algorithm.

Diffusion-limited aggregation models can be made much more efficient by allowing the random walkers to take long steps when they are far from the cluster. This approach has been used in the past [421] to simulate processes such as steady-state heat conduction that can be expressed in terms of the Laplace equation. This method is based on the idea that if the random walker

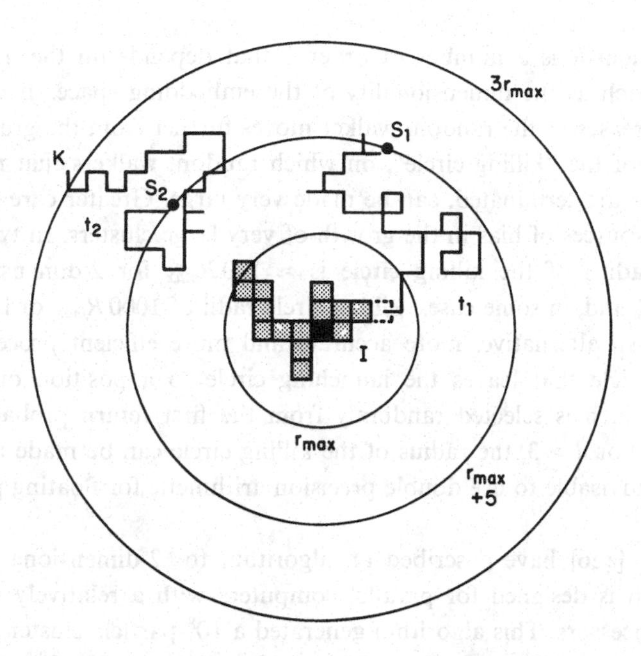

Figure 3.6 A simple, square lattice DLA model. The initial seed or growth site is shown in black. Sites that have already been added to the growing cluster are shaded. Two typical random walk trajectories, originating at random points on the launching circle with a radius of $r_{max} + 5$ lattice units are shown, where r_{max} is the maximum cluster radius measured from the initial growth site. Trajectory t_1 starts from point S_1 and eventually reaches the unoccupied perimeter site, with dashed borders, which is filled to represent the growth process. Trajectory t_2, starting from point S_2, reaches the killing circle with a radius of $3\,r_{max}$, before reaching the unoccupied perimeter, and is terminated.

can be surrounded by an empty region with a simple shape, then its random walk path, until it leaves that empty region, can be replaced by a single step to the surface of the empty region. The most simple procedure is to use a hyperspherical region centered on the current position of the random walker. In this case, the trajectory within the hypersphere can be replaced by a single jump to a randomly selected position on its surface. This requires an efficient procedure for determining the smallest distance between the random walker and the cluster [422, 423, 424]. Once this distance has been found, the random walker is allowed to take a step in a randomly selected direction that is one particle diameter shorter than this minimum distance, so that the particle does not reach the cluster in a single long step.

The most efficient algorithms use a hierarchy of "maps" of the cluster to determine the smallest distance between the random walker and the cluster [422, 424]. In practice, a safe estimate for the smallest distance is used. Algorithms of this type can be used to generate clusters containing as many as 10^8 particles, and the computer time required grows only as $s\log(s)^x$, where s is the cluster size and the exponent x is a number of order 1 that depends on the details of the algorithm such as the dimensionality of the embedding space. Because the step length increases as the random walker moves further from the growing cluster, the radius of the "killing circle", on which random walkers that move far from the cluster are terminated, can be made very large. Greater care must be taken to avoid sources of bias in the growth of very large clusters. In typical simulations, the radius of the killing circle is $\approx 100\,R_{max}$ for 2-dimensional DLA clusters [425] and, in some cases, killing circle radii of $1000\,R_{max}$ or larger have been used. An alternative, more accurate and more efficient procedure is to return a particle that leaves the launching circle to a position on the launching circle, which is selected randomly from the first return probability distribution [426]. For $d \geq 3$, the radius of the killing circle can be made much smaller. It is also advisable to use double precision arithmetic for floating point operations.

Kaufman *et al.* [426] have described an algorithm for 2-dimensional, off-lattice DLA, which is designed for parallel computers with a relatively small number of fast processors. This algorithm generated a 10^8-particle cluster in 13 hours on a 32 processor IBM computer. Using this approach, 20 10^8-particle clusters were grown. While the clusters produced in this manner are not DLA clusters from the purist's point of view, the growth process can be started off using a sequential algorithm and large clusters generated using this algorithm can be regarded as DLA clusters for most purposes. However, a cautious approach is advisible because of the subtle scaling properties of DLA clusters.

Important insights into the growth of DLA clusters can be obtained by using color graphics to show the order in which the particles were incorporated into the growing cluster [427]. An example is shown in figure 3.7 (see color plate section). One of the most striking features is that the late arriving green particles do not penetrate into the interior of the cluster. This is necessary if the open fractal structure is to be maintained as the cluster grows. The lack of penetration is a consequence of a strong screening effect. The particles, following random walk paths, are trapped at the sides of the "fjords" leading to the interior and there is a very small probability that a particle, following a random walk path, will reach the interior of the cluster. The screening effect can be described in terms of the screening of a Laplacian field, with absorbing boundary conditions on the surface of the DLA fractal. Other fields, such as flow fields, may be screened in a similar manner, and these screening effects are key ingredients in the development of a better understanding of the physics of fractals. Considerable effort has gone into the development of a more quantitative description of the screening of Laplacian fields in terms of multifractal measures [428], and this topic is discussed in more detail in appendix B. Relatively little work has been done in $d = 3$ and higher-dimensional spaces. However, color-coded cross-sections through 3-dimensional and 4-dimensional DLA clusters reveal a similar strong screening effect, even though these cross-sections are more sparse than 2-dimensional DLA clusters of the same size (length scale L/ϵ).

The acronym "DLA" is sometimes used for any diffusion-limited aggregation process, including cluster–cluster aggregation phenomena. In this book, the term "DLA" is reserved for "particle–cluster" aggregation processes in which a large object is constructed by adding small objects, one at a time.

3.4.4 The Dielectric Breakdown Model

As the name suggests, the dielectric breakdown model was originally developed to obtain a better understanding of dielectric breakdown phenomena in which branched "breakdown patterns" are formed [429]. In this model, it is assumed that the rate of growth of the breakdown pattern is given by

$$V_{\mathbf{x}_s} = f(\nabla_n \Phi_{\mathbf{x}_s}), \qquad (3.22)$$

where $V_{\mathbf{x}_s}$ is the velocity normal to the interface between the breakdown pattern and the surrounding material at position $\mathbf{x}_s$ on the interface and $\nabla_n \Phi_{\mathbf{x}_s}$ is the gradient of the electric potential Φ, measured in the dielectric material at $\mathbf{x}_s$, in a direction normal to the interface. Niemeyer *et al.* [429] simulated the growth of breakdown patterns using a model in which a discretized version of the Laplace equation, $\nabla^2\Phi = 0$, was solved numerically, at each stage in the simulation, with the boundary conditions $\Phi = const.$ ($\Phi = \Phi_\circ$) at all occupied sites used to represent the breakdown region and $\Phi = 0$ on a distant surface enclosing the growing cluster of "breakdown" sites.[5] On a square lattice, the discretized Laplace equation can be written as

$$\Phi(i,j) = [\Phi(i,j-1) + \Phi(i,j+1) + \Phi(i-1,j) + \Phi(i+1,j)]/4, \qquad (3.23)$$

where $\Phi(i,j)$ is the "potential" at the lattice site with coordinates (i,j). After equation 3.23 has been solved, the "growth velocities" $V(i,j)$ are calculated, approximating equation 3.22 by

$$V(i,j) = f(n, |\Phi_\circ - \Phi(i,j)|), \qquad (3.24)$$

where n is the number of occupied sites, representing the breakdown pattern, that are nearest neighbors to the unoccupied perimeter site at coordinates (i,j). It is very often assumed that $V(i,j)$ has the simple homogeneous form

$$V(i,j) = n|\Phi_\circ - \Phi(i,j)|^\eta. \qquad (3.25)$$

5. Because of the linearity of the Laplace equation, equivalent results can be obtained by setting Φ to zero on the occupied sites and setting Φ to any non-zero constant at "infinity".

After the "velocities" have been calculated, one of the active zone sites is picked randomly with selection probabilities $P(i,j)$ given by

$$P(i,j) = V(i,j)/\sum{}' V(k,l), \tag{3.26}$$

where $\sum'$ indicates summation over the entire active zone. The selected active zone site is then filled to represent the growth process. The "growth velocities" $V(i,j)$ must then be recalculated, to proceed to the next growth event. In the dielectric breakdown model, equation 3.23 must be solved numerically, using methods such as over-relaxation [430] or conjugate gradient [431] algorithms. Both methods require large amounts of computer time, and the trade-offs between these two methods are an important practical issue. Relaxation methods, with relatively poor error control, have been used in many simulations. Multi-grid relaxation algorithms [431] may offer substantial advantages, but there seems to be little experience in this direction.

In general, the dielectric breakdown model does not belong to the Eden model universality class because the growth probabilities depend on the global cluster geometry (not just on the local configurations of the growth sites) and the ratio between the largest and smallest growth probabilities (P_{max}/P_{min}) diverges as the cluster size increases. However, for $\eta \rightarrow 0$, in equation 3.25, the dielectric breakdown model does belong to the Eden model universality class, but it does not correspond exactly to any of the simple Eden models described earlier in this chapter because of "trapping" of unfilled sites that are surrounded by the growth process. Similarly, the "dielectric breakdown" version of the diffusion-limited growth walk model described above becomes equivalent to the IGSAW model, described above, in the $\eta \rightarrow 0$ limit.

The dielectric breakdown model is very closely related to the DLA model described above. This was anticipated by Witten and Sander, who pointed out an "electrostatic analogy" between the probability that a random walker reaches a site at position $\mathbf{x}$ on the perimeter of a growing cluster and the growth probability or growth velocity at that site in a Laplacian growth process. Indeed, they commented on the similarity between DLA clusters and lightning. This can be seen by comparing equation 3.23 with the equation

$$P(i,j) = [P(i,j-1) + P(i,j+1) + P(i-1,j) + P(i+1,j)]/4, \tag{3.27}$$

for the probability that a random walker will be found in the site at position (i,j) in terms of the probabilities of finding it in the neighboring sites. This means that a scalar field Φ that obeys the Laplace equation can be simulated by a random walk with the same boundary conditions [421]. The probability $P(i,j)$ in equation 3.27 is then proportional to $\Phi(i,j)$.

For the case in which the growth exponent η in equation 3.25 is unity, the dielectric breakdown model can be replaced by a simple modification of the standard DLA model in which the random walkers are terminated only when they enter an occupied site. After a random walker has been terminated, the previously unoccupied perimeter site (the active zone site from which the random walker stepped onto the cluster) is filled. This model will be called the random walk dielectric breakdown model. A similar approach can be used if the exponent η in equation 3.25 is a small integer (n). In this model [432, 433], the simulation is carried out using the "random walk dielectric breakdown" model just outlined. However, when a random walker steps onto the growing cluster, the previously occupied active zone site is not immediately filled. Instead, $n-1$ random walkers are launched from the last unoccupied site visited by the random walker and, if all of them reach the outer boundary without stepping on the cluster, the unoccupied active zone site is then filled. An "η model" version of DLA can be constructed in a similar way. This algorithm is faster and more accurate than the standard dielectric breakdown model, for $\eta = 2, 3$ and 4. On the other hand, the standard dielectric breakdown model, in which the discretized Laplace equation is solved at each step in the growth process, can be used for non-integer values of the growth exponent η, large values of η or more complex, non-homogeneous forms for the "growth function" ($f(x)$ in equation 3.22). Figure 3.8 shows clusters grown on square and cubic lattices, with $\eta = 2$ and $\eta = 3$, using the random walk DLA model.

In a similar manner, the standard dielectric breakdown model can be modified to simulate DLA local boundary conditions. In this modified dielectric breakdown model, the discretized Laplace equation is solved with the boundary condition $\Phi = 0$ on both the occupied sites and the unoccupied perimeter sites and $\Phi = const.$ at "infinity". The growth velocity $V(i, j)$, in the unoccupied perimeter site at i, j, is given by

$$V(i, j) = [\Phi(i, j-1) + \Phi(i, j+1) + \Phi(i-1, j) + \Phi(i+1, j)]/4. \qquad (3.28)$$

To avoid confusion, the standard dielectric breakdown model will be called Laplacian growth with dielectric breakdown boundary conditions, and the model described in this paragraph will be called Laplacian growth with DLA boundary conditions, to emphasize that the simulations are carried out by numerically solving the discretized Laplace equation.

It is generally believed that the asymptotic scaling properties of the clusters generated by the DLA and dielectric breakdown models are the same. In particular, the exponent β describing the growth of the radius of gyration with increasing cluster size is believed to be the same for both models. In general, it can be anticipated that details in the growth algorithm that depend only on the *local* configuration will not change the long range structure. However, there is no rigorous proof of this conjecture, and there are some indications that it may not always be true [434]. The non-universality of the DLA and dielectric breakdown versions of the diffusion-limited growth walk model, discussed earlier in this chapter (section 3.3.3), may serve as a warning.

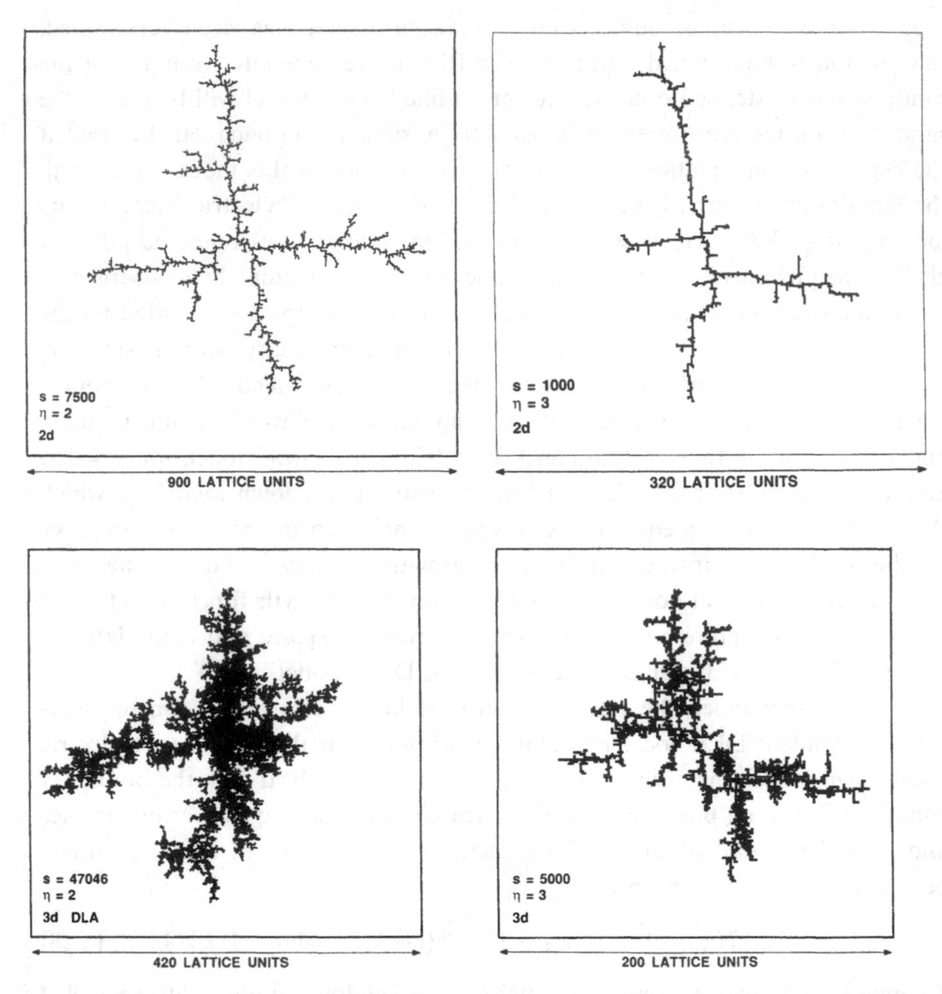

Figure 3.8 The top row shows clusters grown on a square lattice with $\eta = 2$ (left) and $\eta = 3$ (right) using a random walk DLA model. The bottom row shows projections of clusters generated using the corresponding cubic lattice, random walk DLA models, with $\eta = 2$ (left) and $\eta = 3$ (right).

The original simulations of Niemeyer *et al.* [429] were carried out on a quite small scale and may have suffered from incomplete relaxation of the Laplacian field Φ between each successive growth event. Nevertheless, they were able to show, convincingly, that the fractal dimensionality D depends on the exponent η. Values of $2, 1.89 \pm 0.01, 1.75 \pm 0.02$ and ≈ 1.64 were obtained for $\eta = 0, 0.5, 1.0$ and 2.0, respectively [429, 435]. It has been suggested [436] that the structure might become 1-dimensional at a finite value of $\eta = \eta_c$. This idea would be difficult to test numerically, because very accurate values for D would be required and random walk simulations are not very effective for large values of η. At present, this remains an open question [437].

Because of the flexibility of the dielectric breakdown model, a large number of variations on the basic model is possible. Work in this direction has become too extensive to survey here so only a short selection of these models is presented, to illustrate the wide range of possibilities. Other examples will be encountered later in this chapter. A variety of modified dielectric breakdown models has been used to simulate processes such as dendritic solidification. Early examples include the work of Nittmann and Stanley [357] and Family *et al.* [438]. These models may include modifications that enhance the effects of anisotropy, control growth noise and represent the effects of surface tension, time dependent growth conditions and other effects. In most cases, more than one modification to the basic dielectric breakdown model is included, but these models are usually simple enough to provide valuable insight concerning the origins of non-equilibrium morphologies and morphology transitions. Recently, more realistic models have been developed along similar lines. Work in this direction has substantially narrowed the gap between purely "atomistic" models such as DLA and the continuum surface growth models described in chapter 1.

Because the growth probabilities are calculated for all of the sites at each stage in a simulation, models in which more than one site can be filled at a time have been studied. For example, Family *et al.* [439] have explored a model in which all of the unoccupied perimeter sites are examined at each stage in the growth process. The unoccupied perimeter sites are filled with probabilities given by $P_i = E_i/E_{max}$, during each stage in the growth process. Here, E_i is the field (potential gradient) at the ith unoccupied perimeter site and E_{max} is the maximum field for any of the unoccupied perimeter sites. The clusters generated with this model exhibit the effects of the underlying (square) lattice much more strongly than do clusters obtained from the corresponding DLA or dielectric breakdown models. Measurement of the number of occupied sites $N(L)$, within a box of size L centered on the cluster origin, indicates an algebraic relationship between $N(L)$ and L, corresponding to a fractal dimensionality $D(\eta)$ that was found to be in good agreement with the theoretical results of Turkevich and Sher [440], described later in this chapter (section 3.10.2).

Another natural extension of the DLA and dielectric breakdown models is to growth processes in which the growth rates are controlled by a scalar field that obeys the biharmonic equation [441, 442]

$$\nabla^4\Phi' = \nabla^2(\nabla^2\Phi') = 0. \tag{3.29}$$

The discretized version of equation 3.29 is

$$\begin{aligned}\Phi'(i,j) \quad = \quad & (2/5)[\Phi'(i,j-1) + \Phi'(i,j+1) + \Phi'(i-1,j) \\ & + \Phi'(i+1,j)] - (1/10)[\Phi'(i-1,j-1) + \Phi'(i+1,j+1) \\ & + \Phi'(i-1,j+1) + \Phi'(i+1,j-1)] - (1/20)[\Phi'(i,j-2) \\ & + \Phi'(i,j+2) + \Phi'(i-2,j) + \Phi'(i+2,j)].\end{aligned} \tag{3.30}$$

It is clear from equation 3.30 that "biharmonic" growth processes can be simulated using models that are closely related to the dielectric breakdown and DLA models. Clusters grown using this model, with growth probabilities proportional to $|\Phi'(i,j)|$ or $\nabla^2|\Phi'(i,j)|$, generate patterns that are very similar to those obtained using the corresponding DLA or dielectric breakdown models, and it appears that this model can be considered to lie in the DLA universality class [441]. The simulations indicated that the structure of the clusters was more sensitive to the boundary conditions and system size than the Laplacian models. However, this is not a qualitative difference between the Laplacian and biharmonic models. The scaling structure of clusters grown with a fixed value for Φ' at "infinity" and $\Phi' = 0$ on the growing cluster is like that of DLA clusters.

3.4.5 The Scaling Structure of DLA

Witten and Sander [181] first studied the scaling structure of the small DLA clusters generated in their pioneering computer simulations. They measured both the global scaling properties of their aggregates (the dependence of their "mass" or number of particles on their overall size) and the scaling properties associated with the internal mass distribution. Their work showed that, within the uncertainties associated with a small range of length scales and small sample sizes, both the internal structure and the overall mass/length scaling could be described in terms of the same fractal dimensionality.

This first characterization of DLA clusters, grown on 2-dimensional lattices [181], was based on the growth of the radius of gyration $R_g(s)$ with increasing cluster size (number of particles) s and the two-point density–density correlation function or radial distribution function $C(r)$ (chapter 2, section 2.1.3.1). Both of these quantities were found to have the simple algebraic forms

$$< R_g(s) > \sim s^{\beta}, \tag{3.31}$$

for $s \gg 1$ and

$$C(r) \sim r^{-\alpha}, \tag{3.32}$$

on length scales r larger than the particle size (diameter) $d_\circ$ and smaller than the cluster diameter L ($d_\circ \ll r \ll L$).

The observation that the fractal dimensionalities corresponding to the power law relationships in equations 3.31 and 3.32 ($D_\beta = 1/\beta$ and $D_\alpha = d - \alpha$, where d is the Euclidean dimensionality of the embedding space in which the cluster is grown) were equal, within the uncertainties of the measurements at that time, and that small portions taken from different parts of the cluster looked "similar" to each other, led to the idea that DLA clusters were homogeneous self-similar fractals. However, doubts were expressed about the homogeneity because the

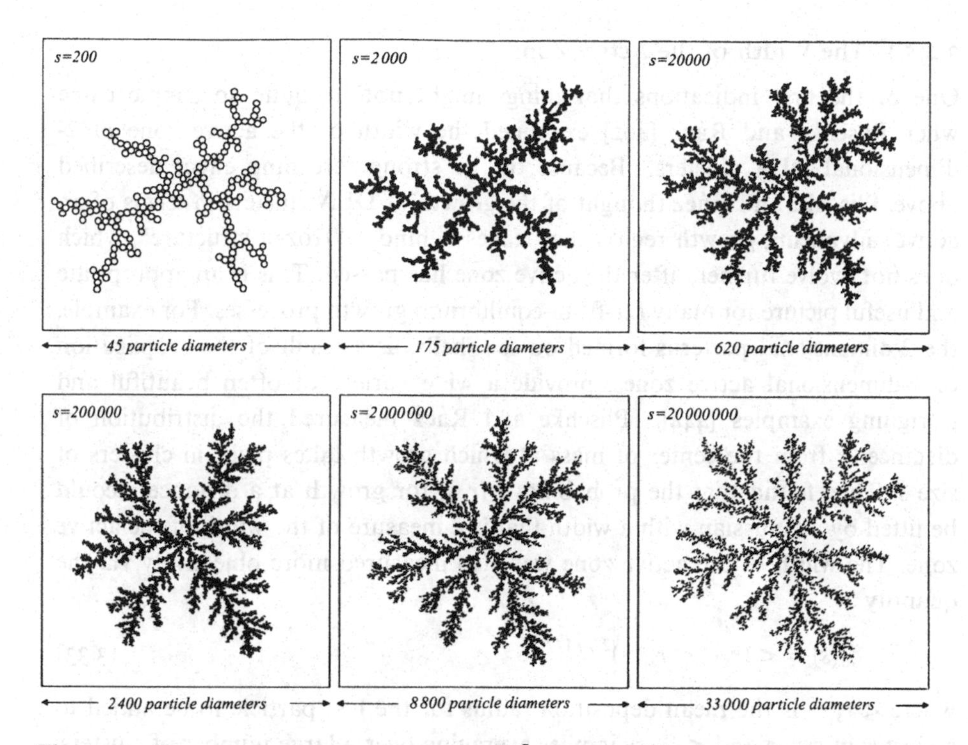

Figure 3.9 Six stages in the growth of a 2×10^7-particle, off-lattice DLA cluster. Each cluster is displayed on a different scale, so that all the clusters appear to have about the same size in the figure. This figure was provided by T. Rage.

region containing the cluster origin can always be identified in a large cluster [443]. This picture of the structure of DLA clusters prevailed for several years. Since that time, a more complex picture of DLA has been emerging. This is illustrated in figure 3.9, which shows six stages in the growth of a large off-lattice 2-dimensional DLA cluster (after 200, 2000, 2×10^4, 2×10^5, 2×10^6 and 2×10^7 particles have been added). While it can be dangerous to draw conclusions from just one simulation like this, it appears that the overall form of the DLA cluster slowly changes as it grows larger. The larger DLA clusters seem to have more arms and to fill space more uniformly (have a smaller lacunarity) than smaller clusters.

In the later parts of this chapter, the structure of DLA clusters and clusters generated by models closely related to DLA will be discussed in terms of a simple self-similar scaling model. This gives valuable insights into the results of computer simulations and experiments. However, the possibility of a more complicated scaling picture and/or the existence of a large transient or large corrections to the asymptotic scaling behavior indicated in this section, and its attendant uncertainties, should be constantly borne in mind.

3.4.5.1 The Width of the Active Zone

One of the first indications that things might not be quite so simple came when Plischke and Rácz [402] examined the width of the active zone of 2-dimensional DLA clusters. Because of the strong screening effect described above, Plischke and Rácz thought of the growth of DLA clusters in terms of an active, advancing growth region that leaves behind a "frozen structure", which does not evolve further, after the active zone has passed. This is an appropriate and useful picture for many far-from-equilibrium growth processes. For example, the 2-dimensional patterns formed on sea shells, as a result of the propagation of 1-dimensional active zones, provide a wide variety of often beautiful and intriguing examples [444]. Plischke and Rácz measured the distribution of distances r from the center of mass at which growth takes place in clusters of size s. They found that the probability $P(r,s)$ for growth at a distance r could be fitted by a Gaussian with a width that is a measure of the width of the active zone. The width of the active zone was also measured more objectively via the quantity

$$\xi_a(s) = < (r_s - < r_s >)^2 >^{1/2}, \tag{3.33}$$

where $< r_s >$ is the mean deposition radius for the last particle to be added to a cluster of size s and $< \ldots >$ implies averaging over a large number of clusters. More generally, the width of the active zone w_a can be defined as

$$w_a^2 = < P_i r_i^2 > - < P_i r_i >^2, \tag{3.34}$$

where P_i is the probability of growth at position i, at a distance r_i, measured from the initial "seed" or the center of mass. Since the probabilities P_i are not readily obtained for DLA clusters, Plischke and Rácz defined the width of the active zone as

$$\tilde{w}_a^2(i) = < r_i(n)^2 - < r_i(n) >^2 > = < (r_i(n) - < r_i(n) >)^2 >, \tag{3.35}$$

where $r_i(n)$ is the deposition radius for the ith particle in the nth cluster and the averages in equation 3.35 are over the N clusters. Plischke and Rácz found that $< r_s >$ grew algebraically with increasing cluster size, with an exponent ν that was indistinguishable from the exponent β (equation 3.31) describing the growth of the radius of gyration ($\nu = 0.584 \pm 0.02$). However, the exponent $\overline{\nu}$, describing the algebraic growth of the width of the active zone $\tilde{w}_a$, was found to be substantially smaller ($\overline{\nu} = 0.48 \pm 0.01$). The width of the active zone, defined in equation 3.35, includes both intercluster and intracluster contributions to the fluctuation in the deposition radius of the ith particle. An alternative procedure that includes only the intracluster contributions is to define the width of the active zone as

$$w'_a = < r_i(n) - r_{i-1}(n) >, \tag{3.36}$$

where $r_{i-1}(n)$ and $r_i(n)$ are measured in the *same* cluster, and the average is over a large number of clusters. In practice, w'_a grows as $w'_a \sim s^{\overline{\nu}'}$, and values obtained for the exponents $\overline{\nu}$ and $\overline{\nu}'$ appear to converge, within statistical uncertainties, for large DLA clusters [445].

3.4.5.2 Angular Correlations

Another indication that 2-dimensional DLA clusters may have a richer scaling structure than the most simple self-similar fractals was revealed by the study of angular correlations by Meakin and Vicsek [446]. They measured the angular correlation function

$$C_R(\theta) = < \rho(\theta_\circ)\rho(\theta_\circ + \theta) >, \tag{3.37}$$

for the distribution of mass in narrow annular regions, with a mean radius R, centered on the cluster origin. For off-lattice clusters, this angular correlation function was found to have the form $C_R(\theta) = \theta^{-\alpha_\theta}$ over a range of about one decade in θ, and a value of 0.41 ± 0.04 was estimated for the exponent α_θ. This indicates that the fractal dimensionality for circular cuts in the interior, frozen, region of DLA clusters has a fractal dimensionality of $D_\theta \approx 0.59$, instead of the value of about 0.71 expected for a self-similar fractal with a dimensionality of $D \approx 1.71$. As the angle θ increases, the angular correlation function reaches a minimum. At still larger angles, $C_R(\theta)$ increases to a maximum, before finally reaching an essentially constant value, corresponding to the average density at a distance R from the center of the cluster, at large angles. For square lattice clusters, this maximum is quite pronounced and lies at $\theta \approx \pi/2$. For the off-lattice clusters, the angular correlation function has a minimum, but the maximum is poorly defined or absent. Similar results were obtained, at about the same time, by Kolb [447], who measured a difference of $\Delta D = D_r - D_\theta = 0.16 \pm 0.05$ between the fractal dimensionalities of circular cuts and cuts along the radius of square lattice DLA clusters containing up to 5000 particles. A combination of lattice anisotropy effects and the small cluster sizes may contribute to the effective exponents measured in this work.

A very similar result, $D_\theta = 0.60 \pm 0.02$, was obtained by Arneodo *et al.* [448] for circular cuts of larger, 10^6-particle, off-lattice DLA clusters [424]. At larger angles, the algebraic decay of $C_R(\theta)$ ceases and a local maximum was found. More recently, Mandelbrot *et al.* [449] have measured a value of 0.65 ± 0.01 for D_θ using off-lattice DLA clusters containing 2×10^7 particles. These results are consistent with measurements of the effective fractal dimensionality in the interior of DLA clusters that give values of $D \approx 1.60$ [448, 450, 451, 452]. Very similar results [453] have also been obtained from 2-dimensional (1 + 1-dimensional) off-lattice diffusion-limited deposition model simulations, described in section 3.7.4, carried out on strips with a width of 8192 lattice units, with periodic boundary conditions in the lateral direction. From the two-point density–density correlation functions of horizontal cuts at different heights, obtained from 50 simulations in which 10^6 particles were deposited in each simulation, a fractal dimensionality of $D_x \approx 0.62$ was obtained.

Erzan *et al.* [454] and Kaufman *et al.* [426] have suggested that cuts, perpendicular to the growth direction, may represent a more fundamental and universal aspect of the structure of DLA clusters and the patterns generated by diffusion-limited deposition than the overall mass/length scaling. In particular "transverse" cuts of clusters, deposits and structures grown with other boundary conditions appear to have the same fractal dimensionality.

3.4.5.3 Mass Distribution and Lacunarity

Figure 3.9 and similar figures [455] suggest that the lacunarity of DLA clusters decreases with increasing size. Meakin and Havlin [456] measured the moments

$$M_n(r) = \sum s^n P(s,r), \tag{3.38}$$

for DLA clusters and other clusters generated using growth models. In equation 3.38, $P(s,r)$ is the probability that a circle of radius r, centered on a particle in a cluster, will contain s particles (for this purpose the particles are treated as point masses). Meakin and Havlin found that the moment ratios,

$$\Lambda_n(r) = M_n(r)/[M_1(r)]^n, \tag{3.39}$$

are independent of r for "simple" statistically self-similar fractals such as clusters generated by 2-dimensional cluster–cluster aggregation models [58], 2-dimensional percolation clusters [457] and 3-dimensional percolation clusters [457]. In contrast to this behavior, $\Lambda_n(r)$ was found to decrease with increasing r for 2-dimensional off-lattice DLA clusters. The difference in this aspect of the structure of clusters generated by off-lattice diffusion-limited cluster–cluster aggregation [58] and off-lattice DLA is illustrated in figure 3.10. They pointed out that this behavior implies that the DLA structure is more uniform on long length scales than it is on short length scales. This observation is consistent with figure 3.9. These moment ratios characterize the lacunarity (chapter 2, section 2.1.2) of the fractal [2, 197].

The analysis of Meakin and Havlin was carried out by selecting a large number of "reference" particles (typically 5000) from the "frozen" central portion of the cluster and measuring $N(s,r)$, the number of particles within a distance r of the reference particle, for $2 \le r \le 20$ particle diameters. The results from 37 50 000-particle clusters were averaged to measure $P(s,r)$. For a self-similar fractal, the distribution $P(s,r)$ can be represented by the scaling form

$$P(s,r) = s^{-1} f(s/r^D), \tag{3.40}$$

and a good data collapse was obtained using self-similar clusters generated by simple growth and aggregation models [456]. This indicates that the moment

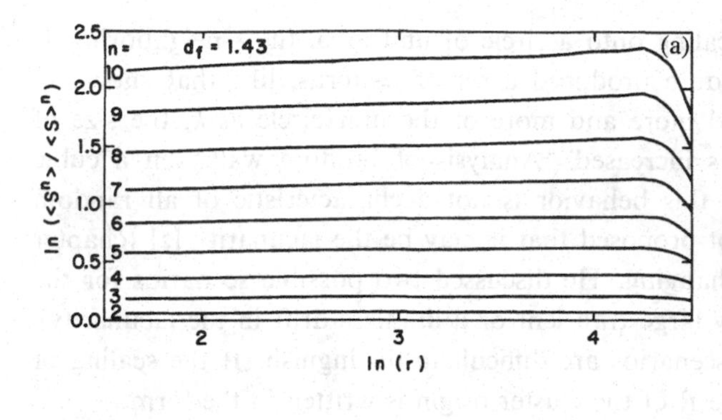

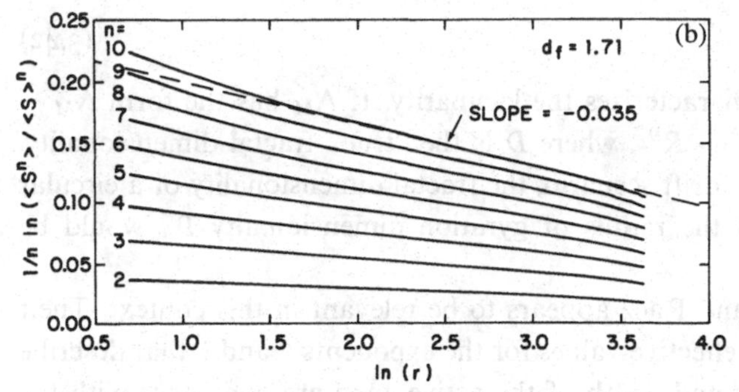

Figure 3.10 Dependence of the moment ratios $\Lambda_n(r) = M_n(r)/[M_1(r)]^n$ or $\Lambda_n(r) = \langle s^n \rangle / \langle s \rangle^n$ on the distance r over which they were measured for clusters generated using a simple 2-dimensional diffusion-limited cluster–cluster aggregation model (a) and 2-dimensional off-lattice DLA model (b).

ratios $\Lambda_n(r)$ are independent of r. If the reference particles in the 2-dimensional DLA clusters were selected randomly from the *entire* cluster, then the distribution $P(s,r)$ could be represented by the scaling form given in equation 3.40 with $D \approx 1.71$ [458], but the data collapse was not as good as that obtained for clusters generated by other models. Similarly, if self-similar clusters are covered with grids with elements of size (length and width) ℓ or randomly placed "balls" of size ℓ, then the corresponding distribution for a self-similar fractal can be represented by the scaling form $N(s,\ell) = s^{-2}g(s/\ell^D)$ or

$$P(s,\ell) = s^{-1}f'(s/\ell^D). \tag{3.41}$$

For the DLA clusters, the best data collapse was obtained with exponents that differed substantially from the values of 1 and D, given in equation 3.41, and the quality of the data collapse was very poor. For 2-dimensional DLA clusters, the scaling functions or lacunarity functions $f(x)$ and $f'(x)$ in equations 3.40 and 3.41 become narrower as r or ℓ increase, indicating a more uniform, less lacunar structure on longer length scales.

Mandelbrot [459] refocused attention on this unresolved problem. Using very large, 3×10^7-particle, 2-dimensional off-lattice clusters, an "ϵ-neighborhood analysis" was carried out on the "dead", non-growing, central portion of the cluster. The regions of the cluster within various distances from the center ($\lambda^k R_0$, where $R_0 = 78$ particle diameters and $\lambda = 1.175$), were scaled onto a circle with a unit area (i.e. all of the coordinates and particles were reduced by a factor of $1/(\pi^{1/2}\lambda^k R_0))$. After scaling onto a circle of unit area, the ϵ-neighborhoods were found.[6] This procedure produced a set of patterns, like that shown in figure 3.11, which covered more and more of the unit circle as k, the size of the unscaled pattern, was increased. Analysis of random walks on a cubic lattice demonstrates that this behavior is not a characteristic of all random fractals [460]. Mandelbrot proposed that it may be the lacunarity [2] (chapter 2, section 2.1.2) that is changing. He discussed two possible scenarios for the observed behavior (a very large transient or a limitless drift in the lacunarity). At this stage, these two scenarios are difficult to distinguish. If the scaling of the mass within a distance R of the cluster origin is written in the form

$$M(R) = \Lambda_M R^D, \tag{3.42}$$

then the amplitude Λ_M characterizes the lacunarity. If Λ_M has the form $\Lambda_M \sim R^{\Delta D}$, then $M(R) \sim R^{D+\Delta D} \sim R^{D^*}$, where D is the "true" fractal dimensionality. According to this limitless drift scenario, the fractal dimensionality of a circular cut would be $D - 1$ and the radius of gyration dimensionality D_β would be $D^* = D + \Delta D$ [449, 459].

The work of Plischke and Rácz appears to be relevant in this context. Their measurement of different effective values for the exponents ν and $\overline{\nu}$ that describe the increase in the radius and width of the active zone are consistent with the scenarios proposed by Mandelbrot. Simulations on a scale larger than those of Plischke and Rácz indicate [445] that $\overline{\nu}$ approaches ν as the cluster size grows. Based on results for clusters of 10^6 particles [424], this process would not be complete for clusters of 3×10^7 particles. Figure 3.12 shows how these exponents depend on the cluster size. This behavior might be taken as evidence for a massive transient scenario.

3.4.5.4 Multiscaling

Although small scale 2-dimensional and 3-dimensional DLA simulations give results consistent with the idea that the density profile $\rho(r, L)$, where $\rho(r, L)$ is the mean density at a distance r from the center of a cluster of size (radius of gyration) L, could be represented in terms of the simple scaling form

$$\rho(r, L) = r^{D-d} f(r/L), \tag{3.43}$$

the results of Plischke and Rácz [402, 403] suggest otherwise. Coniglio and Zannetti [257] suggested that the density profile might be described by the multiscaling form

$$\rho(r, L) = r^{D(r/L)-d} f(r/L) = r^{-\alpha(r/L)} f(r/L), \tag{3.44}$$

to account for the narrowing of the width of the active zone relative to the size of the growing cluster as the cluster size increases. This implies that the

6. The ϵ-neighborhood of a set consists all of those points within a distance ϵ of that set.

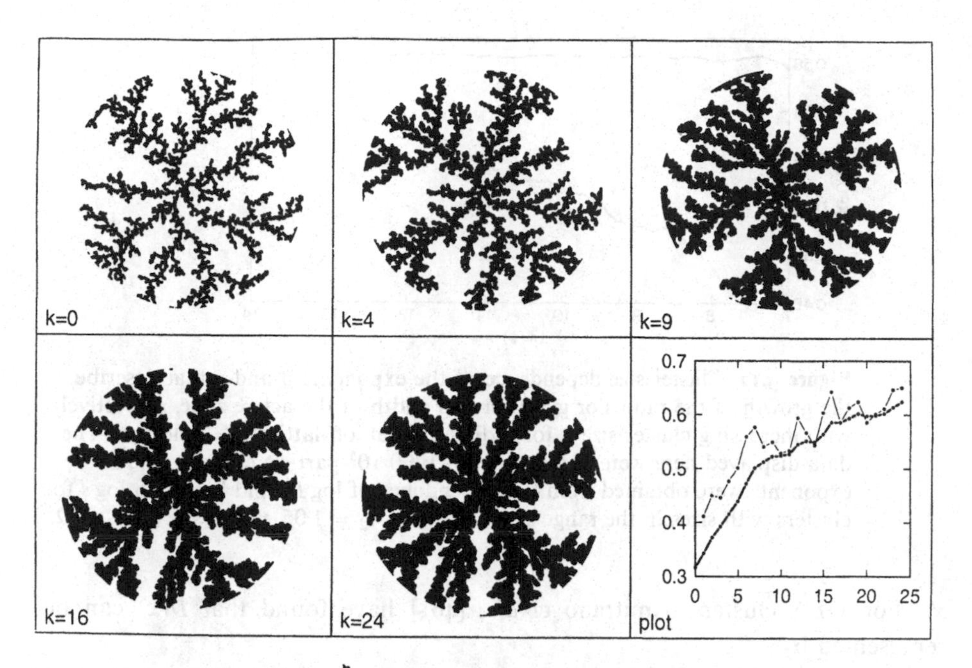

Figure 3.11 The ϵ-neighborhoods for circular regions with radii of $R = 78 \times 1.175^k$ particle diameters taken from the center of a large, off-lattice DLA cluster. The circular regions have been scaled onto a circle of unit area. The ϵ-neighborhoods are shown for $k = 0, 4, 9, 16$ and 24 using a relatively small ϵ. Unlike the behavior found for self-similar fractals, the ϵ-neighborhoods fill a larger fraction of the unit area circles as k increases. This is shown in the lower right part of the figure, where the thick line shows the mean density (area $A(k)$ filled in the circles of unit area) as a function of k. The thin line shows the fraction of the area occupied by the ϵ-neighborhoods for those parts of the cluster lying in the region $78 \times 1.175^{k-1} < R < 78 \times 1.175^k$, in the corresponding outer annulus of the scaled circle. This figure was provided by B. B. Mandelbrot.

cluster has an inhomogeneous structure with a local fractal dimensionality that is a function of r/L. In the case of a growing cluster, the manner in which the density profile evolves as the cluster size increases is of central interest. If the cluster is grown from a size L to λL, then, according to equation 3.44, the density profile becomes

$$\rho(\lambda r, \lambda L, \epsilon) = \lambda^{-\alpha(r/L)} r^{-\alpha(r/L)} f(r/L) \tag{3.45}$$

or

$$\rho(\lambda r, \lambda L, \epsilon) = \lambda^{-\alpha(r/L)} \rho(r, L) = \lambda^{-\alpha_\circ} \lambda^{-\delta\alpha(r/L)} \rho(r, L), \tag{3.46}$$

where ϵ is the particle size and $\epsilon \ll r, \lambda r$. The new (transformed) density profile has the same shape as the original profile only if $\alpha(r/L)$ is a constant ($\delta\alpha(r/L) = 0$). For a growing cluster, like a DLA cluster, $\alpha(x) = \alpha(r/L)$ must increase monotonically, since $D(x)$ must decrease monotonically with increasing

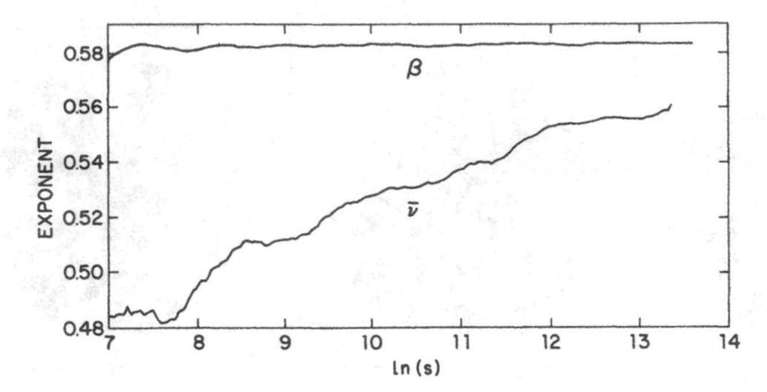

Figure 3.12 Cluster size dependence of the exponents β and $\overline{\nu}$ that describe the growth of the radius or gyration and width of the active zone, respectively, with increasing cluster size s, for 2-dimensional, off-lattice DLA clusters. The data displayed here were obtained from 1000 10^6-particle clusters. The exponents were obtained from the dependence of $\log R_g$ and $\log w_a$ on $\log s$ for clusters with sizes in the range $s_1 - s_2$ with $s_2/s_1 = 1.05$, where $s = (s_1 + s_2)/2$.

x. For DLA clusters, Amitrano *et al.* [461] have found that $D(x)$ can be represented by

$$D(x) = D(0) - ax^{\beta}, \tag{3.47}$$

with $a \approx 1/2$ and $\beta \approx 10$, for small values of x.

Typically, $\alpha(x) = const.$ for $x \ll 1$ and $\alpha(x)$ decreases substantially for $x \gg 1$. This implies that the shape of the density profile will evolve towards a form that decreases more rapidly with increasing x as the cluster size or λ increases. In this case, if $\alpha(x) = \alpha_\circ$ for $x \leq x_\circ$, and $\alpha(x) = \alpha_\circ - k(x - x_\circ)^n$, for $x > x_\circ$, the width of the front will grow as $w_a \sim L(\log L)^{-1/n}$.

Based in part on the results of Meakin and Sander [445], which showed that the exponents ν' and ν converge with increasing cluster size, Coniglio and Zannetti [257] conjectured that

$$(L/w_a)^2 = c(\ln L), \tag{3.48}$$

where w_a is the width of the active zone and L is some characteristic length scale, such as the radius of gyration or mean deposition radius, that measures the cluster size. From this conjecture and the Gaussian form

$$P(r,s) = \frac{1}{(2\pi)^{1/2} w_a} e^{-[(r - <r_s>)^2 / 2w_a^2]}, \tag{3.49}$$

suggested by the results of Plischke and Rácz, for the distribution of deposition radii $P(r,s)$, in clusters of size s, Coniglio and Zannetti concluded that

$$P(r,s) = \frac{(c \ln L(s))^{1/2}}{(2\pi)^{1/2} L(s)} L(s)^{-(c/2)[(r/L(s)) - 1]^2}, \tag{3.50}$$

which has a multiscaling form.[7] Coniglio and Zannetti also showed that the data of Plischke and Rácz [402] could be fit quite well by equation 3.50, with $L(s) \sim s^{\beta} \sim s^{1/D_{\beta}}$ ($\beta = 0.585$) and $c = 6.46$. Later, Amitrano *et al.* [259] showed that the density profiles obtained from 200 10^5-particle, 2-dimensional, off-lattice clusters could be scaled much better using multiscaling than the standard scaling form given in equation 3.43. Density profiles were measured at 20 stages during the growth process, and $D(r/L)$ was obtained from the dependence of the density $\rho(r, L)$ on r at constant r/L ($\rho(r, L)_{r/L=x} = r^{D(x)-d}$). Ossadnik [260] has come to similar conclusions based on data from 100 10^6-particle, 2-dimensional, off-lattice clusters and a single cluster of 5×10^7 particles. Unfortunately, the statistical uncertainties appear to be large for the 5×10^7-particle cluster.

It has recently been suggested, on the basis of numerical studies with large clusters [463, 260], that the width of the active zone grows with increasing cluster size according to

$$w_a(s) \sim s^{1/D}/(\log s)^{1/2} \sim R_g/(\log s)^{1/2}, \tag{3.51}$$

which is consistent with equation 3.48.

3.4.5.5 Some Comments

The large scale simulations described earlier in this section and the scaling models that they have spawned point to a growth scenario that appears to be untenable [445]. The multiscaling and continuously drifting lacunarity models imply that as the cluster grows larger and larger it can be scaled onto a unit circle that it fills more and more homogeneously. At the same time, the active zone becomes concentrated onto an increasingly perfect circle. This is inconsistent with the well established Mullins–Sekerka instability, described in appendix A (section A.1), which indicates that the circular shape is unstable with respect to the growth of shape perturbations under Laplacian growth conditions. Although the linear stability analysis of Mullins and Sekerka does not demonstrate how shape perturbations grow to finite amplitudes, a plethora of numerical studies and theoretical work demonstrates that the circular shape is completely unstable. In fact, DLA appears to be the most dramatic manifestation of the Mullins–Sekerka instability. Even logarithmic differences between the rates of growth of the cluster size and active zone width would eventually lead to homogeneous, circular clusters. This simple argument strongly favors the "massive transient" scenario. It appears that this unprecedented behavior must come about as a result of a subtle interplay between the Mullins–Sekerka instability, the uncontrolled randomness generated by the random walkers and the discrete

7. More recent studies indicate that the deposition probability grows algebraically with increasing distance $R(r) \sim r^a$ for small values of r. The exponent a has an effective value of about 8 [462, 463], but the data can also be fitted quite well by the scaling form $P(r, s) = (r^{a'}/w_a(s)) f[(r - r_a(s))/w_a(s)]$ with $a' \approx 2$ [463], where $r_a(s)$ is the radius of the active zone and $r_a(s) \sim R_g(s)$.

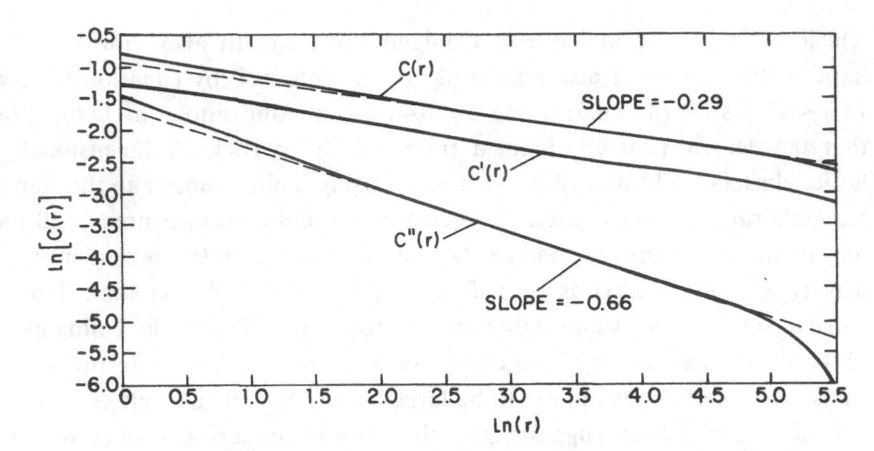

Figure 3.13 Comparison of the two-point density–density correlation function $C(r)$ and the orientational correlation function $C_v(r)$ defined in equation 3.52 ($C''(r)$ in the figure). The correlation function $C'(r)$ is defined as $C'_v(r) = << |\mathbf{u}(\mathbf{r}_\circ) \cdot \mathbf{u}(\mathbf{r}_\circ + \mathbf{r})| >>_{|\mathbf{r}|=r}$. This correlation function differs from the standard orientational correlation function $C_v(r)$, because the magnitude of $\mathbf{u}(\mathbf{r}_\circ) \cdot \mathbf{u}(\mathbf{r}_\circ + \mathbf{r})$ is used.

nature of the particles. Perhaps these interactions enhance the instability during the early stages of growth and the massive transient is a consequence of a crossover from this early time behavior to an asymptotic self-similar growth process.

3.4.5.6 Vector Correlations

The structure of off-lattice DLA clusters has also been characterized using a vector correlation function calculated from the directions of the bonds connecting contacting particles [423]. After each particle had been added to the growing cluster, the vector $\mathbf{u}(\mathbf{r})$ from the center of the contacted particle in the cluster, at position $\mathbf{r}$, to the center of the newly added particle was determined. Data from 54 50 000-particle, 2-dimensional, off-lattice clusters were used to calculate the correlation function

$$C_v(r) = << \mathbf{u}(\mathbf{r}_\circ) \cdot \mathbf{u}(\mathbf{r}_\circ + \mathbf{r}) >>_{|\mathbf{r}|=r} . \tag{3.52}$$

This correlation function, shown in figure 3.13, has the form $C_v(r) \sim r^{-\alpha_v}$, with $\alpha_v \approx 0.66$, for length scales in the range $\epsilon \ll r \ll L$. Since the ordinary density–density correlation function has the form $C(r) \sim r^{-\alpha}$, with $\alpha \approx 0.29$, it is apparent that the correlation of the bond orientations decays as $C_\theta(r) \sim r^{-\alpha_\theta}$, with $\alpha_\theta \approx 0.37$.

More recently, Hegger and Grassberger [464] have measured the dependence of the mean angle of contact ϕ_n (the angle between the bond connecting the nth particle to the growing deposit and the normal to the substrate) in 2-dimensional (1 + 1-dimensional), off-lattice diffusion-limited deposition. From 100 runs with

10^6 particles per run and a system width of 20 480 particle diameters, they found that $< \cos\phi_n > \sim n^{-0.24\pm0.01}$. Since the mean deposition height $< h_n >$ grows as $< h_n \sim n^{1/(1-\alpha)} >$ [465], and $\alpha \approx 0.29$ for 2-dimensional, off-lattice DLA, this result implies that $< \cos\phi_n > \sim h_n^{0.24/0.71} \sim h_n^{0.34}$, which is in quite good agreement with the earlier results obtained from off-lattice DLA clusters by Meakin [423].

Lam *et al.* [269] have measured the distribution $P(\theta, s)$ of angles θ between the direction of growth (the vector from the contacted particle to the contacting particle) and the direction of the vector from the origin to the contacted particle, as a function of the cluster size s. They analyzed the distribution in terms of the cosine series

$$P(\theta, s) = (1/2\pi) + \sum_n a_n(s) cos(n\theta). \quad (3.53)$$

The coefficients were found to decay algebraically with increasing cluster size s ($a_n(s) \sim s^{-\gamma_n}$), with $\gamma_1 = 0.0997 \pm 0.0003$ and $\gamma_2 = 0.67 \pm 0.03$.

These three studies demonstrate that DLA is locally isotropic on sufficiently long length scales. It seems likely that DLA clusters grown in higher-dimensionality spaces will have a similar asymptotic local isotropic nature, but this idea remains to be tested.

3.4.5.7 Minimum Path Dimensionality

The fractal dimensionality D_{min} of the shortest path between pairs of points on the cluster is an aspect of the geometric scaling of DLA, which has important implications for understanding the physical properties of systems with the DLA structure. For $d = 2, 3$ and 4 it has been shown numerically that $D_{min} \approx 1$ [466]. This indicates that there is no upper critical embedding space dimensionality d_c at which the minimum path dimensionality D_{min} reaches the asymptotic value of 2, and simple mean field theories give exact values for the fractal dimensionality. For systems that can be described in terms of a simple mean field model, above the critical dimensionality d_c of the embedding space [366, 467, 468] (such as lattice animals, percolation clusters, cluster–cluster aggregates and linear polymers) $D_{min} = 2$ for $d \geq d_c$, since the minimum length path between a pair of points is a random walk. In contrast to the behavior found for DLA clusters, D_{min} increases with increasing embedding space dimensionality if $d < d_c$, for cluster–cluster aggregation. The amount of mass within a minimum distance ℓ, measured on the structure, is also an important quantity. This leads to a new fractal dimensionality D_ℓ, called the spreading dimensionality [467], defined by $M_\ell \sim \ell^{D_\ell}$. It is apparent, from this definition, that the spreading dimensionality D_ℓ is given by $D_\ell = D/D_{min}$.

3.4.6 Other Aspects of DLA

Arneodo *et al.* [469] have studied the average "density" distribution $\rho(\mathbf{x})$ or $\rho(x, y)$ of the patterns obtained from a large number of square lattice DLA simulations in a channel, with reflecting lateral boundaries. They defined the width of the distribution w by

$$w(y) = (1/\rho_{max}) \int_0^{L_x} \rho(x, y) dx, \tag{3.54}$$

where L_x is the channel width and ρ_{max} is the maximum density at the center of the "cell". If those sites with densities greater than the density $\rho(w/2)$ at a distance of $w/2$ from the center (longitudinal axis) of the cell are displayed, then a pattern very similar to that of a stable Saffman–Taylor [470] viscous finger emerges. This is shown in figure 3.14. Here, w is the mean width of the distribution $\rho(\mathbf{x})$, and $\rho(w/2)$ is measured in that region where the density profile across the lattice has a constant shape. In addition, the finger width obtained in this manner is equal to $1/2$ of the channel width, within the uncertainty of the simulations. This suggests that the "average" structure for DLA in a channel is a solution to the Saffman–Taylor problem (appendix A, section A.2). The procedure used to set the threshold for $\rho(\mathbf{x})$ seems to be somewhat arbitrary, and a different width would be obtained with a different threshold. This may correspond to the selection of a different member of the infinite family of Saffman–Taylor solutions for viscous fingering in a Hele-Shaw cell. However, the procedure used by Arneodo *et al.* to select the threshold is supported by its success in other geometries. Simulations have also been carried out with reflecting boundary conditions, at the straight walls, in sector shaped cells with diverging and converging flows. For these boundary conditions, averages over a large number of DLA simulations also lead to shapes similar to those obtained from analytical solutions [471] to the corresponding Saffman–Taylor problem.

In diffusion-limited aggregation, the growth is concentrated onto the outer regions of the cluster, and the complex structure of the "active zone" may be approximated by a smooth surface. With these boundary conditions, this smooth surface appears to have the same geometry as a Saffman–Taylor "finger" and its growth can be described by the same equations. The observation that the overall (average) shape of DLA clusters may be understood in terms of the corresponding stable growth process might prove to be an important step towards the development of a general theoretical understanding. However, it does not seem likely that this will lead to important insights concerning the complex internal structure of DLA patterns.

Arneodo *et al.* [469] also studied the effects of noise reduction on DLA growth on a square lattice in a channel. Noise reduction enhances the effects of lattice anisotropy, as is described in more detail later in this chapter. As the noise reduction parameter m is increased, the DLA pattern becomes narrower.

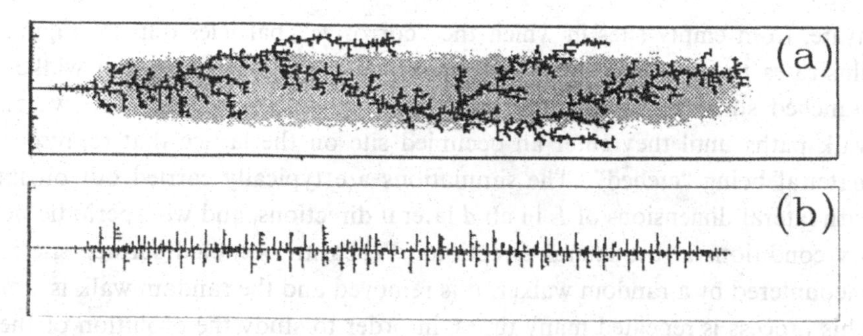

Figure 3.14 DLA clusters grown in a channel with a width of 64 lattice units. Part (a) shows a "finger" grown without noise reduction ($m = 1$), and part (b) shows a cluster grown with a noise reduction parameter of $m = 8$. The shaded region shown in part (a) corresponds to an average over 250 clusters and shows those regions in which the mean occupation density is greater than $\rho(w/2)$. The shape of this area is like that of a Saffman–Taylor finger. This figure was provided by A. Arneodo.

The numerical results [469] indicate that the relative width ($\lambda = w/L_x$) can be represented by the scaling form

$$\lambda = f[1/(m^{3/2}L_x)], \tag{3.55}$$

where the scaling function $f(x)$ has the form $f(x) = 1/2$ for $x \gg x_c$ and $f(x) = x^{1/2}$ for $x \ll x_c$, where x_c is a finite critical value for x.

DLA clusters have been used extensively in computer models designed to explore the physics and chemistry of fractals [57]. This choice is well motivated because the DLA model is representative of a large class of patterns of scientific and practical importance. However, the complex, incompletely understood scaling properties of finite DLA clusters exacerbate the difficulty of interpreting the simulation results. The use of square and cubic lattice DLA clusters in many studies is particularly unfortunate. Not only do these clusters inherit the complexity of off-lattice DLA, they also suffer from additional complexity associated with the crossover from the noise dominated, small size regime to the lattice anisotropy dominated, large size regime. For clusters of a practical size, neither the shape of the cluster envelope, nor the scaling of the overall cluster dimension L with size s, remain stable as the clusters grows.

3.4.7 Diffusion-Limited Annihilation

Diffusion-limited annihilation models are similar to DLA models, but the random walkers are used to remove or annihilate filled sites on a lattice, instead of to add sites. In the most simple version of this model [472], the simulation is carried out on a $d + 1$-dimensional lattice, with an initially smooth, flat d-dimensional surface or interface separating filled sites, representing the substrate, from empty sites in which the "corrosive" particles diffuse. Figure 3.15 illustrates the $1+1$-dimensional version of this model. Random walkers are launched sufficiently far above the d-dimensional surface and follow random walk paths until they enter an occupied site on the lattice that represents the material being "etched". The simulations are typically carried out on lattices with lateral dimensions of L in all d lateral directions, and with periodic boundary conditions in the lateral directions. When an occupied surface site is first encountered by a random walker, it is removed and the random walk is stopped. This process is repeated many times, in order to study the evolution of the surface. Since the most exposed or least screened sites are preferentially removed, a quite smooth interface is generated. The decay of shape perturbations can also be demonstrated in the simulations. This model may be relevant to electropolishing [473] processes in which diffusion of acceptor molecules to the anode is the rate limiting step [474, 475, 476]. This model has also been called the "anti-DLA" model, and it has been used to simulate the displacement of a low viscosity fluid by a high viscosity fluid in a porous medium [477]. A related model called "internal DLA", in which the cluster is "eaten" from the inside by random walkers starting from the origin, has been described by Lawler [391].

In the $1+1$-dimensional case (diffusion-limited annihilation of a flat surface on a square lattice strip) the surface width grows as $(\log s)^{1/2}$ or $(\log t)^{1/2}$, where t is the "time" and s is the number of removed sites. For the $2+1$-dimensional case (diffusion-limited annihilation of a flat surface on a cubic lattice) the width of the surface roughness quickly saturates to a small, constant value. Krug and Meakin [478] have shown that a small periodic perturbation of a flat surface with a wave vector $\mathbf{q}$ in the $d+1$-dimensional space, perpendicular to the direction of motion of the interface, decays, with a decay constant $k(\mathbf{q})$ given by

$$k(\mathbf{q}) = V|\mathbf{q}|, \tag{3.56}$$

where V is the mean velocity of the interface. This form for the decay of perturbations of a smooth surface shape has been observed in $1+1$-dimensional computer simulations [355, 472]. This leads immediately to the stochastic differential equation

$$\partial h_{\mathbf{q}}(t)/\partial t = -k(\mathbf{q})h_{\mathbf{q}}(t) + \eta_{\mathbf{q}}(t), \tag{3.57}$$

for the motion of the Fourier components $h_{\mathbf{q}}(t)$ of the surface roughness $h(\mathbf{x}, t)$. The Fourier components $\eta_{\mathbf{q}}(t)$ of the Gaussian, uncorrelated noise are given by

$$< \eta_{\mathbf{q}}(t)\eta_{\mathbf{q}'}(t') > = (\Delta/L^d)\delta(\mathbf{q}+\mathbf{q}')\delta(t-t'), \tag{3.58}$$

where $\Delta = V\ell_\eta^{d+1}$ and ℓ_η is a length scale associated with the randomness. This linear equation can be solved quite easily [478]. For the $1+1$-dimensional case,

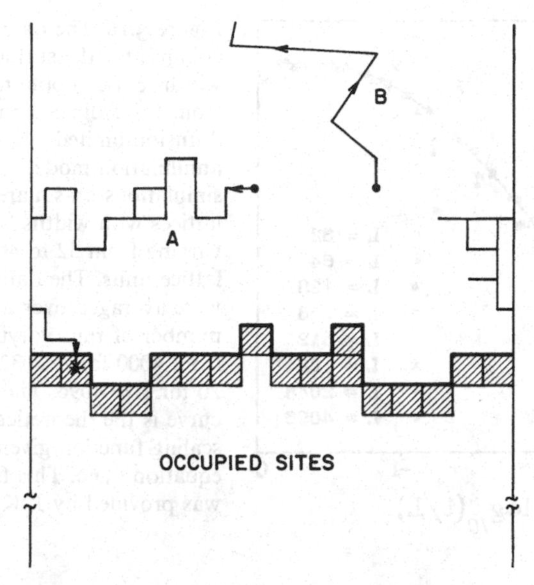

Figure 3.15 A simple 1 + 1-dimensional, square lattice, diffusion-limited annihilation model. The filled active zone sites, which can be removed in the next stage in the simulation, are shaded. The random walkers are launched from randomly selected positions on the "horizontal" launching surface, above the interface. Trajectory *A* eventually moves the random walker onto a filled active zone site, which is then removed. Trajectory *B* moves the random walker away from the interface. If it reaches the "killing surface", far from the interface, it will be terminated and a new random walker will be started from a randomly selected position on the launching surface, a few lattice units above the interface. Trajectory *B* also illustrates the long, off-lattice steps, far from the interface.

the surface width $\xi_\perp(L,t)$ measured over the lateral width L at time t is given by

$$\xi_\perp^2(L,t) = \frac{\Delta}{2\pi V}[\ln(L/\epsilon) - f(Vt/L)], \tag{3.59}$$

where ϵ is the lattice cut-off length scale and $f(x)$ is a scaling function with the form

$$f(x) = -\ln(1 - e^{-4\pi x}). \tag{3.60}$$

These theoretical results were confirmed by the good data collapse obtained by plotting $\xi_\perp^2(L,t) - \xi_\perp^2(L,\infty)$ as a function of t/L. The scaling function obtained from the data collapse was in good agreement with the theoretical scaling function given in equation 3.60. The data collapse and the theoretical scaling function are shown in figure 3.16.

Tang [355, 479] has used a model of this type to simulate the Saffman–Taylor process (appendix A, section A.2). At the start of a simulation, the sites in the central region in a square lattice strip of width L are filled (those

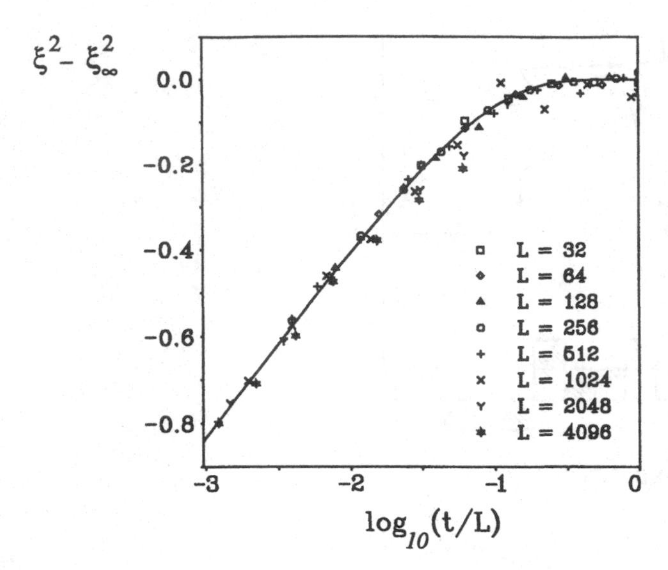

Figure 3.16 The data collapse for the surface widths $\xi_\perp(L,t)$ obtained from $1+1$-dimensional, diffusion-limited annihilation model simulations on square lattices with widths varying from 32 to 4096 lattice units. The data were averaged over a number of runs varying from 4000 for $L = 32$ to 20 for $L = 4096$. The curve is the theoretical scaling function given in equation 3.60. This figure was provided by J. Krug.

sites with x coordinates in the range $(1-f)L/2$ to $(1+f)L/2$, where f is the fraction of the strip that is filled). This central region represents the fluid "finger" in the Saffman–Taylor process. At the start of the simulation, the "finger" does not reach to the $+y$ end of the lattice and has an arbitrary shape at its end. The random walkers are launched from positions with randomly selected x coordinates, beyond the end of the finger of occupied sites, and reflecting boundary conditions are used to represent the lateral walls of the Hele-Shaw cell. In Tang's simulations, a noise reduction procedure (similar to that described below for DLA) was used. Simulations were carried out for various values of f in the range $0.25 < f < 0.75$, and the resulting tip shapes were found to be in excellent agreement with analytic solutions of the Saffman–Taylor problem (appendix A, section A.2), even for the quite narrow channels ($L = 64$ lattice sites) used in these simulations. This model simulated the "time reversed" Saffman–Taylor process. Because of the time reversal, perturbations to the steady-state solution decay rather than grow, so that the shape of the fluid–fluid interface corresponding to the penetration of a stable finger of non-viscous fluid into a viscous fluid can be obtained.

3.5 Percolation and Invasion Percolation

Percolation models [350, 480] have been used extensively to develop a better understanding of the structure and properties of multiphase media. Simple percolation models provide some of the best established examples of statistically self-similar fractals and have played an important role in the development of concepts such as scaling, universality, fractal subsets and the use of scaling exponents, such as the fracton dimensionality, to describe dynamical, transport, mechanical and optical properties. Although percolation clusters can be generated by non-equilibrium growth algorithms, it has been shown [481] that the percolation problem corresponds to the q-state Potts model in the $q \rightarrow 1$ limit. The nearest-neighbor Potts model Hamiltonian can be written as

$$\mathscr{H} = J \sum\nolimits_{nn}' \delta(\sigma_i, \sigma_j), \tag{3.61}$$

where the "spin" σ_i at the ith site can take on q values, J is a positive coupling constant and $\sum'_{nn}$ indicates summation over nearest spin pairs, avoiding multiple counting. The $q = 2$ Potts model is equivalent to the Ising model. This provides an important link to equilibrium statistical mechanics that is absent for most non-equilibrium growth models. Consequently, percolation clusters are much better understood than, for example, DLA clusters. Exact exponents are known for the 2-dimensional Potts models [216], and these can be related to the fractal dimensionalities of the critical percolation cluster and its subsets, such as the hull and red bonds [482] described below.

A square lattice percolation cluster can be generated by randomly filling the sites of a square lattice of size $L \times L$. As the lattice is filled, the clusters of connected sites are identified (nearest-neighbor filled sites belong to the same cluster, but filled sites that cannot be connected to each other by a path consisting of nearest-neighbor steps connecting only filled sites belong to different clusters). At first, only small clusters are found but, as the fraction of filled sites (p) increases, the clusters become larger and larger until one of the clusters becomes large enough to "span" the lattice (connect one side to the opposite side). In the limit $L \rightarrow \infty$, an infinite spanning cluster is formed when p reaches the critical value p_c, known as the percolation threshold probability. At this point, the infinite spanning cluster is a self-similar fractal with a fractal dimensionality of exactly 91/48. It is important to recognize that at the percolation threshold, where $p = p_c$, a large fraction of the lattice sites are filled. Consequently, the entire "percolating" system is uniform on all but quite short length scales. A scattering "experiment" will not "see" the fractal structure of these clusters, since they interpenetrate to form an almost uniform structure. It is necessary to separate the interpenetrating clusters or label them according to their connectivity to "see" their fractal character. In systems such as polymer gels, the clusters can be separated by dilution and their structures may be studied by techniques such as small-angle light and x-ray scattering. However, dissolution in a good solvent will remove the screening of the self-avoiding interaction that is present in a dense system of interpenetrating molecules. This causes the clusters (sol molecules) to swell, and they have the fractal dimensionality of lattice animals, rather than that of percolation clusters [358].[8]

It is only after the infinite percolating cluster has been identified, via its connectivity, that a self-similar fractal emerges. In the limit $L \to \infty$, this fractal percolation cluster occupies only a negligible fraction of the lattice sites. At the percolation threshold, there will be many other large clusters besides the "infinite" cluster. The size distribution for these clusters is given by

$$N_s \sim s^{-\tau}, \tag{3.62}$$

where N_s is the number of clusters of size s occupied sites. These clusters have the same fractal structure as the infinite cluster on length scales ℓ in the range $1 \ll \ell \ll s^{1/D}$ lattice units.

If a region of size (length scale) ℓ is selected randomly in a system at the percolation threshold $p = p_c$, it is very unlikely that the region will be spanned by the infinite cluster. However, the probability that it will be spanned by part of a smaller cluster is quite high. Recent work [483, 484, 485] has shown that the probability of spanning a region of size ℓ at $p = p_c$ converges to a value of 1/2 at large ℓ, for several 2-dimensional site and bond percolation models. Cardy [486] has obtained a theoretical expression for the spanning probability for rectangular regions of any aspect ratio. The results were expressed in terms of hypergeometric functions and appear to be in excellent agreement with simulation results. Stauffer *et al.* [487] have shown that the spanning probability is about 0.42 for 3-dimensional site and bond percolation.

Many versions of the percolation problem have been studied. The fractal dimensionality of the infinite cluster at $p = p_c$ and the exponent τ in equation 3.62 are universal (they are independent of details such as the lattice structure and connectivity rules used to identify clusters, but do depend on the dimensionality of the embedding space). On the other hand, the location of the percolation threshold p_c is sensitive to model details. In some cases, p_c is known exactly. For example, p_c is 1/2 for the square lattice bond percolation model, in which bonds are randomly added to an empty square lattice until a spanning path of bonds has been formed.[9] In most cases, only numerical estimates are available. For example, the most accurate estimates of p_c for the square lattice site percolation model described above are 0.5927464 ± 0.0000005 and 0.5927460 ± 0.0000005 [485]. These two estimates for p_c were obtained using different methods and are in excellent agreement with each other.

Many aspects of "percolation" models near to the percolation threshold can be described using scaling concepts, in terms of a fractal "blob" model. The

8. This does not seem possible in the 2-dimensional case since $D_a D_{min} < D_p$, where D_a, D_{min} and D_p are the animal, minimum path and percolation dimensionalities, respectively. However, for $d = 3$, $D_a D_{min} > D_p$.

9. Alternatively, bonds can be removed randomly from a square network of bonds until there is no spanning path.

basic idea is that, near to the percolation threshold, there is a characteristic correlation length ξ that diverges algebraically as the percolation threshold is approached:

$$\xi = A_{\pm}(|p - p_c|)^{-\nu}. \tag{3.63}$$

The exponent ν has a universal value, which is the same above and below the percolation threshold. For $d = 2$, the percolation correlation length exponent has a value of exactly 4/3 [488]. For $d = 3$, this exponent is not yet known exactly, but a value of about 0.88 has been established, within about ± 0.01, from large scale computer simulations [489]. The simulation results have been summarized by Nakayama *et al.* [490]. The amplitudes $A_{\pm}$ are not universal, but the ratio A_{+}/A_{-} does have a universal value [491] for all regular lattices with the same dimensionality d.

Below the percolation threshold all the clusters are of finite size, and the characteristic length ξ describes the size of the largest clusters. The cluster size distribution has the form

$$N_s = s^{-\tau} f(s/s^*), \tag{3.64}$$

where the function $f(x)$ has the form $f(x) = const.$ for $x \ll 1$ and $f(x)$ decays faster than any power of x for $x \gg 1$. The characteristic cluster cut-off size s^* is given by $s^* \sim \xi^D$ (the large clusters are fractal "blobs" of size ξ). For all $p < p_c$, the most extended clusters in an infinite system, with characteristic sizes L_c much larger than ξ, appear to have the scaling properties of lattice animals [365]. The cluster size distribution exponent τ is given by [353]

$$\tau = (D + d)/D. \tag{3.65}$$

Above the percolation threshold, the "infinite cluster" forms a continuous network. This network can be described as a packing of fractal blobs of size ξ. The mean density is the average blob density

$$\rho \sim \xi^{D-d}. \tag{3.66}$$

Above the percolation threshold, the probability $P(p)$ that a given site or bond will belong to the infinite cluster grows as

$$P(p) \sim (p - p_c)^{\beta} \tag{3.67}$$

for small values of $p - p_c$. It follows from equation 3.66 that

$$\beta = \nu(D - d), \tag{3.68}$$

or

$$D = d - (\beta/\nu). \tag{3.69}$$

This is a hyperscaling relationship (it depends explicitly on the Euclidean dimensionality d and is true for all d). For $d = 2$, the fractal dimensionality, $D = 91/48$, can be obtained from the exactly known values for β (5/36) [492] and ν (4/3, see above). For $d = 3$, computer simulations give $\nu \approx 0.88$, $\beta \approx 0.42$ and $D \approx 2.525$ [489, 493, 494]. The uncertainties in these exponents are of the order of ± 0.01, and they are consistent with equation 3.68. The characteristic size of the largest "holes" in the network is given by $s^* = \xi^d$, and the size distribution of these holes is given by

$$N_s = s^{-\tau^+} f^+(s/s^*). \tag{3.70}$$

The function $f^+(x)$ has the same "cut-off" form as $f(x)$ in equation 3.64, and the exponent τ^+ is given by the general relationship

$$\tau^+ = (D + d)/d, \tag{3.71}$$

which is equivalent to equation 2.49.

Another important exponent relates the mean cluster size S, defined as

$$S = \sum s^2 N_s / \sum s N_s, \tag{3.72}$$

to $p - p_c$. Below the percolation threshold, $S \sim |p - p_c|^\gamma$, and above the percolation threshold, $S' \sim (p - p_c)^\gamma$, where the contribution of the infinite or spanning cluster is removed from the sums in equation 3.72. The exponent γ has a value of exactly 43/18 for $d = 2$ and about 1.795 for $d = 3$. This simple scaling describes the dependence of S and S' on $p - p_c$ only in the $p \to p_c$ limit.

A variety of fractal subsets can be identified in an infinite percolation cluster, at $p = p_c$. These subsets are often important in understanding the properties of systems that can be described in terms of percolation clusters. Similar subsets are important for other fractals, for similar reasons. One of these subsets is the set of "red bonds" [495] that has a fractal dimensionality D_{rb} given by

$$D_{rb} = 1/\nu, \tag{3.73}$$

where ν is the correlation length exponent [482]. These red bonds (or sites) are the bonds that, if removed individually, would sever the path between two distant points at opposite sides of the cluster. Similarly, the "backbone" between two distant points on a cluster can be defined as all of the sites (or bonds) that would carry current between the two points, if they were maintained at a different potential. The backbone is the union of all non-reentrant paths or self-avoiding walks between the two distant points [496]. The fractal dimensionality of the backbone is about 1.645 for $d = 2$ and about 1.74 for $d = 3$ [497, 484]. It is important to recognize that both the "red" bonds and the backbone are defined with respect to a pair of points on the fractal. The fraction of bonds that belong to the backbone connecting two opposite ends of an infinite cluster above the percolation threshold grows as $|p - p_c|^{\beta_b}$, where the exponent β_b has a value of about 0.5 for $d = 2$ [498] and about 1.1 for $d = 3$ [498].

The backbone can be described in terms of a chain of "blobs",[10] connected together by red bonds [499]. If a critical percolation cluster ($\xi \rightarrow \infty$) is examined on a length scale ℓ, by extracting regions of size ℓ^d and constructing their backbones, then the blob size distribution can be represented by the scaling form [497]

$$N_s(\ell) = s^{-\tau} f(s/\ell^{D_{bb}}), \tag{3.74}$$

where D_{bb} is the fractal dimensionality of the backbone, $\tau = 1 + (d/D_{bb})$ and $f(x)$ has the usual cut-off form ($f(x) = const.$ for $x \ll 1$ and $f(x)$ decays faster than any power of x for $x \gg 1$). As the length ℓ is increased, the small blobs on one length scale ℓ_o may be linked together to form larger blobs on length scales greater than ℓ_o, as red bonds identified on the ℓ_o length scale become part of loops on longer length scales and lose their identity as red bonds. As larger length scales are approached, the number of small blobs (in a unit volume) decreases to form larger blobs and the scaling form given in equation 3.74 is preserved. Consequently, the evolution of the blob size distribution, as ℓ is increased, is similar to the evolution of the cluster size distribution in an aggregating system in which large clusters are formed from smaller clusters that were present at an earlier stage [58].

The definitions of the subsets do not suggest very practical ways of extracting them from a fractal pattern grown on a lattice or obtained from a digitized image. The minimum path between the two points P_1 and P_2 can be found by starting with one of the points (lattice sites) P_1 and giving it the label "1". All of its nearest neighbors are then given the label "2" and the unlabeled neighbors of the sites labeled "2" are given the label "3", etc., until the point P_2 is reached. This process is often called "burning". The process is then repeated by first giving point P_2 the label "1′"; the nearest neighbors of P_2 that were given a lower label than that given to P_2 in the first burning or labeling stage are then given the label "2′". In the next stage, the nearest neighbors of sites labeled "2′" that were given a lower label in the first "burning" stage are then given the label "3′", etc. When the point P_1 is reached, all of the sites with "primed" labels will form the union of minimum paths or "elastic backbones" [498]. Figure 3.17(a) shows the elastic backbone between two points in a small cluster that has been identified in this manner. In the first stage of labeling or burning, each "attempt" to label a previously labeled or burned site identifies a site that forms a node in a network of loops. The sites $\{P_j^{(n)}\}$, from which attempts to label a previously labeled site were made, are the sites at which

10. These blobs can have any size in a critical percolation cluster; they should not be confused with the blobs of size $\xi \sim |p - p_c|^{-\nu}$ in a non-critical percolation cluster.

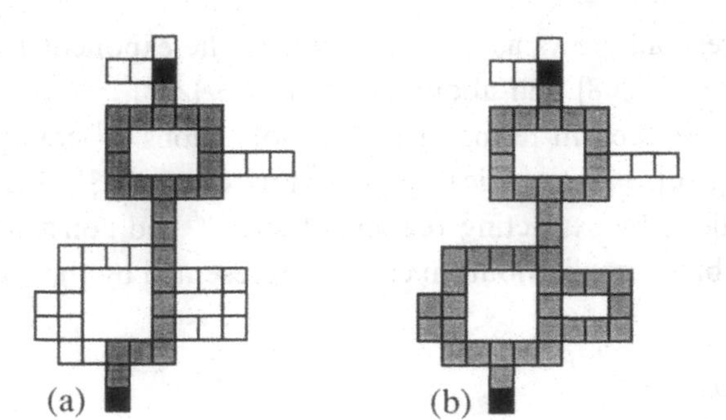

Figure 3.17 The elastic backbone (a) and the backbone (b) between two points on a small cluster. The points are indicated by blackened lattice sites and the backbones are indicated by gray lattice sites.

loops were closed in the first stage of burning, and they can be identified during the first burning.

To identify the ordinary backbone, the elastic backbone is used as the initial configuration and the backbone is "grown" from the elastic backbone, which is a subset of the backbone, by adding loops. A new burning stage is started, using all of the loop closing sites $\{P_j^{(n)}\}$ found in the first stage as origins. This third and final stage of "burning" then proceeds in the same way as the second stage, in the direction of decreasing values of the labels assigned during the first stage. During this final stage, sites in the elastic backbone (or sites added to the elastic backbone to form the growing backbone at an earlier part of the third stage) cannot be burned. If the burning process, starting from one of the sites $P_i^{(n)}$ in $\{P_j^{(n)}\}$, reaches the "growing backbone" at two or more locations, then the sites that have been burned from the site $P_i^{(n)}$ are part of the backbone. They are added to the growing backbone and cannot be used at later stages in the search process. If the sites found during the burning process from one of the sites $P_i^{(n)}$ in $\{P_j^{(n)}\}$ contact the "growing backbone" at only one location, then these sites are not part of the backbone since a walk entering this region from the elastic backbone cannot return without entering previously entered sites. This procedure is repeated for all of the origins $\{P_j^{(n)}\}$. Figure 3.17(b) shows the backbone between a pair of sites in a small cluster.

The "surfaces" of percolation clusters are important in models for catalysis, dissolution of binary alloys, the migration of non-wetting fluids in porous media and many other applications. The perimeter, hull and external perimeter of 2-dimensional square lattice site percolation clusters have different fractal dimensionalities. The occupied and unoccupied perimeters have the same fractal dimensionality ($D = 91/48$) as the entire cluster. The hull has a fractal dimensionality of exactly $D_{hull} = 7/4$ [216, 500], and the external perimeter has a fractal dimensionality of $D_{ex} = 4/3$ [216, 501, 502]. Figure 3.18 shows the occupied hull and occupied external perimeter of a small "percolation" cluster.

In the case of site percolation on a triangular lattice, $D_{ex} = D_{hull} = 7/4$ [503]. For $d = 3$, essentially all the sites are on the "surface" and $D_{hull} = D_{ex} =$

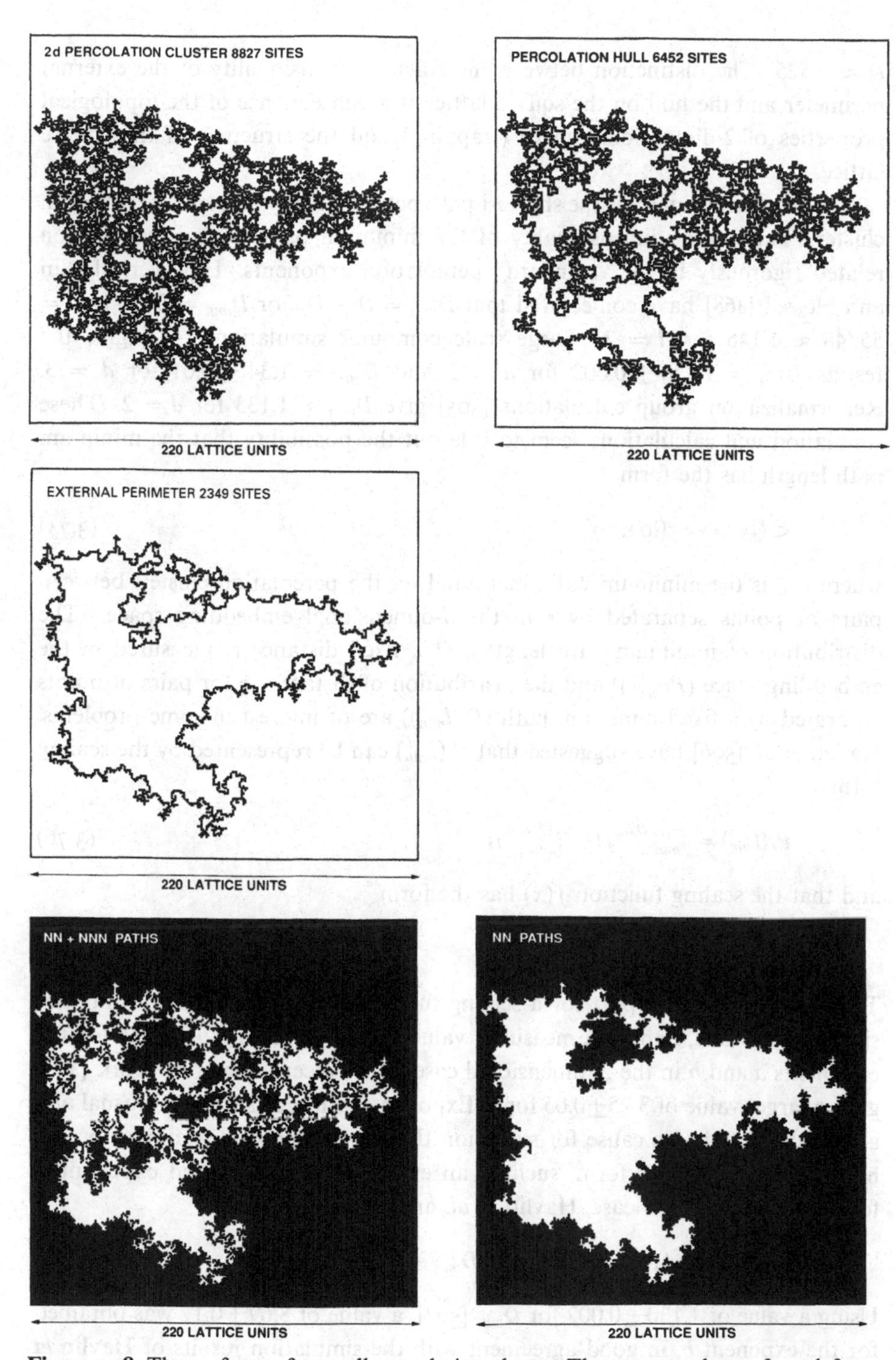

Figure 3.18 The surfaces of a small percolation cluster. The top row shows, from left to right, the cluster of 8827 sites and the occupied hull of 6452 sites. The middle panel shows the occupied external perimeter of 2349 sites. The bottom row shows all of the paths consisting of steps between nearest-neighbor and next-nearest-neighbor unoccupied sites that were used to find the hull, and all of the paths consisting of steps between nearest-neighbor unoccupied sites that were used to find the external perimeter.

$D \approx 2.525$. The distinction between the fractal dimensionality of the external perimeter and the hull on the square lattice is a consequence of the topological properties of 2-dimensional space (trapping) and the structure of the square lattice.

Another fractal subset is the shortest path between two points on a percolation cluster. The fractal dimensionality of the minimum path, D_{min}, has not been related rigorously to the "standard" percolation exponents. However, Havlin and Nossal [468] have conjectured that $D_{min} = D - D_{rb}$ or $D_{min} = D - (1/\nu) = 55/48 \approx 1.146$ for $d = 2$. Large scale computer simulations [504] give the results $D_{min} = 1.130 \pm 0.002$ for $d = 2$ and $D_{min} = 1.34 \pm 0.01$ for $d = 3$. Renormalization group calculations [505] give $D_{min} \approx 1.133$ for $d = 2$. These simulation and calculations seem to rule out the possibility that the minimum path length has the form

$$< l_{min} > \sim r(\log(r))^x, \tag{3.75}$$

where l_{min} is the minimum path, measured on the percolation cluster, between pairs of points separated by r in the d-dimensional embedding space. The distribution of minimum path lengths, at a fixed distance r, measured in the embedding space ($P_{l_{min}}(r)$) and the distribution of distances r for pairs of points separated by a fixed minimum path ($P_r(l_{min})$) are of interest in some problems. Havlin *et al.* [506] have suggested that $P_r(l_{min})$ can be represented by the scaling form

$$P_r(l_{min}) = l_{min}^{-1/D_{min}} f(r/(l_{min}^{1/D_{min}})) \tag{3.76}$$

and that the scaling function $f(x)$ has the form

$$f(x) \sim x^a e^{-cx^b}. \tag{3.77}$$

They found strong support for a scaling function of this form using computer simulation results, and they measured values of 2.5 ± 0.3 and 9.8 ± 0.5 for the exponents a and b in the 2-dimensional case. More recent numerical work [457] gave a larger value of 3.25 ± 0.05 for a. Exponents as large as 9.8 are unusual and are often a legitimate cause for suspicion that the asymptotic relationship may have a non-power law form, such as an exponential or stretched exponential form. However, in this case, Havlin *et al.* argued that

$$b = 1/[1 - (1/D_{min})] = D_{min}/(D_{min} - 1). \tag{3.78}$$

Using a value of 1.130 ± 0.002 for D_{min} [504], a value of 8.69 ± 0.12 was obtained for the exponent b, in good agreement with the simulation results of Havlin *et al.* In a later numerical study, Neumann and Havlin [457] extended this work to 3-dimensional percolation clusters and obtained a more accurate estimate of the exponent b ($b = 9.8 \pm 0.02$) for 2-dimensional percolation. This and a similar discrepancy between the values obtained for b and $D_{min}/(D_{min}-1)$ from the 3-dimensional simulations caused them to express doubts about equation 3.78.

Hovi and Aharony [505] have used an approximate real-space renormalization group method, which gives accurate results for a variety of problems, including square lattice bond percolation, to study the distribution function $P_{l_{min}}(r)$ of the minimum path distances l_{min} for fixed distances r measured in the embedding space. These distributions can be represented by the scaling form

$$P_{l_{min}}(r) = 1/l_{min} g(l_{min}/r^{D_{min}}). \qquad (3.79)$$

The results of Hovi and Aharony indicated that the scaling function $g(x)$ had the form

$$g(x) \sim x^{a'} e^{-c'x^{b'}}, \qquad (3.80)$$

with $a' = D_{min}/(D_{min}-1)$ and $b' = -1/(D_{min}-1)$ for $x \ll 1$. In the $x \gg 1$ limit, the scaling function was found to have the form

$$g(x) \sim e^{-\overline{c}x^{\overline{b'}}}, \qquad (3.81)$$

with $\overline{b'} = 2/(2-D_{min})$. Similar results were obtained for the distributions of self-avoiding walk lengths and cluster masses, except that D_{min} must be replaced by D_{SAW} or D in equation 3.79 and the expressions for the exponents a', b' and $\overline{b'}$, where D_{SAW} is the fractal dimensionality of a self-avoiding walk on the percolation cluster.

Based on these results, Hovi and Aharony deduced that the scaling function $f(x)$ in equation 3.77 should decay as e^{-cx^b} for $x \gg 1$, with $b = D_{min}/(D_{min}-1)$, in accordance with the earlier results of Havlin *et al.* [506]. They also proposed that the scaling function $f(x)$ should decay as $f(x) \sim e^{-\tilde{c}x^{\tilde{b}}}$, with $\tilde{b} = -2D_{min}/(2-D_{min})$ for $x \ll 1$. This is very different from the power law form obtained from the numerical studies of Havlin *et al.* Equation 3.77, obtained from the numerical work, implies that $f(x) \sim x^a$ for $x \ll 1$. However, a scaling function with the form given in equation 3.77 could still provide an accurate description of the distribution of distances $P_r(\ell_{min})$ for intermediate values of x.

Havlin *et al.* [507] have defined the skeleton of a percolation cluster (or other object) as the set of all shortest paths connecting a selected site to all the sites at a minimum path distance L from the selected site. Numerical studies indicate that the number of skeleton sites $M(\ell)$, within a minimum path distance ℓ from the selected site, scale as $M(\ell) \sim \ell$ (for $\ell \ll L$) for 2-dimensional percolation clusters, and theoretical arguments indicate that $M(\ell) \sim \ell$ for $d \geq d_c = 6$ [507]. It appears that $M(\ell) \sim \ell$ for all d, so that the skeleton is topologically linear and the fractal dimensionality of the skeleton is equal to D_{min}.

A new cluster can be created by randomly selecting a position (site) as an origin on an "infinite" percolation cluster and extracting all of those sites that lie within a selected minimum path distance L from the origin. This cluster can be considered to consist of concentric shells lying at minimum path distances of $m = 1, 2, \ldots, L-2, L-1, L$ from the randomly selected origin. The number of sites contained in the first m shells is given by

$$s(m) \sim m^{D/D_{min}} \sim m^{D_L}, \tag{3.82}$$

and the number of sites in the mth shell is given by

$$B(m) \sim m^{D_L - 1}. \tag{3.83}$$

The exponent D_L is the "spreading dimensionality" [467], discussed earlier in this chapter (section 3.4.5.7). 2-Dimensional computer simulations, with a square lattice bond percolation model (for which the threshold of $p_c = 1/2$ is known exactly), lead to the estimate $D_L = 1.675 \pm 0.0005$ [508]. This should be compared with the value of $D_L = 1.678 \pm 0.002$ obtained from the relationship $D_\ell = D/D_{min}$, with $D = 91/48$ and $D_{min} = 1.130 \pm 0.002$. Havlin and Nossal also define the quantity D_w^L that describes the average minimum path distance ℓ traveled by a random walker in time t ($\ell \sim t^{1/D_w^L}$).

It has been conjectured that the fractal dimensionalities of the subsets of 2-dimensional percolation clusters satisfy the conformal invariance equation [509]

$$D = (100 - n^2)/48, \tag{3.84}$$

where n is an integer. All exactly known fractal dimensionalities for 2-dimensional percolation (D = 91/48, 7/4, 4/3 and 3/4) are given by equation 3.84 with n = 3, 4, 6 and 8, respectively. For the backbone, there is no exact value, but computer simulations [497] give $D = 1.62 \pm 0.02$. The nearest conformal dimensionality (n = 5) is 25/16, or about 1.56. For the minimum path dimensionality, the discrepancy between the conformal invariance conjecture and computer simulations is also much larger than the reported numerical uncertainties. The best simulation result is $D_{min} = 1.130 \pm 0.002$, and the closest conformal dimensionality (n = 7) is 17/16, or 1.0625. It seems unlikely that the uncertainties in values obtained for the backbone fractal dimensionality and the minimum path fractal dimensionality are large enough for the simulation results to be consistent with the conformal invariance conjecture.

Lattice based percolation models are convenient for numerical studies but continuum percolation models are usually more realistic. There are many continuum percolation models. In one of the most simple of these, sometimes called the Swiss-cheese model, d-dimensional hyperspherical voids are removed randomly from a d-dimensional conducting medium until a percolation threshold is reached at which the material is no longer connected. In this model, the hyperspherical voids may be monodisperse or may be selected from a distribution of sizes, providing that the distribution is not too broad. Alternatively, the percolation problem associated with the formation of an infinite spanning path through the connected hyperspheres may be studied.

Another important continuum model is the random potential model [245, 510]. In this model, the connectivity between those parts of a random surface $h(\mathbf{x})$ that lie above or below a constant height h_o is used to define a percolation threshold. Here, $h(\mathbf{x})$ is a single-valued function of $\mathbf{x}$ that fluctuates about a 2-dimensional flat surface. For $h_o < h_c$, those parts of the surface with $h(\mathbf{x}) > h_c$ percolate, while for $h_o > h_c$ those parts of the surface with $h(\mathbf{x}) < h_c$ percolate. This may be regarded in terms of a land surface that is filled to a constant level h_o with water. For $h_o < h_c$, the landscape consists of a continuous land mass filled with lakes, while for $h_o > h_c$ a continuous sea with islands appears (there are, of course, islands within the lakes and lakes within the islands ...). There is a transition, at $h = h_c$, from continuous land to continuous sea. For a surface that is statistically invariant to inversion about the mean surface height $h_c = < h >$, the percolation threshold is found at $p = p_c = 1/2$, where p is the fraction of the surface above (or below) h_c. For $d > 2$, continuous (spanning) "land masses" and "oceans" can exist simultaneously, and there are two percolation thresholds. This model has played an important role in establishing a relationship between the fractal dimensionality of contours and the Hurst exponent of the corresponding surfaces [511] (chapter 2, section 2.7.3.1).

Rosso [512] has characterized the surface of 2-dimensional continuum percolation clusters, generated by randomly depositing discs of equal size, that are allowed to overlap, onto a plane. The occupied "hull" can be defined as all of those discs that can be reached from outside the cluster by paths (followed by particles of zero size) that do not intersect any discs. In the work of Rosso, a constant disc concentration gradient was used in a strip with periodic boundary conditions in the lateral direction. The surface accessible to discs of different sizes, from the low concentration side, was studied after the "finite" clusters were removed. A transition from $D = 1.75 \pm 0.02$ to $D = 1.35 \pm 0.02$ was observed as the size of the probing particle used to find the accessible surface, relative to the size of the particles used to construct the percolation cluster, was increased. These fractal dimensionalities are in good agreement with results from lattice model simulations, which indicate that $D = 7/4$ (the hull dimensionality) and $D = 4/3$ (the external perimeter dimensionality) for probe particles that are small and large compared with the cluster particles, respectively. In addition, these simulations led to a relatively accurate estimate for the 2-dimensional continuum percolation threshold ($p_c = 0.6766 \pm 0.0005$).

Kolb [513] has shown that, for the continuum percolation model, the shape of the surface that can be reached by discs of finite radius δr without contacting the cluster can be described in terms of a crossover scaling function with the form

$$\mathscr{S}(R, \delta r) = R^{D_{hull}} f(R/R^*(\delta r)). \tag{3.85}$$

Here, $\mathscr{S}(R, \delta r)$ is the length of the surface contained in a region of size (spatial extent) R that can be found by "probe particles" with a radius of δr. Alternatively, $\mathscr{S}(R, \delta r)$ is the length of the surface of the cluster, after it has been fattened by a distance of δr that can be found by paths starting outside the region occupied by the cluster that do not intersect the fattened cluster. The characteristic crossover length R^* is given by $R^*(\delta r) \sim \delta r^{-\nu}$, where ν is the percolation correlation length exponent. The function $f(x)$ in equation 3.85 has the form $f(x) = const.$ for $x \ll 1$ and $f(x) = x^{D_{ex}-D_{hull}}$ for $x \gg 1$, where D_{ex} is the fractal dimensionality of the external perimeter, described earlier. This study, and that of Rosso [512], support the idea that there are only two fractal dimensionalities associated with the surfaces of 2-dimensional percolation clusters [216].

In the gradient percolation model, the site occupation probability varies linearly ($p_A = p_\circ + gh$, $p_B = 1 - p_A$) in the "h" direction, parallel to the gradient. This model can be used to obtain quite accurate values for the percolation threshold. For this purpose, the occupied hull or frontiers of the "infinite" "A" cluster is found. This hull contains N_A sites and is at an average position h_A in the direction of the gradient. All of the "B" sites that are adjacent to the A hull are then identified (there are N_B of them and they have an average position h_B). The percolation threshold is then given by $p_c = p_A(< h >) = p_\circ + g < h >$, where

$$< h > = \frac{N_A h_A + N_B h_B}{N_A + N_B}, \tag{3.86}$$

and this estimate for p_c can be averaged over many simulations carried out using several values for g and extrapolated to the $g \to 0$ limit. A second estimate for p_c can be obtained by reversing the roles of the A and B sites by first identifying the occupied hull of the infinite B cluster and then following the same procedure. Using this approach, Rosso *et al.* [514] obtained the estimate that $p_c = 0.592802 \pm 0.000010$. It turns out that this estimate differs from the best current value for p_c obtained using similar ideas by about 0.000055.

A percolation cluster encloses unoccupied holes of all sizes from single sites up to holes with sizes s^* of the order of the size of the entire cluster ($s^* \sim L^d$). The size distribution can be written in the form

$$N_s = N_\circ(L) s^{-\tau} f(s/s^*), \tag{3.87}$$

and the total area occupied by the enclosed holes is given by

$$\mathscr{A} = N_\circ(L) \sum_{s=1}^{s=\infty} s s^{-\tau} f(s/s^*) \approx N_\circ(L) \int_{s=0}^{s=\infty} s^{1-\tau} f(s/s^*) ds. \tag{3.88}$$

Assuming that $f(x) \approx 1$ for $x \ll 1$ and that $f(x)$ decays faster than any power of x for $x \gg 1$, it follows from equation 3.88 that $\mathscr{A} \approx N_\circ(L)(s^*)^{2-\tau}$ if $\tau < 2$.

Since small holes are present everywhere in the cluster, $N_\circ(L) \sim L^D$ and the cut-off size s^* is given by $s^* \approx L^d$, it follows that[11]

$$\mathscr{A} \sim L^D(s^*)^{2-\tau} \sim L^{D+d(2-\tau)}. \tag{3.89}$$

The total area occupied by all of the holes is given by $\mathscr{A} \sim L^d$, and this area must be the same as that given in equation 3.89. Consequently, $d = D(2-\tau)$ and the size distribution exponent is given by

$$\tau = (D+d)/d. \tag{3.90}$$

This simple argument demonstrates the validity of equation 2.49 and equation 3.71. For 2-dimensional percolation clusters, $D = 91/48$ and $\tau = 187/96$. For 2-dimensional invasion percolation clusters with trapping, $D \approx 1.82$, and it follows from equation 2.49 that $\tau \approx 1.91$, in good agreement with computer simulation results [515].

The hull of a 2-dimensional percolation cluster encloses a distribution of empty holes (clusters of empty sites). The size distribution for these holes has the form $N_s \sim s^{-\tau_h}$ with $\tau_h = (D_h + d)/d = 15/8$, from equation 2.49 or equation 3.71. The perimeter lengths ℓ_p of these holes also have a power law form $N_\ell \sim \ell^{-\tau_\ell}$. Since the holes consist of Euclidean spaces having fractal perimeters with a fractal dimensionality of D_x, the fractal dimensionality of the external perimeter, the perimeter length ℓ is related to its size s by $\ell \sim s^{D_x/d}$. Using this relationship and equation 2.84 to calculate the perimeter length distribution from the hole size distribution, it follows that

$$\tau_\ell = (d/D_x)(D_h/d) + 1 = (D_h/D_x) + 1. \tag{3.91}$$

Similarly, the distribution of the perimeter lengths of all of the holes in the entire cluster is a power law with an exponent given by $\tau_\ell = (D/D_x) + 1$ [516]. The larger value of τ_ℓ for the holes trapped by the entire cluster can be regarded as being a consequence of the much larger number of small holes trapped in the whole cluster.

If the external perimeter sites are filled, then the hull of the percolation cluster can be decomposed into loops that contact the external perimeter. Kolb and Rosso [517] investigated the distribution of the lengths ℓ of these loops. Assuming that the loop length distribution has the form

$$N_\ell \sim \ell^{-\tau'_\ell} f(\ell/\ell^*), \tag{3.92}$$

11. A more general approach is to express equation 3.88 in terms of the argument of the scaling function by making the substitution $x = s/s^*$. Equation 3.88 then becomes $\mathscr{A} = N_\circ(L)(s^*)^{2-\tau} \int_{x=0}^{x=\infty} x^{1-\tau} f(x)dx$. If the integral $\int_0^\infty x^{1-\tau} f(x)dx$ is finite, $\mathscr{A} \sim N_\circ(L)(s^*)^{2-\tau}$.

then the total length of the hull $\mathscr{L}_h$ for a cluster spanning a length L can be written as

$$\mathscr{L}_h = \sum_{\ell=\ell_{min}}^{\infty} \ell N_\ell. \tag{3.93}$$

In the asymptotic limit ($L \rightarrow \infty$), equation 3.93 can be replaced by

$$\mathscr{L}_h \sim L^{D_h} \sim N_\circ \int_0^\infty \ell \cdot \ell^{-\tau'_\ell} f(\ell/\ell^*) d\ell. \tag{3.94}$$

Since $N_\circ$ (the number of small loops) is proportional to the length of the external perimeter (L^{D_x}), and the function $f(x)$ decays faster than any power of x for $x \gg 1$, equation 3.94 indicates that, if $\tau'_\ell < 2$,

$$L^{D_h} \sim L^{D_x} \ell^{*^{2-\tau'_\ell}}. \tag{3.95}$$

The length ℓ^* in equations 3.92 and 3.95 is the length of the largest loop, and $\ell^* \sim L^{D_h}$ if the hull is a self-similar fractal. This means that equation 3.95 can be replaced by

$$L^{D_h} \sim L^{D_x} L^{D_h(2-\tau'_\ell)}, \tag{3.96}$$

so that $D_h = D_x + D_h(2 - \tau'_\ell)$ or [517]

$$\tau'_\ell = 1 + (D_x/D_h) = 37/21. \tag{3.97}$$

In the most simple lattice percolation models, it is assumed that the lattice sites (or bonds) are occupied or empty with probabilities that are independent of their environment. For many physical applications, this assumption is too simple and the site occupation densities are correlated. These correlations may be characterized by correlation functions such as

$$C_\theta(\mathbf{r}) = <\theta(\mathbf{r}_\circ)\theta(\mathbf{r}_\circ + \mathbf{r})> - p^2, \tag{3.98}$$

where $\theta(\mathbf{r})$ is 1 if the site at $\mathbf{r}$ is occupied and 0 if it is unoccupied. In equation 3.98, p is the average occupation probability ($p = <\theta>$). Harris [518] showed that the percolation exponents do not depend on short range correlations that decay faster than any power of r, at large r. If $C_\theta(\mathbf{r})$ decays faster than $r^{-2/\nu}$, with increasing r ($r = |\mathbf{r}|$), the percolation exponents are also unchanged [518]. This result is based on the idea that the fluctuations in $|p-p_c|$ in a region of size ξ (volume of ξ^d) should be smaller than $|p-p_c|$ to remain in the standard percolation universality class. For ordinary percolation, the number of occupied sites (or bonds) in a region of size ξ or volume $V \sim \xi^d$ is pV, and the fluctuations in that volume will be proportional to $V^{1/2}$. Consequently, the fluctuation in $|p - p_c|$ within the correlation volume V will be given by $\delta|p - p_c| \approx \xi^{-d/2}$ or $\delta|p - p_c| \approx |p - p_c|^{\nu d/2}$. Since $\delta|p - p_c|$ must remain smaller than $|p-p_c|$ as the percolation threshold is approached, νd must be greater than 2.

If the correlation function $C_\theta(\mathbf{r})$ in equation 3.98 has the form $C_\theta(\mathbf{r}) \sim |\mathbf{r}|^{-2\zeta}$, then the fluctuation of the average occupation probability in a region of size ξ is given by $\delta|p-p_c| \approx \xi^{-\zeta}$, and the requirement that the fluctuations in $|p-p_c|$ are smaller than $|p-p_c|$ leads to the result

$$\zeta > 1/\nu. \tag{3.99}$$

This indicates that correlations that fall off more rapidly than $|\mathbf{r}|^{-2/\nu}$ are not sufficient to change the universal percolation exponents. On the other hand, more slowly decaying correlations bring about an essential change in the scaling properties. Similar arguments can be applied to a wide range of related problems [519, 520].

3.5.1 Growth Models for Percolation

The percolation threshold p_c is known exactly for a few lattices and is known quite accurately for several others. This allows several quite simple methods to be used to generate "incipient infinite percolation clusters" for the purpose of simulating physical and chemical processes on percolating systems at the threshold. In practice, the percolation threshold is known with sufficient accuracy that the correlation length ξ will exceed the system size, for almost any realistic simulation, if a fraction p_{cs} of the lattice sites is randomly filled, where p_{cs} is the "best" value for p_c obtained from numerical studies. It is often more convenient to grow a "percolation cluster" to a particular size than to fill a lattice at random and select a large cluster of filled sites. This can be done quite easily [362, 363] by modifying version A of the Eden model, described above (section 3.4.1). Each time an unoccupied surface site is selected according to the Eden model rules, it is filled with probability p and "blocked" with probability $1-p$. Sites that have been blocked are not filled and cannot be filled at subsequent stages in the growth process. If the blocking probability is set to $1-p_c$, where p_c is the critical percolation probability, this algorithm will generate a "percolation cluster".

In many applications the required information can be obtained from the hull or the external perimeter of the percolation cluster. These "surfaces" can be generated directly, without generating the entire cluster, using random walks on a randomly blocked lattice or network. This can be done with the hull generating walk [521, 522] on a square lattice or the honeycomb lattice IGSAW model [384] described above (in section 3.3.2). An efficient algorithm for finding the hull of a percolation cluster, without generating the cluster [521, 522], is closely related to the percolation cluster growth algorithm and the "hull walk" algorithm for finding the hull of a percolation cluster, both described above. The hull generating walk follows the rules of the "hull walk" described in section 3.1. As the walk proceeds, newly encountered sites are randomly determined to be occupied (part of the percolation cluster) with probability p, or vacant (not part of the percolation cluster) with probability $1-p$. Since the starting point will not, in general, be at the largest or smallest x or y coordinate of the percolation cluster, the hull generating walk does not stop until it passes the origin *in the same direction as the first step.* This hull generating walk approach can also be used to obtain the hulls of percolation clusters in an occupation probability gradient and to obtain very accurate numerical estimates of percolation threshold values [522]. The percolation threshold can be obtained from the relationship $p_c = N_{occ}/N_{tot}$, where N_{occ} is the number of sites occupied by the walk and N_{tot} is the number of occupied sites plus the number of blocked sites generated by the walk. A second way of measuring the percolation threshold p_c is via the relationship $p_c = p(\overline{y})$, where $\overline{y}$ is the average y coordinate for the occupied sites in the linear gradient. Ziff and Sapoval [522] estimated a value of 0.592745 ± 0.000002 for p_c on a square lattice, based on simulations in which 2.6×10^{11} occupied plus blocked sites were generated. For the triangular lattice, a value of 0.652704 was obtained using 1.2×10^{10} occupied plus blocked sites, compared with the exact value of $p_c = 1 - 2\sin(\pi/18) = 0.6527036\ldots$ [523].

Keefer and Schaefer [524, 525] have developed a closely related "poisoned Eden model" for the reaction limited growth of one-, two-, three- and four-coordinate monomers on a square lattice. This model was motivated by the formation of colloidal particles from partially hydrolyzed silicon alkoxide solutions. If the fraction of high coordination number monomers is high, porous structures that are uniform on all but quite short length scales and that have self-affine surfaces are formed. These structures most probably belong to the Eden model universality class. If the fraction of high coordination number monomers is low, the surface may become completely poisoned and growth will cease. The simulation results of Keefer and Schaefer indicate that the effective fractal dimensionality increases with the average (unblocked) monomer coordination number, but also depends on the variance of the coordination number. It is not clear if these effective fractal dimensionalities represent asymptotic fractal dimensionalities or are a consequence of crossovers between a few simple universality classes such as the percolation, self-avoiding random walk and Eden growth classes.

Sintes *et al.* [526] have developed a more realistic version of this model for simulating the growth of particles from partially hydrolyzed silicon alkoxide solutions. 3-Dimensional simulations were carried out on a face-centered cubic lattice using a growth algorithm that more realistically represents the steric constraints on the polymerization process. The results obtained from this model were qualitatively similar to those obtained from the square lattice simulations of Keefer and Schaefer. The scattering structure functions $S(k)$ obtained from these simulations appear to be consistent with scattering experiments on the hydrolysis of silicon tetraethoxide/water mixtures of different compositions that

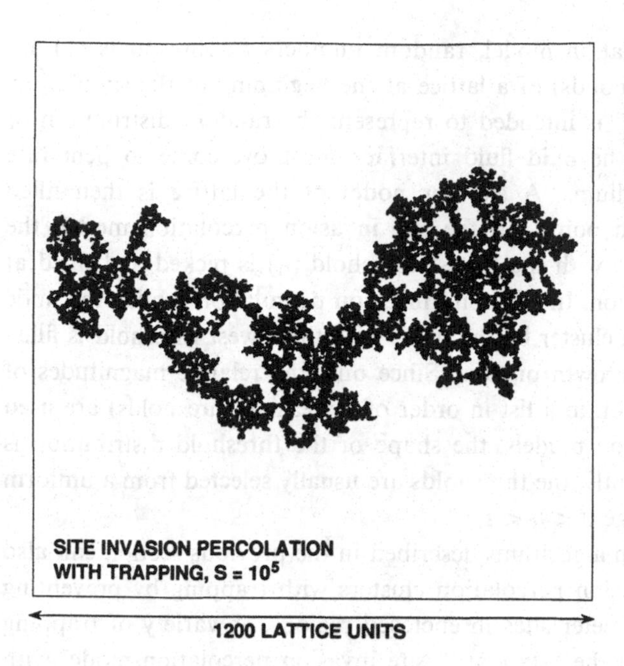

Figure 3.19 A cluster of 10^5 sites grown on a square lattice using the site invasion percolation model. The fractal dimensionality is about 1.82.

generated a mixture of $Si(C_2H_5O)_{4-n}(OH)_n$ monomers with a composition that depends on the $Si(C_2H_5O)_4$/water ratio.

3.5.2 Invasion Percolation

The invasion percolation model was originally developed to study the slow displacement of a wetting fluid by a non-wetting fluid in a porous medium [213, 214, 215], and this is still its main application. A variety of important processes, including the drying of soils, wood and ceramic powders [527], the secondary migration[12] of oil [222, 221, 528] and the redistribution of fluids in oil reservoirs, can be described quite well by the invasion percolation model, or simple extensions of this model, under appropriate limiting conditions. The model is based on the idea that, for very slow displacement rates, when the effects of fluid viscosity can be ignored, the fluid–fluid displacement process will be dominated by capillary effects and the fluid–fluid interface will move via the pore "throat" with the lowest capillary pressure, determined by its geometry and wetting characteristics. In practice, things are much more complicated, even for quite simple laboratory micromodels [529], but the invasion percolation model leads to invasion patterns that closely resemble those seen in experiments [530, 218]. A cluster of 10^5 sites, grown on a square lattice, using the site invasion percolation model with trapping, is shown in figure 3.19.

12. Migration from mature source rocks, through the intervening porous, sedimentary rocks to a trapping formation or reservoir.

In the invasion percolation model, random numbers or thresholds (t_i) are assigned to the sites (or bonds) of a lattice at the beginning of the simulation. This "quenched disorder" is intended to represent the random distribution of capillary pressures that the fluid–fluid interface must overcome to penetrate through the porous medium. A site (or node) of the lattice is then filled to represent the injection point. In the site invasion percolation model, the unoccupied perimeter site with the lowest threshold (t_i) is picked and filled at each stage of the simulation. In the bond invasion percolation model, the node connected to the growing cluster by the bond with the lowest threshold is filled during each step in the growth process. Since only the relative magnitudes of the thresholds (the position in a list in order of increasing thresholds) are used in the invasion percolation models, the shape of the threshold distribution is not important. Consequently, the thresholds are usually selected from a uniform distribution over the range $0 < t_i < 1$.

The percolation growth algorithms described in the previous section can also be used to generate invasion percolation clusters with trapping by preventing growth in all of the perimeter sites in enclosed regions. A variety of trapping rules can be used but, in the "standard" site invasion percolation model with trapping, growth is allowed only in sites that belong to the unoccupied external perimeter. Unoccupied sites that do not lie on the unoccupied external perimeter are considered to be "trapped".

Cao and Wong [531] used a square lattice percolation growth algorithm to study growth from a line. In these simulations, active and blocking sites were added randomly, with probabilities of p and $1 - p$, to the active zone. Only those unoccupied sites that were connected, via bonds between nearest-neighbor sites, to a filled, unblocked site and to the "infinite cluster" of unoccupied sites surrounding the growing cluster were considered to be part of the growth zone. In other words, the active zone consists of those sites of the unoccupied external perimeter of the cluster of active and blocked sites that are nearest neighbors to at least one occupied active site. Simulations were carried out at $p = p_c$, and other values of p, where the percolation threshold probability p_c has the same value as site percolation on a square lattice.

If trapping by both filled sites and blocked sites is included, then four surfaces can be defined: the external perimeter of the cluster of filled *and* blocked sites with a fractal dimensionality of 4/3; the hull of the cluster of filled *and* blocked sites with a fractal dimensionality indistinguishable from that of an ordinary percolation cluster hull ($D = 7/4$); the hull of the cluster of filled sites with a fractal dimensionality that is also indistinguishable from that of an ordinary percolation cluster hull ($D = 7/4$); and the external perimeter of the cluster of filled sites with a fractal dimensionality of about 1.52 [531]. The fractal dimensionalities characterize the self-similar surfaces generated with $p = p_c$. The external perimeter of the cluster of filled sites appears to be a fractal with quite different scaling properties than those of the other percolation "surfaces". The effective fractal dimensionality of the external perimeter of the filled site cluster is a non-universal exponent that depends on model details. For example, a value of about 1.67 was obtained if the surface of the growing cluster was blocked with dimers, represented by a pair of nearest-neighbor sites, instead of monomers. This non-universality seems surprising. The results were obtained from quite large scale simulations on square lattices, with sizes up to $L = 3000$ lattice units. However, the dependence of $\log N(r)$ on $\log(r)$, where $N(r)$ is the number of sites within a distance r of an occupied site, shows evidence of curvature for the external perimeter of the cluster of filled sites, and additional work may be needed before these results can be accepted without reservation.

Clusters generated using the square lattice bond invasion percolation model with trapping (growth is allowed only in sites that are connected to regions outside of the cluster via a path of bonds between unoccupied nearest-neighbor sites) appear to have a fractal dimensionality of $D \approx 1.82$, which is the same as that for square lattice site invasion percolation with trapping. For this model, computer simulations indicate that both the hull and external perimeter have a fractal dimensionality of $D \approx 4/3$. This is attributed to the greater probability of growth in unoccupied sites with two or more occupied nearest neighbors, which fills in more "corners" and blocks paths that would otherwise connect trapped regions to the exterior via next-nearest-neighbor steps between unoccupied sites [515].

The problem of invasion percolation on fractal substrates is not an entirely artificial one. Fractures play a dominant role in the transport of fluids through most types of rock, and under some circumstances the parts of the fracture network that are the most important in fluid flow may be approximated by a percolating network near to the percolation threshold. Under some conditions, the pore space may also be near a percolation threshold [532]. However, in most cases, a simple percolation model will not provide a realistic model for the distribution of fracture apertures or pore space. An infinite percolation cluster, at the percolation threshold, contains an infinite number of "singly connected" bonds or sites (bonds that will result in disconnection of the fluid entry and exit points, preventing the fluid from flowing if they are removed). The "invading fluid" must pass through all of these bonds or sites that are between the injection site(s) and the "exit" site(s). Some of these bonds or sites will have very high thresholds, and, in the absence of trapping, a finite fraction of the sites on the percolating substrate will be filled and the fractal dimensionality of the invaded region will be the same as that of the substrate ($D = 91/48$ for $d = 2$). For invasion percolation with trapping, corresponding to incompressible fluids, the invading fluid cannot enter dead ends and is confined to the backbone of the percolation cluster. Consequently, the process can be investigated by first isolating the backbone of the percolation cluster and then studying invasion percolation on the backbone.

Using computer simulations, Paredes and Octavio [533] have found that $D = 1.81 \pm 0.01$ for $p > p_c$ and $D = 1.37 \pm 0.01$ for $p = p_c$, for invasion percolation *with trapping* on a percolation cluster, where p is the occupation probability used in the generation of the substrate. The value of $D = 1.81 \pm 0.01$ for $p > p_c$ is the same as the fractal dimensionality for invasion percolation with trapping on a 2-dimensional square lattice. The simulation results are consistent with the idea that there is a crossover from $D \approx 1.37$ on short length scales ($\lambda \ll \xi$) to $D \approx 1.81$ on long length scales ($\lambda \gg \xi$), where ξ is the substrate correlation length. At the percolation threshold, a value of 1.20 was measured for the fractal dimensionality of the backbone of the invasion percolation pattern. This is reasonably close to the minimum path fractal dimensionality for percolation clusters ($D = 1.130 \pm 0.002$ [504]; see above). In the absence of trapping, a fractal dimensionality of about 1.89 was obtained for all $p \geq p_c$, in accordance with the simple argument given above.

Dias and Wilkinson [219] have included trapping effects in the standard square lattice and cubic lattice percolation models. The simulation starts with an empty lattice, and sites are selected randomly and filled as in the most simple percolation models. The new sites need not be on the perimeter of previously occupied sites. However, as the simulation progresses, only sites that belong to the "infinite" cluster of unoccupied sites are selected and filled. Growth is not allowed in the "trapped" unoccupied sites that have become isolated from the infinite cluster of unoccupied sites surrounding the growing cluster. Eventually, a cluster of filled sites spans the lattice. Unlike invasion percolation with trapping, the fractal dimensionality of clusters generated on a square lattice, using this percolation with trapping model, appears to be the same as that for ordinary percolation.

3.5.3 Diffusion Fronts and the Effect of Gradients

The blurring of an initially sharp interface due to diffusion is a common phenomenon in areas such as solid state chemistry and metallurgy.

A simple 2-dimensional model for a diffusion front can be constructed by labeling the sites in a square lattice of width L_x and length L_y "A" for those sites with $h \leq L_y/2$ and "B" for those sites with $h > L_y/2$, where h represents the "height" or y coordinate. If pairs of nearest-neighboring sites are selected at random, and their labels are interchanged, the initially smooth interface between the A and B regions will evolve into a diffusion front.[13] This model neglects all interactions between the "particles" represented by the labeled lattice sites. Alternatively, the lattice can be filled with "A" sites, and a source of "B" sites at a constant concentration ($p_B = 1$, where p_B is the fraction of "B" sites) can be

13. The algorithm can be made more efficient by maintaining a list of "interface" sites and selecting only these interface sites and their nearest neighbors for exchange.

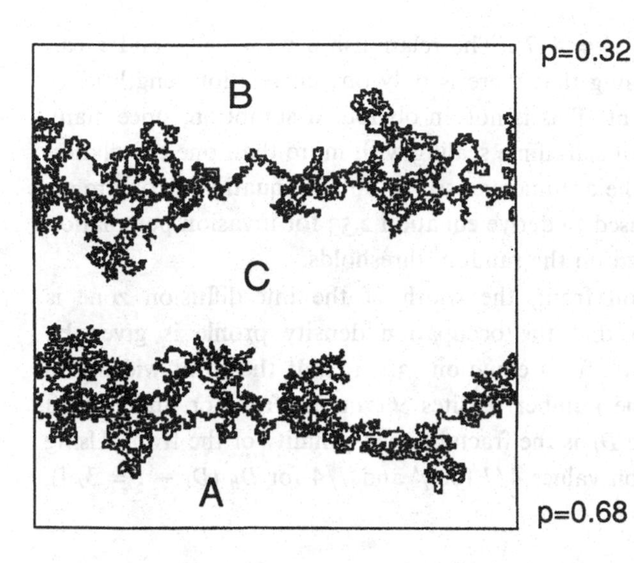

Figure 3.20 Cluster hulls obtained using a constant occupation probability gradient model to represent diffusion fronts on a square lattice. The fraction of "A" sites, p_A, increases linearly from top to bottom, while the fraction of "B" sites, $\rho_B = 1 - \rho_A$, decreases. Regions A and B contain parts of the "infinite" connected clusters of A and B sites, respectively. Region C contains "finite" A and B clusters that are not attached to regions A and B. This figure was provided by B. Sapoval.

maintained at $h = 0$. In either case, the A and B concentration profiles can be obtained by solving the classical diffusion equation. For example, in the latter case, the density profile for the B sites is given by

$$p_B(h,t) = \ erfc\{h/[2(\mathscr{D}t)^{1/2}]\} = 1 - (2/\pi^{1/2}) \int_o^{h/[2(\mathscr{D}t)^{1/2}]} e^{-z^2} dz, \quad (3.100)$$

where $\mathscr{D}$ is the diffusion constant and $erfc$ is the associated error function. As is the case for ordinary percolation, a much more interesting structure is seen if connectivity is taken into account [500, 534, 535]. A quite similar structure can be generated by randomly labeling sites on a lattice with an imposed linear gradient in the occupation densities ($p_B(h)$ and $p_A(h) = 1 - p_B(h)$). For example, figure 3.20 shows the occupied hulls ($D = 7/4$) of the largest ("infinite") A and B clusters obtained from a square lattice simulation carried out in this way. The intermediate region labeled "C" consists of interpenetrating A and B clusters that do not contact their respective main clusters.

The "interfaces" shown in figure 3.20 are fractals on short scales ($1 \ll \lambda \ll \xi$) and are flat on long length scales ($\lambda \gg \xi$). They can be described in terms of a string of fractal blobs of size ξ. The correlation length ξ is related to the occupation probability gradient ∇p by

$$\xi \sim (\nabla p)^{-\gamma'}, \quad (3.101)$$

where

$$\gamma' = \nu/(1+\nu). \quad (3.102)$$

Here, ν is the correlation length exponent from ordinary percolation theory

($\nu = 4/3$ for $d = 2$, so that $\gamma' = 4/7$). The relationship between γ' and ν can easily be obtained by assuming that there is only one correlation length associated with the diffusion front. This is not an obvious assumption, since many propagating interfaces exhibit self-affine scaling with more than one correlation length (chapters 2 and 5). The arguments used to obtain equation 3.102 are essentially the same as those used to derive equation 2.53 for invasion percolation with a gradient superimposed on the random thresholds.

In the case of a diffusion front, the width of the interdiffusion zone is proportional to $(\mathscr{D}t)^{1/2}$, so that the occupation density profile is given by $\nabla p \sim (\mathscr{D}t)^{-1/2}$, and it follows, from equation 3.101, that the front width ξ is proportional to $(\mathscr{D}t)^{\gamma'/2}$. The number of sites $\mathscr{S}_h$ in the front or hull is then proportional to ξ^{D_h-1}, where D_h is the fractal dimensionality of the front. Using the 2-dimensional percolation values 4/7 for γ' and 7/4 for D_h ($D_h - 1 = 3/4$), the result

$$\mathscr{S}_h \sim L_x(\mathscr{D}t)^{3/14} \tag{3.103}$$

is obtained.

The situation is a little more complex in three dimensions [536]. For 3-dimensional site percolation on a cubic lattice, the percolation threshold is found at $p_c = 0.311605 \pm 0.000010$ [489], so that both the A and B sites percolate for $p_c \leq p_A(h) \leq 1 - p_c$. In this range, the A and B clusters interpenetrate forming an entangled network (figure 3.21). Consequently, the hulls of both the A and B clusters extend through the entire interpenetration region in which $p_c \leq p_A(h) \leq 1 - p_c$ and $p_c \leq p_B(h) \leq 1 - p_c$. Those parts of these hulls that extend beyond the region where $p_c \leq p_A(h), p_B(h) \leq 1 - p_c$ (to occupation densities smaller than p_c) correspond to the 2-dimensional gradient percolation fronts and will be referred to as the external frontier (of the corresponding infinite cluster), in the 3-dimensional case. The relationship between the external frontier width ξ and the gradient ∇p is still given by equations 3.101 and 3.102. Because trapping is short ranged in three dimensions, the external frontiers of the two connected "infinite" clusters (and the adjacent parts of the "internal" frontiers, within a distance of about ξ from the plane where $p = p_c$) are expected to have a fractal blob structure with fractal dimensionalities of $D \approx 2.525$ (that of a 3-dimensional percolation cluster) on length scales up to the width of the fronts. This has been confirmed by computer simulations [537].

Another quantity measured by Gouyet *et al.* [537] is the density of hull or "frontier" sites ρ_h in the plane in which $p = p_c$, where p is the density of *all* of the occupied sites. This is a quantity that can easily be calculated using the fractal blob approach. The density ρ of sites in the intersection of a plane with a D dimensional self-similar fractal blob of size ξ in 3-dimensional space

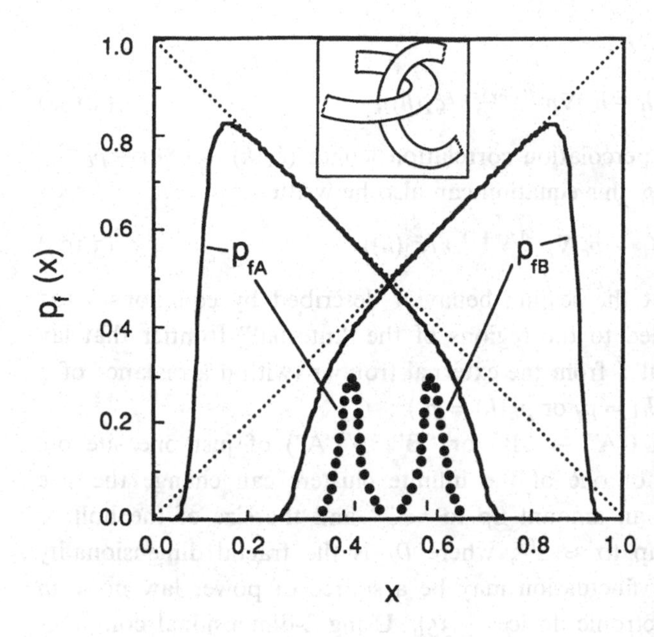

Figure 3.21 Distribution of cluster hull sites obtained using a constant occupation probability gradient model to represent diffusion fronts on a cubic lattice. Here, p_{fA} and p_{fB} are the frontier (hull) densities for the "infinite" connected clusters of A and B sites, respectively. For comparison, the dots show p_{fA} and p_{fB} for the corresponding square lattice model. The frontiers overlap over a considerable range of compositions for the 3-dimensional case, but do not overlap in two dimensions. This figure was provided by B. Sapoval.

is given by $\rho \sim \xi^{D-1}/\xi^2$, or, more generally, $\rho \sim \xi^{D-d}$ (equation 2.5). Since $\xi \sim \nabla p^{-\nu/(1+\nu)}$, it follows that the density of hull sites is given by

$$\rho_h \sim \nabla p^{(d-D_h)\nu/(1+\nu)}, \tag{3.104}$$

where D_h is the fractal dimensionality of the hull. For the $d = 3$ case, where $D_h = D \approx 2.525$ and $\nu \approx 0.88$, $\rho_h \sim \nabla p^{\approx 0.23}$, which is in good agreement with the numerical results of Gouyet *et al.* [537] ($\rho_h \sim \nabla p^{\approx 0.23 \pm 0.01}$). It is reasonable to suppose that the density profile for the frontier ($d = 2$) or external frontier ($d = 3$) can be represented by the scaling form

$$\rho_h(h) \sim \nabla p^{(d-D_h)\nu/(1+\nu)} f((h-h_c)/\xi). \tag{3.105}$$

A similar scaling form

$$\rho'(h) \sim \nabla p^{(d-D)\nu/(1+\nu)} f'((h-h_c)/\xi) \tag{3.106}$$

can be used for the density of sites in the infinite cluster, where h_c is the height at which $p = p_c$. Gouyet *et al.* [537] proposed an equivalent scaling form and obtained a good data collapse for $\rho'(h)$ using results from 3-dimensional simulations, but the data collapse for $\rho_h(h)$ was not as good. The relatively poor data collapse was attributed to larger corrections to scaling. There is no reason to believe that equation 3.105 fails in the asymptotic, large system size limit. Gouyet *et al.* also demonstrated that the density–density correlation function for a planar cut at a height h is given by

$$C(r,h) = r^{D-d} f((h-h_c)/\xi, r/\xi_2(h)) \tag{3.107}$$

or

$$C(r,h) = r^{D-d} f'((h-h_c)\nabla p^{\nu/(1+\nu)}, r/\xi_2(h)), \tag{3.108}$$

where ξ_2 is the ordinary percolation correlation length ($\xi_2(h) \sim |p(h) - p_c|^{-\nu}$). Since $h - h_c \sim (p - p_c)/\nabla p$, this equation can also be written as

$$C(r,h) = r^{D-d} f''((p-p_c)\nabla p^{-1/(\nu+1)}, r/\xi_2(h)). \tag{3.109}$$

Gouyet *et al.* found that the scaling behavior described by equations 3.105 to 3.109 could be extended to the regions of the "internal" frontier that lay within a distance of about ξ from the external frontier (within a distance of ξ from the plane where $p_A(h) = p_c$ or $p_B(h) = p_c$).

A change in the label ("A" → "B" or "B" → "A") of just one site on, or adjacent to, the hull of one of the infinite clusters can change the size of the infinite cluster by an amount up to $\approx \xi^D$ and the size of the hull or frontier by an amount up to $\approx \xi^{D_h}$, where D_h is the fractal dimensionality of the hull. This type of fluctuation may be a source of power law noise in some inhomogeneous electronic devices [535]. Using 2-dimensional computer simulations, Sapoval *et al.* [535] found that in a time interval δt the size of the front changed by an amount proportional to $\delta t^{\approx 0.4}$ ($< \delta \mathcal{S}_h(t)^2 >^{1/2} \sim t^{\approx 0.4}$, where $\delta \mathcal{S}_h(t)$ is the change in the number of sites in the front or hull during a time interval δt). This can be interpreted in terms of a "signal" with a Hurst exponent of $H \approx 0.4$, and the corresponding power spectrum $P(\omega)$ would scale as $P(\omega) \sim \omega^{-(1+2H)} \sim \omega^{\approx -1.8}$. The measured exponent of 0.4 is not much different from the asymptotic value of $H = 0.5$ that would be expected for a distribution of uncorrelated events with finite moments ($\delta \mathcal{S}_h \leq c\xi^{D_h}$, where c is a constant of order 1). A subsequent scaling theory analysis [538] showed that

$$< \delta \mathcal{S}_h(t)^2 > \sim A_1 t \tag{3.110}$$

on short time scales ($t \ll t_c$) and

$$< \delta \mathcal{S}_h(t)^2 > \sim A_2 \tag{3.111}$$

for $t \gg t_c$. The amplitude A_2 in equation 3.111 is simply the product of the number of blobs of size ξ and the square of the blob size ($A_2 \approx \xi^{2D_h}(L/\xi)^{(d-1)}$ or $A_2 \approx L^{(d-1)}\nabla p^{-[(\nu/(\nu+1))(2D_h-d+1)]}$). Estimation of the amplitude A_1 requires a more intricate scaling argument, and it is given by $A_1 \approx L^{(d-1)}\nabla p^{-[1+(\nu/(\nu+1))(2D_h-d)]}$. It follows from these values for A_1 and A_2, together with equations 3.110 and 3.111, that the crossover time t_c scales as $t_c \sim \nabla p^{1/(1+\nu)}$. These results were confirmed by 2-dimensional computer simulations [538].

Kolb *et al.* [539] have studied the effects of attractive and repulsive interactions on diffusion fronts using computer simulations. In their 2-dimensional simulations, diffusion into an initially empty region was studied with a fixed concentration (fraction of filled sites) of $\rho = 1$, along one edge of the lattice ($x = 0$).

Multiple-site occupancy was forbidden, and only short range (nearest-neighbor) interactions were included. A standard "heat bath" Monte Carlo algorithm (chapter 1, section 1.9) was used with nearest-neighbor site label exchanges. For weak attractive or repulsive interactions, the concentration profiles $\rho(x,t)$ could be represented by the scaling form

$$\rho(x,t) = f_E(x/X(t)), \tag{3.112}$$

where $X(t)$ is the mean position in the x direction at time t and $X(t) \sim t^{1/2}$. The shape of the scaling function $f_E(y)$ was found to depend on the reduced interaction energy E, where $E = E_a/k_BT$ and E_a is the interaction energy between each adjacent pair of sites, labeled to represent the diffusing species. There appeared to be a transition between strong and weak attractive interaction behaviors. For strong attractive interactions, the density profile took on a quite different, non-scaling, behavior. For those parts of the profile with intermediate densities, the shape of the profile, expressed as $\rho(x - x_{1/2})$, where $x_{1/2}$ is the distance from the edge of the lattice to the position at which $<\rho(x)> = 1/2$, evolved towards a time independent form. However, the profile continued to spread for large and small values of $|x - x_{1/2}|$. The transition appeared to take place at $E = E_c = -(1 + 2^{1/2})$, which is the critical point interaction energy for a corresponding lattice gas on a square lattice. For $|E| > |E_c|$, the spreading process slowed down, and the simulations suggested that $< x > \sim t^{1/z}$ with $z > 2$. However, the approach to asymptotic behavior was very slow for $|E| \geq |E_c|$, and the asymptotic behavior was not reached in the simulations. This suggests that there might be an asymptotic scaling regime for the concentration profiles that was inaccessible in the simulations. The behavior with large repulsive interactions appeared to be more complex and is not as well understood. Kolb *et al.* used the techniques described earlier in this section to measure the percolation threshold as a function of the reduced interaction energy.

3.5.4 Directed Percolation

In ordinary, 2-dimensional percolation, the sites of a lattice of size L ($L \rightarrow \infty$) are filled randomly until a path consisting of steps between neighboring filled sites (sites on an occupied cluster) span the lattice. If the process is defined on a square lattice, then a cluster is considered to span the lattice when any path on the cluster consisting of steps between nearest-neighbor occupied sites in any of the four possible directions connects opposite edges of the lattice. In $1 + 1$-dimensional directed percolation, only paths consisting of steps in the positive x and positive y directions are considered. Clearly, this will increase the percolation threshold, and there will be two correlation lengths $\xi_\perp$ and $\xi_\parallel$ that are related to the fraction of filled sites (p) by

$$\xi_\perp \sim |p - p_c|^{-\nu_\perp} \tag{3.113}$$

and

$$\xi_\parallel \sim |p - p_c|^{-\nu_\parallel}, \tag{3.114}$$

instead of the single correlation length found in ordinary percolation. The correlation lengths $\xi_\perp$ and $\xi_\parallel$ are measured perpendicular and parallel to the "preferred" direction (parallel and perpendicular to the (11) direction). The exponents $\nu_\perp$ and $\nu_\parallel$ have values of 1.097 ± 0.001 and 1.733 ± 0.001, respectively [540]. (At the time of writing, the best values for these exponents, obtained from a series expansion analysis, were $\nu_\perp = 1.09684 \pm 0.00002$ and $\nu_\parallel = 1.73383 \pm 0.00003$ [541].) The spanning path on the infinite (directed) percolation cluster is then a self-affine fractal with a Hurst exponent given by $H = \nu_\perp/\nu_\parallel$, or $H \approx 0.63$. A variety of models, more or less closely related to directed percolation [542], has been invoked to explain the "anomalously" high roughness exponents found in some interface propagation experiments (chapter 5).

The work of Seiden and Schulman [14, 15] on galactic structure is an interesting application of directed percolation. It helps to illustrate the broad range of applications that simple growth models have found. Very similar models have been used in areas such as biology and ecology. In this $2+1$-dimensional model, the two lateral coordinates are used to represent the distribution of stars, and the additional coordinate, along which the process is directed, is time. The idea behind the model is that stars are created in regions of relatively high gas density (molecular clouds with densities in the range $10^2 - 10^4$ atoms cm^{-3}). The clouds have a typical size of a few hundred parsec (1 parsec $\approx$ 3.2 light years), comparable with the thickness of a typical galaxy. A typical disc galaxy has a much larger diameter, ranging from about 10^3 to 10^6 parsecs. Near the center of the galaxy, there is a bulge with little or no star formation. The stars in a molecular cloud are formed in clusters and have a quite wide range of sizes (masses of 0.1 $M_\odot$ to 50 $M_\odot$, where $M_\odot$ is the mass of our own sun). The larger of these stars are relatively short lived ($\approx 10^7$ years) and provide enormous amounts of energy during their short lives. They play an important role in the formation of molecular clouds via their stellar winds, ionization shock fronts and supernova shock waves. These processes disrupt the local molecular cloud and help to assemble new molecular clouds in neighboring regions. Without these shock waves to assemble molecular clouds, star formation would not take place, or would take place only very slowly.

In the model of Seiden and Schulman [14, 15], the galaxy is represented by a 2-dimensional array of cells or sites, in the form a disc consisting of concentric annuli, as shown in figure 3.22. The sites in this array represent regions with a

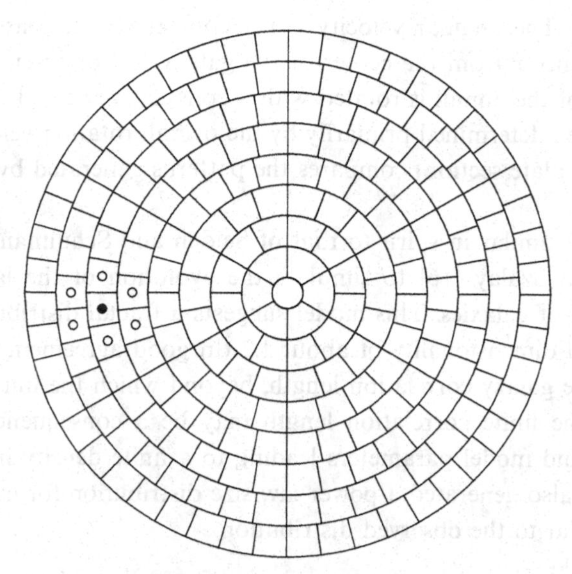

Figure 3.22 The 2-dimensional array of equal area cells used in the galaxy model of Seiden and Schulman. The filled circle indicates an active region of star formation. In the next step in the simulation, it may induce star formation in the contiguous cells (cells that share a common boundary) labeled with open circles. At later stages, the neighboring cells in adjacent rings will change, because of the differential angular velocities. This figure was provided by P. E. Seiden.

size of a few hundred parsecs. Full sites correspond to regions in which stars are being created and empty sites represent "inactive" regions. The time scale is also discretized in units corresponding to $\approx 10^7$ years, so that the evolution of the galaxy is represented by a series of 2-dimensional concentric annular arrays corresponding to the times $t_n, t_{n+1}, \ldots$. The evolution of the system in this model is determined by the local rules relating the configuration of the disc-like array at time t_{n+1} to that at time t_n. Sites in the t_{n+1} array are filled with probability p if they are adjacent to a filled site in the t_n array. Adjacent sites are sites that share part of a common boundary.

To represent the star-formation process more accurately, the basic directed percolation model must be modified. After a region has become active, the gas in that region will have become heated and depleted. There will be only a small probability of that region becoming active again, for a period of about 10^8 years or about ten time steps in the model. This phenomenon is taken into account by reducing the reactivation probability in a previously active cell by an amount $1 - e^{-t/\tau} = 1 - e^{-m/\tau}$, where τ is the decay time for conversion of "inactive gas" into "active gas" and m is the number of time steps since the site was last active. The probability for activation of a site that is adjacent to an active site is then given by $p = p_\circ(1 - e^{-m/\tau})$, where τ is the decay time, in units of the model time scale. The angular velocity is *not* constant, it decreases as $1/r$ with increasing distance r from the center of the galaxy. To represent this *non-rigid* rotation, each of the annuli is rotated with a constant velocity V. The shape of the galaxy is then determined primarily by the overall rotation velocity V. Figure 3.23 (see color plate section) compares the patterns generated by this model with real galaxies.

A model which is very similar in spirit to that of Seiden and Schulman was developed by Vicsek and Szalay [16] to simulate the evolution of the larger scale, fractal distribution of galaxies. This model suggests a fractal distribution of galaxies with a fractal dimensionality of about 1.2 (in good agreement with observations) and a finite galaxy correlation length, beyond which the universe is uniform. However, the finite correlation length may be a consequence of the use of a finite grid and model parameters leading to a finite density in the simulations. This model also generated a power law size distribution for galaxy clusters that is very similar to the observed distribution.

3.5.5 The Screened Growth Model

The problems encountered in developing a theory for DLA arise from the difficulty of solving the Laplace equation with moving, fractal boundary conditions. The non-local screening of the Laplacian field in DLA suggests that it should be possible to generate fractal structures, and perhaps contribute to the understanding of DLA, by studying other growth models that incorporate non-local screening in a more simple manner. The development of the screened growth model [408] was motivated by these ideas. Like lattice DLA and Eden models, the screened growth model is a surface growth model in which the growth process is represented by the successive filling of unoccupied perimeter sites. The screening effect at the ith potential growth site is represented in terms of multiplicative contributions from each of the occupied sites, so that the probability that the ith site will be filled during the next step, in a cluster that already has s filled sites, is given by $P_i \propto \Pi_{j=1}^{j=s} \sigma_{i,j}$, where $\sigma_{i,j}$, the screening "contribution" of filled site j to the growth rate or probability at unoccupied perimeter site i, is given by $\sigma_{i,j} = exp(-A/R_{ij}^{\varepsilon})$. Here, R_{ij} is the distance between growth site i and occupied site j, so that

$$P_i = \Pi_{j=1}^{j=s} exp[-A/(R_{ij})^{\varepsilon}] / \sum_{k=1}^{k=N} \Pi_{j=1}^{j=s} exp[-A/(R_{kj})^{\varepsilon}]. \quad (3.115)$$

In equation 3.115, N is the number of unoccupied perimeter sites and A is a constant.

The screened growth model is a simplified version of a finite screening range cluster growth model developed by Rikvold [543] in which the screening contributions are given by $\sigma_{i,j} = exp[(-A/(R_{ij})^{\epsilon})exp(-R_{ij}/\xi)]$. The parameter ξ is

Figure 1.10 Patterned ground in Svalbard, Norwegian Arctic. Nearly circular areas of fine-grained soil, 2–3 m across, are surrounded by gravel ridges about 0.2m high. These patterns are created by seasonal freeze/thaw cycles in which the fine-grained soils move in a convection-like manner. More than 4×10^4 years ago this area was a wave cut platform covered by a 1–2 m thick layer of coarse, mixed beach sediments. This figure was provided by B. Hallet.

These plates are available for download in color from www.cambridge.org/9780521452533

facing page 242

Figure 3.7 This color figure illustrates the way in which a 2-dimensional DLA cluster was constructed by the addition of successive particles. The continuous color scale ranging from white through blue to red and finally to green indicates the order in which the particles were added to the cluster.

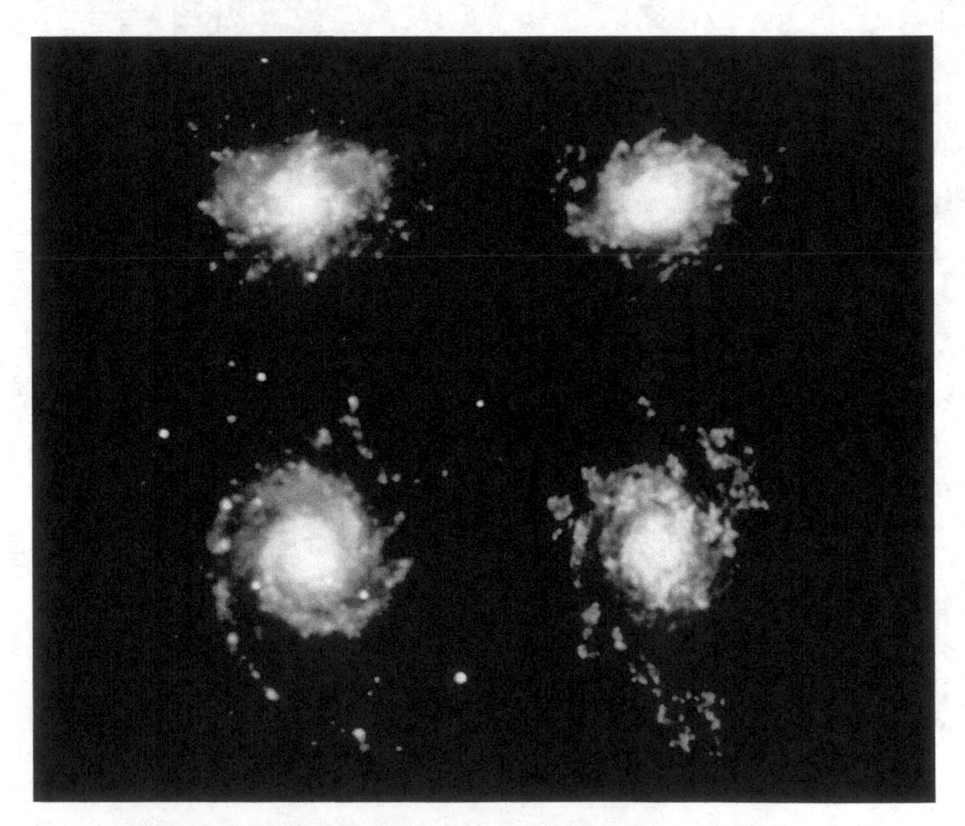

Figure 3.23 The left-hand part of this figure shows digitized images of galaxies NGC-7793 (top) and NGC-628 (bottom), while the right-hand side shows patterns generated by the directed percolation galaxy model of Seiden and Schulman. This figure was provided by P. E. Seiden.

These plates are available for download in color from www.cambridge.org/9780521452533

Figure 4.12 Branched patterns formed by the invasion of relatively warm water through holes in "black ice" into a layer of overlying slush. After this event, the slush froze to form "white ice", and the melted regions formed black ice dendrites within the white ice layer. This figure was provided by C. A. Knight.

These plates are available for download in color from www.cambridge.org/9780521452533

Figure 5.7 Rough surfaces generated by the 2 + 1-dimensional single-step, solid-on-solid model. Part (a) shows the surface after a mean height $< h >$ of 5000 lattice units has been reached, and part (b) shows the surface at $< h > = 10\,000$, in the same simulation. The simulations were carried out with $L_x = L_y = 1024$, but the surfaces were reduced to 512×512 points for display purposes. This figure was provided by T. Johnsen.

These plates are available for download in color from www.cambridge.org/9780521452533

a correlation length, so that, for length scales greater than ξ, the cluster has a uniform density. The asymptotic scaling properties are the same as those of Eden model clusters and the Rikvold model belongs to the same universality class as the Eden model [408]. At each stage in the growth process, a perimeter site is selected at random, with probabilities given by equation 3.115, and filled. After a new site has been occupied, the new unoccupied perimeter sites must be found and the growth probabilities for all of the unoccupied perimeter sites must be updated (for the "old" sites) or calculated (for the "new" sites). Because of this, it is not possible to grow very large clusters with this model. The computer time required grows as s^2, where s is the number of occupied sites in the cluster. However, an advantage of this model is that all of the growth probabilities are accurately known at all stages in the growth process. For this reason, the screened growth model has been valuable in developing concepts related to growth probability distributions ([408, 544, 545], appendix B, section B.3).

The clusters generated by this model are branched, random fractals, much like DLA clusters. Typical examples are shown in figure 3.24. Computer simulations indicate that the fractal dimensionality is equal to the screening exponent ε, for clusters grown on 2-dimensional and 3-dimensional lattices. This simple relationship between D and ε is supported by theoretical arguments [408, 546, 547], which indicate that $D = \varepsilon$ for all $d \geq 2$, if $1 \leq \varepsilon \leq d$. The parameter A, in equation 3.115, acts much like the sticking probability in DLA. A small value for A will lead to clusters that are compact on short length scales $\ell \ll \xi(A)$ and fractal on long length scales $\ell \gg \xi(A)$, where $\ell(A)$ is a correlation length that depends on the parameter A in equation 3.115. Large A values lead to clusters that have a more "stringy" local structure.

3.5.6 Faceted Growth Models

Simple models for the growth of rough surfaces frequently result in mean surface growth velocities that depend on the orientation of the coarse-grained "reference" surface with respect to the underlying lattice. Dahr [548] showed theoretically that the growth velocity along the axes exceeds the velocity in the "diagonal" directions for the $d \geq 54$ Eden models. Computer simulations show that growth velocity anisotropy and the resulting shape anisotropy remains for dimensionalities as low as 2. For $d = 2$, the growth velocity anisotropy is small for Eden model clusters [549, 550, 551] and Ising model clusters in the Eden regime [552], but it is sufficient to exacerbate the problem of measuring the roughness exponent and the dependence of the width of the roughness on the growth time or cluster size, using cluster growth algorithms. This is one of the reasons why strip geometry simulations, rather than cluster growth simulations, are almost invariably used to measure surface scaling properties.

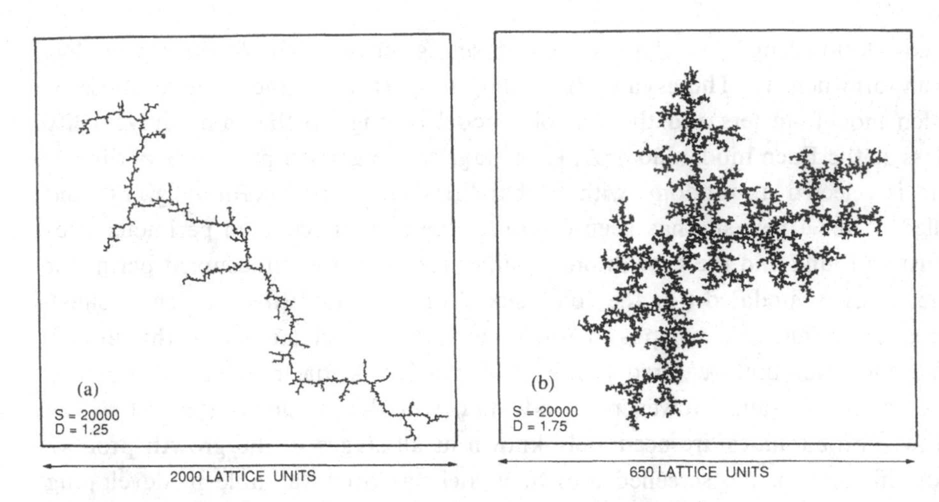

Figure 3.24 2-Dimensional, 20 000-site screened growth model clusters grown on a square lattice, with $A = 1.0$. Parts (a) and (b) show clusters generated with values of 1.25 and 1.75, respectively, for $\varepsilon = D$.

In the standard Eden model, described earlier in this chapter, one of the unoccupied perimeter sites is picked randomly and filled. If, instead, all of the perimeter sites are filled at each stage in the growth process, a regular, square faceted shape with faces in the (11) and the three other symmetrically equivalent directions will appear. Richardson [553] found that, as the fraction p of external perimeter sites filled at each growth stage was increased, a transition took place from an essentially round form with a rough surface to the square shape with smooth surfaces.[14] The simulations reported by Richardson were carried out on a quite small scale. It is difficult to tell whether or not the transition is continuous, but there is some indication of the existence of a faceted shape for $p > p_c$, where p_c is the transition threshold. More recent theoretical work [554, 555] confirmed the existence of a faceting transition at the directed percolation threshold $p_c \approx 0.705489$. A similar transition, from an almost round shape to a faceted shape, is found as the noise reduction parameter m, defined above, is increased in the Eden model [556, 295]. This is illustrated in figure 3.25. In this case, the transition is continuous, and intermediate shapes with curved faces and rounded corners are generated if small values of the noise reduction parameter are used. As figure 3.25 shows, the intermediate shapes soon reach an asymptotic form, as the cluster size is increased. Batchelor and Henry [557] have observed a similar continuous transition for the noise reduced

14. In the original Richardson model, the bonds joining the external perimeter sites to the occupied sites were found and turned "on" with probability p and off with probability $1 - p$. The external perimeter sites connected to occupied sites in the cluster by "on" bonds were then filled.

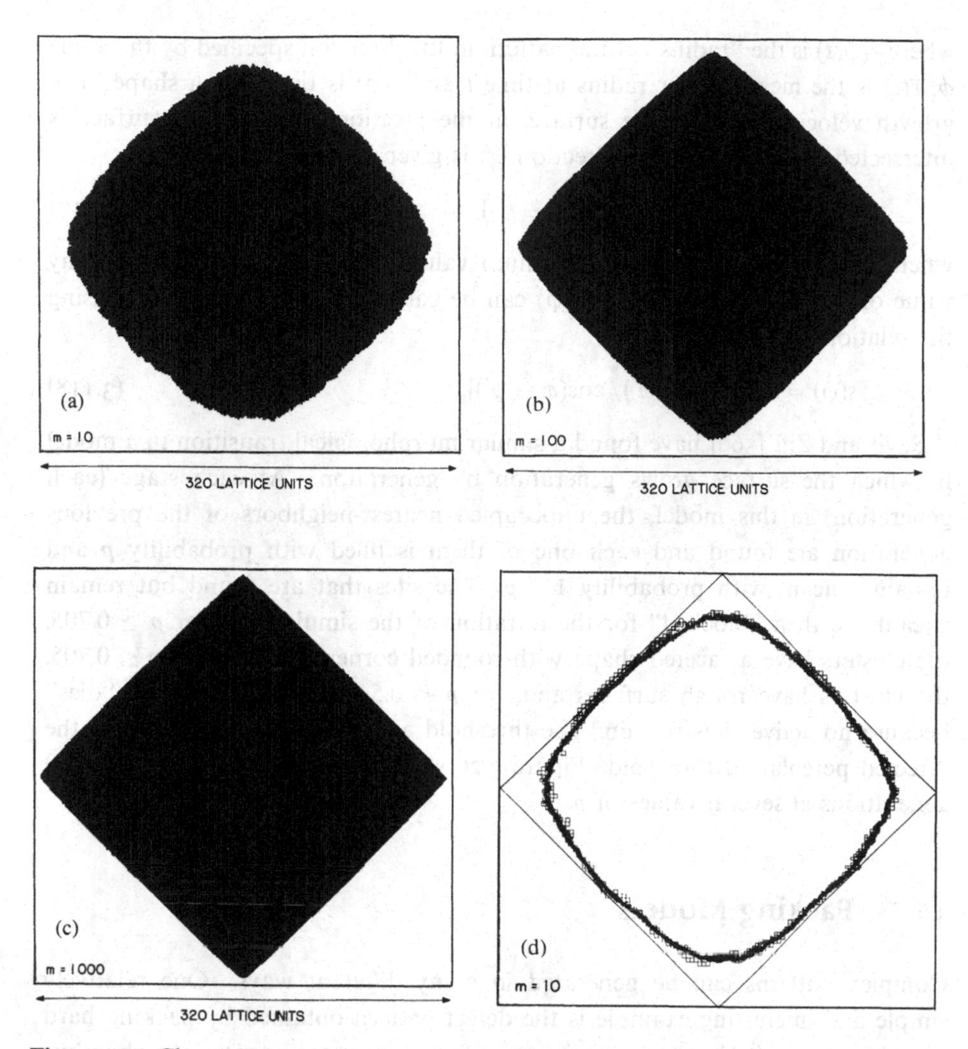

Figure 3.25 Clusters generated using a noise reduced, square lattice Eden model. Parts (a), (b) and (c) show clusters grown with noise reduction parameters of $m = 10$, $m = 100$ and $m = 1000$, respectively. The superposition of the unoccupied perimeters, or active zones, for a cluster grown with a noise reduction parameter of $m = 10$, after reaching sizes of $s = 5 \times 10^3, 10^4, 2 \times 10^4, 4 \times 10^4, 8 \times 10^4, 1.6 \times 10^5, 3.2 \times 10^5$ and 5×10^5, shown in part (d), demonstrates that a limiting shape is reached at a quite early stage during the growth process.

cubic lattice Eden model, from an almost spherical shape, for $m = 1$, to a bipyramid for $m \to \infty$.

The direction dependence of the growth velocity can be obtained from the shapes of compact clusters with convex perimeters using the Wulff [558] construction [559]. The convex perimeter shape can be represented by

$$r(\phi, t) = \overline{r}(t)s(\phi), \tag{3.116}$$

where $r(\phi, t)$ is the "radius" of the pattern in the direction specified by the angle ϕ, $\bar{r}(t)$ is the mean cluster radius at time t and $s(\phi)$ is the pattern shape. The growth velocity $V(\phi')$ of the surface, at the position in which the surface is intersected by a vector in the direction ϕ', is given by

$$V(\phi') = max\{\phi\}[s(\phi)\cos(\phi - \phi')], \tag{3.117}$$

where $max\{\phi\}f(\phi)$ means the maximum value of the function $f(\phi)$, for any value of ϕ. Similarly, the shape $s(\phi)$ can be calculated from the velocity using the relationship

$$s(\phi) = min\{\phi'\}[V(\phi')/\cos(\phi - \phi')]. \tag{3.118}$$

Savit and Ziff [560] have found a similar morphological transition in a model in which the surface grows generation by generation. At each stage (each generation) in this model, the unoccupied nearest-neighbors of the previous generation are found and each one of them is filled with probability p and remains vacant with probability $1 - p$. The sites that are found but remain vacant are then "blocked" for the duration of the simulation. For $p \geq 0.705$, the clusters have a faceted shape with rounded corners. For $0.54 \leq p \leq 0.705$, the clusters have rough surfaces and, for $p < 0.54$, the cluster always "dies" because no active sites remain. The threshold at $p \approx 0.705$ corresponds to the directed percolation threshold. Figure 3.26 shows clusters generated after 250 generations at several values of p.

3.6 Packing Models

Complex patterns can be generated in many different ways. One relatively simple and interesting example is the defect pattern obtained by packing hard discs about an initial pattern consisting of four discs at the corners of a rhombus [561]. In the model of Onoda and Toner, the packing is constructed by adding new discs in positions that contact two discs, already in the packing. Each new disc is added in the "contacting" position closest to the center of the rhombus, with "ties" being broken by random selection. In most cases, the center of the newly added disc and the two discs that it contacts form an equilateral triangle that becomes part of a close-packed network. However, in some cases, the new disc and its two contacts form two sides of a rhombus, and it is these rhombuses that form a fractal network in the otherwise regular close-packed structure. Figure 3.27 shows typical defect networks generated by this model. Onoda and Toner [561] were able to calculate exactly the fractal dimensionality of the defect network.

Figure 3.28 compares the fractal dimensionality obtained from an analysis of the defect network generated in simulations of packing around different

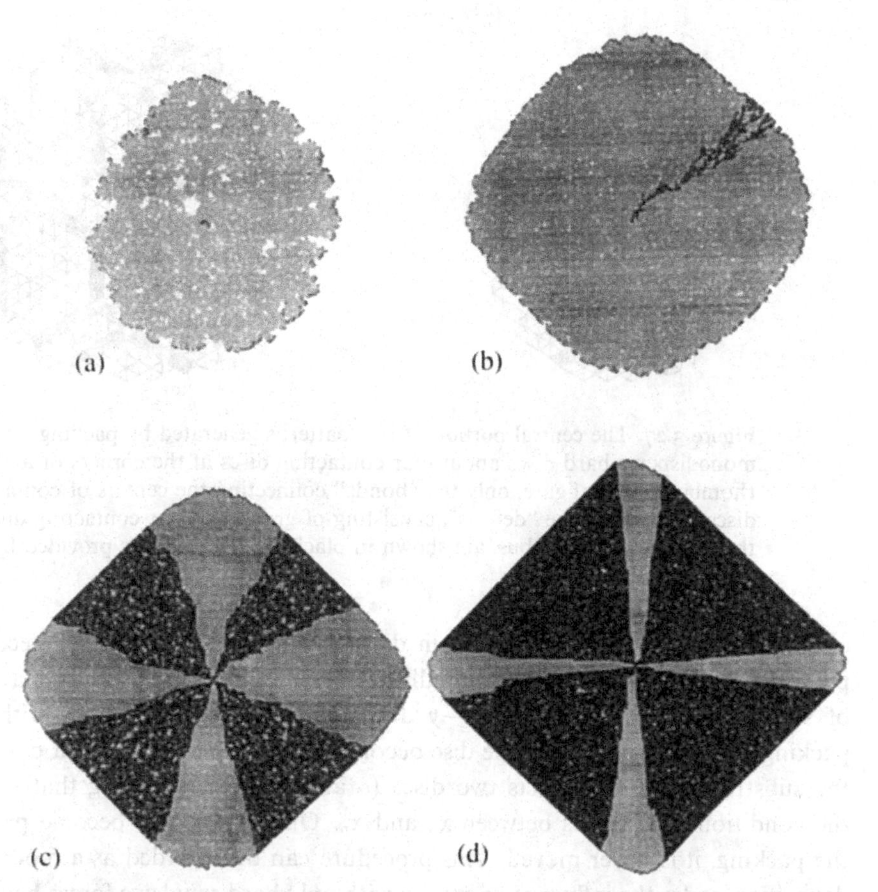

Figure 3.26 Clusters generated using the faceted growth model of Savit and Ziff. The clusters are shown after 250 generations. The "active" unoccupied perimeter sites and the sites that are filled, at the facets, along directed paths from the origin are shown in black. These sites are directed percolation clusters and they are self-affine fractals at the directed percolation threshold, $p = p_c \approx 0.705$. The other occupied sites are shown in gray. Parts (a), (b), (c) and (d) show clusters with $p = 0.6$, 0.7, 0.8 and 0.9, respectively. This figure was provided by R. M. Ziff.

rhombus shapes with the theoretical predictions of Onoda and Toner. Unlike most other models, the fractal dimensionality varies continuously with a model parameter, the shape of the initial rhombus.

A similar defect network pattern is formed if monodisperse discs are deposited, one at a time, onto a horizontal substrate. In this model, the discs are deposited at randomly selected lateral (x) coordinates, x_n for the nth particle. The initial position of a deposited particle is $(x_n, y_\circ)$, where $y_\circ$ is the highest y coordinate at which the disc contacts either the horizontal substrate at $y = 0$ or a disc that is already part of the packing. If the disc contacts the substrate, where the

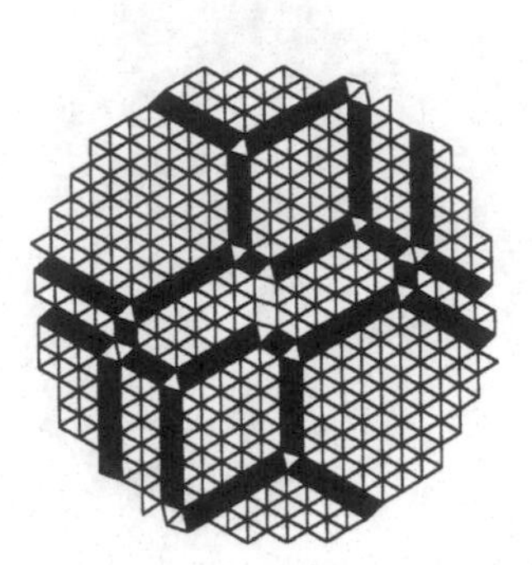
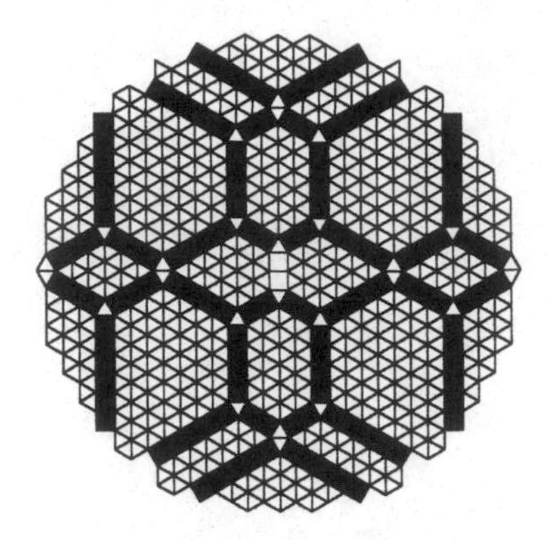

Figure 3.27 The central portion of two patterns generated by packing monodisperse hard discs about four contacting discs at the corners of a rhombus. In this figure, only the "bonds" connecting the centers of contacting discs are shown. The "defects", consisting of groups of four contacting discs at the corners of a rhombus, are shown in black. This figure was provided by J. Toner.

coordinate of its center is $(x_n, 1/2)$ in disc diameters, it immediately becomes part of the packing. If the disc initially contacts the deposit, it follows a path of steepest descent, towards the $-y$ direction, maintaining contact with the packing. This path stops and the disc becomes part of the packing if it contacts the substrate or if it contacts two discs (o and p) in the packing that satisfy the condition that x_n lies between x_o and x_p. Once a disc has become part of the packing, it is never moved. This procedure can be regarded as a model for deposition under the influence of gravity with only hard repulsive forces between a "moving" particle and the deposit and strong "sticking" between non-moving particles and the deposit. In such a packing, each deposited particle contacts two previously deposited particles in the substrate, so that the average coordination number is exactly 4. Most of the discs in the deposit have a coordination number of 4; those that do not can be considered to be topological defects in a packing that is otherwise composed of four-coordinate discs. Figure 3.29 shows a small part of a large packing generated by this model near to the substrate. Figure 3.30 shows a much larger region in which only the defects are identified.

Large scale computer simulations [562] indicate that the decay of the defect density $\rho(y)$, with increasing height, can be described by the power law

$$\rho(y) \sim y^{-\alpha} \tag{3.119}$$

at large heights. The exponent α has a value of about 1.92. The simulations are not accurate enough to rule out a value of 2.0 for α, but $\alpha = 1.92$ would correspond to a fractal dimensionality of 0.08. In the asymptotic $y \to \infty$ and $\rho(y) \to 0$ limit, there are no defects, and the network of "bonds" between contacting discs forms a tiling of the plane with rhombuses of various shapes.

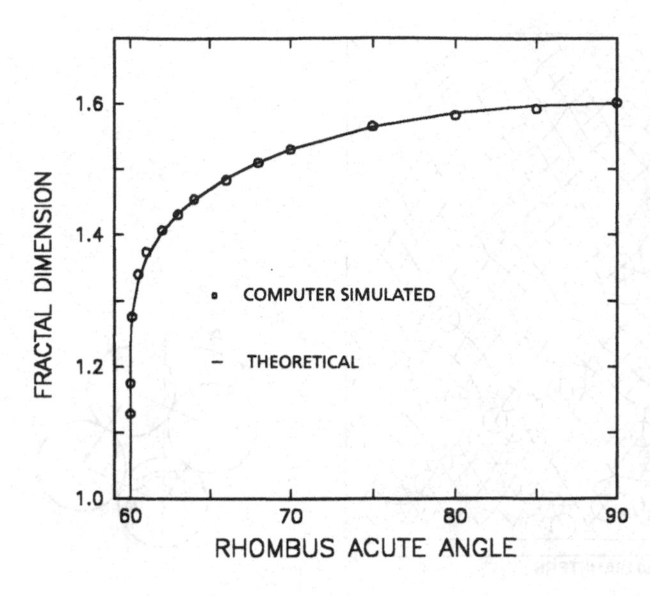

Figure 3.28 The dependence of the fractal dimensionality of the defect pattern generated by the model illustrated in figure 3.27 on the shape of the central rhombus, expressed in terms of the acute angle between adjacent sides. This figure compares the results obtained by analyzing more extensive versions of patterns like those shown in figure 3.27, indicated by small open circles, and the theoretical dependence of D on θ, represented by the curve. This figure was provided by J. Toner.

If the packing is carried out in a system of finite lateral extent L, with periodic boundary conditions in the lateral y direction, or if deposition in a channel of finite width L with hard walls is simulated, then the packing becomes asymptotically periodic [563, 564]. In the asymptotic (large y or large "time") limit, the defect density decays exponentially [563],

$$< \rho_L(y) > \sim e^{-k_L y}, \tag{3.120}$$

where $< \rho_L(y) >$ is the average of $\rho(y)$ over a large number of packings in systems of width L. It appears that $k_L \approx 1/L$, where L and y are measured in disc diameters. Although this was not addressed in the simulations, it can be conjectured that the defect density profile can be represented by

$$< \rho_L(y) > \sim y^{-\alpha} e^{-cy/L}, \tag{3.121}$$

where c is a constant of order 1.

In the large height, defect free limit, the inclination angle θ of the bond between the centers of contacting discs is confined to the range $30° \leq \theta \leq 60°$, measured from the normal to the surface. Angles outside this range can be considered to be geometrical defects and they are always associated with the topological defects. "Flat" defects with $\theta > 60°$ do not propagate as the heap grows, since they are soon capped by the formation of triangular defects. Consequently, the distribution of defects is controlled by the sharp defects with $\theta < 30°$. Defects are formed whenever a continuation of the defect free packing of rhombuses would generate a rhombus with an angle ϕ given by $\phi = \theta_1 + \theta_2 < 60°$. Such a configuration would require overlap of the discs. This

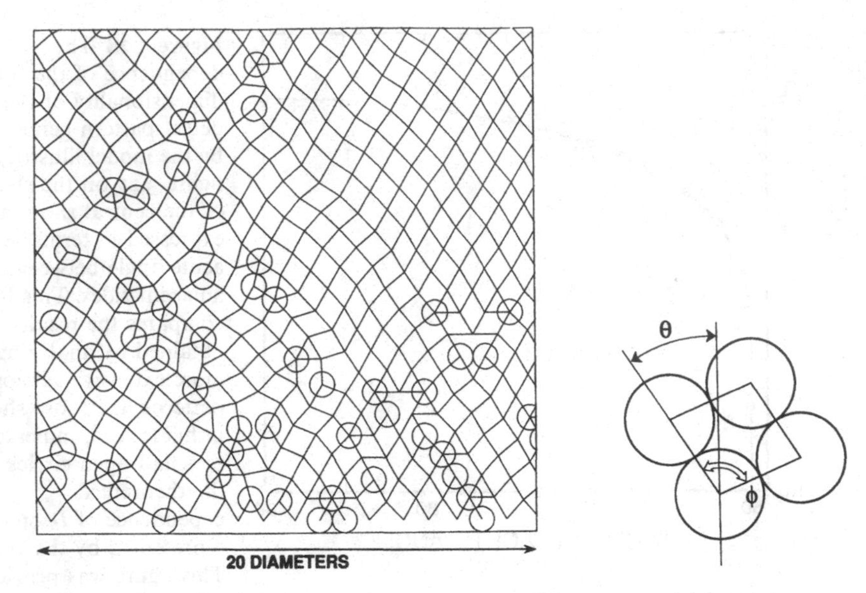

Figure 3.29 A small part of a packing obtained from the model for the random deposition of discs onto a 2-dimensional heap of discs supported by a flat substrate. This figure shows the "bonds" connecting the centers of contacting discs and identifies the "defect particles" that have coordination numbers (number of contacting neighbors) other than 4. The right-hand side of the figure shows how the angles θ and ϕ used to describe the bond orientations, as well as the shapes and orientations of the rhombuses, are defined.

cannot happen, and a defect is formed instead. The formation of defects depends on both the strength, $30° - \theta$, and the density of sharp defects. Using a mean field theory based on these ideas, Socolar [565] has shown that the decrease in density and strength of sharp defects contribute equally to the decreasing density of topological defects with increasing height. This theory predicts that the exponent α in equation 3.119 should have a value of 2.0. This mean field theory value for α is consistent with the simulation results.

3.7 Growth Models Related to DLA

Since the introduction of the DLA model [181], many other models, more or less closely related to DLA, have appeared. This proliferation of models has been motivated by several different objectives. Some of this work has been carried out in connection with experiments that lead to DLA-like patterns, but that differ in substantial ways from the patterns obtained from standard DLA simulations. Very often, theoretical considerations and/or experimental observations indicate ways in which the basic DLA model can be modified to provide a better representation of experimental growth processes. In other

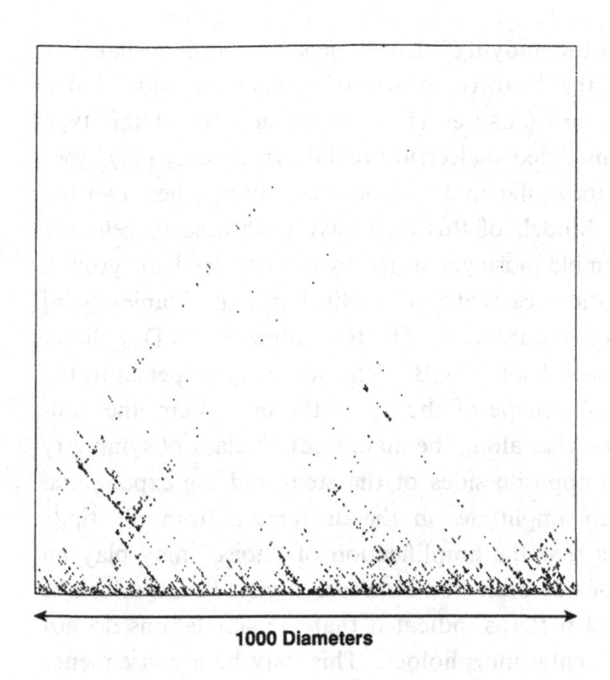

Figure 3.30 A relatively large deposit of discs obtained using the random deposition model, with steepest descent relocation to a local minimum, illustrated in figure 3.29. In this figure, only the defects are shown.

cases, frustration with the slow progress towards a full understanding of the DLA model has led to the development of simplified DLA models, with the idea that these new models might be more theoretically tractable, and that insights developed during the theoretical analysis of these models might be applicable to the more difficult, general DLA problem. In many cases, models have been developed in a more exploratory spirit and have been directed towards the more general issue of the relationships between cluster geometry and growth algorithms. In most cases, the results have had the most impact when the modifications have been as simple as possible.

Even if consideration is restricted to simple modifications, many possible perturbations to the simple DLA model have been explored. These include changes in the particle trajectories, nature and dimensionality of the substrate, boundary conditions, sticking rules and particle concentration. If the "perturbations" are homogeneous, then no new length scales are expected to appear, and it can be anticipated that the structure generated by the new model will exhibit a simple algebraic relationship between the cluster radius of gyration and the cluster mass or number of particles. If the perturbation is not homogeneous, then it can be expected to induce new characteristic length scales in the patterns generated by the model. Very often these new length scales are directly related to characteristic length scales associated with the perturbation. However, the current understanding of DLA does not allow reliable theoretical predictions to be made, and unexpected simulation results should not be rejected on theoretical grounds alone.

One of the main motivations for studying DLA models is the insight that they provide into processes such as the dendritic growth of solids from undercooled melts and supersaturated solutions (chapter 1). Most simulations of this type have been carried out using modified dielectric breakdown models [357, 566] because of the flexibility that they offer and the focus on issues other than the asymptotic scaling properties. Models of this type have been used to generate patterns that quite closely resemble those generated by natural dendritic growth processes. For example, simulations carried out by Nittmann and Stanley [566] successfully reproduce some of the characteristic features observed by Dougherty *et al.* [567, 568] during the growth of NH_4Br dendrites from supersaturated solutions, including the parabolic shape of the tip of the main stem, the non-periodic distribution of side branches along the main stem, the lack of symmetry between the side branches on opposite sides of the stem and the exponential dependence of the side-branch amplitude on the distance z from the tip.[15] These results support the idea that the amplification of "noise" may play an important role in the growth of side branches. However, a visual comparison of the experimental and simulated patterns indicated that the simulations do not completely capture the experimental morphology. This may be a consequence of a wide disparity (by a factor of $\approx 10^{12}$) between the number of molecules in the real dendrites and the number of occupied lattice sites in the simulations, as well as the use of a 2-dimensional model to represent a pseudo 2-dimensional process.

The DLA and dielectric breakdown models for dendritic growth correspond to growth in the quasi-static, low Peclet number, infinite diffusion length limit. More realistic, and necessarily more elaborate, models have been developed [569] that bridge the gap between DLA and dielectric breakdown models on the one hand and numerical solutions of the equations for transport and interface evolution on the other.

A variety of DLA-based models have been developed to simulate specific processes in detail. In many cases, the connection between the model and the process(es) that they are intended to simulate is tenuous and/or poorly justified. However, in other cases, the models are well grounded physically and have had considerable success. For example, DLA-like models have frequently been used to model electrochemical deposition. In most cases, serious attention has not been paid to the physics and electrochemistry of this growth processes. At the time of writing more realistic models were being developed. For example, the work of Roberts and Knackstedt [570] provides an example of a study in which a DLA-like model was developed with the purpose of capturing a more

15. The exponential growth of the side-branch amplitude was observed only near to the tip. For larger values of z, the side-branch amplitude was much smaller than that predicted by an extrapolation of the exponential growth with increasing z observed at small values of z to larger values. The experimental data was also consistent with the idea that the side-branch amplitude grew as $e^{s^{1/2}}$, where s is the arc length, measured from the tip.

realistic representation of the physico-chemical mechanisms of electrochemical deposition processes. However, this is a very complex process (chapter 4, section 4.1.1), and it is easy to generate the "right" patterns for the wrong reasons.

3.7.1 Homogeneous Perturbations

The DLA model can be modified without destroying the homogeneous nature of the Laplacian field by replacing the random walks in DLA by Levy flight [2] or Levy walk particle trajectories [372], described earlier in this chapter (section 3.3). The Eden, ballistic aggregation and DLA models are aggregation models in which the particles follow paths with fractal dimensionalities of 0, 1 and 2, respectively. The use of Levy flights and walks leads to a continuous family of models in which the Eden, ballistic aggregation and DLA models have special places. As the fractal dimensionality D_w of the particle trajectories increases, the particle trajectories become less penetrating and screening effects become more important. The screening is strong enough to generate fractal clusters with $D < d$, only if $D_w > 1$.

In a Levy walk, the "particle" can be considered actually to follow linear segments with a distribution of lengths Δ given by equation 3.9, and, in a DLA-like model, the particle sticks to the first particle in the cluster that it encounters in its path. In a Levy flight, the particle "hops" between the end of the segments, and, in a DLA model, it can stick to the growing cluster only at the end point of each step. A random trajectory (Levy flight or Levy walk)[16] constructed with a distribution of step lengths, given by equation 3.9, has a fractal dimensionality of f if $1.0 \leq f \leq 2$. Computer simulations have been carried out using both square lattice and off-lattice Levy walk particle trajectories [372]. For $f = 2$, the fractal dimensionalities D_α, equation 3.32, and D_β, equation 3.31, for clusters generated by these models have an effective value of about 1.75. These appear to be significantly larger than the value of about 1.71 obtained from DLA with random walk trajectories. Clusters generated with $f = 1.25, 1.5, 1.75$ and 2.0 are shown in figure 3.31. As the parameter f, which describes the step length distribution, is decreased below 2.0, the fractal dimensionality of the clusters appears to increase continuously, until at $f = 0$ an effective fractal dimensionality of about 1.93 was obtained for clusters containing 25 000 particles. A value of zero for f corresponds to ballistic (linear) particle trajectories, and the asymptotic fractal dimensionality is 2.0. Simulations have also been carried out using a corresponding Levy flight

16. The total scale of a Levy flight or Levy walk is dominated by the longest step if $1.0 \leq f \leq 2.0$. However, the total path length $< \sum_{i=1}^{i=n} \Delta_i >$ of a "typical" Levy walk is proportional to the number of steps n.

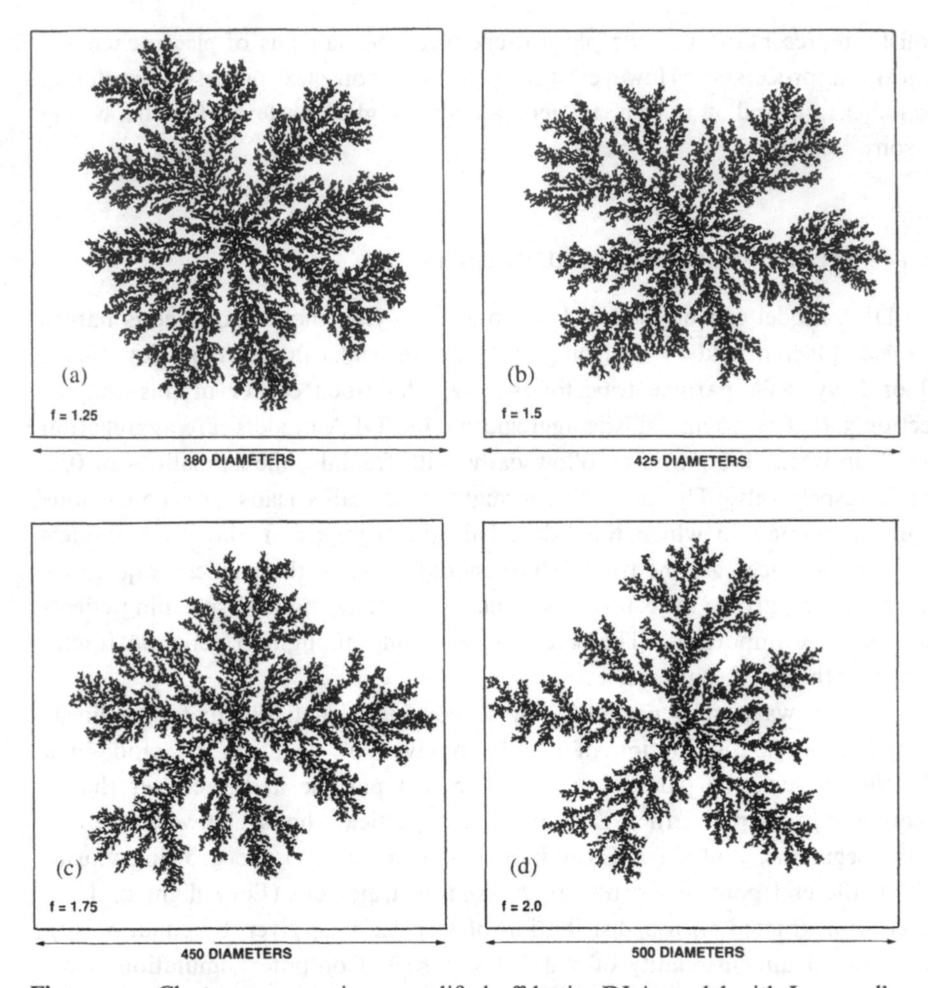

Figure 3.31 Clusters grown using a modified off-lattice DLA model with Levy walk trajectories. Parts (a), (b), (c) and (d) show clusters generated with values of 1.25, 1.5, 1.75 and 2.0 for the parameter f (equation 3.9) defining the step length distribution in the Levy walk.

model. Similar behavior is observed. In this case, $f = 0$ corresponds to the Eden model ($D = d = 2$).

Another example of a homogeneous modification of the DLA model is to replace the Laplacian growth probability $\mu(\mathbf{x})$ by a radially biased growth probability of the form

$$P(\mathbf{x}) \sim \mu(\mathbf{x})|\mathbf{x}|^{\phi}, \tag{3.122}$$

where $\mu(\mathbf{x})$ is the harmonic measure at position $\mathbf{x}$ on the surface of the growing cluster [571]. Clusters can be generated using this model by following the ordinary DLA rules until the random walker "contacts" the cluster at position

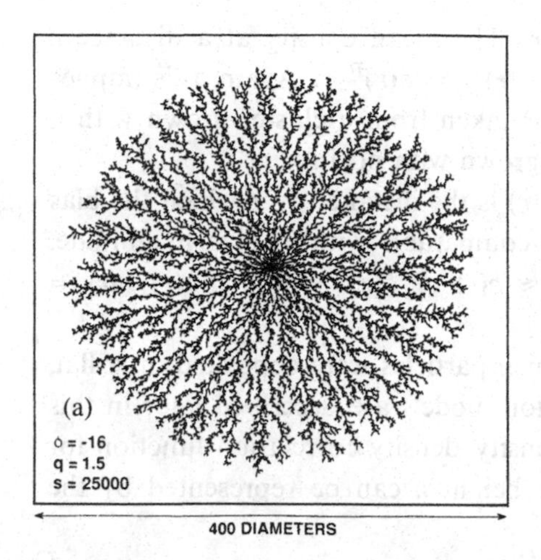

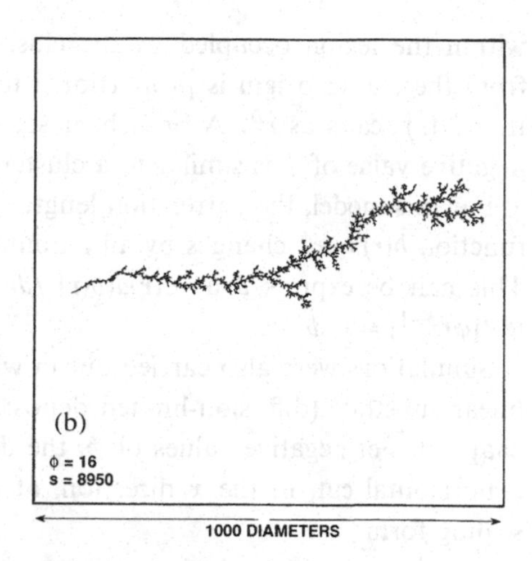

Figure 3.32 Clusters grown using the radially biased DLA model. Parts (a) and (b) show clusters grown with $\phi = -16$ and $\phi = 16$, respectively, where ϕ is the exponent in equation 3.122.

x. At this point, the random walker is added to the cluster with a probability proportional to $|\mathbf{x}|^{\phi}$ or is killed without addition. Figure 3.32 shows clusters generated using this model [571]. Although the clusters look very different from DLA clusters, the growth of the radius of gyration can be represented by $R_g \sim s^{\beta}$, and the exponent β has the same value that is found using the standard DLA model.

To characterize the overall shape of these clusters, the inertial tensors **I** were diagonalized and the eigenvalue ratios $R_{\Lambda} = \Lambda_{max}/\Lambda_{min}$ were measured, as a function of cluster size. Garick *et al.* [572] concluded that, for off-lattice DLA, the eigenvalue ratio R_{Λ} slowly approached an asymptotic limiting value of 1.0. For negative values of ϕ, R_{Λ} also appears to approach an asymptotic value of 1.0, and the rate of convergence to this value grows with increasing ϕ. For positive values of ϕ, the eigenvalue ratio R_{Λ} appears to approach a finite value, greater than 1.0, which increases with increasing $|\phi|$. The cluster structures can be described in terms of a fractal blob model with a DLA-like structure on short length scales (within the blob size ξ). For positive values of ϕ, the blob size ξ is proportional to r/ϕ, where r is the distance from the center. The blobs form a directed chain away from the cluster origin. This blob model implies that the mass per unit length $m(r)$ is proportional to $\xi(r)^{D-1}$, so that $m(r) \sim r^{D-1}$, where D is the fractal dimensionality and the number of particles within a distance r from the origin $N(r)$ scales as r^D, as it does in ordinary DLA.

A similar fractal blob model can be used to describe clusters grown with negative values for the exponent ϕ. The correlation length or blob size ξ is proportional to $r/|\phi|$. In this case, the blobs are packed together to fill the space within the region occupied by the cluster. The mean density at a distance r from the cluster origin is proportional to $\xi(r)^{-\alpha} = \xi(r)^{D-d}$. Again, this implies that $N(r)$ scales as r^D. A branch or sector taken from a cluster grown with a negative value of ϕ is similar to a cluster grown with positive ϕ.

For this model, the correlation length $\xi(r)$ is the distance over which the bias function $b(r) = r^{\phi}$ changes by an amount comparable with its own magnitude. This can be expressed as $\xi(r)d(b(r))/dr \approx b(r)$ or $\xi(r) \approx b(r)/[d(b(r))/dr] = r^{\phi}/[\phi r^{\phi-1}] = r/\phi$.

Simulations were also carried out in which particles were deposited on a flat, linear substrate (diffusion-limited deposition models are described later in this chapter). For negative values of ϕ, the density–density correlation function for a horizontal cut, in the x direction, at a height h can be represented by the scaling form

$$C(x,h) = A(h)\xi_h^{-\alpha} f(x/\xi_h^{\alpha/\nu}), \tag{3.123}$$

where $\xi_h = h/|\phi|$, $\alpha \approx 1/3$, $\nu \approx 1/2$ and $A(h)$ is a height dependent amplitude, related too the lacunarity. The scaling function $f(y)$ has the form $f(y) \sim y^{-\nu}$ on short length scales ($y \ll 1$) and $f(y) = const.$ for $y \gg 1$. This suggests a self-affine fractal blob model with a correlation length $\xi_\perp(h) = \xi_h \approx h/\phi$ in a direction perpendicular to the substrate and a correlation length of $\xi_\| \sim \xi_\perp^{\alpha/\nu}$ parallel to the substrate. Consequently, equation 3.123 can be written as

$$C(x,h) = A(h)\xi_\|^{-\nu} f(x/\xi_\|(h)). \tag{3.124}$$

The fractal dimensionality $D_\|$ of a horizontal cut is $1-\nu \approx 1/2$, and the fractal dimensionality $D_\perp$ of a vertical cut is $1-\alpha \approx 2/3$. This implies that the mean density at a height h is given by $\rho(h) \sim \xi_\|(h)^{D_\|-1} \sim \xi_\perp(h)^{-\alpha} \sim (h/\phi)^{-\alpha}$ with $\alpha \approx 1/3$. This is in good agreement with the observed density profile, which has the form $\rho(h) = A(\phi)r^{-\alpha}$. The amplitude $A(\phi)$ increases as $|\phi|$ increases, and the simulation data are consistent with the idea that $A(\phi) \sim |\phi|^{\alpha}$. This is consistent with the diffusion-limited deposition model simulations of Meakin and Family [573] and the strip geometry simulations of Evertsz [574], using a dielectric breakdown model. Both studies indicate that the deposit structure has self-affine scaling properties.

3.7.2 Inhomogeneous Perturbations

An example of an inhomogeneous perturbation is the addition of a drift to the random walk paths [575, 576] that carry the random walkers to the growing cluster. In this case, it can be imagined that the particle trajectories start at $x = -\infty$ and that the drift is in the x direction. This can be implemented, in a square lattice model, by selecting steps in the $+x, -x+y$ and $-y$ directions with probabilities of $(1/4)+\delta, (1/4)-\delta, 1/4$ and $1/4$, respectively. For $|\delta \ll 1|$, this will generate trajectories that are 2-dimensional on short length scales and 1-dimensional on long length scales. The crossover will occur after N^* steps, where $N^*\delta \approx (N^*)^{1/2}$ or $N^* \sim \delta^{-2}$, and the distance moved after N steps can be expressed in terms of the scaling form

$$< R(N) >= N^{1/2} f(N/N^*), \tag{3.125}$$

where $f(x) = 1$ for $x \ll 1$ and $f(x) \sim x^{1/2}$ for $x \gg 1$. Again, it is useful to think of the random walk in terms of a fractal blob model. The characteristic crossover length scale ξ is given by $\xi \approx R(N^*) \approx (N^*)^{1/2} \approx \delta^{-1}$. On short length scales, within a blob of size ξ ($\ell \ll \xi$), the path is like an unperturbed random walk. On long length scales, where $\ell \gg \xi$, the path can be thought of in terms of a string of blobs forming a self-affine ($H = 1/2$) directed walk in the drift direction. The crossover in the particle trajectories is reflected in the aggregate structure [575]. On short length scales $\lambda \ll \xi$, the structure is like that of a DLA cluster ($D \approx 1.715$), but on long length scales the cluster is like a ballistic aggregation cluster ($D = 2.0$). The density–density correlation function has the form

$$C(r) = r^{-\alpha} f(r/\xi), \tag{3.126}$$

where α is the codimensionality of DLA. The function $f(x)$ has the form $f(x) = const.$ for $x \ll 1$ and $f(x) \sim x^{\alpha}$, so that $C(r)$ has a constant value proportional to $\xi^{-\alpha}$, for $x \gg 1$.

Based on a real-space renormalization group analysis, Nagatani [577, 578] has proposed the quite different relationship,

$$\xi \sim [Pe/(1 + Pe)]^{-1/\alpha}, \tag{3.127}$$

between the crossover length ξ and the drift bias or Péclet number Pe. For small Péclet numbers, this implies that

$$\xi \sim Pe^{-1/\alpha} \sim \delta^{-1/\alpha} \sim \delta^{1/(D-d)}. \tag{3.128}$$

Diffusion-limited aggregation models with drift towards [579] and away [580] from the center of the cluster have also been explored. For an outward bias, there is a crossover from an ordinary DLA-like structure on short length scales to a "stringy" structure, with a fractal dimensionality of $D \approx 1$ on long scales. The dependence of the crossover length scale ξ on the bias strength was not explored quantitatively. Off-lattice simulations in which the growing clusters are allowed to undergo rotational diffusion [581] or the clusters rotate continuously with a constant angular velocity [579] have also been carried out. In the latter case, a single spiral arm dominates the resulting pattern. The spiral arm is quite compact because the particle trajectory, with respect to the rotating spiral arm, is a ballistic path on all but quite small length scales. Very similar patterns [582] are generated by a ballistic particle–cluster aggregation model in which particles are "fired" one at a time, at a constant rate of one per time step, from a randomly selected direction, towards the origin with velocity selected randomly from a uniform distribution ($10 < |v| < 50$ diameters per time step). At the start of the simulation, a massive particle ($m_o = 10^6$ or 10^{12}) with a large angular momentum is placed at the origin. The "low mass" $m = 1$ particles aggregate with the growing cluster with conservation of momentum and angular momentum. For a sufficiently large initial angular momentum, the dominant arm forms a logarithmic spiral.

Biased random walk DLA models have been used to represent the effects of fluid flow on electrodeposition experiments [583]. However, this is probably a poor representation of these complex experiments. In the work of Nagatani and Saugues [583], the effects of drift towards a linear substrate in which the particles are deposited was simulated using a non-normal drift direction. In the case of a horizontal drift, a crossover from diffusion-limited deposition to ballistic deposition at grazing incidence (chapter 5, section 5.7.1) was seen.

More realistic simulations by Warren *et al.* [584], in which particles are carried along low Reynolds number fluid-flow streamlines towards a growing cluster, or aggregate, lead to very different structures. In the model of Warren *et al.*, the forces on each of the particles were calculated under Stokes flow conditions. The magnitude of the force $\mathbf{f}_i$ that the ith particle exerts on the fluid is proportional to the probability that the next particle will be added to that particle and the direction of the force $\mathbf{f}$ determines the direction of the bond between the new particle and the cluster. In practice, the flow field was recalculated after a small number of particles had been added to the cluster (so that the shape of the cluster does not change too much before the forces are recalculated), but not after the addition of every particle. Simulations were carried out in which the orientation of the growing cluster with respect to the far field flow remained fixed and in which the flow direction varied randomly with respect to the cluster orientation, each time the flow field was recalculated. A slightly smaller effective fractal dimensionality was obtained for the fixed orientation simulations, but the difference was within the statistical uncertainties. In any event, fractal dimensionalities of $D \approx 1.6$ and $D \approx 2.5$ were obtained from 2-dimensional and 3-dimensional simulations, respectively. These values for D, which are essentially the same as the fractal dimensionalities of the corresponding 2-dimensional and 3-dimensional DLA clusters, reflect the similarity in the screening of the penetration of flow fields and Laplacian fields into fractal aggregates. In fact, the simulation results are consistent with the idea that diffusion-limited aggregation and "convection-limited aggregation" might belong to the same universality class.

Many natural processes can be better represented by growing clusters from a "sea" of randomly walking particles that initially occupy a fraction ρ of the lattice sites [585, 586] than by the DLA model, which implies that $\rho \to 0$. A model of this type was proposed by Rosenstock and Marquardt [587] for the growth of silver aggregates on the surface of silver halide crystallites. However, computer simulations were not carried out. An irregular cluster geometry was anticipated, but the clusters were represented by a simple shape to facilitate a theoretical study of the growth kinetics. A computer simulation might have led to a different perspective on this problem and a quite different theoretical approach. Such a simulation can be carried out on a lattice of size $L_x = L_y = L$ by filling (labeling) a site in the middle of the lattice to represent the origin of the cluster and then filling ρL^2 randomly selected sites, avoiding multiple selection of the same site, to represent the random walkers. The random walkers are then selected randomly and moved to a randomly selected nearest-neighbor site. In most cases, this is done only if it would not result in multiple occupancy of the site, but such details have little effect on the final outcome. If a random walker enters an unoccupied site on the perimeter of the growing cluster, it becomes part of the cluster. The simulation can be carried out with a fixed number of "particles" (random walkers or cluster sites), or the concentration of filled sites at the boundaries of the lattice can be maintained at a constant value of ρ. The clusters generated by this model also have a DLA-like structure on short length scales $\epsilon \ll \lambda \ll \xi$ and are uniform on longer length scales. In this case, the correlation length or crossover length ξ is determined by the requirement that the average density in the final cluster and the original "sea" of particles must be the same. The length ξ is the diffusion length (chapter 1, section 1.4.2), and it characterizes the width of the depleted region near to the growing pattern. This means that the cluster density–density correlation function must have the form $C(r) \sim r^{-\alpha}$ for length scales in the range $\epsilon \ll \lambda \ll \xi$, and must have a constant value on larger length scales $C(r > \xi) = \rho$. Since $C(\epsilon) \approx 1$, it follows that $\xi \approx \epsilon\rho^{-1/\alpha}$, where α is the codimensionality, $d - D$.

Clusters that have a DLA-like structure on short length scales $\epsilon \ll \lambda \ll \xi$ and are uniform on longer length scales can be obtained using a model that is closely related to the DLA model [588, 589]. In this model particles are launched from randomly selected positions on a surface that encloses the growing cluster (a closed loop in 2-dimensional simulations). Each point on the launching surface is at the same minimum distance ξ from the cluster, so that the launching surface is the boundary of the ξ-neighborhood of the cluster. The distance ξ between the launching surface and the cluster represents the diffusion length in the growth process. If a random walker enters an unoccupied perimeter site (in a lattice model), then the random walk is terminated and that site is filled, to represent the growth process. If a random walker returns to the launching surface, it is stopped and a new random walker is launched from a randomly selected position on the launching surface. The launching surface must be changed continuously to maintain a constant shortest distance ξ between all points on the launching surface and the cluster.

In many real growth processes, the scalar field Φ that controls the growth is "screened", so that the Laplace equation for Φ must be modified to represent the more local nature of the growth process. In some cases, it is appropriate to replace the Laplace equation by the Poisson equation

$$(\nabla^2 - k^2)\Phi = 0. \tag{3.129}$$

This equation can be thought of in terms of random walkers. The second term on the left-hand side of equation 3.129 represents a sink for random walkers, and, on average, the random walker will move of the order of k^{-2} steps, or by a distance of $\approx k^{-1}$, before it is annihilated. Similarly, if equation 3.129 is changed to $(\nabla^2 + k^2)\Phi = 0$, creation of new random walkers from previously existing random walkers is implied. It might be thought that equation 3.129 simply introduces a characteristic length scale $\xi = k^{-1}$ into the problem and that the structure can be described in terms of a packing of fractal blobs, with a size of ξ. Such a structure can be generated using the model described in the preceding paragraph. However, things are not that simple. One approach to the algorithmic modeling of equation 3.129 would be to launch random walkers from randomly selected sites on the unoccupied perimeter of a growing aggregate. At each step in the random walk, the random walker could be killed with a probability of k^2. If the random walker was killed before it returned to the perimeter of the aggregate, the perimeter surface at which the growth process started could be filled to represent the growth process. This would generate a pattern that would initially grow with a low density, somewhat like that expected from the fractal blob model. However, as the aggregate pattern grew it would become more and more dense, behind the advancing interface. Unlike the Laplace equation, equation 3.129 is not invariant to the transformation $\Phi \rightarrow \Phi + const.$ (it is not gauge invariant). Consequently, more attention must be given to the boundary conditions. The scenario outlined above places the outer boundary at "infinity" and assumes a fixed value for Φ ($\Phi \neq 0$) on the growing surface.

Louis *et al.* [590] have carried out square lattice dielectric breakdown model simulations based on equation 3.129, with a growth probability in the ith perimeter site proportional to $|E_i|$, where E_i is the corresponding field or potential gradient. The simulations are carried out on finite strips of width L with fixed "potentials" Φ_b and Φ_t at the bottom and top of the strip, respectively. If $\Phi_t > \Phi_b$, then a transition can be found from a "stable" dense pattern at heights h smaller than h^* to an unstable branched pattern for $h > h^*$. The possibility of this occurring can be seen by solving the 1-dimensional version of equation 3.129 for the field at a flat aggregate surface at height h. This leads to the expression

$$E(h) = \frac{\Phi_t \cosh[\xi(H-h)] - \Phi_b}{\sinh[\xi(H-h)]}\xi, \tag{3.130}$$

for the field at the interface, after the growth process has reached a height of h in a cell with a total height of H ($H - h$ is the remaining gap between the "electrodes"). Here, $\xi = k^{-1}$, and $E(h) = 0$ if $\Phi_b/\Phi_t = \cosh[\xi(H - h)]$ and this determines the transition height h^*. For $h < h^*$, the growth process slows down as $h \to h^*$ and the pattern is filled in to form a dense compact structure. Once the growth has been forced past the height h^*, it becomes unstable on all length scales, like DLA, and a single branch will grow from the compact $h < h^*$ regime. In terms of random walks the transition takes place when the "negative" random walkers from the electrode at a height H dominate the "positive" random walkers released from the growing pattern at the surface of the pattern. In practice, random walk simulations would not generate patterns like those obtained by Louis *et al.* because the field calculated in the model of Louis *et al.* could only be obtained by averaging over an enormous number of random walkers.

A variety of patterns and transitions can be generated using this type of model [590, 591], and a richer morphology is predicted for antiscreening models, where $(\nabla^2 + k^2)\Phi = 0$. In this case, the field may vanish at more than one stage during the growth process, leading to a number of growth transitions. If 2-dimensional simulations are carried out with a point inner "electrode", from which the growth originates, and a circular, square or triangular outer "electrode", then the boundary of the dense region from which a single tenuous branch grows has a shape very similar to that of the outer boundary or electrode [592]. This model illustrates that dramatic morphological transitions, similar to the Hecker effect seen in electrochemical deposition experiments (chapter 4, section 4.2.1), may take place as a result of changing conditions at the advancing interface. However, because of the change in the sign of the potential gradient at the active zone during the transition, implying a change in sign in the charges of the ions deposited at the interface, this model does not explain the Hecker effect. In addition, the screening length, relative to the cell size, is very much smaller in typical experiments than in the simulations. Under these conditions, the transition would not take place in the observed position, more or less midway between the cathode and anode of the cell. The arrival of a "chemical" front at the growing cathode (chapter 4, section 4.2.1) provides a much more convincing explanation of the Hecker effect.

One of the most simple modifications of the DLA model is to introduce a sticking probability [419, 593] P_σ so that, when a random walker reaches an unoccupied perimeter site (or attempts to move onto an occupied site if dielectric breakdown boundary conditions are used), it "sticks" with probability P_σ and continues on its random walk, without entering an occupied site, with probability $1 - P_\sigma$ [419, 593, 594, 595]. This modification weakens the screening of the penetration of random walkers into the interior of the cluster and generates a cluster that is compact on short length scales ($\ell \ll \xi$) but has the same scaling properties as ordinary DLA on long length scales ($\ell \gg \xi$) [594, 593, 419], where ξ is a correlation length that increases as the sticking probability decreases.

Witten and Sander [419] argued that, in the continuum limit, the normal surface growth velocity $V_n(\mathbf{x}_s)$ is related to the normal component of the field gradient $\nabla_n \Phi(\mathbf{x}_s)$ by

$$V_n(\mathbf{x}_s) \sim aP_\sigma \Phi(\mathbf{x}_s) \sim a^2 \nabla_n \Phi(\mathbf{x}_s), \tag{3.131}$$

where P_σ is the sticking probability, $\Phi(\mathbf{x}_s)$ is the magnitude of the field in the unoccupied sites adjacent to the surface and a is the lattice constant. This implies that there is a characteristic length scale $\xi = a/P_\sigma$ which plays a role similar to the capillary length described in chapter 1, section 1.4.2. Based on a simple linear stability analysis, Witten and Sander proposed that the structure should be uniform on length scales ℓ smaller than ξ and DLA-like on longer scales. They found evidence in favor of this idea for quite small clusters with sticking probabilities in the range $0.1 \leq P_\sigma \leq 1.0$.

Simulations carried out using a triangular lattice indicate that the crossover length ξ is related to the sticking probability by

$$\xi \sim P_\sigma^{-\nu_X}, \tag{3.132}$$

where the exponent ν_X has a value of about 1.2 [595, 596]. A theoretical value of 0.85 was found [578] using a real-space renormalization approach based on a resistor network representation of the DLA process and a minimum energy dissipation principle [578]. In the limit $P_\sigma \to 0$, the DLA model becomes equivalent to an Eden model, with "trapping" of enclosed voids, and a compact but porous structure (with a density ρ in the range $0.7 < \rho < 0.71$ [597] for growth on a square lattice) with a self-affine surface is generated. For $d \geq 3$, a similar crossover takes place. In the $P_\sigma \to 0$, Eden model limit, it is much more "difficult" to trap voids if $d > 2$, and the clusters will have densities much closer to 1.

Based on a linear stability analysis of an almost flat surface, Halsey and Leibig [598] proposed that the crossover from Eden-like to DLA-like growth should take place when the cluster size s has reached the critical number $s_c \sim (L_w/\epsilon)^2 \propto [(1/P_\sigma) - 1]^2$, where L_w is a characteristic length scale called the Wagner length by Halsey and Leibig. For length scales longer than L_w, the surface instability is like the Mullins–Sekerka instability, discussed in appendix A. On shorter length scales, the surface is marginally unstable corresponding to Eden growth. This leads to the scaling form

$$s = (R_g/\epsilon)^2 f([(1/P_\sigma) - 1]/R_g^{1/\mu}), \tag{3.133}$$

with $\mu = 1$. Halsey and Leibig obtained a better data collapse with $\mu = 1.25$ and concluded that $\mu = 1.25 \pm 0.25$, corresponding to a value of $1/\mu$ or 0.8 for the exponent ν_X in equation 3.132.

Similar models, with sticking probabilities that depend on the local structure [419], can be used to generate almost completely dense patterns [357, 566, 599, 600, 601]. This is illustrated in figure 3.33, which shows clusters generated on a six-coordinate triangular lattice, with sticking probabilities for unoccupied perimeter sites with z occupied nearest-neighbors given by [602]

$$P_\sigma = r^{6-z}, \tag{3.134}$$

where $r < 1$. In some cases [601], the structure of the patterns have been interpreted in terms of a fractal dimensionality $D(r)$ that depends on the selectivity parameter r. However, it seems most likely that the patterns can be described much better in terms of a crossover from $D = 2$ on short length scales (scales less than $\xi(r)$), to a fractal dimensionality of about 1.7, the characteristic DLA value, on length scales longer than $\xi(r)$.

Batchelor *et al.* [603] have generated patterns with a very similar model (with $P_\sigma = r^{5-z}$) in the zero noise limit ($m \to \infty$, where m is the noise reduction parameter discussed in section 3.1). The implementation of zero noise growth models is discussed later in this chapter (in section 3.8). The simulations were carried out by numerically solving the Laplace equation on a triangular lattice, with the boundary condition $\Phi(\mathbf{x}) = 0$ on occupied sites (nodes) and $\Phi(\mathbf{x}) = 1$ on a distant boundary. This requires a modification of the discretized Laplace equation, for the unoccupied perimeter sites and their nearest neighbors, to represent the effects of the local sticking probabilities. For these sites,

$$\Phi(\mathbf{x}) = \sum_{\{nn\}} (1 - P_\sigma(\mathbf{x}'))\Phi(\mathbf{x}')/(6 - z(\mathbf{x}')), \tag{3.135}$$

where $z(\mathbf{x}')$ is the number of occupied nearest-neighbors of the site at position $\mathbf{x}'$, which is one of the nearest neighbors of the unoccupied perimeter site at $\mathbf{x}$. In equation 3.135 P_σ is zero for sites that are not part of the external perimeter, and this equation reduces to the standard discretized Laplace equation, $\Phi(\mathbf{x}) = (1/6)\sum_{\{nn\}} \Phi(\mathbf{x}')$, for sites that are neither part of the unoccupied perimeter nor nearest neighbors to a site in the unoccupied perimeter.

The growth probabilities are given by

$$k(\mathbf{x}) = P_\sigma(\mathbf{x})\Phi(\mathbf{x}), \tag{3.136}$$

where $P_\sigma(\mathbf{x})$ and $\Phi(\mathbf{x})$ are the sticking probability and the value of the Laplacian field Φ in the unoccupied perimeter site at position $\mathbf{x}$, respectively. In this noiseless model, the growth proceeds by stages in which symmetrically equivalent sites are filled. After 100 growth stages, clusters generated without site-selective sticking ($r = 1$) have six equivalent, densely side-branched arms. As the sticking

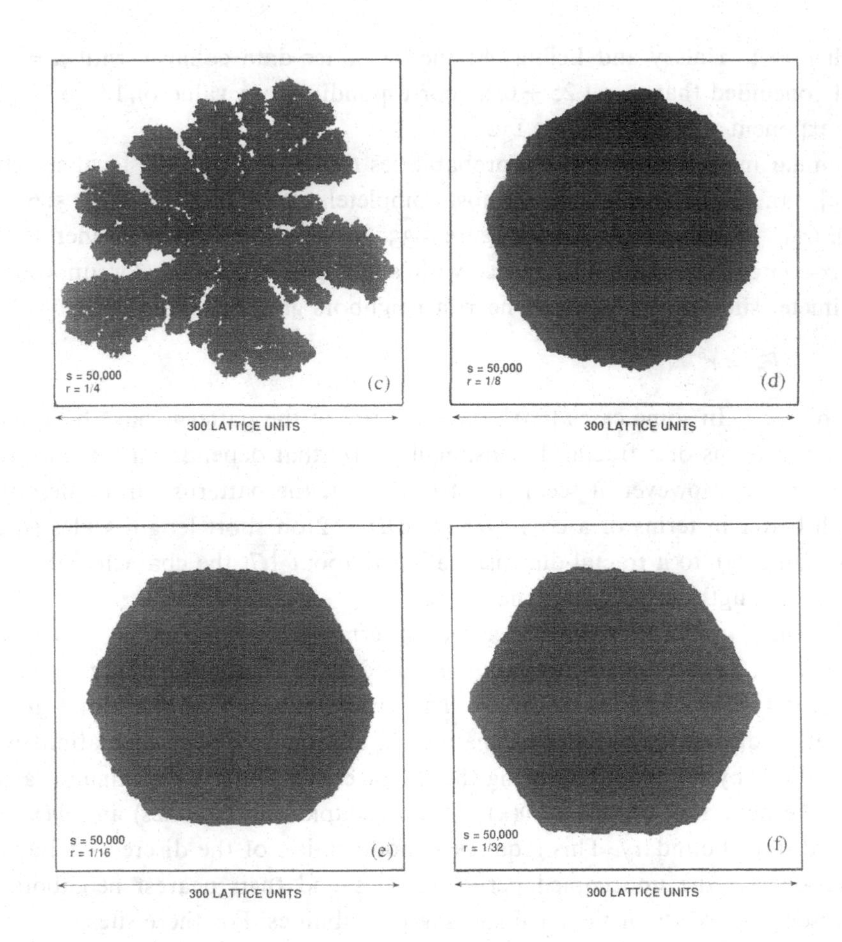

Figure 3.33 Clusters grown on the cells of a six-coordinate lattice using a DLA model with site-selective sticking probabilities. Parts (a) – (f) show 50 000-site clusters grown with selectivity parameters (r in equation 3.134) of 1, 1/2, 1/4, 1/8, 1/16 and 1/32, respectively.

selectivity parameter r is reduced, and the sticking becomes more selective, the side branching disappears and the arms of the cluster become broader. As the sticking selectivity parameter r is progressively reduced, this broadening continues via a series of six-armed patterns with a smooth transition to a compact hexagonal pattern, at small values of the sticking parameter r. The patterns grown using this model become more complex as the growth process continues. The sequence of patterns depends on the cluster sizes, and their asymptotic shapes are not known. Similar effects probably occur in the related models described below. In addition, simulations of this type can be exquisitely sensitive to the boundary conditions and other model details [604].

Simulations were also carried out using a noiseless, $m \to \infty$ DLA model in which the random walkers are always terminated in the first unoccupied perimeter site that they reach, but in which the growth probability, after the first unoccupied perimeter site has been reached, is given by $P_\sigma^* = r_\beta^{5-z}$. In the Laplacian growth version of this model, the Laplace equation is solved with the boundary conditions $\Phi(\mathbf{x}) = 0$ on occupied sites and $\Phi(\mathbf{x}) = 1$ on a distant boundary. The growth rates are given by

$$k(\mathbf{x}) = P_\sigma^*(\mathbf{x}) \left[\sum_{\{nn\}} \Phi(\mathbf{x}') \right] /6, \tag{3.137}$$

where $P_\sigma^*(\mathbf{x})$ and $k(\mathbf{x})$ are the growth probability after contact and the growth rate in the site at position $\mathbf{x}$, respectively, and the sum is over the nearest-neighbor sites with Laplacian fields of $\Phi(\mathbf{x}')$. The patterns were examined after 200 growth stages. For clusters of this size, this model shows a transition from a densely side-branched, six-armed pattern, in the noiseless DLA limit, to a compact, six-armed pattern, as r is decreased. As r is decreased further, a 12-armed pattern is formed, and this is replaced by another six-armed pattern rotated by 30° with respect to the initial (large r) six-armed pattern, at smaller values of r. As r is decreased still further, the arms merge, and eventually a new pattern with six compact arms, in the original orientation, is formed. At very low values of r, the pattern looks like a flower with six petals.

Batchelor and Henry [605, 606] have also investigated a square lattice version of the Laplacian DLA model, with sticking probabilities given by

$$P_\sigma = r^{3-z} \tag{3.138}$$

(a square lattice version of equation 3.134), in the noiseless, $m \to \infty$ limit. For large values of r, the growth process is like the $m \to \infty$ limit of the noise reduced DLA model and a "cross" oriented along the lattice axes is formed. As r is reduced, a transition takes place to a pattern in which tip splitting is observed in each of the four arms. At $r = 1/2$, a pattern like that obtained from the noise reduced dielectric breakdown model (see section 3.8) is formed. As r is further reduced, a transition to a compact square takes place.

Batchelor and Henry [605, 606] have also investigated a square lattice Laplacian DLA model that corresponds to termination of the random walkers in the first encountered unoccupied perimeter site and growth with a probability given by

$$P_\sigma^* = r_\beta^{3-z}. \tag{3.139}$$

For small values of r_β, growth in the "diagonal" directions is favored, but for large values of r_β, four arms grow along the axes of the lattice. The transition between these two limits is surprisingly complex [606], as figure 3.34 illustrates.

Decreasing the sticking selectivity parameter r in equation 3.134 is equivalent to increasing the "surface tension". The rich variety of patterns and morphology transitions generated by these and related models reflects the results of the interplay between anisotropy and surface tension found in real dendritic growth processes and theoretical models for dendritic growth (chapter 1, section 1.4). In the presence of anisotropy, classical dendrites are formed. In general, increasing the "surface tension" via the parameter r in equation 3.134 leads to patterns with thicker "arms" that evolve via tip-splitting processes and become compact, in the large surface tension limit. Decreasing the sticking selectivity parameter r suppresses side branching because of the relatively low probability of nucleating a new side branch and the much larger rate of lateral growth compared to nucleation, once a side branch has been nucleated. This is a purely non-equilibrium, kinetic effect. However, the results are in accordance with the intuitive idea that side branching is suppressed because of the large surface energy associated with side branching.

Many other versions of the DLA model have been used to explore pattern formation. Processes such as surface diffusion [607, 608] are important in a variety of growth phenomena of practical interest, such as the growth of dendritic "islands" in molecular beam epitaxy (chapter 4, section 4.1.3). Some applications of this type of model will be discussed elsewhere [57].

Clusters grown on a square lattice with a quenched multifractal distribution of sticking probabilities, like those illustrated in figure 3.35, have a quite inhomogeneous appearance, but the exponent β that describes the growth of the radius of gyration appears to be unchanged [609].

Many models which can be thought of in terms of the effects of inhomogeneous perturbations on DLA have been described in the literature. In many cases, the clusters generated by these models have been analyzed in terms of a fractal dimensionality that varies continuously with the "strength" of the perturbation. A more careful analysis, sometimes requiring much larger scale simulations, almost invariably shows that the data can be represented much better in terms of a crossover with a crossover length that depends on the strength

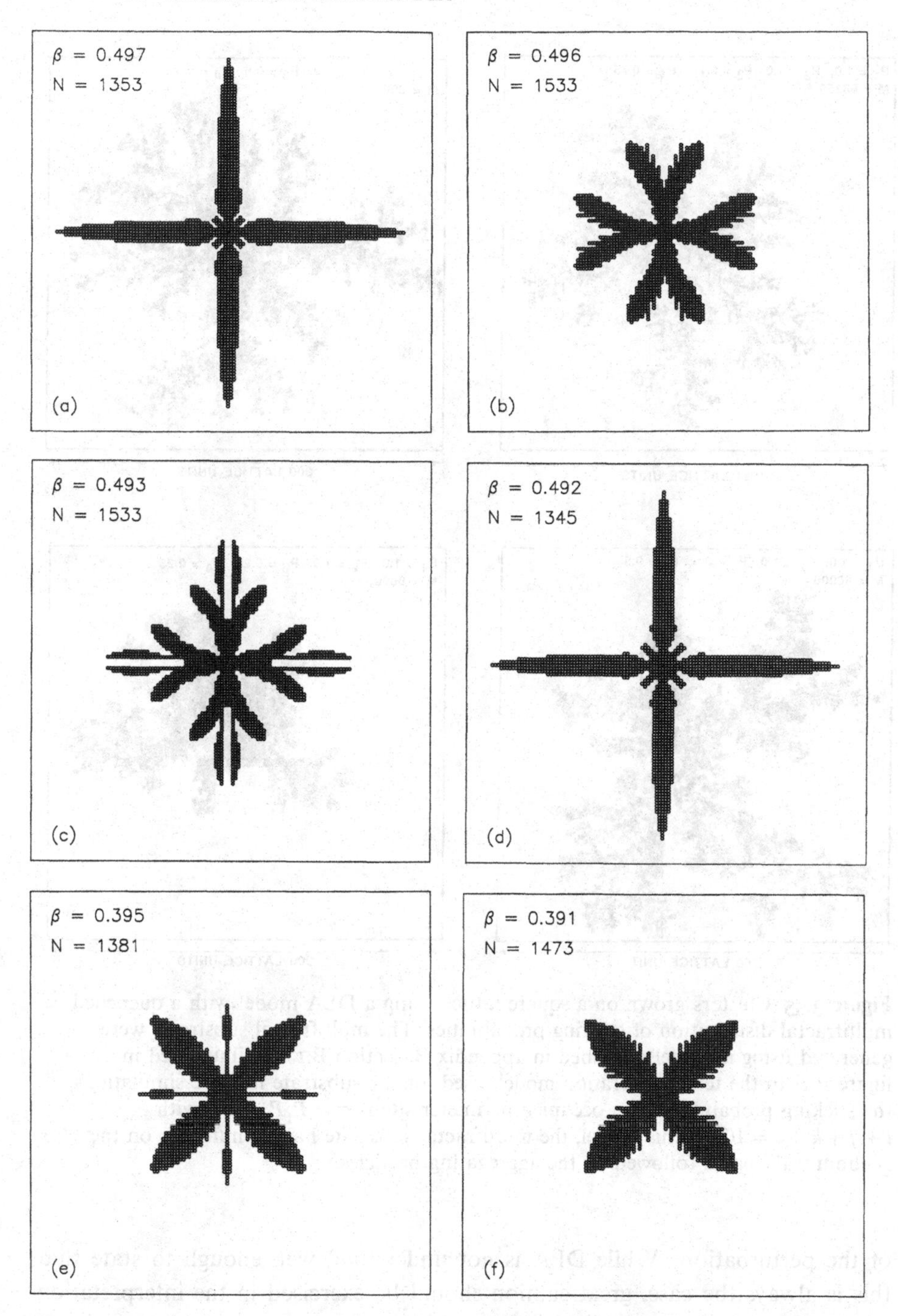

Figure 3.34 Clusters grown using a zero-noise-limit square lattice Laplacian DLA model in which random walkers are terminated in the first unoccupied perimeter that they encounter and contribute to the growth process with a probability $P(z)$ given by $P(z) \sim r_\beta^{3-z}$, where z is the site coordination number. Parts (a) to (f) show the patterns formed as r_β (β in the figures) is decreased from 0.497 in part (a) to 0.391 in part (f). This figure was provided by M. T. Batchelor.

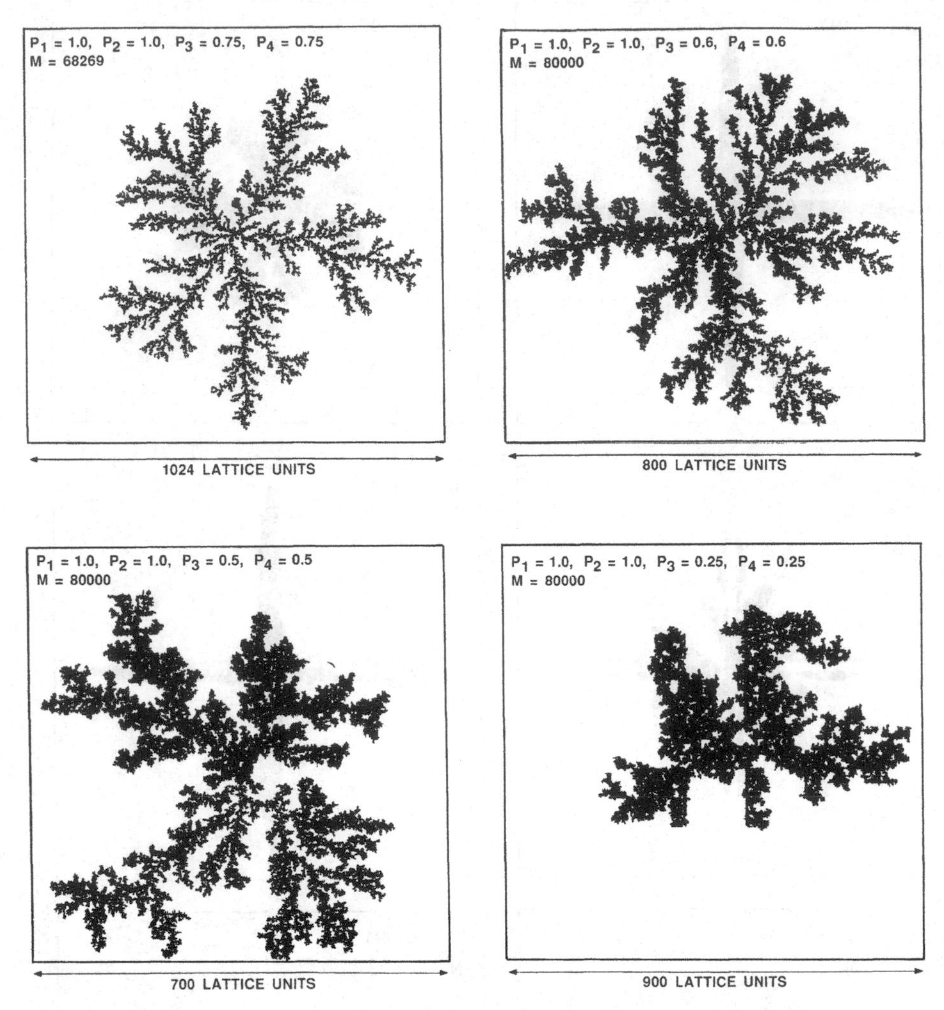

Figure 3.35 Clusters grown on a square lattice using a DLA model with a quenched multifractal distribution of sticking probabilities. The multifractal substrates were generated using a model described in appendix B, section B.1 and illustrated in figure B.2. In the tenth generation model, used for the substrate in these simulations, the sticking probability in an occupied perimeter site is $\sigma = P_1^i P_2^j P_3^k P_4^l$, with $i + j + k + l = 10$. In this model, the multifractal substrate has no influence on the random walk paths followed by the aggregating particles.

of the perturbation. While DLA is not understood well enough to state that this is always the case, great caution should be exercised in the interpretation of simulations (or experiments) with inhomogeneous perturbations in terms of continuously varying fractal dimensionalities.

One of the main applications of the DLA model has been to the process of fluid–fluid displacement or viscous fingering [610]. In this context, the DLA model corresponds to the displacement of a viscous fluid by a non-viscous fluid, with an infinite viscosity ratio, in a homogeneous porous medium or Hele-Shaw cell. The DLA model corresponds to the limit in which there is no interfacial tension between the two fluids and the pressure field in the fluids obeys the Darcy–Laplace equation. The effects of surface tension can often be neglected if the displacement process is sufficiently fast. However, for very fast flows, the linear Darcy equation breaks down and the scaling properties of the displacement pattern may be different from that of DLA [611]. Because of the efficiency of random walk algorithms, and the ease with which the computer programs can be written, this is an attractive approach to the simulation of fluid–fluid displacement and other moving boundary problems. The status of research in this direction will be discussed elsewhere [57].

In practice, the surface tension or surface energy associated with fluid–fluid interfaces plays an important role in pattern formation and must be included in realistic models. The most elegant solution to this problem [171, 612], within the context of the DLA model, is to launch random walkers, with probabilities given by

$$P(\mathbf{r}) = A\mathscr{K}(\mathbf{r}) + B, \tag{3.140}$$

from the lattice sites representing the fluid–fluid interface and kill them when they return to the interface. In equation 3.140, $\mathscr{K}(\mathbf{r})$ is a local "discretized" curvature calculated from the configuration of filled sites near to the position $\mathbf{r}$ on the interface. This is combined with noise reduction [355] by keeping track of the score on each site, which is increased by 1 for an absorption event and decreased by 1 for a "launching". If the score reaches a value of m, the interface advances by one lattice unit, and if it reaches $-m$ the interface recedes. The "viscous fingering" patterns generated using this approach are in good agreement with experimental patterns [612].

3.7.3 Attractive Interaction Model

A simple particle–cluster aggregation model which generates clusters that are difficult to distinguish from DLA clusters has been introduced by Block *et al.* [613, 614]. The particles are added, one at a time, to the growing cluster, starting from "infinity". After the particles have been launched from a randomly selected point, far from the cluster, they follow deterministic paths to the cluster. At all times, the particles move in the direction of a force resulting from attractive interactions with all of the particles in the cluster. For a cluster of N particles, this force is given by

$$\mathbf{F} = \sum_{i=1}^{i=N} (|\Delta\mathbf{r}_i|)^{-(\zeta+1)} \Delta\mathbf{r}_i. \tag{3.141}$$

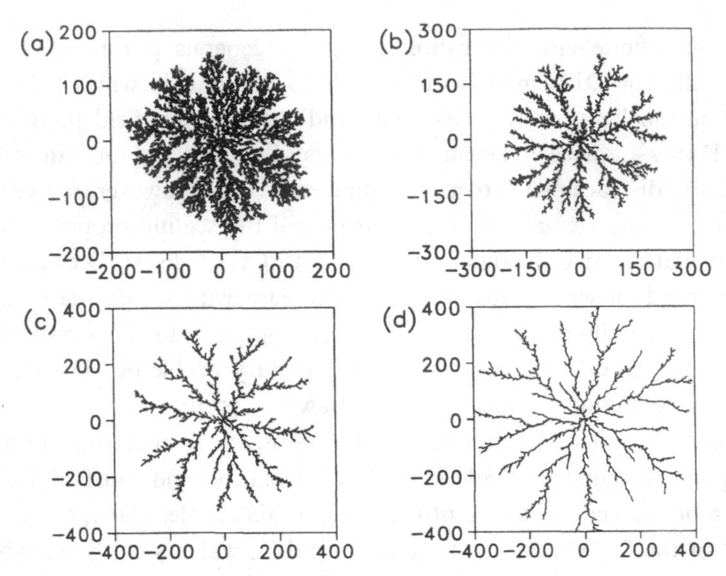

Figure 3.36 Clusters generated using the attractive interaction model. Parts (a), (b), (c) and (d) show 12 000-particle clusters generated with interaction exponents of 2, 3, 4 and 5, respectively. This figure was provided by A. Block.

Here, $\Delta\mathbf{r}_i = \mathbf{r}_i - \mathbf{R}$, where $\mathbf{r}_i$ is the position of the center of the ith particle in the cluster and $\mathbf{R}$ is the position of the "moving particle". In practice, the particle is moved by a small distance in the direction of the force $\mathbf{F}$

$$\mathbf{R}' = \mathbf{R} + (\mathbf{F}\delta_x/|\mathbf{F}|), \tag{3.142}$$

where $\mathbf{R}'$ is the new position and δ_x is a sufficiently small number. After the particle has moved to the new position, the force $\mathbf{F}$ is recalculated and another move is executed. This process continues until the particle comes into contact with the growing cluster and is incorporated into the cluster at the first contact position. Particles are added repeatedly to the growing cluster in this manner until a sufficiently large cluster has been generated. Figure 3.36 shows clusters that have been generated using this algorithm, with interaction exponents ζ of 2, 3, 4 and 5. An exponent, ζ, of 2.7, in equation 3.141, generates clusters that bear a striking resemblance to DLA clusters. In addition, the distribution of growth probabilities is similar to that found for DLA growth, including the existence of a "phase transition" in the multifractal growth probability measure (appendix B, section B.3).

At present, it is not known if this is just a coincidence or if there is a fundamental relationship between this model and DLA. If such a connection does exist, then this may provide an important route towards the development of a better understanding of DLA. However, it is clear that there is no simple equivalence between this model and DLA. This can be seen by considering a circle with a large "mass" inside it. For the DLA process, each point on the circle is equally accessible to random walkers and the growth probability will be everywhere the same. For the attractive interaction model, the particles will be attracted towards the mass inside the circle, so that the growth probability distribution will depend on the mass distribution inside the circle.

To model real growth processes, both interactions (attractive and/or repulsive) and the effects of thermal fluctuations will have to be taken into account, and the particles will follow random walk paths that are biased by interactions with the cluster. This could be done by means of Langevin dynamics simulations.

Hurd [615, 616] has carried out cluster growth simulations using a Monte Carlo model with attractive and repulsive interactions. In this model, a Boltzmann factor $P(\mathbf{r}) = e^{-V(\mathbf{r})/k_B t}$, where $V(\mathbf{r})$ is the potential at the site at position $\mathbf{r}$, is associated with each of the unoccupied lattice sites. The Boltzmann factors $P(\mathbf{r})$ then determine the trajectories followed by the particles, as they are added to the growing clusters. The probability that a random walker will move into a site at position $\mathbf{r}$ that is a nearest neighbor to its current position is given by $P(\mathbf{r})/\sum_{\{nn\}} P(\mathbf{r}')$, where $\{nn\}$ is the set of nearest neighbors. Long range repulsive interactions enhance the screening effect, which leads to the formation of more tenuous clusters and destabilizes the essentially isometric shape of the DLA cluster [617] into an elongated form with two major "arms". Long range attractive interactions reduce the probability that random walkers will be captured by the most exposed parts of the cluster, and may lead to the formation of more compact clusters. The simulations of Hurd [615, 616] were carried out using "Coulomb" ($V(r) \sim 1/r$), "screened Coulomb" ($V(r) \sim e^{-kr}/r$) and "dipole" ($V(r) \sim 1/r^3$) attractive and repulsive pair potentials.

Quite different effects are seen in modified DLA models if the interactions have a short range nature, like the screened Coulomb interaction. The structure is not changed on long length scales, but on short length scales attractive interactions result in a more open, tenuous structure, while repulsive interactions lead to a more compact form. Short range repulsive interactions can be represented by a small sticking probability, and short range attractive interactions can be represented by "capture" of particles that wander within a distance ξ_c, the "capture length", of some part of the growing cluster [618]. In the capture process, a particle reaching within a distance ξ_c from the cluster is immediately transferred to the nearest contacting position on the cluster and becomes part of the cluster.

Pastor-Satorras and Rubi [619] have carried out 2-dimensional and 3-dimensional off-lattice DLA simulations with dipolar interactions between the aggregating particles and the growing cluster. The randomly walking particles are started, with randomly oriented dipole moments, far from the cluster. This is necessary for simulations with long range interactions because random walkers starting far from the cluster do not enter a circle enclosing the cluster for the first time with equal probabilities at all positions on the circle. The path of an aggregating particle is obtained using standard, finite temperature, Metropolis Monte Carlo methods [188] (chapter 1, section 1.9), with trials in which the particle attempts to move by one particle diameter in a randomly selected direction. After each successful step, the orientation of the moving particle dipole moment is rotated to the direction of the local field (due to the dipole moments on the growing cluster) at its current position. At very large temperatures, the model converges to the standard DLA limit. At very low temperatures, the dipole interactions direct the particle towards the cluster. Under these conditions the algorithm is relatively fast, and it was possible to grow clusters containing up to 5000 particles. At intermediate temperatures, the algorithm becomes quite time consuming, and it was possible to grow clusters up to a size of only 1000 particles.

The simulation results were interpreted in terms of a fractal dimensionality that varied continuously from a low temperature limit of $D \approx 1.13$ to a high temperature (DLA) limit of $D \approx 1.71$. However, these results could also be interpreted in terms of a crossover from $D \approx 1.13$, on short length scales, to $D \approx 1.71$, on long length scales, with a crossover length that depends on the reduced temperature T_r ($T_r = \epsilon^3 k_B T/\mu^2$, where μ is the dipole moment strength and ϵ is the particle diameter). From the 3-dimensional simulations, a fractal dimensionality of $D = 2.46 \pm 0.04$ was obtained in the high temperature (DLA) limit, in which $T_r > \approx 10^{-2}$, and in the low temperature limit $D = 1.37 \pm 0.03$. The vector correlations were also characterized via the correlation function $C_v(r)$, defined in equation 3.52. The vector correlation function $C_v(r)$ was found to decrease algebraically with increasing distance r but with a larger exponent than the density–density correlation function $C(r)$. For low reduced temperatures ($T_r < \approx 10^{-2}$), $C_v(r) \sim r^{-1}$. Similar behavior has been found for the bond vector correlation function, described earlier in this chapter (section 3.4.5.6), in ordinary 2-dimensional, off-lattice DLA.

3.7.4 Growth on Fibers and Surfaces

In an important class of processes, material is deposited onto an essentially smooth surface or fiber to form a deposit. In many cases, the deposit is quite dense, but the surface of the deposit may have a self-affine or self-similar geometry. Such processes are discussed in more detail in chapter 5. Similar processes can also lead to the growth of much more open, highly branched structures. Patterns of this type can be generated using a simple modification of the DLA algorithm, described above. In general, particles are deposited onto a d-dimensional surface or substrate from a $d+n$-dimensional space and such processes can be referred to as $d+n$-dimensional processes. By far the most important processes of this type are $d+1$-dimensional with $d = 1$ or $d = 2$.

In the $d+1$-dimensional diffusion-limited deposition model [465], particles are added, one at a time, to the growing deposit via random walk trajectories that originate far "above" the deposit. The random walkers stop when they reach a site that is adjacent to either the d-dimensional surface or the growing deposit (figure 3.37). The corresponding unoccupied perimeter site is then filled to represent the growth of the deposit on the surface. Periodic boundary conditions are used in the lateral direction(s). Simulations have been carried out using the $1+1$-dimensional (deposition from a plane onto a line), $2+1$-dimensional (deposition from 3-dimensional space onto a plane) and $1+2$-dimensional (deposition from 3-dimensional space onto a line) versions of this model [465]. The deposits consist of a "forest" of "trees" that do not contact each other (figure 3.38). Shielding or screening of the smaller trees by their larger neighbors inhibits their growth, and the low probability of a random walker diffusing through the forest of trees almost completely prevents new growth on the substrate after the early stages of growth. These simulations indicated that the density correlations inside the deposit had the same power law form as in the corresponding $d+n$-dimensional DLA clusters. In particular, the mean density $\rho(h)$ at a distance h from the substrate had the form

$$\rho(h) \sim h^{-\alpha_{(d+n)}} \sim h^{D_{(d+n)}-(d+n)}, \tag{3.143}$$

where $\alpha_{(d+n)}$ is the codimensionality for $d+n$-dimensional diffusion-limited deposition and $D_{(d+n)}$ is the corresponding fractal dimensionality.

Computer simulations indicate that, providing the lateral system size L is much greater than the height or depth of the growing deposit, the distribution of tree sizes N_s can be described quite well by the form

$$N_s \sim s^{-\tau} f(s/s^*), \tag{3.144}$$

first proposed by Rácz and Vicsek [620], where N_s is the number of clusters of size s and τ is the size distribution exponent. The cut-off function $f(x)$ has the form $f(x) = const.$ for $x \ll 1$ and $f(x)$ decreases faster than any power of x, as x increases, for $x \gg 1$. Here, s^* is the characteristic cut-off size that is determined by the deposit height (h^*), if this height is much less than the strip width L.

The exponent τ can be obtained if it assumed that equations 3.143 and 3.144 describe the scaling properties of the deposit. The power law density profile $\rho(h)$, in equation 3.143, will have a strong cut-off near the "top" of the deposit and can be written in the form

$$\rho(h) \sim h^{-\alpha} g(h/h^*), \tag{3.145}$$

where the cut-off function $g(x)$ has properties similar to $f(x)$ in equation 3.144.

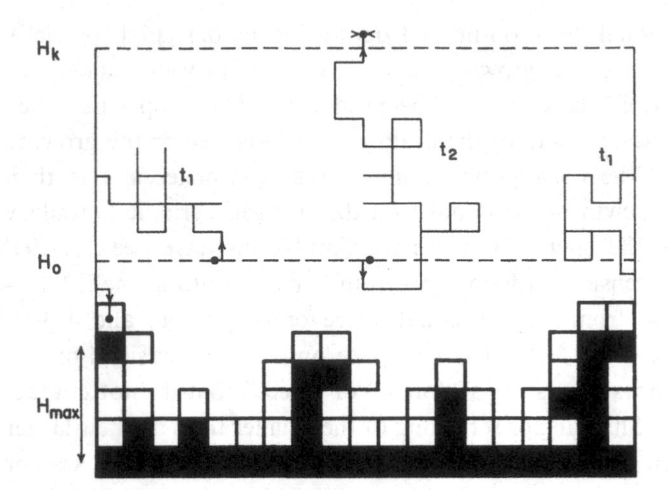

Figure 3.37 A simple square lattice model for diffusion-limited deposition onto a line. The previously filled sites are indicated in black and the unoccupied perimeter sites, in which growth can take place, are indicated by empty squares with borders. Trajectory t_1 eventually brings the particle (random walker) to a site in the active zone, and that site will be filled, to represent the growth process. This trajectory also illustrates the periodic boundary conditions used in the model. Trajectory t_2 carries the random walker far from the growing deposit. This trajectory is killed when it reaches the "killing surface" at a height of H_k. A new random walker will then be launched from a randomly selected position on the launching surface, at a height of H_o.

Consequently, the mass (number of particles or sites) contained in a deposit that has grown to a height of h^* is given by

$$M \sim h^{*^{n-\alpha}}, \tag{3.146}$$

if $\alpha < n$, which is always true for diffusion-limited deposition. The mass of the deposit can also be expressed in terms of the cluster size distribution

$$M = \sum sN_s. \tag{3.147}$$

Using the form for the distribution given in equation 3.144, the expression

$$M \sim s^{*^{(2-\tau)}} \tag{3.148}$$

is obtained, for the case $\tau < 2$. Assuming that the individual clusters are self-similar fractals that can be characterized by the same fractal dimensionality ($D = D_{d+n}$) as the entire deposit, equation 3.148 can be replaced by

$$M \sim h^{*^{(2-\tau)D}}. \tag{3.149}$$

Comparing the expressions for the deposit mass in equations 3.148 and 3.146, it is apparent that

$$(2-\tau)D = n - \alpha \tag{3.150}$$

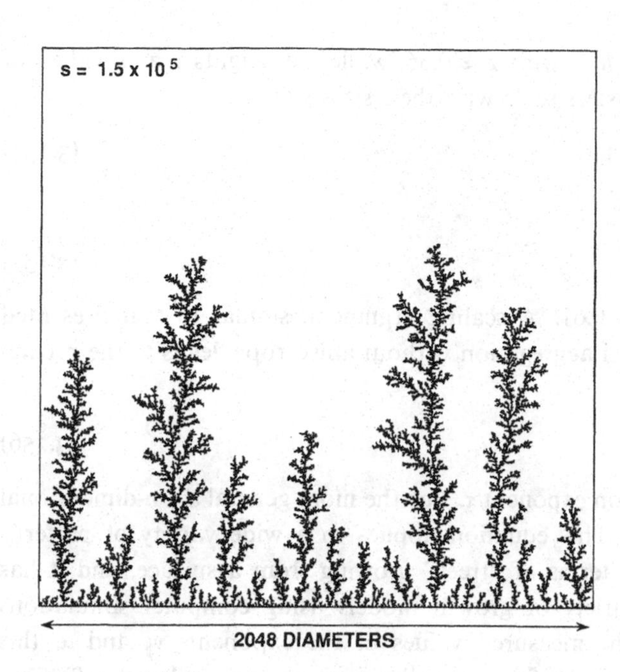

Figure 3.38
A 1 + 1-dimensional pattern generated using the model illustrated in figure 3.37.

or

$$(2-\tau)D = D - d + n, \tag{3.151}$$

from which the result

$$\tau = 1 + (d/D) \tag{3.152}$$

is obtained, for $n = 1$. An expression equivalent to equation 3.152 was first obtained by Rácz and Vicsek [620] using similar arguments. Equation 3.152 indicates that $\tau = (1/D) + 1$ (≈ 1.585), for the 1 + 1-dimensional case, and $\tau = (2/D) + 1$ (≈ 1.8), for the 2 + 1-dimensional case. Simulation results [621] are in quite good agreement with these predicted values for the size distribution exponent τ.

Equation 3.145 implies that the mass of the deposit M is related to the "height" h^* by equation 3.146 and the height profile, after a mass M has been deposited, can be represented by the scaling form

$$\rho(h, M) \sim h^{-\alpha} g(h/M^{1/(n-\alpha)}). \tag{3.153}$$

As is the case for the growth of DLA clusters on a square lattice, noise reduction enhances the effects of lattice anisotropy on diffusion-limited surface growth models [622]. The forest of trees looks more like a forest of "conifers" than the "deciduous" forest shown in figure 3.38. For large noise reduction parameters, the density profile, measured for patterns generated in 1 + 1-dimensional simulations, has the form $\rho \sim h^{-\alpha}$, with $\alpha \approx 0.36$, while the heights y and widths of the individual trees or clusters scale with their sizes s as

$$y \sim s^{\nu_\parallel} \tag{3.154}$$

and

$$x \sim s^{\nu_\perp}, \tag{3.155}$$

with $\nu_\parallel \approx 0.65$ and $\nu_\perp \sim 0.61$. A scaling argument, similar to that presented above, for diffusion-limited aggregation without anisotropy, leads to the prediction [623] that

$$\tau = 2 - \nu_\parallel(1 - \alpha), \tag{3.156}$$

for the tree size distribution exponent τ. For the more general $d+n$-dimensional case, $\tau = 2 - \nu_\parallel(n - \alpha)$. This equation applies to a wide variety of patterns that can be described in terms of "trees" growing from a surface, and it has been confirmed for a variety of growth models using computer simulations [623, 624, 625]. With the measured values of the exponents $\nu_\parallel$ and α, this leads to the estimate $\tau \approx 1.58$ for $1+1$-dimensional noise-reduced diffusion-limited deposition. This estimate is consistent, but not in good agreement, with the measured value of $\tau \approx 1.64$. The discrepancy may arise because the simulations were not carried out on a large enough scale to reach, and accurately characterize, the asymptotic scaling regime, or because a more complex scaling model is required.

If the deposit is considered to be made up of a compact packing of fractal blobs with a self-similar structure on length scales in the range $\epsilon < \ell < \xi(h)$, then the blob size $\xi(h)$ will be of the order of the mean "tree width" at a height of h, so that $\xi(h) \sim h^{\nu_\perp/\nu_\parallel}$, and the density at this height will be given by

$$\rho(h) \sim h^{(\nu_\perp/\nu_\parallel)[D-(d+n)]}. \tag{3.157}$$

The increase in the size of a surviving cluster with increasing height is given by

$$< ds/dh > \approx \rho(h)\xi_h^{d+n-1} \sim h^{(\nu_\perp/\nu_\parallel)(D-1)}. \tag{3.158}$$

This equation indicates that the size s (number of sites or particles) of a cluster of height h is given by $s \sim h^{[(\nu_\perp/\nu_\parallel)(D-1)+1]}$. Since the cluster size is also related to the height by $s \sim h^{1/\nu_\parallel}$, it follows that $(\nu_\perp/\nu_\parallel)(D-1)+1 = 1/\nu_\parallel$, or [576]

$$D_{eff} = 1 + \frac{1-\nu_\parallel}{\nu_\perp}, \tag{3.159}$$

for all d and n. If this model is accepted, then an effective fractal dimensionality of $D_{eff} \approx 1.57$ is obtained from the $1+1$-dimensional noise reduced diffusion-limited deposition model simulations.

Although the $L \gg h^*$ limit, where L is the substrate size, is the most important from an experimental point of view, the $h^* \gg L$ limit is also of interest in the context of Hele-Shaw cell and other experiments. Computer simulations on a square lattice [573] indicate that, in this limit, the average density $\overline{\rho}$ is related to the strip width by

$$\overline{\rho} \sim L^{-\alpha} \sim L^{-0.340\pm0.002}. \tag{3.160}$$

If the growth of a "tree" or a cluster from a substrate is described by equations 3.154 and 3.155, then the height of the cluster in a strip of width L might be expected to grow according to the scaling form

$$h = s^{\nu_\parallel} f(s/L^{1/\nu_\perp}) = L^{\nu_\parallel/\nu_\perp} f'(s/L^{1/\nu_\perp}) \tag{3.161}$$

or

$$h = L^{\nu_\parallel/\nu_\perp} L^{-1/\nu_\perp} s g(s/L^{1/\nu_\perp}). \tag{3.162}$$

In the large height limit, $h \sim s$, so that $g(x) = const.$ for $x \gg 1$ and the deposit density ρ_∞ is given by

$$\rho_\infty \sim s/hL \sim L^{[(1/\nu_\perp)-1-(\nu_\parallel/\nu_\perp)]}. \tag{3.163}$$

The large height limit density is also expected to be given by

$$\rho \sim L^{-\alpha} \sim L^{D-2}. \tag{3.164}$$

Comparing equations 3.163 and 3.164, it is apparent that

$$D - 2 = (1/\nu_\perp) - 1 - (\nu_\parallel/\nu_\perp), \tag{3.165}$$

or

$$D = 1 + [(1 - \nu_\parallel)/\nu_\perp], \tag{3.166}$$

in accordance with equation 3.159, and $\alpha = 2 - D = 1 - [(1 - \nu_\parallel)/\nu_\perp]$. Equation 3.166 was derived for the $1 + 1$-dimensional case. However, the arguments used to derive this equation can easily be extended to $d + 1$-dimensional growth processes, and it can be shown that equation 3.166 applies for all d. If $\nu_\perp = \nu_\parallel = \nu$, equation 3.166 indicates that $D = 1/\nu$. Using values of 0.65 and 0.61 for $\nu_\parallel$ and $\nu_\perp$, respectively, from the $1 + 1$-dimensional, noise reduced diffusion-limited deposition simulations [622], a value of ≈ 0.43 is predicted for the exponent α, which is not in very good agreement with the result $\alpha \approx 0.36$ obtained directly fron the density profile measured using the same noise-reduced DLA model simulations. If, instead, values of $\approx 2/3$ and $\approx 1/2$ (predicted for the asymptotic, large cluster size, limit by the wedge growth model [440, 626] described below) are used for $\nu_\parallel$ and $\nu_\perp$, then a value of $1/3$ is predicted for α. These values for $\nu_\parallel$ and $\nu_\perp$ are supported by the good data collapse obtained using the scaling form

$$\overline{h} = L^{-4/3} f(s/L^2), \tag{3.167}$$

for 1+1-dimensional diffusion-limited deposition without noise reduction, where $\bar{h}$ is the mean deposit height [573].

Perhaps the firmest conclusion from the numerical studies of the $1+1$-dimensional version of the diffusion-limited deposition model (without noise reduction) is that, for $L \gg h$, the density decreases as $\rho_L \sim L^{-\alpha}$, with $\alpha \approx 1/3$. The simulations of Meakin and Family [573] and Evertsz [574] also strongly indicate that there is a correlation length in the lateral direction $\xi_{\|}(h)$ that grows as $\xi_{\|}(h) \sim h^{\zeta}$ with $\zeta \approx 3/4$. The mean density at a height h is given by

$$\rho(h) \sim \xi_{\|}(h)^{D_{\|}-1} \sim h^{-\alpha}, \tag{3.168}$$

if the horizontal cut can be described in terms of a self-similar fractal blob model with a fractal dimensionality of $D_{\|}$ over length scales in the range $\epsilon < \ell < \xi_{\|}(h)$. It follows from equation 3.168 that $\zeta(1-D_{\|}) = \alpha$, or $D_{\|} \approx 5/9 \approx 0.556$, if the values of $1/3$ for α and $3/4$ for ζ are accepted. This value for D is reasonably close to the value of $D \approx 0.6$ found for circular cuts of DLA clusters [446, 448] (section 3.4.5.2).

Recent, larger scale, off-lattice, $1+1$-dimensional simulations [453] indicate that the two-point density–density correlation function $C(r,h)$ for horizontal cuts at a height h, in wide strips, can be represented quite well by the scaling form

$$C(r,h) = h^{-a} f(r/h^{\zeta}), \tag{3.169}$$

where the scaling function $f(x)$ has the form $f(x) \sim x^{-\alpha_{\|}}$, with $\alpha_{\|} \approx 0.38$ for $x \ll 1$. At larger values of x, $f(x)$ decreases more rapidly with increasing x, reaching a minimum at $x \approx 1$. For $x > 1$, $f(x)$ rises to a maximum before decreasing slightly to a constant value for large x. The exponent a has a value of about 0.27, and the effective value for the exponent ζ increases slowly in the range $0.80 \leq \zeta \leq 0.85$ with increasing height. It may approach a value of 1.0 at very large heights. The effective values measured for the exponent ζ, describing the growth of the lateral correlation length, are larger than the value of ≈ 0.75 estimated on the basis of smaller scale square lattice model simulation results.

The density $\rho(h,L)$ in strips with a width of L can be described quite well by the similar scaling form

$$\rho(h,L) = L^{-\alpha} g(h/L^{\gamma}), \tag{3.170}$$

where the scaling function $g(x)$ has the form $g(x) \sim x^{-\alpha_{\perp}}$, with $\alpha_{\perp} \approx 0.28$ for small x, and crosses over quite quickly to a constant value at large x. Values of $\alpha \approx 0.33$ and $\gamma \approx 1.20$ were measured for the scaling exponents in equation 3.170.

The form for $\rho(h,L)$ implies that the exponent α relating the deposit density to the strip width L in the large h limit should be related to the exponents $\alpha_\perp$ and γ by

$$\alpha = \gamma\alpha_\perp, \tag{3.171}$$

and this relationship is satisfied quite well by the values obtained for α, $\alpha_\perp$ and γ. However, the similar exponent scaling relationship

$$a = \zeta\alpha_\parallel, \tag{3.172}$$

expected for the exponents associated with equation 3.170, is not satisfied very well by the values determined from the simulation data.

It is also reasonable to expect that the density will stop decreasing with increasing height when the correlation length in the lateral direction, which grows as h^ζ according to equation 3.169, reaches the strip width L. This will happen at a height of $L^{1/\zeta}$. According to equation 3.170, the density stops decreasing at a height h^* given by $h^* \sim L^\gamma$, which implies that

$$\zeta = 1/\gamma. \tag{3.173}$$

The overall situation appears to be similar to that found in 2-dimensional off-lattice DLA clusters. The decrease in density with increasing distance from the substrate appears to have the same form, and essentially the same exponent, as the decrease in density as a function of distance from the origin in a DLA cluster. This could be interpreted in terms of a fractal dimensionality of $D \approx 1.72$. In addition, the fractal dimensionality D_x of a horizontal cut in the $1+1$-dimensional deposit seems to be the same as, or similar to, the fractal dimensionality of a circular cut in a DLA cluster. This could be interpreted in terms of a fractal dimensionality of $D \approx 1.60$ for the cluster or deposit. A value less than 1 for the exponent ζ would be consistent with an active zone thickness that grows less than linearly with increasing height. These results raise the same unresolved issues concerning the nature of an appropriate asymptotic scaling model and the nature of the approach to asymptotic behavior, as do results obtained from the $1+1$-dimensional, off-lattice DLA cluster growth model.

3.7.5 Simplified DLA Models

Structures that are much more simple than those generated by the $d+n$-dimensional diffusion-limited deposition models described above can be obtained using a model in which growth is allowed to occur only in the direction away from the substrate. Models of this type generate a "forest of needles", or "field of grass", with a broad distribution of needle heights [627]. They capture the essential aspects of competitive growth in diffusion-limited aggregation and are

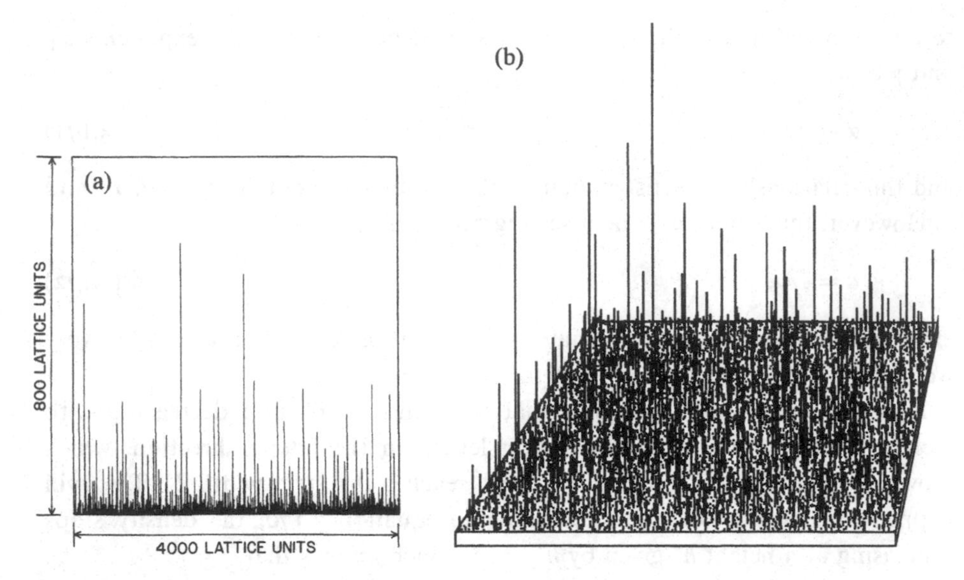

Figure 3.39 Patterns obtained from (a) 1 + 1-dimensional and (b) 2 + 1-dimensional diffusion-limited needle growth models. The 1 + 1-dimensional pattern was generated using a substrate of $L = 4000$ lattice units with reflecting boundary conditions on the sides of the needles. The 2 + 1-dimensional pattern was obtained using a substrate of 256 × 256 lattice units. The 2 + 1-dimensional figure was provided by T. Rage.

much more amenable to theoretical analysis. However, they have not, so far, contributed substantially to a deeper understanding of the full DLA problem.

Figure 3.39 shows the patterns obtained from 1 + 1- and 2 + 1-dimensional simulations. The distribution of needle heights can be described by the scaling form

$$N_h \sim h^{-\tau} f(h/h^*), \quad (3.174)$$

where the characteristic maximum height h^* grows as a power of the total mass (number of sites s) in the deposit ($h^* \sim s^{1/(2-\tau)}$), only if $\tau < 2$. If $\tau \geq 2$, the tallest needles will not make a significant contribution to the total mass and a "runaway" growth process will take place.

Similarly, the density distribution $\rho(h)$ can be represented by

$$\rho(h) \sim h^{-\alpha} g(h/h^*). \quad (3.175)$$

Computer simulations, with reflecting boundary conditions at the sides of the growing columns, indicate that the functions $f(x)$ and $g(x)$ in equations 3.174 and 3.175 both have a simple exponential form (e^{-kx}). The exponents τ and α must be related by $\tau = 1 + \alpha$, but the values obtained from relatively small scale simulations, $\tau \approx 1.47$ and $\alpha \approx 0.83$, do not satisfy this relationship very well. Rossi [628] obtained values of ≈ 0.7 and ≈ 1.7 for the exponents α and τ, respectively, using the same model. He also obtained a theoretical estimate of 0.75 for α. Subsequent computer simulations and analytical work indicated that the early simulation results were quite inaccurate. Rossi [629] obtained more accurate estimates for the exponents ($\alpha \approx \tau - 1 \approx 0.85$) and extended the simulation studies to growth from 2- and 3-dimensional surfaces. The numerical results from this work were $\alpha_{2+1} \approx 1.00$, $\tau_{2+1} \approx 2.05$ and $\alpha_{3+1} \approx 1.00$, $\tau_{3+1} \approx 2.18$.

The theoretical model of Rossi focuses attention on the competition between the N tallest needles in a cell of width ℓ, where N is a small integer. Ignoring the shorter needles, which are considered to be screened, this competition can be described in terms of the height increase required to screen the growth of the shortest $N-1$ needles to the extent that their effect on the growth of the remaining needle in the cell of width ℓ can be neglected. An alternative description of the growth process could be based on the height increase required to screen the growth of the shortest of the N needles until its effect on the growth of the $N-1$ remaining needles in the cell of width could be neglected. Unlike some of the shadowing models for needle growth discussed in chapter 5, section 5.7.3, there is no natural division between "screened" and "unscreened" needles.

To maintain the number N of competing needles at a constant value, the cell width ℓ must grow to maintain the self-similarity of the growth process. The competition between the absorbing (weakly screened) needles was analyzed using periodic boundary conditions to represent the effect of neighboring cells. The reliability of the small N approximation is not well established; it breaks down in high dimensionality spaces where the screening is weakened, and the adsorbtion of walkers becomes a collective process in which the background of shorter needles plays an important role. This approximation is weakest for the reflecting boundary condition process in which the screening of the Laplacian field by the tallest needles is least effective.

The growth can be thought of in terms of a hierarchical process in which the cell size ℓ is increased, and the analysis proceeds to a larger cell size, to preserve the number of absorbing needles when the lowest absorbing needles become sufficiently screened. In the $N = 2$, two-absorber version of this model, the exponent α is given by $\alpha_2 = \log 2/\log[\Delta h(2\ell)/\Delta h(\ell)]$, where $\Delta h(\ell)$ is the height increment of the tallest needle in a cell of width ℓ at the stage at which the shorter needle has become screened. In general, $\alpha_N = \log N/\log[\Delta h(N\ell)/\Delta h(\ell)] = \log N/\log(\Delta h_{n+1}/\Delta h_n)$ at the stage where N is the number of needles and Δh_n is the height increment required for the tallest needle to screen the $N-1$ other needles during the nth stage. If α_N converges to a constant value as N is increased (as the cell width containing N needles increases), then the theoretical model indicates scaling with an exponent α_∞.[17]

17. The calculation of α could also be based on an estimation of the height increment needed to screen M out of the N needles. In this case, $\alpha_{N,M} = \log[N/(N-M)]/\log[\Delta h([N/(N-M)]\ell)/\Delta h(\ell)]$.

The results of this theoretical approach depend on the screening criterion and the model used for the probability $P(\ell, \delta)$ that the randomly walking particles will stick to the tallest needle in a cell of size ℓ, when the height difference between the needles is δ. Rossi considered the case in which $P(\ell, \delta)$ has the form

$$P(\ell, \delta) = [1 + C(2\delta/\ell)^a]/2. \tag{3.176}$$

This equation implies that

$$d\delta/dt = C(2\delta/\ell)^a + \eta, \tag{3.177}$$

where η represents the fluctuations in the deposition process. In this process, the fluctuations are important only in the early stages and set the conditions for the subsequent deterministic, asymptotic growth stage [630]. Rossi obtained values of 1.0, 0.75 and 2/3 for the density profile exponent for $a = 1/3$, 2 and 3, respectively, if the shorter needle was considered to be screened when $\delta = \ell/2$. Rossi argued that $a = 2$ – corresponding to adsorbtion of a random walker by the needle that is closest to the point at which a random walker starting at "infinity" first crosses a horizontal line at the height of the tip of the tallest needle (a point selected at random on the line) – is a good approximation to the Laplacian needle growth process.

Rossi [629] studied the competitive growth process in cells containing more than two absorbing needles using Monte Carlo simulations in the spirit of the Monte Carlo renormalization group method[631, 632].[18] A fixed small number N of columns or needles with absorbing tips and reflecting sides was used in the Monte Carlo simulations. The value of about 0.75 estimated for α was insensitive to, but not completely independent of, the parameters N, ℓ and the screening criterion, providing that it was not too weak or too strong.

Several other theoretical investigations have been carried out to calculate the density profile exponent α and other characteristics of the patterns generated by this model. A value of 1.0 has been obtained for α using a continuum approximation that neglects fluctuations [633, 634]. The mean field equations [633, 634] for the growth of the density profile $\rho(h,t)$, with reflecting boundary conditions on the sides of the needles, consist of the equation

$$\partial\rho(h,t)/\partial t = -\Phi(h,t)\partial\rho(h,t)/\partial h \tag{3.178}$$

for the growth in the field $\Phi(h,t)$, and the continuity equation is

$$[\partial\rho(h,t)/\partial t] + [\partial\Phi(h,t)/\partial t] = \partial^2\Phi(h,t)/\partial h^2. \tag{3.179}$$

Here, $\Phi(h,t)$ represents the mean density of walkers or the Laplacian field,

18. In Monte Carlo renormalization group analysis, the linearized renormalization group transformation matrix **T** (chapter 2, section 2.5.1) is calculated from spin correlation functions, via Monte Carlo simulations at the critical point [632].

Kassner [634] has shown that the quasistatic assumption is inaccurate so that the "$\partial\Phi(h,t)/\partial t$" term in equation 3.179 cannot be neglected. This breakdown of the quasistatic approximation is related to the ever-increasing growth rate of the longest needles [635]. Assuming a solution of the form $\rho(h,t) = \rho_\circ + h^{-\alpha} f(h/t^\gamma)$, it can be shown that $\alpha = 1$ and $\gamma = 1/2$ [634], within the mean field approximation.

In a more recent theoretical analysis of this problem [630], the two-absorber approach of Rossi [628, 629] was used in conjunction with conformal mapping [636] and scaling arguments. This conformal mapping approach indicated that, in the case of growth from a line in 2-dimensional space, the exponent a in equation 3.176 has a value of 1 and

$$\rho(h) \sim (\log h/h), \tag{3.180}$$

for absorbing boundary conditions on the sides of the needles. This theoretical result, for $d = 1$, is consistent with computer simulations [630]. Scaling arguments and computer simulations indicated that $\rho(h) \sim (\log h/h)$ for the $1+1$-dimensional model with reflecting boundary conditions as well. For $d = 2$ (deposition onto a plane from 3-dimensional space), an algebraic density decay of the form $\rho(h) \sim h^{-3/2}$ was predicted for the absorbing boundary condition process. This prediction has not yet been adequately tested by computer simulations.

The mean field scaling theory is based on the idea that, for a regular array of needles of equal height protruding from a flat d-dimensional substrate in a $d+1$-dimensional space, the number of absorbing sites $N_a(\ell)$ in a region of size ℓ or volume ℓ^{d+1} scales as $N_a(\ell) \sim (\ell/\xi)^d$, for reflecting boundary conditions, and scales as $N_a(\ell) \sim \ell(\ell/\xi)^d$, for absorbing boundary conditions, where ξ is the characteristic lateral spacing between adjacent needles. The adsorption length ℓ_a is the length at which the product of the number of absorbing sites ($N_a(\ell) \sim (\ell/\xi)^d$ or $N_a(\ell) \sim \ell(\ell/\xi)^d$) and the number of steps in the random walk, which scales as ℓ_a^2, is equal to the volume ($\ell_a^{(d+1)}$). This implies that the absorption length ℓ_a is given by

$$\ell_a = \begin{array}{ll} \xi^{d/2} & \text{for absorbing boundary conditions} \\ \xi^d & \text{for reflecting boundary conditions ,} \end{array} \tag{3.181}$$

except for $d = 1$, where the random walker must move by at least a distance ξ before being absorbed. If the penetration depth is much greater than the needle spacing, then a mean field theory can be used [629, 633, 634], leading to the profile

$$\rho(h) \sim h^{-1}, \tag{3.182}$$

for reflecting boundary conditions, and

$$\rho(h) \sim h^{-2}, \tag{3.183}$$

for the absorbing boundary condition process, which is a better approximation to diffusion-limited deposition [635, 637]. Mean field behavior is expected when $d \geq d_c$, with $d_c = 2$ for reflecting boundary conditions and $d_c = 3$ for absorbing boundary conditions, in accordance with computer simulations [629]. A more complete exposition of competitive needle growth mediated by a Laplacian field has been provided by Krug [637].

Dogterom and Liebler [638, 639] have modified the 2+1-dimensional branchless DLA model by including a finite concentration of walkers, reversibility and monomer degradation, to simulate the assembly of linear protein fibers.

In an attempt to bridge the gap between the needle growth model and diffusion-limited deposition, Devillard and Stanley [640] studied a square lattice model in which only one generation of side branching was allowed (linear, needle shaped side branches were allowed to grow in the lateral direction from each main stem or "trunk"). Sticking was allowed only at the tips of the trunk, the sides of the trunk or the tips of the first order side branches. A density profile of the form $\rho(h) \sim h^{-\alpha}$, with $\alpha = 0.34 \pm 0.04$, was found. This value for α is very close to the value obtained from $1+1$-dimensional, square lattice, diffusion-limited deposition model simulations, with no side branching restrictions. The "tree" size distribution was also measured, and a value of 1.61 ± 0.02 was obtained for the size distribution exponent τ. This value for τ is in good agreement with the the value calculated from equation 3.152, if it is assumed that $D = 2 - \alpha$ ($\tau = 1 + [1/(2 - \alpha)]) \approx 1.60$). However, the individual trees may have self-affine scaling properties and the exponents τ and α may be related via equation 3.156. The needle growth model and this model can be regarded as the first two members of a series of models that approach the square lattice DLA limit from the needle growth limit. It appears that convergence to the DLA limit is rapid. The patterns generated by the restricted side branching model of Devillard and Stanley resemble quite closely those generated by the noise reduced diffusion-limited deposition model.

Graff and Sander [635] have analyzed a growth model in which a needle approximation is used to calculate the increase in mass of the trees in a diffusion-limited deposition process, but the trees are considered to have a fractal dimensionality of D_t. Consequently, the rate of height increase $\partial h/\partial t$ is related to the rate of mass increase $\partial m/\partial t$ by $\partial h/\partial t = h^{D_t - 1}\partial m/\partial t$. This controls the runaway increase in the height of the tallest needles or trees and provides a more realistic representation of the diffusion-limited deposition process. In this model, the flux of random walkers along an entire needle, not just the flux at the tip, is used to calculate the growth rate. The $1+1$-dimensional version of this model reproduces most aspects of the corresponding diffusion-limited deposition model, including the self-affine scaling of the deposit [573, 574]. A mean field theory for this model gave results that were consistent with simulations carried out using the corresponding diffusion-limited deposition model simulations. Krug [637] has argued that this approach must break down for $d > d_c$, where d_c is the critical dimensionality of the absorbing needle model, because the needle forest is no longer able to mimic the strong screening that is characteristic of DLA in high-dimensionality spaces.

Additional information on the mean field approach to Laplacian needle growth and its relationship with the full DLA problem can be found in reviews by Krug [637, 641].

The growth of DLA clusters has also been idealized by the growth of n needle shaped "spokes" from a central point or "hub" [642, 643]. Derrida and Hakim [436] used a conformal mapping approach to study the growth of n needles of zero width in a Laplacian field, with a tip velocity proportional to the field gradient at the tip and with the boundary conditions $\Phi = 1$ at infinity and $\Phi = 0$ on the surfaces of the needles. For the case of n needles growing in n equivalent directions, with angles of $2\pi/n$ between each pair of adjacent needles, they showed that patterns with more than six arms are unstable with respect to growth in fluctuations in the needle lengths, a result which was obtained earlier by Ball [644]. A similar analysis of the growth of two "arms" separated by an angle θ shows that for $\theta > \pi(1 - (2/3)^{1/2})$, or $\theta > 33.0306\ldots$ degrees, the two needles grow to essentially the same length [436]. For smaller angles, the symmetry is broken and the length ratio between the two needles becomes unequal. These theoretical predictions appear to be confirmed by computer simulations carried out using absorbing boundary conditions on the sides of the needles [645]. The symmetry breaking found in this model also occurs in related models [646, 647] for Laplacian needle growth.

Derrida and Hakim also extended their analysis to explore how the stability of the needle armed patterns depends on the growth exponent η. As might be expected intuitively, the number of arms that can grow to form a stable symmetric pattern decreases with increasing η. They found that $n = 5$ is unstable for $\eta > 10/7$, $n = 4$ is unstable for $\eta > 2$ and $n = 3$ is unstable for $\eta > 6$. The general stability condition can be expressed as $\eta < 4/(n - 2)$ for even n and $\eta < 4n/(n^2 - 2n - 1)$ for odd n.

3.8 Noise Reduction and Deterministic Models

The main focus of this chapter is on random growth models. This is appropriate since "randomness" plays an important role in most natural phenomena. However, randomness enters these models in an essentially uncontrolled fashion, and they are too random to represent many natural processes. For example, it was realized by Witten and Sander [181] that the DLA model might provide insight

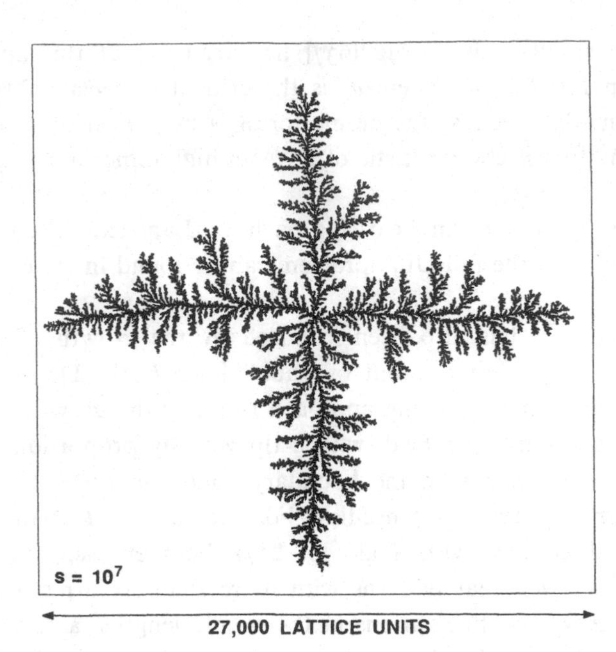

Figure 3.40 A cluster of 10^7 sites grown on a square lattice with the efficient DLA model of Brady and Ball [422].

into the dendritic morphology associated with many crystal growth processes, but the structures generated by simple DLA models are much too disorderly for this purpose. If DLA clusters are grown to a large size on a square lattice, then a more organized, large length scale structure does begin to appear, as the cluster of 10^7 sites shown in figure 3.40 indicates. Very similar patterns can be generated using much less computer time, because they consist of much fewer particles or sites, using the corresponding noise reduced DLA model. However, for very large noise reduction the random walk DLA approach becomes inefficient.

Figure 3.41 shows 50 000-site square lattice DLA clusters grown using the noise reduction approach described earlier in this chapters with noise reduction parameters of 2, 3, 10 and 30. The clusters shown in figures 3.41(a) and 3.41(b) were generated using noise reduction parameters of $m = 2$ and $m = 3$, respectively. They look very much like square lattice DLA clusters grown to a very much larger size, without noise reduction [648] (see figure 3.40). This supports the idea that noise reduction allows the asymptotic structure associated with growth models to be explored, without growing extremely large clusters. Noise reduction has also been used to enhance the effects of lattice anisotropy in 3-dimensional (cubic lattice) and 4-dimensional [649] (hypercubic lattice) simulations. Figure 3.42 shows a 43 253-site cubic lattice cluster generated with a noise reduction parameter of $m = 30$.

The value of noise reduction as a means of approaching the asymptotic limit is uncertain. Large noise reduction parameters enhance the effects of lattice

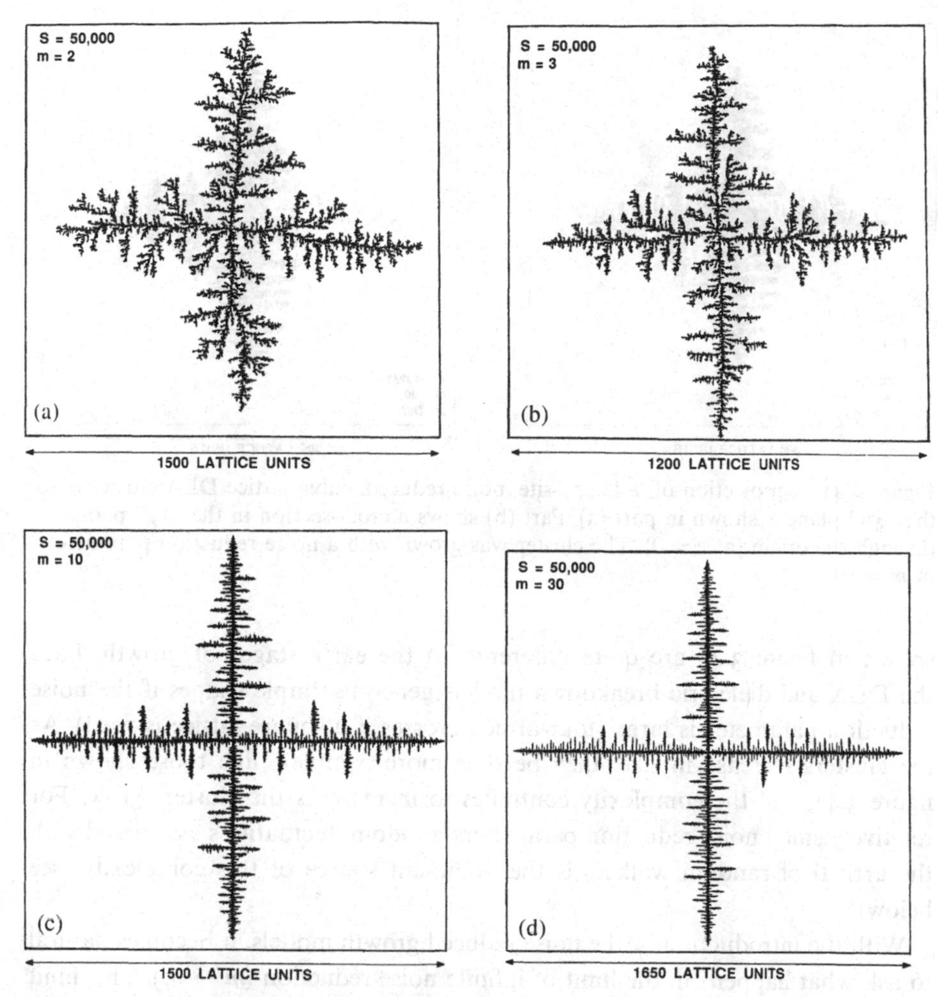

Figure 3.41 Clusters of 50 000 sites grown on a square lattice using a noise reduced algorithm with noise reduction parameters m of 2, 3, 10 and 30 in parts (a), (b), (c) and (d), respectively. The clusters in parts (a) and (b) should be compared with the 10^7-site DLA cluster grown without noise reduction ($m = 1$) in figure 3.40.

anisotropy, but they also suppress the development of ramified structures by side branching or tip splitting. This leads to the idea [604] that there might be an optimum value for m. As the noise reduction parameter m is increased, the patterns seem to become more sensitive to the details in the growth rules. This is intuitively reasonable if randomness is playing a smaller role. Figure 3.43(a) shows a cluster grown using the square lattice DLA model, with a noise reduction parameter of $m = 10\,000$, and figure 3.43(b) shows a cluster generated with the random walk dielectric breakdown model, also with $m = 10\,000$. While clusters generated using these two models with $m = 1$ can be difficult to distinguish (the dielectric breakdown model cluster is locally more compact), the two clusters

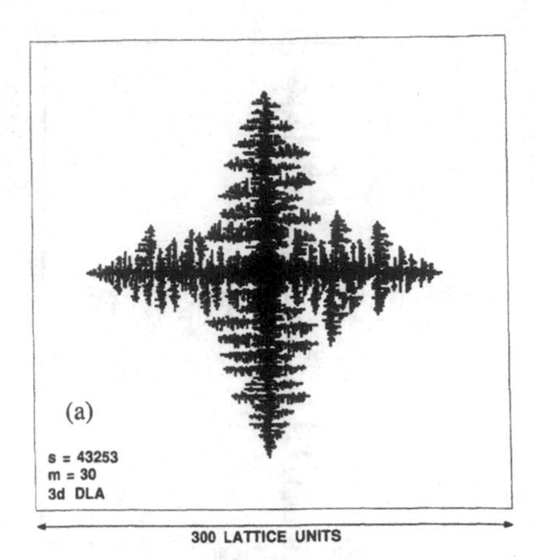

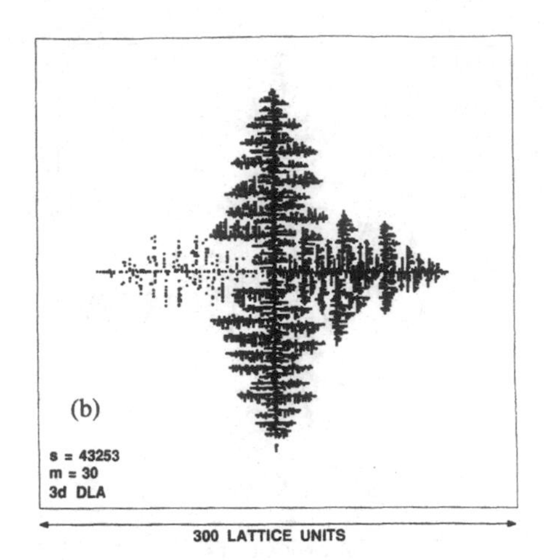

Figure 3.42 A projection of a 43 253-site, noise reduced, cubic lattice DLA cluster onto the "xy" plane is shown in part (a). Part (b) shows a cross-section in the "xy" plane through the origin (at $z = 0$). The cluster was grown with a noise reduction parameter of $m = 30$.

shown in figure 3.43 are quite different. In the early stages of growth, both the DLA and dielectric breakdown models generate simple shapes if the noise reduction parameter is large (four-armed crosses if a square lattice is used). As the clusters increase in size, they become more complex, like those shown in figure 3.43, and the complexity continues to increase as the clusters grow. For relatively small noise reduction parameters, random fluctuations associated with the arrival of random walkers is the dominant source of this complexity (see below).

With the introduction of the noise reduced growth models, it becomes natural to ask what happens in the limit of infinite noise reduction ($m \to \infty$). This limit cannot be approached using random walk models. Instead, the growth rates must be calculated by numerically solving the Laplace equation, with $\Phi = 0$ at occupied lattice sites and $\Phi = 1$ at "infinity". As in the noise reduced DLA model, a score σ_i is associated with all of the unoccupied perimeter sites. New unoccupied perimeter sites, generated by growth events, are assigned a zero score. However, pre-existing unoccupied perimeter sites that neighbor newly filled sites retain their previous scores. At each stage in the growth process, the quantities δ_i given by

$$\delta_i = (1 - \sigma_i)/R_i, \tag{3.184}$$

where R_i is the growth rate for the ith site, are calculated. The sites with the lowest value for δ_i (there are four or eight of them by symmetry for the square lattice) are filled and the scores associated with all the other perimeter sites are incremented, so that $\sigma_i \to \sigma_i + \delta_{min} R_i$ (here δ_{min} is the minimum value for δ_i).

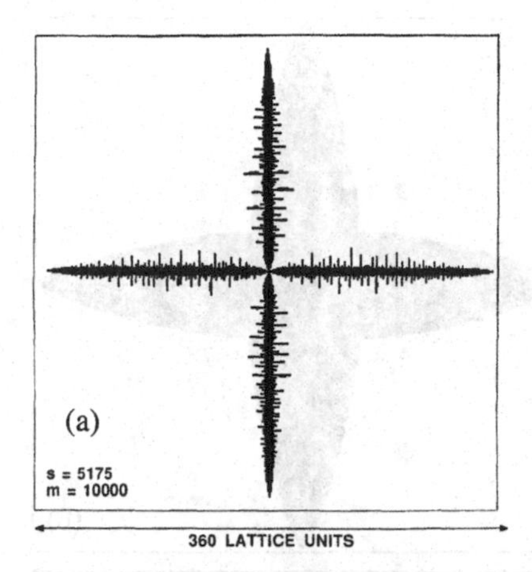

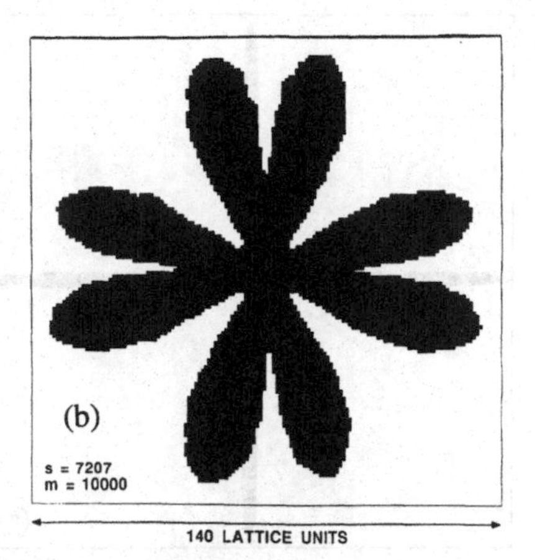

Figure 3.43 Clusters of similar size grown on a square lattice with a noise reduction parameter of $m = 10\,000$. The cluster in part (a) was grown using a DLA model, whereas the cluster in part (b) was generated using a corresponding random walk dielectric breakdown model.

The patterns obtained using this approach [605] are in excellent agreement with patterns generated using standard, Monte Carlo DLA models with very large noise reduction parameters [650].

Some DLA model simulations indicate that, even in the $m \to \infty$ limit, complex patterns eventually emerge [603]. However, a careful investigation by Kassner [604] has shown that for Laplacian growth on a square lattice with DLA boundary conditions, side branching is a consequence of the anisotropy that arises when the discrete Laplace equation is solved numerically with a "circular" outer boundary. As the radius of this outer boundary is increased, side branching first appears at larger and larger cluster sizes. It appears that similar effects may compromise the reliability of other noiseless growth model simulations.

Other growth models have been studied in the infinite noise reduction limit. For example, figure 3.44 shows clusters of about 25 000 sites grown on a square lattice with "infinite noise reduction", using the screened growth model [651]. The patterns that are formed are symmetric and have four compact arms, like square lattice DLA model clusters with infinite noise reduction. The screened growth model has several advantages over Laplacian growth models. There is no outer boundary, the model is more efficient than the Laplacian growth model (larger clusters can be grown) and the growth probabilities can be calculated with greater accuracy because the algorithm does not require the solution of the discretized Laplace equation or a similar equation. However, the DLA model is of much more intrinsic interest, and the numerical problems could probably be reduced via a Green's function approach [165].

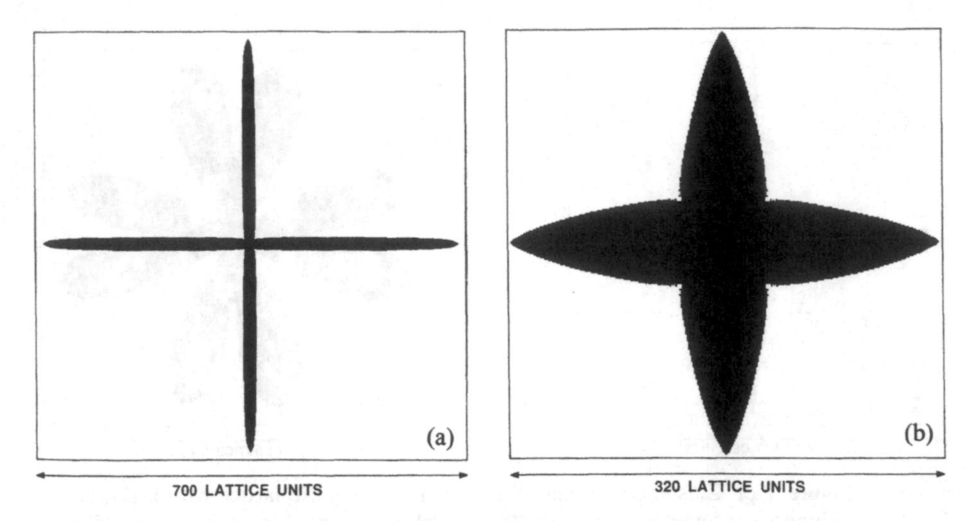

Figure 3.44 Clusters of about 25 000 sites grown on a square lattice using an infinite noise reduction screened growth model. Part (a) shows a cluster grown with $A = 0.1$ and $\epsilon = 1.25$ in the screening function given by equation 3.115. For the cluster shown in part (b), $A = 0.1$ and $\epsilon = 1.75$.

It is apparent from figure 3.43 that the square lattice DLA cluster takes on a more complex form by a process of "side branching", while the dielectric breakdown model cluster grows by "tip splitting". Similar behavior [603] has been found in triangular lattice Laplacian growth model simulations with DLA model and dielectric breakdown model boundary conditions in the infinite noise reduction limit. Simulations with DLA model boundary conditions indicate that the growth pattern evolves from a pattern with six needle shaped arms into a pattern of six needles with side branches. For simulations with dielectric breakdown model boundary conditions, the growth pattern evolves from an almost circular shape via tip splitting into a pattern with six-fold symmetry and 12 arms. Since the dielectric breakdown model favors growth in sites that have a relatively high coordination number, it can be considered to correspond to random growth with a weak surface tension. It appears that this small surface tension, possibly combined with a weakening of the lattice anisotropy effects, is sufficient to bring about a transition from growth via side branching to growth via tip splitting.

A variety of other approaches has been used to control the fluctuations in dielectric breakdown and related models. For example, in the "maximum-likelihood model" [572, 652], the maximum growth probability P_{max} (normalized so that $\sum_i P_i = 1$) is calculated and the $n = \lfloor(1/P_{max}) + (1/2)\rfloor$ unoccupied perimeter sites, with the highest growth probabilities, are then filled after n has been increased ("overfill" model) or decreased ("underfill" model) to satisfy the symmetry of the lattice. Here, "$\lfloor x \rfloor$" stands for the integer part of x. Both the overfill and underfill versions of this model give complex patterns that have

Figure 3.45 A cluster of 5077 sites grown on a square lattice using the deterministic overfill model. This cluster was obtained with a fixed potential on the sites that best approximate an unbroken circle with a radius of 125 lattice units, centered on the cluster origin, and a zero potential on the occupied sites. The dots on the periphery identify the unoccupied perimeter sites with potential gradients that are more than $10^{-3.5}$ times the maximum potential gradient. This figure was provided by R. Richter.

scaling properties similar to those associated with the DLA model. Figure 3.45 shows a cluster of 5077 sites generated using the overfill model. For clusters generated on a square lattice, the envelope is a well defined square (rotated 45° with respect to the lattice). This means that the theoretical arguments of Turkevich and Scher [440], discussed below, can be applied, and a fractal dimensionality of 5/3 should describe the dependence of the cluster size on its mass (number of occupied sites). For this model, the cluster envelope has a simple shape and the width of the active zone, as defined by Plischke and Rácz [402], grows in the same manner as the radius of gyration, since it is determined by the anisotropy of the square envelope shape.

3.8.1 Lattice Structure Effects

The structure of the lattice has a large effect on the interplay between anisotropy and noise reduction. In general, the effects of anisotropy are largest if there are fewer distinct growth directions. This is illustrated in figure 3.46, which shows a relatively small DLA cluster grown on a triangular lattice in which the random walkers can move in any of the six possible directions, but bonds can be formed between the sites in the cluster and the unoccupied perimeter sites that are entered by the random walkers in only three of the six possible directions (at angles of 60°, 180° and 300° with respect to the $+y$ direction). If the random walker enters an unoccupied perimeter site from which a bond to a

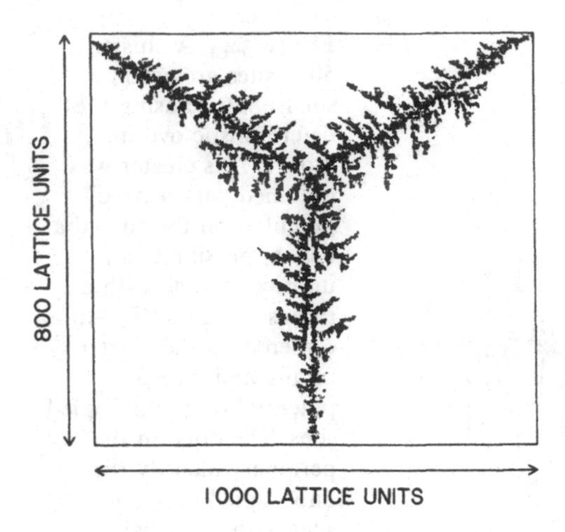

Figure 3.46 A cluster of 40 000 sites grown on a triangular lattice using a DLA model in which random walkers move in all of the six possible directions between adjacent cells or nodes, but sticking (bond formation) is allowed in only three of the six possible directions.

nearest-neighbor filled site cannot be formed, in one of the three symmetrically disposed bonding directions, then the random walker continues to a randomly selected *unoccupied* nearest-neighbor site. Even without noise reduction, the anisotropy of the cluster is quite pronounced. Clusters generated using a DLA model in which the random walkers can move in only three of the six possible directions, but sticking occurs when the random walkers enter any unoccupied perimeter site, allowing bonds to be formed in all six possible directions, are much less anisotropic. This demonstrates that the restriction in the growth directions is much more important than the restriction in the number of step directions in the random walks [653].

A more systematic study of the effects of the number of growth directions can be carried out using the "semi-lattice" model [653] shown in figure 3.47. In this model, the particles consist of monodisperse discs. The discs are added to the growing cluster via completely off-lattice random walk trajectories. When the randomly walking particle contacts a particle in the cluster, the contacting particle is rotated about the contacted particle until the vector from the contacted to the contacting particle points in the nearest of n symmetrically disposed directions. In the noise reduced version of this model, counters are associated with each of the n directions, for each particle. When the score, associated with a particular particle and direction, reaches the threshold value of m, a particle is added in the corresponding position and the counters associated with the new particle are set to zero.

Figure 3.48 shows clusters grown using the semi-lattice model with a noise reduction parameter of $m = 100$ and $n = 3$ to $n = 8$ growth directions. For $n = 3, 4$ and 5, the clusters have n distinct arms. For $n = 6$, the results depend on the model details. For the triangular lattice model, six distinct arms grow, but for the semi-lattice model, as is illustrated in figure 3.48, this is not the

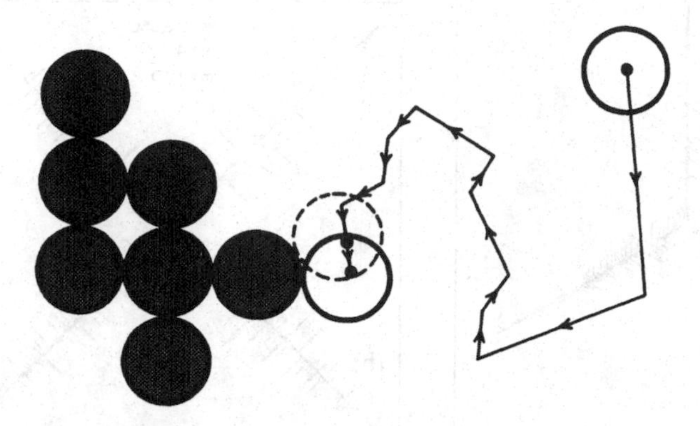

Figure 3.47 A semi-lattice model for the growth of DLA clusters with n symmetrically disposed growth directions. The added particles follow off-lattice trajectories until they contact the cluster (dotted circle). The vector from the contacted particle, in the cluster, to the contacting particle is then rotated to the nearest of the n directions. This moves the particle to the position indicated by the full circle, at the end of the trajectory, where it is permanently incorporated into the cluster. The $n = 4$ case is illustrated in this figure.

case. For $n \geq 7$, there is no obvious way of determining the number of arms in the cluster, and the clusters become more like off-lattice DLA clusters as the number of arms increases. These results are consistent with the theoretical work of Ball [644] and Derrida and Hakim [436], which showed that patterns with more than six needle arms lose their six-fold character under Laplacian growth.

Square lattice DLA model simulations [648, 654] and simulations with closely related models [357] indicate that the aspect ratio of the cluster arms reaches a constant asymptotic value as the cluster size increases. This suggests that the internal structure of the cluster can be characterized by the exponent β that describes the growth of the radius of gyration. If this is the case, then the internal structure has a fractal dimensionality that is significantly smaller than that of an off-lattice DLA cluster. The effective fractal dimensionality D_β and the "inside" angle at the arm tips decrease as the number of arms or growth directions is decreased. The decrease in the fractal dimensionality with decreasing tip angle is in accordance with the theoretical model of Turkevich and Scher (described below, in section 3.10.2).

The effects of anisotropy can be reduced by growing clusters on a triangular lattice or by allowing random walkers to stick when they reach unoccupied nearest-neighbor (*nn*) and next-nearest-neighbor (*nnn*) sites, with different sticking probabilities at the *nn* and *nnn* sites to "tune out" the anisotropy. The effective anisotropy of square lattice DLA can also be controlled by partial transfer of the particle between *nn* and *nnn* sites, after the active zone has been reached. It is sometimes convenient to construct DLA clusters

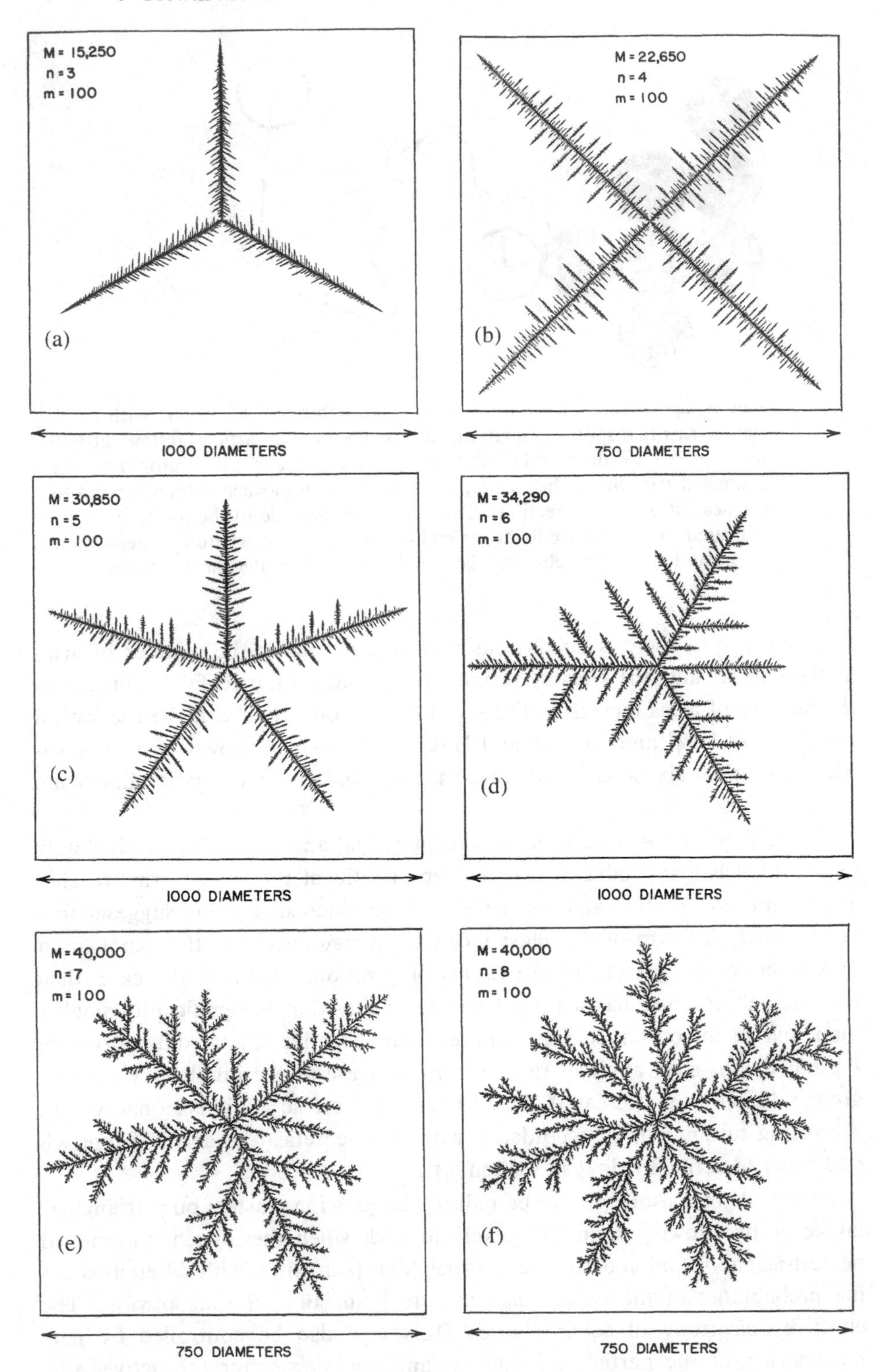

Figure 3.48 Clusters containing from 15 250 to 40 000 particles grown using the 2-dimensional semi-lattice DLA model with a noise reduction parameter of $m = 100$. Clusters with $n = 3$ to $n = 8$ growth directions are shown in parts (a) – (f).

on a lattice, but for most applications it is better to use off-lattice aggregates. There are several other ways in which the lattice anisotropy can be controlled. One approach is based on the semi-lattice DLA model shown in figure 3.47. Instead of rotating the bond between the added (contacting) and contacted particle to the closest of n discrete directions, the direction of the newly formed bonds can be merely biased towards these directions.

Another approach [655] is to start with a regular lattice, displace the points at the centers of the lattice sites and construct the Voronoi tessellation [99, 107] (chapter 1, section 1.3.2) for the displaced points. Moukarzel [655] has used this approach to simulate infinite noise reduction Laplacian growth models with DLA and dielectric breakdown model boundary conditions. In this model, the discretized Laplace equation is solved via the Kirchhoff equations for an equivalent conducting network. The "conductivities" between adjacent cells i and j are given by $\sigma_{ij} \sim \ell_{ij}/r_{ij}$, where ℓ_{ij} is the length of the common edge and r_{ij} is the distance between the cell centers. Moukarzel [655] showed that even an infinitesimal amount of disorder of this type will remove the effects of anisotropy, with DLA boundary conditions, while, for the dielectric breakdown model boundary conditions, there is a slow transition from orderly to disorderly patterns as the displacements are increased. Infinitesimal displacements have a large effect on the topology of the lattice obtained after the Voronoi construction and add new cells to the active zone for the DLA boundary condition models. Consequently, it is not possible to approach the small anisotropy limit for the DLA models using these lattices with non-zero displacements. However, for dielectric breakdown model boundary conditions, the conductivity σ_{ij} between cells sharing short edges is small, and it approaches zero in the small anisotropy limit.

The anisotropy can also be controlled and enhanced by restricting the growth of the pattern to sites with coordinates $(m'i, m'j)$, where m', i and j are integers [556, 656]. The constant m' plays a role similar to that of the noise reduction parameter m in noise reduced DLA. In the model of Matsushita and Kondo [656], the random walkers are allowed to step onto any lattice site. When a random walker reaches an unoccupied perimeter site it is terminated, but growth occurs only in sites with coordinates $(m'i, m'j)$.

3.9 Models with Quenched Disorder

The interplay between anisotropy and random fluctuations plays an important role in DLA models. In the standard DLA model, the disorder is said to be "annealed" because there are no "time" correlations (apart from the position of the moving boundary, the addition of a random walker to the growing cluster is not related in any way to the paths followed by the random walks generated at earlier stages of the simulation). Very similar patterns can be generated using quenched disorder models in which the disorder is associated with the substrate on which the growth takes place and does not vary during the simulation. An example of a model of this type would be to modify the standard dielectric breakdown model by assigning random thresholds t_i to all of the bonds in the lattice at the beginning of a simulation. The simulation would then proceed by filling the unoccupied perimeter site connected to the growing cluster by the bond with the maximum value of $\delta\Phi_i/t_i$, where $\delta\Phi_i$ is the difference in the scalar field across the ith bond. This model has been studied by Family *et al.* [657].

Hansen *et al.* [658] have suggested that a distribution of thresholds $\{t\}$ for the quenched disorder, Laplacian growth model can be obtained from the corresponding annealed disorder model. In the annealed model simulation, each of the bonds has the same conductivity and is initially given a threshold of $t_i(0) = 1$. Each time a bond (bond j) is picked in the annealed disorder simulation, all of the bonds that have not previously been occupied are updated according to the algorithm

$$t_i(n) = max\left[t_i(o), \frac{\delta\Phi_i}{\delta\Phi_j}t_j(o)\right], \tag{3.185}$$

where $t_i(n)$ is the new threshold for the ith bond and $t_i(o)$ is the old threshold for the ith bond. Using these bond thresholds, the quenched disorder (deterministic) model will generate *exactly* the same cluster as the annealed disorder model used to generate the cluster. This procedure does not determine all of the thresholds. Only limits are found for the bonds that were not used in the annealed model simulation. Using this approach, Hansen *et al.* found that the probability density $P(t)$ for the distribution of thresholds that are overcome during the growth of a 2-dimensional dielectric breakdown model pattern, using this preconditioned, deterministic, quenched disorder Laplacian growth model, could be described by the power law form

$$P(t) \sim t^{-(1+\gamma)} \tag{3.186}$$

over a significant range of thresholds. They obtained values of 0.07, 0.16, 0.43 and 0.46 for the exponent γ in equation 3.186 from simulations with $\eta = 1, 2, 3$ and 4, respectively, where η is the growth probability exponent (equation 3.25).

A variety of quenched disorder DLA models have been developed to represent processes such as fluid–fluid displacement in inhomogeneous porous media [398, 610, 659, 660] and pattern formation on fractal substrates [427, 434]. This work will be described elsewhere [57].

3.9.1 Growth in High-Dimensionality Spaces

In comparison with the enormous amount of numerical work on 2-dimensional models, 3-dimensional models and simulations with $d > 3$ have been comparatively neglected. This is unfortunate since 3-dimensional processes are of obvious practical importance, and the results from 2-dimensional simulations do not necessarily provide a reliable guide as to what will happen in 3-dimensional space. The topological properties of 2-dimensional and 3-dimensional space are quite different. In particular, the possibility of forming loops, which can enclose large regions in 2-dimensional space, is absent in three dimensions. Loops do not appear to play an important role in 2-dimensional DLA models, but large regions of space are almost enclosed by adjacent branches, resulting in strong screening of the Laplacian field in the almost enclosed regions that play an important role in the growth process. The study of processes such as DLA in higher, $d > 3$, dimensionality spaces is also important, for theoretical reasons. A complete theory for DLA should predict the scaling structure in spaces of all d, and comparisons between theory and simulations for $d > 3$ provide a valuable check on theoretical results. Many growth processes become simple in spaces of high enough dimensionality ($d > d_c$), and the analytical results obtained for $d > d_c$ provide a basis for developing an understanding in lower-dimensional spaces, but this is not the case for DLA.

It is particularly unfortunate that more work has not been done on 3-dimensional DLA. It is now possible to grow quite large DLA clusters using both cubic lattice and off-lattice algorithms [424]. Figure 3.49 shows large clusters grown using both models. The large cubic lattice model cluster shown in figure 3.49 shows evidence of the effects of lattice anisotropy, and these effects become more apparent if noise reduction is used, as is illustrated in figure 3.42.

DLA simulations have been carried out in spaces and lattices with dimensionalities of up to eight [424]. The fractal dimensionalities (D_β, obtained from the growth of the cluster radius of gyration) are $D = 1.715 \pm 0.004$ for $d = 2$, 2.49 ± 0.01 for $d = 3$ and $D = 3.40 \pm 0.02$ for $d = 4$. Effective fractal dimensionalities of about 4.33, 5.4, 6.6 and 7.5 were obtained for $d = 5$, 6, 7 and 8, respectively [424]. As d increases, the cluster sizes become smaller as a result of practical difficulties, and the range of length scales L/ϵ becomes much smaller. For $d \geq 5$, the radius of gyration of the largest clusters generated so far is less than ten particle diameters, and there are large uncertainties in the values estimated for D_β.

Figure 3.50 shows the dependence of the exponents β and $\overline{\nu}$ that describe the growth of the cluster radius of gyration and width of the active zone on cluster size for 3-dimensional off-lattice DLA clusters. This figure indicates that the scaling structure of 3-dimensional off-lattice DLA clusters has the same complexities as that of the 2-dimensional clusters. Consequently, it

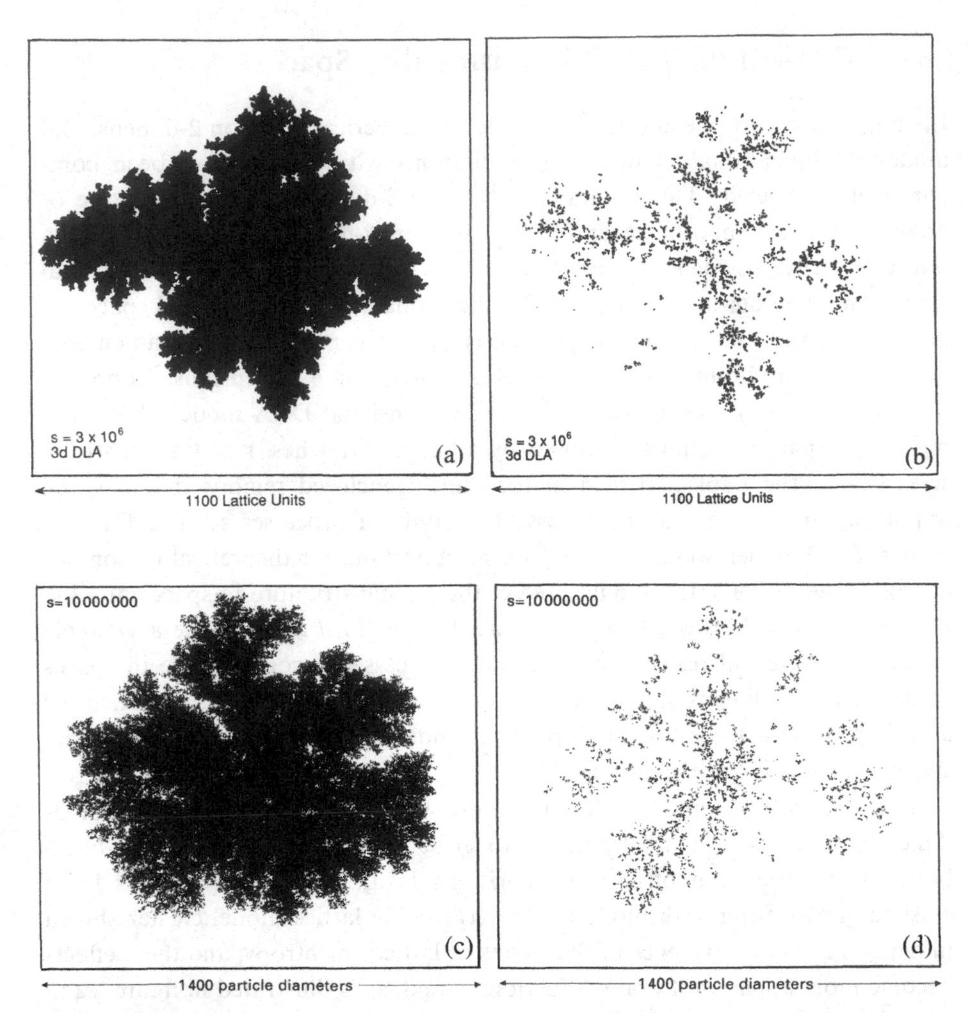

Figure 3.49 Projections (parts (a) and (c)) and planar cuts ((b) and (d)) through the origin for large cubic lattice and 3-dimensional off-lattice DLA clusters. The lattice model cluster, on the top row, contains 3×10^6 sites and the off-lattice cluster consists of 10^7 identical spheres. For the off-lattice cluster, only 5×10^5 of the 10^7 particles are shown. The cut through the origin intersected 14 313 particles and the intersected particles are represented by dots in the lower right corner. The off-lattice cluster was generated by T. Rage.

appears that this behavior is not related to the unique properties of random walks in two dimensions or the formation of large regions that are almost enclosed by the arms of 2-dimensional DLA clusters that result in a completely different distribution of *small* growth probabilities in the 2-dimensional and 3-dimensional cases. There is a clear need for more work on 3-dimensional DLA.

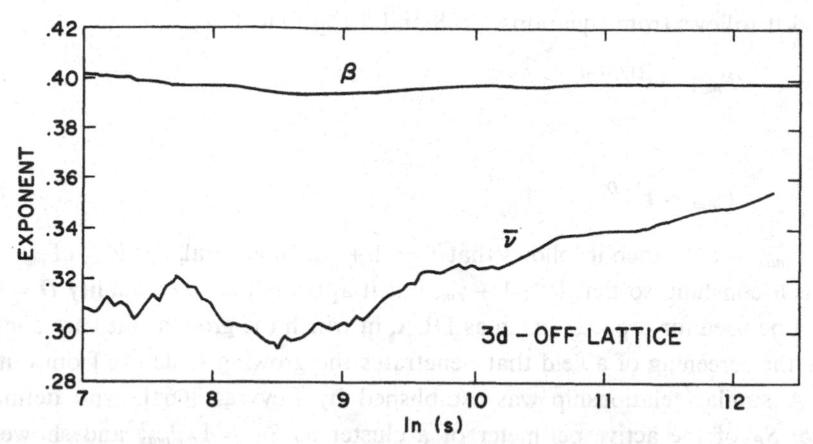

Figure 3.50 Cluster size dependence of the effective exponents β and $\bar{\nu}$ that describe the growth of the cluster radius of gyration and width of the active zone, with increasing cluster size s for 3-dimensional, off-lattice DLA clusters. The data displayed here were obtained from 86 3.5×10^5-particle clusters. The exponents were obtained from the dependence of $\log R_g$ and $\log w_a$ on $\log s$ for clusters with sizes in the range $s_1 - s_2$, with $s_2/s_1 = 1.05$, where $s = (s_1 + s_2)/2$.

3.10 Theoretical Methods

There are no generally successful or routine theoretical methods that can be used to predict the outcome of simple growth algorithms. In some cases, important results can be obtained quite easily, but in other cases considerable theoretical effort has led to little progress. From a mathematical point of view, there are very few rigorous results for DLA, and these results [391] (bounds on the fractal dimensionality such as $D > (d+1)/2$ and $D \leq d$) are of no value in physics.

Although there are no general theoretical approaches that can be applied to non-equilibrium growth processes, there are a few simple ideas that are of general value [661, 662]. For example, if it can be assumed that the number of particles or sites s and radius r of a growing cluster are related by

$$s \sim r^D, \tag{3.187}$$

then it follows that

$$dr/ds \sim s^{(1/D)-1}. \tag{3.188}$$

For many cluster growth processes, the most exposed part of the cluster will have the largest growth probability, and, on average, when growth does occur at this position, the cluster radius will increase by an amount of the order of one lattice unit or particle diameter. This means that

$$dr/ds \sim P_{max}, \tag{3.189}$$

and it follows from equations 3.188 and 3.189 that [662]

$$P_{max} \sim s^{(1/D)-1} \tag{3.190}$$

or

$$P_{max} \sim r^{1-D}. \tag{3.191}$$

If $P_{max} \sim r^{-\gamma_m}$, then it follows that $D = 1+\gamma_m$. In general, $dr/ds \leq cP_{max}$, where c is a constant, so that $D \geq 1+\gamma_m$, but it appears that the equality $D = 1+\gamma_m$ can be used for processes such as DLA, in which the growth rates are controlled by the screening of a field that penetrates the growing structure from outside.

A similar relationship was established by Leyvraz [662], who defined the size S_P of the active perimeter of a cluster as $S_P = 1/P_{max}$ and showed that S_P cannot grow faster than R^{D-1}, where R is a length, such as the radius of gyration, characteristic of the overall size of the cluster.

One of the most important classes of growth processes is particle–cluster aggregation, in which particles are added, one at a time, to a growing cluster or aggregate via trajectories with a fractal dimensionality D_w. Although it is difficult to obtain any rigorous results for this class of models, Ball and Witten [661] showed that it was possible to obtain a useful lower limit on the fractal dimensionality using the very simple argument that follows.

If a particle following a trajectory with a fractal dimensionality of D_w enters the region occupied by a growing cluster of radius r, then the average number of contacts between the particle and the cluster is given by

$$N \sim r^{D_w+D-d} \tag{3.192}$$

Since D_w+D-d is always positive, the particle always contacts the aggregate and the aggregate behaves like an absorbing hyperspherical surface with a radius of order r. The rate at which particles will enter the region occupied by the cluster and be trapped is given by[19]

$$R_t \sim r^{(d-D_w)}, \tag{3.193}$$

and it follows from equations 3.191 and 3.193 that the maximum capture rate or growth rate for any particle in the cluster is given by

$$R_{cm} \sim r^{1-D+d-D_w}. \tag{3.194}$$

19. After a time t (or after t steps), a particle following a trajectory with a fractal dimensionality D_w will have moved through a distance $\mathscr{R}(t)$, which scales as $\mathscr{R}(t) \sim t^{1/D_w}$. This means that all of the particles within a volume proportional to $(\mathscr{R}(t)^d \sim t^{d/D_w}$ will have traveled far enough to reach an absorbing hypersphere of radius r. The probability P_a that a typical trajectory will actually have entered the absorbing region of size r is given by $P_a \sim (r/\mathscr{R})^{d-D_w}$ (if $d > D_w$), and the total number of adsorbtion events N_a scales as $N_a \sim \mathscr{R}(t)^d[r/\mathscr{R}(t)]^{d-D_w} \sim r^{(d-D_w)}R(t)^{D_w} \sim r^{(d-D_w)}t$. Consequently, the rate of adsorbtion, R_t, scales as $R_t \sim r^{(d-D_w)}$.

The growth rate cannot be greater than that of a completely unscreened single particle, so that

$$1 - D + d - D_w \leq 0 \tag{3.195}$$

or [661]

$$D \geq d - D_w + 1. \tag{3.196}$$

For the Eden, ballistic and DLA models, $D_w = 0, 1$ and two, respectively. Consequently, equation 3.196 implies that $D = d$ for the Eden and ballistic aggregation models and that $D > d - 1$ (for all d) for the DLA model.

Because of the wide range of applications, the aesthetic qualities of DLA clusters and the idea that it should be possible to understand such a simple process in rigorous quantitative terms, many attempts have been made to develop a better understanding of DLA. Progress has been made, but a theory that would allow exact values for any of the important scaling exponents to be calculated (even in principle) still seems to be quite far away. Many of the theoretical methods that have been applied to this problem are based on assumptions concerning the asymptotic scaling structure. Several theoretical approaches lead to estimates for the fractal dimensionality that are in remarkably good agreement with the measured values for D_β obtained from the dependence of the radius of gyration on the number of particles in a cluster. However, as is discussed in some detail earlier in this chapter, considerable uncertainty remains concerning the interpretation of the simulation results. It is now apparent that the long range anisotropy associated with the lattice structure [422] can have an important effect on the asymptotic scaling properties. Consequently, care should be taken to distinguish between theories for off-lattice DLA and theories that are based on a lattice and may implicitly include the effects of long range anisotropy.

There are few exact results on which a theory of DLA can be built. Makarov [663] has shown that the information dimensionality $D(q = 1)$ (see chapter 2 and appendix A) of the growth probability measure is 1 for $d = 2$. It has been speculated that $D(q = 1) = d - 1$ for all d, but this does not appear to be true for all self-similar fractals. In addition, Halsey [664] has shown that

$$D = \tau(3) + 2 - d, \tag{3.197}$$

where $\tau(3)$ is one of the partition function exponents defined in chapter 2, or

$$D = 2D(q = 3) + 2 - d. \tag{3.198}$$

At present, these appear to be the only reliable, exact results.

3.10.1 Mean Field Theories

Mean field theories for fractal growth are based on the assumption that the strongly fluctuating, but strongly correlated, density $\rho(\mathbf{r}, t)$, in a random fractal, can be replaced by a "smeared out" radial density function $\rho(r, t) = <\rho(\mathbf{r}, t)>_{|\mathbf{r}|=\mathbf{r}}$, where $\rho(\mathbf{r}, t)$ is the cluster density at position $\mathbf{r}$ with respect to the cluster seed or origin. This is a drastic approximation that takes away the essential fractal nature of the DLA internal structure. For this reason alone, mean field theories cannot be expected to provide a completely satisfying understanding of DLA.

In the mean field description of DLA, growth will take place in the region where the density profile $\rho(\mathbf{r}, t)$ and the field Φ describing the distribution of random walkers overlap. Consequently, the Laplace equation is replaced by

$$\nabla^2\Phi \sim \rho\Phi, \tag{3.199}$$

and

$$d\rho/dt \sim \rho\Phi. \tag{3.200}$$

However, this does not account for growth of the cluster into previously unoccupied regions. In DLA, the growth occurs on the unoccupied perimeter, not on the cluster. This means that the growth process should be formulated in terms of the density of unoccupied perimeter sites $P(\mathbf{r})$. Based on a coarse-graining approximation (averaging over regions of size ℓ, where $\epsilon \ll \ell \ll L$), Witten and Sander [419] and Nauenberg *et al.* [665, 666, 667] proposed the continuum approximation

$$P(\mathbf{r}) = \rho(\mathbf{r}) + a^2\nabla^2\rho(\mathbf{r}) + \text{higher order terms}, \tag{3.201}$$

with the corresponding continuum growth equations

$$\partial\rho/\partial t = \Phi P(\mathbf{r}) = \Phi(\rho + a^2\nabla^2\rho) \tag{3.202}$$

and

$$\nabla^2\Phi = \Phi(\rho + a^2\nabla^2\rho). \tag{3.203}$$

In equations 3.201 to 3.203 $a^2\nabla^2$ is the lattice Laplacian and a is the lattice constant. Approximate analytical solutions were obtained for these equations in the long time limit, and the equations were solved numerically for $d = 3$. The main result is that $D = d - 1$ for $d > 2$. This appears to be a promising start to a theory for DLA, and Ball *et al.* [667] suggested ways in which this mean field model could be improved. Unfortunately, this mean field approach has not been developed along the lines suggested by Ball *et al.* and this direction of investigation seems to be inactive at present.

It has been suggested [668] that equation 3.203 should be replaced by

$$d\rho/dt = \nabla^2\Phi = \Phi(\rho^\gamma + a^2\nabla^2\rho), \tag{3.204}$$

with $\gamma > 1$ to represent the fact that there must be at least one occupied site in a region of volume ℓ^d for growth to take place in that region. This cutoff can be represented by a growth rate that decreases rapidly to zero (faster than linearly) with decreasing "aggregate" density ρ as $\rho \to 0$. Brener *et al.* [668] have simulated the mean field growth process by solving the 2-dimensional Poisson equation $\nabla^2\Phi - \Phi(\rho^\gamma + a^2\nabla^2\rho) = 0$ numerically, using Hele-Shaw cell boundary conditions to obtain Φ and $\partial\rho/\partial t$. At each stage, the density field was incremented using the solution to the Poisson equation and the Poisson equation was solved again. Numerical solutions were obtained for several values of the exponent γ. The density contour patterns generated by the numerical solution of equation 3.204 are like stable Saffman–Taylor fingers (section 3.4.6 and appendix A, section A.2). The density profiles and density contours are in good agreement with those obtained by Arneodo *et al.* from averaging a large number of DLA simulations, with the same boundary conditions [469]; see section 3.4.6, above. The results are not sensitive to the value used for the exponent γ. Brener *et al.* [668] showed how equation 3.204 generates an effective surface tension. If the Laplacian in equations 3.203 and 3.204 is replaced by $a(\partial^2\rho/\partial x^2)+b(\partial^2\rho/\partial y^2)$, to represent the effects of anisotropy, Saffman–Taylor fingers with different widths can be obtained. This work gives important insights into the *average* properties of DLA patterns, but it cannot be used to study the essential internal structure (density fluctuations on all scales). However, measurement of the mean density $\overline{\rho}$, far behind the finger "tip", as a function of the channel width w or the lattice constant a indicates algebraic relationships with exponents that are insensitive to γ, in the range $1.5 < \gamma < 2$. The dependence of $\overline{\rho}$ on w corresponds to a fractal dimensionality of 0.64 for a transverse cut and suggests that the mean field calculations might be used to estimate a fractal dimensionality.

Tu *et al.* [669] have extended this mean field theory to diffusion-limited growth under non-adiabatic, non-zero Peclet number regime conditions in which the field Φ has a finite relaxation time. Growth under these conditions can be described by

$$d\rho/dt = \Phi(\rho^\gamma + a^2\nabla^2\rho) \tag{3.205}$$

and

$$d\Phi/dt = \mathscr{D}\nabla^2\Phi - d\rho/dt, \tag{3.206}$$

where $\mathscr{D}$ is the diffusion coefficient, with the boundary conditions $\Phi = \Phi_\infty$ and $\rho = 0$, far from the growing pattern. In this model, Φ_∞ can be interpreted as the ambient concentration of the sea of random walkers in which the growth process takes place. Simulations, based on these mean field equations, with $\gamma > 1$, revealed a transition from a "dendritic" pattern with a concave envelope to a "dense radial" convex envelope pattern, as Φ_∞ was increased beyond a critical value, $\Phi_\infty^{(c)}(\gamma)$. In these simulations, a term of the form $\varepsilon[\partial^4\rho/\partial x^4 + \partial^4\rho/\partial y^4]$ was added to the Laplacian term in equation 3.204 to represent the effects of anisotropy. There was no evidence for a discontinuity in $\partial V/\partial\Phi_\infty$ when Φ_∞ reached the critical value $\Phi_\infty = \Phi_\infty^{(c)}(\varepsilon, \gamma)$ at which the transition occurred, where V is the growth velocity. This result appears to contrast with the simulation results obtained by Shochet *et al.* [68, 156, 157], described in chapter 1, section 1.3. A similar, continuous transition was found using a related model [670] in which equation 3.205 was replaced by a growth equation based on detailed balance controlled by a free energy functional that describes both the interfacial energy and the chemical potential difference between the two phases.

Inspired by the success of Flory theory for self-avoiding random walks [375] and related problems, a variety of other, Flory type, mean field theories for DLA have been developed. For example, Muthukumar [671] has obtained the relationship

$$D = (d^2 + 1)/(d + 1) \tag{3.207}$$

using a mean field theory. This is a remarkable result; it gives the correct result ($D = d - 1$) in the limits $d = 1$, satisfies the limit $D \geq d - 1$ for $d \to \infty$, given in equation 3.196, and is in excellent agreement with simulation results for all values of d, for which reliable results are available [182, 424, 593], except for $d = 2$. For the $d = 2$ case, there are sufficient uncertainties concerning the asymptotic scaling structure that the value for D predicted by equation 3.207 may eventually prove to be, in some sense, correct. Unfortunately, the theoretical arguments used by Muthukumar to derive equation 3.207 are quite opaque and have not been widely accepted.

The same result, given in equation 3.207, was obtained by Tokuyama and Kawasaki [672], using a quite different mean field theory, closely related to the Flory mean field theory for polymers [375]. They also extended their analysis from the growth of clusters from a point to growth from lines and surfaces. Again, the mean field theory results were in good agreement with available simulation results [465].

Somewhat different results were obtained by Hentschel [673], using an approach that is similar to that of Tokuyama and Kawasaki but incorporating different assumptions about the structure of DLA. The dependence of the fractal dimensionality on both the dimensionality of the particle trajectories (D_w) and the dimensionality of the embedding space was explored, leading to the result

$$D(d, D_w) = \begin{matrix} [4D_w - d(2D_w - 4) + 5d^2]/[5D_w - 4 + 5d] & \text{for } d < d_c \\ [4D_w - 4d + d^2]/[D_w - 4 + d] & \text{for } d > d_c, \end{matrix} \tag{3.208}$$

where the critical dimensionality d_c of the embedding space, above which only the "entropic" term in the "free energy" is important and the aggregate becomes like a non-interacting string of blobs, is given by

$$d_c^2 - (4 - D_w)d_c - 8D_w = 0. \tag{3.209}$$

These results are applicable only if the fractal dimensionality of the particle trajectories satisfies $1 < D_w < d$. For $D_w = 1$, $D = d$, in accordance with simulation results and more reliable theoretical results. However, the values calculated for large d, from the expression for $d > d_c$ given in equation 3.208, violate equation 3.196, which is generally regarded to be reliable. This mean field theory does not give a definite result for $D_w = d = 2$, but for $d = 2$ and $D_w \rightarrow 2$ a fractal dimensionality of 1.75 is obtained.

Another mean field theory has been proposed by Honda *et al.* [674], leading to the result

$$D = (d^2 + D_w - 1)/(d + D_w - 1), \tag{3.210}$$

which simplifies to that given in equation 3.207 for $D_w = 2$. This approach has been extended to models in which the growth probability exponent η in equation 3.25 is not equal to 1 [432], leading to the result

$$D = (d^2 + \eta(D_w - 1))/(d + \eta(D_w - 1)). \tag{3.211}$$

Despite the extremely good overall agreement between these mean field theory predictions, simulation results and the limited results from more rigorous theoretical methods, they must be regarded with suspicion.

A mean field approximation is clearly a poor representation of any fractal that has fluctuations on all length scales. In the mean field approach, these fluctuations are "smoothed away". While the fractal dimensionality can be obtained from the mean density $< \rho(r) >$ (where $< \rho(r) >$ is the mean density at a distance r from the growth site), it is not clear what is lost in the averaging process. These mean field theories also incorporate additional assumptions about the scaling and/or topological structure of DLA clusters that may not be valid. In most cases, the mean field analysis is based on a screening length or penetration length, and the relationship between this length and the particle density (a crucial part of these theories) is not well justified or explained and is not the same in all of the mean field theories. In some respects, the situation seems to be similar to that for "polymers" or self-avoiding random walks. However, for the case of self-avoiding random walks, the basic assumptions are more transparent and physically easier to justify within the framework of a mean field approximation. The Flory mean field theory gives very good results for the fractal dimensionality of self-avoiding walks, but it is recognized that the Flory mean field theory is fundamentally unsound. It is not understood why such "poor" theories do so well, for either DLA or self-avoiding walks.

At this stage, equations 3.207 and 3.211 must be regarded as good empirical approximations for DLA.

3.10.2 Wedge Growth Theories

If the dependence of the maximum growth probability on the cluster size (number of particles or sites s) is given by

$$P_{max} \sim s^{-\gamma}, \tag{3.212}$$

then the fractal dimensionality is given by

$$D = 1/(1 - \gamma) \tag{3.213}$$

for the broad class of growth processes for which equation 3.190 applies. One approach that has enjoyed considerable success is to estimate the exponent γ from an idealized model for the shape of the cluster [440, 626]. For example, the cluster can often be thought of as being made up of wedges with a characteristic angle θ (figure 3.51). The distribution of growth probabilities near the tip of the wedge can be calculated by solving the Laplace equation for a scalar field with absorbing boundary conditions on the wedge surface and a fixed value at infinity using conformal mapping [636]. This shows that the growth probability at a distance r from the wedge tip is given by

$$P(r) \sim r^{\chi-1}, \tag{3.214}$$

where

$$\chi = \pi/(2\pi - \theta). \tag{3.215}$$

The maximum growth probability is then given by

$$P_{max} \approx \int_0^{\epsilon} r^{\chi-1} / \int_0^{L} r^{\chi-1}, \tag{3.216}$$

where ϵ is the inner cut-off length scale (the lattice constant or particle size) and L is a length characteristic of the overall cluster size. This implies that

$$P_{max} \sim (\epsilon/L)^{\chi}, \tag{3.217}$$

or $P_{max} \sim L^{-\chi}$, since ϵ is constant. If the growing structure can be considered to be a self-similar fractal, then $P_{max} \sim s^{-\chi/D}$, so that $\gamma = \chi/D$ and

$$D = 1/[1 - (\chi/D)], \tag{3.218}$$

or

$$D = 1 + \chi = (3\pi - \theta)/(2\pi - \theta). \tag{3.219}$$

This theoretical approach was developed by Turkevich and Scher [440], who applied it to DLA using the assumption that the "wedge" angle θ could be

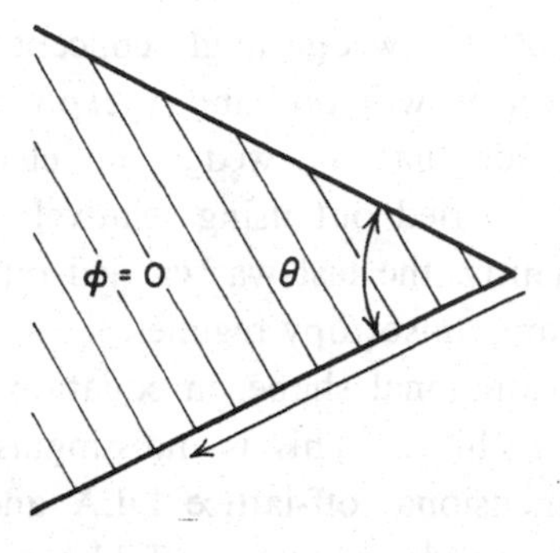

Figure 3.51 Idealized wedge shape used to represent the structure of the most exposed tips in DLA model clusters. The angle θ and distance r from the tip are shown.

obtained from the structure of the underlying lattice ($\theta = \pi/2$ for a square lattice, $\theta = 2\pi/3$ for a triangular or honeycomb lattice, for example).

Unfortunately, this idea is not supported by computer simulation results. For example, while clusters with a relatively modest size of $\approx 10^5$ sites, grown on a square lattice, do have envelopes that approximate the purported square shape, the cluster eventually develops four quite distinct arms, like the arms of the cluster shown in figure 3.40, and it is apparent that the effective wedge angle θ is much smaller than $\pi/2$. For $\theta = 0$, a fractal dimensionality of $D = 3/2$ is predicted by this approach. The exponent β describing the growth of the radius of gyration with the cluster size should therefore have a value of 2/3. Measurements using noise reduced DLA clusters grown using the semi-lattice model [653] with four growth directions indicate that θ has an asymptotic value of about $32 \pm 2°$ [675] and the exponent β has a value slightly smaller than 2/3, in accordance with the observed wedge angle and equation 3.219. For clusters with $n = 3$, 5 and 6 arms, the effective tip angles are $16 \pm 1°$, $36 \pm 4°$ and $60 \pm 10°$, respectively [675]. The theoretical approach of Turkevich and Scher can be extended to higher-dimensional embedding spaces [676] by calculating the strength of the growth probability distribution associated with a cone ($d = 3$) or "hypercone" ($d > 3$). For large d, the fractal dimensionality predicted in this manner for a right angled hypercone approaches the lower limit of $d - 1$ given in equation 3.196 .

These results can easily be generalized for the 2-dimensional dielectric breakdown model with $\eta \neq 1$ to give

$$D = 1/\{1 - [(\eta\chi - \eta + 1)/D]\}, \tag{3.220}$$

which gives $D = 2$ for $\eta = 0$, and more generally

$$D = 2 + \eta\chi - \eta \tag{3.221}$$

or

$$D = 2 - [\eta(\pi - \theta)/(2\pi - \theta)]. \tag{3.222}$$

Evertsz *et al.* [677] have attempted to test the wedge angle concept using dielectric breakdown model simulations with growth probability exponents η in the range $0.25 \leq \eta \leq 2.0$. They conclude that the wedge model is not generally valid, but these simulations were carried out using relatively small clusters, without noise reduction. Consequently, the test was carried out in a slow crossover from weak anisotropy to strong anisotropy regimes.

An angle of $3\pi/5$, corresponding to a pentagonal shape, in equation 3.219 gives a fractal dimensionality of $12/7 = 1.714\ldots$. This is indistinguishable from the measured value of D_β for 2-dimensional off-lattice DLA and has led to considerable speculation about the possible role of five-fold symmetry in DLA. Idealized geometries have been used in attempts to obtain a better understanding of other aspects of DLA. In particular, a variety of simple models has been proposed for the shapes of the "fjords" between cluster arms in order to address the problem of the distribution of small growth probabilities (appendix B, section B.3).

Ball [644] has analyzed the stability of 2-dimensional clusters consisting of alternating short and long arms under diffusion-limited growth conditions. Assuming that the cluster arms preserve their shapes and scaling properties under growth, Ball showed that the relative growth rate for the short and long arms is given by

$$dR_+/dR_- = [R_+/R_-]^{(D-1)(m-1)}, \tag{3.223}$$

where $2m$ is the number of arms and D is the fractal dimensionality, defined in terms of equations 3.212 and 3.213. This means that the cluster will be stable with respect to fluctuations in the arm length if

$$(D-1)(m-1) < 1, \tag{3.224}$$

so that the maximum number of arms n_{max} is given by

$$((n_{max}/2) - 1)(D-1) = 1. \tag{3.225}$$

This implies that for off-lattice DLA, with $D \approx 1.71$, the maximum number of arms is about 4.9, and for needle armed clusters, for which the mass/length scaling is like that of a fractal with $D = 1.5$, $n_{max} = 6$. It is not clear, based on computer simulations, how many arms there are in a large off-lattice DLA cluster. Although algorithms can easily be designed to determine the "number of arms" in DLA clusters, the significance of the results is not clear, and visual inspection of figures such as figure 3.9 suggests that the number of arms increases slowly with increasing cluster size. The decrease in the width of the active zone and the slow decrease in the lacunarity found in 2-dimensional DLA clusters, as the cluster size increases, supports the impression that the number of arms increases with increasing cluster size, and this increase may be unlimited. However, the results of Ball [644] and of Derrida and Hakim [436] suggest that the number of arms in a 2-dimensional DLA cluster cannot become much larger than ≈ 6, and are more compatible with the "massive transient" scenario than the "continuous drift" scenario.

If it is assumed that the fractal dimensionality is related to the "wedge angle" by equation 3.219 and that the effective wedge angle is defined by the n-edged regular polygon envelope of the n-armed cluster, then

$$D = 1 + [n/(n+2)]. \tag{3.226}$$

If n, in equation 3.226, is equal to n_{max} in equation 3.225, then

$$[(n^2/2) - n]/[n+2] = 1, \tag{3.227}$$

which has a positive solution $n = 2(1 + 2^{1/2})$ corresponding to a fractal dimensionality $D = 1 + (2^{1/2}/2) = 1.707\ldots$. This result [644] can be thought of in terms of a non-integer number of arms or a wedge angle of $\pi - (\pi/2^{1/2})$. The value obtained for D in this manner is in good agreement with computer simulations for off-lattice DLA, but the geometric interpretation of the effective wedge angle or non-integer number of arms has not been established.

While DLA itself has proven to be quite intractable, progress has been made on a number of more or less closely related problems. Much of this progress is based on the idea of Turkevich and Scher [440] that insight into the growth of DLA clusters could be developed by considering the distribution of Laplacian field gradients (the growth probability distribution) for objects that had a regular shape related to that of the lattice. For example, Szép and Lugosi [678] analyzed the problem of the growth from a point of n symmetrically disposed needles. They included the lateral growth of each needle as well as its elongation. The elongation of the needles is much faster than their lateral growth, so that their aspect ratio continues to increase as they grow. Based on this observation, the distribution of growth probabilities (surface growth velocities) along the needles can be approximated quite well by ignoring the width of the needles. The zero needle width growth probability was obtained by solving the Laplace equation using conformal mapping onto a circle. The conformal map was then solved numerically in order to grow patterns with n arms, such as those shown in figure 3.52. The growth of the radius of gyration was found to increase algebraically with increasing size (pattern area) with exponents (β_n) that approaches a value of 2/3 with increasing size for all values of n. Based on the growth near the tip, Szép and Lugosi were able to show that the shapes of the arms of the pattern could be described by the scaling form

$$w(kx, kr) = k^{1/2} w(x, r), \tag{3.228}$$

where $w(x,r)$ is the width of a needle of length r at distance x from its tip. This means that the arm length grows as $s^{2/3}$, while the width grows as $s^{1/3}$, where s is the size or area of the pattern. Zero noise limit, cubic lattice DLA

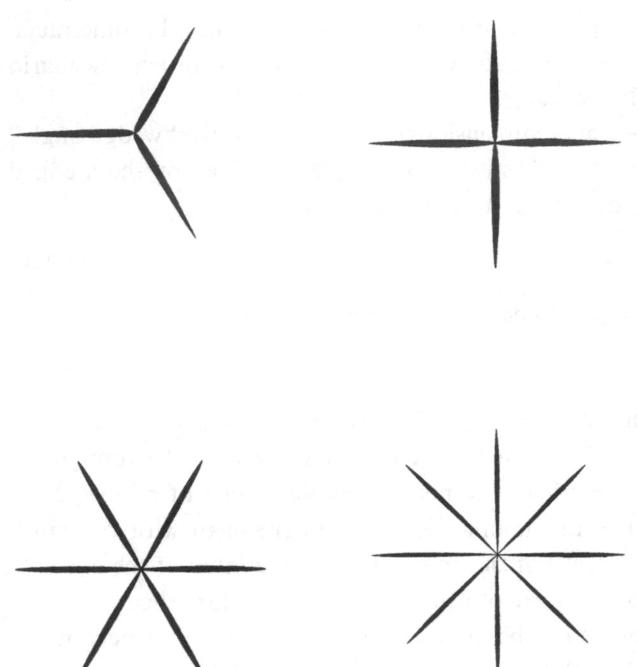

Figure 3.52 Patterns with needle shaped arms grown using a Laplacian growth model. The figure shows patterns grown with $n = 3$, four, six and eight symmetric arms. As the patterns grow, the aspect ratio in each arm increases. Each pattern was grown to the same size (arm length) of 740, starting with n zero width arms of length 1. The arm shapes were calculated assuming that their tips have the same constant curvature. The change in the arm shapes and the narrowing of the arms with increasing n reflects the increase in the mutual screening. This figure was provided by J. Szép.

simulations generate clusters with six planar leaf-like arms and the arm length grows as $s^{1/1.7}$ [679], so that the width grows as $s^{1-(1/1.7)}$ or $s^{\approx 0.41}$.

A quite similar approach was taken by Eckmann *et al.* [680, 681] to develop a theoretical model for the structure of noise reduced DLA clusters. This analysis was applied specifically to the noise reduced, 2-dimensional, semi-lattice DLA model, illustrated in figure 3.47 and described above. At first, the cluster grows into a cross with n compact arms. The theoretical model indicates that the cluster will begin to grow side branches when it reaches a size $s^*(m)$ given by

$$s^*(m) \sim (\log m)^3, \tag{3.229}$$

and this prediction was confirmed by computer simulations.

During the first stage of growth, the width of the compact cluster arms or the mass per unit length $M(x, r)$ for an arm of length r at a distance x from the tip is given, within the "needle" approximation, by equation 3.228 or

$$M(x, r) = x^{1/2}F(x/r) = r^{1/2}G(x/r). \tag{3.230}$$

In the later stages of growth, the cluster arms evolve towards an asymptotic shape characterized by a constant length to width (aspect) ratio that does not grow with increasing cluster size [654, 648]. The shape of the cluster arms becomes invariant to isotropic rescaling. This is illustrated in figure 3.53, which shows cluster arms of two different sizes grown with four growth directions and a noise reduction parameter of 300, on length scales proportional to $s^{2/3}$, where s is the cluster size. This figure indicates that the asymptotic shape is reached

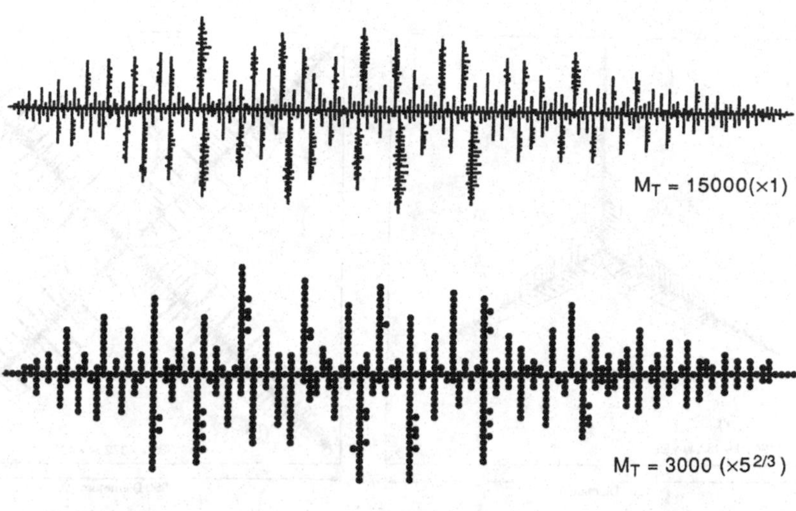

Figure 3.53 Rescaling of the cluster geometry for DLA clusters generated with four growth directions ($n = 4$) and a noise reduction parameter of $m = 300$. One arm of the cluster is shown after the cluster has grown to a size of $s = 3000$ particles (bottom) and 15 000 particles (top). The arms are shown on length scales proportional to $s^{2/3}$, where s is the cluster size. The tips of the arms are on the right-hand side.

at relatively small cluster sizes s. If the scaling of the arm lengths and widths with cluster size can be characterized by a fractal dimensionality of D, then, in the later stages of growth, the width of the arms at a distance r from the tip is given by $w(x,r) \sim M(x,r)^{1/(D-1)}$. In the "needle" limit, $D = 3/2$ and

$$w(x,r) \sim M^2(x,r) = xF^2(x/r) = rG^2(x/r). \tag{3.231}$$

The function $F(x/r)$ in equation 3.231, which describes the shapes of the cluster arms, depends on the number of arms n in the cluster. It is given by

$$F(x/r) = \frac{n}{\pi} \int_{x/r}^{1} \{y^{[(n/2)-(5/2)]}\}/\{(1-y^n)^{1/2}\}dy. \tag{3.232}$$

Figure 3.54 shows a comparison between clusters simulated with $n = 3, 4$ and 5 and the cluster shapes calculated from the shape function

$$w(x,r) = cxF^2(x/r), \tag{3.233}$$

where the function $F(x/r)$ is given by equation 3.232 and c is an adjustable parameter.

The theory can be improved [681] by using the shape function for the cluster arm envelope to recalculate the mass distribution $M(x,r)$ along the cluster arms. This is the first step in an iterative procedure, and the shape function can be obtained as the fixed point of the corresponding functional equation. In this

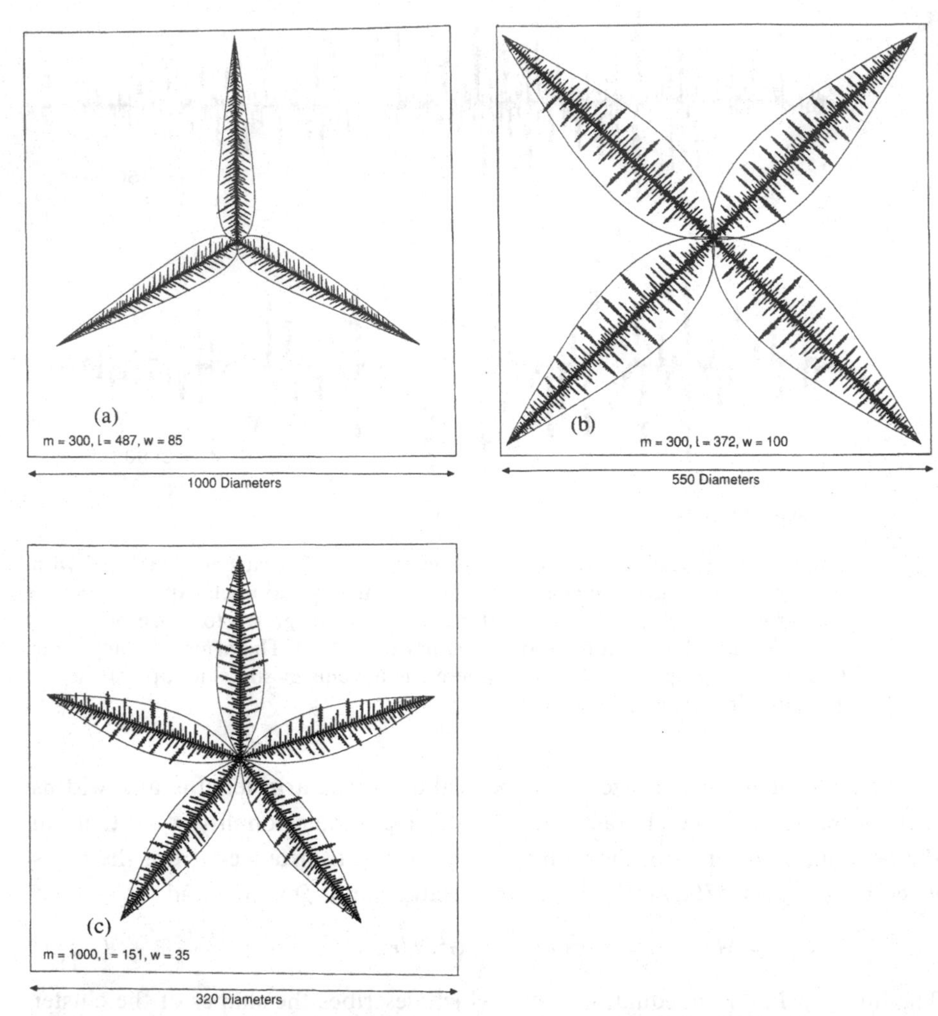

Figure 3.54 A comparison between the theoretical arm shapes for DLA clusters grown with noise reduction and a discrete number of growth directions. Parts (a), (b) and (c) show the results for $n = 3, 4$ and 5, respectively. The cluster envelopes were calculated from equations 3.232 and 3.233.

manner, it can be shown [681] that needle shaped arms with a zero tip angle are not fixed points. This approach to DLA with long range anisotropy also lends support to the idea that lattice DLA and noise reduced DLA, on the same lattice, are in the same universality class and that the asymptotic arm shapes are independent of the noise reduction parameter.

The conformal mapping theory also predicts the shapes of the cluster tips. For large noise reduction parameters, the ends of the cluster arms consist of a short linear array of particles with no decoration. The length of this "exposed tip" is predicted to be approximately 6, 4 and three (particles) for $n = 3$, 4 and

5 arms, respectively. This prediction is in good agreement with the simulation results.

Ball *et al.* [675] have used a similar approach to address the problem of the distribution of side branches in DLA with anisotropy. Their analysis was based on the idea that the side-branching pattern is generated by a growth process that is marginally stable in the sense that

$$\delta\ell/\ell = \delta V/V, \tag{3.234}$$

where δV is the change in the side arm growth velocity if its length is changed by $\delta\ell$. In this theoretical model, the side branches, as well as the main arms (the envelope of the side branches), are represented by wedges, and the side arms form a saw-like pattern along the sides of the main branches, with the teeth or side branches oriented at angle ϕ with respect to the axis of the main branch. The sizes of the teeth of the saw increase with distance from the end of the saw (cluster tip) and the saw-tooth pattern is self-similar about the tip. The tips of the side branches lie on a wedge, with the tip of the main branch as an apex, at distances proportional to λ^m from the apex, where m is an integer and $\lambda < 1$. The stability of this pattern was analyzed with respect to a perturbation that alternately increased and decreased the side arm lengths along the sides of the main arms. Using conformal mapping the strength of the perturbation $\delta\ell/\ell$ and the response $\delta V/V$ were calculated in terms of the exponent λ that characterizes the spacing between the side branches, the exponent χ (equation 3.214) characterizing the strength of the growth probability singularity at the side-branch tips, the main-branch tip angle and the angle ϕ between the side branches and the main branch.

The exponent χ (equation 3.215), for the side branches, is not an independent parameter in the model. To maintain the overall wedge-like shape of the cluster arm envelope, with its characteristic tip angle, the exponent χ describing the growth probability singularity must be the same for the tips of the side branches and the tip of the main branch. In this model, the exponent χ is not given by equation 3.215, where θ is the envelope wedge angle, because the wedge is not solid and the effective wedge angle θ_{eff} is smaller than the envelope angle.

The model allows the parameters ϕ, λ and the wedge envelope angle θ to be calculated if one of them is known, via the marginal stability hypothesis and the requirement that the shape must be preserved during growth. Ball *et al.* used this theoretical model to calculate the side-branching structure of cluster with three, four and five symmetrically disposed arms from the side-branching angle ϕ which is determined by the anisotropy ($\phi = 2\pi/n$, where n is the number of arms or growth directions). The results were found to be consistent with the results obtained from the noise reduced, semi-lattice DLA model simulations described earlier in this chapter. However, the simulations did not provide a very accurate estimate for λ and do not provide very strong support for this aspect of the calculations.

The growth of fluid–fluid displacement patterns in a Hele-Shaw cell is closely related to the DLA problem. Almgren *et al.* [682] have used a boundary integral method to calculate the growth of Hele-Shaw cell patterns generated by the displacement of a viscous fluid by a non-viscous fluid, at a constant injection rate, with an anisotropic surface tension of the form

$$\Gamma = \Gamma_o(1 - \epsilon \cos m\theta), \tag{3.235}$$

so that the scalar harmonic field on the surface is given by

$$\Phi(\mathbf{x}_s) = \Gamma_o(1 - \epsilon \cos m\theta)\mathscr{K}(\mathbf{x}_s), \tag{3.236}$$

where $\Phi(\mathbf{x}_s)$ is the value of the scalar field inside the viscous fluid at position $\mathbf{x}_s$, at arc length s, on the interface and $\mathscr{K}(\mathbf{x}_s)$ is the curvature at $\mathbf{x}_s$ (chapter 1, section 1.4). The simulations were carried out with $m = 4$. For large values of the anisotropy factor ϵ, the patterns are dominated by four compact arms with a large aspect ratio. The lengths of the arms grow as $t^{3/5}$ and their widths grow as $t^{2/5}$, where t is the time. This means that the shape of the arms can be described by a scaling form, similar to that given in equation 3.228,

$$w(kx, kr) = k^{2/3}w(x, r), \tag{3.237}$$

where $w(x, r)$ is the width of an arm of length r at distance x from its tip. This scaling is shown in figure 3.55. Since the arm length grows as $t^{3/5}$, measurement of the growth of the radius of gyration as a function of the displaced area A will lead to the result $R_g \sim A^{3/5}$, which can be interpreted as a "fractal dimensionality" of 5/3. Dougherty and Chen [683] have determined that the width $w(z)$ a distance z from the tip grows as $w(z) \sim z^{0.74 \pm 0.6}$, for 3-dimensional NH_4Cl dendrites grown from supersaturated glycerin/water solutions.

Kessler *et al.* [165] and Sander *et al.* [684] had earlier obtained numerical solutions for essentially the same noiseless, continuum interface dynamics model for Laplacian growth as Almgren *et al.* The work of Kessler *et al.* was carried out in a dendritic solidification context. They found that the area $\mathscr{A}$ of the pattern and the total length $\mathscr{S}$ of the interface grew algebraically with increasing time, with effective exponents that depended on model parameters, such as the symmetry. For a pattern with four-fold symmetry they found that $\mathscr{A} \sim t^a$ and $\mathscr{S} \sim t^b$, with the effective exponents $a \approx 0.98$ and $b \approx 0.65$. Sander *et al.* carried out similar simulations with no surface tension anisotropy ($\epsilon = 0$). In this work, the evolution of the interface was calculated by inverting the equation

$$1 + \frac{1}{4\pi}\int \mathscr{K}(\mathbf{x}'_s)\frac{\partial G(\mathbf{x}_s, \mathbf{x}'_s)}{\partial n'}d\mathbf{x}'_s = \int V_n(\mathbf{x}'_s)G(\mathbf{x}_s, \mathbf{x}'_s)d\mathbf{x}'_s. \tag{3.238}$$

Here, $G(\mathbf{x}_s, \mathbf{x}'_s)$ is the Green's function for the 2-dimensional Laplace equation

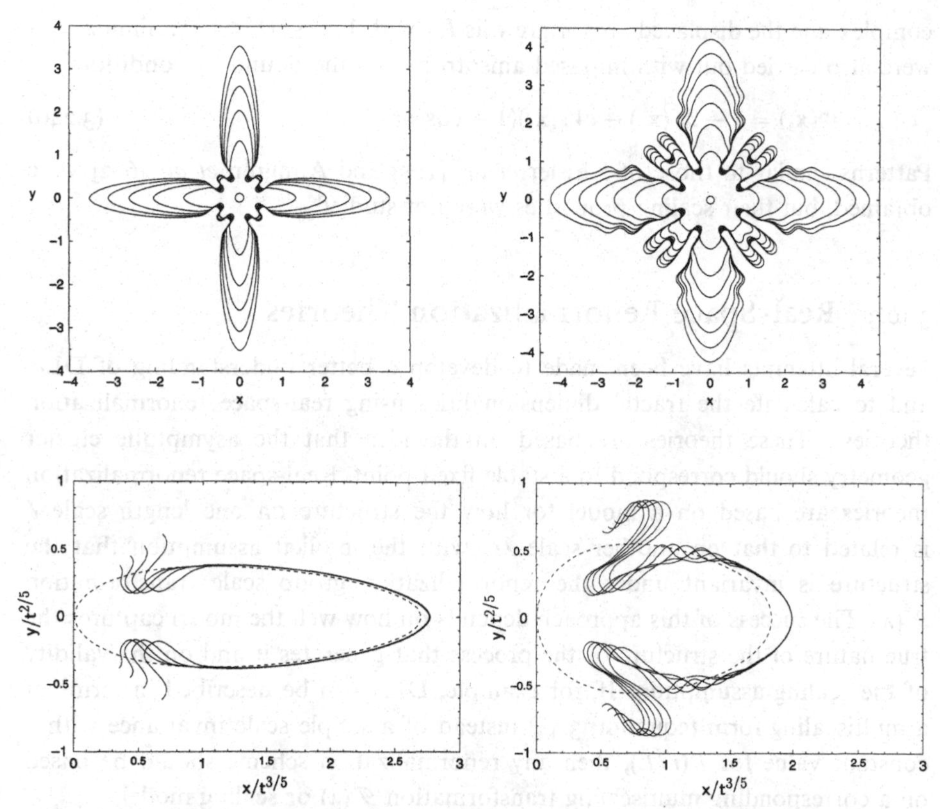

Figure 3.55 Scaling of the displacement patterns calculated using a boundary integral model for an anisotropic Hele-Shaw cell. The upper row shows several stages in simulations carried out using high ($\epsilon = 0.5$) and low ($\epsilon = 0.1$) anisotropies on the left-hand and right-hand sides, respectively. The lower part shows how the patterns can be scaled. This figure was provided by V. Hakim.

[165] and $V_n(\mathbf{x}_s) = -\mathbf{n}(\mathbf{x}_s) \cdot \nabla\Phi(\mathbf{x}_s)/(4\pi)$ is the normal velocity at $(\mathbf{x}_s)$, where $\mathbf{n}(\mathbf{x}_s)$ is a unit vector, in a direction normal to the interface. Sander *et al.* started with an interface of the form $r(\theta) = r_\circ(1 + \delta\cos(m\theta))$ and obtained shapes that preserved the initial m-fold symmetry. Simulations were also carried out with the modified integro-differential equation

$$1 + \frac{1}{4\pi}\int [\mathscr{K}(\mathbf{x}'_s)]^N \frac{\partial G(\mathbf{x}_s, \mathbf{x}'_s)}{\partial n'} d\mathbf{x}'_s = \int V_n(\mathbf{x}'_s) G(\mathbf{x}_s, \mathbf{x}'_s) d\mathbf{x}'_s, \tag{3.239}$$

in order to approach the DLA limit (large values for the exponent N allow small curvatures to develop but prevent very large curvatures from developing). As the exponent N in equation 3.239 was increased, the patterns became more complex and the displaced area A grew as R_G^D with $1.72 \leq D \leq 1.75$. Simulations were also carried out with imposed anisotropy, via the boundary condition[20]

$$\Phi(\mathbf{x}_s) = 1 - \mathscr{K}(\mathbf{x}_s) - cV_n(\mathbf{x}_s)(1 - \cos m\theta). \tag{3.240}$$

Patterns similar to those of Kessler *et al.* [165] and Almgren *et al.* [682] were obtained, but their scaling properties were not studied.

3.10.3 Real-Space Renormalization Theories

Several attempts have been made to develop a better understanding of DLA and to calculate the fractal dimensionalities using real-space renormalization theories. These theories are based on the idea that the asymptotic cluster geometry should correspond to a stable fixed point. Real-space renormalization theories are based on a model for how the structure on one length scale ℓ is related to that on another scale $\lambda\ell$, with the implicit assumption that the structure is invariant under the renormalization group scale transformation $\mathscr{T}(\lambda)$. The success of this approach depends on how well the model captures the true nature of the structure or the process that generates it and on the validity of the scaling assumption. If, for example, DLA can be described in terms of a multiscaling form (equation 3.44) instead of a simple scale invariance with a constant value for $D(r/L)$, then any renormalization scheme should be based on a corresponding multiscaling transformation $\mathscr{T}(\lambda)$ or scaling model.

Gould *et al.* [685] obtained fractal dimensionalities of 1.71 and 1.67 for square lattice DLA using cells of size 2×2 and 3×3, respectively. This approach is based on an enumeration of all the possible ways of generating spanning clusters in the cell. A quite different point of view has been proposed by Halsey and Leibig [596]. They suggest that the growth can be described in terms of "escape" from an unstable fixed point on all length scales.

Wang *et al.* [686] have used a kinetic renormalization approach to DLA that leads to the result $D = 1.727$ and $D = 2.494$, for the square and cubic lattices, respectively. This renormalization model can also be used to estimate the exponents that describe the multifractal distribution of growth probabilities [687] (chapter 2, section 2.8 and appendix B). Wang and Huang [688] have extended the work of Wang *et al.* [686] to $d > 3$. They obtain $D = 1.737$ for $d = 2$, $D = 2.515$ for $d = 3$ and results that are consistent with simulation results for $d > 3$. They also find that $D \to d - 1$ for $d \to \infty$, in agreement with the limit of Ball and Witten [661].

Nagatani [689] has made extensive use of renormalization group models for DLA and related growth processes. In these renormalization group models,

20. Equations 3.236 and 3.240 are not incompatible. Almgren *et al.* [682] defined the scalar potential Φ so that $V_n(\mathbf{x}_s) = \mathbf{n}(\mathbf{x}_s) \cdot \nabla\Phi(\mathbf{x}_s)$ and used far field boundary conditions that were different from those used by Sander *et al.* [684].

the DLA cluster is represented in terms of a network of superconducting links embedded in a normal conducting mesh. His approach is based on the idea that the pattern growth is governed by a minimum energy dissipation principle. The energy dissipation Q is given by

$$Q = V_\circ \sum_{i \in \mathcal{S}} \sigma_i V_i, \tag{3.241}$$

where $V_\circ$ is the total potential drop across the system, σ_i is the conductivity of the ith bond, V_i is the potential drop across the ith bond and the summation is over the surface $\mathcal{S}$ of the cluster. In terms of an equilibrium system, the energy dissipation Q plays the role of the Hamiltonian and the surface conductivity has the role of a coupling constant. Using a small cell (2×2) renormalization scheme, Nagatani calculated a fractal dimensionality of 1.628 for 2-dimensional DLA. A number of other real-space renormalization group schemes for DLA have been proposed and investigated.

Real-space renormalization group theories have also been used to calculate the multifractal "spectrum" $D_f(\alpha)$ (chapter 2, section 2.8, and appendix B) for the distribution of growth probabilities on the active zone of DLA clusters. They all appear to assume that this distribution can be described in terms of a multifractal scaling model and fail to account for the breakdown of this model for very small growth probabilities. However, the distribution of these very small growth probabilities in real space and probability space can be regarded as a *consequence* of the growth process, and these distributions have no effect on the way in which the pattern evolves.

A problem that is common to all real-space renormalization theories is that they are based on renormalization of DLA on a lattice for which the effects of anisotropy are not well understood, while the most interesting problem, from both a theoretical and experimental point of view, is off-lattice DLA. Very often, the results from real-space renormalization group calculations based on lattice models are compared with the results from off-lattice DLA simulations, and they give results that are in much better agreement with the fractal dimensionality ($D_\beta \approx 1.71$ for $d = 2$) measured for off-lattice DLA than the fractal dimensionality obtained from lattice model DLA simulations. In addition, the real-space renormalization and related approaches *assume* that the DLA clusters are self-similar.

Pietronero *et al.* [690] have introduced a "fixed-scale transformation" approach to DLA that is based on the scale invariance of the Laplace equation. In this approach, the structure of a d-dimensional cluster is characterized in terms of its intersection with a flat $d-1$-dimensional surface perpendicular to the growth direction. The fixed scale transformation method is well suited to diffusion-limited deposition on a flat surface. At the most simple level, this model is based on the idea that the "horizontal" cut can be subdivided on finer and finer scales. If the cut is divided into 2^n equal boxes at the nth level, then at the $n+1$st level each occupied box will be replaced by either one occupied box and one unoccupied box with probability C_1 or by two full boxes with probability C_2. The fractal dimensionality of the cut is then given by

$$D_1 = \log(C_1 + 2C_2)/\log 2. \tag{3.242}$$

The probabilities C_1 and C_2 provide a coarse-grained picture of the structure at any level of subdivision. In the fixed-scale transformation analysis, attention is focused on the way in which a pair distribution of type 1, consisting of one filled and one empty subcell, or type 2, consisting of two filled cells, is followed by a pair distribution of type 1 or 2, *at a fixed level of subdivision or fixed scale,* as the height increases inside the "frozen" region, well behind the advancing active zone, where no more growth takes place. The evolution of the distribution (C_1, C_2) is expressed by the iterative equation

$$\begin{pmatrix} C_1^{(k)} \\ C_2^{(k)} \end{pmatrix} = \begin{pmatrix} M_{11} & M_{12} \\ M_{21} & M_{22} \end{pmatrix} \begin{pmatrix} C_1^{(k-1)} \\ C_2^{(k-1)} \end{pmatrix}, \tag{3.243}$$

where M_{ij} describes the probabilities that a distribution or configuration of type j will follow a distribution of type i, in the direction of growth, and $M_{ii} + M_{ij} = 1$. This equation expresses the statistical invariance of the structure to translation in the "vertical" direction (inside the frozen zone). The fixed point of equation 3.243 gives the values of C_1 and C_2 needed to calculate the fractal dimensionality D_1. The matrix elements M_{ij} are obtained from the potential field calculated using the discretized Laplace equation for the configurations that arise in the model with $\phi = 0$ on occupied cells with appropriately chosen lateral boundary conditions. In general, the elements of $\mathbf{M}$ can be calculated via lattice path integrals for the possible growth processes. These "integrals" should extend close enough to the asymptotic $t \rightarrow \infty$ limit, where t is "time", corresponding to the situation in which the active zone is far away. It is at this point that this becomes a model for DLA. The lateral boundary conditions are important in the fixed-scale transformation approach. In practice, $\mathbf{M}$ is a properly weighted average over the matrices $\mathbf{M}_i$ calculated for all possible boundary conditions (surrounding configurations). The incorporation of "fluctuations" in this manner appears to be substantially advantageous over standard real-space renormalization group analysis. These fluctuations depend on the distribution of gaps (unoccupied regions) in the horizontal cuts, which can also be expressed in terms of C_1 and C_2 (or D_1). The most simple fixed-scale transformation provides remarkably good results, but these are still only rough estimates for D_1. The accuracy of the results can be improved in several ways. One approach is to enlarge the configuration space

$\{C\} = (C_1, C_2, C_3, \ldots, C_n)$. An important example is the addition of the empty configuration C_3 to the full configuration C_2 and the partly full configuration C_1, in the two-cell model. This can be considered to represent the effects of lattice paths that enter the two-cell system from the side in a model based on more than two cells. These paths allow empty configurations to be followed by non-empty configurations, and they may be considered to represent sideways growth into empty regions. In principle, a larger configuration basis $\{C_n\}$ could be constructed using more than two cells. However, the number of lattice paths needed to ensure convergence grows rapidly with the number of cells [454].

The fixed-scale transformation has been used with considerable success for a wide variety of growth models [454]. It can be systematically improved and can give excellent results, if the basic scaling assumption is correct. For 1 + 1-dimensional DLA, fractal dimensionalities of 1.55, 1.64 and 1.62 were obtained for closed boundary conditions, open and closed boundary conditions and a three configuration (C_1, C_2, C_3) calculation, respectively, where C_3 represents the possibility of an empty pair of cells. The results obtained for 3-dimensional DLA, $D = 2.54$, and the Levy flight DLA model described above, are in excellent agreement with simulation results.

Recently, Cafiero *et al.* [691] have included noise reduction in the fixed-scale transformation analysis of DLA. They showed that the fixed point is attractive and that the fixed-point value m^* of the noise reduction parameter m is small ($m^* \approx 2.4$). This indicates that the screening effect, which is essential to the growth of a fractal structure, is effective on all length scales. It is interesting, in this context, that the large square lattice DLA cluster shown in figure 3.40 has an appearance similar to the clusters generated with noise reduction parameters of 2 and 3 in figures 3.41(a) and (b), but is unlike the clusters generated with noise reduction parameters of 10 and 30 in figures 3.41(c) and (d). However, this observation could be misleading, since the shapes of the clusters in these figures may not be near enough to their asymptotic limits.

The general approach taken in the fixed-scale transformation analysis is similar to that used in the real-space renormalization group method, and it suffers from some of the same disadvantages, including the assumption of self-similarity.

3.10.4 Other Approaches

Halsey and Leibig [596, 692] have proposed a hierarchical binary tree model for DLA. This model supposes that no more than two tributary segments or sub-branches join at each node in the tree. The tree can be considered to be composed of linear segments corresponding to the bonds between contacting particles. A binary tree model corresponds to a growth process in which at most two particles can be added to each particle in the cluster. It seems intuitively obvious that a DLA-like growth model in which particles become "non-sticky" after two particles have contacted and been added to them will have the same scaling properties as ordinary, off-lattice DLA. Certainly, computer simulation results obtained from DLA-like models with short range attractive interactions and models with small sticking probabilities (models in which it is either more common or rarer to find nodes with more than two tributary segments, respectively) indicate that the scaling properties are not sensitive to this aspect of the tree topology. Hentschel [693] has carried out simulations using a DLA model in which random walkers that contact particles with two tributary branches continue on their random walk paths, without crossing over the cluster. Hentschel stated that

> Binary DLA is a fractal in the same universality class as normal DLA (in fact its fractal dimension appears to be even closer to $D = 5/3$ than normal DLA)....

Hentschel's simulations were carried out on a quite small scale, and there is no reason for believing that the asymptotic fractal dimensionality of binary DLA may be closer to $D = 5/3$ than normal DLA.

The theoretical model of Halsey and Leibig focuses on the sizes (number of particles) s_1 and s_2 and the growth probabilities p_1 and p_2 for each of the branching nodes in the tree (s_1 and s_2 are the sizes of the sub-branches and their growth probabilities are p_1 and p_2). In this respect, their approach is similar to the two-needle competitive growth model used by Rossi [628, 629] to study diffusion-limited needle growth from a surface. It is convenient to work with the normalized growth probabilities x and tributrary sizes y defined as

$$x = \frac{p_1}{p_1 + p_2} \tag{3.244}$$

and

$$y = \frac{s_1}{s_1 + s_2}. \tag{3.245}$$

It follows from the definitions of x and y in equations 3.244 and 3.245 that the equation of motion for y can be written as

$$dy/d(\ln s_b) = (ds_1/ds_b) - (s_1/s_b) = x - y, \tag{3.246}$$

where $s_b = s_1 + s_2$. Halsey and Leibig assumed that a similar equation

$$dx/d(\ln s_b) = f(x, y) \tag{3.247}$$

describes the evolution of x. Symmetry requires that $f(1 - x, 1 - y) = -f(x, y)$. There are three obvious fixed points for the dynamics described by equations 3.244 and 3.245 ($(x, y) = (0, 0)$, $(1, 1)$ and $(1/2, 1/2)$). DLA simulations strongly indicate that, in the asymptotic limit, one of the two sub-branches takes all of the particles and all of the growth probability, so that the fixed point at $(1/2, 1/2)$ must be unstable. Halsey and Leibig concentrated on the scenario in which there is one stable and one unstable direction at the $(1/2, 1/2)$ fixed point. They described the evolution of a branch in terms of trajectories on the (x, y) plane that start near the unstable fixed point at $(1/2, 1/2)$ and converge onto the unstable manifold. The actual dynamics was approximated by trajectories starting out on the unstable manifold, near the $(1/2, 1/2)$ fixed point at $s_b \approx 1$. As the cluster evolves, the growth of the branch can be described in terms of motion along the unstable manifold, towards a stable fixed point. In most cases, the trajectory will die because of screening from other branches in the cluster.

Halsey and Leibig carried out computer simulations to follow the trajectories in the (x, y) plane. The growth probabilities were calculated by representing the shapes of the off-lattice DLA clusters on a square lattice and calculating the growth probabilities by numerically relaxing the discretized, square lattice Laplace equation (equation 3.23). The approach used by Halsey and Leibig allows significant parts of trajectories in the (x, y) plane to be followed. In general, these trajectories move away from the fixed point at $(1/2, 1/2)$ towards the attractive fixed points at $(0, 0)$ and $(1, 1)$. They appear to fluctuate about an average trajectory with an infinite slope dy/dx at the stable fixed points and a slope of 0.61 ± 0.17 at the unstable $(1/2, 1/2)$ fixed point.

Within the framework proposed by Halsey and Leibig, the scaling properties of DLA can be calculated from the trajectory $y(x)$ describing the evolution of a branch point. This indicates that the structure of DLA clusters is determined not by convergence to a stable fixed point but by the way in which the trajectories diverge from an unstable fixed point. Near to the fixed point at $x = 1/2$, $y = 1/2$, $y(x)$ has the form $(y - (1/2)) = (x - (1/2))/c$, where c is a constant that depends on the details of the growth process. In one specific example, it was assumed that this linear form for $y(x)$ persisted over the full range $0 < x < 1$. At $x = 0$ or $x = 1$, one of the branches continues to grow while the other stops growing entirely. This continued growth of the "winning" branch can be described by a trajectory following a path at $x = 0$ or $x = 1$, towards the corresponding fixed point at $(0, 0)$ or $(1, 1)$. In this model, everything is determined by the parameter c. The actual trajectories found in a simulation were qualitatively similar but quantitatively different from this primitive model for $y(x)$. This model allows both the fractal scaling properties of the tree and the distribution of growth probabilities to be calculated. The value of the parameter c was determined by requiring that

$$(d - 1)/D = d\sigma(q)/dq|_{q=1}, \tag{3.248}$$

where D is the fractal dimensionality, and

$$\sigma(q) = \frac{\ln < \sum_i P_i^q >}{\ln s} = \tau_a(q)/D. \tag{3.249}$$

In equation 3.249, $\{P_i\}$ are the growth probabilities and the indices $\tau_a(q)$ are the multifractal scaling exponents describing the dependence of the annealed moments, or annealed average partition function, (appendix B, section B.3.1) of the growth probabilities on the overall cluster length scale L. Equation 3.248 makes this general model for branched growth into a theory for DLA. It expresses the idea that the information dimension D_I for the harmonic measure or the distribution of growth probabilities is given by $D_I = d - 1$. This a rigorous and exact result for $d = 2$. Exceptions are known for $d > 2$, but this is expected to be at least a very good approximation for DLA. Once c has been determined, the fractal dimensionality $D(d)$ can be calculated. Numerically, the results obtained from this simplified model were in good agreement with the empirical equation 3.207. The model also predicts that the maximum growth probability should decrease as L^{1-D}. The decrease of the minimum growth probability P_{min} with increasing cluster size is sensitive to the shape of $f(x, y)$ near the attractive fixed points at $(0,0)$ and $(1,1)$. This implies that there is no close association between the fractal dimensionality of the cluster and the way in which P_{min} depends on the cluster size, since the fractal dimensionality is determined by the escape from the fixed point at $(1/2, 1/2)$, rather the way in which the attractive fixed points are approached.

The formation of these ideas into a theory for DLA rests on several assumptions, including a hyperbolic form for the unstable fixed point. In a subsequent development of this approach, Halsey [692] derived the hyperbolic form using a renormalized mean field approximation and estimated a fractal dimensionality of $D = 1.66$.

Another possible route to a theory for DLA was pioneered by Shraiman and Bensimon [694]. They showed how conformal mapping techniques can be used to transform the interface growth equations

$$V_n(\mathbf{s}(\mathbf{x})) = -\nabla_n \Phi(\mathbf{x}_s) \tag{3.250}$$

and

$$\nabla^2 \phi = 0 \tag{3.251}$$

with $\Phi(\mathbf{x}_s) = 0$ and $\Phi(\infty) = 1$ into a many body equation of motion described by ordinary differential equations for the critical points (poles and zeros) within the unit disc. In equations 3.250 and 3.251, $\mathbf{x}_s$ is a position on the evolving surface and $\nabla_n \Phi(\mathbf{x}_s)$ is the component of the gradient of Φ normal to the surface.

A serious problem with this approach is that the equations of motion bring the quasi-particles to the unit circle, at the edge of the unit disc, in finite time, corresponding to the formation of cusp singularities in the corresponding physical boundary that is related to the unit circle by the conformal map.

Blumenfeld [695] has shown that the equation of motion for the quasi-particles can be expressed in terms of a Hamiltonian and that under some conditions the Hamiltonian is integrable. Blumenfeld [695, 696] has further proposed that this approach can be extended to a continuous density of poles and zeros. The problem of the emergence of singularities as the critical points approach the unit circle can be controlled in several ways. One of these is to add surface tension to the problem [697]. In terms of the quasi-particle Hamiltonian, this corresponds to the addition of an appropriate repulsive interaction between the quasi-particles and the unit circle that prevents them from reaching the unit circle. This repulsive interaction diverges as the quasi-particles approach the unit circle. Another key ingredient of DLA is the presence of noise resulting from the random deposition process. The quasi-particle Hamiltonian formalism allows noise to be introduced in a natural and direct manner.

Blumenfeld has shown how a power law probability density for the curvature may arise, indicating that a fractal structure, for the physical interface, may be a consequence of this theoretical approach rather than an assumption. Another intriguing aspect of this theoretical model is that creation and annihilation of the quasi-particles may take place, and that the splitting of zeroes ($Z \rightarrow 2Z + P$, where Z is a zero and P is a pole) may correspond to processes such as side branching or tip splitting [698]. This theory appears to offer a hope that a "first principles" theory for DLA may eventually be developed. At the present stage, this idea appears to offer a lot of promise, but it faces substantial difficulties before a theory of DLA can hopefully emerge. Despite the advantages of a Hamiltonian formulation, it is not yet clear if this attractive theoretical approach will provide a path towards a better understanding or merely transform one formidable theoretical problem into another. Field theoretical methods have been used [699, 700, 701] in an attempt to develop a theory for DLA. While it is possible to write down formal equations for the action [700], with non-local interactions, this approach has, so far, encountered very serious obstacles.

A number of workers have suggested that off-lattice DLA clusters have five arms, and Argoul *et al.* [451, 452] and Kuhn [702] have carried out a wavelet transformation analysis[21] of DLA clusters, and have interpreted

21. Wavelet analysis [703] is similar to Fourier analysis. It is based on an orthonormal basis set that can be generated by dilation and translation from a single wavelet function $\Psi(x)$, which decays rapidly with increasing $|x|$ and has a zero integral ($\int_{-\infty}^{\infty} \Psi(x)dx = 0$). In 1-dimensional systems the members of the basis set have the form $\Psi_{jk}(x) = 2^{-j/2}\Psi(2^j x - k)$, for the most simple binary dilation case. Unlike a Fourier series, a wavelet series is defined on the entire d-dimensional space. The wavelet basis functions differ from the sine and cosine functions used in Fourier analysis in that they are localized and can be used to represent inhomogeneous distributions such as multifractal fields [704]. The wavelet transform of a function $f(\mathbf{x})$ representing the distribution of a measure is given by $F(\mathbf{X}, \lambda) = \int \Psi[(\mathbf{x} - \mathbf{X})/\lambda]f(\mathbf{x})d\mathbf{x}$, where $\mathbf{X}$ is the translation and λ is the dilation.

their results in terms of a preferential screening angle of $\pi/5$. Argoul *et al.* have taken this approach further by using the wavelet transformation approach to analyze the intersections between large off-lattice DLA clusters and circles of radius r centered on the origin of the cluster, inside the internal "frozen" structure of the cluster. If the local maxima of the wavelet transformation are plotted as a function of the "resolution" of the transformation for a Cantor set, then a tree-like structure emerges that reveals the hierarchical nature of the Cantor set and represents the construction of the Cantor set in the way described in chapter 2, section 2.1. The fractal dimensionality is then the given by the branching ratio $r_B = 2$ and the scale factor $r_L = 3$ (length ratio or resolution ratio) between successive generations ($D = \log r_B / \log r_L$).

For the intersections between DLA clusters and circles, the trees generated by the wavelet transformation analysis are less well defined. However, Arneodo *et al.* [448] measured a scale factor ratio of $r_L = 2.2 \pm 0.02$ using 23 10^6-particle off-lattice DLA clusters. This value for r_L, with a fractal dimensionality of 1.60 ± 0.02 [450, 451, 452] (0.60 ± 0.02 for the azimuthal intersection), gives a value of about $2.2^{0.6}$, or about 1.61, for the branching ratio r_B. Arneodo *et al.* [448] pointed out that this value for r_B is close to the golden mean $(1 + 5^{1/2})/2 = 1.61803\ldots$ However, the uncertainty in both D [705] and r_L are probably underestimated, and the numerical evidence supporting the idea of a golden mean branching ratio is weak. In particular, this work does not take into account the possibility of a very slow approach to the true asymptotic behavior (strong corrections to scaling) or the possibility of a more complex scaling model such as those discussed earlier in this chapter. Arneodo *et al.* [448] also related the structure of the "tree" generated by the wavelet transformation to the Fibonacci sequence defined by the iterative rule

$$F_n = F_{n-1} + F_{n-2}, \tag{3.252}$$

with $F_0 = 1$ and $F_1 = 1$. In the limit $n \to \infty$, the ratio F_n/F_{n-1} approaches the golden mean. They showed that the number of maxima in each "generation" of the wavelet transformation can be represented quite well by the Fibonacci numbers F_n. 2-Dimensional wavelet transformations were also used to illustrate a quite striking relationship between the internal branching structure of DLA and the Fibonacci sequence.

At present, is not possible to assess the significance of these observations. The numerical results are not compelling, and it is not clear how a theory for DLA can be based on these ideas. At this stage, the relationship between DLA and Fibonacci sequences must be regarded as speculative.

3.11 Additional Information

The application of computer simulations to far-from-equilibrium growth processes has been the subject of a number of conference proceedings and reviews. In addition, the book entitled *Fractal growth phenomena* by Vicsek [706] covers a range of topics similar to those discussed in this book. Bunde and Havlin [698] have edited an attractive collection of original reviews, with an emphasis on percolation theory and fractal growth, under the title *Fractals and disordered systems. Introduction to percolation theory* by Stauffer and Aharony [220] provides an account of recent advances in percolation theory.

Chapter 4

Experimental Studies

While much of the work on non-equilibrium growth has been concerned with the development of simple models and attempts to understand the consequences of these models, the main objective of this work has been to obtain a better understanding of natural phenomena. Consequently, considerable effort has been devoted to experimental studies of simple growth processes. In many cases, this work has entailed detailed comparisons between experiments, theory and computer simulations. Since experiments are usually more expensive and require more planning and preparation, they have often lagged behind simulations or theoretical work. However, simple experiments, often carried out by scientists lacking extensive experimental experience, have made important contributions to the development of this area and have stimulated additional simulations and theoretical work. Most pattern-formation experiments are relatively inexpensive. While equipment such as a synchrotron radiation source or a high resolution CCD camera may be a valuable asset in some experiments, most investigations have been performed with much more modest equipment. In this chapter, the results of some experimental studies will be presented in the context of the models described in chapter 3. Laboratory and natural growth processes related to DLA are emphasized. Experiments on the growth of self-affine rough surfaces are described in chapter 5.

In this chapter, no attempt is made to provide a comprehensive survey of the experimental study of disorderly growth processes. Instead, experiments in which serious attempts were made to characterize quantitatively the geometry of patterns formed under far-from-equilibrium conditions in terms of fractal geometry or scaling are emphasized. Emphasis is also placed on processes such as the growth of densely branched patterns, which have been the subject of computer modeling and/or theoretical work. The fascinating area of biological growth processes has been completely neglected. This subject will be discussed elsewhere [57].

4.1 DLA Processes

Although the complex, fractal patterns generated by the diffusion-limited aggregation (DLA) model were quite surprising, a wide variety of experimental realizations were soon found, and it quickly became apparent that DLA-like patterns are common in nature. This can be regarded as a consequence of the central role played by the Laplace equation in physical processes. Whenever a "random" growth process is controlled by a scalar field that obeys the Laplace equation, with the appropriate boundary conditions, then a DLA-like pattern can be expected to emerge. In principle, a more regular pattern will arise in an environment with strong anisotropy and weak disorder. However, in most cases, there is no long range anisotropy, since the materials are amorphous or polycrystalline and/or the environment is strongly disordered. DLA-based models have been used to study the evolution of a wide range of dendritic patterns, including the form of urban centers [18, 420, 19]. In many of these applications, the study of DLA models was initially motivated by the similarities between the observed morphologies and DLA patterns. Only at a later stage was a serious attempt made to identify the field obeying the Laplace equation and to develop more realistic models based on an understanding of the underlying processes.

One of the most familiar examples of DLA-like patterns is the manganese oxide/iron oxide "pseudo-fossils" (MnO_2 dendrites) that are found at the joint surfaces in sedimentary rocks such as Solnhofen chalks [708]. An example is shown in figure 4.1. The origin of these patterns has not been unambiguously established. They are evidently associated with the transport of Fe^{2+} and Mn^{2+} ions along the joint surfaces. Van Damme [709] and Garcia Ruiz *et al.* [710] have suggested that they are "fossilized viscous fingers", but others have suggested that they are formed by diffusion-limited precipitation reactions [711]. In either case, the relationship with DLA is clear. Garcia Ruiz *et al.* [710] have carried out experiments to support their idea of "self-fed" viscous fingering. Three 5 mm thick glass sheets were stacked with layers of toothpaste, consisting mainly of colloidal silica aggregates in water, between them. The stack was then immersed in a vessel containing colloidal γFeOOH and colloidal γMnOOH particles. The stack of glass sheets, separated by toothpaste, was then fractured with a hammer. After a few minutes, dendritic patterns were observed in the thin layers of toothpaste that separated the glass sheets. Similar patterns were obtained in an experiment in which the toothpaste layer between the glass sheets

Figure 4.1 A "MnO_2" pseudo-fossil that has grown along a bedding plane in Solnhofen chalk. This figure was provided by A. Fowler.

was replaced by clay/BaO which was immersed in a solution containing Fe^{2+} and Mn^{2+} ions. The DLA-like structures are one among a variety of patterns that are formed by iron and manganese in the bedding planes of sedimentary rocks. The origins of these patterns have not been thoroughly studied, despite their familiarity. The presence of MnO_2 dendrites overlying feather impressions has been cited as evidence for the veracity of *Archaeopteryx* [712] fossils found near Solnhofen.

4.1.1 Electrochemical Deposition

One of the most extensively studied pattern-formation processes, in the post-DLA era, has been the growth of metallic forms by electrochemical deposition. Depending on the material being deposited and the electrochemical conditions, a wide variety of patterns including whiskers, polyhedra and platelets [713] can be generated. Most studies have been carried out using Zn, Cu or Ag as the deposited metal. However, similar patterns have also been generated by the electrochemical polymerization of pyrrole [714] and the electrodeposition of lead [715], cobalt [716], iron and nickel. It is becoming increasingly apparent that electrochemical deposition is a very complex process that, in most systems, is not well understood. From this point of view, electrochemical deposition is a poor experimental model for the study of pattern-formation processes. However, it compensates for this weakness by presenting a very rich pattern-formation phenomenology.

The first studies of the geometric scaling properties of electrodeposited metals were carried out by Brady and Ball [717] and Matsushita *et al.* [718]. In the experiments of Matsushita *et al.*, a layer of n-butyl acetate was floated on a thin (4 mm deep) layer of 2 M zinc sulfate in a 20 cm inside diameter glass vessel. A 3 mm thick, 17 cm diameter circular zinc anode, with a height of about 2.5 cm, was placed on the bottom of the glass container, a carbon (graphite pencil lead) cathode was then inserted through the n-butyl acetate layer, into the center of the electrolyte layer, and zinc metal was deposited by applying a dc potential difference of about 5 V. Figure 4.2 shows a typical zinc dendrite that was deposited on the cathode in one of these experiments. For reasons that are not well understood, the zinc dendrites grow along the interface between the n-butyl acetate and electrolyte layers. A similar phenomenon has been observed in the electrochemical growth of Ag from aqueous silver nitrate solutions with a polymer liquid crystal Langmuir monolayer [720]. In this case, it was suggested that the growth along the Langmuir monolayer was a consequence of a negatively charged monolayer, of higher conductivity, adjacent to the polymer layer. However, it is difficult to understand the growth of zinc dendrites along an n-butyl acetate/water interface on this basis, since ions would be repelled from the interface because of the relatively low dielectric constant of the organic fluid. In any event, the aspect ratio of the electrolyte used in the experiments of Matsushita *et al.* was large, and the "cell" was considered to be essentially 2 dimensional. The growth of the dendrites was recorded using a camera and the images were digitized. An effective fractal dimensionality of 1.66 ± 0.03 was obtained from the two-point density–density correlation function $C(r)$, defined in chapter 2, section 2.1.3.1, for patterns grown at applied voltages of less than 8 V. At higher voltages, the effective fractal dimensionality increased with increasing voltage.

Quasi 2-dimensional electrodeposition experiments can be carried out under better controlled conditions using thin cells. This is an approach that was pioneered simultaneously by Grier *et al.* [721] and Sawada *et al.* [722]. This work led to the discovery of densely branched radial patterns as well as DLA-like patterns and a variety of other growth forms.

The formation of DLA-like patterns during electrochemical processes is not limited to metallic systems. For example, Kaufman *et al.* [714, 723] have obtained DLA-like structures by the electrochemical oxidation of pyrrole, to form polypyrrole, in a 0.01 inch gap cell with gold electrodes and an electrolyte consisting of 0.1 M silver tossylate (silver p-toluenesulfonate) and 0.1 M pyrrole in acetonitrile. The cathode consisted of a 2000 Å thick gold ring and the anode was a 0.05 inch diameter gold wire, which was inserted through a hole in one of the glass cell walls and cut flush with the cell wall. A fractal dimensionality

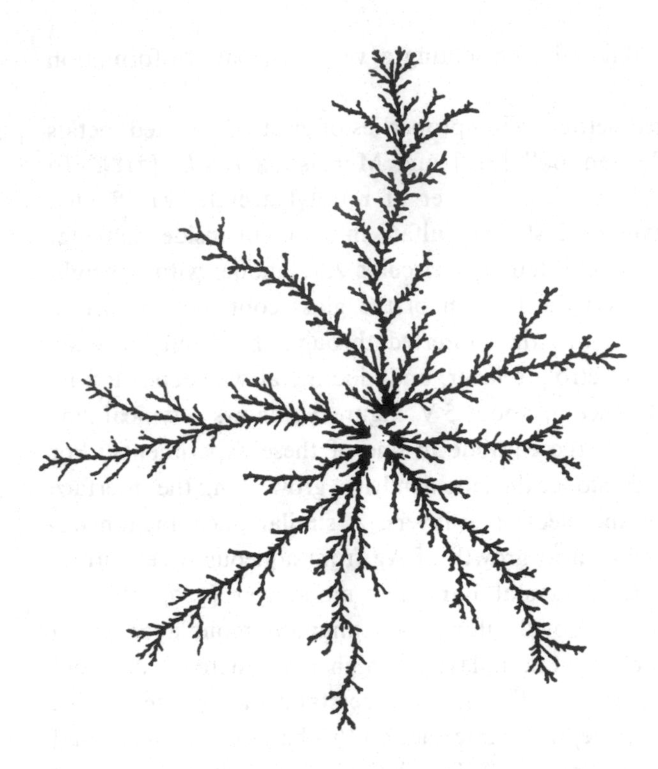

Figure 4.2 A zinc dendrite grown by electrodeposition. This figure was provided by M. Matsushita.

of $D = 1.74 \pm 0.03$ was measured for the dendritic polypyrrole pattern formed at an applied potential difference of 6 V. Kaufman *et al.* were able to show that the width of the active zone (chapter 3, section 3.4) grew linearly with the pattern radius. Polypyrrole is a mechanically durable, semi-conducting, polymeric material. These properties allowed the resistance R to be measured as a function of the distance r between contact points on the dendritic structure. They found that $R \sim r^{0.96 \pm 0.04}$, consistent with a minimum path dimensionality D_{min} of 1 for these loopless structures. It follows from the relationship between diffusion or random walks and conductance that [724, 725]

$$\mu = D - D_w. \tag{4.1}$$

In this equation, the conductance exponent μ is defined through the relationship $g(\ell) \sim \ell^{\mu}$, where $g(\ell)$ is the conductance between two points separated by a distance ℓ, measured in the embedding space, and D_w is the fractal dimensionality of a random walk on the fractal. This relationship applies equally well to the backbone of the fractal, defined in chapter 3, section 3.5, and the whole fractal. This implies that $D_w = D_{wb} - D_b + D$, where D_b is the fractal dimensionality of the backbone and D_{wb} is the fractal dimensionality of a random walk on the backbone. Since the conductivity measurements indicate that $D_{wb} - D_b \approx 0.96$, it follows that $D_w \approx 0.96 + 1.74 \approx 2.7$, and the fracton or spectral dimensionality

[726, 727] is given by $D_s = 2D/D_w \approx 3.48/2.7 \approx 1.29$. This is in accordance with measurements of the spectral dimensionality for 2-dimensional DLA clusters [728], using random walk simulations, which give $D_s = 1.3 \pm 0.1$.

Quite striking DLA-like patterns were obtained by Fujii *et al.* [729] using an electrochemical cell consisting of a nickel plate anode, an electrolyte and a polypyrrole film cathode with an area of about 1cm × 2cm, that was grown electrochemically on a glass plate coated with conducting In/SnO_2. The electrolyte consisted of 0.05 mmol tossylate, 0.5 mmol acetonitrile/1% water and 0.1 mol pyrrole monomer. The cell was operated by applying a potential difference of 3 V for 3 hours. At this stage, the coated glass plate was the anode of the cell where pyrrole was electrochemically oxidized to form a polypyrrole film. Half of the electrolyte was then removed and a potential of −3 V was applied for 1.5 hours to undope the bottom part of the polypyrrole film adjacent to the electrode. The surface of the undoped film was then examined using an electron microscope, and DLA-like random dendrites were observed growing along the surface of the film. A fractal dimensionality of $D \approx 1.69$ was measured from the two-point density–density correlation function. These dendrites were believed to be composed primarily of nickel, but their composition was not determined. During the undoping process, the nickel cations migrated towards the polypyrrole-coated conducting In/SnO_2 cathode. Nickel was deposited preferentially on regions of high local conductivity, and the conductivity remained high in the parts of the polypyrrole film covered by the nickel dendrites. As the process proceeded, the anionic dopant was removed from the uncovered parts of the doped polypyrrole film, which became less conductive. At this stage, the nickel ions were believed to drift towards the polypyrrole–electrolyte interface and diffuse on this interface until they reached and were deposited on a growing nickel dendrite. The formation of dendrites can adversely affect the performance of batteries and other devices.

A variety of polypyrrole patterns including DLA-like, needle-like and leaf-like (lobed) patterns have been generated under constant current conditions using a similar geometry and similar electrolyte [730]. The nature of the pattern depended on the current, the electrolyte composition and particularly the pyrrole concentration. Under constant current conditions, the nature of the pattern may change during the course of an experiment due to a change in the effective voltage. Such morphological transitions were not found to be characteristic of constant voltage experiments.

A variety of processes can contribute to pattern formation by electrochemical deposition. At large screening electrolyte[1] concentrations, in an appropriate potential range, the diffusion of cations may be the dominant transport process leading to growth. However, diffusion-limited growth is a slow process, and in

1. A screening electrolyte is an electrolyte that is electrochemically inert under the experimental conditions and screens the electrostatic field.

many experiments the growth velocities observed in the formation of fractal patterns (as high as 1 mm s^{-1} [731]) cannot be accounted for in terms of a diffusion-limited growth mechanism. In the absence of convection, the transport of the ions in a simple salt solution is described by the equations

$$\partial c^+/\partial t = -\nabla j^+, \tag{4.2}$$

$$\partial c^-/\partial t = -\nabla j^-, \tag{4.3}$$

$$j^+ = -\mathscr{D}^+\nabla c^+ + \mathscr{M}^+ c^+ E \tag{4.4}$$

and

$$j^- = -\mathscr{D}^-\nabla c^- - \mathscr{M}^- c^- E, \tag{4.5}$$

where E is the electric field, $\mathscr{D}^\pm$ are the ion diffusion coefficients, $\mathscr{M}^\pm$ are the ion mobilities, which are related to the diffusion coefficients by $\mathscr{M}^\pm = \mathscr{D}^\pm e|z^\pm|/k_B T$, $c^\pm$ are the concentrations and $j^\pm$ are the ion fluxes. For the transfer of cations from an anode to a cathode made of the same metal, via a dilute electrolyte, the propagation of a growth front (in the x direction) can be understood in terms of the steady-state transport equations, which can be reduced to

$$(\mathscr{D}^\pm d^2c^\pm/dx^2) \pm [\mathscr{M}^\pm d(c^\pm dV/dx)/dx] = 0 \tag{4.6}$$

and

$$d^2V/dx^2 = -e(z^+c^+ + z^-c^-)/\epsilon\epsilon_\circ, \tag{4.7}$$

obtained from equations 4.2–4.5. Here, $\epsilon_\circ$ is the vacuum permittivity, ϵ is the dielectric constant of the solvent (usually water), $ez^\pm$ are the ion charges, including the sign of the charge, and V is the electrostatic potential. The effective dielectric constant of the solvent in aqueous electrolytes is somewhat smaller than that of pure water, but this would be important only in accurate quantitative comparisons between theory and experiments. Chazalviel [731] has analyzed these steady-state transport equations and has numerically solved the 1-dimensional time dependent ion transport equations, obtained from equations 4.2–4.5. The picture that emerges is that the cations and anions are depleted in a "space charge" region in the vicinity of the cathode. The potential, cation concentration and anion concentration decrease almost linearly from the anode to the "boundary" of the space charge region, within the "quasi-neutral" region. Most of the potential drop takes place across the space charge region. Within the quasi-neutral region, $|V_\circ - V(x)| \leq k_B T/e$, where $V_\circ$ is the potential at the anode, $V(x)$ is the potential at position x and $-e$ is the electron charge. The potentials are measured relative to the cathode potential of $V_c = 0$. According to the analysis of Chazalviel, the width w of the space charge region is given by

$$w \approx L(\lambda_D/\lambda_b)^{2/3} \approx 10\ \mu\text{m}. \tag{4.8}$$

In this equation, L is the distance between the electrodes, λ_b is the backstream diffusion length ($\lambda_b \sim \mathscr{D}^+/\mathscr{M}^+V_\circ L$) and λ_D is the Debye screening length of the electrolyte given by $\lambda_D = [\epsilon\epsilon_\circ k_B T/(e^2 c_\circ(\sum_i n_i z_i^2))]^{1/2}$, where $c_\circ$ is the nominal electrolyte concentration and $n_i c_\circ$ is the concentration of the ith ion in the electrolyte. Chazalviel argued that a voltage drop of ≈ 10 V across a length w of $\approx 10\ \mu$m does not occur in practice, and avoidance of this large potential gradient is the source of instability in the growth front. The unphysically large potential gradients may be avoided if convection (including electroconvection, and electro-osmosis [732]) takes place. If convection is inhibited, the advancing growth front will become unstable due to a Mullins–Sekerka-like instability (appendix A, section A.1). Fluctuations in an otherwise straight or flat growth front will be amplified. This will take place well before the system reaches the steady state described by equations 4.6 and 4.7.

Fleury *et al.* [733, 734, 735] have demonstrated that an electroconvection process, driven by charges near the tips of the branches, takes place during the formation of DLA-like patterns in thin cells. If no supporting or screening electrolyte is present, the force $d\mathbf{f}$ acting on a small volume of fluid $d\Omega$ is given by

$$d\mathbf{f} = [z^+\rho^+ + z^-\rho^-]e\mathbf{E}d\Omega, \tag{4.9}$$

where ρ^+ and ρ^- are the ion densities, $\mathbf{E}$ is the electric field and $-e$ is the electron charge. Since most of the electrolyte is electrically neutral, this force acts primarily on the excess of cations near to the cathode. This electroconvective force is strongest near the most exposed tips of the growing structure and drives the fluid into the branched pattern. At low ionic strengths the electroconvective process is the dominant transport mechanism that determines the concentration fields in the electrochemical cell. Under other conditions buoyancy driven convection may be dominant, and electro-osmotic convection can also be important [732].

The 1 + 1-dimensional electrochemical deposition process has been studied theoretically by Fleury *et al.* [733, 734]. The dendrite forest was represented by an array of needles, like the teeth on a comb. The low Reynolds number fluid flow problem was solved with a force at each needle tip, directed towards the needle tips. The picture that emerged was that of an arch between each adjacent pair of needle tips. The arches separate a high electrolyte concentration region, above the arches, from a depleted region, between the needles, under the arches. A vortex pair forms at each tip beneath the arches. In the absence of diffusion, the arches would form a sharp interface. In experiments, this interface is blurred by diffusion. When tip splitting takes place, a new vortex pair is formed. If a branch dies by trailing too far behind the active front, its vortex pair dies. The concentration of electrolyte in the "funnel" formed by the arches above each needle tip is the same as that of the bulk electrolyte.

Fleury *et al.* were able to observe the convective motion using small oil droplets, with diameters of about 1.0 μm, as tracers. The convective motion drives the fluid towards the most actively growing tips. This may be, at least in part, responsible for the sensitivity of the growth morphology to small impurity concentrations. This electroconvective motion completely dominates other transport processes in the vicinity of the growth front. It is strong enough to deform the branches of the growing pattern and allows the pattern to grow much more rapidly than it would if the growth was diffusion-limited. Electroconvective processes are also important in the growth of dense radial patterns [735, 736]. Considerable motion in the tenuous deposit branches can be seen during electrochemical deposition in a thin cell. This motion appears to be a direct consequence of the electrolyte convection, and may also play an important role in the pattern-formation process. The electrically driven flow funnels electrolyte towards the most exposed tips and appears to generate a mechanical tip-splitting instability by pushing apart the delicate, highly ramified tip structure. Electroconvection, migration in the potential gradient and diffusion all contribute to ion transport and the pattern-formation process. Fleury *et al.* [735] have suggested that the transition from DLA-like to dense radial patterns may be related to the displacement of growing branches during the electrochemical deposition process and the interaction between these displacements and the cation concentration field. Since hydrodynamic, mechanical and electrochemical phenomena all appear to play an important role in electrochemical growth, and since these phenomena are strongly coupled to each other, it may be some time before a quantitative understanding of the formation of these patterns is achieved. It is quite remarkable that simple models can reproduce the observed morphologies so well. This may be a consequence of an underlying universality.

The fluid motion during electrodeposition has been studied by Huth *et al.* [732] by recording snapshots in a thin cell containing 10^8 neutrally buoyant particles cm^{-3} with diameters of 1 to 3 μm. Figure 4.3(a) shows the particle paths obtained during a 6 s period, using a dilute electrolyte to enhance the role of electroconvection. Figure 4.3(b) shows the fluid motion obtained from an electroconvection model. The observed and simulated fluid motions are in good qualitative agreement; however, the measured electroconvective velocities were found to be much smaller than those predicted by the theoretical, model of Fleury *et al.* [734].

Wang *et al.* [737] have studied the electrodeposition of Fe from $FeSO_4$ solutions. They observed a new morphology, in which the deposit forms a network with many loops. This pattern was formed only at low pH. A dense radial pattern was formed at higher pH values. Wang *et al.* suggested that the formation of this class of patterns could also be qualitatively understood

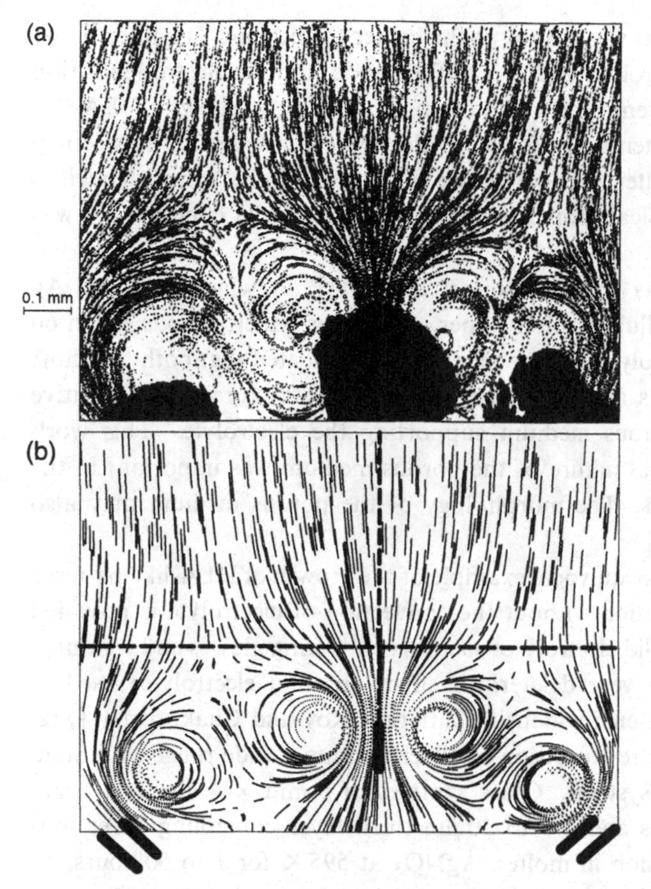

Figure 4.3 A comparison between observed and simulated particle tracer patterns during electrochemical deposition. Part (a) shows the experimental particle path streaks during electrochemical deposition of Cu from 0.01 M $CuSO_4$ in a 25 mm × 25 mm × 0.1 mm cell at a current density of 4 mA cm^{-2}. Part (b) shows the particle path streaks simulated using an electroconvection model. The solid bars show the positions and orientations of the branch tips at the beginning and end of the 6 s recording period. This figure was provided by H. Swinney.

in terms of electroconvection and its effects on the ion concentration fields. It appears that, in these experiments, the electroconvection was driven mainly via the H_3O^+ ions. If the H_3O^+ concentration is low, as a result of either a high initial pH or electrochemical H_3O^+ ion depletion, the electroconvection is weakened and a dense branching morphology develops. The formation of H_2 bubbles was observed in the low pH experiments. These bubbles may play an important role in the pattern-formation process, and their presence added an additional complication.

Although it is apparent that electroconvection plays a key role in some electrochemical deposition processes, it does not seem to be essential for the formation of DLA-like patterns. For example, Hibbert and Melrose [738, 739, 740] have generated DLA-like patterns by the electrodeposition of copper using an electrolyte consisting of filter paper dampened with copper sulfate solutions in 1.0 M sulfuric acid. The electrolytic cell consisted of the damp filter paper electrolyte lying on a polymethylmethacrylate sheet, a copper ring anode with an internal diameter of 8 cm, which was clamped on top of the filter paper, and a copper wire cathode that was placed at the center of the ring. The filter paper was connected to an electrolyte reservoir at the anode, and the electrodeposition was carried out under potentiostatic conditions. Figure 4.4(a) shows a DLA-like pattern obtained at a potential of 0.5 V. In this system, convection is strongly suppressed. Since the filter supported the otherwise fragile metal dendrites, the measurement of physical properties such as the electrical conductivity was facilitated [739].

Subsequent experiments by Atchison *et al.* [731], in which Cu, Zn, Ag and Pd were grown in cellulosic filter paper, glass-fiber paper, porous nylon 66 membranes and porous polypropylene membranes, indicated that both the short length scale morphologies and the large length scale structures were sensitive to the nature of the porous medium supporting the electrolyte. This work indicated that the chemical nature of the porous medium was important in the pattern-formation process. The morphology of the porous medium may also play a role.

A more dramatic and unequivocal example of the growth of DLA-like patterns by electrochemical deposition without the convection of electrolyte is provided by electrodeposition in solid electrolytes such as conducting glasses and ceramics. This process can have a very deleterious effect on solid electrolytes used in applications such as high-energy-density batteries. Roy and Chakravorty [732] have investigated the electrodeposition of silver in quite a wide range of lithium and sodium oxide glass systems. Glass samples of 5 mm $\times$ 4 mm $\times$ 1.5 mm were cut from cast pieces and a 5 to 30 μm thick layer containing silver ions was produced by immersion in molten $AgNO_3$ at 595 K for 1 to 50 hours. In the electrochemical growth experiments, conducting silver paint electrodes were applied to one face of the ion exchanged samples and dc potential differences in the range 1 to 50 V were applied for periods of a few minutes to half an hour. The electrodeposited silver was studied by optical and electron microscopy. Because of the high aspect ratio of the Ag^+ doped layer between the electrodes, the electrodeposition process can be considered to be 2 dimensional.[2] DLA-like patterns were formed, and fractal dimensionalities in the range $1.52 \pm 0.04 \leq D \leq 1.89 \pm 0.02$ were measured for different materials and ion exchange conditions. In general, higher fractal dimensionalities were found for samples in which Li^+ was replaced by Ag^+ than for those in which Na^+ was replaced by Ag^+. Roy and Chakravorty [732] discussed their results in terms of the "η" Laplacian growth model, described in chapter 3, section 3.4.4, in which $\mathbf{V}_n(\mathbf{x}_s) \sim [|\nabla\Phi(\mathbf{x}_s)|]^\eta$, where $\mathbf{V}_n(\mathbf{x}_s)$ is the growth velocity at position $\mathbf{x}_s$ on the surface and $\Phi(\mathbf{x}_s)$ is the potential. Although such a model can give fractal dimensionalities in the range measured by Roy and Chakravorty, these authors did not provide a satisfactory explanation of the origin of the values of η that are not equal to unity. An

2. The silver ion conductivity is much larger than either the sodium or lithium ion conductivities.

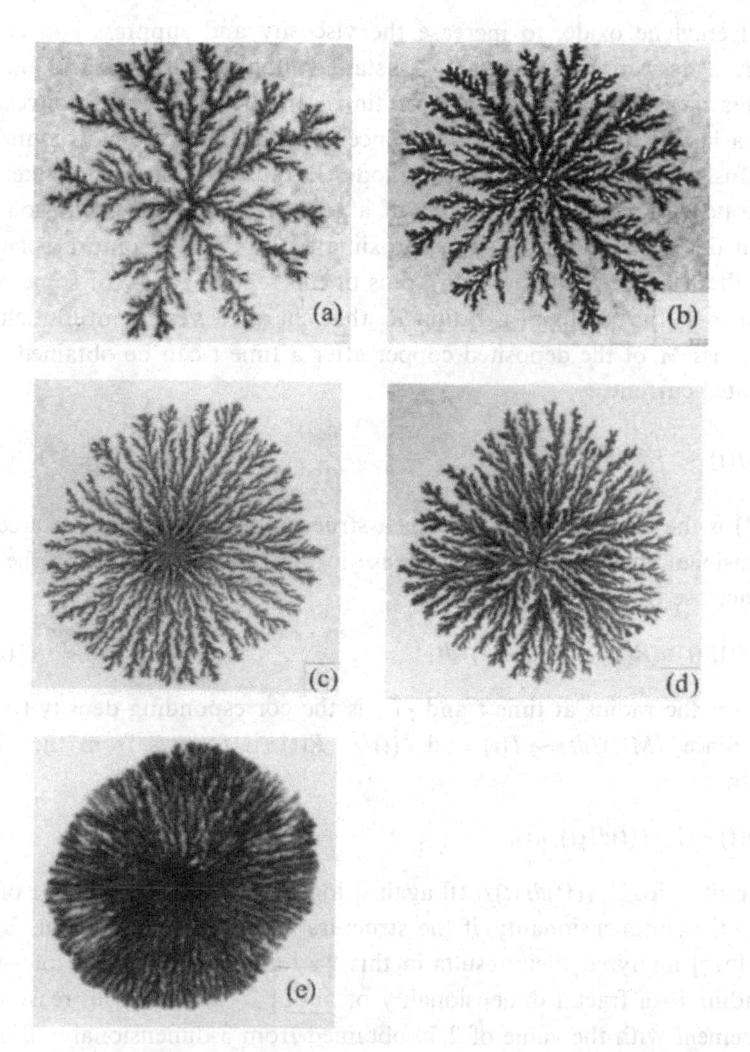

Figure 4.4 Electrodeposition patterns obtained using an electrolyte consisting of filter paper dampened by an aqueous 1.0 M sulfuric acid/0.75 M $CuSO_4$ solution. Part (a) shows the DLA-like pattern obtained at a potential of 0.5 V. Parts (b)–(e) show the transition from DLA to dense radial patterns as the voltage is increased (0.7, 0.8, 1.6 and 3.0 V in parts (b), (c), (d) and (e), respectively). This figure was provided by J. R. Melrose.

explanation, suggested by Roy and Chakravorty, in terms of the random walk η model of Matsushita *et al.* [432] (chapter 3, section 3.4.4), with the emission of randomly walking silver atoms after each Ag^+ ion is deposited, does not seem to be realistic.

The 3-dimensional electrodeposition of copper onto the exposed tips of otherwise insulated copper wires was studied by Brady and Ball [717]. They used solutions of copper sulfate, with sodium sulfate as the screening electrolyte, and added polyethylene oxide, to increase the viscosity and suppress convection. The experiments were carried out at constant voltages of 0.3 – 0.8 V, and the current was measured as a function of time. In the absence of convection, and with a large screening electrolyte concentration, the current is controlled by the diffusion of Cu^{2+} ions to the cathode. Because the random walkers do not penetrate into the interior regions of a growing cluster, the diffusion flux (current in the experiment) can be approximated by the flux onto a spherical surface with a radius R equal to the radius of the "active zone", or some other distance such as the radius of gyration R_g that characterizes the overall cluster size. The mass M of the deposited copper after a time t can be obtained from the integrated current

$$M(t) \sim \int_0^t I(t')dt', \tag{4.10}$$

where $I(t')$ is the current at time t'. For a structure growing out from a center in 3-dimensional space, the rate of increase in the radius is related to the rate of mass increase by

$$\rho(t)R(t)^2 dR(t)/dt \sim dM(t)/dt, \tag{4.11}$$

where $R(t)$ is the radius at time t and $\rho(t)$ is the corresponding density ($\rho(t) = \rho(R(t))$). Since $dM(t)/dt \sim I(t)$ and $I(t) \sim R(t)$, it follows from the above equation that

$$\rho(t) \sim 1/(I(t)dI(t)/dt), \tag{4.12}$$

so that a plot of $\log[1/(I(t)dI(t)/dt)]$ against $\log I(t)$ should have a slope of $-\alpha$, where α is the codimensionality, if the structure is a self-similar fractal. Brady and Ball [717] analyzed their results in this way and found a slope of –0.57, corresponding to a fractal dimensionality of about 2.43. This is in reasonably good agreement with the value of 2.50 obtained from 3-dimensional, off-lattice DLA model simulations [424].

Electron micrographs and extrapolation of the density–radius relationship to the density of copper indicated an inner cut-off length of about 40 nm. Consequently, it appears that the fractal scaling may extend over four full orders of magnitude ($L/\epsilon \approx 10^4$). However, the copper wire cathode had a diameter of 12.5 or 25 μm and the deposition current was not measured over a correspondingly wide range. Despite these pioneering studies of Brady and Ball, there has been very little experimental work on 3-dimensional DLA-like growth processes.

More recently, the electrodeposition of silver onto a spherical (0.076 cm diameter) Pt cathode from an electrolyte consisting of 0.50 M Na_2SO_4/0.010 M H_2SO_4/0.0050 M Ag_2SO_4 in aqueous agarose was studied under potentiostatic conditions (–200 mV with respect to a saturated calomel electrode) by

Carro *et al.* [733, 734]. The agarose concentration was varied in the range $0.00033 - 0.005\,\mathrm{g\,ml^{-1}}$, to cover both the sol and gel regimes. In agarose free electrolytes, the silver deposit has a fractal dimensionality of 3 with a rough surface, $D_s \approx 2.5$. The approach used to measure the surface fractal dimensionality D_s was based on the relationship $A \sim Q^{D_s/3}$, where A is the surface area and Q is the deposit volume. The surface area A was measured via the underpotential deposition [745, 746] of Pb or Cd at sweep rates of 0.010 or $0.020\,\mathrm{V\,s^{-1}}$ in 10^{-2} M $Pb(CH_3COO)_2$/0.5 M $NaClO_4$/10^{-2} M $HClO_4$ or 10^{-2} M $CdSO_4$/0.5M $NaClO_4$/10^{-2}M H_2SO_4 aqueous electrolytes, respectively. In these voltagrammetric studies, the amount of charge required to electrodeposit and remove a monolayer of Pb or Cd was measured, and the area was estimated from these measurements [747].

In underpotential deposition, a metal cation (Pb^{2+} in this case) is deposited onto a substrate using electrochemical conditions under which the substrate (Ag in this case) is inert. The measurements are carried out using a linear potential sweep, which starts at a value that is anodic with respect to the Pb^{2+} deposition potential. As the potential is swept towards more negative values, a monolayer of metal is deposited on the substrate, before bulk, multilayer deposition appears. The monolayer deposition leads to a peak in the current/voltage curve with an area that corresponds to the amount of deposited metal in one monolayer. Because of the electronegativity difference between the deposited metal and the substrate, partial electron transfer takes place, leading to an increase in the bond energy, which allows the first layer to be deposited at an underpotential [746]. Fine structure can arise because of different underpotentials on different crystal faces. Fine structure in the current voltage curve can also arise for underpotential deposition on single crystal faces because of the formation of more than one phase in the monolayer of deposited metal atoms and transitions between them [748]. Bewick and Thomas [748] have reported that a single layer of Pb is formed on Ag(111) and Ag(100) faces in the underpotential regime but that on Ag(110) a second layer of Pb starts to form before the onset of bulk deposition. These complications do not appear to compromise seriously the value of this approach in the measurement of the volume dependence of the surface areas.

The mass fractal dimensionality was estimated from the dependence of the deposit mass (from the electrodeposition charge) on its radius. The effective fractal dimensionality measured in this way decreased with increasing agarose concentration and reached a value of about 2.50 at $0.005\,\mathrm{g\,cm^{-3}}$ agarose. After a characteristic size had been reached ($r > \xi$, where r is the deposit radius and $\xi \approx 0.3$ cm), the deposit density stopped decreasing and became constant at a value of about $2.8 \times 10^{-3}\,\mathrm{g\,cm^{-3}}$. This behavior suggests a crossover from a DLA-like structure on short length scales to a dense radial pattern on longer length scales, at later times. However, the structures were not characterized in detail. At low agarose concentrations, convection played an important role, and the overall shape of the deposit became cone-like, with the tip of the cone pointing downwards.

While most electrochemical deposition growth experiments have been carried out using a radial Hele-Shaw cell geometry, experiments using a linear electrode have also been popular [450, 749, 750]. These experiments frequently generate patterns that look like "a forest of trees" and can be simulated quite well using a 2-dimensional model for diffusion-limited deposition on a line [465]. Figure 4.5 compares zinc dendrites that were grown on a linear cathode in a narrow cell and a DLA model simulation with the corresponding boundary conditions. Matushita *et al.* [749] measured the 2-dimensional tree size distribution. They found that the cumulative size distribution had a power law form over about a decade in the tree sizes s with an exponent τ_c of about 0.54. This corresponds to a size distribution exponent τ with a value of about 1.54, in good agreement with the value of about $1 + (1/D) \approx 1.585$ expected for a self-similar DLA deposit (chapter 3) and the value of 1.55 ± 0.05 measured for τ using diffusion-limited deposition simulations [621]. In these experiments, the rms deposit height or thickness h_2 was observed to grow algebraically with increasing time or deposit mass m ($h_2 \sim m^a$, where the exponent a had a value of $1/(0.70 \pm 0.06)$). This can be understood in terms of a density profile of the form $\rho(h) \sim h^{-\alpha} f(h/t^{1/(1-\alpha)})$, where the scaling function $f(x)$ has the form $f(x) = const.$ for $x \ll 1$ and $f(x)$ decays faster than any power of x for $x \gg 1$. Here, $\alpha = 1 - 1/a$ is the codimensionality, so that the deposit can be considered to be a fractal with the same density correlations that are found in 2-dimensional DLA, characterized by a dimensionality of about 1.7. In view of the complexities of electrochemical deposition and uncertainties in the scaling model for 1+1-dimensional diffusion-limited deposition, this facile interpretation of the experiments of Matushita *et al.* should be regarded with suspicion.

The effects of high magnetic fields (up to 8 T) on the electrodeposition of silver and lead have been studied by Mogi *et al.* [715, 751]. Starting with conditions that lead to the growth of DLA-like patterns (with no magnetic field), a transition to a densely branched ($D \approx 2$) radial growth pattern was observed for the silver dendrites as the magnetic field was increased. In the case of the lead dendrites, a similar but less dramatic change was found (the effective fractal dimensionality increased from about 1.58 to about 1.66). In both cases, the branches showed a slight curvature, all in the same sense. Mogi *et al.* suggested that these effects on the morphology may be a consequence of magneto-hydrodynamic phenomena. However, a detailed understanding of the growth of these patterns has not been reached.

Microscopic DLA-like patterns have also been observed, but not analyzed, by Bedekar *et al.* [752] in thin films of poly(o-anisidene) that were grown electrochemically from aqueous ortho-anisidene solutions on fluorine doped SnO_2

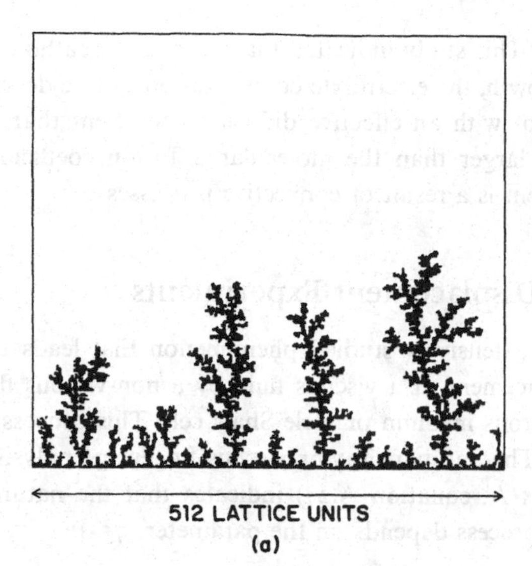

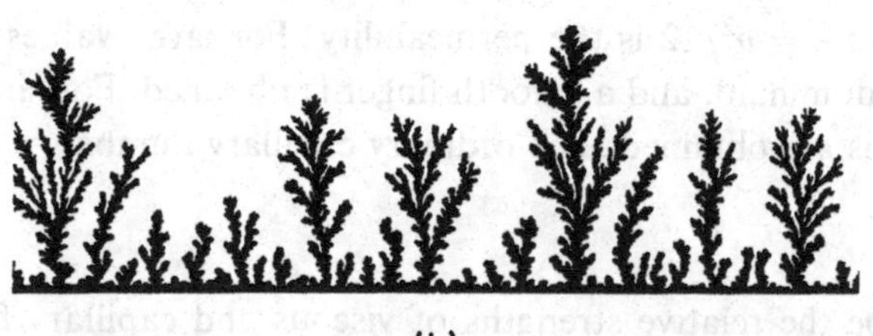

Figure 4.5 A comparison between a "forest" of zinc dendrites deposited on a graphite electrode (b) and a corresponding 2-dimensional DLA simulation (a). Part (b) was provided by M. Matsushita.

electrodes with HCl as a "supporting" electrolyte. A variety of morphologies were obtained. The most DLA-like were obtained using low monomer concentrations (0.1 M) with relatively high electrode conductivities (1560 $(\Omega\ \text{cm})^{-1}$).

Kuhn *et al.* [753] have generated patterns that look very much like DLA clusters by electroless deposition of silver on copper in cells with a small gap ($\approx 50\ \mu$m), to suppress convection. In these experiments, a small piece of thin, 99.99% pure Cu foil was inserted into the center of a circular, glass walled cell filled with 0.05 M $AgNO_3$. The cell was connected to a reservoir containing the same electrolyte. The similarity between the Ag dendrites grown in these experiments and off-lattice DLA clusters was established by measuring their fractal dimensionality and by wavelet transform analysis.

At the time of writing, interoferometric experiments are beginning to provide important information concerning the dynamics of the electrolyte concentration field during the growth of fractal electrodeposits [702]. For example, Argoul *et al.* [744] have analyzed the evolution of interference fringes during the electrodeposition of Zn from an aqueous $ZnSO_4$ electrolyte in a thin cell with parallel, linear electrodes. This study indicated that near to the cathode, during the early stages of the growth, the electrolyte concentration can be described by the diffusion equation, but with an effective diffusion coefficient that is about one order of magnitude larger than the molecular diffusion coefficient. The enhanced diffusion coeffient is a result of convective processes.

4.1.2 Fluid–Fluid Displacement Experiments

An important and quite extensively studied phenomenon that leads to DLA-like patterns is the displacement of a viscous fluid by a non-viscous fluid in a pseudo 2-dimensional porous medium or Hele-Shaw cell. This process is often call "viscous fingering". The Saffman–Taylor viscous fingering analysis for the Hele-Shaw cell (appendix A, equation A.27) indicates that the nature of the fluid–fluid displacement process depends on the parameter [755]

$$B = (\Gamma b^2)/(12\mu V w^2) = (\Gamma/\mu V)(k/w^2) = (\Gamma(b/w)^2)/12\mu V, \tag{4.13}$$

where Γ is the interfacial tension, b is the separation between the inside faces of the cell, w is the cell width, μ is the viscosity of the more viscous fluid, V is the "finger" velocity and $k = b^2/12$ is the permeability. For large values of B, the capillary forces are dominant, and a smooth finger is obtained. For small B, highly branched patterns are obtained. The ordinary capillary number

$$C_a = \mu V/\Gamma \tag{4.14}$$

is often used to describe the relative strengths of viscous and capillary forces, and the control parameter B in equation 4.13 can be written in the form $B = 1/[12(w/b)^2 C_a]$. For fluid–fluid displacement in a porous medium, a pore level capillary number Ca_p can be defined as

$$Ca_p = V\mu a^2/k\Gamma, \tag{4.15}$$

where a is a length characterizing the pore diameter and k is the permeability of the porous medium. In this case, $1/Ca_p$ in equation 4.15 plays a role similar to B in equation 4.13. In a series of geometrically similar materials, the pore sizes scale linearly with the particle sizes ϵ, and it is often convenient to define the pore level capillary number as $Ca_p = V\mu\epsilon^2/k\Gamma$.

Fluid–fluid displacement phenomena that lead to DLA-like patterns have been studied almost as extensively as electrochemical deposition. In both conventional Hele-Shaw cell experiments and experiments in which the cell is filled with a porous medium, the field obeying the Laplace equation is the pressure field in the viscous fluid that is displaced by a very much less viscous fluid. In the porous media experiments, the disorder is the quenched disorder of the porous medium itself. In the conventional Hele-Shaw cell experiments, the origin and role of the disorder are not as clear. The residual roughness (chapter 5) of the smooth glass walls, wetting heterogeneity due to impurities, dust and gas bubbles may be important in some experiments.

An early example of the generation of a fractal, DLA-like pattern is provided by the work of Måløy *et al.* [756] on the displacement of liquid epoxy or glycerin/water by air in a Hele-Shaw cell containing a monolayer of randomly placed glass beads. The 2-dimensional porous medium was prepared by randomly depositing 1.0 or 1.6 mm diameter glass spheres on a circular polymethylmethacrylate plate that had been coated with a thin layer of liquid epoxy. After the epoxy had hardened, the excess glass beads were removed and the monolayer of beads was covered by a second polymethylmethacrylate plate that had also been coated with a thin layer of liquid epoxy. The resulting 2-dimensional porous medium had a porosity of about 0.45. The fluid–fluid displacement experiments were carried out by first filling all of the interstices between the glass beads and the enclosing plates with viscous fluid. The "non-viscous" fluid was then injected through a small hole in the middle of one of the two circular plates. At high capillary numbers (large injection rates) DLA-like patterns were formed. A typical example is shown in figure 4.6. A fractal dimensionality of 1.62 ± 0.04 was found by measuring the amount of fluid (number of pixels in a digitized image) $N(r,n)$ within a distance r of the injection point at several stages n during the experiment. Måløy *et al.* found that the function $N(r)$ could be represented quite well by the simple scaling form

$$N(r,n) = N_\circ(n)(r/R_g(n))^D f(r/R_g(n)), \tag{4.16}$$

where $R_g(n)$ is the radius of gyration of the pixel image at the nth stage in the displacement process and $N_\circ(n)$ is the number of pixels in the image of the displaced region. The fractal dimensionality measured in this manner was smaller than that found in off-lattice DLA model simulations.

Stokes *et al.* [757] have studied the displacement of more viscous wetting fluids by non-wetting fluids and thc displacement of more viscous non-wetting fluids by wetting fluids in a 30 cm long × 7 cm wide × 0.15 cm Hele-Shaw cell filled with an unconsolidated packing of glass beads with a small diameter dispersion. The permeability k of the porous medium was controlled via the bead diameter ϵ ($k \sim \epsilon^2$). The "wetting" fluids were glycerin/water mixtures and the "non-wetting" fluids were oils. In all experiments, the viscosity ratio μ_2/μ_1 was greater than 100. In most cases, the displacement patterns could be seen quite easily because of the much larger refractive index difference between water/glycerin and glass than between the glass and the oils used in these experiments. If necessary, dye was added to the displacing fluid to enhance the optical contrast. In the case of displacement of a more viscous fluid by a less

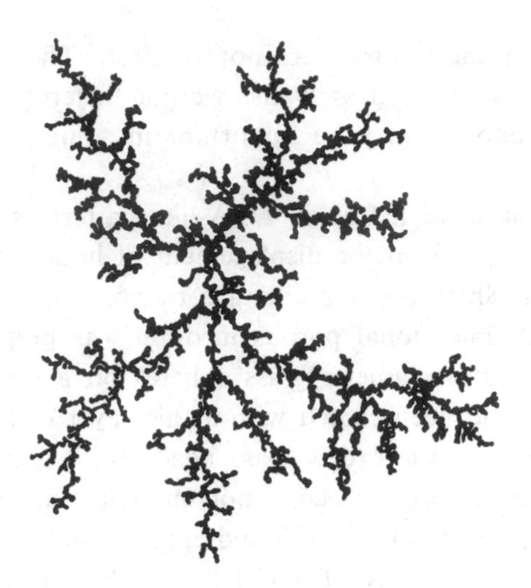

Figure 4.6 A DLA-like pattern formed by the displacement of liquid epoxy by air in a 40 cm diameter, 2-dimensional porous medium consisting of a disordered monolayer of glass spheres confined to a narrow gap between two transparent, circular plates. The pattern was formed after 19.1 s at a constant pressure difference of 20 mbar across the cell. This figure was provided by J. Feder.

viscous, more wetting, fluid, a branched pattern was generated, and the finger width w was found to scale as

$$w \sim (k/Ca_p)^{1/2}. \tag{4.17}$$

In all cases, the finger width was substantially larger than the length scales of order ϵ (the size of the glass beads) associated with the relatively homogeneous porous medium and much smaller than the channel width.

The patterns generated by the displacement of a more viscous, more wetting fluid by a less wetting fluid were quite different. The finger width was approximately equal to the particle size ϵ and independent of Ca. The number of fingers across the channel was observed to increase with Ca. Stokes *et al.* commented that DLA-like patterns were found only in the displacement of a wetting fluid by a non-wetting fluid. At low capillary numbers, an invasion percolation-like pattern will be generated, so that DLA is expected only at high capillary numbers, with a less wetting displacing fluid.

In order to reduce the possible stabilizing effect of interfacial energies, Daccord *et al.* [758] used water and viscous solutions of polymers in water as the two fluids. The experiments were carried out in a radial Hele-Shaw cell consisting of two glass plates each with a diameter of 1 m and a thickness of 2 cm. The glass plates were separated by a distance of 0.5 mm using spacers. The (dyed) water was injected through a hole in the center of one of the glass sheets. The fractal dimensionality of the pattern was measured in several ways by analyzing digitized pictures. A fractal dimensionality of 1.70 ± 0.05 was obtained for viscosity ratios in the range $10^{-4} \leq \mu_1/\mu_2 \leq 10^{-2}$. A typical displacement pattern is shown in figure 4.7.

Figure 4.7 A "viscous fingering" pattern obtained by the injection of a low viscosity Newtonian fluid (colored water) into a viscous non-Newtonian fluid (aqueous polymer solution) in a radial Hele-Shaw cell. This figure was provided by H. E. Stanley.

Although the patterns appear to have a branched structure, like that of DLA, and have essentially the same scaling properties as 2-dimensional DLA clusters, they clearly do not have the *same* structure as off-lattice DLA clusters. There has been relatively little effort devoted to the development of ways of distinguishing quantitatively between patterns with the same scaling properties. The work of Blumenfeld and Ball [199], discussed briefly in chapter 2, section 2.1.2, is a notable exception. Daccord *et al.* [758] examined the distribution of tip-splitting angles. They found the distribution to be narrower for patterns like that shown in figure 4.7 than for DLA clusters. Similar experiments were carried out using the standard, narrow channel Hele-Shaw cell configuration [759]. An effective fractal dimensionality of $D \approx 1.4$ was measured for the fluid–fluid interface fractal dimensionality using a "walking divider" approach (chapter 2, section 2.7.3.6), and an even smaller fractal dimensionality was obtained by measuring the amount of injected fluid inside square boxes of size ℓ, which were centered on each pixel in the digitized image of the branched displacement pattern, as a function of the side length ℓ. Similar results were obtained from corresponding DLA simulations of growth on a square lattice strip. Since the range of length scales was quite small, in both the experiments and the simulations, the effective fractal dimensionality obtained in this manner may be seriously affected by finite-size effects.

The fluids used in the experiments of Daccord *et al.* [758] were quite strongly non-Newtonian, and the flow velocity **V** in the Hele-Shaw cell was related to the pressure gradient by

$$\mathbf{V} \sim -(\nabla P)|(\nabla P)|^{m-1}. \tag{4.18}$$

It may seem surprising that DLA-like patterns corresponding to a growth probability exponent of $\eta = 1$ (see chapter 3, section 3.4.4) were generated with fluids for which $m = 2.5$ to 10.0. However, the increased growth velocity at the most exposed advancing tips of the pattern, relative to the more strongly screened inner parts, implied by equation 4.18, is offset by the relatively smaller pressure gradients at the tips of the pattern. Daccord *et al.* showed that these two effects cancel out for the simple geometry of a circular "bubble" growing in a Hele-Shaw cell. However, simulations of fluid–fluid displacement for highly non-Newtonian fluids do not appear to have been carried out. It would be interesting to see if such simulations would account for some of the differences between the patterns shown in figure 4.7 and the off-lattice DLA clusters shown in chapter 3, section 3.4.5.

In the growth of a DLA-like pattern, in a radial Hele-Shaw cell, the arms of the cluster are "confined" by their neighbors. This situation can be simplified by carrying out viscous fingering experiments using Hele-Shaw cells shaped like disc sectors with an interior angle θ. However, this does not allow the effective number of arms to increase as the displacement pattern grows. Thomé *et al.* [760] showed that the parameter B in equation 4.13 can be replaced by

$$B' = [\Gamma(b/\theta r)^2]/(12\mu V), \tag{4.19}$$

in sector shaped cells, where r is the distance that the "finger" has advanced from the origin of the sector shaped cell. This equation indicates that B' decreases as the finger advances, at constant velocity, constant pressure or constant injection rate, so that the displacement process becomes progressively less stable. To maintain a constant value for B', the fluid injection must be controlled so that $V \sim r^{-2}$. Only under these conditions can the shape of the displacement pattern be preserved during growth.

The problem of the displacement of a viscous fluid by a non-viscous fluid in a circular geometry (radial Hele-Shaw cell) has not been solved analytically, even in the 2-dimensional Laplacian approximation. However, this phenomenon can be partially understood in terms of the "sector displacement problem" [761]. In the unstable regime, in which highly branched patterns are formed, the growth of a "finger tip" takes place in the presence of neighboring fingers. Consequently, the growth process can be thought of as taking place in a sector whose walls are the mid-lines bisecting the regions between adjacent fingers.

Long range correlated anisotropy plays a similar role in fluid–fluid displacement and other pattern-formation processes such as dendritic growth and

DLA. The effects of anisotropy in fluid–fluid displacement have been studied by etching a regular network of channels in one face of a 2-dimensional Hele-Shaw cell [762, 763, 764]. Ben-Jacob *et al.* [762] studied the displacement of 96% glycerin/H_2O in a 25 cm radius Hele-Shaw cell. Grooves, with a depth of $b_1 = 0.015$ inch, a width of 0.030 inch and an edge to edge separation of 0.030 inch were engraved into the bottom plate in the form of a triangular lattice. The anisotropy was controlled via the spacing b_o between the two plates. The experiments were carried out with anisotropies, $a = b_1/b_o$, in the range $0.1 \leq a \leq 1$, and injection pressures of $10^{-3} \leq P \leq 10^{-1}$ atm. At low injection pressures, faceted growth was observed, and the interface advanced one lattice spacing at a time via kink propagation. As the applied pressure was increased, a transition from faceted growth to the formation of disorderly patterns with tip splitting was found. The pressure at which this transition took place increased with increasing anisotropy. At still larger pressures, a second transition to dendritic pattern formation via side branching took place. In this case, the transition took place at lower pressures for larger anisotropies. These experiments demonstrated that the *local* geometrical and boundary layer models, described in chapter 1, section 1.5.2, could provide insight into important aspects of *non-local* growth phenomena.

Chen and Wilkinson [763] carried out experiments in which oil with a viscosity of 1 cP was injected into glycerin with a viscosity of $\approx$ 1200 cP, in a 2-dimensional network of "channels". The interfacial tension is about 20 dyne cm^{-1} for this pair of fluids. The experiments were carried out at a constant flow rate of 1.4×10^{-4} cm^3 s^{-1}. In one experiment, a random network was generated by placing discs with a diameter of 250 μm at random on a plane and etching the region between them to form a network of channels. The channel network surrounding the approximately circular disc shaped "islands" was then enclosed to form a random 2-dimensional channel network or porous medium. The fluid–fluid displacement patterns formed in this network were very similar to off-lattice DLA patterns. If an array of channels was etched in the form of the bonds of a square lattice with channel dimensions of 500 μm long $\times$ 250 μm wide $\times$ 35 μm deep, a pattern like that generated by a square lattice DLA simulation, with a large noise reduction parameter, such as those shown in chapter 3, section 3.8, was formed. This pattern had four distinct arms with compact cores supporting an array of short side branches. Experiments were also carried out using a channel network with channels 500 μm long $\times$ 44 μm deep and widths that varied in the range 80 μm $\leq w \leq$ 730 μm. In these experiments, the displacement patterns resembled those generated by a square lattice DLA simulation with an intermediate noise reduction parameter. Chen and Wilkinson [763] were also able to generate very similar patterns using a tube network model simulation.

In a later series of experiments, Chen [764] used very similar fluids in a Hele-Shaw cell consisting of parallel glass plates with sizes of 10 cm × 10 cm × 0.55 cm separated by gaps b of $75 \pm 2\,\mu$m or $199 \pm 2\,\mu$m. In each experiment, the inner surface of one of the glass plates was etched to form either a 180 × 180 square array of channels with lengths of 500 μm or a "random disc" array similar to that used by Chen and Wilkinson. Three etched patterns were used: a regular pattern of channels with widths of 250 μm and depths of 25 μm; a "random width" array with channel widths of $250 \leq w \leq 500\,\mu$m; and the random disc array. In the random disc array, channels with a depth of 22 μm separated randomly placed discs with diameters of approximately 250 μm. The average channel width in the "random disc" array was about 250 μm. Figure 4.8 shows some of the patterns generated in these experiments. Parts (a) and (b) illustrate the effects of surface tension. In figure 4.8(a) the fluids were miscible and the effective surface tension was small. In part (b) the conditions were the same as in part (a) (including the fluid viscosities) except that the water in part (a) has been replaced by oil in part (b) with a density of 0.74 g cm^{-3} and a surface tension of $\approx$ 20 dyne cm^{-1} with respect to the viscous fluid (glycerin, with a viscosity of $\approx$ 1050 cP). The tip-splitting instability is apparent in part figure 4.8(b). Parts (c), (d) and (e) show the dominating effects of four-fold anisotropy. In parts (c) and (e) the side-branching instability can be clearly seen. In the presence of quenched disorder, both the miscible, part (f), and immiscible displacement processes produced randomly branched DLA-like patterns at the flow rates used in these experiments.

In most Hele-Shaw cell experiments, the separation between the plates is kept fixed and fluid is injected into the cell. An alternative procedure [765] is to open the cell about one of its edges like a book. This process also generates complex branched patterns and may provide a useful model for "mode I" fracture [57] and, with more imagination, directional solidification [762].

4.1.3 Thin Films and Interfaces

Patterns that resemble DLA clusters are frequently seen in thin film deposits (see references [766, 767, 768, 769, 770, 771, 772, 773, 774, 775, 776, 777, 778, 779, 780, 781], for example). In most cases, the lateral extent of the patterns is many times the film thickness, so that the system can be thought of as being 2 dimensional. For example, in the experiments of Elam *et al.* [768], with sputter deposited $NbGe_2$, the film thickness was in the 2000 – 5000 Å range while the DLA-like dendrites with an effective fractal dimensionality of about 1.7 extended over more than 0.1 mm (10^6 Å). Figure 4.9 shows the dendritic patterns obtained from two thin film deposition experiments [768, 767]. Very often, the mechanism(s) responsible for the formation of these patterns are complex and poorly understood. However, the scanning tunneling microscopy

Figure 4.8 Patterns generated by the displacement of glycerin by less viscous fluids. Parts (a) and (b) show patterns generated using smooth cell walls, injection rates of 1.4×10^{-4} cm^3 s^{-1} and a gap of $b = 75$ μm. For the miscible displacement, shown in part (a), the displacing fluid was dyed water, and for the immiscible displacement, shown in part (b), the displacing fluid was dyed oil with a viscosity of 1 cP. Parts (c) and (d) show patterns from experiments carried out using the same fluids in a Hele-Shaw cell in which a 120×120 square network of 25 μm deep $\times$ 125 μm wide $\times$ 500 μm long channels was etched into one inside wall of the cell. A displacement rate of 7.8×10^{-3} cm^3 s^{-1} and a gap of $b = 199$ μm were used. The miscible displacement is shown in part (c). Parts (e) and (f) show patterns generated in miscible displacement experiments with a gap of $b = 75$ μm. In part (e), the displacement rate was 5.4×10^{-4} cm^3 s^{-1} and a 180×180 square network of 18 μm deep $\times$ 250 to 500 μm wide $\times$ 500 μm long channels was used. In part (f), the displacement rate was 5.6×10^{-3} cm^3 s^{-1} and a random pattern of raised "circular" dots with a height of 18 μm and a diameter of 250 μm separated by an average width of ≈ 250 μm was etched into one of the inside walls of the cell. This figure was provided by J-D. Chen.

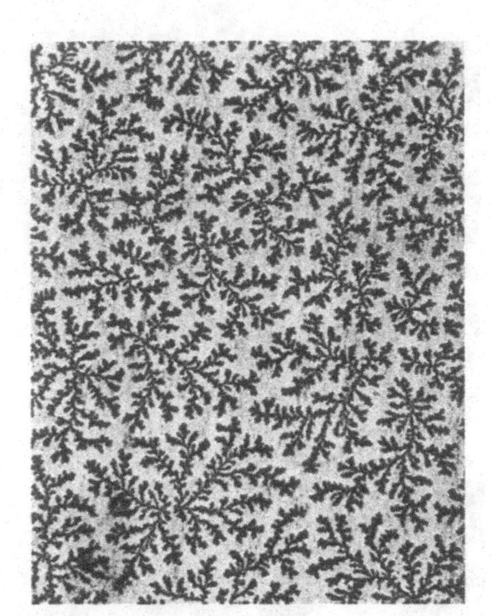

Figure 4.9 DLA-like patterns grown in thin films. The right-hand side shows an optical micrograph, provided by W. T. Elam, of a "$NbGe_2$" film sputter deposited onto a sapphire substrate. The left-hand side shows a transmission electron micrograph of a $GeSe_2$ film, provided by T. Vicsek. The $GeSe_2$ film was grown to a thickness of ≈ 700 Å by vapor deposition onto a thin carbon layer, supported on a copper grid, at a rate of ≈ 8 Å min^{-1}. In both micrographs, the clusters or dendrites extend over a distance on the order of $\approx 10 - 20\ \mu\text{m}$.

(STM) study of the growth of Au clusters on Ru(0001) surfaces [782] seems to be a particularly well defined example of diffusion controlled growth. In this study, sub-monolayer doses of gold were deposited onto a cold (room temperature) ruthenium substrate using vapor deposition from a gold-coated tungsten filament. The experiments were carried out in an ultra-high vacuum chamber with a Au flux of 0.2 to 2.0 monolayers min^{-1}. The STM images were recorded as a function of the mean coverage. A fractal dimensionality of 1.72 ± 0.07 was measured by analyzing the STM images to obtain the dependence of the radius of gyration $R_g(r)$ of that part of the cluster within a distance r measured from the "center" or nucleation site on the covered area $A(r)$, in the same region. More recently, a variety of experiments in which DLA-like patterns appear during sub-monolayer deposition on atomically smooth surfaces have been described, and such experiments have become the subject of detailed computer modeling.

The patterns observed by Hwang *et al.* [782] appear very similar to those generated by DLA-like models with multiple nucleation sites and continual addition of random walkers [783]. This model provides a satisfactory explanation for several prominent aspects of this pattern-formation process, including the formation of DLA-like patterns at the early stages of growth and the way in which a cluster avoids growing into its neighbors until a very late stage is reached. Figure 4.10 shows some patterns generated using this model. Hwang *et al.* also studied the annealing of the dendrites at 650 and 1100 K. Similar patterns have been reported for sub-monolayer deposition of Ag on Pt(111) at 110 K [784]. These experiments were simulated by Jensen *et al.* [785], using a model in which random walkers are continually added to the surface. In this model, particles and clusters are allowed to move with a diffusion coefficient $\mathscr{D}(s)$, given by $\mathscr{D}(s) = \mathscr{D}_\circ s^{-\gamma}$, and contacting clusters are irreversibly combined on contact. This is a cluster–cluster aggregation model with continuous particle injection [58]. However, in the limit of low particle fluxes and large values for the diffusion coefficient exponent γ, it becomes similar to the DLA-based model of Witten and Meakin [783]. Simulations with $\gamma = 1$ and $\phi/\mathscr{D}_\circ = 10^{-3}$, where ϕ is the injected particle flux, give patterns very much like those shown by Röder *et al.* [784].

Zhang *et al.* [608] have pointed out that the formation of DLA-like patterns, as a result of sub-monolayer metal-on-metal deposition, has been observed almost exclusively on substrates with a six-coordinate, triangular structure or systems in which the adatoms form a structure with triangular geometry. They attributed this to the long distances that an atom may have to diffuse along the edge of a compact pattern growing on a square lattice, before a position with a coordination number of 2 or greater is encountered. By contrast, on a triangular lattice, a site with a coordination number of 2 or greater is found almost immediately. For a triangular lattice, a transition from a fractal, DLA-like pattern to a compact pattern will take place only if the temperature is high enough to mobilize two-coordinate atoms at the edge of the growing pattern. To form large (fractal or compact) clusters, the hopping rate $k_\circ$ for single atoms on the surface must be much greater than the deposition rate $k_\downarrow$.

Zhang *et al.* [608] identified two DLA-like pattern growth regimes. In the low temperature regime ($T < T_1$, regime I), the hopping rate k_1, for motion of atoms attached to the growing clusters by a single bond along the perimeter of the cluster, is small, so that new atoms will arrive from outside of the region occupied by an adatom on the edge of the cluster at a rate greater than the hopping rate. In this regime, DLA-like patterns with arms that are only one or two atoms thick, like ordinary DLA model clusters, are formed. In the second, intermediate temperature, fractal growth regime ($T_1 < T < T_2$, regime II) atoms that are connected to the growing cluster by only one bond hop rapidly, but the hopping rate k_2 for atoms connected by two bonds is slower than the rate of arrival of new atoms in the vicinity of a two-coordinate atom. In this regime, DLA-like patterns are also formed, but the arms are several atoms or more thick. In the third, high temperature regime, the two-coordinate atoms hop rapidly and a compact structure appears. All three regimes appear if the sub-monolayer grows on a triangular lattice, but for a square lattice the second regime is absent, and a direct transition from regime I to regime III takes place

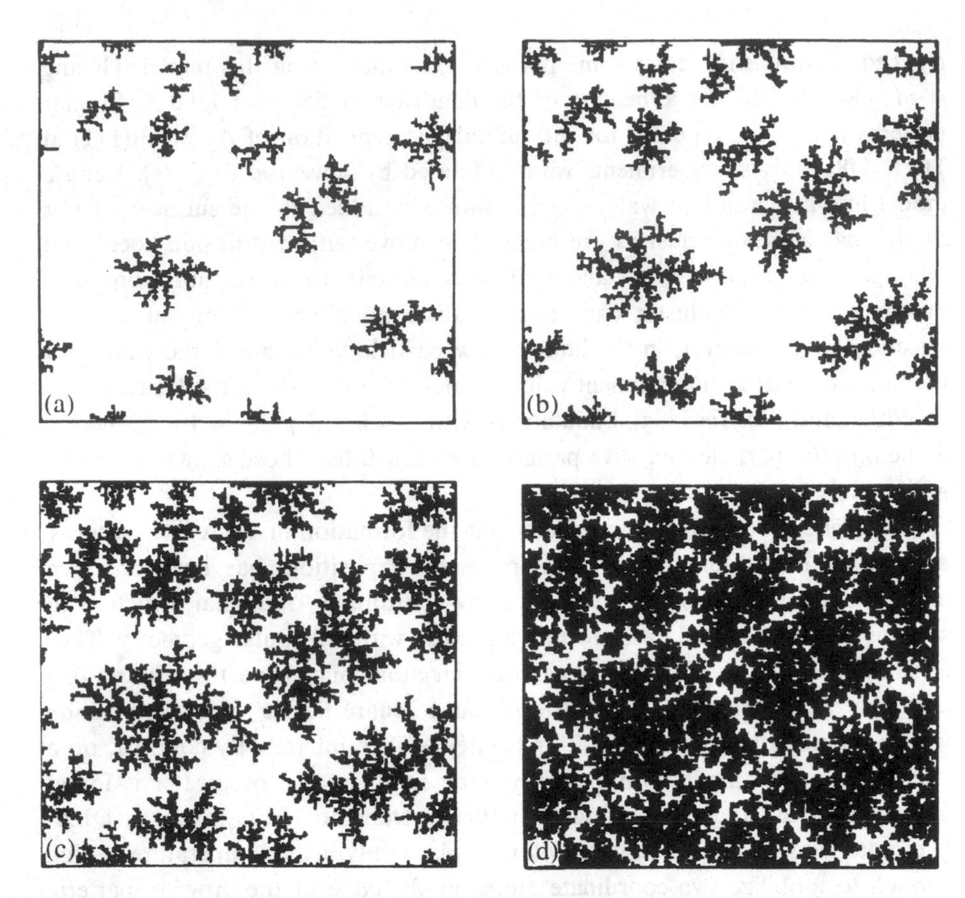

Figure 4.10 Random dendrites obtained from a 2-dimensional simulation in which particles were added randomly, one at a time, to a 200 × 200-site square lattice that initially contained 16 randomly placed nucleation sites. The randomly added particles followed random walk paths until they reached a site with one or more occupied nearest neighbors (nucleation site or previously added particle). Parts (a), (b), (c) and (d) show the patterns formed after 2500, 5000, 10 000 and 20 000 particles were added.

at the lower transition temperature T_1 . The transition temperature T_1 may lie below the range covered in an experiment, or $k_\circ/k_\downarrow$ may be too small so that a large number of small clusters are formed that do not grow to a size that is larger than the arm width. In either event, DLA-like patterns would not be observed on a square lattice.

Hohage *et al.* [776] have studied the growth of randomly branched Pt adatom islands formed by the slow deposition of Pt atoms onto Pt(111) surfaces using scanning tunneling microscopy (STM) and Monte Carlo simulations. After deposition, the samples were cooled rapidly to 20 K to avoid further changes during STM imaging. Hohage *et al.* found evidence for a new growth regime at temperatures between 150 K and 180 K in which growth occurs preferentially in the $< \overline{1}\overline{1}2 >$, $< \overline{1}2\overline{1} >$ and $< 2\overline{1}\overline{1} >$ directions in the Pt(111) surface.

This growth regime lies between regime I and regime II. The clusters, or islands, formed in this intermediate regime were quite small, but the anisotropy was quite distinct, and the clusters resembled small clusters generated by DLA models in which the growth was restricted to three symmetrically disposed directions (chapter 3, section 3.8.1). A detailed comparison between the experiments and simulations indicated that, in the intermediate regime, adatom jumps between singly coordinated positions (with respect to the growing island) and two-coordinate positions take place preferentially in the direction in which the coordination number of the transition state is the highest, with respect to the underlying Pt(111) substrate.

Röder *et al.* [787] have described a scanning tunneling microscopy study of the growth of Ag clusters on Pt(111) and Ag(111) during vapor phase epitaxy. On both substrates DLA-like patterns were formed over quite a wide range of temperatures, and the cluster branch thickened with increasing temperature. Röder *et al.* attributed this behavior to a DLA-like growth process with thermally activated edge diffusion. The temperature dependence of the branch widths could be represented by an Arrhenius form

$$I^{1/2}w = Ae^{-E_a/k_BT}, \tag{4.20}$$

where I is the deposition flux, w is the arm width measured from the STM images and E_a is an "activation energy". Based on a simple stability analysis, Röder *et al.* concluded that the arm width w is given by $w \approx (\mathscr{D}_e/I)^{1/2}$, where $\mathscr{D}_e$ is the edge diffusion constant. This means that the edge diffusion constant can be estimated from $\mathscr{D}_e \approx w^2 I$, and E_a, in equation 4.20, is 1/2 of the activation energy for edge diffusion. Using the temperature dependence of the branch widths, measured in the experiments, activation energies of 125 ± 10 meV and 65 ± 10 meV were obtained for Ag/Pt(111) and Ag/Ag(111), respectively. These activation energies represent an average over the environments along the edge of the DLA-like clusters. The corresponding activation energies for diffusion on flat (111) surfaces are $E_{\mathscr{D}} \approx 157$ meV and $E_{\mathscr{D}} \approx 100$ meV for Ag/Pt(111) and Ag/Ag(111), respectively. Consequently, it might appear that the diffusion along the edges is faster than the diffusion along the substrate and that compact Ag clusters should be formed. However, the measurements of Röder *et al.* indicate that the kinetic "attempt frequencies" are substantially lower for edge diffusion than for diffusion on the substrate. Consequently, $\mathscr{D}_e \ll \mathscr{D}_s$, where $\mathscr{D}_s$ is the diffusion coefficient for Ag atoms on the bare substrate, and these experiments provide a good illustration of regime II behavior.

DLA-based models have been developed to study the formation of dendritic patterns in sub-monolayers of diffusing atoms. In these models, particles are deposited onto a planar surface. They move on the surface via random walk trajectories until stable "nuclei" of two (or more) adjacent particles are formed [788, 789, 790, 791]. The various models of this type differ in the way in which they treat atoms that are deposited on top of the growing dendrites. Patterns that resemble DLA clusters are then formed by the sticking of random walking particles to the stationary nuclei. Amar *et al.* [788] identified four scaling regimes, which they called "the low coverage nucleation regime", "the intermediate coverage regime", "the aggregation scaling regime" and the "coalescence and percolation regime". These scaling regimes will be discussed elsewhere [57].

Brune *et al.* [781] have studied the growth of Ag dendritic patterns formed by the ultra-high vacuum deposition of Ag atoms on Pt(111) at a temperature of 80 − 130 K using a scanning tunneling microscope. The dendrites are one atom thick and the arms are only approximately two atoms wide. If the Ag atom flux was increased, the cluster shape changed from a pattern similar to that of an off-lattice DLA cluster, with a measured fractal dimensionality of 1.76 ± 0.07, to a pattern that resembled a DLA cluster grown with three preferred local growth directions (chapter 3, section 3.8.1). The fractal dimensionality of $D = 1.77 \pm 0.05$, obtained from a box counting analysis, was essentially the same as that obtained under low flux conditions. Brune *et al.* offered an explanation of this remarkable transition in terms of a model that requires two diffusing Ag atoms to meet at the tips, oriented in the preferred direction. However, the mechanism proposed by Hohage *et al.* [776], which does not require two diffusing Ag atoms to meet at the tips, may play an important role in this system as well.

Fractal, DLA-like dendrites have also been found on the surface of a dislocation-free silicon bicrystal grown by horizontal pulling from a silicon melt [792]. The dendrites were found to have overall sizes ranging from a few micrometers to several tens of micrometers, and fractal dimensionalities in the range $1.6 \leq D \leq 1.9$ were measured, depending on the growth conditions. Electron diffraction studies demonstrated that the dendrites have the same composition and orientation as the underlying Si substrate. The DLA-like structure was attributed by Cheng [792] to a layer-by-layer growth mechanism. Scanning tunneling microscopy indicated that the dendrites have a height of several hundred angstroms. The random dendrites appear to be part of a growing interface that leaves behind completely dense Si with a (211) surface. It appears that the DLA-like patterns are formed via nucleation and preferential growth in the lateral directions.

A quite spectacular example of the formation of DLA-like patterns in thin films is provided by the growth of silicon-rich dendrites during the annealing of thin films of aluminum deposited on SiO_2 [793]. The silicon rich dendrites are formed via the reduction of SiO_2 by Al. This is an important process because of its deleterious effects on electronic devices with Al/SiO_2 interfaces.

Another relatively well defined example of "multiple DLA" patterns is provided by the work of Yasui and Matsushita [794] on the dendritic growth of NH_4Cl on the surface of thin layers of agar gel from which water was allowed to evaporate at $20 \pm 0.2\,°C$. The initial concentrations of NH_4Cl and gel were low (0.5% NH_4Cl, 0.8% agar). Similar patterns have been observed in polyethyleneoxide/NH_4I that was cast into the form of a thin film from solution in methanol to which Al_2O_3 was added. After several days, iodine dendrites with structures similar to DLA patterns were observed [795]. The role played by the Al_2O_3 particles is not clear, but they may provide nucleation sites and/or enhance the quenched disorder.

DLA-like patterns have been observed in a variety of essentially 2-dimensional systems such as lipid monolayers on water in a Langmuir trough. These include L-α-dimyristoylphosphatidylethanolamine lipid monolayers on water near the 2-dimensional fluid–solid phase transition [796, 797] and ethyl octadecanoate at the "liquid-condensed"/"liquid-expanded" (LC/LE) phase transition [798]. Patterns that look much more like DLA clusters have been obtained by Wang *et al.* [799] in the LC/LE coexistence regime of R-(-)-N,N-dihexadecyl-(2-[1-(1-imidazolyl)-propyl]amine), which is chiral. As the coexistence regime was reached, DLA-like LC patterns appeared, embedded in the LE phase. Because of the chiral nature of the lipid, the arms of the cluster were distinctly curved. Under the local heating effect of the microscope used to observe the patterns, the arms thickened and eventually became faceted. In an extension of this work [800], pattern formation in monolayers of N,N-dihexadecyl-3-(1-imidazolyl)-propylamine, which is not chiral, were also studied. This work revealed that illumination resulted in decomposition of the fluorescent dye, and that this decomposition played an important role in the growth of the ramified LC domains.

In the experiments of Miller *et al.* [796, 797], a fluorescent dye was incorporated into the phospholipid layer. The dye had a low solubility in the dense phase, and the concentration of the dye into the low density phase made the growth of the solid domains visible. In addition, the dye played an important role in the phase transition. The transition pressure increased with dye concentration, and the diffusion of the dye appeared to control the growth of the "crystalline" dendrites. Since a high dye concentration increased the transition pressure, and growth occurred preferentially at thosc positions on the growing crystalline domain at which the dye concentration was lowest. The latent heat of crystallization probably had a negligible effect because of the large heat capacity and thermal conductivity of the underlying aqueous substrate. In these experiments, the dendritic domains consisted of single crystals. An effective fractal dimensionality of $D \approx 1.5$ was measured, in the early stages of domain growth, following a rapid surface area decrease to bring about crystallization. This value for D is substantially smaller than the value of $D \approx 1.715$ associated with 2-dimensional DLA. This apparent discrepancy may be due, at least in part, to the limited range of length scales that was accessible in these experiments. Eventually, the domains coarsened in response to edge tension. Patterns that resemble DLA clusters have also been seen in fluorescence microscopy studies of phospholipid monolayers at hydrocarbon/water interfaces.

Dietrich *et al.* [801] have studied pattern formation in two closely related phospholipids (1-hexadecyl-2-(2-tetradecylpalmitoyl)glycero-3-phosphocholine, or C1, and 2-hexadecyl-1-(2-tetradecylpalmitoyl)glycero-3-phosphocholine, or C2). Domain-formation was observed in the monolayers using the fluorescence microscopy method. The domain-formation process was followed after compression from the LE to the LE/LC phase coexistence range. In the case of C1, circular domains appeared, but these became unstable and a branched pattern evolved via tip splitting. The behavior observed in the C2 monolayers was quite different. Compact domains with hexagonal symmetry soon appeared and developed via tip growth and side branching into hexagonal dendrites. The lipid C2 packs to a higher density than C1 (a minimum molecular area of $54\,\text{Å}^2$ compared with $59\,\text{Å}^2$) and forms a much more anisotropic structure. Dietrich *et al.* suggested that the transition observed in the C2 monolayers may be a LE/solid-condensed (SC) transition, while the transition observed for C1 is more like a conventional LE/LC transition.

Experiments have also been carried out in which the domains have a cholesteric character. In these experiments, a small amount of cholesterol was added to the lipid [802], or a chiral phospholipid such as dipalmitoylphosphatidylcholine was used [803]. The 2-dimensional solid domains have a random dendritic DLA-like structure, but the dendrites have a structure that is "curved" on all length scales. It seems likely that these structures could be represented by a simple modification of the DLA model.

4.1.4 Dissolution, Melting and Erosion of Porous Media

The experiments of Daccord *et al.* [804, 805, 806] on the dissolution of plaster of Paris ($CaSO_4 \cdot 0.5H_2O$) provide some of the most spectacular examples of DLA-like patterns. In these experiments, water was slowly injected into water-saturated plaster of Paris, which is a porous material with a small solubility in water. Despite the relatively small solubility of $CaSO_4$, water in contact with plaster of Paris soon becomes saturated with $CaSO_4$ and will dissolve no more $CaSO_4$. Consequently, the dissolution of the plaster of Paris takes place near to the water/plaster of Paris interface, and the rate of dissolution (the velocity of the interface) at position $\mathbf{x}$ on the interface is proportional to the flux density of water through the interface at $\mathbf{x}$, if the injection rate is neither too fast nor too slow. The flow of water through the interface is controlled by the pressure field inside the porous plaster of Paris, since there is essentially no pressure drop inside the cavity, etched by the water, in the porous medium.

Slow flow inside a porous medium can be described by Darcy's equation,

$$\mathbf{V}(\mathbf{x}) = -m\nabla\Phi(\mathbf{x}), \tag{4.21}$$

where $m = k/\mu$ is the mobility, $\mathbf{V}(\mathbf{x})$ is the fluid velocity at $\mathbf{x}$ and $\Phi(\mathbf{x})$ is the hydraulic potential ($\Phi(\mathbf{x}) = P + \rho gh$, where P is the fluid pressure, ρ is the fluid density, g is the acceleration due to gravity and h is the distance in the direction of the gravitational field). For incompressible fluids, $\nabla.\mathbf{V} = 0$, so that Darcy's law implies that $\nabla^2\Phi = 0$, and the hydraulic potential field driving the fluid through the porous medium obeys the Laplace equation inside the porous medium. Consequently, the experiment can be described in terms of the random propagation of an interface controlled by a field that obeys the Laplace equation (the pressure field inside the porous plaster of Paris), and the dissolution process can be represented quite well by the DLA model.

Experiments were carried out with several different boundary conditions, including injection through a small hole in a "radial" Hele-Shaw cell consisting of two parallel circular glass sheets, which confined a thin layer of plaster of Paris, with a diameter of 250 mm and a separation (porous layer thickness) of 1 mm. The patterns produced in these experiments closely resembled those generated in corresponding DLA model simulations [465]. Figure 4.11 shows the dissolution pattern obtained by injecting water at a rate of 48 cm^3 hour^{-1} into plaster of Paris with a porosity of 60% and a permeability of 60 μm^2, where 1 μm^2 = 1.013 25 darcy. After the water had been injected, the plaster of Paris was dried and Wood's metal was injected into the cavity etched into the plaster, and the remaining plaster of Paris was then dissolved. Figure 4.11 shows one of these Wood's metal castings. At lower injection rates, the branches were broader and less ramified.

The work of Daccord *et al.* was partially motivated by applications in enhanced oil recovery (a processes called "wormholing"). Extension of these laboratory models and the DLA model to represent this process better will be discussed elsewhere [57].

A phenomenon closely related to the dissolution of porous materials has been described by Knight [807]. This is the formation of branched patterns (figure 4.12, see color plate section) in "white ice" when snow is deposited on top of a layer of clear "black ice" on a lake. This phenomenon is quite common. I have observed these patterns many times on ponds in Eastern Pennsylvania. The overall size of the patterns ranges from $\approx$ 20 cm to $>$ 10 m, and these branched patterns often emanate from the edge of the pond. If sufficient snow is deposited on the ice, water may flood the surface of the ice, forming a layer of slush that later freezes. The pattern consists of branches of clear ice within the layer of white ice (frozen slush). The pattern originates at a hole in the underlying layer of black ice. Relatively warm water flows through this hole into the layer of slush on the ice. The warm water melts some of the slush, but is soon cooled to the

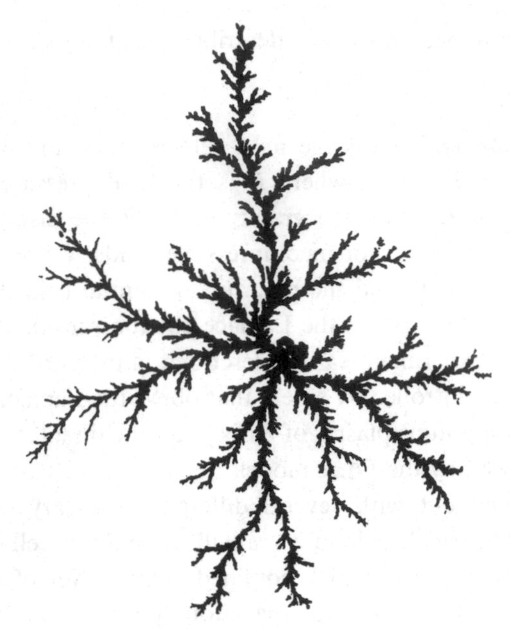

Figure 4.11 A Wood's metal casting of a hole etched into a 1 mm thick layer of plaster of Paris, in a radial Hele-Shaw cell. This figure was provided by G. Daccord, and is reproduced with the permission of Dowell-Schlumberger.

freezing point because of the large latent heat of melting of water. The cooled water penetrates through the slush layer. This causes no further melting, but the flow of water through the slush, which can be described by Darcy's equation, controls the flow of warm water to the water/slush interface.[3] Consequently, all the ingredients of DLA (a random growth process controlled by a scalar field that obeys the Laplace equation) are present. Others have suggested that these patterns are related to cracking of the ice [808]. This may take place under some circumstances. However, the generation of similar patterns in laboratory experiments, and experiments in which water slightly above the freezing point was poured onto lakes with snow covered ice [809], lends strong support to the "flow controlled melting" scenario.

Certain "river" systems, particularly drainage channels in dry climates, have a quite strong resemblance to 2-dimensional DLA clusters. In some cases, there is reason to believe that this resemblance may be more than superficial. The importance of mechanisms such as "piping" and "sapping" in the initiation and extension of streams has been recognized for some time [810, 811, 812, 813, 814]. In piping and sapping, accelerated erosion takes place when ground water flow emerges to generate overland flow. The ground water flow can be described in terms of the Darcy equation. This process is similar to the dissolution and

3. The pressure gradient is related to a gradient in the water depth, in the slush. In principle this depth variation will require corrections to the 2-dimensional Laplace equation. In some cases, the flow-limited melting may take place in a natural Hele-Shaw cell consisting of a layer of snow trapped between two layers of ice. In this event, no correction would be required.

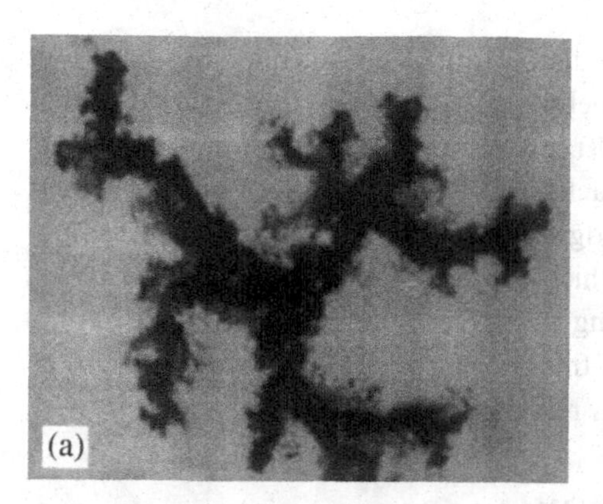

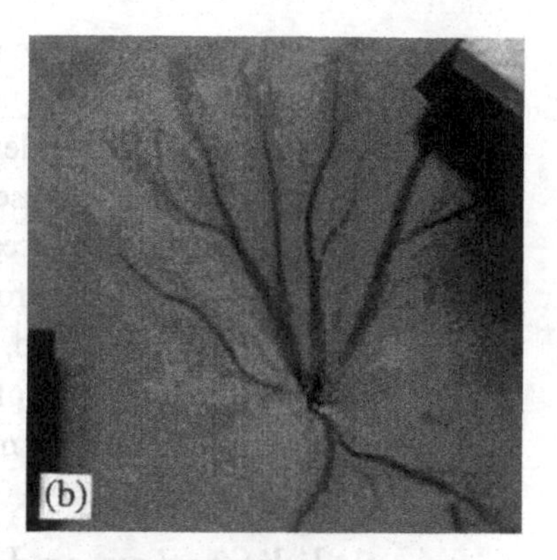

Figure 4.13 Branched patterns formed by erosion of unconsolidated materials in a Hele-Shaw cell. Part (a) shows erosion channels obtained using Fontainebleau sand, and part (b) shows a pattern obtained in a similar experiment using glass spheres with diameters in the 5 μm to 60 μm range. These figures were provided by P. Mills and P. Cerasi.

melting processes discussed above, except that the direction of the water flow is reversed. If the rate of erosion at the water saturated soil/water interface is proportional to the flux density, then a DLA model could be a good starting point for simulating these phenomena. In practice, the dependence of the rate of erosion on the flux density is likely to be more complex, but may be approximated by a power law, suggesting that an "η" model (chapter 3, section 3.4.4) might be used.

Mills *et al.* [815] have carried out experiments in which water/glycerin was withdrawn from the center of a radial Hele-Shaw cell filled with Fontainebleau sand saturated with the water/glycerin mixture. The gap between the plates of the Hele-Shaw cell was about 300 μm, and the diameters of the sand grains were in the range 100 μm to 300 μm. Experiments were carried out using a range of pumping schedules. At low flow rates, the fluid flowed through the cell with little or no erosion. At a threshold flow rate $Q_c^{(1)}$, localized erosion began and an unstable erosion front propagated through the unconsolidated porous medium. At a second threshold flow rate $Q_c^{(2)}$, fluidization took place with motion of the entire porous medium. Branched erosion channels were formed under quite a wide range of conditions with fractal dimensionalities of $D \approx 1.70$ (figure 4.13(a)), consistent with the fractal dimensionality of 2-dimensional off-lattice DLA. Under some conditions, less well defined channels with lower fractal dimensionalities were obtained. Similar experiments were carried out using glass beads in place of the sand (P. Cerasi, private communication, 1995). Branched patterns like that shown in figure 4.13(b) were also formed at intermediate flow rates, but the morphology seemed to be quite different from that obtained with sand.

Processes that involve melting and/or dissolution of porous materials resulting in the concentration of flow onto a network of channels are believed to be important in the Earth's mantle [816, 817] and in metamorphic processes. For example, Hart [818] has proposed a fractal tree model for transport of magma from the mantle to the mid-oceanic ridges and island "hot spots". According to this model, the channels through which the melt is transported coalesce as shallow depths are approached, forming an inverted tree. It appears that the physico-chemical processes involved in this process are quite complex, and it is not clear if a simple DLA-like model is relevant to this process.

4.1.5 Solidification and Crystallization

The relationship between DLA and dendritic solidification was pointed out by Witten and Sander in their seminal paper on diffusion-limited aggregation [181]. In most cases, dendritic solidification is a complex process involving convection, thermal diffusion, molecular diffusion, interfacial phenomena and other processes (chapter 1, section 1.4). However, the DLA model does provide a paradigm for understanding some aspects of random dendritic solidification and crystallization. For example, Honjo *et al.* [819] have studied the growth of NH_4Cl crystals from supersaturated solutions of NH_4Cl in a thin ($\approx 5\,\mu$m) layer between two parallel glass sheets. When the surfaces of both glass sheets are smooth, the inherent anisotropy of NH_4Cl dominates and a "snowflake"-like pattern grows [819]. Figure 4.14 shows a "regular" NH_4Cl dendrite grown under these conditions in a 5 μm thick cell. Honjo *et al.* also carried out experiments in which the inner surface of one of the glass cell walls was roughened, with a correlation length $\xi_{\parallel}$ of about 7.5 μm. Under these conditions, the fluctuations associated with the rough surface are strong enough to dominate the lattice anisotropy of the NH_4Cl and a disorderly pattern, like that shown in the lower part of figure 4.14, was formed. The mass (area in a digitized picture) of the pattern was measured as a function of the radius of gyration. Although the range of length scales between the finger width and the overall size of the pattern was small, an effective fractal dimensionality of 1.671 ± 0.002 was obtained. This is in quite good agreement with the fractal dimensionality of about 1.715 measured for from 2-dimensional DLA model computer simulations, without lattice anisotropy.

This was an important experiment, since it elucidated the role played by quenched disorder in the formation of DLA-like patterns in laboratory experiments and in nature. In a later study, Ohta and Honjo [820] analyzed photographs of the growing pattern at regular intervals to measure the distribution of growth rates. They found that this distribution was like the "multifractal" distribution found for DLA from computer simulations (see appendix B, section B.3). A similar demonstration was provided by the experiments of

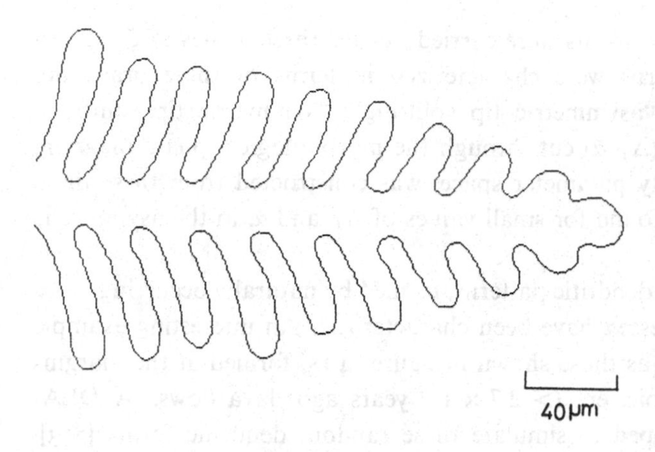

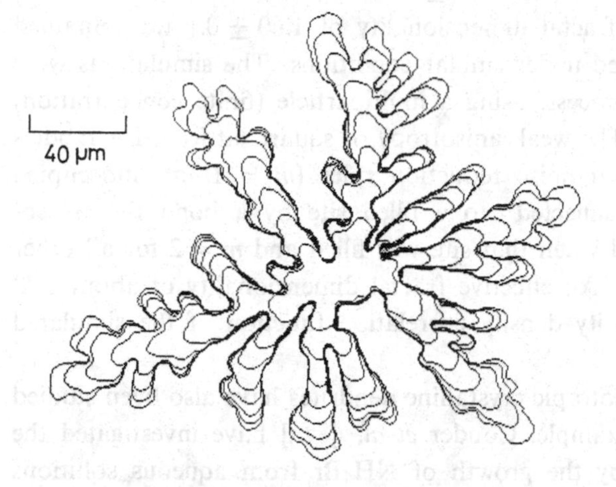

Figure 4.14 Dendrites grown from a supersaturated aqueous solution of NH_4Cl. The top part of the figure shows a dendrite grown in a smooth walled thin (5 μm) cell. The lower part shows the pattern obtained using a rough walled thin (5 μm) cell. The characteristic length associated with the surface roughness (of one of the cell walls) was about (7.5 μm). This figure was provided by M. Matsushita.

Chen and Wilkinson on viscous fingering in regular and random etched channel networks [763]. These experiments are described earlier in this chapter (section 4.1.2).

A relationship between long range orientational order and dendritic morphology was also found by Grier *et al.* [821] in their studies of electrochemical deposition of Zn and other metals (Cd, Pb and Ag). Under fast growth conditions, relatively regular dendrites with long range orientational and positional ordering were found. At intermediate growth rates, more disorderly DLA-like patterns with diffuse electron scattering patterns, indicating correlation lengths of less than 100 Å, were found.

Honjo *et al.* [822] carried out a more systematic series of experiments on the effects of quenched disorder on dendritic growth using undercooled succinonitrile, instead of supersaturated ammonium chloride. The results of these experiments were described in terms of the control parameters Δ_T (the supercooling) and $\bar{\alpha} = h/\xi_{\|}$, where $h = 14$ μm is the cell gap thickness and $\xi_{\|}$ is the characteristic distance between scratches in a randomly scratched interior cell surface. Experiments were carried out for three values of $\xi_{\parallel}$ (10, 16 and 34 μm). The patterns were characterized in terms of three branching mechanisms designated "asymmetric tip splitting", "symmetric tip splitting" and "tip oscillation". A $(\Delta_T, \bar{\alpha})$ cut through the morphological phase diagram, in the high-dimensionality parameter space, was constructed from these data. DLA-like patterns were found for small values of Δ_T and $\bar{\alpha}$, in the asymmetric tip-splitting regime.

A variety of disorderly dendritic patterns, formed by naturally occurring, non-equilibrium growth processes, have been characterized. An interesting example is olivine dendrites, such as those shown in figure 4.15, formed at the margins of archaean, or archeozoic, era ($> 2.7 \times 10^9$ years ago) lava flows. A DLA-based model was developed to simulate these random dendritic forms [823]. An effective fractal dimensionality of 1.80 ± 0.05 was measured for the olivine dendrites, and an effective fractal dimensionality of 1.60 ± 0.1 was obtained for pyroxine dendrites formed under similar conditions. The simulations were carried out to model this process, using a multiparticle (finite concentration) square lattice DLA model. The weak anisotropy of square lattice DLA models was enhanced by using local noise reduction rules ($m = 1$ for unoccupied perimeter sites that are "connected" to a filled site by a bond that is co-linear with the bond formed when that site was filled and $m = 2$ for all other unoccupied perimeter sites). An effective fractal dimensionality of about 1.73 was obtained from the density–density correlation function of the simulated dendrites.

Less disorderly, more anisotropic crystalline dendrites have also been studied and related to DLA. For example, Couder *et al.* [824] have investigated the dendritic patterns formed by the growth of NH_4Br from aqueous solutions in a Hele-Shaw cell geometry with a gap of 30 μm. In these experiments, thermal diffusion is much faster than mass diffusion, so the growth of the dendrites is limited by the diffusion of NH_4Br. The dendrites appear similar to those generated by noise reduced, square lattice DLA models. A box counting analysis of images of the dendrites, far from the tip, gave an effective fractal dimensionality of 1.58 ± 0.03, similar to the fractal dimensionality found for noise reduced DLA from the dependence of the radius of gyration on the number of particles or sites. In addition, the area of the dendrites within a distance x measured from the dendrite tips was found to scale as $x^{1.5}$ over three decades in x. The length of the lateral side branches at a distance x from the tips was proportional to $x^{1/2}$, so that the envelope of the dendrite had a parabolic shape like that of the tip of an advancing compact dendrite. This is reminiscent of the relationship between DLA patterns generated in a strip with reflecting lateral boundaries and the Saffman–Taylor family of solutions for the viscous fingering problem (chapter 3, section 3.4.6 and appendix A, section A.2).

Figure 4.15 Olivine dendrites formed at the margins of an archaean era lava flow. This figure was provided by A. D. Fowler.

4.1.6 Dielectric Breakdown

Dielectric breakdown is another complex phenomenon that can lead to fractal patterns that are very similar to DLA clusters. Dielectric breakdown processes are frequently observed as highly branched patterns formed in solids, liquids [825] and gases. A quite striking example is the formation of "arcing defects" during the commercial coating of metal tools by titanium nitride using arc sources [826]. A more important example, from a commercial point of view, is the growth of "water trees" in buried electrical cables with polyolefin insulation. The water trees are formed by a "dielectric breakdown" process in the presence of water. A wide variety of morphologies has been observed, including highly ramified structures that resemble DLA clusters and dense branching patterns that resemble those discussed later in this chapter. A number of studies indicated that the size L of water trees grew as $L \sim t^a$, where t is time. A wide range of values, $0.2 \leq a \leq 0.9$ [827], has been reported for the exponent a. There is no consensus concerning the mechanism of water treeing. It appears that it is not simply a dielectric breakdown process, and, in some cases, a crossover from water treeing to electrical treeing (dielectric breakdown) may take place as the pattern grows. Both mechanical and chemical processes may be important. One prominent idea is that electrical forces drive the high dielectric constant water through the low dielectric constant polymer [828] in a process that is more like viscous fingering than dielectric breakdown.

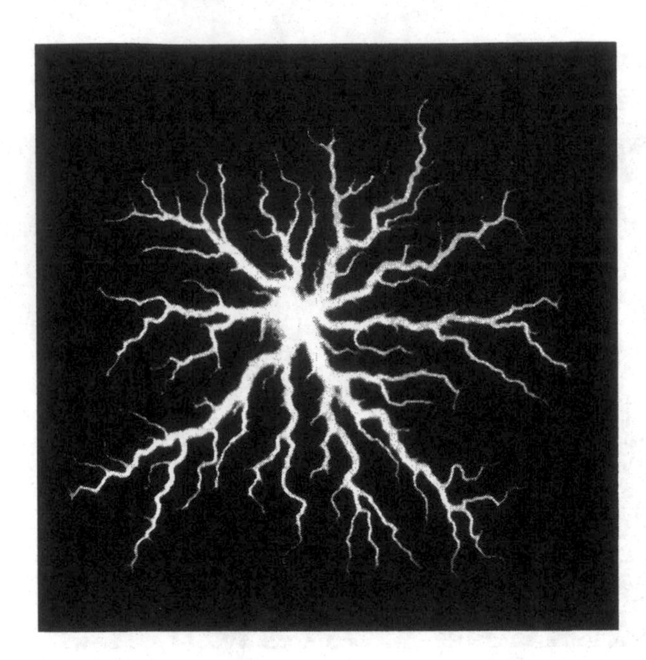

Figure 4.16 A time integrated photograph of a surface discharge or Lichtenberg figure generated on a 2 mm thick glass plate in 300 kPa SF_6 using a 30 kV applied potential pulse for 1 μs. This figure was provided by L. Niemeyer.

Lightning provides a familiar large scale example of dielectric breakdown. Apart from the branched discharge patterns in the air, branched patterns can also be observed in the surface of golf courses [829]. Branched structures called fulgurites, which are made of fused sand, are often formed when lightning strikes sandy soil [829]. Quite intricate branched patterns have been observed in experimental electrical discharges in wet and dry sand [830].

In dielectric breakdown, it is clear that the electric field that obeys the Laplace equation has something to do with the pattern-formation process. One of the most simple dielectric breakdown processes is the formation of Lichtenberg figures in gases such as SF_6 [831, 832]. Figure 4.16 shows a discharge pattern obtained using a 2 mm diameter electrode rod that contacted a glass sheet (lying in a plane perpendicular to the axis of the electrode) with a conductive coating, in 300 kPa SF_6. The side of the glass sheet opposite to the electrode was grounded. A fractal dimensionality of about 1.7 was obtained [429] from digitized images of similar patterns by measuring the number of branches $N_b(r)$ intersected by circles of radius r centered on the origin of the discharge. The corresponding fractal dimensionality was obtained using the relationship $N_b(r) \sim r^{D-1}$.

4.1.7 Growth Probability Distributions

Because of the theoretical interest in the distribution of growth probabilities on the surface of a DLA cluster and the possible insights that this may lead to, there has been considerable interest in measuring this distribution in terms of a multifractal measure (chapter 2, section 2.8, and appendix B). In principle, the distribution of growth rates or probabilities can be measured via the increments in the surface of the growing pattern at times t and $t + \delta t$, for sufficiently small values of the time increment δt. This has been accomplished by Måløy *et al.* [833] for DLA-like viscous fingers in a 2-dimensional porous medium. The $D_f(\alpha)$ curve (chapter 2, section 2.8) was obtained using the histogram approach (chapter 2, section 2.8 and appendix B, section B.2) by defining the Holder exponent α as

$$\alpha = \frac{-\ln \mu}{\ln(L/\epsilon)}, \tag{4.22}$$

where ϵ is the resolution, and the radius of gyration R_g was used to measure the overall cluster size L. Equation 4.22 provides a way of calculating the index α for a region on the surface of size ϵ in which the growth probability measure (the normalized growth probability) is μ. The fractal dimensionality $D_f(\alpha)$ of the corresponding subset was calculated from

$$D_f(\alpha) = \frac{\ln(\rho(\alpha) - \ln c)}{\ln(L/\epsilon)}, \tag{4.23}$$

where $\rho(\alpha)$ is the probability density that describes the distribution of the measure over the subsets ($\rho(\alpha')\delta\alpha'$ is the probability that the index α lies in the range $\alpha' - (\delta\alpha'/2) < \alpha < \alpha' + (\delta\alpha'/2)$). In equation 4.23, the constant c is a correction to scaling chosen so that the maximum value of $D_f(\alpha)$ would be 1.0 for all the data sets (different values of L/ϵ).

A $D_f(\alpha)$ curve was also obtained by measuring the moments Z_q of the growth probability measure and calculating $D_f(\alpha)$ by Legendre transformation of the function τ_q, where τ_q is the scaling exponent that describes how Z_q scales with L/ϵ (chapter 2, section 2.8, and appendix B). The $D_f(\alpha)$ curve obtained in this manner was in quite good agreement with the $D_f(\alpha)$ curve obtained using the histogram approach (appendix B, section B.2). However, the statistics are necessarily poor.

Other investigators have digitized images of DLA-like patterns and *calculated* the harmonic measure (the distribution of gradients on the surface of the pattern for a scalar field that obeys the Laplace equation, with absorbing boundaries on the pattern). This gives the distribution of growth probabilities if and only if the pattern evolution is indeed a Laplacian growth process. This type of analysis has been carried out for viscous fingering [834], random crystal growth [820] and electrochemical deposition [835] patterns. The growth increment approach to measuring the distribution of growth velocities in experimental studies [820, 833] is not sensitive enough to measure the small growth probabilities in the strongly screened part of the growing surface. Consequently, the distribution of growth probabilities measured in experiments [820, 833] is very different from that obtained by numerical methods [820, 834, 835]. This may account, at least in part, for the difference between the $D_f(\alpha)$ curves obtained in the experimental work of Måløy *et al.* and those obtained by solving the Laplace equation. There may be an essential difference between the harmonic measure calculated by solving the Laplace equation with absorbing boundary conditions and the distribution of growth increments. Even for very small growth increments the mutual screening between the growth increments may play an important role. Numerical studies [836] indicate that the fractal dimensionality of the "old growth/new growth" interface[4] is approximately 1.0. This implies that the maximim value for $D_f(\alpha)$ is ≈ 1.0 for the growth increment measure. For the harmonic measure, $D_1 = D_f(1) = 1$ [663], and the maximum in $D_f(\alpha)$ is, in general, greater than unity.

Other studies have been carried out in which experimental patterns that are very different from DLA clusters have been characterized via the $D_f(\alpha)$ curve for the harmonic measure. In general, this does not appear to be a very useful or direct way of characterizing structure, but it may be useful in investigations of the physical and chemical behavior of complex structures.

A comparison between the distribution of growth probabilities measured during the electrochemical deposition of openly branched zinc dendrites and the distribution calculated by solving the Laplace equation using a digitized image of the growing patterns has been described by Mach *et al.* [837]. Deposition of zinc from a 0.4 M $ZnSO_4$ solution was carried out using a line electrode in a cell with a gap of 70 μm between the parallel walls at a potential difference of 6 V. Good agreement was found between the left, high growth probability, side of $D_f(\alpha)$ calculated from the experimental data and the left side of $D_f(\alpha)$ obtained by solving the Laplace equation. A similar agreement, with a different $D_f(\alpha)$, was found for similar experiments carried out using 0.02 M $ZnSO_4$ and a potential difference of 40 V, to generate a densely branched pattern.

4.2 Dense Branching Morphology

Many processes lead to patterns that have a superficial resemblance to DLA clusters but do not share their fractal scaling properties. In two dimensions, these patterns often consist of a randomly branched structure with an essentially circular boundary. Unlike DLA patterns, the branches fill the space uniformly on length scales greater than a characteristic correlation length ξ. Similar patterns are frequently observed in surface growth phenomena [838] and fluid–fluid displacement experiments [838]. The "spherulitic" morphology, which has been studied in semi-crystalline polymers for many years, also appears to be a dense

4. The old growth/new growth interface consists of all of those places where a cluster of size s is contacted by particles that are added after the cluster reaches a size s as the cluster is grown to a much larger size S, where $S/s \gg 1$ and the cluster of size s is completely screened from further addition of particles.

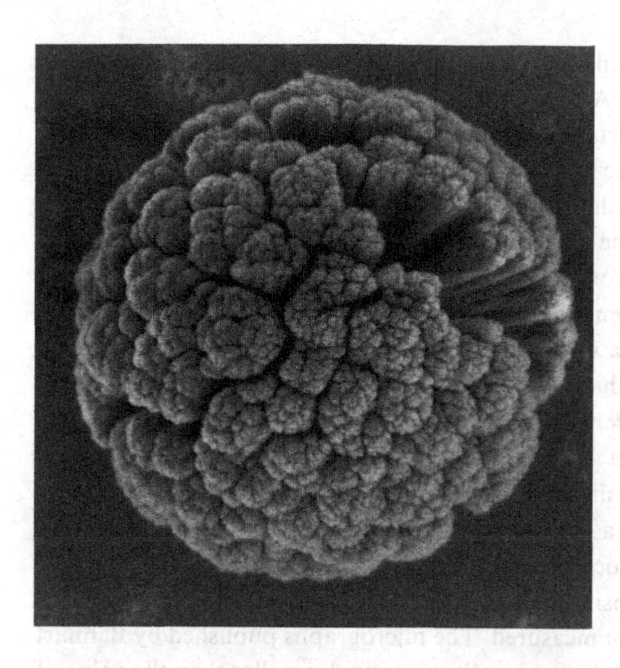

Figure 4.17 An electron micrograph of a carbon particle aggregate formed in a dusty 15 kHz helium plasma. The columnar internal structure and rough surface are clearly revealed in this figure. The aggregate had a diameter of about 3 μm. The aggregating particles were of molecular size and may have been generated by sputtering and/or plasma chemistry. This figure was provided by P. D. Haaland.

branching morphology. A quite spectacular example of a dense radial morphology is provided by the work of Haaland *et al.* [839, 840, 841] on dust particle aggregates formed in dusty plasmas. Aggregates of carbon particles were formed in helium or argon plasmas generated at pressures of $\approx$ 1 Torr using carbon electrodes and frequencies in the tens of kilohertz range. Similar structures were formed using different electrode materials and frequencies. The aggregates were trapped by electrostatic forces and could be studied *in situ*. Dust samples were collected and examined using high resolution, low energy, field emission electron microscopy. An example is shown in figure 4.17, which shows "spherulitic" internal structure and a "cauliflower-like", possibly self-affine (chapters 2, section 2.7, and 5), surface structure. Garscadden *et al.* [840] proposed a regular "midpoint displacement" model, for the surface geometry. This model is like the random midpoint displacement models, described in chapter 2, section 2.7.1, that are used to generate statistically self-affine fractal surfaces.

The growth of carbon filaments on insulating materials, such as boron nitride crystals, under irradiation in an electron microscope [842, 843] provides a related and equally dramatic example of a 3-dimensional densely branched morphology. These structures were formed from neutral and ionized hydrocarbon molecules that were present in the microscope, which was used without cryoshields. In both cases, the structure can be attributed to a modified ballistic aggregation mechanism [839] (chapter 3, section 3.4.2). The trajectories of particles will be modified by the electrostatic field surrounding the aggregates. Positively charged particles will be directed inwards towards the negatively charged aggregates or deposit. The trajectories followed by the neutral hydrocarbon molecules will be determined by their polarizabilities, dipole moments and the charge distribution on the growing aggregate. At the pressures used in these experiments, the mean free path of the molecules is much larger than the aggregate size, and the effects of collisions on the particle trajectories are small. Mechanical effects (radial stresses) associated with the electrostatic repulsion of the negative surface charge may be important. The formation of an insulating hydrocarbon layer on the rapidly growing tips, before conversion to conducting, amorphous carbon, may modify the electric potential near the tips, and the diffusion of adsorbed hydrocarbon may also play a role in the growth process.

In these experiments, Banhart [831, 832] found that sparsely branched dendrites were formed at low deposition rates ($\approx 2 \times 10^4$ carbon atoms s^{-1} for a dendrite with a size of about 1 μm). At a deposition rate of $\approx 3 \times 10^5$ carbon atoms s^{-1}, the carbon deposits had a densely branched morphology, and at $\approx 5 \times 10^7$ carbon atoms s^{-1} a compact structure was formed. The growth rate was controlled via the hydrocarbon pressure and the electron beam intensity. The sparsely branched deposits appeared to have a fractal structure, but the fractal dimensionality was not measured. The micrographs published by Banhart [832] indicate that the fractal dimensionality was much smaller than the value of $D \approx 2.5$ expected for 3-dimensional DLA. The transition from a branched fractal structure at low growth rates to a densely branched morphology at higher growth rates is similar to the behavior found in the growth of bacterial cell colonies, electrochemical deposition experiments and other growth processes. In addition, morphology transitions, similar to the Hecker effect in electrochemical deposition, were observed in some of the densely branched structures. Because of the complexity of these processess, the significance of the morphological similarities is not clear.

Similar morphologies were found in the earlier experiments of Sharma [833] in which Cu and Zn targets were laser ablated in atmospheres of oxygen, wet air, wet argon or hydrogen, at pressures near to 1 atmosphere, leading to the growth of dendrites on a thin Cu wire, which was separated from the target by a distance of 3–5 mm and was held at a potential of 200–400 V with respect to the ablation target (very similar patterns were obtained with positive or negative potential differences). If O_2 or H_2 atmospheres were used, the dendrites were formed from particles of the corresponding oxide or hydride. In wet atmospheres the composition was uncertain, but the dendrites may have contained a mixture of oxide, hydroxide and hydride. The ablation was carried out using 100–150 mJ focused pulses from a 303 nm XeCl excimer laser, with a pulse repetition rate of about 8 Hz. The ablation process resulted in the formation of small non-conductive particles in the cold background gas, near the target. The electrically neutral, dipolar particles generated by this process migrated in the electric field and stuck to the dendrites growing from the wire. The resulting deposits were very weak and could be fragmented by increasing the potential difference to

$\approx$ 600 V. This appears to be a complex process in which mechanical effects probably play an important role. No attempt was made to characterize the morphologies in a quantitative manner.

The formation of branched structures by the aggregation of aerocolloidal systems in electric fields is a common phenomenon that has been known for many years [845]. However, the quantitative study of the morphologies formed by these processes is only just beginning.

4.2.1 Electrochemical Deposition

Dense radial patterns can be grown electrochemically [721, 722, 738, 740, 846], under conditions that are similar to those leading to DLA patterns (except that the applied voltage or current is higher and the growth velocity is larger). Similar morphologies can be obtained using long parallel electrodes [750]. In most experiments, dense radial patterns were obtained without the addition of an inert screening electrolyte. Figure 4.4 shows the transition from DLA-like to dense radial patterns in the experiments of Hibbert and Melrose [738, 740].

There appears to be a continuous increase in the fractal dimensionality from that of DLA ($D \approx 1.71$) at small voltages to $D = 2$ at high voltages. However, a crossover with a voltage dependent crossover length ξ seems more likely. These two possible scenarios cannot easily be distinguished using the results from these experiments. It is not immediately obvious why dense radial patterns are stable under these conditions. Grier *et al.* [847] have proposed that the pattern is stabilized by the electrical potential drop across the filamentary pattern. They carried out a perturbative linear stability analysis, like that of Mullins and Sekerka [149, 159], discussed in appendix A, section A.1, to show that this stabilizing mechanism is realistic, even for a quite good conductor.

Inside the dense radial structure, the electrolyte is depleted and the current can flow only in the radial direction, along the filaments. This anisotropy plays an important role in the stability analysis. If the conductivity in the growing metallic structure was isotropic, then a circular envelope would be unstable, in the quasi-stationary, long diffusion length limit, unless the resistivity of the metallic deposit ρ_d was greater than that of the electrolyte ρ_e. The anisotropic conductivity leads to the stable growth of a circular deposit for values of $m' = \rho_d^r/\rho_e$ larger than $m'_{min} \approx 1/(1 + \ln(r_\circ/r_1))$, where ρ_d^r is the resistivity of the metallic deposit in the radial direction, ρ_e is the resistivity of the electrolyte, $r_\circ$ is the radius of the anode and r_1 is the radius of the cathode. It follows that for a parallel electrode configuration $m'_{min} = 1$ [847]. For isotropic conduction, the conductivity ratio plays the same role as the viscosity ratio in fluid–fluid displacement,[5] and $m'_{min} = 1$ in the circular cell, radial growth geometry. The

5. The ratio $m = \mu(2)/\mu(1)$ is usually used to describe fluid–fluid displacement processes, where $\mu(1)$ is the viscosity of the injected fluid and $\mu(2)$ is the viscosity of the displaced fluid. The restivity ratio m' plays a role similar to $1/m$.

stability of the circular envelope will be further enhanced if the diffusion length $\xi_{\mathscr{D}}$ is shorter than the size of the cell, and stable growth becomes possible at quite small values of m' ($\leq 10^{-3}$, in typical cases) [847]. Correlations between the growth of different regions on the interface separated by distances greater than $\xi_{\mathscr{D}}$ are quenched. This can explain the uniform nature of the growing pattern on scales ℓ greater than $\ell \approx \xi_{\mathscr{D}}$. However, the envelope of the pattern could still become unstable if the conductivity of the dense radial pattern was too large. In the small diffusion length limit, the electrochemical potential of the cation is described, to a good approximation, by the diffusion equation, and it is the diffusion equation, not the Laplace equation, that must be solved between the growth front and the anode. Grier *et al.* [847] found that stable growth can take place for quite small, but not zero, values of m' if the diffusion length is small. A densely branched pattern can also grow between parallel electrodes if the diffusion length is small. In this case, a forest of densely packed trees advances behind the smooth "interface" between depleted and undepleted electrolyte.

In subsequent work, Grier and Mueth [837] studied this model in the quasi-static, large diffusion length limit, using both linear stability analysis and computer simulations. This study indicated that, for a large conductivity anisotropy ratio ($\rho^r_d/\rho^a_d \approx 10^{-4}$, where ρ^a_d is the azimuthal resistivity in the direction parallel to the growth front), the dense radial pattern will be stable with respect to low mode number perturbations if the resistivity ratio m' is larger than about 1/5. If m' lies in the range $\approx 1 > m' > \approx 1/5$, the pattern may be stable to low mode number perturbations, but unstable to high mode number perturbations, leading to preservation of the overall circular envelope, while allowing branching to occur on smaller scales. In a more recent extension of this work, Lin and Grier [838] showed that dense radial growth could occur in 3-dimensional systems with resistivity ratios as low as $m' \approx 0.01$.

On the other hand, Fleury *et al.* [735] proposed that the transition to a dense radial pattern is related to the increased importance of electroconvection at higher growth velocities. They suggested that the stronger electroconvection induces stronger fluctuations which suppress the mechanically induced tip-splitting mechanism, leading to the rapid dominance of one branch with the other forming a "spur" on the side of the dominant branch. Although the use of a filter paper support does not completely suppress fluid motion, it seems unlikely that this scenario can apply to the transition from DLA-like to dense radial morphologies, with increasing potential, found by Hibbert and Melrose [738] using filter paper wetted with $CuSO_4/H_2SO_4$ solutions (figure 4.4). It is apparent that electroconvection does play an important role in many, if not most, electrochemical deposition processes and that the effects of electroconvection on ion transport and/or the mechanical effects of electroconvection must be considered in the analysis of electrodeposition experiments.

Barkey *et al.* [846, 850] have observed an electrolyte depleted boundary layer surrounding dense radial aggregates grown without a screening electrolyte. The outer surface of this layer forms a smooth envelope surrounding the growing dense radial pattern. This suggests that the growth of dense radial pattern may be similar to the growth of DLA clusters in a finite concentration of random walkers and the growth of Al and Ge dense radial patterns from an amorphous Ge/Al alloy [851], described later in this section. In the case of widely separated branches, the boundary layer observed by Barkey *et al.* penetrates between the branches. In this respect the shape of the boundary between the concentrated and depleted electrolyte regions is like that found for growth of DLA clusters from a "sea" of particles at a non-vanishing concentration, and is quite different from that observed by Fleury *et al.* for electroconvection driven growth, where the boundary bulges outwards between the growing tips to form "arches".

In these experiments, the growth velocity is essentially constant throughout the experiment. The advance of the dense radial pattern maintains the narrow width of the depletion zone. This width is the diffusion length discussed in chapter 1, section 1.4.2, and is given by $\xi_{\mathscr{D}} \approx \mathscr{D}/V$, where $\mathscr{D}$ is the electrolyte diffusion coefficient and V is the growth velocity. Barkey *et al.* proposed a mechanism for stabilization of the circular boundary of the dense radial pattern that is similar to that which stabilizes finite density DLA and does not require the dense radial pattern to have a significant electrical resistance.

A computer model that incorporates similar ideas has recently been described by Erlebacher *et al.* [852]. In this multiparticle, off-lattice DLA cluster growth model, the particles follow random walk paths in the screened region, far from the growing aggregate. In the essentially unscreened depletion zone, the particles (representing metal cations) follow migration controlled paths. In the model, the particles follow straight line paths to the nearest position at which they can contact the cluster when they move within a distance ξ from the aggregate. In this respect, the model is similar to the "attractive interaction DLA model" of Meakin [618]. In the model of Erlebacher *et al.*, particles are continuously injected into the system on a circle that represents the anode. The number of particles injected at each iteration of the simulation is given by $J = jL(p)$, where $L(p)$ is the length of the perimeter of the depleted zone at a distance ξ from the growing cluster. Every particle is moved at each iteration of the algorithm, and, since the particles that contact the cluster during a particular iteration are not added until that iteration ends, particles in the cluster may overlap (but particles cannot be added to the strongly screened interior regions). A quite wide variety of patterns can be generated by varying the model parameters ξ and j. Dense radial patterns appear for large values of the parameter j that controls the concentration of random walkers. For small values of j, randomly branched, DLA-like patterns are formed.

The stabilization mechanism in this model is very different than that in the "anisotropic resistivity" model proposed by Grier *et al.* [847]. In the model of Erlebacher *et al.* [852] the circular envelope appears to be stabilized as a result of the microscopic dynamics of the model. The overlap of the particles at the most exposed tips of the growing pattern supresses their rate of growth and stabilizes the growth envelope. Erlebacher *et al.* suggested that the overlap might represent 3-dimensional effects in the quasi 2-dimensional growth process. However, this mechanism seems unphysical and does not provide a satisfactory explanation of the stability of the circular envelope.

Barkey *et al.* have also suggested that the large concentration gradient near the outer boundary of the depletion envelope may behave like a fluid–fluid interface with a non-zero interfacial energy and the associated capillarity effects [853], and that these "capillarity" effects may play an important role in stabilization of the dense radial pattern [853]. However, the possible magnitude of this effect was not quantitatively assessed, and it appears likely that it is too small to play an important role in the growth process.[6]

A phenomenon that has been known for many years but has recently been rediscovered and called the "Hecker effect" is often seen in dense patterns grown by electrochemical deposition. The Hecker effect is an abrupt change in morphology (see figure 4.18, for example) that takes place when the growing pattern reaches a constant distance from the anode [713, 857, 858, 859]. This effect is most striking when the anode has a non-circular shape. It has been suggested that the Hecker effect provides evidence that the Laplace field or the electrical potential plays an important role in the growth process [860]. However, it is now generally accepted that the Hecker transition occurs when a chemical front from the dissolving anode reaches the advancing "active zone" [861, 862], and, as is illustrated in figure 4.18, it has been suggested that such morphological transitions can actually be used to measure ionic mobilities [863] if the effects of convection can be supressed. In the presence of an alkali metal salt, the morphological transition takes place at a distance ℓ_c, from the cathode, given by $\ell_c = \ell_t \mathcal{M}_a/(\mathcal{M}_a + \mathcal{M}_m)$, where $\mathcal{M}_a$ is the anion mobility, $\mathcal{M}_m$ is the alkali metal ion mobility and ℓ_t is the distance between the parallel electrodes

6. Assuming that the free energy functional $F(\mathrm{x})$ has the form $F(\mathrm{x}) = f(\phi(\mathrm{x})) + c|\nabla\phi(\mathrm{x})|^2$ where ϕ is the volume fraction of one of the two components, it can be shown [854, 855, 856] that the interfacial surface tension Γ is given by

$$\Gamma = 2c \int_{-\infty}^{\infty} c|\nabla\phi(x)|^2 dx. \tag{4.25}$$

Here, $\phi(x)$ is the volume fraction at a distance from a flat interface. It follows from equation 4.25 that the surface tension will be inversely proportional to the width of the interface. Although equation 4.25 cannot be applied to systems in which $1/|\nabla\phi(x)|$ is of the order of molecular sizes, the surface tension calculated from equation 4.25 reaches values that are similar to those associated with the interfaces between immiscible fluids only for interface widths of molecular dimensions. Consequently, the effective interfacial tension is very small, except for very narrow interfaces.

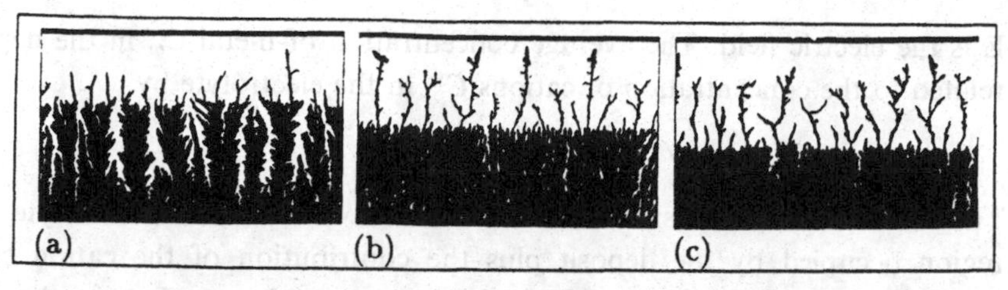

Figure 4.18 An illustration of the "Hecker effect" in the electrochemical deposition of Zn from 0.05 M $ZnSO_4$. In parts (a), (b) and (c), the electrolyte also contained 2.5×10^{-4} M Li_2SO_4, Na_2SO_4 and K_2SO_4, respectively. The transition from a dense pattern to a sparse dendritic pattern takes place when the growth front encounters the alkali metal ion depleted zone moving down from the anode (top). The position of the transition reflects the relative mobilities ($\mathscr{M}_{Li^+} < \mathscr{M}_{Na^+} < \mathscr{M}_{K^+}$). This figure was provided by F. Argoul.

in the thin cell. Similar transitions can take place as a result of convective transport [732] instead of migration of ions in the electric field.

In general, the evolution of the concentration field due to ion transport in an electrochemical cell can be written as

$$\partial c_{\pm}/\partial t = -\nabla \mathbf{j}^{\pm} = \nabla \cdot (c^{\pm}\mathscr{M}^{\pm}\nabla\phi + \mathscr{D}^{\pm}\nabla c^{\pm}) + \text{convective terms} \quad (4.24)$$

for a binary electrolyte, where, $\mathbf{j}^{\pm}, c^{\pm}, \mathscr{M}^{\pm}, z^{\pm}$ and $\mathscr{D}^{\pm}$ are the flux, concentration, mobility, charge and diffusion constant of the cation and anion. The interplay between the different terms on the right-hand side of equation 4.24 leads to the rich morphological behavior observed in the experiments. This scenario can be further complicated by material dependent effects such as surface "poisoning". Kuhn and Argoul [864, 865] have shown that the morphology of zinc, copper and tin electrodeposits can be quite sensitive to "impurities" such as dissolved O_2 (or H_2O_2) or small concentrations of alkali metal cations. This may help to explain why different experimental groups have had difficulty reproducing each other's results. Similarly, the work of Atchison *et al.* [731] demonstrated the important effect of complexing agents, such as cyanide and ethylenediaminetetra-acetate (EDTA) that can form negatively charged, metal containing anions such as $[Cu(CN)_4]^{2-}$ and $[CuEDTA]^{2-}$, on the morphology of electrodeposited metals. In many experiments, there does not appear to have been any attempt to control dissolved oxygen, CO_2 or other impurities.

To maintain electroneutrality and to prevent the generation of a large space charge near the cathode, the growth of the densely branched patterns must match the *anion* [866] front with a velocity $\mathbf{V} = -\mathscr{M}^{-}\mathbf{E}_b$ (towards the anode), where $\mathbf{E}_b$ is the electric field in the bulk electrolyte. In a 2-dimensional cell, the growth front of the densely branched pattern can be approximated by a straight line that moves with a velocity $\mathbf{V}^{-}(\mathbf{x}) = -\mathscr{M}^{-}\mathbf{E}(\mathbf{x})$, towards the anode, where

E is the electric field. The average concentration of metal C_m in the deposit is related to the concentration of cations C^+ in the electrolyte by

$$C_m = C^+[1 + (\mathcal{M}^+/\mathcal{M}^-)]. \tag{4.26}$$

The metal in the deposit contains contributions from the electrolyte in the region occupied by the deposit plus the contribution of the cation current. The impurity cations released from the anode travel toward the cathode with a velocity $\mathbf{V}_i^+(\mathbf{x}) = \mathcal{M}_i^+\mathbf{E}(\mathbf{x})$. Fleury *et al.* [862] have calculated the electric field $\mathbf{E}(\mathbf{x})$ between a circular boundary representing the front of the growing densely branched pattern and a square boundary representing a square anode. This was used in a simulation of the Hecker effect. At each stage in the model of Fleury *et al.*, the circular boundary of the densely branched pattern is advanced by a small amount, and the impurity front at position **x** is advanced by an amount $\mathcal{M}_i^+\mathbf{E}(\mathbf{x})\delta t$ $(\mathbf{x} \rightarrow \mathbf{x} + \mathcal{M}_i^+\mathbf{E}(\mathbf{x})\delta t)$. The points at which the impurity front first encounters the growth front is recorded, and these points form the simulated Hecker transition boundary. This boundary does not have a simple square shape. Instead, it is extended toward the corners of the square anode, in agreement with experimental observations.

In the experiments of Fleury *et al.* [862], on the electrodeposition of copper from copper sulfate solutions with copper cathodes and brass anodes, two Hecker transitions were observed. In the case of the brass anodes, these transitions were attributed to the separate arrival of H_3O^+ and Zn^{2+} cation fronts. In the case of copper anodes, the two Hecker transitions were attributed to H_3O^+ and an unidentified impurity (or impurities) with a mobility similar to that of Zn^{2+}. Because of the high mobility of H_3O^+, it generated a Hecker transition nearer to the cathode than to the anode.

Based on an experimental study of the electrodeposition of Zn from aqueous $ZnSO_4$ electrolytes with parallel Zn electrodes that act as spacers in a thin cell, Zik and Moses [856] have concluded that the Hecker transition takes place when the original electrolyte is depleted. However, this observation is consistent with the idea that the transition is caused by the arrival of impurity ions from the anode at the growth front if the transport properties of the impurity and Zn^{2+} are similar. These experiments were carried out under constant current conditions with careful control of the temperature and the electrolyte composition (including dissolved gases). After the Hecker transition had taken place, the growth velocity did not change, despite the much lower deposit density. This indicates that the electrodeposition current efficiency was low after the Hecker transition, and that electrochemical processes other than metal ion deposition were taking place at the cathode.

4.2.2 Thin Films

The crystallization of amorphous Al/Ge thin films has been studied extensively by Deutscher, Lereah and others [838, 851, 868, 869, 870]. This system provides another example of a dense branching morphology. In a typical experiment, a thin film of amorphous Al_xGe_{1-x} ($0.4 < x < 0.55$) was prepared by the simultaneous electron bombardment evaporation of Al and Ge onto coated microscope slides at room temperature with a background pressure of $\approx 10^{-6}$ Torr. The vapor deposition process was continued at a rate of about 4 Å s^{-1}, until a film thickness of about 550 Å was reached. Because of the geometry of the experiment, the composition varied in the range $0.4 < x < 0.55$, along the slide, so that samples with different compositions could be obtained by cutting the slide. After the slide had been cut into pieces, the Al_xGe_{1-x} film was removed by dissolving the soluble coating on the glass slide, and the film samples were mounted on copper or nickel grids. The films were then heated and observed in a transmission electron microscope.

If the Al_xGe_{1-x} thin films were heated to a temperature of 230 °C, or higher, phase separation accompanied by the nucleation and growth of crystalline regions took place. The crystalline regions consisted of an Al-rich single crystal matrix, with about 10% Ge, surrounding a polycrystalline Ge core with a dense branching morphology (figure 4.19). The distance between the tips of the Ge core and the envelope of the Al-rich region grew during the early stages, but soon reached a constant value. In this respect, the growth of the dense branching pattern is like that of a "DLA" cluster in a sea of particles at a non-zero density [57, 586, 585]. The perimeter of the Al-rich region advanced at a constant velocity V that grew exponentially with increasing temperature [851]. The growth of the velocity V with increasing temperature correlated strongly with the growth of $\mathcal{D}/\xi$, where $\mathcal{D}$ is the diffusion coefficient of Ge in Al and ξ is the distance between the perimeter of the Al-rich region and the Ge dendrites growing inside. In the growth of this dense radial pattern, the Ge atoms diffused from the amorphous Al_xGe_{1-x} region through the Al single crystal, where the Ge diffusion coefficient was high, to the polycrystalline, dense radial Ge pattern. As the Al crystal grew, Ge was expelled into the amorphous Al_xGe_{1-x} region. The overall composition inside the region enclosed by the advancing Al crystallization front was the same as that of the amorphous Al_xGe_{1-x} film. The stable propagation of the Al crystallization front determined the overall shape of the pattern, and the internal length scale was determined by the competition between nucleation and growth of the Ge crystallites [870]. Under some conditions, patterns that looked more like DLA clusters could be seen [869].

The dense branching Ge patterns formed from amorphous Al_xGe_{1-x} films have been studied quantitatively [871]. Although there were some loops, the patterns could be represented by trees. These trees were analyzed by measuring

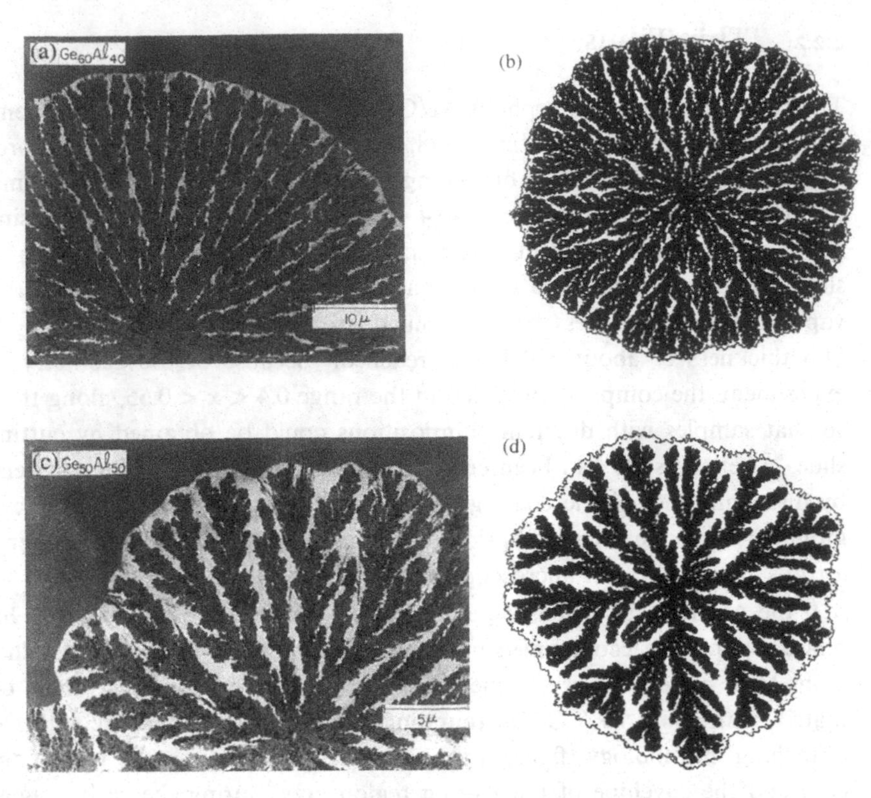

Figure 4.19 Densely branched patterns grown in amorphous Al_xGe_{1-x} thin films. The white regions contain crystalline aluminum and the dark dendrites within the white regions correspond to germanium rich regions. The left column shows experiments and the right column shows patterns obtained from the "DLA in a box" model, described in the text. This figure was provided by Y. Lereah.

the characteristic lengths $\xi_\perp(r)$ (the distance between adjacent branch segments, in a direction perpendicular to the growth direction, determined from the number of intersections with a circle of radius r centered on the initial growth site) and $\xi_\parallel(r)$ (the distance between nodes along the main branch, or the distance between the points at which adjacent side branches emerge from the main branch). The angle between branches measured at the nodes was also used to characterize the patterns. The characteristic length scales were found to decrease with increasing growth velocity V. Data recorded for growth velocities covering a range of about two decades (3×10^{-3} μm s^{-1} to 3×10^{-1} μm s^{-1}) indicated that $\xi_\parallel(r) \sim V^{a_\parallel}$, $\xi_\perp(r) \sim V^{a_\perp}$ with $a_\parallel \approx 1/6$ and $a_\perp \approx 1/4$. (Except for small values of r, near to the origin, $\xi_\parallel(r)$ and $\xi_\perp(r)$ are essentially independent of r.) In addition, the angles between the branches were found to increase essentially linearly with increasing Al content, in the original films. The growth of these dense branching patterns was simulated by using a model that combined the Eden growth model, for the advancing Al envelope, and the DLA model, for the growth of the branched Ge pattern inside the Al envelope. In this model, the sites on a 2-dimensional lattice are labeled at the start of a simulation to represent randomly distributed Al and Ge with probabilities of x and $1 - x$, respectively. An Eden growth process is then started at the center of the lattice to represent the growth of the Al envelope. As the "Al" envelope advances into the "Al/Ge" region "Ge" particles (labeled sites) are released and follow a random path walk, inside the Eden growth cluster, until they contact the growing Ge cluster. The random walkers that represent Ge "atoms" are stopped and become part of the growing Ge cluster when they reach the origin or reach an unoccupied perimeter site on the Ge cluster. At each step in the simulation, all of the free Ge sites inside the Al Eden cluster move by one lattice unit (and stick if they reach the perimeter of the Ge cluster), and a fraction f of the sites on the external perimeter of the Al Eden cluster are added to the Eden cluster. The parameter f corresponds to the ratio between the growth velocity V and the diffusion coefficient of Ge in Al. Figure 4.19 shows electron micrographs of typical Al/Ge densely branched patterns and patterns generated using the computer model.

To better represent the experiments, the basic model described above was modified to include sticking probabilities that depend on the local interface curvature [599, 872] and a small modification to the Eden growth rules to account for the non-zero solubility of Ge in Al. The patterns generated by this model are very similar to those seen in the experiments. Patterns similar to the Ge/Al dense radial patterns have been studied in other systems such as poly(ϵ-caprolactone)/polystyrene thin films [873]. Similar patterns can also be generated using the DLA-like model, described in chapter 3, section 3.7.2, in which particles are launched from a surface at a constant distance from the growing cluster corresponding to a fixed diffusion length.

4.2.3 Fluid–Fluid Displacement

The question of the asymptotic behavior of "Saffman–Taylor" displacement in a radial geometry remains unresolved. Ben-Jacob *et al.* [838] found a dense radial morphology when air was injected at constant pressures, in the 50 ± 2 mm Hg to 150 ± 2 mm Hg range, into 1200 cP glycerin confined in gaps of $0.4 - 0.8$ mm between 3/4 inch thick, 23 inch diameter polymethylmethacrylate sheets that were flat to a tolerance of about 0.005 inch. Rauseo *et al.* [874] studied the displacement of 24 cP paraffin oil by nitrogen in a radial Hele-Shaw cell constructed from 1/4 inch thick float glass. In these experiments, the interfacial tension Γ was 50 dyne cm^{-1} and the nitrogen was injected at a constant rate. In typical experiments, an algebraic relationship was found between the displaced area and the radius of gyration, corresponding to a fractal dimensionality of about 1.79. However, the patterns exhibited only one level of tip splitting.

Similar, somewhat more complex, patterns generated by the displacement of glycerin by air in a 600 mm diameter Hele-Shaw cell constructed from 13 mm thick glass discs were reported earlier by Paterson [875]. Paterson also studied the "inward fingering" process that takes place when glycerin is withdrawn from the center of the cell and air is allowed to enter from the edges.

Patterns that more closely resemble DLA clusters were obtained in the experiments of Couder [876] in which the cell was constructed from 1.5 cm thick glass plates. In these experiments, the displacement was carried out by allowing air to enter at a pressure of 1 atmosphere, through a hole in the middle of the upper glass plate, while the viscous fluid was evacuated from the open edge of the cell. Rauseo *et al.* [874] and Couder [761, 876] commented on the sensitivity to experimental conditions and imperfections. It appears that DLA-like patterns may only appear close to the limit of infinite viscosity ratio, perfect wetting by the viscous fluid and rigid cell walls. Couder [761, 876] has suggested that if these conditions are not met, then a crossover to a dense radial pattern will take place.

May and Maher [877] studied the displacement of heavy paraffin oil by nitrogen in a cell consisting of a 1 inch thick, 24 inch diameter top plate on a 22inch diameter base plate with a $b = 1.5$mm gap. The flexion of the glass plates was estimated to be no greater than 0.020 mm, at the highest pressure used in these experiments. The experiments were characterized by the dimensionless control parameter

$$C = j\mathcal{M}/b\Gamma, \tag{4.27}$$

where $\mathcal{M}$ is the mobility Γ is the surface tension and j is the injection rate, measured in terms of the area A of fluid displaced in the Hele-Shaw cell. To reach control parameter values larger than about $C = 35$, lead bars were placed on the top plate. A fractal dimensionality was estimated from the relationship between A and the radius of gyration of the displacement pattern. For small values of the control parameter, $D \approx 2.0$. As C was increased, the fractal dimensionality decreased, and was almost constant, with a value close to the DLA value of 1.715, for control parameters in the range $10 \leq C \leq 40$. At $C \approx 40$, the fractal dimensionality increased quite rapidly to a second plateau of about 1.79 ± 0.04, which extended to at least $C = 75$, the limits of the experiment.

A linear stability analysis (appendix A, section A.1) indicates that displacement of a viscous fluid by a less viscous fluid should eventually lead to an instability in the fluid–fluid interface for any viscosity ratio m. However, the linear stability analysis cannot predict how the pattern will evolve. If m is finite, the DLA screening effect is weakened and the interior parts of the displacement continue to evolve as a result of pressure gradients in the more viscous fluid, induced by flow of the less viscous fluid and the surface tension. Even in the

$m \to \infty$ limit the interior will continue to evolve in response to surface tension acting through the curvature of the interface. Consequently these interior regions can become quite compact. Similar effects may also cause the "inactive", non-growing parts of the pattern to fragment into strings of bubbles. This process would preserve the overall mass/length, or area/length scaling, providing that the bubbles did not move after they were formed .

Based on numerical solutions of the 2-dimensional equations of motion for the fluid–fluid interface during displacement of a viscous fluid by a non-viscous fluid ($m = \infty$) in a Hele-Shaw cell, with circular boundary conditions, Jasnow and Yeung [878] concluded that the displacement patterns are asymptotically 2-dimensional, under constant injection rate conditions. The surface energy induced reorganization of the interior of the pattern can be clearly seen in the simulations. These simulations suggest a crossover from an initially compact pattern via an intermediate scaling regime, in which the radius of the pattern grows as $R_g \sim A^{\beta}$, with $\beta \approx 0.57$ ($D_{eff} \approx 1.75$), where A is the displaced area, to an "asymptotic" regime in which $R_g \sim A^{0.515}$ ($D_{eff} \approx 1.94$). The effective value of the exponent β tends to smaller values as the displaced area increases, suggesting an asymptotic value of 0.5 or $D_{\beta} = 2$. While these simulations are not conclusive, they do quite strongly suggest that the asymptotic displacement patterns are uniform on length scales greater than a characteristic scale ξ. DLA-like scaling of the radius of gyration may be found on intermediate length scales in the range $R_{\circ} \ll \ell \ll \xi$, where $R_{\circ}$ is the radius of the initial interface. Based on the numerical results and a simple scaling model, Jasnow and Yeung suggest that the pattern size $R \approx R_g$ can be represented by the scaling form

$$R = (\Delta A)^{1/2} f(L_{\circ}/(\Delta A)^{1/2}, R_{\circ}/L_{\circ}), \tag{4.28}$$

where $\Delta A = A(t) - A(0)$ and where the scaling function $f(x, y)$ has the form

$$f(x,y) = \begin{array}{ll} c_1 & \text{for fixed } y \ (0 < y < \infty), \text{ as } x \to 0 \\ c_2(xy)^{-a} & \text{for } y \to \infty,\ xy \to 0 \\ c_3 xy & \text{for } xy \to \infty, \end{array} \tag{4.29}$$

where c_1, c_2 and c_3 are constants. Equation 4.29 describes three distinct regimes: the first line on the right-hand side of equation 4.29 corresponds to a uniform ($D = 2$) asymptotic pattern; the second line corresponds to a fractal structure ($D = 2/(1 + a)$); and the third line ensures that $R = R_{\circ}$ at early times. Consequently, a DLA-like pattern can be found when $L_{\circ} \ll R_{\circ}$. Here, $L_{\circ} = \xi_c \mathcal{M}/j$, where ξ_c is the capillary length, $\mathcal{M}$ is the mobility determined by the fluid viscosity and the internal width of the Hele-Shaw cell and j is the injection rate.

Highly branched viscous fingering patterns can be found outside the laboratory. They are quite commonly observed as a form of failure in laminated glass products in which layers of glass are separated by thin layers of a soft polymer such as polyvinyl butyral [879]. A fine example (patterns in the windows of the Los Lebreros Hotel, Seville) has been described by Garcia-Ruiz [880]. Such patterns are frequently generated as a result of stresses induced by heating and cooling.

Shochet and Ben-Jacob [74] have studied the transition from a dense branched morphology, in which the pattern evolves via tip splitting, to an anisotropic dendritic pattern, in which the pattern grows by a side-branching mechanism. This study was carried out using a model that combined the features of DLA and continuum growth models. In this case, Shochet and Ben-Jacob were able to show that the growth velocity V_b for the dense branched morphology pattern, with a convex "envelope", scales as $(\Delta\mu)^3$, while the growth velocity V_d for the anisotropic dendritic morphology, with a concave "envelope", scales as $(\Delta\mu)^{3/2}$, where $\Delta\mu$ is the difference between the chemical potential in the "solid" and that in the "liquid" at infinity. They found that a sharp transition takes place when $V_b(\Delta\mu) = V_d(\Delta\mu)$ at $\Delta\mu = \Delta\mu^*$, so that a dendritic pattern emerges when $\Delta\mu < \Delta\mu^*$ and a densely branched morphology is found when $\Delta\mu > \Delta\mu^*$. This provides strong support for the existence of sharp morphological transitions and for the idea of a "fastest growing morphology" selection principle. Transitions from a branched, fractal, DLA-like morphology at low driving forces to dense branched morphology under large driving force, high growth velocity conditions have been found in a variety of experimental systems.

Quite different results [881] were obtained using the "mean field" DLA model of Brener *et al.* [668] (chapter 3, section 3.10.1) at a non-zero particle density. In these simulations, a transition from a concave pattern envelope, interpreted as being representative of dendritic growth, to a convex pattern envelope (dense branching-like morphology), without a discontinuity in the velocity slope, was observed as the undercooling parameter was increased, while the anisotropy parameter remined constant. It is apparent that it will be some time before a general understanding of morphological transitions emerges. This will be an important direction for future studies.

4.2.4 Spherulites

It appears that spherulitic growth in bulk polymers and many other materials, including natural minerals and organic compounds, is a 3-dimensional example of the dense branching morphology [882]. Polymer spherulites [883] typically consist of a radiating array of anisotropic crystallites, forming branches at small angles that are not related to the crystalline structure. This branching process continues as the spherulite grows, leaving behind an almost space filling structure that is uniform on all length scales larger than the characteristic length ξ. On a detailed level, a very wide variety of spherulitic structures can be found in polymeric systems. However, the spherulitic morphology is an almost universal characteristic of polymers. The spherulitic morphology in polymers is attributed to the growth of a polycrystalline branched structure with the rejection of "impurity". The velocity of growth is given by $V \sim \delta\mu$, where $\delta\mu$ is the chemical potential difference for the crystallizing component across the rejected impurity layer. The dimensionless concentration field U_s at the spherulite surface is given by

$$U_s = \Delta - \xi_{c\phi}\mathscr{K} - (V_n/k), \tag{4.30}$$

where Δ is the dimensionless supersaturation, $\xi_{c\phi}$ is the capillary length, $\mathscr{K}$ is the curvature, V_n is the growth velocity in a direction normal to the interface and k is a kinetic parameter (chapter 1, section 1.4.2). Goldenfeld [882] has argued that if the radius of the pattern is sufficiently large, the growth of the interface is like that of a planar front and that this front will travel with a velocity V_n given by

$$V_n = (\Delta - 1)k. \tag{4.31}$$

In the early stages, the diffusion length $\xi_{\mathscr{D}}$ (chapter 1, section 1.4.2) will be larger than the spherulite radius $R(t)$, and $R(t)$ is expected to grow as $t^{1/2}$. At later times, $R(t) \gg \xi_{\mathscr{D}}$ and $R(t) \sim t$. Goldenfeld also pointed out that the growth process is kinetically controlled and the Mullins–Sekerka kinetic instability (appendix A, section A.1) results in a pattern with a characteristic morphology length ξ_m given by $\xi_m = 2\pi(\xi_{c\phi}\xi_{\mathscr{D}})^{1/2}$.

4.3 Percolation

Multiphase systems that are near to the percolation threshold are very common and are important in very many areas of science and technology.

In most cases, percolation can be studied more easily and more directly in 2-dimensional than in 3-dimensional systems. For example, Voss *et al.* [884] have analyzed electron micrographs of thin layers of gold that were vapor deposited onto 30 nm thick amorphous Si_3N_4 "windows" supported on a Si wafer frame. By moving a shutter in front of the substrate during the deposition, a small gradient in the deposited film thickness was created, so that the film thickness varied from 6 nm (insulating) at one side to 10 nm (conducting) at the other. Near to the insulator/conductor percolation transition, the film does not cover the surface uniformly (figure 4.20). Instead, the gold forms ramified clusters with a thickness of about 10 nm and a characteristic length scale of about the same magnitude. The clusters are compact on small length scales $\lambda \leq \epsilon \approx 10$ nm and ramified on longer length scales. Images, like those shown in figure 4.20, were digitized and the gold clusters (connected regions of gold) were found. The percolation threshold was found at a coverage of 0.74 ± 0.01. The gold clusters were found to have a fractal dimensionality of about 1.9, on length scales in the range $\epsilon < \ell < L$, where L is the cluster size (diameter), in good agreement with the theoretical value of 91/48 for 2-dimensional percolation. The number of clusters with sizes (areas) larger than s^* ($N(s > s^*)$) was measured as a function of s^*. Although Voss *et al.* did not analyze a large enough area to reduce statistical uncertainties to a very small level, their results were consistent with a size distribution of the form $N(s > s^*) \sim (s^*)^{-(\tau-1)} f(s^*/s_m^*)$ with $\tau = (D + d)/D = 187/91 = 2.0549\ldots$, for the finite clusters. Here, s_m^* is the characteristic cut-off size. The cut-off function $f(x)$ is constant for $x \ll 1$ and decays faster than any power of x for $x \gg 1$.

Kapitulnik and Deutscher [885] have carried out similar studies in which lead was deposited onto amorphous germanium. In this work, a value of 1.9 ± 0.02 was obtained for the fractal dimensionality of the infinite cluster, on length scales in the range $\epsilon < \ell < \xi$, where ξ is the correlation length. In addition, a value of 1.65 ± 0.05 was obtained for D_{bb}, the fractal dimensionality of the backbone, and the exponent τ for the power law size distribution of the finite clusters was found to have a value of 2.1 ± 0.2. All of these results are in satisfactory agreement with 2-dimensional percolation theory.

More recently, Jensen *et al.* [886] studied thin films of antimony deposited onto several substrates in a residual vacuum of 10^{-4}Pa. A copper grid was coated with 5nm of αC and 10nm of SiO_2 for transmission electron microscopy (TEM). The deposition process was stopped just before the conductivity threshold was reached. Under these conditions, the antimony film was too thin to crystallize. The antimony formed a stable amorphous film with a few crystallized "embryos". The film was then irradiated with an electron beam, from the high resolution TEM used in this work, in the vicinity of one of the embryos. This converted one of the amorphous connected clusters to the crystalline phase. The crystalline and amorphous materials could be readily distinguished in the electron microscope. The crystallized clusters were digitized and analyzed. The clusters were found to have fractal dimensionalities of 1.90 ± 0.01. Fractal dimensionalities of 1.70 ± 0.05 and 1.35 ± 0.1 were obtained for the hull and external perimeter, respectively. These results can be compared with the exact values of 1.75 and 4/3 for 2-dimensional critical percolation clusters (chapter 3, section 3.5). The major advantage of this method is that the crystallization process reveals the connectivity.

An annealed percolation model [887] has been proposed for the structure of percolating thin films such as those described above. This model is similar to the zero temperature phase separation model [888], an Ising model with conserved dynamics [889, 890, 891, 892]. In this model, the sites of a lattice are filled randomly, with probabilities p_A and $p_B = 1 - p_A$, to represent the two components in a binary mixture. The system is then annealed at "zero temperature" by randomly selecting adjacent sites with opposite labels (A or B) and exchanging the labels if the exchange will lower the energy

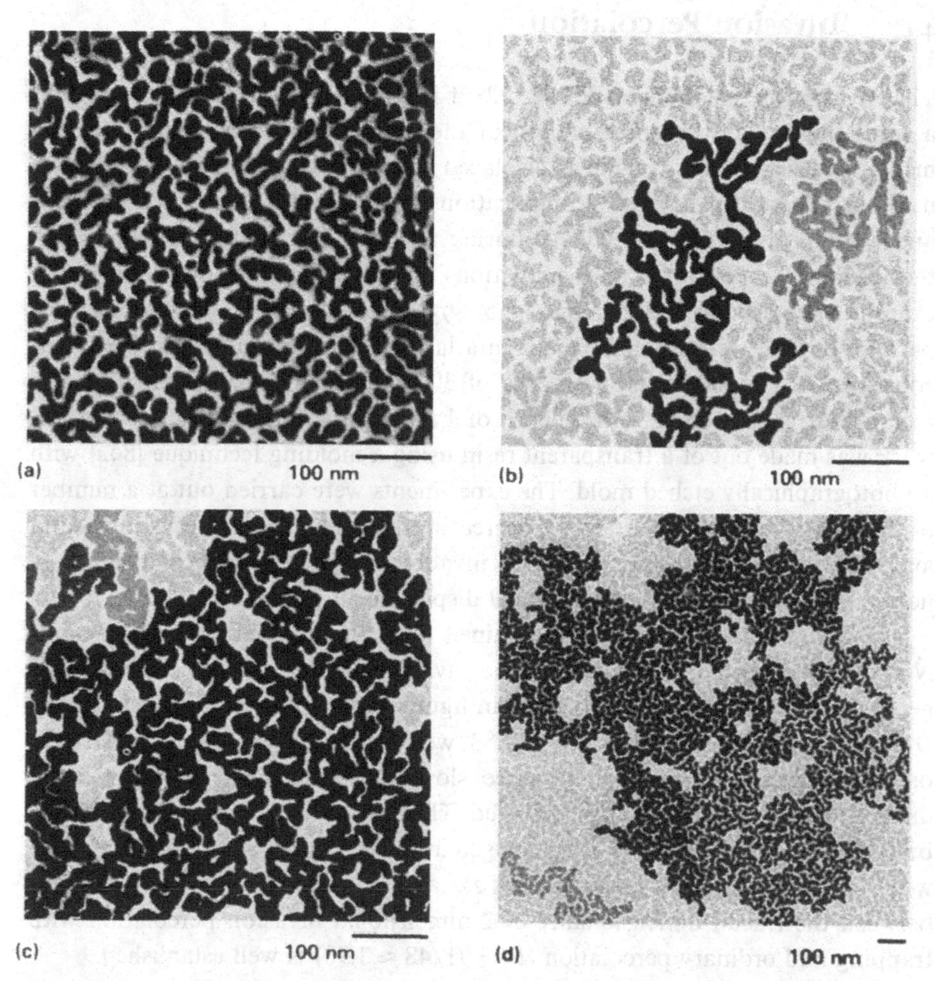

Figure 4.20 A digitized representation of a transmission electron micrograph of a thin layer of gold deposited via electron beam evaporation onto a Si_3N_4 substrate held at room temperature. Part (a) shows a digitized image at a coverage of $p = 0.64$, and the largest clusters in part (a) are shown in part (b) with darker shades indicating larger clusters. Part (c) shows clusters with $p = 0.64$ and part (d) shows clusters with $p = 0.72$, near the percolation threshold, at a 3× lower magnification. This figure was provided by R. F. Voss.

of the system. The process is continued until no more exchanges are possible. In the original model, an Ising Hamiltonian with the nearest-neighbor coupling constants $J_{AA} = J_{BB} = J$ and $J_{AB} = 0$ was used. In the work of Wollman *et al.* [887], next-nearest-neighbor interactions were included with $J_{nnn} = J_{nn}/3$. In accordance with universality ideas, the percolation threshold p_c is changed by "annealing" but the characteristic percolation exponents are not.

4.4 Invasion Percolation

The invasion percolation model described in chapter 3, section 3.5.2 provides a surprisingly realistic representation of the patterns generated during experiments in which a wetting fluid is displaced by a non-wetting fluid in a porous medium. There can be many complications (such as film flow, viscous effects and wetting properties that change during an experiment), but, in some cases, the results of experiments and simulations can be difficult to distinguish. For example, Lenormand and Zarcone [530, 893] have studied the displacement of paraffin oil (the wetting fluid) by air in a large 2-dimensional channel network consisting of 250 000 ducts in an area of 300 mm × 300 mm. The ducts had a rectangular cross-section with a depth of 1 mm and a variable width. The network was made out of a transparent resin using a molding technique [894] with a photographically etched mold. The experiments were carried out at a number of different displacement rates, by decreasing the pressure in the wetting fluid and allowing the non-wetting fluid to invade through one edge of the square network. Figure 4.21(a) shows a typical displacement pattern.

The fractal dimensionality was obtained from the number of invaded ducts $N(r)$ within a distance r of one of the invaded ducts, near the center of the network. Some of the data are shown in figure 4.21(b). Fractal dimensionalities D lying in the range $1.80 \leq D \leq 1.83$ were obtained over a 20-fold range of displacement rates. For the three slowest displacement rates, a fractal dimensionality of 1.83 ± 0.02 was reported. This is in excellent agreement with the best simulation results ($D \approx 1.82$) for square lattice bond invasion percolation with trapping (chapter 3, section 3.5.2), and it appears that the difference between the fractal dimensionality of 2-dimensional invasion percolation with trapping and ordinary percolation ($D = 91/48 \approx 1.98$) is well established.

The geometry of invasion fronts has also been investigated experimentally in 3-dimensional systems. For example, Clément *et al.* [895] have carried out experiments in which Wood's metal was injected into a non-consolidated packing of crushed glass. The crushed glass with a relatively narrow size distribution (mesh sizes between 200 μm and 225 μm) was deposited into a tube with a height of 15 cm and an inside diameter of 10 cm. The tube was then evacuated and liquid Wood's metal was injected slowly from the bottom, so that the invasion front was gravity stabilized. The experiment was carried out at low injection rates. A typical experiment took about 48 hours. At the end of an experiment, the system was allowed to cool in order to freeze the Wood's metal, with a melting point of about 70 °C. The free glass particles were then removed and the Wood's metal with embedded glass particles was potted in epoxy. Cross-sections were obtained by cutting the potted Wood's metal in the horizontal and vertical directions, machining the cuts and then polishing them. The machined and polished cuts were then photographed and the photographs were

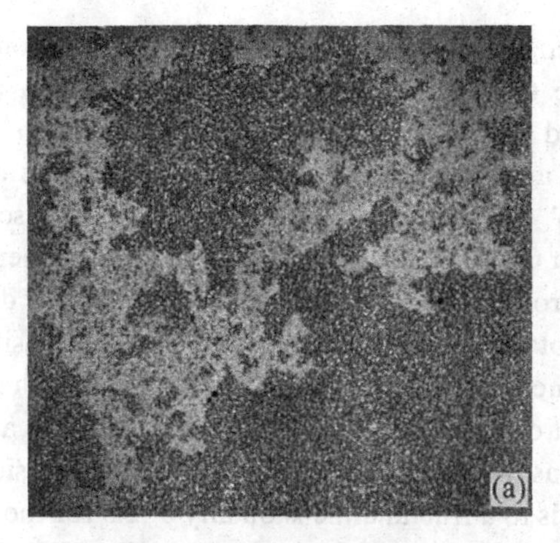

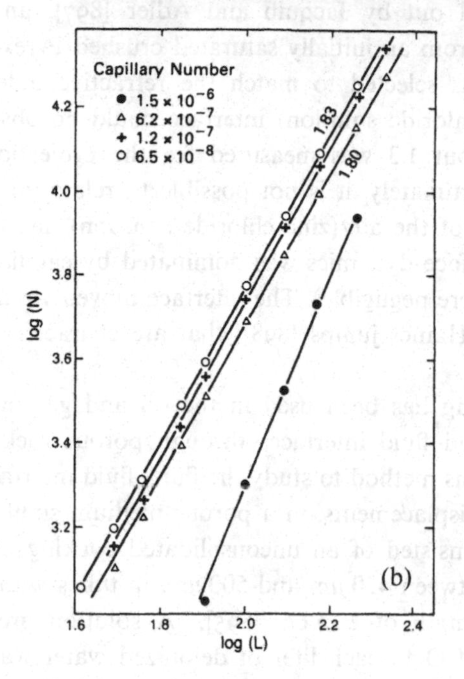

Figure 4.21 Part (a) shows a pattern generated by the slow displacement of a wetting fluid (paraffin oil) by a non-wetting fluid (air) in a 2-dimensional channel network. Part (b) shows the dependence of the number of filled ducts (N), in square regions of size $L \times L$, on the length scale (L) obtained from experiments like that shown in the part (a). The data for the three lowest capillary numbers indicate a fractal dimensionality of 1.80 – 1.83. For the highest capillary number, an effective fractal dimensionality greater than 2 was obtained, revealing a possible weakness in the method used to measure D. This figure was provided by R. Lenormand.

digitized. A fractal dimensionality of about 1.65 was obtained by measuring the two-point density–density correlation function $C(r)$ for the horizontal sections. This corresponds to a fractal dimensionality of $D \approx 2.65$ compared with a value of about 2.52 expected from invasion percolation models and the idea that the fractal dimensionality for 3-dimensional invasion percolation, with or without trapping, should be the same as ordinary 3-dimensional percolation. The correlation function was found to have the form $C(r) \sim r^{-\alpha}$ on short length scales ($\epsilon < \ell < \xi$) and was almost constant for length scales in the range $\xi < \ell < L$, where ϵ is the glass grain size and L is the system size. Here ξ can be interpreted as the characteristic size of the fractal blobs in the interface.

The "fractal" range of length scales ($\epsilon < \ell < \xi$) extended over only about one decade, and this may be the reason for the difference between the measured fractal dimensionality and that expected for invasion percolation.

In a later study [896], a parallelepiped was cut from the core of the sample and milled from the "top" to obtain many horizontal cuts, each separated from the previous cut by about 0.2 mm. This would allow a full, 3-dimensional image or map of the invasion front to be created. The effective fractal dimensionality of the horizontal cuts, obtained from the two-point density–density correlation functions, was found to increase with increasing cut height. A value of about 1.5 was found for the fractal dimensionality of the lower cuts, with a high Wood's metal saturation, by measuring the Wood's metal density–density correlation function. This corresponds to a fractal dimensionality of 2.5 for the full structure.

In a similar study, carried out by Jacquin and Adler [897], an aqueous $ZnCl_2$ solution was drained from an initially saturated crushed Pyrex packing. The $ZnCl_2$ concentration was selected to match the refractive index of the glass, so that the air/(zinc chloride solution) interface could be observed. A fractal dimensionality of about 1.3 was measured for the projection of the interface onto a plane. Unfortunately, it is not possible to relate this quantity to the fractal dimensionality of the air/(zinc chloride solution) interface in 3-dimensional space. The interface dynamics was dominated by capillary effects and gravity (viscous forces were negligible). The interface moved via a series of pronounced "jumps", called Haines jumps [898], that are characteristic of an invasion percolation process.

Magnetic resonance imaging has been used in the oil and gas industry to study the propagation of fluid–fluid interfaces through porous rock samples. Chen *et al.* [899] have used this method to study the fluid–fluid interface during gravity stabilized oil/water displacements, in a porous medium similar to that used by Clément *et al.* It consisted of an unconsolidated packing of crushed glass beads sieved to sizes between 420 μm and 500 μm, in tubes with an inner diameter of 6.3 cm and a length of 23.5 cm [895]. A solution prepared by adding 44.77 g of $NiSO_4 \cdot 6H_2O$ to each liter of deionized water was used to reduce the proton relaxation time in the water. In this way, the H_2O signal was suppressed in the spin echo detection and only the oil was observed. The non-wetting phase was Soltrol 100, with a density of 0.737 g cm^{-3} and a viscosity of 1.02 cP. The interfacial tension was $\Gamma = 38.8$ dyne cm^{-1}. Both stabilized drainage (displacement of water by oil) and imbibition (displacement of oil by water) were studied. The magnetic resonance imaging technique generated 256×256 pixel images of horizontal cross-sections through the vertical packings. The pixel "intensity" was proportional to the amount of oil in regions of size 0.45 mm $\times$ 0.45 mm $\times$ 1.7 mm, in the vertical direction. Consequently, these images could be used to estimate the fractal dimensionality of the horizontal cross-section by measuring the amount of oil $M(r)$ in occupied regions of size r. In this manner, effective fractal dimensionalities in the range 1.2 to 1.8 were obtained for the horizintal cuts, corresponding to effective fractal dimensionalities in the range 2.2 to 2.8 for the full 3-dimensional structure. The effective values for D depended on the height of the cross-section, in much the same way as in the experiments of Clément *et al.* [896]. It is apparent that the resolution and limited range of length scales do not permit an accurate value for D to be estimated from these experiments.

The effects of gravity play an important role in all of the 3-dimensional experiments. These experiments can be simulated using 3-dimensional versions of the gradient invasion percolation models discussed in chapter 2, section 2.3, and they can be described quantitatively using fractal blob models and scaling models, like those discussed in chapter 2 (section 2.3) and chapter 3 (section 3.5.3).

Shaw [527] has shown that the drying of a confined pseudo 2-dimensional porous medium can be regarded as an immiscible displacement process. Drying experiments were carried out using a homogeneous packing of 0.5 μm diameter silica spheres confined in a gap of 15–20 μm between two glass slides with an area of 2.5 cm $\times$ 4.0 cm. The two slides formed the walls of a thin cell. After the cell was packed with silica spheres and filled with distilled water, three of its sides were sealed with epoxy, and evaporation was allowed to take place via the unsealed edge of the cell. Since the silica spheres had almost the same refractive index as water, the propagation of the air/water interface into the cell could be easily observed and recorded using an optical microscope. Digitized images were analyzed; the region "invaded" by air was found to have a fractal dimensionality of about 1.89, and the interface between the "leading edge" of the air and water was found to have a fractal dimensionality of about 1.38. In this process, liquid regions that did not appear to be connected to the drying front or the open edge of the cell were in fact connected via crevices at particle contacts and filled pore spaces, so that trapping did not occur. This explains why the observed fractal dimensionality was closer to the value of 91/48 associated with invasion percolation without trapping than the value of about 1.82 found with trapping. There was a flow of air through the large gaps between the glass spheres and a counter flow of water to the open edge of the cell, where most of the water evaporation took place. Because of the counter flow, via crevices at particle contacts and filled pore spaces, the drying process was much faster with, than without, the silica sphere packing. Similar phenomena are important in commercial drying processes and are of considerable economic importance.

On large length scales, the interface was stable, so that the fractal scaling extended over length scales ℓ in the range $\epsilon < \ell < \xi$, where ϵ was approximately the sphere diameter and ξ was the correlation length for the invasion front. The correlation length ξ decreased as the drying rate increased and the drying front became more stable. The experiments of Shaw indicated that the correlation length ξ was related to the mean velocity of the invasion front by a simple power law

$$\xi \sim V_f^{-\gamma}, \tag{4.32}$$

and a value of 0.48 ± 0.1 was measured for the exponent γ. If it is assumed that the pressure gradient across the interface inside the aqueous phase is constant and proportional to the front velocity, then the relationship $\xi \sim V_f^{-\nu/(\nu+1)}$ can be derived [219, 527], where ν is the correlation length exponent for percolation (chapter 3, section 3.5).

As a result of capillary effects, a quite large negative pressure was generated in the water that filled the interstices between the silica spheres. If an air bubble is trapped in the porous medium, or is generated from the air dissolved in the water, a fluid–fluid displacement process can occur in which air displaces water in the pseudo 2-dimensional porous medium. Figure 4.22 shows a displacement pattern obtained in this way. It seems likely that similar processes will occur naturally during the drying of rocks and soils

In an interesting extension of the scope of application of percolation models Balázs [900] has used the percolation growth model with trapping (like the model described in chapter 3, sections 3.5.1 and 3.5.2) to simulate the open-circuit corrosion of a 31 nm thick aluminum film, deposited on a flat glass substrate. In the experiments, aluminum coated glass slides were immersed in a 0.5 mM $Fe_2(SO_4)_3$/5 mM HCl/10 mM Na_2SO_4/5 mM NaCl aqueous electrolyte. The pits grew via the advance of a corrosion front that left isolated islands of aluminum behind, as it propagated into the uncorroded film. The growth of individual pits in the aluminum film was recorded using an optical microscope with a CCD camera, and the pit perimeters were characterized using box counting analysis of the digitized images recorded by the CCD camera. The dependence of the number of occupied boxes on the box size could be represented by a power law over a box size range of about two decades, and a fractal dimensionality of $D = 1.33 \pm 0.01$ was obtained, in excellent agreement with the fractal dimensionality of 4/3 expected for a 2-dimensional percolation process.

4.5 Displacement in Complex Fluids

There is a natural tendency to focus attention on processes in simple systems that generate simple and/or familiar patterns. Even in simple Newtonian fluids, complex patterns that are not described at all well by the models discussed in chapter 3 can be formed. In rheologically complex fluids, displacement processes can lead to a very wide variety of patterns, some of which are neither simple

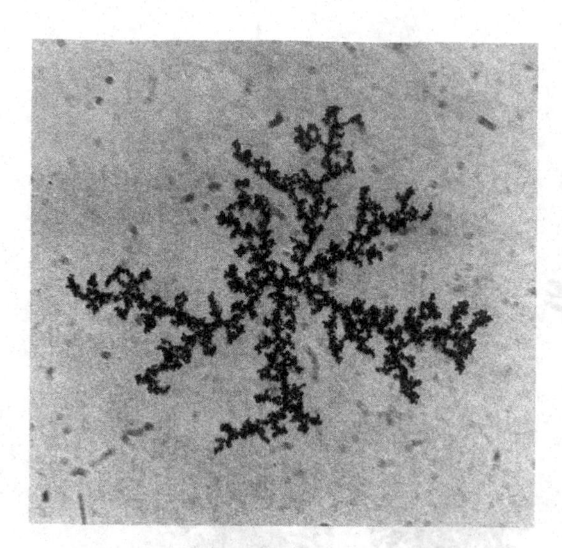

Figure 4.22 A DLA-like displacement pattern produced during the drying of a thin layer of silica spheres contained between two parallel sheets of glass to simulate a pseudo 2-dimensional porous medium. This figure was provided by T. M. Shaw.

nor familiar. However, these processes are of great practical importance and intrinsic interest.

4.5.1 Polymer Solutions

The complexities that can be found in even quite simple fluids are well illustrated by the work of Zhao and Maher [901] in which the displacement of non-viscous fluids (polyethylene oxide solutions) by a Newtonian fluid (water) was studied in a small (4.8 cm radius, 0.4 ± 0.02 or 0.08 ± 0.04 mm gap) radial Hele-Shaw cell. These are miscible fluids, so that the interfacial tension does not introduce a large inner cut-off length. Figure 4.23 illustrates the variety of patterns found in this system.

4.5.2 Colloidal Systems

Colloidal systems can exhibit a variety of rheological properties, including shear thinning, shear thickening and critical stresses that can result in unique applications. Clays constitute one of the most important classes of colloidal materials and have been studied extensively. For example, the crossover from stable Saffman–Taylor fingering to DLA-like displacement patterns has been studied by Van Damme *et al.* [902, 903] for the case of the displacement of a non-Newtonian colloidal fluid (an aqueous clay dispersion) by a low viscosity Newtonian fluid (air) in a Hele-Shaw cell. The experimental conditions were characterized by the control parameter B', given by $1/B' = 12(w/b)^2 C'_a$, where w is the cell width, b is the internal spacing between the cell walls and C'_a is the capillary number given by $C'_a = \mu' V/\Gamma$. Here, μ' is the ratio between the shear stress and the shear strain rate (the "non-Newtonian viscosity"), V is the tip velocity of the invading "finger" and Γ is the interfacial surface tension. For

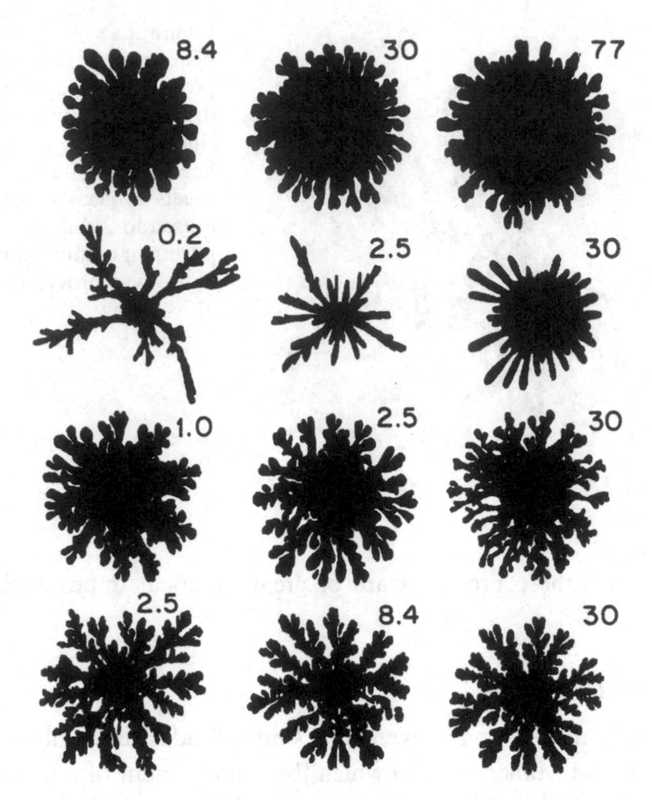

Figure 4.23 Patterns generated by the displacement of non-Newtonian polyethyleneoxide solutions by dyed water in a circular Hele-Shaw cell. In each row, the injection rate increases from left to right. The concentration of polyethyleneoxide increases from top to bottom (0.3 wt%, 0.8%, 5% and 10%, respectively). High molecular weight polymer ($\approx 5 \times 10^6$ daltons) was used in the low concentration experiments shown in the top two rows. Lower molecular weight polymer ($\approx 3 \times 10^5$ daltons) was used in the high concentration experiments shown in the bottom two rows. The injection rate (given in milliliters per minute) is shown by each pattern. This figure was provided by J. V. Maher.

Newtonian fluids, this definition of B' is the same as that given for B in equation 4.13. Figure 4.24 shows the displacement patterns obtained at ten values of the control parameter. The large range of control parameter values was obtained by tuning both the clay particle concentration and the air injection pressure. The smectite clays used in this work are shear thinning and have a non-zero yield stress σ_y. The relationship between the shear stress σ_s and the shear strain rate $\dot{\gamma}$ can be represented quite well by [904, 905]

$$\sigma_s = \sigma_y + b\dot{\gamma}^n, \tag{4.33}$$

with a shear thinning exponent n smaller than unity.

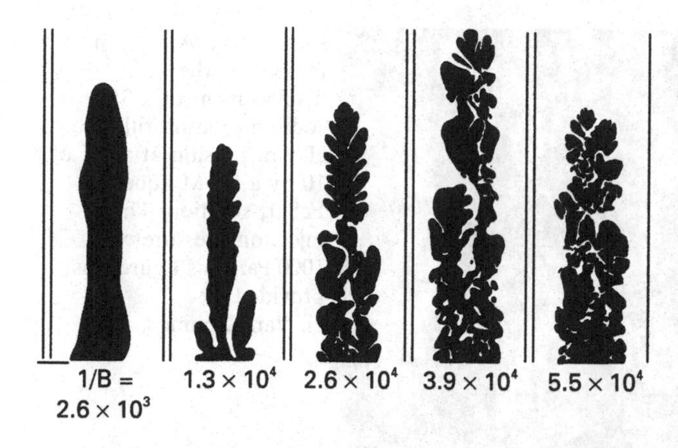

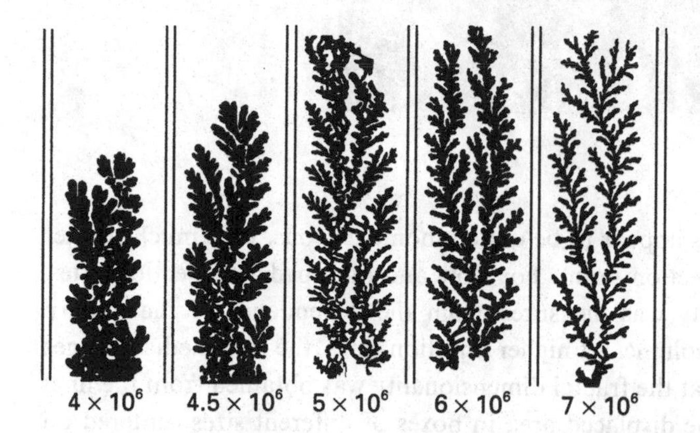

Figure 4.24 Viscous fingering patterns obtained by the injection of air into aqueous clay dispersions in a Hele-Shaw cell. This figure shows the continuous transition from a stable, Saffman–Taylor-like pattern (upper left corner) to an unstable, DLA-like pattern (lower right corner) as the control parameter $1/B'$ was increased through more than four orders of magnitude. This figure was provided by H. Van Damme.

At higher concentrations, clay dispersions also exhibit viscoelastic behavior that can be characterized by the Deborah number De given by

$$De = t_x/t_f, \tag{4.34}$$

where t_x is the characteristic relaxation time for the fluid, which can often be interpreted as a structural relaxation time, and t_f is a characteristic time for the flow process ($1/\dot{\gamma}$, for example). For $De \ll 1$, the dispersion behaves as a viscous fluid, and for $De \gg 1$ the behavior is like that of an elastic solid. The relatively complex rheological behavior leads to a wide variety of pattern-formation phenomena but complicates quantitative interpretation.

Van Damme *et al.* [906] measured effective fractal dimensionalities ranging from 1.3, at very small tip velocities, to 2.0, at high tip velocities, when water was injected into Bentonite clay pastes, with a range of water contents, in a radial Hele-Shaw cell. In these experiments, a constant injection pressure was used. The interface between the injected water and the clay remained sharp during the displacement process. Consequently, the experiments can be described in terms of immiscible displacement with a negligible surface tension. In most

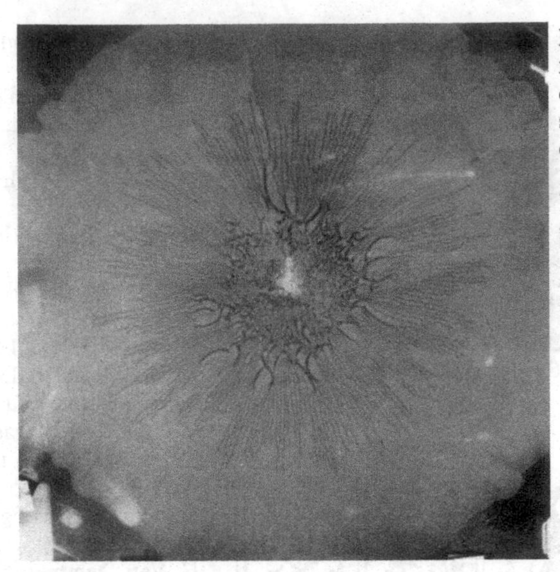

Figure 4.25 A pattern formed by the displacement of a 3% sodium montmorillonite clay dispersion at a pH of 10 by a 0.2 M aqueous $FeSO_4$ solution. The injection pressure was 1000 Pa. This figure was provided by H. Van Damme.

experiments, the patterns appear to be quite inhomogeneous, with much thicker branches near to the injection point. For slow and intermediate injection rates, the fractal dimensionality was measured from the dependence of the pattern radius on the displaced volume. At higher injection rates, the intermediate stages were not recorded, so that the fractal dimensionality was obtained from the final pattern by measuring the displaced area in boxes of different sizes centered on the injection point or by measuring the number of branches $N(\ell)$ that intersect the boundaries of the boxes of size ℓ. In view of the inhomogeneity, the fractal dimensionalities obtained using these approaches may be quite different, and the measurement of a single scaling index does not provide a very satisfactory characterization of these patterns.

Quite different patterns were obtained by Alsac *et al.* [907] on injecting a deaerated solution of ferrous sulfate (typically at a concentration of about 0.2 M) into a sodium montmorillonite clay dispersion raised to a pH of 10 by addition of NaOH. In this system, $Fe(OH)_3$ was precipitated at the interface between the two fluids. The $Fe(OH)_3$ was formed by the reaction of Fe(II) with OH^- in the presence of oxygen dissolved in the clay. Without the precipitation process, DLA-like viscous fingering patterns would be formed. Instead, the complex pattern shown in figure 4.25, consisting of an inner part with a dense network of connected fingers and an outer dense sparsely branched radial pattern, was formed. Although these experiments were not carried out to simulate the behavior of a specific natural phenomenon, it is clear that processes that combine fluid displacement and precipitation are common and thus may be important in areas such as geophysics and geochemistry.

Van Damme *et al.* [908] have also used the injection of water into aqueous clay pastes (Wyoming bentonite, N.L. Industries) to study the effects of the clay viscoelasticity, cell thickness and injection pressure P on the finger width ξ_w of the displacement patterns. They found that $\xi_w \sim \sigma_y^{0.25}$, where σ_y is the yield stress of the clay. In addition, the finger width ξ was found (in common with many other fluid displacement experiments) to increase linearly with the interior cell thickness b, at constant clay concentration and injection pressure. In addition, the experiments indicated that the finger width decreased slightly with increasing injection pressure. However, the experimental data were not accurate enough or extensive enough to establish an empirical relationship between ξ_w and P. Van Damme *et al.* [908] pointed out that their experimental results are not in agreement with the relationship

$$\xi_w = \frac{bP^{1/2}}{(12G)^{1/2}} \tag{4.35}$$

predicted by theoretical models [909, 875] for non-Newtonian fluids with a shear modulus G. Based on the flow equation (equation 4.33)

$$V \sim -\nabla P |\nabla P|^{(1/n)-1}, \tag{4.36}$$

or $|V| \sim |\nabla P|^{1/n}$, Van Damme *et al.* proposed the relationship

$$\xi_w^{(n+1)/n} = \frac{b^2(P - P_\circ)^{1/n}}{12G}, \tag{4.37}$$

where $P_\circ$ is the flow pressure threshold. For the clays used in these experiments, the values of the exponent n in equations 4.33 and 4.36 were found to lie in the range $0.42 \le n \le 0.6$.

4.5.3 Foams

Like clays, foams exhibit unique "history" dependent rheological properties that combine some of the characteristics of both solids (a static shear modulus) and liquids (viscous flow at high shear stress). Park and Durian [910, 911] have studied the displacement of commercial shaving cream foam by nitrogen in radial Hele-Shaw cells, with a range of sizes and plate separations. Figure 4.26 shows displacement patterns obtained at two quite different displacement rates with a gap (plate separation) of $b = 0.104$ cm. In these experiments, the N_2 was injected into circular foam samples prepared by injecting foam into the cell through its central hole until a foam filled region with a radius of R was formed. The experiments were characterized by a typical shear strain rate $\dot{\gamma}$ (the ratio of the finger velocity to the finger width). A morphology transition from a low shear rate pattern (figure 4.26(b)) to a high shear rate pattern (figure 4.26(a)) was found at $\dot{\gamma} \approx 9\ \mathrm{s}^{-1}$. The finger width ξ_w was found to scale as $(R/P)^{1/2}$,

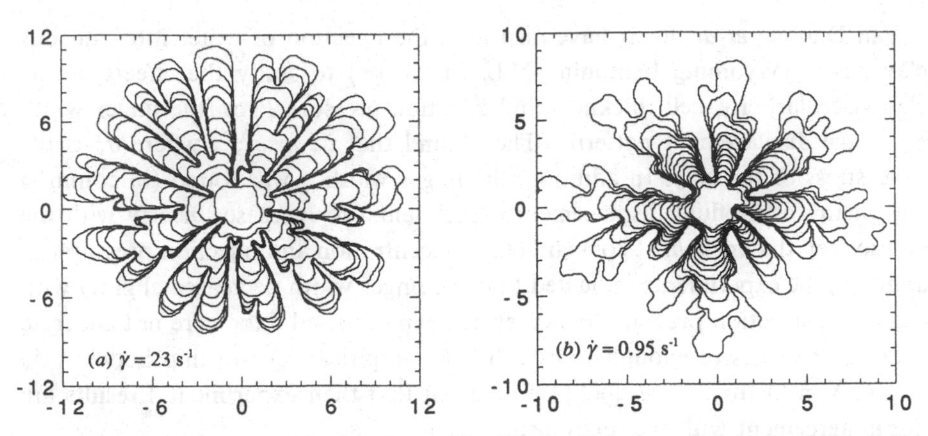

Figure 4.26 The growth of patterns formed by the displacement of foam by N_2 in radial Hele-Shaw cells. Part (a) shows the growing interface at time intervals of 1/60 s for an experiment carried out with a gap of $b = 0.104$ cm and a driving pressure of 10 kPa (the shear strain rate was 23 s^{-1}). Part (b) shows the interface at intervals of 0.5 s for an experiment carried out with a gap of $b = 0.034$ cm and a driving pressure of 6 kPa (the shear strain rate was 0.95 s^{-1}). The initial radii of the foam samples were 13.6 cm and 13.0 cm in parts (a) and (b), respectively. This figure was provided by D. J. Durian.

where P is the constant driving pressure. Park and Durian showed that the finger width could be related to the experimental parameters by

$$\xi_w = (\Gamma R/P)^{1/2}[c + (b/2L_b)] \tag{4.38}$$

or

$$\xi_w(P/\Gamma R)^{1/2} = [c + (b/2L_b)], \tag{4.39}$$

where c is a constant and Γ is the surface tension. Equations 4.38 and 4.39 are based on a linear stability analysis for the fastest growing wavelength λ (chapter 1, section 1.4, and appendix A, section A.1), which gives

$$\lambda = (\Gamma R/P)^{1/2}[\pi(12)^{1/2} + (b/2L_b)]. \tag{4.40}$$

The experimental finger widths can be represented quite well by equations 4.38 to 4.40 with $c = \pi(12)^{1/2}$. The "$b/2L_b$" term in equations 4.39 and 4.40 was obtained empirically.

Effective fractal dimensionalities of $D \approx 7/4$ were measured for both high and low shear strain rate patterns. This illustrates the limitation of the fractal dimensionality alone as a means of describing and distinguishing between quite different patterns.

4.5.4 Fractal Systems

From a practical point of view, fluid–fluid displacement processes taking place in complex porous media such as soils, rocks and biological tissues are of more importance than fluid–fluid displacement phenomena taking place in simple environments. There is a growing realization [57] that the patterns formed during fluid–fluid displacement in natural porous media such as those found in oil reservoirs and aquifers are, to a large extent, dominated by heterogeneities on all scales. Experiments carried out in the laboratory, using random but essentially homogeneous porous media, may provide a poor basis for understanding similar processes in systems that are both random and (extremely) inhomogeneous. A study of fluid–fluid displacement in relatively simple self-similar fractal channel networks provides insight into related phenomena in more complex systems. If the fluids are incompressible, then the pattern formed by the injected fluid is restricted to the "conducting backbone" of the fractal substrate.

For example, Oxaal *et al.* [912] have studied the displacement of a high viscosity, wetting fluid (glycerol) by a low viscosity, non-wetting fluid (air) in a channel network in the form of a 2-dimensional percolating network near to the percolation threshold. The experiments were carried out at both slow and fast injection rates. The displacement patterns obtained at slow injection rates were more compact than those obtained at high injection rates, but the differences were relatively small. This small difference is a consequence of the low order of ramification[7] of the percolation cluster. The fluid must flow through the "red bonds" (chapter 3, section 3.5) to reach the outside, and both patterns are confined to the backbone(s) between the injection point and the exit point(s) where the channel network reaches the edge of the cell. The fast displacement experiments were simulated using a modified dielectric breakdown model, and the slow displacement experiments were modeled using a modified Eden model in which growth was confined to perimeter sites of the growing cluster, on the percolation cluster substrate. As regions of the "defending" fluid become trapped by the growing cluster they are removed from the list of active surface sites (unoccupied perimeter sites) that can be occupied at later stages. The simulations and experiments agree very well in both the slow displacement and fast displacement regimes, but this is, at least in part, a consequence of the restricted geometry of the backbone and the low order of ramification.

Oxaal [913] has also carried out experiments to study immiscible displacement of dyed glycerin/water by air in a 2-dimensional channel network with two types of channels. The experimental model was constructed from a computer generated pattern using photosetting technology with 50% narrow (width w_1) low permeability channels and 50% wide (width w_2) high permeability channels, randomly distributed on the bonds of a 140×140 square lattice array. This means that the distribution of high permeability channels, on the bonds of a square lattice, was at its percolation threshold. The permeability ratio between the two types of channels was given by $R = k_1/k_2 = (w_1/w_2)^4 \approx 4 \times 10^{-4}$ (the

7. The order of ramification of a set is the smallest number of points that must be removed to isolate all the points in a finite (bounded) subset from an infinite set.

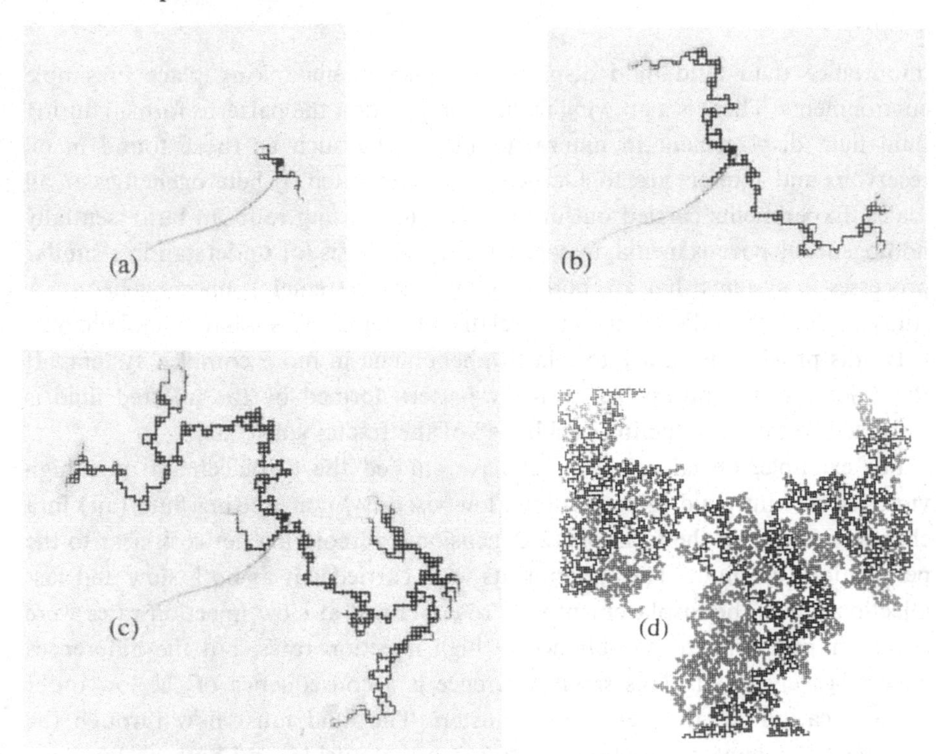

Figure 4.27 Three stages in the miscible, essentially incompressible displacement of glycerin/water by dyed glycerin/water in a 2-dimensional channel network model of a percolation cluster. Part (d) shows a larger scale view of the percolation cluster used to construct the experimental model. The skeleton between the injection point and the perimeter of the cell is shown in black, and the rest of the percolation cluster is represented by open squares. Part (c) was obtained after 337 s, at an injection rate of 0.0648 ml hour^{-1}. This figure was provided by M. Døvle.

channel widths and depths were approximately equal). The experiments were in agreement with patterns generated using a modified DLA model [660] and with the idea that the structure should exhibit a crossover from the structure of DLA on an "infinite" percolation cluster at the percolation threshold ($D \approx 1.30$ [427, 912]) on short length scales ($\epsilon < \ell < \xi(R)$) to ordinary square lattice DLA on longer length scales, $\xi(R) < \ell < L$. Computer simulations suggest that $\xi(R) \sim R^{-a}$, with $a \approx 0.25$ [660]. An effective fractal dimensionality of $D \approx 1.5$ between these limiting values was measured. Related experiments and simulations were carried out on networks for which the fraction p of wide channels was varied near the percolation threshold at $p_c = 1/2$ [914].

Similar effects were seen in miscible displacements carried out using the same type of substrates [915]. Figure 4.27 shows three stages in an experiment in which glycerin/10% water was displaced by the same fluid with $< 0.01\%$ nigrosin dye, in a 2-dimensional percolation channel network.

In general, the displacement patterns are not as strongly confined in most porous media as they are on an incipient infinite percolation cluster. Nevertheless, the heterogeneity plays an important, and often dominating, influence in many fluid–fluid displacement processes, and these simple experiments provide a valuable insight into the origins of this effect.

4.6 Other 2-Dimensional Patterns

A variety of other patterns has been observed in electrochemical deposition and other processes. These include more regular "dendritic" patterns that resemble those produced by crystallization processes, very tenuous "stringy" patterns and compact patterns with rough surfaces. The results of the more extensive studies have frequently been presented in terms of "morphological phase diagrams" [721, 722, 750]. However, the electrochemical deposition is complex, and these phase diagrams should be regarded as cuts through a much higher-dimensionality morphology diagram. In general, different experimental series can be expected to generate different cuts, so that results from these different experiments may be difficult to compare. Additional difficulties arise if the patterns are not homogeneous. This is often a consequence of changing conditions during the growth process. In most, if not all, experiments, some of the important experimental conditions are not controlled. For example, the important role played by small impurity concentrations is often overlooked. To reduce the ambiguities arising from pattern inhomogeneity, Sawada *et al.* [722] based their classification on the nature of the pattern a fixed distance (1.5 cm) from the cathode.

A quite striking example of the formation of a variety of complex fractal and non-fractal patterns in a relatively simple system is provided by the work of Schwartz *et al.* [916] on the growth of octadecyltrichlorosilane (OTS) submonolayers on steam treated mica surfaces. Although atomically smooth cleaved mica has a high energy surface, it is not suitable for the growth of OTS monolayers. The steam treatment of the cleavage plane surfaces makes them suitable for the adsorption experiments. This may be a consequence of the formation of surface hydroxyl groups. The treated mica surfaces were exposed for different time periods to 0.5 mM OTS solutions in bicyclohexyl and studied using atomic force microscopy. After exposure to the OTS solution, the mica discs were rapidly immersed in stirred chloroform to remove any unadsorbed OTS.

Both the entire patterns and the individual clusters (figure 4.28) obtained after short exposure times (low coverages) appear to be fractals over a substantial range of length scales. Fractal dimensionalities of $D \approx 1.68$ and $D \approx 1.79$ were measured for OTS aggregates after 3.1 and 6.6 s of exposure, respectively, using a box counting analysis (chapter 2, section 2.1.3.4). The structures of the aggregates were simulated using a modified DLA model with multiple growth sites, similar to that illustrated in figure 4.10. Particles were deposited randomly onto a square lattice, with multiple growth sites to represent adsorption from the solution. Any particle on the lattice that was not part of the aggregate moved to adjacent unoccupied lattice sites with the same diffusion constant. Unlike the simulations illustrated in figure 4.10, these simulations were not carried out in the $k_{\circ} \gg k_{\downarrow}$ limit, where $k_{\circ}$ is the hopping rate for adsorbed molecules and $k_{\downarrow}$ is the deposition rate. The effective fractal dimensionality increased from $D \approx 1.60$ for very low coverages to $D \approx 1.80$ for coverages of $> 30\%$. The simulation results should be interpreted in terms of a crossover from the fractal dimensionality of (square lattice) DLA at low coverages to $D = 2$ at high coverages, instead of a fractal dimensionality that varies continuously with the coverage (chapter 3, section 3.7.2). In contrast to most studies in which an effective fractal dimensionality is seen to vary with conditions, Schwartz *et al.* [916] *did*, correctly, interpret their results in terms of a crossover. In general terms, the agreement between the simulations and experiments was quite good. However, the simulations do not capture the thickening of the arms of the cluster (figure 4.28) as the exposure time is increased. This may be a consequence of restructuring by mechanisms such as surface diffusion along the margins of the aggregate while it is exposed to the solution. The simulations are in quite good agreement with the observed kinetics, but there is an initial period of slow growth in the simulations in which particles are deposited but have not formed aggregates. This initial slow growth period is not seen in the experiments. The simulations do not address the interesting larger scale structure that is evident in figure 4.28.

A variety of quite different fractal patterns have also been studied in thin films. For example, Cui *et al.* [917] observed the formation of ramified patterns during phase separation of Bi/Fe films. The 30–40 nm thick, 60% Bi/40% Fe films were formed by codeposition, and were placed on Mo transmission electron microscopy (TEM) grids. The films were then thermally annealed in a vacuum or aged at room temperature and examined by TEM. The films consisted of a distribution of Bi-rich clusters with a small cut-off length scale ϵ_1 embedded in an Fe-rich phase. These Bi-rich clusters had a fractal dimensionality of 1.88 ± 0.05. This is close to the fractal dimensionality of 2-dimensional percolation, and the Bi-rich clusters appear to form a pattern corresponding to a system just below the percolation threshold. This pattern formed a background for much larger Bi-rich clusters with a larger cut-off length, ϵ_2. These clusters resembled closely those formed in 2-dimensional cluster–cluster aggregation model simulations [58], but the fractal dimensionality of 1.83 ± 0.05, reported by Cui *et al.*, is much larger than that

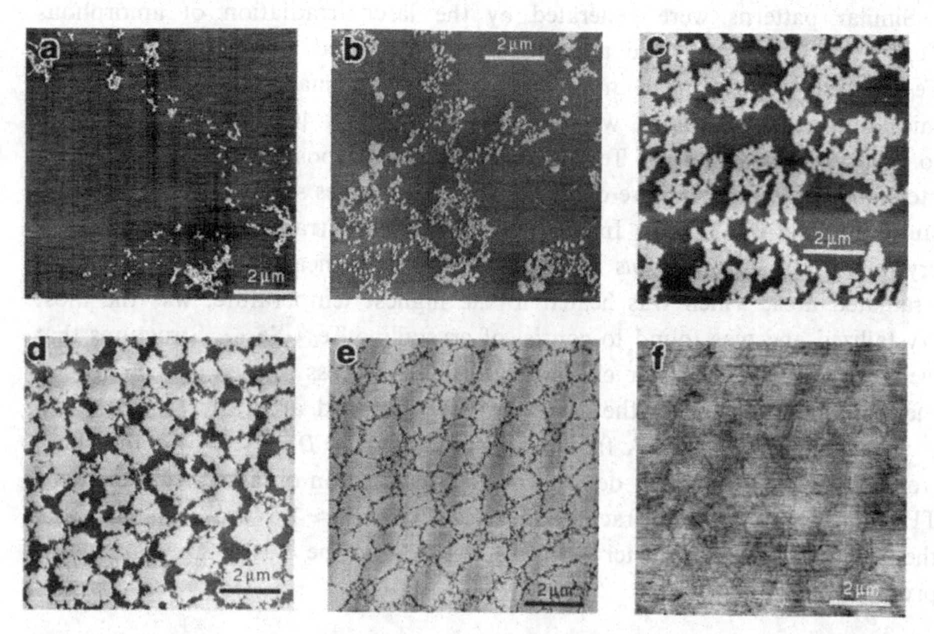

Figure 4.28 10 μm × 10 μm atomic force microscopy images of steam treated mica surfaces exposed to octadecyltrichlorosilane solutions for different times. The exposure times were 0.4, 2, 10, 38, 66 and 304 s in parts (a), (b), (c), (d), (e) and (f), respectively. The white parts of the images represent the regions covered by octadecyltrichlorosilane, in the sub-monolayers. This figure was provided by D. K. Schwartz.

associated with the 2-dimensional diffusion-limited cluster–cluster aggregation model or the 2-dimensional reaction-limited cluster–cluster aggregation model [58].

Fractal clusters that resemble patterns formed by cluster–cluster aggregation [58] were also studied by Carrière *et al.* [918] in gold films deposited on amorphous Al_2O_3 supported on carbon-coated mica substrates. The gold was sputter deposited onto 4 nm thick amorphous Al_2O_3 layers, at a low deposition rate with a substrate temperature of 300 K. The films were then irradiated by a pulsed argon ion laser with a wavelength of λ = 514.5 nm. After irradiation, the Al_2O_3/Au films were floated off and captured on copper TEM grids. After deposition, the gold formed a distribution of drop-like particles with a more or less Gaussian size distribution and a mean diameter of 2.5 nm. After three days in air at room temperature, the mean droplet diameter had increased to about 8.3 nm. The fractal clusters were formed from these gold particles as a result of the laser irradiation (the clusters formed at the edges of the irradiated areas). Fractal dimensionalities of 1.5 – 1.55 were measured via the density–density correlation functions of digitized TEM images.

Similar patterns were generated by the laser irradiation of amorphous Te/Se/Br thin films. The thin films were prepared by rf sputtering of a $Te_{0.3}Se_{0.5}Br_{0.2}$ target in an argon atmosphere onto quartz and carbon-coated mica substrates. The films, with thicknesses of about 40 nm, were determined to have a composition of $Te_{0.45}Se_{0.46}Br_{0.09}$. The deposited films were irradiated, and TEM images were obtained using techniques similar to those used to study the gold films [918]. In this system, the laser irradiation brought about crystallization. After 300 μs irradiation, the region near to the center of the irradiated area, which was heated to the highest temperature, was the most crystallized and was found to consist of crystalline $Te_{0.55}Se_{0.45}$. Structures that were attributed to a cluster–cluster aggregation process formed at the edges of the irradiated area or in the center of the irradiated area after shorter irradiation times ($\approx$ 75 μs). A fractal dimensionality of $D \approx 1.53$ was measured from the two-point density–density correlation function obtained from digitized TEM micrographs, and a fractal dimensionality of $D \approx 1.57$ was obtained from the dependence of the cluster radius of gyration on the cluster size (number of pixels).

4.7 Additional Information

Compared with the large volume of work published on computer modeling, there are few, if any, books or long reviews devoted to the experimental aspects of fractal growth. *The fractal approach to heterogeneous chemistry* [919], edited by Avnir, contains several contributed chapters that are concerned mainly with experiments. Several books, including *Fractals and chaos in geology and geophysics* by Turcotte [920], *Fractal models in the earth sciences* by Korvin [921] and *Fractal geometry and its uses in the earth sciences* [922] edited by Barton and LaPointe are concerned primarily with the applications of fractals to natural phenomena. Relatively few books concerned with applications in the life sciences have appeared. *Fractal physiology* by Bassingthwaighte, Liebowitch and West [413] is a significant exception. Feder's book [327], *Fractals*, is oriented toward the applications of fractal geometry to experiments and natural phenomena.

Chapter 5

The Growth of Surfaces and Interfaces

The structure of many systems can be described in terms of more or less uniform and continuous homogeneous regions separated by interfaces at which physical and/or chemical properties change abruptly. Growing surfaces can evolve into many forms. Flat, faceted, cusped and disorderly surfaces are familiar forms, but surfaces may also develop grooves, solid or hollow whiskers, platelets, dendrites, scales, spirals and other, often intriguing and complex, structures [923]. The problem of understanding the physical processes that control these morphologies poses a major challenge with important practical implications. The kinetic aspects of surface growth is also of great scientific and practical interest.

It has been realized, for a long time, that many processes of considerable practical importance take place at or near disorderly interfaces which cannot be adequately described in terms of simple Euclidean shapes. In addition, transport of materials to and across irregular surfaces and interfaces is important in many applications such as heterogeneous catalysis, electrochemistry and biology. Nevertheless, most fundamental experimental studies have been carried out using specially prepared "smooth" surfaces, and most theoretical models have been concerned with idealized planar interfaces. One of the reasons for this situation has been that the difficulty of describing rough surfaces in quantitative terms has discouraged both experimental and theoretical work. With the wide dissemination of the concepts of fractal geometry [2] and other ideas of modern statistical physics such as scaling and universality, processes occurring near to random rough surfaces, as well as the nature of the surfaces themselves, are now the subject of a broad range of experimental and theoretical studies. This has become an exciting, rapidly developing area of physics that has a wide range of important implications for other areas of science.

A clearer understanding of the growth of surfaces is also important to our understanding of the internal structure of a wide range of objects. Very often, growth processes can be described in terms of the propagation of a growth front or "active zone" (chapter 3, section 3.4) which leaves behind a structure that does not change. For some simple growth models, this is an exact picture and, at least in principle, the internal structure can be understood completely in these terms. In other cases, the picture of growth in terms of a propagating active zone leaving behind a "frozen" structure, that may provide a record of its passage, is at least a good approximation. The growth of diffusion-limited aggregation (DLA) clusters [181] (chapter 3) is an important example.

Computer simulations have played a major role in the development of a better understanding of the formation of rough surfaces under both equilibrium and non-equilibrium conditions. In many cases, computer simulations provide a better means of testing theoretical ideas than experimental studies. In computer simulations, the system can be precisely controlled and unanticipated phenomena can generally be avoided. In addition, computer simulations can be used to explore conditions that are not physically accessible and systems that do not correspond to physical reality, but may be of considerable theoretical interest.

Rough surfaces and interfaces can frequently be described quite well in terms of the concepts of fractal geometry [2] (chapter 2). Many rough surfaces appear to exhibit self-affine scaling [2, 326] over a significant range of length scales. In addition, the scaling properties associated with a surface or interface generated by one process may be very similar to the scaling properties associated with a superficially quite different process. These scaling and universality properties have quite naturally attracted the attention of statistical physicists.

In many important systems, an essentially smooth, flat surface evolves into a rough surface. This takes place in processes such as vapor deposition, corrosion, erosion, wear, and electrodeposition. Very often, the growth of the surface roughness can be described in terms of the correlation lengths $\xi_\perp$ and $\xi_\parallel$, discussed in chapter 2, section 2.7. Here, the perpendicular correlation length $\xi_\perp$ describes the "width" of the surface and the parallel correlation length $\xi_\parallel$ describes the lateral distance over which the surface "height" fluctuations are correlated. The correlation length $\xi_\perp$ is usually defined by $\xi_\perp = w(\ell \gg \xi_\parallel) = \xi_\perp^{(q=2)}$, where

$$\xi_\perp^{(q)} = < |h_i - \overline{h}|^q >^{1/q} . \tag{5.1}$$

In equation 5.1, h_i is the height of the surface in the ith region of size ϵ and $\overline{h}$ is the mean value of the h_i. The heights are measured with respect to the initial ($t = 0$, where t is the growth time) surface or with respect to a flat surface or

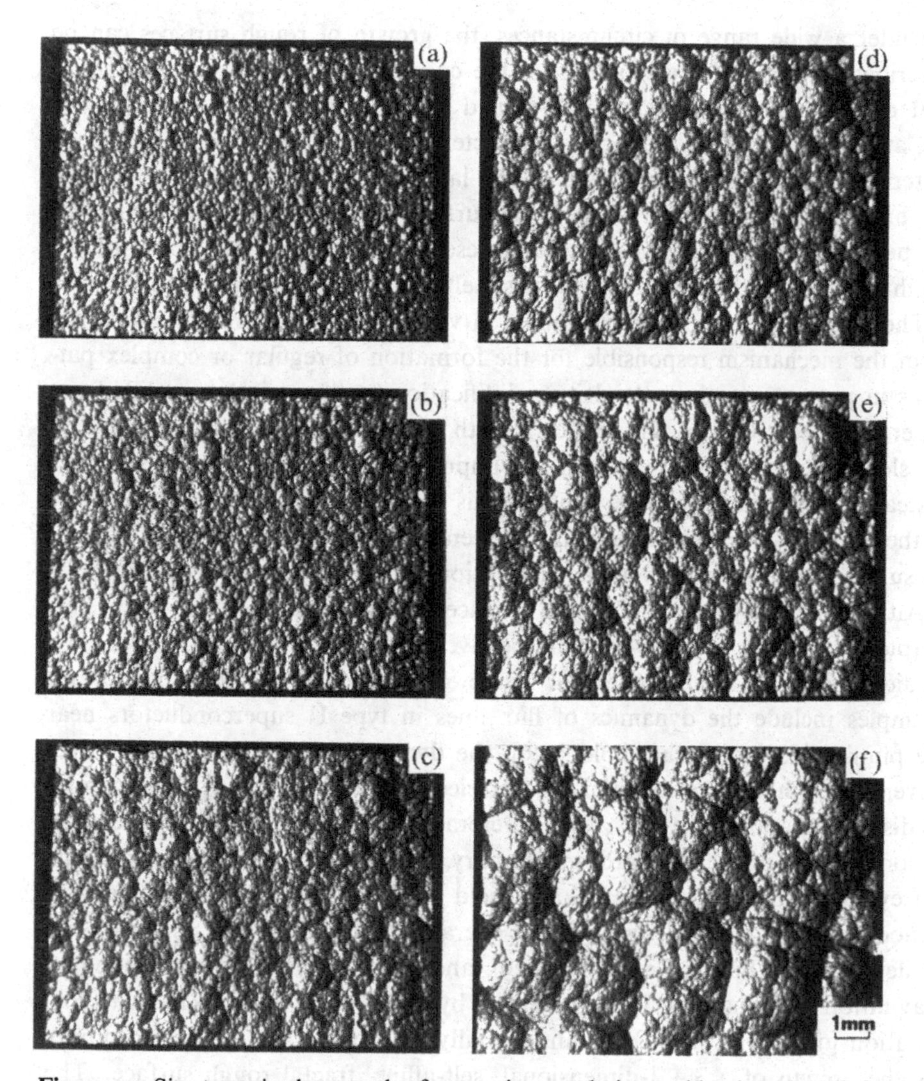

Figure 5.1 Six stages in the growth of a massive pyrolytic graphite deposit. All six parts are shown at the same magnification. The film thicknesses are approximately 1, 4, 11, 14, 20 and 45 mm, in parts (a)–(f), respectively. This figure was provided by R. Messier.

smooth, coarse-grained surface that approximates the rough surface. In most cases, $\xi_\perp^{(2)} \approx \xi_\perp^{(q)} \approx \xi_\perp^{(q')}$, for all positive values of q and q'.

Figure 5.1 shows several stages in the growth of a thick pyrolytic graphite deposit [924]. This figure shows quite clearly the growth of the correlation lengths $\xi_\perp(t)$ and $\xi_\parallel(t)$. The figure also suggests that the surface height fluctuations at different stages in the growth process might be related (statistically) by suitable rescaling of the lengths, in the directions perpendicular and parallel to the plane of the surface.

Under a wide range of circumstances, the growth of rough surfaces can be described in terms of the evolution of the coarse-grained surface height $h(\mathbf{x}, t)$ that can be assumed to be a single-valued function of the lateral coordinates $(\mathbf{x})$, at all times t. Other important characteristics of surfaces can be described in terms of single-valued functions of the lateral position $\mathbf{x}$. Examples include the brightness and color of images, the surface chemical composition and the temperature. These quantities can be represented in terms of functions like the height field $h(\mathbf{x})$, which may also possess self-affine scaling properties [925].

The origins of the surface roughness in vapor deposition are quite different from the mechanism responsible for the formation of regular or complex patterns in processes such as dendritic solidification or the growth of solids from supersaturated solutions. In surface growth, the destabilizing effects related to the slow diffusion of heat, material and impurities in the fluid phase are absent. Instead, the disorderly surface roughness is created by processes such as noise in the deposition process, shadowing (screening), growth velocities that depend on surface inclination, irreversible adsorption and restricted surface diffusion.

Although the dynamics of rough surfaces is the focus of attention in this chapter, a variety of other systems behave in a similar manner and can be studied using the same theoretical framework as rough surfaces. Important examples include the dynamics of flux lines in type II superconductors near the pinning/depinning threshold [926], the dynamics of charge density waves driven through impure materials by an electric field [927] and the dynamics of dislocations and polymers [928]. Berera and Fang [17] have argued that, in some stages of the standard, inflationary, "big bang", model for cosmology, the evolution of the peculiar velocity field (velocity fluctuations relative to a "smooth background" describing the overall expansion of the universe) can be described by Burgers' equation [929], and the growth of the corresponding gravitation potential can be represented by the Kardar–Parisi–Zhang (KPZ) equation [930] (with non-Gaussian, spatially and temporally correlated noise), for the growth of a $3 + 1$-dimensional, self-affine, fractal rough surface. The KPZ equation is discussed later in this chapter.

5.1 The Structure and Growth of Rough Surfaces

The quantitative description of rough surfaces and interfaces has been an important challenge for many years. In a review, Nowicki [931] described 32 parameters and functions that have been used to characterize rough surfaces. Similarly, Klinkenberg [932] has listed 24 morphometric parameters used to characterize the Earth's surface. More recently, it has been realized that fractal geometry and scaling concepts can considerably simplify this task for a quite wide range of systems. A similar approach can also be used as a basis for understanding the growth of rough surfaces. In the progress of science, the ability to describe phenomena in precise, quantitative terms has frequently led to important advances in understanding. This certainly seems to be true in the case of surface growth.

It is important to classify phenomena so that the task of describing and understanding them can be reduced to a reasonable magnitude. Most models for the growth of single-valued self-affine surfaces $h(\mathbf{x}, t)$ can be described in terms of stochastic differential equations, or Langevin equations, and it now appears that a classification scheme based on the concept of universality and the study of Langevin equations for the evolution of rough surfaces and interfaces is taking form. In quite a large number of cases, these equations can be solved analytically to obtain exact values for scaling exponents and sometimes for the amplitudes as well. This is in marked contrast to the situation described for DLA growth processes in chapter 3. Consequently, theoretical advances are a much more important aspect of this chapter than they are for chapter 3. The use of self-affine fractal geometry to describe structures such as surfaces (structures with different scaling properties in different directions) has been described in chapter 2, section 2.7.

In keeping with the theme of emphasizing results that can be obtained using simple scaling theory, without the use of intricate mathematical arguments, the powerful renormalization group methods that are needed to analyze non-linear surface growth equations are not discussed in this book. This is a quite specialized approach that has led to important advances towards the development of a better understanding of surface growth phenomena. However, some of the most important and basic surface growth equations have not yielded to this approach. The more theoretically inclined reader can find information about these methods in the literature [933, 934].

5.1.1 Basic Surface Growth Equations

The growth of the surface or interface is frequently driven by a random process, such as the deposition of particles onto a cold surface. The most simple description for the growth of a rough surface, under these conditions, is provided by the equation

$$\partial h(\mathbf{x}, t)/\partial t = \eta(\mathbf{x}, t), \tag{5.2}$$

where $\eta(\mathbf{x}, t)$ represents a random process controlling the growth of fluctuations in the surface height. In the moving coordinate system, in which $< h(\mathbf{x}, t) >= 0$, the term $\eta(\mathbf{x}, t)$ can often be considered to be a random, uncorrelated Gaussian process with

$$< \eta(\mathbf{x}, t) > = 0 \tag{5.3}$$

and

$$< \eta(\mathbf{x}, t)\eta(\mathbf{x}', t') > = 2\mathscr{D}\delta(\mathbf{x} - \mathbf{x}')\delta(t - t'). \quad (5.4)$$

Equation 5.2 does not provide a realistic description of most surface growth phenomena because it neglects lateral correlations in the growth process, which lead to a smoother surface profile.

Edwards and Wilkinson [935] carried out a theoretical investigation of the growth of rough surfaces via random deposition. They assumed that the flux of particles onto the growing surface was weak, so that correlations between depositing particles could be ignored, and that once a particle had settled on the surface, under the influence of gravity, it did not move when other particles were added above it. These are essentially the same assumptions that are made in simple ballistic deposition models in which the deposited particles follow a path of steepest descent on the surface, after contact with the surface [936, 937, 938] (section 5.2).

Edwards and Wilkinson [935] showed that the growth of the surface, as a result of this process, could be described by the Langevin equation

$$\partial h(\mathbf{x}, t)/\partial t = a\nabla_d^2 h(\mathbf{x}, t) + \eta(\mathbf{x}, t), \quad (5.5)$$

known as the Edwards–Wilkinson or EW equation, where ∇_d^2 is the d-dimensional Laplacian.[1] The term $\eta(\mathbf{x}, t)$ represents the noise driving the growth of the interface, and $\eta(\mathbf{x}, t)$ is assumed to satisfy equations 5.3 and 5.4. Since equation 5.5 is linear, the Fourier amplitudes of the evolving surface can be obtained by Fourier transforming equation 5.5. Using this approach, Edwards and Wilkinson [935] showed that in finite, 2+1-dimensional systems, with a lateral length scale L,

$$\xi_\perp \sim (\log(t))^{1/2}, \quad (5.6)$$

at "short" times t and

$$\xi_\perp \sim (\log(L))^{1/2}, \quad (5.7)$$

at "long" times. Similarly, for the $1 + 1$-dimensional case, it can be shown [939] that the roughness exponent α for the self-affine surface is $\alpha = 1/2$ and that the correlation lengths $\xi_\perp$ and $\xi_\parallel$ grow algebraically ($\xi_\perp \sim t^\beta$ and $\xi_\parallel \sim t^{1/z}$) with the exponents $\beta = 1/4$ and $1/z = \beta/\alpha = 1/2$. In this case, the Hurst exponent H is equal to the roughness exponent α. The definition of the exponents α, β, z and H, as well as the basic properties of self-affine fractals, can be found in chapter 2, section 2.7.2.

1. In the remainder of this chapter, operators like ∇_d will be replaced by ∇; it should be understood that ∇ is the d-dimensional gradient operator.

The exponents α and β can also be obtained for the EW equation (equation 5.5) by exploiting the idea that the growth process is invariant to the transformation

$$\mathbf{x} \rightarrow \lambda \mathbf{x}, \quad (5.8)$$

$$h \rightarrow \lambda^{\alpha} h \quad (5.9)$$

and

$$t \rightarrow \lambda^{z} t. \quad (5.10)$$

The rescaled, or transformed, EW equation has the form

$$\lambda^{\alpha - z} \partial h(\mathbf{x}, t)/\partial t = a\lambda^{\alpha-2}\nabla^2 h(\mathbf{x}, t) + \lambda^{-(d+z)/2}\eta(\mathbf{x}, t). \quad (5.11)$$

The noise term is rescaled by $\lambda^{-(d+z)/2}$ because the uncorrelated noise integrated over a "volume" proportional to $\lambda^d \lambda^z$ is given by $\eta'(\mathbf{x}, t) \sim \lambda^{(d+z)/2}$, so that the noise in the rescaled coordinate system has an amplitude proportional to $\lambda^{(d+z)/2}/\lambda^{(d+z)} \sim \lambda^{-(d+z)/2}$. Equation 5.11 can be written as

$$\partial h(\mathbf{x}, t)/\partial t = a\lambda^{z-2}\nabla^2 h(\mathbf{x}, t) + \lambda^{(z-d-2\alpha)/2}\eta(\mathbf{x}, t). \quad (5.12)$$

Since the rescaled equation should be independent of λ, equation 5.12 implies that $z = 2$ and $\alpha = (2 - d)/2$. If the coefficient a in equation 5.5 is negative, the growth process becomes unstable, the surface is not self-affine and this approach cannot be used.

Kardar *et al.* [287] demonstrated that equation 5.5 does not provide an adequate description of most surface growth phenomena. They proposed that the non-linear Langevin equation

$$\partial h(\mathbf{x}, t)/\partial t = a\nabla^2 h(\mathbf{x}, t) + b(\nabla h(\mathbf{x}, t))^2 + \eta(\mathbf{x}, t) \quad (5.13)$$

is a more satisfactory basis for understanding a variety of surface growth processes. The noise $\eta(\mathbf{x}, t)$ was assumed to satisfy equations 5.3 and 5.4. Equation 5.13 provides a full description (in the long wavelength or hydrodynamic limit) of the growth of the surface fluctuations for a variety of surface growth models, including the ballistic deposition and Eden growth models, and most probably a variety of real surface growth processes as well. However, the KPZ equation (equation 5.13) has proven to be a much less universal equation for real surface growth phenomena than was initially expected on the basis of computer simulations with simple models and theoretical optimism. There is, as yet, no unambiguous experimental demonstration of KPZ growth.

Because of the non-linear term $b(\nabla h(\mathbf{x}, t))^2$ in equation 5.13, the surface growth exponents cannot be determined directly from invariance to the transformation given in equations 5.8, 5.9 and 5.10. For the $1 + 1$-dimensional case, Kardar *et al.* showed that the growth of the surface roughness described by equation 5.13

can be characterized by the exponents $\beta = 1/3$ and $\alpha = 1/2$. For the $2+1$-dimensional case, exact values for these exponents have not yet been obtained. However, large scale computer simulations indicate that $\alpha \approx 2/5$ and $\beta \approx 1/4$. The scaling relationship $\alpha + (\alpha/\beta) = 2$ given in equation 2.183, chapter 2, is satisfied exactly. A range of values for β and α have been obtained by Schwartz and Edwards [940] ($\alpha \approx 0.31$), Bouchaud and Cates [941] ($\alpha \approx 0.26$) and Tu [942] ($\alpha \approx 0.385$), using theoretical methods that give exact results in the $1+1$-dimensional case. The lateral correlation length $\xi_\parallel$ grows as $\xi_\parallel \sim t^{1/z} \sim t^{\beta/\alpha}$. For the $1+1$-dimensional case, $1/z = 2/3$, and for the $2+1$-dimensional case, $1/z \approx 0.625$. In both cases, $1/z > 1/2$, and the growth of $\xi_\parallel$ is often said to be "superdiffusive" for this reason.

The non-linear term $b(\nabla h(\mathbf{x},t))^2$ in equation 5.13 may have different origins in different models [944] or physical processes. For example, in the Eden model this term arises primarily because the velocity $(\partial h/\partial t)$ in the vertical direction of an inclined surface (growing with a constant velocity in a direction parallel to the local normal to the surface) increases with increasing angle of inclination. In the ballistic deposition model, the non-linear term is a result of the fact that the deposit density decreases, and hence the growth velocity increases, with the angle of inclination. By itself, the non-linear term $b(\nabla h(\mathbf{x},t))^2$ would convert a surface consisting of "rounded hills" into a surface of "plateaus" separated by narrow, steep-sided "canyons".

5.1.2 Surface Diffusion

The term "$a\nabla^2 h(\mathbf{x},t)$" on the right-hand side of equation 5.13 does not describe the effects of surface diffusion. It can arise as a result of desorption of adsorbed atoms, a process that is related to the Gibbs–Thompson effect, that plays an important role in dendritic growth in dense fluids (chapter 1, section 1.4). This term can also arise as a result of lateral correlations in the growth process.

In the absence of other processes, the evolution due to surface diffusion of the curved surface of a solid with a uniform density can be described by the equation

$$\partial h(\mathbf{x},t)/\partial t = -\nabla \mathbf{j}_s(\mathbf{x}), \tag{5.14}$$

where $\nabla \mathbf{j}_s(\mathbf{x})$ is the divergence of the d-dimensional surface flux density or current density $\mathbf{j}_s$. Since $\mathbf{j}_s(\mathbf{x}) \sim -\nabla\mu(\mathbf{x})$, where $\mu(\mathbf{x})$ is the chemical potential at $\mathbf{x}$, the surface diffusion transports material from high chemical potential, convex parts of the surface, where the curvature $\mathscr{K}(\mathbf{x})$ is positive, to regions of low chemical potential, where the surface is concave. The chemical potential is given by

$$\mu(\mathbf{x}) - \mu_\circ \sim \mathscr{K}(\mathbf{x}), \tag{5.15}$$

where μ_o is the chemical potential at a flat interface. In the hydrodynamic, long length scale limit, the curvature $\mathcal{K}(\mathbf{x})$ of the surface is given by

$$\mathcal{K}(\mathbf{x}) = -\nabla^2 h(\mathbf{x}). \tag{5.16}$$

Consequently, the evolution of a surface under surface diffusion can be represented by the equation [945, 946, 947]

$$\partial h(\mathbf{x},t)/\partial t = -c\nabla^4 h(\mathbf{x},t) = -\tilde{\mathcal{D}}_s \nabla^4 h(\mathbf{x},t). \tag{5.17}$$

A more detailed analysis indicates that $c = \tilde{\mathcal{D}}_s = \mathcal{D}_s \Gamma v_a^2 \nu / k_b T$ [946, 947], where $\mathcal{D}_s$ is the surface diffusion coefficient, Γ is the surface free energy density, v_a is the atomic volume, ν is the number of atoms per unit area of surface, k_B is the Boltzmann constant and T is the temperature. It follows from equation 5.17 that the growth of a surface subject to random deposition and surface diffusion can be described by [948, 949, 950]

$$\partial h(\mathbf{x},t)/\partial t = -c\nabla^4 h(\mathbf{x},t) + \eta(\mathbf{x},t). \tag{5.18}$$

This equation is often called the Mullins–Herring equation and will be referred to as the "MH equation" in this chapter. Equations 5.17 and 5.18 are appropriate only if there are no other contributions such as those responsible for the linear $\nabla^2 h(\mathbf{x},t)$ or non-linear $(\nabla h(\mathbf{x},t))^2$ terms in the KPZ equation.

The derivation of equation 5.18 neglects the fact that the path length between two points $(\mathbf{x}_1, h(\mathbf{x}_1))$ and $(\mathbf{x}_2, h(\mathbf{x}_2))$ measured along the surface is longer than the distance $|\mathbf{x}_2 - \mathbf{x}_1|$ measured in the lateral coordinate system. Consequently, equation 5.18 is correct only in the small slope, $\nabla h(\mathbf{x},t) \rightarrow 0$, limit. In general, surface diffusion generates additional terms in the Langevin equation of motion. The growth of the surface height and the velocity of the surface in the direction perpendicular to the local surface $V_n(\mathbf{x})$ is given by

$$\partial h(\mathbf{x},t)/\partial t = g^{1/2} V_n(\mathbf{x},t), \tag{5.19}$$

where $g = 1 + (\nabla h(\mathbf{x}))^2$ is the local surface area density (the amount of surface per unit area in the lateral coordinate system). The growth of the surface height $h(\mathbf{x},t)$ as a result of transport on the surface can then be expressed as

$$\partial h(\mathbf{x},t)/\partial t = g^{1/2} V_n(\mathbf{x},t) = -g^{1/2} \nabla_s \mathbf{j}_s(\mathbf{x},t) = -\nabla \mathbf{j}_s, \tag{5.20}$$

where ∇_s and $\mathbf{j}_s$ are defined in the local coordinate system on the surface.

In the case of volume conserving transport by surface diffusion, the surface flux $\mathbf{j}_s$ is given by

$$\mathbf{j}_s(\mathbf{x},t) \sim -\partial\mu(\mathbf{x},t)/\partial\mathcal{S}(\mathbf{x},t) \sim -\nabla_s \mu(\mathbf{x},t) \sim -g^{-1/2} \nabla\mu(\mathbf{x},t), \tag{5.21}$$

where μ is the chemical potential and $\mathcal{S}(\mathbf{x},t)$ is the position in the local "surface" coordinate system, at position $\mathbf{x}$ in the lateral coordinate system. In the $1+1$-dimensional case, $\mathscr{S}$ is the arc length s. It follows from equations 5.20 and 5.21 that

$$\partial h(\mathbf{x},t)/\partial t \sim \nabla g^{-1/2}\nabla\mu(\mathbf{x},t). \tag{5.22}$$

The dependence of the chemical potential on the surface curvature $\mathscr{K}$ is given by equation 5.15. The curvature $\mathscr{K}$ can be expressed in terms of the height field $h(\mathbf{x},t)$ as $\mathscr{K} = -g^{-3/2}\nabla^2 h(\mathbf{x},t)$. Consequently, the rate of growth of the surface, in a direction perpendicular to the coarse-grained reference surface, is given by [946]

$$\partial h(\mathbf{x},t)/\partial t = -\tilde{\mathscr{D}}_s\nabla g^{-1/2}\nabla g^{-3/2}\nabla^2 h(\mathbf{x},t), \tag{5.23}$$

where $\tilde{\mathscr{D}}_s$ is defined below equation 5.17. In the $2+1$-dimensional case, equation 5.23 can be replaced by

$$\partial h(\mathbf{x},t)/\partial t = g^{1/2}\tilde{\mathscr{D}}_s\nabla_{LB}^2\mathscr{K}, \tag{5.24}$$

where $\mathscr{K}$ is the curvature and ∇_{LB}^2 is the Laplacian in the local surface coordinate system, called the Laplace–Beltrami operator, given by [951]

$$(\nabla_{LB}^2)_{ij} = g^{-1/2}\frac{\partial}{\partial x_i}\{g^{1/2}[\delta_{ij} - g^{-1}(\partial h/\partial x_i)(\partial h/\partial x_j)]\}\partial/\partial x_j. \tag{5.25}$$

Consequently,

$$\partial h(\mathbf{x},t)/\partial t = -g^{1/2}\tilde{\mathscr{D}}_s\nabla_{LB}^2 g^{-3/2}\nabla^2 h(\mathbf{x},t), \tag{5.26}$$

which simplifies to equation 5.23 for $d=1$. Krug [637] has shown how this equation can be generalized to include growth under non-equilibrium conditions, with a deposition flux.

The noise $\eta_{\mathscr{D}}(\mathbf{x},t)$ associated with surface diffusion can be expressed as $\eta_{\mathscr{D}}(\mathbf{x},t) = -\nabla_s\mathbf{j}(\mathbf{x},t)$ where $\mathbf{j}(\mathbf{x},t)$ is a random surface flux density. This noise is uncorrelated and is given by the fluctuation–dissipation relationship

$$\eta_{\mathscr{D}}(\mathbf{x},t)\eta_{\mathscr{D}}(\mathbf{x}',t') = -2\tilde{\mathscr{D}}_s\nabla_s^2\delta(\mathbf{x}-\mathbf{x}')\delta(t-t'). \tag{5.27}$$

If equation 5.24 is linearized, equation 5.17 is obtained, and if the effects of diffusion noise are included the equation of motion for surface diffusion can be written as

$$\partial h(\mathbf{x},t)/\partial t = -\tilde{\mathscr{D}}_s\nabla^4 h(\mathbf{x},t) + \eta_{\mathscr{D}}(\mathbf{x},t), \tag{5.28}$$

where the diffusion noise $\eta_{\mathscr{D}}$ is given by equation 5.27. In general, the noise $\eta(\mathbf{x},t)$ will have contributions from both the unconserved deposition noise $\eta_{\varphi}(\mathbf{x},t)$, which satisfies equation 5.4, and conserved diffusion noise $\eta_{\mathscr{D}}(\mathbf{x},t)$, which satisfies equation 5.27. If the noise in equation 5.18 has a purely depositional origin, then invariance under the transformation given in equations 5.8 to 5.10 gives the results

$$z = 4 \tag{5.29}$$

and

$$\alpha = (4 - d)/2. \tag{5.30}$$

The effects of the rescaling transformation given in equations 5.8 to 5.10 on purely conserved diffusional noise $\eta_{\mathscr{D}}(\mathbf{x}, t)$ is given by[2]

$$\eta'_{\mathscr{D}}(\mathbf{x}, t) = \lambda^{-(d+2+z)/2}\eta_{\mathscr{D}}(\mathbf{x}, t). \tag{5.31}$$

Consequently, the rescaled Langevin equation of motion (linearized Langevin equation) can be written as

$$\partial h(\mathbf{x}, t)/\partial t = -c\lambda^{z-4}\nabla^4 h(\mathbf{x}, t) + \lambda^{(z-d-2\alpha-2)/2}\eta(\mathbf{x}, t), \tag{5.32}$$

from which it follows that

$$z = 4 \tag{5.33}$$

and

$$\alpha = (2 - d)/2. \tag{5.34}$$

If conserved (surface diffusion) and non-conserved (depositional) noise are both present, the conserved noise decreases more rapidly than the non-conserved noise under the transformation given in equations 5.8 to 5.10, as the long length scale limit, $\lambda \to \infty$, is approached. The asymptotic behavior will be the same if both types of noise are present or if just depositional noise is present. Consequently, "$\eta(\mathbf{x}, t)$" can be assumed to represent depositional noise unless explicitly stated otherwise in this chapter.

5.1.3 Universality Classes

Equations 5.2, 5.5, 5.13 and 5.18 define universality classes for surface growth models and experiments. Several different approaches can be taken to determine to which universality class a model belongs. The most straightforward is to measure the exponents α, H and β and then to compare them with values obtained analytically or numerically, using simple, well understood models. This method may not be reliable because of a slow approach to the asymptotic behavior. In addition, processes with the same exponents may not belong

2. This result can be obtained using arguments similar to those used to establish the rescaling relationship $\eta'(\mathbf{x}, t) = \lambda^{-(d+z)/2}\eta(\mathbf{x}, t)$ for unconserved depositional noise. In the case of conserved diffusion noise, the random surface currents $j_s(\mathbf{x}, t)$ rescale in the same way as random deposition noise and the diffusion noise $\eta_{\mathscr{D}}(\mathbf{x}, t) = \nabla j_s(\mathbf{x}, t)$ rescales as $\eta'_{\mathscr{D}}(\mathbf{x}, t) = \lambda^{-(d+z+2)/2}\eta_{\mathscr{D}}(\mathbf{x}, t)$.

to the same universality class. For example, $1+1$-dimensional lattice gas simulations of the roughening of immiscible fluid–fluid interfaces [952] lead to the same exponents, $\beta = 1/3$ and $\alpha = 1/2$, as the $1+1$-dimensional KPZ equation for surface growth, but there is no obvious physical or mathematical relationship between these two processes. In other cases, the presence or absence of specific terms in the stochastic differential equation corresponding to the model may be determined theoretically using symmetry, physical arguments or other approaches discussed below.

An important symmetry is invariance to the transformation $\mathbf{x} \to -\mathbf{x}$. This symmetry excludes terms involving odd powers of ∇ such as $\nabla h(\mathbf{x},t)$ and $\nabla^3 h(\mathbf{x},t)$. In general, surface growth is not invariant to inversion about the d-dimensional coarse-grained surface plane ($h \to -h$). In some cases, such as the EW equation, the continuum limit surface growth equation is invariant under $h \to -h$. This symmetry forbids terms such as $[\nabla h(\mathbf{x},t)]^2$. The absence of invariance under $h \to -h$ implies that the direction of propagation is important and that the growth is a genuinely non-equilibrium process. The use of symmetry arguments requires the correct identification of all the relevant symmetries. This is generally based on the assumption that the symmetry of the Langevin equation of motion is the same as that of the microscopic processes in a computer model or physical model. In practice, important symmetries are often overlooked.

Very often, terms that are allowed by symmetry have no influence on the asymptotic ($\ell \to \infty$, $t \to \infty$) scaling behavior. For example, $\nabla^2 h(\mathbf{x},t) \to \lambda^{\alpha-2}\nabla^2 h(\mathbf{x},t)$ and $\nabla^4 h(\mathbf{x},t) \to \lambda^{\alpha-4}\nabla^2 h(\mathbf{x},t)$ under the affine transformation $h(\mathbf{x}) \equiv \lambda^{-\alpha} h(\lambda\mathbf{x})$. Consequently, the term $\nabla^4 h(\mathbf{x},t)$ approaches zero more rapidly than $\nabla^2 h(\mathbf{x},t)$, in the long length scale limit, and a term of the form $-c\nabla^4 h(\mathbf{x},t)$ can be ignored if a term of the form $a\nabla^2 h(\mathbf{x},t)$ is also present. However, the crossover to the asymptotic behavior may be difficult to reach if $c \gg a$. Similarly, a term of the form $e\nabla^2 h(\mathbf{x},t)[\nabla h(\mathbf{x},t)]^2$ is "irrelevant", in a scaling sense, if a term of the form $b\nabla h(\mathbf{x},t)^2$ is also present. In other cases, the mean growth velocity can be measured as a function of surface inclination [953], to demonstrate the presence or absence of terms involving powers of $\nabla h(\mathbf{x},t)$.

Lam and Sander [954] have developed a general approach for determining the form of the evolution equation from simulation results. They pointed out that this equation can be written in the general form

$$\partial h(\mathbf{x},t)/\partial t = \mathbf{a} \cdot \mathbf{Q}(x,t) + \eta(\mathbf{x},t), \tag{5.35}$$

where $\mathbf{a}$ is a vector of coefficients ($a, b, \ldots$ in equations such as equation 5.13) and $\mathbf{Q}(x,t)$ is a vector containing the various derivatives of $h(\mathbf{x},t)$ and powers of these derivatives. The determination of the coefficients $\{a\}$ is then based on measuring the vector $\mathbf{Q}(x,t)$ and the increments $\Delta h(\mathbf{x},t)/\Delta t$ using a discretized time scale, with time increments of Δt, on a coarse-grained growing surface with a characteristic length scale λ. The calculation of the coefficients is then based on minimization of the quantity

$$\mathscr{D}' = \frac{1}{N}\sum_{1=1}^{1=N}[(\Delta h(\mathbf{x},t)/\Delta t) - \mathbf{a}\cdot\mathbf{Q}(x,t)]^2 \tag{5.36}$$

with respect to **a**. For simple examples, Lam and Sander found that the coefficients $\mathbf{a}(\lambda)$ were independent of the scale of coarse graining λ, except for small values of λ. Data from several simulations can be averaged. Coefficients a_n that have very small values are judged to correspond to terms that are absent from the Langevin equation.

This approach seems to work well for simple cases. It is not known if it offers substantial advantages in cases where there is a long transient pseudo-scaling regime before the asymptotic behavior becomes apparent. This type of behavior can occur if a "relevant" term in the Langevin equation of motion has a small coefficient. In such cases, it might be difficult to determine if the small coefficient is zero or not. However, Sander and Ji [955] have used this method to study the slow approach of $1+1$-dimensional surfaces generated by the deterministic Kuramoto–Sivashinsky equation (described below) to a limiting dynamics that can be described by the stochastic KPZ equation. So far, this method has not been used extensively, but it may prove to be valuable in the analysis of experimental data where other, mainly theoretical, approaches may be difficult or impossible.

Another approach which has been used to study a variety of processes is to measure the exponential decay of perturbations as a function of the perturbation wavelengths λ or wave number k [956]. This approach will succeed only for linear Langevin equations. It has been used effectively by Kessler and Orr [957] to demonstrate the presence of a term of the form $a\nabla^2 h(\mathbf{x},t)$ in simple models for molecular beam epitaxy.

Rácz *et al.* [958] and Vvedensky *et al.* [959] have described a different approach to the derivation of Langevin equations of motion for models of processes such as molecular beam epitaxy, in which the Langevin equation is derived from a master equation for the dynamics. In this approach, a lattice Langevin equation is obtained, for each of the column heights, from the master equation. The lattice Langevin equation is then regularized by replacing non-analytical quantities by analytical forms and retaining only the leading-order terms to obtain a continuum Langevin equation for the surface. This approach has the advantage that the sources of the terms in the Langevin equation can be identified and their magnitudes related to the microscopic dynamics. In some cases, terms that are allowed by symmetry may be absent, and this approach allows them to be identified. In the work of Vvedensky *et al.*, the microscopic dynamics was based on a solid-on-solid model combined with Arrhenius rate equations that satisfy detailed balance, while Rácz *et al.* also studied models in which the microscopic dynamics violated detailed balance. Krug *et al.* [960] have recently given a more rigorous derivation of the Langevin equation for surface diffusion using an activated hopping (Arrhenius kinetics) solid-on-solid model.

Another important way of determining the Langevin equation that corresponds to a growth algorithm is to measure the dependence of the growth velocity and/or the net surface flux $\mathbf{j}_s$ on the inclination of the surface [961]. This fruitful approach will be discussed later in this chapter.

For the case of $1+1$-dimensional "KPZ" growth processes, the value of $1/2$ obtained for the roughness exponent α is a consequence of the symmetry of the growth dynamics of 1-dimensional interfaces (Galilean invariance) and a fluctuation–dissipation relationship [962]. Since there is only one independent exponent for "KPZ" interfaces, this requires an "exponent universality" for $1+1$-dimensional KPZ growth processes. However, the universality of the KPZ equation goes beyond this exponent universality [963]. The height difference correlation functions can be written as

$$C_q(r) = A_q r^{H_q}. \tag{5.37}$$

In this equation, the amplitude A_q is given by $A_q = c_q(\mathscr{D}/a)^{1/2}$, where $\mathscr{D}$ is defined in terms of the noise in equation 5.4 and c_q is a universal constant. Similarly, the time dependence of the surface width $\xi_\perp^{(q)}$ (equation 5.1) is given by

$$\xi_\perp^{(q)} = B_q t^{\beta_q}, \tag{5.38}$$

where $B_q = b_q(|b|(\mathscr{D}/a)^2)^{1/3}$ and b_q is also a universal constant [290, 291]. Amar and Family [291] have predicted universal scaling functions and amplitude scaling functions for $2+1$-dimensional KPZ surface growth based-on-scaling analysis and have used mode coupling theory [964] to calculate these quantities. They have also demonstrated these universalities using a variety of $2+1$-dimensional surface growth models.

Amar and Family [965] have also studied surface growth with the Langevin equation

$$\partial h(x,t) = a\nabla^2 h(x,t) + b|[\nabla h(x,t)]|^\gamma + \eta(x,t). \tag{5.39}$$

By numerically integrating the $1+1$-dimensional equation, they found KPZ-like behavior for a wide range of values for the exponent γ ($1/2 \le \gamma \le 4$) in the above equation. These studies and similar work on the $1+1$-dimensional Kuramoto–Sivashinsky equation (described below) demonstrate that observation of KPZ exponents does not imply an underlying KPZ equation for the surface growth process. Instead, the surface growth process may belong to a much larger class of equations that "renormalize" to the KPZ equation in the hydrodynamic limit, or at least have KPZ exponents in the hydrodynamic limit.

5.1.4 Exponent Scaling Relationships

An important property of the KPZ equation (equation 5.13) is its invariance to the "Galilean" transformation

$$h \to h + \delta\mathbf{r} \cdot \mathbf{x} \tag{5.40}$$

and

$$\mathbf{x} \to \mathbf{x} + b\delta\mathbf{r}t, \tag{5.41}$$

where $|\delta\mathbf{r}| \to 0$. This corresponds to tilting the surface by a small angle. This symmetry leads to the scaling relation

$$\alpha + z = 2, \tag{5.42}$$

expressed in equation 2.183.

The motivation for use of the term "Galilean" transformation to describe equations 5.40 and 5.41 is not very transparent. However, it reflects the theoretical interests of the physicists who pioneered the use of the KPZ equation for surface growth phenomena and acknowledges an important link with other areas of physics. The KPZ equation is related to the Burgers' equation [929]

$$\partial\mathbf{V}/\partial t = a\nabla^2\mathbf{V} - b\mathbf{V} \cdot \nabla\mathbf{V} - \nabla\eta(\mathbf{x}, t), \tag{5.43}$$

for a vorticity free velocity field $\mathbf{V}$, via the transformation

$$\mathbf{V} = -\nabla h(\mathbf{x}, t). \tag{5.44}$$

The "noisy" Burgers' equation is used to describe vorticity-free fluid dynamics [966] and is invariant to the Galilean transformation

$$\mathbf{V}(\mathbf{x}, t) - \mathbf{V}_\circ \to \mathbf{V}'((\mathbf{x} - \mathbf{V}_\circ t), t). \tag{5.45}$$

It follows from equation 5.44 that the classical Galilean transformation of the dynamics of a fluid described by Burgers' equation to a moving coordinate system corresponds to transformation of the dynamics of a surface growth process described by the KPZ equation to a rotated coordinate system (to a tilted surface).

A simple, physically motivated derivation of equation 5.42 can be found in the review of Krug and Spohn [555]. They point out that a growing rough surface at time t will contain "bulges" of all sizes up to bulges with a breadth of $\xi_\parallel(t)$ and a height of $\xi_\perp(t)$. Equation 5.42 was then obtained by focusing attention on the growth of one of the largest bulges. Assuming that the direction of growth is normal to the surface, and idealizing the side of the bulge by a plane with a small inclination of $\xi_\perp(t)/\xi_\parallel(t)$, the rate of growth of the bulge width $\xi_\parallel(t)$ is proportional to the horizontal projection of the normal growth velocity, which is proportional to the angle of inclination. This can be expressed as

$$d(\xi_\parallel)/dt \sim \xi_\perp/\xi_\parallel \sim \xi_\parallel^{\alpha-1}. \tag{5.46}$$

Consequently, $\xi_{\parallel}(t) \sim t^{1/(2-\alpha)}$, so that $1/(2-\alpha) = 1/z$ (from equation 2.165) or $\alpha + z = 2$. This derivation of equation 5.42 rests on the assumption that the direction of growth is normal to the surface. This is not generally true, but Krug and Spohn [555] showed that equation 5.46 can be generalized to

$$d(\xi_{\parallel})/dt \sim (\xi_{\perp}/\xi_{\parallel})^{a-1}, \tag{5.47}$$

so that

$$z = a + \alpha(1-a), \tag{5.48}$$

where a is the exponent that describes the dependence of the growth velocity v on $\nabla h(\mathbf{x})$ ($v = \partial h/\partial t = v(0) + C|\nabla h|^a$, where C is a constant).

Application of the transformation given in equations 5.8, 5.9 and 5.10 to the KPZ equation gives

$$\partial h(\mathbf{x},t)/\partial t = a\lambda^{z-2}\nabla^2 h(\mathbf{x},t) + b\lambda^{\alpha+z-2}\nabla h(\mathbf{x},t)^2 + \lambda^{(z-d-2\alpha)/2}\eta(\mathbf{x},t). \tag{5.49}$$

Unfortunately, the invariance of this rescaled equation to the transformation cannot be used to calculate the exponents α and z, because it is not possible to find values for z and α that will allow equation 5.49 to be written in a form that does not depend on λ. However, it turns out that the coefficient b of the non-linear term is constant, leading immediately to equation 5.42.

Other surface growth equations are invariant to similar transformations, leading to different exponent scaling relationships. For example, Wolf and Villain [967] proposed that

$$2\alpha = z - d \tag{5.50}$$

for any process described by the surface diffusion equation 5.14 for $d \le d_c$, where d_c is the critical dimensionality, beyond which $\alpha = 0$. Here, the exponent α is defined by the relationship $\xi_{\perp} \sim \xi_{\parallel}^{\alpha}$, and $\alpha = H$ in most cases. The derivation of equation 5.50 is based on the idea that the fluctuation in the amount of material deposited into a region of the surface of size $\xi_{\parallel}$ in a small fixed time interval is proportional to $\xi_{\parallel}^{d/2}$, and that the corresponding height fluctuation in the same region is proportional to $\xi_{\parallel}^{-d/2}$. Since these fluctuations are statistically independent of earlier fluctuations, the increase in $\xi_{\perp}^2$ will be proportional to $\xi_{\parallel}^{-d}$, or

$$d\xi_{\perp}^2/dt \sim \xi_{\parallel}^{-d} \sim t^{-d/z}. \tag{5.51}$$

It follows from this equation and the relationship $\xi_{\perp} \sim t^{\beta} \sim t^{\alpha/z}$ that $d\xi_{\perp}^2/dt \sim t^{(2\alpha/z)-1}$. The terms in equation 5.51 must scale with the same power of t, and this requires that $(2\alpha/z) - 1 = -d/z$, or $2\alpha = z - d$. This simple argument does not work for processes described by the KPZ equation because the growth velocity fluctuations depend on the surface slope and there are correlations between fluctuations at different times, mediated by the slope.

Sun *et al.* [968] have studied the 1+1-dimensional surface growth equation

$$\partial h(\mathbf{x},t)/\partial t = -\nabla^2[c\nabla^2 h(\mathbf{x},t) + e(\nabla h(\mathbf{x},t))^2] + \eta(\mathbf{x},t). \quad (5.52)$$

In this model, it is assumed that the surface evolution is a process that conserves the volume of the deposit, so that the noise $\eta(\mathbf{x},t)$ has the conservative form $\eta(\mathbf{x},t) = \eta_{\mathscr{D}}(\mathbf{x},t)$, where

$$< \eta_{\mathscr{D}}(\mathbf{x},t)\eta_{\mathscr{D}}(\mathbf{x}',t') > = -\nabla^2\delta^d(\mathbf{x}-\mathbf{x}')\delta(t-t'). \quad (5.53)$$

This equation is invariant to the transformation

$$h \rightarrow h + \mathbf{r}\cdot\mathbf{x} \quad (5.54)$$

and

$$\mathbf{x} \rightarrow \mathbf{x} - bt\mathbf{r}\nabla^2, \quad (5.55)$$

where $\mathbf{r}$ is a constant vector. This symmetry leads to the exact exponent scaling relationship

$$\alpha + z = 4. \quad (5.56)$$

Sun *et al.* developed a 1+1-dimensional model to simulate equation 5.52, and this model gave the effective exponents $\beta = 0.091 \pm 0.002$ and $\alpha = 0.35 \pm 0.03$, in good agreement with the theoretical values for β ($\beta = 1/11$), obtained from a renormalization group analysis, and $\alpha = 1/3$, from this value for β with equation 5.56. However, Rácz *et al.* [969] have shown that this model is invariant under the transformation $h(\mathbf{x},t) \rightarrow -h(\mathbf{x},t)$ and cannot generate the $\nabla^2(\nabla h(\mathbf{x},t))^2$ term. Consequently, the apparent agreement between the simulations and theoretical results remains to be explained. The noise $\eta(\mathbf{x},t)$ plays an important role in the growth of rough surfaces, and the scaling properties of the evolving surface can be changed by introducing correlations into the "noise" $\eta(\mathbf{x},t)$ [970] or by replacing the Gaussian noise with non-Gaussian noise [971]. Models of this nature are described below.

5.1.5 The Kuramoto–Sivashinsky Equation

The Kuramoto–Sivashinsky equation

$$\partial h(\mathbf{x},t)/\partial t = -\nabla^2 h(\mathbf{x},t) - \nabla^4 h(\mathbf{x},t) + (\nabla h(\mathbf{x},t))^2 \quad (5.57)$$

has been used to describe the dynamics of flame fronts and other processes involving the propagation of interfaces [139, 140, 972]. On short length scales, a cusped pattern is formed, but on long length scales, the interface generated by this equation is a disorderly, self-affine fractal. Yakhot [973] suggested that the scaling properties of the Kuramoto–Sivashinsky equation should be the same, in the hydrodynamic limit, as those for the stochastic KPZ equation, and a similar conclusion was reached by Zaleski [974]. This idea is supported by more recent theoretical work [963]. Asymptotic KPZ-like behavior has been confirmed numerically by Sneppen *et al.* [975] for the $1+1$-dimensional case. A similar conclusion has been reached by Hayot *et al.* [976]. There is a long intermediate scaling regime that makes it difficult to reach the asymptotic regime, and this led to the opposite conclusion in earlier numerical work. A detailed understanding of the way in which the deterministic, $1+1$-dimensional Kuramoto–Sivashinsky equation becomes equivalent to the stochastic KPZ equation, in the hydrodynamic limit, has been developed by L'vov and Procaccia [977].

A quite different conclusion has been reached by Procaccia *et al.* [978] for the $2+1$-dimensional Kuramoto–Sivashinsky equation. Computer simulations carried out by Procaccia *et al.* indicated that the roughness exponent α is 0 and that the behavior is more like that associated with the $2+1$-dimensional EW equation (equation 5.5) than the KPZ equation. This conclusion is supported by a theoretical analysis of the Kuramoto–Sivashinsky equation. It has now been shown by L'vov *et al.* [977, 979, 980] that the Kuramoto–Sivashinsky equation is equivalent to the KPZ equation for $d < 2$, where the non-linear term $(\nabla h(\mathbf{x}, t))^2$ is dominant in both equations, but that at $d = 2$, where the linear terms remain relevant, the solutions of these equations may differ. For $d > 2$, the dynamical exponent z has a value of 2 for the Kuramoto–Sivashinsky equation in the strong coupling regime. While the behavior of the $2+1$-dimensional Kuramoto–Sivashinsky and KPZ models remains controversial [981, 982, 983] and more work must be done, there now appears to be a good understanding of the relationship between the Kuramoto–Sivashinsky and KPZ equations.

Anisotropic versions of the Kuramoto–Sivashinsky equation, with and without added shot noise, have been used to describe processes such as sputter erosion and epitaxial growth [984, 985]. Some anisotropic Kuramoto–Sivashinsky equations generate ripple patterns similar to those observed in some sputter erosion experiments. These ripple patterns become unstable at late times.

5.2 Simple Models

Simple algorithmic models have played a central role in the development of an increased understanding of surface growth phenomena. This enterprise began more than 30 years ago, when digital computers were in their infancy. With the increased power, ease of use and availability of computers, combined with a decreased cost in real terms, this is still a rapidly growing area. For some of the most simple surface growth models, it is now possible to deposit 10^{12} (approximately $N_\circ^{1/2}$, where $N_\circ$ is the Avogadro number) sites in a single simulation. The development of simple models for disorderly growth phenomena has stimulated considerable theoretical work, which has now advanced much further, in the case of the surface growth models described in this section, than in the case of DLA.

5.2.1 Eden Growth Models

The Eden model was one of the earliest growth models studied on the computer [348, 349]. This model generates compact structures with self-affine surfaces. For the Eden model, the width of the active zone ξ_a, described in chapter 3, section 3.4.5, provides a measure of the surface width. Plischke and Rácz [402, 403] found that the dependence of $\xi_a(s)$ on the cluster size s could be represented by the algebraic form $\xi_a(s) \sim s^{\overline{\nu}}$, with a value of 0.18 ± 0.03 for the exponent $\overline{\nu}$. Since the cluster is compact, the mean radius $\overline{r}(s)$, or the radius of the active zone, grows with an exponent of 1/2, so that $\xi_a(s) \sim (\overline{r}(s))^{2\overline{\nu}}$ or $\xi_a(\overline{r}) \sim \overline{r}^{\beta}$ with $\beta \approx 0.36$. Subsequent work demonstrated that it is difficult to obtain reliable values for the exponent β in this fashion because the clusters do not have perfectly circular shapes (chapter 3, sections 3.4.1 and 3.5.6). The most reliable values for β are obtained from simulations of growth from a line with periodic boundary conditions in the lateral directions [288]. However, the work of Plischke and Rácz did indicate that the surfaces of Eden model clusters have interesting scaling properties, despite their rather boring internal structure.

In many surface growth models, accurate measurement of the scaling exponents α, β and H can be compromised by a large, non-scaling, intrinsic width ($\langle w_i \rangle$ equation 2.182, chapter 2). Noise reduced growth models, like the noise reduced Eden model [295, 296, 300], described in chapter 3, section 3.5.6, have been used extensively to reduce the intrinsic width in surface growth simulations. It is widely believed that noise reduction accelerates the approach to the asymptotic scaling limit, without changing the scaling exponents. However, this idea has not been rigorously justified for all growth models, and must be used cautiously. It appears that noise reduction does not change the universality of ballistic deposition and Eden growth models. This can now be understood, since noise reduction changes the coefficients of the terms in the KPZ equation that describes the growth of the surfaces, but does not change the basic structure of the Langevin equation. Care in the selection of the noise reduction parameter (m) must be exercised, since a very large value of m will substantially reduce the magnitude (amplitude) of $\xi_\perp$ and the simulations will require a large amount of computer time, proportional to m, to fill a particular number of sites. This will make it more difficult to measure the roughness exponent α, the Hurst exponent H and the exponent β that describes the algebraic growth of $\xi_\perp$.

Using the noise reduced Eden model, Wolf and Kertész [296] obtained values of 0.33 ± 0.01 and 0.24 ± 0.02 for the roughness exponent α from 2+1-dimensional and 3 + 1-dimensional Eden model simulations, respectively. They found that the exponents α and z satisfied the theoretical exponent scaling relationship

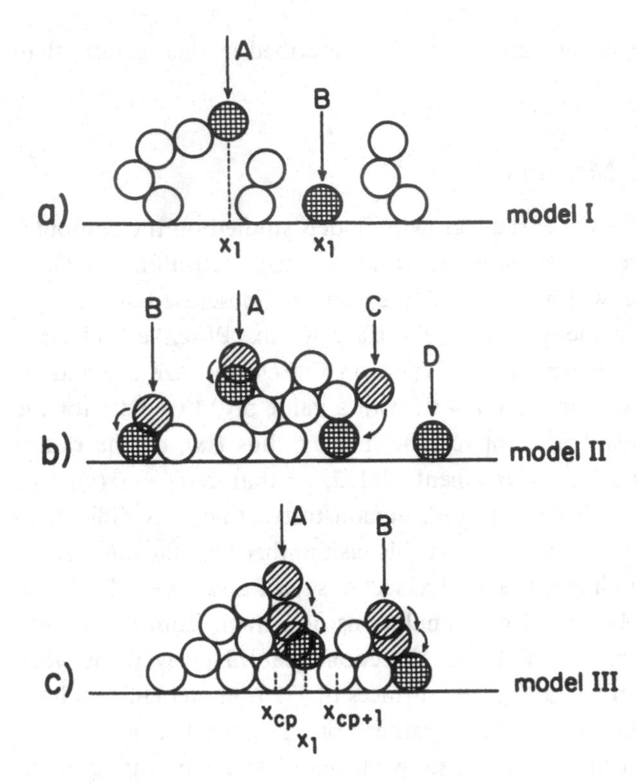

Figure 5.2
1 + 1-Dimensional ballistic deposition models. Part (a) shows the simplest version of this model, with sticking at the position of first contact. Parts (b) and (c) show models in which the deposited particles follow paths of steepest descent until they simultaneously contact two particles in the deposit (b) or reach a local minimum on the surface of the deposit (c). In all cases, the particles become part of the growing deposit at their position of first contact with the substrate, if the substrate is reached. This figure was provided by R. Jullien.

$\alpha + z = 2$, given in equation 5.42, quite well, and that $\beta = 0.24 \pm 0.02$ for $d = 2$ and $\beta = 0.146 \pm 0.025$ for $d = 3$. Devillard and Stanley [986] have obtained the result $\alpha = 0.4 \pm 0.06$ for the 2 + 1-dimensional case, without noise reduction. This leads to a value of 0.25 ± 0.05 for β from equation 5.42.

5.2.2 Ballistic Deposition Models

Off-lattice ballistic deposition models are based on the ballistic aggregation model described in chapter 3, section 3.4.2. In the most simple of these off lattice models, particles (discs for the 1 + 1-dimensional case and spheres for the 2 + 1-dimensional case) are deposited, one at a time, via randomly selected vertical trajectories, onto a horizontal substrate. The deposited particles become part of the growing deposit as soon as they either contact a particle previously incorporated into the growing deposit or contact the substrate. Figure 5.2(a) illustrates the 1 + 1-dimensional version of this model, and figure 5.3 shows a small deposit generated using this model.

The first simulations of ballistic deposition were carried out by M. J. Vold [414, 417, 418, 987], who was interested in the structure of sediments formed by the deposition of small particles from a colloidal dispersion. These were very small scale simulations in which, at most, 160 particles were deposited, and

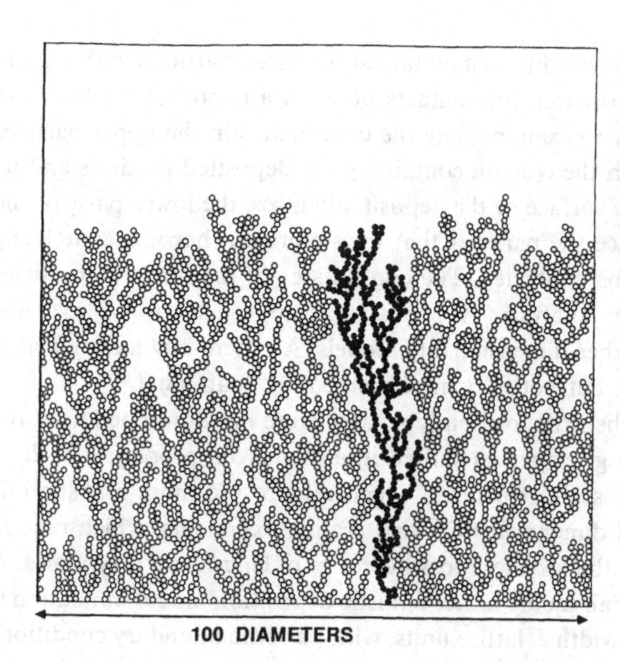

Figure 5.3 A small, $1+1$-dimensional deposit grown using the off-lattice ballistic deposition model illustrated in figure 5.2(a). A "tree" of connected particles is identified by filled circles. The growth of the upper surface (highest occupied point at each lateral position) belongs to the KPZ universality class.

Vold did not use periodic boundary conditions to reduce finite-size effects. A density of about 0.125 was obtained for the central "core" region of the deposit. Much larger scale simulations ($\approx 10^8$ particles) carried out using an essentially identical model gave a packing density of 0.1465 ± 0.003 [938]. This result is in good agreement with independent simulations [988]. Vold studied a variety of modifications to the basic ballistic deposition model, in order to represent effects such as attractive interactions and restructuring after the initial contact between a sphere and the deposit had occured. Some of these models have been revisited by others, in recent years, to explore both the internal structure and surface geometry of granular deposits. Vold [418] also carried out simulations with non-spherical particles.

During the 1970s, several $1+1$-dimensional and $2+1$-dimensional off-lattice ballistic deposition models were developed to represent the growth of thin films deposited at low pressures [989, 990, 991, 992]. The simplest ballistic deposition models lead to structures that have unrealistically low densities. Consequently, simple restructuring mechanisms were included in most of these models. Much of the work carried out during this period was concerned with the "columnar" morphology that is characteristic of many deposited structures, particularly those grown at non-normal angles of incidence.

In order to explore the asymptotic scaling structure associated with off-lattice ballistic deposition, it is important to design efficient algorithms. In the $1+1$-dimensional case, this can be accomplished by subdividing the deposit into columns with a width of one disc diameter. A list of particle coordinates, ordered according to their heights, is maintained for those particles with centers lying in each column. To search for contacts between a deposited particle and the deposit it is necessary to examine only the coordinates in the upper parts of those lists associated with the column containing the deposited particles and its nearest neighbors. As the surface of the deposit advances, the lower parts of the lists can be removed, since the particles that they represent become completely screened and will never be contacted. The coordinates of the removed particles can be stored for further analysis. A very similar approach can be used for 2+1-dimensional and higher-dimensionality models. A potentially more efficient algorithm was described, but not fully implemented, by Joag [993].

Lattice models for ballistic deposition [292, 465] were not developed until the growing interest in scaling properties shifted attention from properties such as density and short length scale structures, which require the more realistic off-lattice models, to fractal dimensionalities and scaling exponents, which require simple models to explore the broadest possible range of length and time scales. A standard, 1+1-dimensional, square lattice ballistic deposition model simulation is carried out on a strip of width L lattice units, with periodic boundary conditions in the lateral (x) direction. Columns of the lattice are selected randomly and, after the ith column has been selected, the site at height $h_i' = max(h_{i-1}, h_i+1, h_{i+1})$, in the ith column, is filled. Here, h_i is the height (y coordinate) of the highest occupied site in the ith column. The simulation is started by filling all of the L_x lattice sites at height $h = 0$. This model is fast and requires little storage, since only the height of the highest filled site in each column needs to be stored to continue the growth process. Because growth can occur by "particles" sticking to the sides as well as the tops of previously deposited parts of the structure, fluctuations in the surface height propagate in the lateral as well as in the vertical direction. This sidewards spreading is responsible for the growth of horizontal correlations and the establishment of the horizontal correlation length $\xi_\parallel$. In this case, the horizontal correlations spread the height fluctuations in the horizontal direction(s) and reduce the fluctuations in the vertical direction, so that $\beta < 1/2$. This model is illustrated in figure 5.4. In some cases, it is more convenient to use the height of the active zone h_i^a in each column. In this version of the model, a column of the lattice (column i) is selected at random. The active zone site in that column is then "filled", so that the new active zone heights in columns $i-1, i$ and $i+1$ are given by

$$(h_i^a)' = h_i^a + 1, \tag{5.58}$$

$$(h_{i-1}^a)' = max(h_{i-1}^a, h_i^a) \tag{5.59}$$

and

$$(h_{i+1}^a)' = max(h_{i+1}^a, h_i^a). \tag{5.60}$$

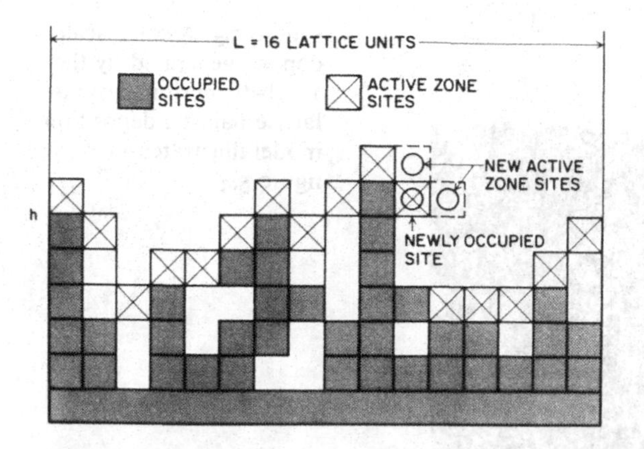

Figure 5.4 The $1+1$-dimensional, square lattice ballistic deposition model. The previously occupied sites are shaded and the active zone sites are labeled "×". If the active zone site labeled "⊗" is filled, then the site labeled "O" will become part of the new active zone and the active zone sites at lower levels in the columns occupied by these sites will cease to be part of the active zone.

Very similar models can be used to simulate ballistic deposition in $d+1$ dimensions, with $d > 1$.

Figure 5.5 shows a "deposit" generated by a small scale, $1+1$-dimensional, square lattice ballistic deposition model simulation. It is apparent from this figure, and from a more careful analysis of deposits generated using much larger scale simulations, that the structure has a quite low average density but is uniform on all but quite small length scales. These features are related to the formation of quite large "steps" between the maximum occupied heights of adjacent columns. Computer simulations show that the distribution of step heights $N(\delta h)$, where $N(\delta h)$ is the number of steps of size δh lattice units and $\delta h = |h_i - h_{i-1}|$, has the asymptotic form

$$N(\delta h) = e^{-k\delta h}, \tag{5.61}$$

where the "decay constant" k has a value of about 0.39 [288]. Because of the exponentially decaying form of $N(\delta h)$, the large steps do not change the asymptotic scaling properties associated with the growing rough surface. However, they do lead to a quite large value for the intrinsic width (w_i in equation 2.182) and the concomitant large corrections to scaling.

In the $1+1$-dimensional, single-step model [288], the simulation is started by occupying columns on a square lattice with coordinates $2i$, where i is an integer, to a height of one lattice unit, while the columns with lateral position $2i+1$ have initial heights of zero. The simulation proceeds by randomly selecting columns with heights lower than those of their two nearest neighbors and increasing their heights by *two* lattice units. This procedure ensures that the height difference between adjacent columns is always $\pm$ one lattice unit. Consequently, there are no large "steps" on the surface of the deposit, and corrections to the asymptotic scaling properties are minimized. This model is illustrated in figure 5.6.

In the $2+1$-dimensional, cubic lattice version of this model, the initial configuration consists of a checkerboard-like pattern, with "black" columns

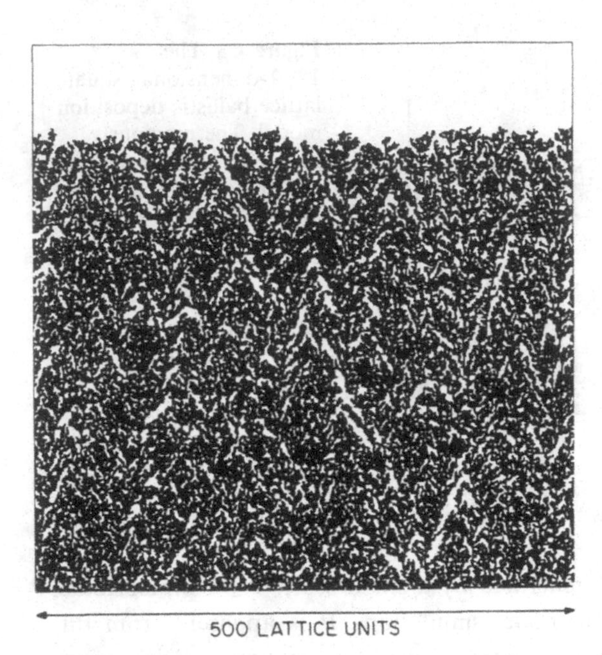

Figure 5.5 A small scale deposit generated by the $1+1$-dimensional, square lattice ballistic deposition model illustrated in figure 5.4.

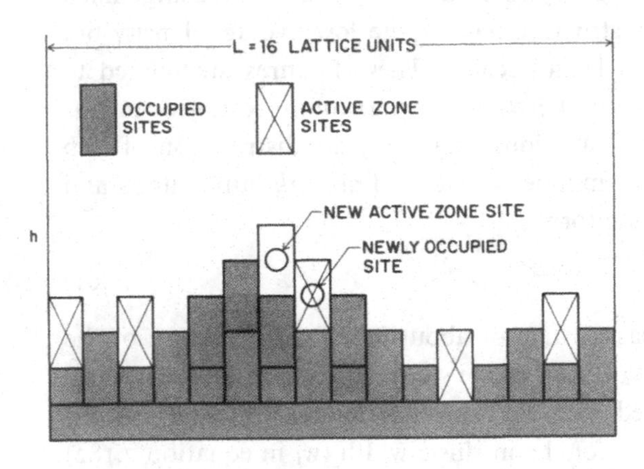

Figure 5.6 The $1+1$-dimensional "single-step" ballistic deposition model. The previously occupied sites are shaded and the active zone sites, which can be filled during the next stage in the growth process, are labeled "×". If the active zone site labeled "⊗" is filled, then the site labeled "O" will become part of the new active zone.

having a height of one lattice unit and "white" columns having a height of zero ($h(i,j)=1$ if $i+j$ is even and $h(i,j)=0$ if $i+j$ is odd). Columns for which all four nearest neighbors have a greater height are selected randomly and their heights are increased by two lattice units. The efficiency of these simulations can be improved by maintaining, and continuously updating, a list of "active zone" sites (sites that can be filled according to the algorithm) and randomly selecting sites (columns) from this list. Figure 5.7 (see color plate section) shows rough surfaces generated using this model.

The single-step model is an example of a "solid-on-solid" model, discussed below. In these models, the deposit density does not depend on the inclination of the surface. Instead, the non-linear $b\nabla(h(\mathbf{x},t))^2$ term in the KPZ equation

(equation 5.13) arises because the density of active sites in which growth may occur depends on the inclination of the surface.

A similar model, called the restricted step height solid-on-solid, ballistic deposition model, was introduced by Kim and Kosterlitz [297]. In the restricted step height model, columns of a hypercubic lattice are selected at random and their heights are increased by one lattice unit. However, if a growth event would generate a "step" between the heights of adjacent columns greater than a fixed amount, chosen at the start of the simulation, the "violating" growth event does not take place and a new column is randomly selected.

Park *et al.* [994] have studied a $1+1$-dimensional, restricted step height, ballistic deposition model in which particles are deposited onto an inclined surface using the standard square lattice ballistic deposition model rules, with the restriction $h_{i+1} - h_i \geq 0$. At the start of a simulation, the surface forms a staircase with $h_{i+1} - h_i = 1$ for all i. The growth of the fluctuations about the mean inclined surface can be described in terms of the exponents $\alpha = 1/2$ and $\beta = 1/2$. Thus this model does not lie in either the KPZ or EW universality classes.

5.2.3 Solid-on-Solid Models

In the standard ballistic deposition models a porous structure that is uniform on all but very short length scales, with a self-affine surface, is generated. In solid-on-solid models [995], new sites are always grown on top of old sites, so that there are no overhangs in the advancing surface, or holes in the structure that is left behind the advancing active zone, at any stage in the growth process. These models are more amenable to a detailed theoretical analysis, and provide a more realistic representation of processes such as molecular beam epitaxy, than the standard ballistic deposition models. The most simple of these models is the independent column model [939, 996], which can be represented by equations 5.2 to 5.4. In this model, columns of the lattice are selected at random and their heights are incremented by one lattice unit. For this model, the exponent β in equation 2.163 has a value of $1/2$ for all d, but there are no correlations in the lateral direction(s). Consequently, the surface width $w(\ell, t)$, which usually depends on the width ℓ over which it is measured or the system size L, is a function of time only ($w(\ell_1, t) \equiv w(\ell_2, t)$). If, instead, the height of the lowest columns at positions $i-1$, i or $i+1$ is incremented when the ith column is selected, in the $1+1$-dimensional case, then the exponent β is found to have a value of exactly $1/4$ and the exponent α has a value of $1/2$ [939, 996]. In this model, the column that is to be incremented can be selected randomly from those columns at positions $i-1$, i or $i+1$ with the lowest heights, if two or more of them have equal lowest heights. As the exponents suggest, this model belongs to the EW universality class, described by equation 5.5. Similar results were obtained from simulations in which the column to be incremented was chosen randomly from the columns at positions $i-n$, $i-(n-1)$, ..., i, ..., $i+(n-1)$, $i+n$, where n is any small integer.

Figure 5.8 shows two equivalent, simple examples of 1+1-dimensional ballistic deposition models that also belong to the EW universality class. When a "particle" is deposited on the growing surface, it follows a downward path on the surface until a local minimum is found. In the model shown in figure 5.8(a), the deposited particles are represented by two vertically oriented nearest-neighbor filled sites on a square lattice. The model is similar to the process of building a wall with vertical bricks. If one or more of the columns that are the nearest neighbors to the column on which a particle has been randomly deposited have lower heights, then the particle is transferred to the top of a lower nearest neighbor and ties are broken by random selection. This transfer process continues until a local minimum is reached, and the particle becomes part of the growing deposit. The transfer process leads to a surface current given by $\mathbf{j}(\mathbf{x}) \sim -\nabla h(\mathbf{x})$ for small slopes $|\nabla h(\mathbf{x})|$, and the contribution of this current to the surface growth process can be written in the form $\partial h(\mathbf{x},t)/\partial t \sim -\nabla \mathbf{j} \sim \nabla^2 h(\mathbf{x})$ in the surface growth equation.

Chan and Liang [997] have introduced a 1 + 1-dimensional model for "self-organized surfaces" in which columns grow independently, as in the models of Weeks *et al.* [996] and Family [939], described above. However, if the height of one of the columns exceeds that of one of its nearest neighbors by Δh lattice units, then $N = \Delta h - 1$ sites "topple" from the "tall" column onto N of its nearest neighbors, so that, if $h(i) - h(i+1) \geq \Delta h$, then $h(i) \to h(i) - N$ and $h(j) \to h(j) + 1$ for $j = i+1$ to $i+N$. Similarly, if $h(i) - h(i-1) \geq \Delta h$, then $h(i) \to h(i) - N$ and $h(j) \to h(j) + 1$ for $j = i - N$ to $i - 1$. If both $h(i) - h(i-1) \geq \Delta h$ and $h(i) - h(i+1) \geq \Delta h$, then the direction in which the top part of the tall column will topple is selected randomly. Here, Δh is a small integer that is a parameter in the model. Chan and Liang [997] did not carry out particularly large scale simulations. They measured a roughness exponent of about 0.47 and a value of about 0.2 for β. From this it appears likely that this model belongs to the EW universality class defined by equation 5.5, with $H = \alpha = 1/2$ and $\beta = 1/4$, for the 1 + 1-dimensional case. Krug and Socolar [998] have argued persuasively that the model of Chan and Liang should belong to the EW universality class. In particular, they point out that the model does not generate a non-linear $(\nabla h(\mathbf{x},t))^2$ term.

In the $d+1$-dimensional "hypercube stacking model" [999, 1000], $d+1$-dimensional hypercubes are added to or removed from a lattice subject to solid-on-solid and step height restrictions, along a selected direction on the hypercubic lattice. For example, if the (111) direction is selected, only the (001), (010) and (100) faces of the cube may be exposed, and the model consists of randomly adding and removing cubes that allow only these faces to be

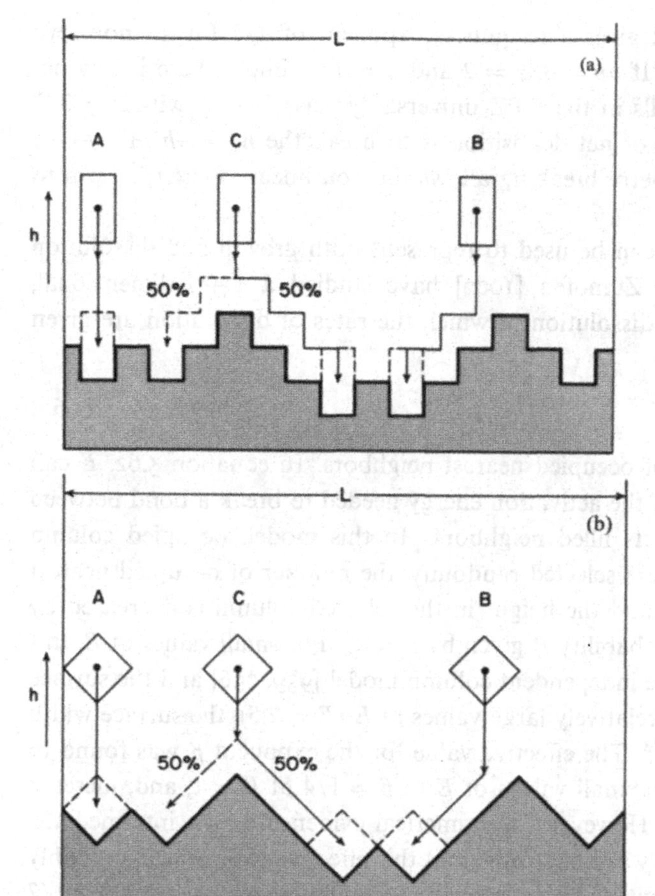

Figure 5.8 Two simple solid-on-solid models that belong to the $1+1$-dimensional EW universality class.

exposed. The $1+1$-dimensional, pure deposition version is equivalent to the "single-step" model described above, but for $d \geq 2$ both the substrate structure and the step height restrictions are different. The hypercubic stacking model is simple enough to allow large scale simulations and, in the pure deposition mode, belongs to the KPZ universality class. Values of 0.240 ± 0.001 and 0.180 ± 0.005 have been obtained for the exponent β from very large scale $2+1$-dimensional and $3+1$-dimensional simulations, respectively [1000]. These results appear to rule out the conjecture that $\beta(d) = 1/(d+2)$ [297]. But it is not unusual for the uncertainties in the exponents obtained from computer simulations to be underestimated by a factor of 10 or more. In general, both the growth rate p^+ and evaporation rate p^- can be varied. In the limit $p^+ = p^-$, the non-linear term in the corresponding KPZ equation disappears, and the strength of this term can be adjusted by tuning $p^+ - p^-$. In the $1+1$-dimensional case, sites may be added to the bottoms of "valleys", as in the "single-step" model, or removed from "hill" tops. The restriction to step heights of ± 1 is maintained at all times. This model gives a roughness exponent of 1/2 for all non-zero values of $\delta p = p^+ - p^-$. If $\delta p = 0$, $z = 2$ and $\beta = 1/4$, but if there is any net deposition, the model falls in the KPZ universality class [1001], with $z = 3/2$ and $\beta = 1/3$. The effect of net deposition is to break the $h \to -h$ or $t \to -t$ symmetry, and the symmetry breaking allows the non-linear $[\nabla h(\mathbf{x}, t)]^2$ term to appear.

Solid-on-solid models can be used to represent both growth and dissolution processes. Poupart and Zumofen [1002] have studied a 1 + 1-dimensional, solid-on-solid model for dissolution in which the rates of dissolution are given by

$$k_n \sim e^{-nE} = CK^n, \tag{5.62}$$

where n is the number of occupied nearest neighbors. In equation 5.62, E can be thought of in terms of the activation energy needed to break a bond between a filled site and one of its filled neighbors. In this model, occupied column tops on a square lattice are selected randomly, the number of occupied nearest neighbors is determined and the height in the selected column is decreased by one lattice unit with a probability P given by $P = k_n$. For small values of E, this model is equivalent to the independent column model [939, 996] and the surface width grows as $t^{1/2}$. For relatively large values of E ($E = 2.5$), the surface width was found to grow as $t^{1/3}$. The effective value for the exponent β was found to decrease from $\beta \approx 1/2$ at small values of E to $\beta \approx 1/4$ at $E \approx 1$, and, increase to $\beta \approx 1/3$ for large E. However, the numerical evidence for an intermediate, $\xi_\perp \sim t^{1/4}$, scaling regime was not strong, and this effective exponent is probably part of the crossover from $t^{1/2}$ to $t^{1/3}$ growth. A roughness exponent of $\alpha \approx 1/2$ was found for all E.

A 2+1-dimensional version of this model was used by Fernandes *et al.* [1003] as a model for anodic dissolution. In this model, the dissolution rates are given by

$$k_n \sim min[1, e^{(V-nE)}] \sim min[1, CK^n], \tag{5.63}$$

where the parameter V accounts for the biasing of the reaction rates due to the potential. In this model, the asymptotic width $\xi_\perp(\infty)$ scales as $L^{0.28\pm0.2}$, where L is the lateral system size. The dependence of the asymptotic width on the model parameters L and V can be represented by $\xi(t = \infty, L, V) = L^{0.28\pm0.2} f(V)$, where $f(V)$ depends linearly on V, but with quite different slopes for $V < V_c$ and $V > V_c$, where $V_c \approx 4.0$ in units where $k_B T = 1$.

5.2.4 The Polynuclear Growth Model

The polynuclear growth model [1004, 1005, 1006] is a solid-on-solid growth model in which "islands" grow laterally on an initially smooth surface. In this model, the islands all have the same thickness, and they grow at a constant velocity. New islands are nucleated at a constant rate on either the original substrate or on an island that has already started to grow. Islands grow only in the lateral direction(s) and stop growing when all of their external perimeter sites are occupied by other islands at the same level. Because the islands grow at the same velocities, overhangs are not generated. A variety of discretized (lattice) versions of this model have been developed [555, 1004, 1007, 1008], and Krug *et al.* [290, 1009] have shown that this model belongs to the KPZ universality class. This result has been confirmed [1010] by numerical solution of a recursion relationship between the coverages in adjacent layers, which leads to the estimates $\beta = 0.33 \pm 0.03$ and $\beta = 0.27 \pm 0.03$ for the $1+1$-dimensional and $2+1$-dimensional models, respectively.

5.2.5 Directed Polymers

In the most simple directed walk model, a "particle" takes a random step in a transverse direction in the d-dimensional lateral coordinate system, each time its position in the longitudinal direction, or time, is incremented by one lattice unit. For example, in a simple $1+1$-dimensional model, the t coordinate is increased by one lattice unit, while the x coordinate changes by 0, $+1$ or -1 during each step. This model is very similar to the discretized Brownian process $B(t)$ described in chapter 2, section 2.7. In general,

$$< |\mathbf{x}(t)| > \sim t^{\nu} \sim t^{1/2}, \tag{5.64}$$

for all d. If the directed walk takes place on a lattice with quenched disorder, then the configurations of the directed polymer under the influence of the disordered medium can be treated as an equilibrium statistical mechanics problem. In one version of the directed polymer model [1011], an energy E_i, selected randomly from a uniform distribution over the range $0-1$, is assigned to each of the bonds in the longitudinal direction and an energy γ is assigned to each of the bonds in the transverse direction(s). For this problem, the exponent relating $< |\mathbf{x}(t)| >$ to t, in equation 5.64, changes.

Using a "zero temperature" model, in which the polymer always adopts the lowest energy configuration starting from position $\mathbf{x} = \mathbf{0}$ at time $t = 0$, Kardar and Zhang [1011] found that

$$x(t) \sim t^{\nu} \sim t^{2/3}, \tag{5.65}$$

for the $d = 1$ case. They also found that the energy difference ΔE between the lowest energy path and that constrained to terminate at the point $(\mathbf{x} = \mathbf{0}, t)$ grew algebraically with increasing path length

$$< \Delta E(t) > \sim t^{\omega} \sim t^{1/3}, \tag{5.66}$$

where $< \ldots >$ indicates averaging over a large number of trials with different quenched, random disorder. These values for the 1+1-dimensional directed walk have also been derived by Perlsman and Schwartz [1012], using a theoretical approach based on an ultrametric tree model for directed walks.

Kardar and Zhang [1011] found that the evolution (with increasing path length t) of the overall Boltzmann weight $W(\mathbf{x}, t)$, for all polymers between the points $(\mathbf{0}, 0)$ and $(\mathbf{x}, t)$, could be described in terms of the evolution equation

$$\partial W / \partial t = \gamma \nabla^2 W + \eta_q(\mathbf{x}, t) W, \tag{5.67}$$

where $\eta_q(\mathbf{x}, t)$ is a random process, uncorrelated in space and time, that represents the effects of the quenched disorder. This allowed an important relationship to be established [1011] between the directed polymer model and the KPZ model for surface growth by using the non-linear transformation

$$W(\mathbf{x}, t) = e^{\lambda h(\mathbf{x}, t)/2\gamma}, \tag{5.68}$$

which converts equation 5.67 into

$$\partial h(\mathbf{x}, t)/\partial t = \gamma \nabla^2 h(\mathbf{x}, t) + (\lambda/2)[\nabla h(\mathbf{x}, t)]^2 + (2\gamma/\lambda)\eta_q(\mathbf{x}, t), \tag{5.69}$$

which is the KPZ equation (equation 5.13). This means that the free energy ($\sim \log(W(\mathbf{x}, t))$ in the directed polymer problem plays the same role as the height $h(\mathbf{x}, t)$ in the surface growth problem. Consequently, the exponent ω in equation 5.66 can be identified with β in the surface growth problem, and ν, in equation 5.65, is equivalent to $1/z$.

The choice of whether to work with a surface growth or directed polymer model to simulate a KPZ growth process then becomes a matter of preference or convenience. Kim *et al.* [1013, 1014] have studied a directed walk model in which $|\mathbf{x}(t) - \mathbf{x}(t+1)|$ is 0 or 1 with a bias factor or "bending factor" γ_b against steps with $|\mathbf{x}(t) - \mathbf{x}(t+1)|$ equal to unity. Here, $\mathbf{x}$ is the position of a site in the $d+1$-dimensional transverse coordinate system and t is the position in the direction in which the polymer is directed. The polymer energy at zero temperature (the minimum energy for all possible paths) for a 1+1-dimensional chain ending at (x, t) is given by the recursive equation

$$\begin{aligned} E(x,t) \quad = \quad & min[\{E(x,t-1) + \eta_q(x,t-1)\}, \\ & \{E(x+1,t-1) + \eta_q(x+1,t-1) + \gamma_b\}, \\ & \{E(x-1,t-1) + \eta_q(x-1,t-1) + \gamma_b\}], \end{aligned} \tag{5.70}$$

where $\eta_q(x, t)$ is the random potential assigned to each site (x, t). Similar equations can be used for $d > 1$. The random quenched site potentials $\eta_q(x, t)$

are usually selected from a uniform distribution over the unit interval or from a Gaussian distribution [1013]. The polymer energy $E(\mathbf{x},t)$ then behaves in the same way as the height $h(\mathbf{x},t)$ in a surface growth model, and the results can be expressed directly in terms of $h(\mathbf{x},t)$ [1014, 1015]. The simulations were carried out using lattices of width L with periodic boundary conditions in the lateral "x" direction and were analyzed using the Family–Vicsek scaling form given in equation 2.177, where $w = \xi_\perp = \langle (h - \langle h \rangle)^2 \rangle^{1/2} = \langle (E(t) - \langle E(t) \rangle)^2 \rangle^{1/2} = L^\alpha f(t/L^z)$. In this fashion, Kim *et al.* obtained results similar to those obtained earlier from surface growth models. For the 2 + 1-dimensional model, they obtained a value of 0.24 ± 0.02 for β and about 0.40 for the roughness exponent α.

The close relationship between the Eden growth process and directed polymers has been known for some time (references can be found in a review by Krug and Spohn [555]). It has also been discussed by Roux *et al.* [1016]. The mapping is based on the idea that an unoccupied perimeter site becomes "active" in the Eden growth process when one of its nearest neighbors becomes filled. Consequently, each site in the Eden cluster or deposit has a "parent" (the neighbor that first became filled). This means that each filled site is connected to the origin or substrate by a unique path with a "time" delay τ_i associated with each step in the path or each site in the path (the time delay between a site becoming active and being filled). The "growth time" t_i for the ith site is then the minimum of the sum of the time delays along any path linking that site to the origin. In this optimal path, each site is connected to its parent, starting with the ith site until the origin is reached. This path is not rigorously directed. However, it is directed on large length scales and the Eden model can be modified to allow only paths that are directed at every step, without changing its scaling properties. The total time t_i is analogous to the minimum energy for a polymer starting at the origin and reaching the ith site at position $\mathbf{x}$. The time t_i corresponds to the minimum energy $E(\mathbf{x},t)$ in equation 5.70, and the delays τ', associated with advances in the "forward" direction, correspond to $\eta_q(i-1)$ or $\eta_q(i-1) + \gamma_b$, depending on the direction of the step from the $(i-1)$th site to the ith site. In the case of growth from a flat substrate, the width of the energy distribution corresponds to the width of the distribution of times at which the various parts of the growth front reach a fixed height. A more precise mapping exists between the single-step ballistic deposition model, described above, and the directed polymer process. In the single-step model, the "growth paths" are rigorously directed on all scales. The mapping becomes particularly simple if the single-step ballistic deposition process is carried out using a line in the (11) direction on a square lattice as the substrate. In this case, all the steps in a path are inclined and there is no need to distinguish between steps in the forward and transverse directions. A clear exposition of this mapping has been presented by Krug and Tang [1017].

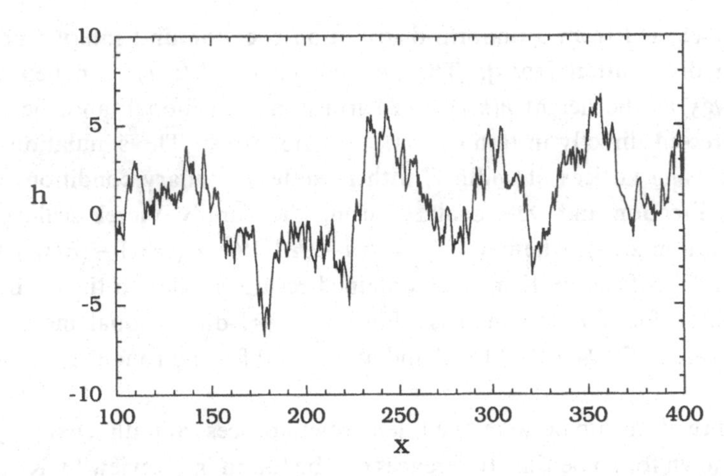

Figure 5.9 Part of a rough "surface" obtained by numerical integration of the $1+1$-dimensional KPZ equation with the parameters $a = 5.0$, $b = 0.65$ and $\mathscr{D} = 0.5$. The "surface" had grown for 100 time units with time increments of $\delta t = 0.005$. The fluctuations about the mean height $< h(t = 100) > = 172$ are shown. The "surface" was represented by 2048 points in a strip of width 1024. This figure was provided by J. Amar and F. Family.

5.2.6 Langevin Dynamics Simulations

An alternative to carrying out a simulation is to numerically integrate the corresponding Langevin equation for the evolution of the surface profile $h(\mathbf{x}, t)$ or to formulate the problem in terms of the Langevin equation alone, and then solve it numerically. This approach has been applied extensively to the $1+1$-dimensional and $2+1$-dimensional KPZ equations [1018]. Figure 5.9 shows the interface obtained by a numerical integration of the $1+1$-dimensional KPZ equation. The values obtained for the exponents α, β and H were in good agreement with large scale lattice model simulations and analytical results.

Numerical integration of the KPZ equation and closely related Langevin equations has been carried out during the course of several other investigations of surface growth models. For example, Honda and Toyoshima [1019] have numerically integrated the Langevin equation

$$\partial\tilde{h}(\mathbf{x}, t)/\partial t = a\nabla^2\tilde{h}(\mathbf{x}, t) + b(\nabla\tilde{h}(\mathbf{x}, t))^2 + k\tilde{h}(\mathbf{x}, t) + \eta(\mathbf{x}, t), \tag{5.71}$$

where $\tilde{h}(\mathbf{x}, t) = h(\mathbf{x}, t) - < h(\mathbf{x}, t) >$ and $\tilde{h}(\mathbf{x}, t)$ is the height in the moving frame. This equation is like the KPZ equation, equation 5.13, except for the third term $(k\tilde{h}(\mathbf{x}, t))$ on the right-hand side. The addition of this term leads to an "unstable" growth process with pronounced cusps in the growing surface. As the surface advances, cusps coalesce and the number of cusps decreases exponentially with a "rate constant" that is proportional to the coefficient k in equation 5.71.

5.2.7 Directed Percolation

A variety of models, more or less closely related to directed percolation (chapter 3, section 3.5.4) [542], have been invoked to explain the "anomalously high" effective roughness exponents found in some interface propagation experiments. Near to the directed percolation threshold, a roughness exponent of $\alpha = \nu_\perp/\nu_\parallel \approx 0.63$ is expected, where $\nu_\perp$ and $\nu_\parallel$ are the directed percolation exponents discussed in chapter 3, section 3.5.4. For example, Buldyrev *et al.* [1020] have interpreted their experiments on $1+1$-dimensional wetting fronts in paper in terms of a simple model that they associate with the directed percolation model.

Tang and Leschhorn [1021] have described a solid-on-solid surface growth model, based on directed percolation. To simulate the advance of the interface, one of the columns with height $h(i)$ at lateral position i is selected at random. If $h(i)$ is greater than either $h(i-1)$ or $h(i+1)$, by at least two units, the height of the lower of the two columns (at lateral position $i+1$ or $i-1$) is increased by one unit. If both of these columns have the same height ($h(i) \geq h(i-1)+2$ and $h(i-1) = h(i+1)$), one of them is selected randomly and its height is incremented. If $h(i) < min[h(i-1), h(i+1)] + 2$, then the height of column i is incremented by one unit, provided that $\eta_q(i, h(i)+1) \leq \mathscr{F}$, where $\eta_q(i, h(i)+1)$ is a random number distributed uniformly over the range $0-1$, to represent the quenched disorder associated with the site at position $(i, h(i)+1)$, and $\mathscr{F}$ is a parameter in the model. It is often useful to think of $\mathscr{F}$ as the force driving an interface through a disordered medium. For this model, the directed percolation threshold is found at $\mathscr{F} = \mathscr{F}_c \approx 0.539$. For $\mathscr{F}$ near to, but smaller than, $\mathscr{F}_c$, effective roughness exponents α somewhat larger than 0.63 were obtained, and a value of about 0.63 was obtained for the exponent β describing the growth of the surface width. The growth of the surface width could be represented in terms of the scaling form

$$\xi_\perp = |\mathscr{F} - \mathscr{F}_c|^{-\nu_\perp} f(t|\mathscr{F} - \mathscr{F}_c|^{\nu_\parallel}), \tag{5.72}$$

where the function $f(x)$ has the form $f(x) \sim x^{\nu_\perp/\nu_\parallel}$ for $x \ll 1$. In the $\mathscr{F} < \mathscr{F}_c$, "pinned" regime, $f(x)$ is constant for $x \gg 1$, and in the $\mathscr{F} > \mathscr{F}_c$, "propagating" regime, $f(x) \sim x^{1/3}$ for $x \gg 1$. The exponent of 1/3 is expected for a process described by equation 5.121 when an interface is forced through a disordered medium, well above the pinning threshold, and the velocity depends on the orientation. Under these circumstances, the quenched disorder acts like the noise driving the growth in a KPZ process. The recognition of the connection between the driven motion of an interface through a medium with quenched disorder and directed percolation [1020, 1021] was a crucial step in the development of a better understanding of a broad range of moving interface processes in disordered media.

The quenched disorder in the "directed percolation" model of Tang and

Leschhorn [1021] plays a completely different role from the noise in random deposition models. In the former case, the quenched disorder pins the interface, retarding its motion, while in the deposition models it is the noise that drives the development of the surface. Quenched disorder models are discussed in more detail later in this chapter (section 5.4).

5.3 Theoretically Motivated Models

Although the simple models described in the previous section have come to be of major importance in statistical physics, most of them were originally developed in other areas to represent processes such as the growth of cell colonies, the sedimentation of colloids and the crystallization of polymers. More recently, a much wider variety of models has been developed to explore issues that were originally of theoretical interest. Most of these models soon became of more practical interest, as it became apparent that the simple models described above could not explain most experimental results, and serious attempts were made to compare the results of experimental and theoretical studies.

5.3.1 Surface Growth with Weak Non-linearity

The magnitude of the non-linear $b(\nabla h(\mathbf{x},t))^2$ term in the KPZ equation (equation 5.13) depends on the details of the model or growth process. The strength of the non-linearity can be expressed in terms of the reduced coupling constant $\overline{b}$, given by

$$\overline{b} = b(2\mathscr{D}/a^3)^{1/2}, \tag{5.73}$$

where a, b and $\mathscr{D}$ are the coefficients in the KPZ equation (equations 5.13 and 5.4). This can be demonstrated by making the substitutions $h \rightarrow (2\mathscr{D}/a)^{1/2}\overline{h}$ and $t \rightarrow \overline{t}/a$ in equations 5.13 and 5.4, which leads to the transformed Langevin equation

$$\partial\overline{h}(\mathbf{x},\overline{t})/\partial\overline{t} = \nabla^2\overline{h}(\mathbf{x},\overline{t}) + \overline{b}[\nabla\overline{h}(\mathbf{x},\overline{t})]^2 + \overline{\eta}(\mathbf{x},\overline{t}), \tag{5.74}$$

where $\overline{\eta}(\mathbf{x},\overline{t})$ has the form

$$<\overline{\eta}(\mathbf{x},\overline{t})\overline{\eta}(\mathbf{x}',\overline{t})> = \delta(\mathbf{x}-\mathbf{x}')\delta(\overline{t}-\overline{t}'). \tag{5.75}$$

For the $1+1$-dimensional case, $\overline{b}^{-2}$ acts as a horizontal length scale that determines the crossover from the weak coupling or linear regime to strong coupling or non-linear growth regimes, provided that $\overline{b}^{-2}$ is large compared with ϵ, the lattice constant or inner cut-off length scale. A number of models have been introduced that allow the coefficient b of the non-linear term in the KPZ equation to be varied continuously.

For example, Gates and Westcott [1022] have used a generalization of the $1+1$-dimensional, single-step model to represent the growth of lamellar crystals formed by the chain folding of polymers. In this model, the growth probability for a local minimum on the advancing interface depends on its local environment. The growth probabilities are P_-, P_0 or P_+ if the number of growth sites decreases, remains the same or increases, respectively, when the minimum is filled. Gates and Westcott explored the case where

$$P_0 = (P_- + P_+)/2, \tag{5.76}$$

so that the growth probabilities can be written in terms of a single parameter Θ, the reduced inverse temperature in the corresponding Ising model, as

$$P_- = min(1, e^{\Theta}), \tag{5.77}$$

$$P_+ = min(1, e^{-\Theta}), \tag{5.78}$$

and P_0 is obtained from equation 5.76. Krug and Spohn [555] have shown that for this model the coefficient b of the non-linear term in the KPZ equation changes sign at $\Theta = \Theta_c = -2\ln(3)$.

Within the framework of the KPZ equation, the most simple scenario would be for all models to belong to the same universality class, irrespective of the magnitude of the non-linear term, except, of course, for $b = 0$. This has been demonstrated quite convincingly for the 1+1-dimensional models [290, 291]. For the 1+1-dimensional off-lattice ballistic deposition model with steepest descent restructuring of the deposited particles to the nearest local minimum [938], illustrated in figure 5.2(c), a slow crossover from $\beta = 1/4$, characteristic of surface growth without a dependence of the growth velocity on the angle of inclination, to $\beta \approx 1/3$, characteristic of surface growth described by the KPZ equation, was found. Consequently, the asymptotic behavior of this model does not belong to the EW universality class. This is most probably a result of a small coefficient b for the non-linear term in the KPZ equation. The most likely origin for this non-linear term is a dependence of the deposit density on the inclination of the surface. A similar effect is likely to be present in the $2+1$-dimensional off-lattice model for ballistic deposition with steepest descent restructuring, but the value of the coefficient b may be quite small. Consequently, this model, which provided motivation for the EW equation, probably does not lie in the EW universality class. However, there is no direct evidence for a crossover from early stage $2+1$-dimensional EW universality class behavior to late stage $2+1$-dimensional KPZ universality class behavior in the $2+1$-dimensional model. A measurement of the deposit density as a function of the inclination $< \nabla h(\mathbf{x}) >$ would probably be the most sensitive way of determining the asymptotic behavior. A small deposit, generated using the 1+1-dimensional model, is shown in figure 5.10. If the deposition process formed

Figure 5.10 A small scale deposit generated using the $1+1$-dimensional, off-lattice ballistic deposition model, with transfer of deposited discs to a local minimum, illustrated in figure 5.2(c).

a regular, close-packed structure, the surface growth process would belong to the EW universality class. However, the internal structure of the deposit is much more complex; it can be described in terms of a network of rhombuses with an algebraically decreasing density of topological defects with increasing height (chapter 3, section 3.6). In this case, it is reasonable to suppose that the density of the underlying packing is correlated with the inclination of the surface and that this is the origin of the non-linear term in the KPZ equation. This correlation may be amenable to theoretical analysis, but the behavior exhibited by this model will be complicated, during the early stages of growth, by the decreasing concentration of topological defects (chapter 3, section 3.6).

In the $2+1$-dimensional case, the situation is more uncertain. Several models that allow the non-linearity to be varied continuously have been investigated. For example, Yan *et al.* [1023] introduced a ballistic deposition model in which the height at which a new site is deposited in the randomly selected column at position $\mathbf{x}$ is given by

$$\begin{aligned}(\mathbf{x},t+1) &= q\,max[h(\mathbf{x},t)+1,\{h(\mathbf{x}',t)\}] + \\ & [(1-q)/(n_a+1)]\left[h(\mathbf{x},t)+1+\sum_{\{nn\}} h(\mathbf{x}',t)\right]. \qquad (5.79)\end{aligned}$$

Here, $\{h(\mathbf{x}',t)\}$ denotes the heights of the columns that are nearest neighbors to the column at lateral position $\mathbf{x}$ and q is a random variable that takes on the value unity with probability p, and zero with probability $1-p$. In this equation,

$\sum_{\{nn\}}$ implies summation over all the nearest-neighbor columns and there are n_a of them. Yan *et al.* argued that the parameter p is equivalent to the ratio (b/a) of the coefficients in the KPZ equation 5.13, so that the relative strength of the non-linear term can be tuned via p.

Using this model, Yan *et al.* carried out both $2+1$-dimensional and $3+1$-dimensional lattice model simulations. In both cases, they found evidence for distinct "weak coupling" and "strong coupling" or weakly non-linear and strongly non-linear regimes, with a "morphology transition" between them.

A similar result was obtained by Pellegrini and Jullien [1024] using ballistic deposition models with a mixture of "sticky" and "non-sticky" particles. Sticky particles are deposited according to the procedure used in the standard ballistic deposition model, while the non-sticky particles are initially deposited in the unoccupied site at position $(i, h_i + 1)$, where h_i is the height of the highest occupied site in the randomly selected ith column. If the sites in one or more of the nearest-neighbor columns at heights h_i and $h_i + 1$ are both empty, then the deposited particle is transferred at a height of h_i to one of those nearest-neighbor columns and falls vertically until it again reaches a vacant site on top of an occupied site. The column that will allow the particle to fall the farthest is selected and ties are broken by random selection. If possible, the particle then moves in the same manner, following a path of steepest descent, to successively lower and lower positions on the tops of filled sites in nearest-neighbor columns, until no further downward motion is possible. The site at the end of the path of steepest descent is then filled. The sticky particles can create holes, and this can be regarded as the origin of the non-linear $b(\nabla h(\mathbf{x}, t))^2$ term in the corresponding continuum equation.

Amar and Family [1025] had previously carried out very large scale simulations using a 2+1-dimensional, restricted step height, solid-on-solid model in which the growth probabilities are given by

$$P(h_i \to h_i + 1) = \begin{array}{ll} e^{-k\Delta E_i} & \text{if } \Delta E_i > 0 \\ 1 & \text{if } \Delta E_i \leq 0. \end{array} \tag{5.80}$$

Here, $P(h_i \to h_i + 1)$ is the probability of increasing the height in the ith column from h_i to $h_i + 1$ and ΔE_i is the change in "energy" associated with this process (ΔE_i is the change in the surface area measured in square lattice units). Amar and Family found that, as the parameter k was increased from zero, the exponents α and β (equations 2.163 and 2.164) first decreased very slowly with increasing k and then fell rapidly to a value of zero at $k \approx 0.5$. As k was further increased, the exponents α and β then increased rapidly, approaching new asymptotic, large k, values that had effective values ($\alpha \approx 1/4$, $\beta \approx 0.14$) that were smaller than those obtained at small values of k ($\alpha \approx 0.4$, $\beta \approx 1/4$). Both sets of exponents satisfied equation 5.42 quite well. The behavior observed by Amar and Family appears to be different from that found by Yan *et al.*

and by Pellegrini and Jullien. The sharp decrease in α and β at $k \approx 0.5$ is a consequence of the disappearance of the non-linear term in the KPZ equation at this value of k [953, 1026, 1027]. The width of this regime is quite narrow, and the dependence of α and β on k does not suggest the existence of a distinct weakly non-linear regime. The difference between the effective values for the exponents α and β for $k \approx 0$ and $k \gg 1/2$ is not understood. It may be due to finite-size effects, even for these very large scale simulations.

The numerical evidence for separate weakly non-linear and strongly non-linear regimes in the 2+1-dimensional case is not convincing [1028, 1029, 1030, 1031], and the theoretical support for this idea is weak. However, the 2+1-dimensional KPZ equation is still not understood well enough to dismiss the *possibility* of a transition between distinct weak coupling and strong coupling regimes, but a transition of this type seems to be quite unlikely. In the $3+1$-dimensional case, the numerical results are more convincing, and there is theoretical support for the possible existence of a distinct weakly non-linear regime. For example, Cook and Derrida [1032] have obtained bounds on the transition temperature to a high temperature phase in the corresponding 3+1-dimensional directed polymer problem. For the 2+1-dimensional case, only a lower bound on the transition temperature was obtained, leaving open the possible existence of a distinct high temperature, weakly non-linear phase.

Kim *et al.* [1033] carried out $2+1$-dimensional and $3+1$-dimensional directed polymer model simulations in conjunction with calculations using transfer matrix representations of the same models. They found a finite temperature transition from a strong coupling, low temperature phase to a weak coupling, high temperature phase for the $3+1$-dimensional model. For the $2+1$-dimensional model, the asymptotic behavior was found to exhibit strong coupling behavior at all temperatures, but the asymptotic behavior was found only after a polymer length ξ_X, which grew as e^{T^2}, with increasing temperature T. The short polymer length behavior (analogous to the short time behavior in a surface growth process) was found to exhibit weak coupling behavior.

Nattermann and Tang [1034] and Tang *et al.* [1035] have carried out a one-loop dynamic renormalization group analysis of the KPZ equation in the weak coupling regime. This analysis suggests an exponentially slow crossover from linear (EW) behavior on short length scales to non-linear (KPZ) behavior on long length scales for the $2+1$-dimensional case. The crossover length scale depends exponentially on $a^3/b^2\mathscr{D}$ (see equations 5.13 and 5.4). For the $1+1$-dimensional case, the horizontal crossover length $\xi_{\parallel}^*$ is proportional to $a^3/b^2\mathscr{D}$ [291]. For $d > 2$, distinct non-linear (rough) and linear (smooth) regimes are predicted, and a value of $z_c = 2$ is predicted for the dynamic exponent at the transition point. Using arguments based on the mapping between the KPZ equation and the problem of directed polymers in random media, Doty and

Kosterlitz [1036] predicted that the result $z_c = 2$ is an exact, superuniversal result.

5.3.2 Correlated Noise

In the standard KPZ equation (equation 5.13), it is assumed that the noise term $\eta(\mathbf{x}, t)$ is uncorrelated in space and time. This is reasonable for processes such as low pressure vapor deposition onto a cold surface, but in other moving interface problems, such as fluid–fluid displacement in a porous media, important correlations may exist. The problem of surface growth with noise correlations in space and/or time was first considered theoretically by Medina *et al.* [970]. In this work, they assumed that the correlations in the random process, $\eta(\mathbf{x}, t)$, could be represented in terms of the noise spectrum $\mathscr{D}(k, \omega)$ in Fourier space, defined by

$$< \eta(\mathbf{k}, \omega)\eta(\mathbf{k}', \omega') > = 2\mathscr{D}(\mathbf{k}, \omega)\delta^d(\mathbf{k} + \mathbf{k}')\delta(\omega + \omega') \tag{5.81}$$

for a $d + 1$-dimensional process, where the noise spectrum $2\mathscr{D}(\mathbf{k}, \omega)$ has power law singularities of the form

$$2\mathscr{D}(\mathbf{k}, \omega) \sim |\mathbf{k}|^{-2\rho_n}\omega^{-2\theta_n}. \tag{5.82}$$

Medina *et al.* used a dynamic renormalization group analysis [966] of the generalized KPZ equation, with the noise spectrum given by equations 5.81 and 5.82, to calculate the exponents $H = \alpha$, β and z. For the case where the noise has only spatial correlations, they predicted that, above a critical value ρ_{nc} of ρ_n, the surface width growth exponent is given by

$$\beta = (2 + 2\rho_n - d)/(4 - 2\rho_n + d), \tag{5.83}$$

while for $\rho_n < \rho_{nc}$ the white noise exponents are obtained. The value for ρ_{nc} is known exactly only for the $1 + 1$-dimensional process. In this case, $\rho_{nc} = 1/4$, and

$$\alpha = \begin{cases} 1/2 & \text{for } 0 \le \rho_n \le 1/4 \\ 1/2 + \{2[\rho_n - (1/4)]/3\} & \text{for } 1/4 \le \rho_n \le 1 \end{cases} \tag{5.84}$$

and

$$z = \begin{cases} 3/2 & \text{for } 0 \le \rho_n \le 1/4 \\ 3/2 - \{2[\rho_n - (1/4)]/3\} & \text{for } 1/4 \le \rho_n \le 1. \end{cases} \tag{5.85}$$

If the non-linear term is absent from the $d + 1$-dimensional surface growth Langevin equation, then the results

$$\alpha = [(2 - d)/2] + \rho_n + 2\theta_n \tag{5.86}$$

and $\beta = \alpha/2$ for $d < d_c = 2+2\rho_n+4\theta_n$ can be obtained from the invariance of the Langevin equation, with correlated noise, to the rescaling given in equations 5.8

to 5.10 [970]. If both the non-linear term and temporal correlations are present, the results are neither exact nor simple.

A correlated ballistic deposition model was studied by Meakin and Jullien [1037, 1038]. In this model, the position at which the nth particle is deposited ($\mathbf{x}_n$) is given in terms of the (lateral) position at which the previous particle was deposited by

$$\mathbf{x}_n = \mathbf{x}_{n-1} + \delta\mathbf{x}, \tag{5.87}$$

where $\delta\mathbf{x}$ is a d-dimensional vector with a length $\delta x = |\delta\mathbf{x}|$ given by

$$P(\delta x > \Delta X) = (\Delta X)^{-f} \text{ for } \delta x \geq \epsilon \tag{5.88}$$

and $P(\delta x < \epsilon = 0)$. Here, $P(\delta x > \Delta X)$ is the probability that the step length δx is greater than ΔX. Equation 5.88 can also be written as

$$\mathcal{P}(\delta x) \sim (\delta x)^{-(f+1)} \text{ for } \delta x \geq \epsilon, \tag{5.89}$$

where $\mathcal{P}(\delta x)$ is the probability density for the step length distribution. This model corresponds to deposition from the points of a Levy flight in the lateral plane, above the growing surface. A practical way of generating a random distribution of step lengths δx satisfying equation 5.88 is described in chapter 2, section 2.4.

Simulations were carried out for both the 1+1-dimensional and 2+1-dimensional cases, with the normal square lattice ballistic deposition sticking rules and for deposition with "downhill" transfer to an adjacent local minimum (restructuring). The top row in figure 5.11 shows $1+1$-dimensional deposits formed without restructuring with $f = 0.5$ and $f = 1.5$. The bottom row shows $2+1$-dimensional deposits generated without "restructuring" with $f = 1.0$ and $f = 2.0$. Figure 5.12 shows $1+1$-dimensional deposits formed with restructuring for the same two values of f. It is apparent from these figures that the Hurst exponent increases as f increases.

For the $1+1$-dimensional models, the computer simulations of Meakin and Jullien suggest that

$$\alpha = \begin{cases} 1/2 & \text{for } 0 \leq f \leq 1/2 \\ 1/2 + \{[(f - (1/2)]/3\} & \text{for } 1/2 \leq f \leq 2, \end{cases} \tag{5.90}$$

for deposition without restructuring to an adjacent local minimum. With restructuring, the simulation results suggest that

$$\alpha = \begin{cases} (1+f)/2 & \text{for } 0 \leq f \leq 1 \\ 1 & \text{for } 1 \leq f \leq 2. \end{cases} \tag{5.91}$$

The uncertainties for the $2+1$-dimensional simulations are much larger, but the simulation results suggest that α grows from a value of about 0.36 for $f = 0$ to about $1/2$ for values of f in the range $1 \leq f \leq 2$, in the case of deposition

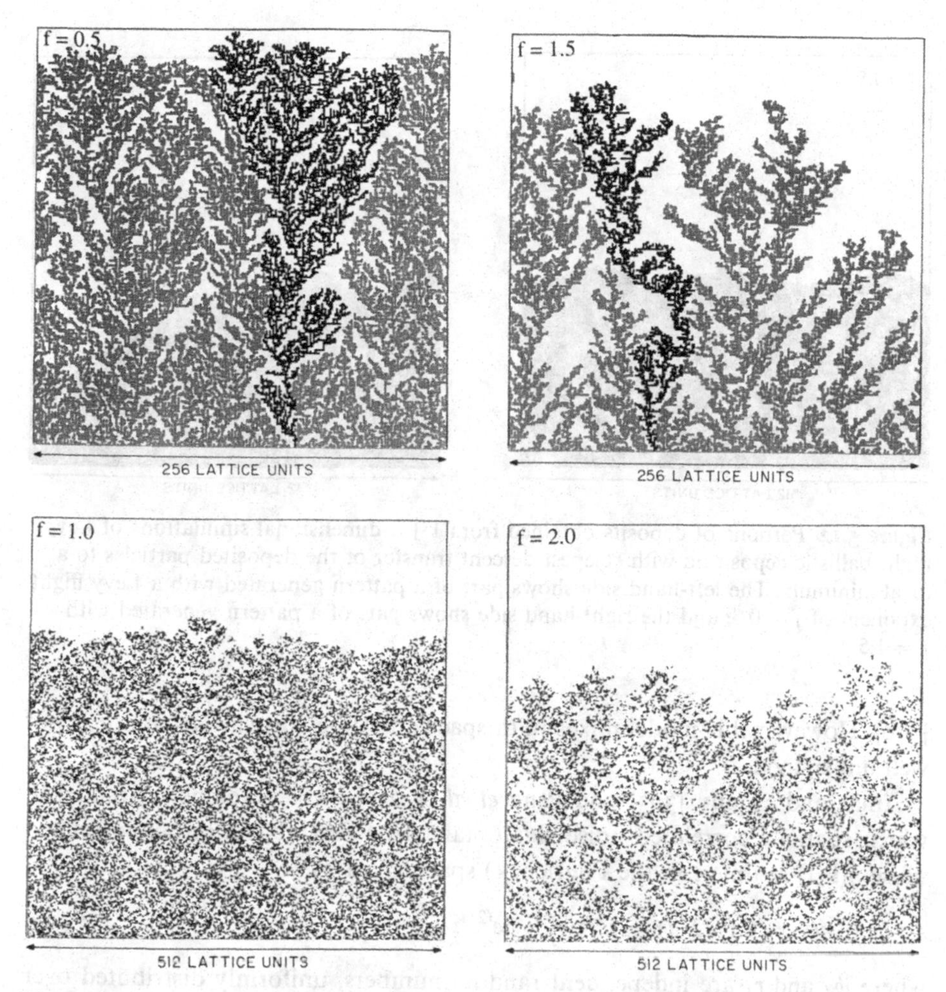

Figure 5.11 Lattice model deposits obtained from Levy flight ballistic deposition models. The top row shows small portions of $1+1$-dimensional simulations with $f = 0.5$ (left) and $f = 1.5$ (right). The bottom row shows vertical cross-sections through $2+1$-dimensional simulations with $f = 1.0$ (left) and $f = 2.0$ (right).

without restructuring. For the $2+1$-dimensional model with restructuring, the simulation results suggest that $\alpha \approx f/2$ for $0 \leq f \leq 2$. Again, the exponent scaling relationship given in equations 2.183 and 5.42 is satisfied quite well for all of the models without restructuring, and the relationship

$$\alpha/\beta = 2 \tag{5.92}$$

is satisfied quite well for the models with restructuring.

Meakin and Jullien argued that the exponent f in equations 5.88 and 5.89 corresponds to the exponent $2\rho_n$ in equation 5.82. If this is correct, then the simulation results were consistent with the theoretical results of Medina *et al.*

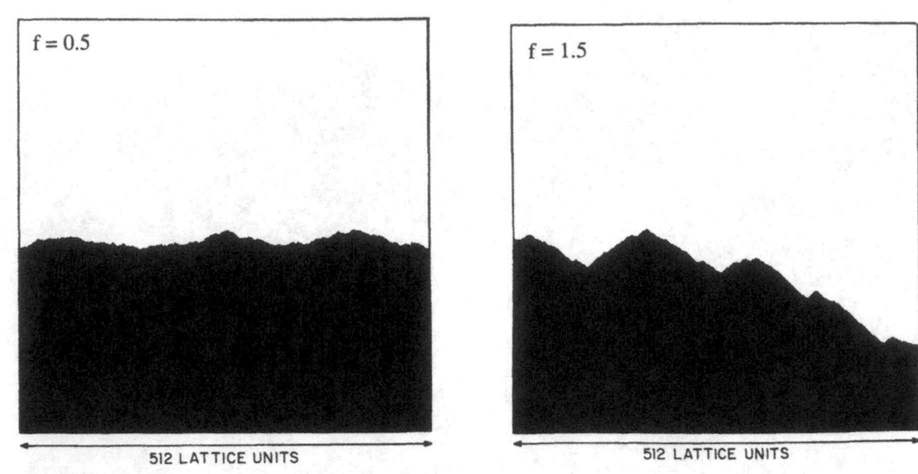

Figure 5.12 Portions of deposits obtained from 1 + 1-dimensional simulations of Levy flight ballistic deposition with steepest descent transfer of the deposited particles to a local minimum. The left-hand side shows part of a pattern generated with a Levy flight exponent of $f = 0.5$, and the right-hand side shows part of a pattern generated with $f = 1.5$.

[970]. However, this model mixes both space and time correlations and it is not well understood.

Amar and Family [305] and Peng *et al.* [1039] have developed a model in which spatially correlated noise was obtained by first generating uncorrelated noise $\eta(\mathbf{k}, t)$ at time t in the Fourier ($\mathbf{k}$) space given by

$$\eta(k_j = 2\pi j/L, t) = j^{-\rho_n}(r_j^{-1/2})e^{(2\pi i\phi_j)}, \tag{5.93}$$

where ϕ_j and r_j are independent random numbers, uniformly distributed over the range 0 to 1. For k = 0, the noise $\eta(0, t)$ was made equal to zero. The noise $\eta(x, t)$ was then obtained by fast Fourier transformation of $\eta(k, t)$. This generates spatially correlated random noise with the form

$$< \eta(x, t), \eta(x', t') > = |x - x'|^{2\rho_n - d}\delta(t - t'). \tag{5.94}$$

Deposition was attempted if $\eta(i, t) > 0$, but not if $\eta(i, t) < 0$. Amar and Family [305] carried out simulations with both standard ballistic deposition rules (equations 5.58 to 5.60) and with restricted step height, solid-on-solid model growth rules. In order to avoid time correlations, a parallel deposition algorithm was used in which deposition was attempted alternately in the sublattice columns with odd or even coordinates at odd or even time steps, respectively. Good agreement with the theoretical results of Medina *et al.* [970] was obtained for both 1 + 1-dimensional models.

Peng *et al.* [1039] have carried out similar simulations. They first generated random, Gaussian, uncorrelated noise $\eta_\circ(\mathbf{x}, t)$, Fourier transformed it to get

$\eta_\circ(\mathbf{k},\omega)$ and back transformed $|\mathbf{k}|^{-\rho_n}\omega^{-\theta_n}\eta_\circ(\mathbf{k},\omega)$ to get the required correlated random noise. Using this correlated noise they integrated the 1 + 1-dimensional KPZ equation numerically, carried out ballistic deposition model simulations and simulated directed polymer growth. Their results did not agree with the theoretical predictions of Medina *et al.*, and the ballistic deposition model simulations did not agree with the two other approaches, except for values of ρ_n close to zero. However, the results from numerical integration of the KPZ equation and the directed polymer simulations were in quite good agreement with the theoretical work of Hentschel and Family [1040], which leads to the expression

$$\beta = 1/(3 - 2\rho_n) \text{ for } 0 \leq \rho_n \leq 1/2, \tag{5.95}$$

for the surface width growth exponent. A third theoretical relationship between β and ρ_n

$$\beta = \begin{array}{ll} 1/3 & \text{for } 0 \leq \rho_n \leq 1/4 \\ (1+2\rho_n)/(5-2\rho_n) & \text{for } 1/4 \leq \rho_n \leq 1 \end{array} \tag{5.96}$$

has been proposed by Zhang [1041]. The Hurst exponent H and the roughness exponent ($\alpha = H$) can be obtained from equations 5.95 or 5.96 and the scaling relationship given in equation 2.183.

Wu *et al.* [1042] have carried out numerical integrations of the 1 + 1-dimensional KPZ equation with spatially correlated noise, using surfaces of size L up to L = 131 072(2^{17}). These simulation results were in satisfactory agreement with equations 5.84 and 5.85 for $0 \leq \rho_n \leq 1/2$. In this work, the noise was simulated by using the series

$$\eta(x) = 2^{1/2} \sum_{n=0}^{n=N-1} A_n cos(k_n x + \phi_n), \tag{5.97}$$

where $A_n = [2S(k_n)\Delta k]^{1/2}$, $k_n = n\Delta k$ and the ϕ_n are random phases uniformly distributed over the range 0 to 2π. Here, $\Delta k = k_u/N$, where k_u is an upper cut-off wave number and $N = L$ is the number of components. The power spectrum $S(k)$ had the form $S(k) = k^{-2\rho_n}$, with $S(0) = 0$. This procedure is essentially equivalent to the Fourier transformation method described above. For the larger values of ρ_n, the measured noise correlation was not the same as the power spectrum exponent used to generate the correlated noise. This discrepancy was attributed to the fact that the noise with $S(k = 0) = 0$ does not represent the singularity in the noise at $k = 0$, particularly for large values of ρ_n. This adds to the uncertainty of these simulations, and others in which the noise is generated in a similar manner. This problem appears to be related to that of generating and characterizing self-affine fractal surfaces, which were discussed in chapter 2, section 2.7. This is an important basic problem that should be resolved as soon as possible since it undermines the reliability of a wide variety of studies concerning rough surfaces. In particular, the work of Yordanov and Nickolaev [317] on the effects of cut-offs on power spectra provides important insights into this problem.

Pang *et al.* [1043] have recently shown how to generate correlated noise that satisfies periodic boundary conditions. For a lattice of width $L = 2N$, the Fourier-space noise is given by

$$\eta(k,t) = (S_{\rho_n}(k))^{1/2}(R_k - 1/2)e^{2\pi\phi_k}, \tag{5.98}$$

where $S_{\rho_n}(k)$ is the power spectrum with the form $S_{\rho_n}(k) \sim k^{-2\rho_n}$. In this equation, R_k and ϕ_k are random numbers between 0 and 1. The power spectrum is given by

$$S_{\rho_n}(k,t) = 1/L \sum_{u=-N}^{u=N-1} g_{\rho_n}(u)e^{-iku}, \tag{5.99}$$

where $u = x - x'$, $g_{\rho_n}(u \neq 0) = |u|^{2\rho_n - 1}$ and $g_{\rho_n}(0) \equiv 2\int_0^{1/2} g_{\rho_n}(u)du = 2\int_0^{1/2} u^{2\rho_n - 1}du$. The real-space noise used in the simulations was obtained by Fourier transformation of $\eta(k,t)$. Pang *et al.* used this discretized noise to carry out simulations using both a directed polymer model and a restricted step height solid-on-solid growth model. However, these simulations have not resolved the uncertainties associated with surface growth driven by correlated noise.

A similar approach has recently been taken by Lam *et al.* [1044]. They used a $1+1$-dimensional ballistic deposition model, instead of a solid-on-solid model, and a "fast fractal Gaussian noise generator" (FFGN) [275] to generate noise with *temporal* correlations. For temporal correlations, Medina *et al.* [970] calculated that the exponents α and β would retain the $1+1$-dimensional KPZ values for $\theta_n \leq 0.167$, and that $\alpha(\theta_n) = 1.69\theta_n + 0.22$, with $\beta = \alpha(\theta_n)(1 + 2\theta_n)/(2\alpha(\theta_n) + 1)$, for $0.167 < \theta_n < 0.5$, using a dynamical renormalization group analysis. The surface height update rules used in this work were

$$h'(x,t) = max[h(x,t) + \eta(x,t), h(x-1,t), h(x+1,t)]. \tag{5.100}$$

This corresponds to ballistic deposition, where $\eta(x,t)$ is the size of the "particle" deposited in column x at time t. Deposition was attempted alternately in either all of the odd columns or all of the even columns, with periodic boundary conditions in the lateral direction. Good agreement with the theoretical results of Medina *et al.* [970] was obtained if the noise was "truncated" in the same way as in the work of Amar and Family [1045], by giving the noise a value of $\eta(x,t) = 1$ if the original FFGN is positive and $\eta(x,t) = 0$ if the original FFGN is negative. In general, they found that the results were sensitive to the details of the FFGN, and much poorer agreement was obtained from simulations in which the heights $h(x,t)$ had continuous values and the updating was carried out using the non-truncated FFGN. A discrepancy, similar to that discussed above, was found between the measured Hurst exponent and that used to generate the FFGN. Although this discrepancy was not large, it adds to the uncertainty associated with this work.

Lam and Family [1046] have studied correlated noise deposition processes described by the continuum equation

$$\partial h(x,t)/\partial t = -K\nabla^4 + \eta(x,t). \tag{5.101}$$

Since this equation is linear, the exponents α, β and z were obtained by a simple scaling analysis based on the invariance of equation 5.101 to the transformation given in equations 5.8 to 5.10, which leads to the rescaled equation

$$\lambda^{\alpha-z}\partial h(x,t)/\partial t = -\lambda^{\alpha-4}K\nabla^4 + \lambda^{\rho_n-(d/2)+z(2\theta_n-1)/2}\eta(x,t) \tag{5.102}$$

or

$$\partial h(x,t)/\partial t = -\lambda^{z-4}K\nabla^4 + \lambda^{\rho_n-(d/2)+[z(2\theta_n+1)/2]-\alpha}\eta(x,t). \tag{5.103}$$

The invariance under rescaling leads directly to the results $z = 4$ and $\alpha = \rho_n + (4-d)/2 + 4\theta_n$, where θ_n and ρ_n are defined in equation 5.82. The theoretical values of these exponents were confirmed by $1+1$-dimensional computer simulations, with θ_n (defined in equation 5.82) equal to zero and by an exact solution of equation 5.101, based on Fourier transformation.

5.3.3 Non-Gaussian Noise

Zhang [1041, 1047] has proposed that the large roughness exponents observed in some experiments, such as those of Rubio *et al.* [1048], Horváth *et al.* [1049, 1050] and Vicsek *et al.* [1051], concerned with 1+1-dimensional surface growth, may be a consequence of non-Gaussian "noise". He carried out simulations on a square lattice in which the interface was advanced according to the discrete evolution equation

$$h(i,t+1) = max[h(i-1,t)+\eta(i-1,t), h(i+1,t)+\eta(i+1,t)], \tag{5.104}$$

where $h(i,t)$ is the height in the ith column at time t and t is an integer. In this equation, $h(i,t)$ and $\eta(i,t)$ are continuous variables.

At each time t, the noise $(\eta(i,t))$ was selected randomly from the power law density distribution

$$\mathscr{P}(\eta) \begin{array}{ll} \sim 1/\eta^{(1+\mu)} & \text{if } \eta > 1 \\ = 0 & \text{if } \eta \leq 1, \end{array} \tag{5.105}$$

and distributed independently among the columns of the lattice. Values for the exponent μ in the range $2 < \mu < \infty$ were considered, and a roughness exponent of 0.75 was obtained for $\mu = 3$.

Similar results were obtained for this model by Amar and Family [1045] and Buldyrev *et al.* [1052]. They also studied several closely related models that appear to belong to the same universality class in the sense that the exponents $\alpha(\mu)$ and $\beta(\mu)$ have the same values for any particular value of μ in the range $2 \leq \mu \leq 7$. In addition to measuring the growth of the surface width and height difference correlation functions, Amar and Family also measured the distribution $\mathscr{P}(\delta h)$ of the height fluctuations, where

$$\delta h(x,t) = h(x,t) - < h(x,t) > . \quad (5.106)$$

They found that $\mathscr{P}(\delta h)$ could be represented quite well by the scaling form

$$\mathscr{P}(\delta h, L) = L^{-\alpha} f(\delta h L^{-\alpha}), \quad (5.107)$$

where $L(\leq \xi_{\|})$ is the lateral system size or lateral scale over which the fluctuations are measured. For $x \gg 1$, the scaling function $f(x)$ in the above equation has the form $f(x) \sim x^{-(\mu+1)}$. It has been suggested that the surfaces generated by this model, and presumably other models in which the growth is dominated by rare events, are examples of multi-affine scaling and require an infinite family of exponents for a full description. However, the scaling form of Amar and Family for $\mathscr{P}(\delta h, L)$ indicates that multi-affine scaling will not be found for moments of δh or $\xi_{\perp}$ smaller than $q = \mu$. The distribution $\mathscr{P}(\delta h)$ was studied theoretically by Havlin *et al.* [1053] by mapping the problem onto a Levy walk. It was found that for small δh, the distribution has a Gaussian shape. The value of 0.75 obtained for α with $\mu = 3$ is in good agreement with the theoretical relationship

$$\alpha = (d+2)/(\mu+1) \quad (5.108)$$

obtained by Zhang [1054] and Krug [1055]. It follows from this result and equation 5.42 that $\beta = (d+2)/(2\mu - d)$.

The expression for α given in equation 5.108 can be obtained from simple scaling arguments by focusing attention on the largest noise fluctuations that occur on given length and time scales [1055]. The average size of the largest event within a region of size δx^d in time δt is given by $\eta_{max}(N) \sim (\delta x^d \delta t)^{1/\mu}$ for large values of $\delta x^d \delta t$. Consequently, the time required for a fluctuation of magnitude δh to appear is $t_f = \delta h^{\mu} \delta x^{-d}$. Once a fluctuation has been created, it will spread laterally, because of the non-linear term in the KPZ equation, with power law noise. If $\delta x \ll \xi_{\|}$, the time t_s required to spread by a distance δx is given by $t_s \sim (\delta x)^2/\delta h$, since $\delta h/\delta t \sim (\delta h/\delta x)^2$. Consequently, the largest fluctuations that contribute to the roughness are those with $t_f = t_s$, so that $\delta h^{\mu} \delta x^{-d} \sim (\delta x)^2/\delta h$ or $\delta h^{\mu+1} \sim \delta x^{2+d}$ and $\alpha = (2+d)/(\mu+1)$ (equation 5.108). This argument applies only if $\mu > d+1$. If this condition is not satisfied, the growth of the entire surface becomes dominated by the single largest event, and the surface can no longer be described in terms of a self-affine scaling model.

Equation 5.108 suggests that, for $\mu \geq \mu_c = 5$, the KPZ exponents should be recovered in the $1+1$-dimensional case. There is some disagreement concerning the exact value of μ_c based on computer simulations [1045, 1056], but most simulations indicate that $\mu_c \approx 7$.

Barabási *et al.* [280] have carried out very similar simulations with the evolution algorithm

$$h(i,t+1) = \quad max[h(i-1,t)+\eta(i-1,t), h(i,t)+\eta(i,t), h(i+1,t)+\eta(i+1,t)]. \tag{5.109}$$

Noise is added to every site, and the heights are then updated according to equation 5.109. For simulations carried out with a noise distribution exponent of $\mu = 3$, a value of 0.75 was obtained for the roughness exponent, in good agreement with equation 5.108 and other simulation results [1047, 1041, 1045, 1052]. The height difference correlation functions obtained from this model exhibit an interesting new crossover. On relatively short length scales ($r \leq \xi^*$), the correlation functions $C_q(r)$ in equation 2.156, chapter 2, have an algebraic form with a corresponding Hurst exponent H_q that decreases with increasing q. On longer length scales, there is crossover from this "multi-affine scaling" behavior to simple self-affine scaling ($H_q \approx 0.75$ for $1 \leq q \leq 9$). At even longer length scales, $r > \xi_{\parallel}(t)$, the correlation functions reach a constant value ($C_q(r > \xi_{\parallel}) \approx \xi_{\perp}$). The factors controlling ξ^* are not well understood, but this crossover length appears to increase with increasing system size L. Arguments based on the idea that ξ^* is dominated by rare events [1054, 1055] indicate that $\xi^* \approx L^{\alpha}$. Although more work is needed, these results appear to be consistent with those of Amar and Family [305].

Lam and Sander [1057, 1058] have shown that equation 5.108 is exact for $d+1 \leq \mu \leq \mu_c$. If $\mu \geq \mu_c$, then the roughness exponent is the same as that for the corresponding KPZ equation with Gaussian noise. The exponents β and α are related by equation 2.183. Since the results of Lam and Sander are based on scaling assumptions and basic symmetries, they can be used to obtain exact results, even in those cases where there are no exact theoretical solutions for the surface growth equation.

The ideas of Zhang [1041, 1047] are supported by experiments carried out by Horváth *et al.* [1050], in which fluctuations in the advancing interface were measured during psuedo 2-dimensional fluid–fluid displacement experiments. In these experiments, glycerin/4% water was injected into an air filled Hele-Shaw cell, with randomly distributed glass beads confined between the two plates. A slow injection, at an average interface velocity of $70-100\mu m s^{-1}$, was carried out to minimize the stabilizing effect of the pressure distribution inside the viscous invading fluid, discussed below.

The deviations $\Delta h(\mathbf{x},t)$ of the interface position from the mean position ($< h(\mathbf{x},t) >$), defined as $\Delta h(\mathbf{x},t) = h(\mathbf{x},t) - \langle h(\mathbf{x},t)\rangle$, were measured after a number of stages during which the interface height $< h(\mathbf{x},t) >$ advanced by a constant amount δh. The noise $\eta(\mathbf{x},t_n)$ was then defined as

$$\eta(\mathbf{x},t_n) = [\Delta h(\mathbf{x},t_{n+1}) - \Delta h(\mathbf{x},t_n)]/\delta h. \qquad (5.110)$$

Because of the possibility of lateral spreading of the height fluctuations, while $< h(\mathbf{x},t) >$ advances by δh, it is not clear if the fluctuations measured in this manner can be compared directly with the noise in the Langevin equation for surface growth. However, Horváth *et al.* found support for their interpretation from computer simulations of the KPZ equation with non-Gaussian noise. The noise distribution for $\eta(\mathbf{x},t) > 0$, averaged over the experiments, could be represented quite well by

$$P(\eta) \sim c\eta^{-(\mu+1)}, \qquad (5.111)$$

with $\mu = 2.67 \pm 0.19$, over about one-and-a-half decades in μ.

Computer simulations using the Zhang model [280, 305, 1047], with this noise distribution exponent, gave a roughness exponent of about 0.8, in good agreement with the exponent measured for the experimental interfaces. Despite the apparent consistency, the origin of the power law noise distribution in the experiments is not understood, and the nature of these interfacial growth phenomena remains an open question.

Csahók and Vicsek [1059] have explored the effects of a broad distribution of particle (disc) sizes on 1 + 1-dimensional, off-lattice, ballistic deposition, with one stage of restructuring [1060]. In their model, a deposited particle follows a path of steepest descent on the surface, after it first contacts the surface, until it contacts a second disc or the substrate. At that point, it stops moving and becomes permanently part of the deposit. The distribution of disc radii is given by

$$\mathscr{P}(r) \sim \begin{array}{ll} r^{-(1+\mu)} & \text{for } r \geq a \\ 0 & \text{for } r < a, \end{array} \qquad (5.112)$$

where the parameter a is the cut-off for the power law distribution of radii sizes. For a size distribution exponent of 3, the results $\beta = 0.51 \pm 0.05$ and $\alpha = 0.77 \pm 0.04$ were obtained. These exponents satisfy the scaling relationship given in equation 5.42 quite well. They are quite similar to the exponents obtained by Zhang [1047], using height increments with the same distribution. This is probably a valid comparison, since a large vertical displacement in the Zhang model will spread rapidly and is more or less equivalent to adding a sphere of the same radius in the model of Csahók and Vicsek [1059].

Vicsek *et al.* [1061] have proposed yet another model, motivated by experiments, that demonstrates the growth of rough surfaces with large values of the roughness exponent. In this model, the evolution of the $1+1$-dimensional surface is assumed to be described by

$$\partial h(x,t)/\partial t = \{a\nabla^2 h(x,t) + b[1+(\nabla(h(x,t))^2]^{1/2}\}\{p+\eta\}, \tag{5.113}$$

where $p > 0$. The noise η has no spatial or temporal correlations, but is not Gaussian, to avoid negative values of $p+\eta$. The term $b[1+(\nabla h(x,t))^2]^{1/2}$ is not simplified to $(\nabla h(x,t))^2$ because $|\nabla h(x,t)|$ is not $\ll 1$ in typical experiments. The simulations were carried out by numerical integration of equation 5.113, with $a = 1$, $b = 0.5$, $p = 10^{-4}$ and η uniformly distributed over the range 0 to 1. A distinct crossover from $\beta \approx 0.65$ at early times to $\beta = 0.28 \pm 0.05$ at later times was found. It appears that, in this limit, this model may belong in the KPZ universality class, and it is expected that $\beta = 1/3$ in the asymptotic limit $t \to \infty$.

5.3.4 Growth on Rough Substrates

Krug and Spohn [555, 1062] have used the Langevin equation

$$\partial h(x,t) = a\nabla^2 h(x,t) + b_\gamma|[\nabla h(x,t)]|^\gamma \tag{5.114}$$

(equation 5.39 without the noisy driving term) to study the smoothing of an initially rough surface that can be characterized by a roughness exponent of $\alpha_\circ$. They argued that a typical surface slope is proportional to $\xi_\perp/\xi_\parallel$, so that the rate of growth of the mean height is given by $d < h(t) > /dt \sim (\xi_\perp/\xi_\parallel)^\gamma \sim \xi_\parallel^{\gamma(\alpha_\circ-1)} \sim t^{\gamma(\alpha_\circ-1)/z}$, and the mean surface height at time t is given by

$$< h(t) > \sim t^{1-\gamma(1-\alpha_\circ)/z}. \tag{5.115}$$

The time required for the surface to become smooth, on length scales up to $\xi_\parallel$, is proportional to $\xi_\parallel^z$, so that the saturation regime is reached when the mean height above the rough substrate reaches a value of

$$< h(t) >^* \sim \xi_\parallel^{z-\gamma(1-\alpha_\circ)}. \tag{5.116}$$

The height at which the surface becomes smooth up to a length scale of $\xi_\parallel$ is also expected to scale as

$$< h(t) >^* \sim \xi_\parallel^{\alpha_\circ}, \tag{5.117}$$

since the non-linear term $b|[\nabla h(x,t)]|^\gamma$ causes smoothing by filling minima in the surface, without raising the height of the maxima, which become flattened by sidewards growth. For large values of the exponent γ in equation 5.114, the first term on the right-hand side dominates, $z = 2$, and it follows immediately, from equations 5.116 and 5.117, that the dynamical exponent z is given by

$$z = min[2, \gamma(1-\alpha_\circ) + \alpha_\circ]. \tag{5.118}$$

The exponent z describes the growth of the lateral correlation length $\xi_{\|} \sim t^{1/z}$, up to which the surface has been smoothed at time t.

This relationship between z and $\alpha_\circ$ was confirmed numerically for the case $\gamma = 1, \alpha_\circ = 1/2$. Equation 5.118 can also be obtained by examining the behavior of the terms in equation 5.114 under the transformation $h \rightarrow \lambda^{\alpha_\circ} h, x \rightarrow \lambda x$ and $t \rightarrow \lambda^z t$, similar to that given in equations 5.8, 5.9 and 5.10 [965]. Equation 5.114 then becomes

$$\partial h(\mathbf{x}, t)/\partial t = a\lambda^{z-2}\nabla^2 h(\mathbf{x}, t) + b_\gamma \lambda^{(\gamma-1)\alpha_\circ + z - \gamma} \nabla h(\mathbf{x}, t)^\gamma. \tag{5.119}$$

If the first term in equation 5.119 is dominant in the large λ limit, then $z = 2$. If the second term is dominant, then $z = \gamma(1 - \alpha_\circ) + \alpha_\circ$. Equation 5.118 has been more extensively tested by Amar and Family [965], leading to very good agreement between the simulation results and equation 5.118.

5.4 Models with Quenched Disorder

A wide range of pattern-formation processes can be thought of in terms of the penetration of a surface or interface through a disordered medium that retards the advance of the interface as a result of "pinning" interactions. In some cases, the pinning forces may be strong enough to stop completely the interface, but in many other cases the interface is forced to advance and overcome all of the pinning forces. Processes of this type can generally be represented better by quenched disorder models that explicitly include the effects of the disordered medium than by the noise driven models described above. The simple "directed percolation" model of Tang and Leschhorn [1021], described earlier in this chapter, section 5.7.2, is a good example of this class of models. Models of this type provide an attractive alternative to the Zhang model [1047] described above, for the penetration of fluid–fluid interfaces, with large values of α, H and β, into porous media.

Although quenched disorder models have received less attention than the noise driven models described above, the study of these processes is progressing in a similar manner. Attempts are being made to classify quenched disorder models in terms of universality classes based on equations of motion such as [1063]

$$dh(\mathbf{x}, t)/dt = a\nabla^2 h(\mathbf{x}, t) + \mathscr{F} + \eta_q(\mathbf{x}, h(\mathbf{x}, t)) \tag{5.120}$$

and [1064]

$$\partial h(\mathbf{x}, t)/\partial t = a\nabla^2 h(\mathbf{x}, t) + b(\nabla h(\mathbf{x}, t))^2 + \mathscr{F} + \eta_q(\mathbf{x}, h(\mathbf{x}, t)), \tag{5.121}$$

where $\mathscr{F}$ is the driving force responsible for the advance of the interface and $\eta_q(\mathbf{x}, h(\mathbf{x}, t))$ represents the quenched disorder or pinning forces. As is the case for the equations representing noise driven growth, the surface morphology and growth exponents depend on the nature of the noise as well as the spatial derivatives of the surface height. Equation 5.120 and 5.121 will be called the quenched disorder Edwards–Wilkinson and quenched disorder Kardar–Parisi–Zhang equations, because of their similarity to the EW and KPZ equations, and they will be represented by the acronyms "QEW" and "QKPZ", respectively.

In general terms, the interface will be trapped, or pinned, by the quenched disorder if the driving force $\mathscr{F}$ is small. Under these conditions, the interface will advance only by a finite distance ξ that can be expected to scale as $\xi \sim (\mathscr{F}_c - \mathscr{F})^{-\nu_\perp}$, where $\mathscr{F}_c$ is the critical driving force needed to make the interface advance through the porous medium, without being arrested by the disorder. As $\mathscr{F} \rightarrow \mathscr{F}_c$ the correlation length ξ diverges. For large driving forces, the interface moves at a constant velocity, and the quenched noise acts like the annealed noise in the EW and KPZ equations [1065]. In some applications, thermal noise and/or noise from external sources may be important, particularly if the driving force is near to $\mathscr{F}_c$. In this event, additional noise terms can be added to the right-hand sides of equations 5.120 and 5.121. Thermal noise and/or external noise may play an important role in depinning interfaces that have become pinned by the quenched noise or disorder. The interplay between quenched disorder and other sources of randomness may be subtle and may depend on the details of the distributions in space and/or time that characterize the noise. At the time of writing, the growth of interfaces in the presence of both quenched disorder and noise had not been extensively investigated. In some models, discussed in this section, the interface is propagated by randomly selecting increments in the column heights, or other local updates that are allowed by the "competition" between the quenched disorder η_q and the driving force $\mathscr{F}$. This introduces uncontrolled annealed disorder. However, this annealed disorder does not influence the balance between the quenched disorder and the driving force, and it does not change the hydrodynamic limit scaling behavior, at least in the $\mathscr{F} \leq \mathscr{F}_c$ regime.

Even in the absence of thermal and/or external noise, there are several possible scaling regimes corresponding to pinned interfaces, critical interfaces and moving interfaces. There is also evidence for a "near-critical" moving interface regime. However, some simulations have been interpreted in terms of a crossover from critical behavior with a large roughness exponent to the "annealed noise regime", where $\alpha = H = 1/2$ for $d = 1$. Most experiments have been carried out in the moving interface, $\mathscr{F} > \mathscr{F}_c$, regime or the purported near-critical moving regime, where the theoretical uncertainties are largest, while most theoretical work has been focused on the pinned, $\mathscr{F} < \mathscr{F}_c$, regime. The quite wide range of Hurst exponents reported in experiments on the penetration of fluids into porous media might reflect crossovers from the critical or near-critical moving regimes to the $\mathscr{F} \gg \mathscr{F}_c$, rapidly moving regime.

The classification of quenched disorder processes is not as fully developed as the classification of noise driven processes, and the conclusions reached at the time of writing were more tentative. However, the experience gained in the study of both equilibrium and non-equilibrium noise driven models is being put to good use, and understanding of this type of interface growth processes is advancing rapidly.

Invasion percolation (chapter 3, section 3.5.2) takes place in the strong disorder limit. Under these conditions, the propagation of the interface through the porous medium is controlled by the local quenched disorder. There are no constraints on overhangs, and a self-similar surface develops. In the large system size ($L \to \infty$) limit, the size of the interface can grow indefinitely. In the case of a non-wetting fluid advancing through a porous medium, the interface dynamics is controlled by the local capillary forces. However, there are no "elastic forces" controlling the interface on longer scales. Under these circumstances, the interface cannot be described in terms of a single-valued function $h(\mathbf{x}, t)$, and the Langevin equation approach is not appropriate.

5.4.1 Models and Simulation Results

Two main approaches to the modeling of the growing interfaces with quenched disorder have been developed: some models have been constructed to simulate processes described by equations such as 5.120 and 5.121 or to solve these equations numerically; other models have been designed to represent specific physical processes or to study the relationships between morphology and growth algorithms. In the latter case, it is not always clear what the appropriate continuum limit growth equation is, or even if the process can be formulated in terms of equations such as such as 5.120 or 5.121.

Jensen and Procaccia [1066] have studied a $1+1$-dimensional, square lattice, quenched disorder, surface growth model with a power law distribution of "pinning strengths". In this model, the distribution of pinning strengths is given by

$$\mathscr{P}[p(\mathbf{x}) = a] = a^{-(\mu+1)}, \tag{5.122}$$

where $\mathscr{P}[p(\mathbf{x}) = a]\delta a$ is the probability that the pinning strength associated with the site at $\mathbf{x}$ lies in the range a to $a + \delta a$. In the model, a "counter" is associated with each site on the advancing interface. The unoccupied perimeter sites are selected randomly and their counters are incremented by unity. When the counter at $\mathbf{x}$ exceeds $p(\mathbf{x})$, the site at $\mathbf{x}$ is filled, provided that this would not generate a step in the advancing interface with a height greater than unity. This model generates self-affine surfaces with exponents α, H and β that depend on

μ. A value of $\mu = 3$ leads to a roughness exponent α of about 0.7. The scaling relationship given in equation 2.183 appears to be satisfied. This model has the advantage that a relatively direct physical interpretation of the pinning strength can be made in terms of the structure of the disordered medium in which the interface is advancing. The disorder plays a different role in this model than in most other quenched disorder models and the random-selection annealed noise *does* compete with the quenched noise. Because all of the pinning forces are eventually overcome, there is no pinning threshold, and this model is much more closely related to the noise driven models. Essentially the same model was studied independently by Tang *et al.* [1067], who showed that it is equivalent to the Zhang [1047] non-Guassian noise model, described above (section 5.3.3).

Quenched disorder models that can be described in terms of the QEW and QKPZ equations (5.120 and 5.121) were studied by Parisi [1064]. For small values of $\mathscr{F}$, smaller than a critical value $\mathscr{F}_c$, the interface becomes pinned by the quenched disorder and the surface width $\xi_\perp$ approaches a constant value as $t \to \infty$ [1063, 1068]. In the $1+1$-dimensional square lattice version of these models, a random variable $\eta_q(i,h)$ is associated with each lattice site (i is the lateral coordinate and h is the height). The noise $\eta_q(i,h)$ is spatially uncorrelated and evenly distributed over the range $-1 < \eta_q < 1$. Two growth algorithms were investigated. In the first algorithm, which was believed to corresponding to the QKPZ equation, the quantities $\mathscr{F}'(i,t)$ defined as

$$\mathscr{F}'(i,t) = \mathscr{F}\{max[h(i+1,t), h(i-1,t)] + 1 - h(i,t)\} \tag{5.123}$$

are calculated, at each stage or time. The interface grows if $\mathscr{F}'(i,t)$ is larger than $\eta_q(i, h(i,t))$, and the new interface height in the ith column is given by

$$h(i,t+1) = max[h(i+1,t), h(i-1,t)] + 1. \tag{5.124}$$

If $\mathscr{F}'(i,t)$ is not larger than $\eta_q(i, h(i,t))$, the interface does not advance in the ith column. This updating procedure is applied alternately to all the odd and all the even columns of the lattice. As Amaral *et al.* [1069] have pointed out, this model generates large steps in the surface, and the surface evolves via the propagation of these steps. If the driving force $\mathscr{F}$ is small, the growth of the interface is concentrated onto a few regions associated with the large steps. This behavior is a consequence of the updating algorithm described by equation 5.124, which raises the height of a column at the foot of a step to a level greater than that of the top of the step. This behavior seems to be quite unrealistic in the context of most physical surface growth processes. It is not clear if the large scale behavior of this model can be described in terms of the QKPZ equation.

In the second algorithm, described by the QEW equation, the evolution of the column heights is given by

$$h(i,t+\epsilon) = h(i,t) + \epsilon[\mathscr{F} + \nabla^2 h(i,t) + \lambda\eta_q(i, h(i,t))], \tag{5.125}$$

where $\nabla^2 h(i,t)$ is the lattice Laplacian ($\nabla^2 h(i,t) = h(i-1,t)+h(i+1,t)-2h(i,t)$). In this model, the heights $h(i,t)$ are not integers. Parisi carried out $1+1$-dimensional simulations based on equation 5.125, with $\epsilon = 0.25$ and $\lambda = 3$ for lattices with widths ranging from $L = 100$ to $L = 1600$ and for times up to $t = 300$. The driving force threshold $\mathscr{F}_c$ with these parameters was found at $\mathscr{F} \approx 1.35$.

For the model described by equations 5.123 and 5.124, the critical value for $\mathscr{F}$ is about 0.1 ($\mathscr{F}_c \approx 0.1$). For values of $\mathscr{F}$ near to, but above, the critical driving force $\mathscr{F}_c$, the interface width $\xi_\perp$ grows as $\xi_\perp \sim t^\beta$, with a characteristic exponent β of about 3/4 up to a time t^* that increases as $\mathscr{F}$ approaches $\mathscr{F}_c$, from above $\mathscr{F}_c$. For simulations carried out using equation 5.125, the value measured for the exponent β was essentially the same for small values of $\mathscr{F} - \mathscr{F}_c$. However, the two models have different exponents: $\beta \approx 1/3$ for the process described by equations 5.123 and 5.124, and $\beta \approx 1/4$ for the model described by equation 5.125 in the large driving force/late time regime where $t \gg t^*(\mathscr{F} - \mathscr{F}_c)$. The behavior of the "QKPZ" model described by equations 5.123 and 5.124, at small values of $\mathscr{F} - \mathscr{F}_c$, suggests that the coefficient b of the non-linear term in the QKPZ equation vanishes as $\mathscr{F} \to \mathscr{F}_c$ for this model, so that the model can be described by the QEW equation in this limit.

Parisi [1064] conjectured that the crossover time t^* diverges as $|\mathscr{F} - \mathscr{F}_c|^{-\gamma}$, with $\gamma = 4/d$ for the $d+1$-dimensional case. He proposed theoretical arguments to support the idea that $\beta = (4-d)/4$, which is certainly consistent with the simulation results for $d = 1$. Simulations were also carried out using $2+1$-dimensional models based on equation 5.125, with $\lambda = 6$ and with lattices of sizes up to $L = 200$ sites. They led to values for the surface width growth exponent β that were consistent with the theoretical value of 1/2. Similarly, simulation carried out using the corresponding $3+1$-dimensional model with $\lambda = 9$ gave $\mathscr{F}_c \approx 1.37$ and a value of about 1/4 for the exponent β, in the critical regime where ($\mathscr{F} \sim \mathscr{F}_c$). This result is also consistent with the theoretical prediction that $\beta = (4-d)/4$. If all sources of uncertainty are taken into account, the values obtained for the surface width growth exponent β by Parisi are also consistent with the results $\beta = 3(4-d)/2(5+d)$ ($\beta = 3/4$, $\beta = 3/7 = 0.42858\ldots$ and $\beta = 3/16 = 0.1875$ for $d = 1, 2$ and 3, respectively) obtained from functional renormalization group analysis [1070, 1071] of the QEW equation.

The model of Buldyrev *et al.* [1020] for the propagation of wetting fronts in porous media is another example of a pinning model. In this $1+1$-dimensional model, a fraction p of the sites on a square lattice are blocked. These sites can be thought of as sites with pinning forces that are too strong to be overcome by the driving force. At each stage in the simulation, one of the unblocked, unoccupied perimeter sites is selected at random and filled. If filling that site creates an overhang, all sites beneath it (all sites in the same column at a lower height, whether they are blocked or not) are also filled, to remove the overhang.

This model generates surfaces with a Hurst exponent of $H = 0.63 \pm 0.02$ and a width growth exponent of $\beta = 0.68 \pm 0.04$ for $p = p_c$, where $p_c \approx 0.47$ is the threshold value for p, above which the interface will not propagate indefinitely in an infinite system.

A quenched disorder model for the motion of an interface through a random medium, that can be described by the QEW equation (equation 5.120) has been introduced by Kessler *et al.* [1065]. The noise $\eta_q(\mathbf{x}, h(\mathbf{x}, t))$ is assumed to be spatially uncorrelated with a mean value of zero. In the $1+1$-dimensional realization of this model, non-zero "pinning forces" $\mathscr{F}(i,j) = \mathscr{F}(\mathbf{x})$ are distributed randomly on a fraction p of the sites on a square lattice, and the pinning force is zero for the other sites. The pinning forces are selected from a Gaussian distribution with a width Δ. Equation 5.120 can be solved numerically with this quenched disorder. Kessler *et al.* showed that, in the large driving force limit ($\mathscr{F} \gg < \eta^2 >^{1/2}$), the QEW equation simplifies to the EW equation and the roughness exponent has the value $\alpha = H = 1/2$. If the interface becomes pinned, $\alpha = 1$ for the $1+1$-dimensional case. Numerical solutions of equation 5.120 indicated that it was possible to obtain effective power law behavior, $\xi_\perp \sim L^\alpha$, with effective exponents in the range $1/2 < \alpha < 1$ over about a decade of length scales L. However, the dependence of $\log \xi_\perp$ on $\log L$ is better represented by a curve with a slope that decreases with increasing L than by a straight line, and it is apparent that this is part of a crossover from $\alpha = 1$ on short length scales to $\alpha = 1/2$ on long length scales. Nattermann *et al.* [1070] have shown that, near to the pinning transition, the Hurst exponent for the $d+1$-dimensional process described by the QEW equation is given by $\alpha = (4-d)/3$ and the dynamical exponent z is $2 - [2(4-d)/9]$, to first order in $\varepsilon = 4 - d$, using a functional renormalization group analysis. This result has been confirmed by Narayan and Fisher [1071], using a slightly different theoretical approach.[3]

Csahók *et al.* [1072, 1073] studied the QKPZ equation numerically. The noise η_q was represented by random numbers on an underlying grid, and these random numbers were uniformly distributed over the range $-1/2 < \eta_q < 1/2$. The discretized version of the QKPZ equation used in the $1+1$-dimensional simulations was

$$\begin{aligned} h(x, t+\delta t) &= \delta t\{h(x-1,t) - 2h(x,t) + h(x+1,t)\} + \delta t\{b[h(x+1,t) \\ &\quad - h(x-1,t)]^2 + \mathscr{F} + \eta_q(x, [h(x,t)])\}, \end{aligned} \tag{5.126}$$

where $h(x,t)$ is a continuous variable and $\eta_q(x, [h(x,t)])$ indicates the value of the noise at the closest grid point to $(x, h(x,t))$. The time increment δt used in these simulations was typically 0.1 or 0.01. This model has a pinning threshold

3. Since the theoretical work of Nattermann *et al.* and Narayan and Fisher was based on an expansion about the critical dimension ($d = 4$), where it appears that $\alpha = H$, it is not clear if the exponent of $(4-d)/3$ should be interpreted as a value for α or a value for H, in those cases in which $\alpha \neq H$. Simulation results appear to support the idea that $H = (4-d)/3$.

at $\mathscr{F} = \mathscr{F}_c \approx 0.05$. Just above this threshold, the surface width $\xi_\perp$ grows as t^β with $\beta = 0.61 \pm 0.06$, and a Hurst exponent of $H \approx 0.71$ was measured from the dependence of the surface width $w(\ell)$ on the length scale ℓ over which it was measured. The values obtained for the exponents β and H are in good agreement with the values of $\beta = 0.6$ and $\alpha = 0.75$ obtained by Kaganovich [1074], using a theoretical approach like that of Hentschel and Family [1040]. Based on a dimensional analysis of the equation of motion of the interface, Csahók *et al.* obtained the theoretical results $\alpha = (4-d)/4$ and $\beta = (4-d)/(4+d)$, so that $\beta = 0.6$ and $\alpha = 0.75$, for $d = 1$. They also predicted an asymptotic crossover to KPZ scaling at a time t^* given by $t^* \sim ((\mathscr{F} - \mathscr{F}_c)^{4+d} b^d/\Delta^2)^{-1/2d}$, where Δ depends on the the noise correlations in the vertical (h) direction and b is the strength of the non-linear term in equation 5.126. Subsequent work, described below, indicates that these simulations should be interpreted in terms of the directed percolation universality class, with $\alpha \approx 0.63$ and $\beta \approx \alpha$. The values of the exponents measured by Csahók *et al.* appear to be consistent with this idea.

Csahók *et al.* [1072, 1073] also carried out numerical simulations using the QEW equation. They found an effective Hurst exponent of $H \approx 0.62$, near to but above the pinning threshold $\mathscr{F} = \mathscr{F}_c$, which increased as $\mathscr{F} \rightarrow \mathscr{F}_c$. This behavior is consistent with the idea that $H = 1$ at $\mathscr{F} = \mathscr{F}_c$.

Leschhorn [1075] has studied a model that is believed to belong to the QEW universality class described by equation 5.120. In this very simple model, the noise $\eta(i, h(i))$ associated with the sites was given a value of $+1$ with probability p and -1 with probability $-p$. In the $1+1$-dimensional model, the entire interface is updated at each "time" step. During each step the interface height in the ith column is incremented by one lattice unit if $v(i) > 0$, where

$$v(i) = h(i-1) + h(i+1) - 2h(i) + \mathscr{G}\eta(i, h(i)). \tag{5.127}$$

Here, the parameter $\mathscr{G}$ represents the strength of the random forces relative to the "elastic forces" acting on the interface. Using large scale simulations, the exponents α and β were found to have values of 1.25 ± 0.01 and 0.88 ± 0.02, respectively, at the pinning threshold ($p_c \approx 0.8004$ for $\mathscr{G} = 1$ and $p_c \approx 0.8748$ for $\mathscr{G} = 2$). The height difference correlation functions were measured, and a good data collapse was obtained using the scaling form $C_2(x) = t^\beta f(x/t^{1/z})$ with $\beta = 0.88 \pm 0.02$ and $z = 1.42 \pm 0.03$. An estimate of the Hurst exponent was not reported, but it appears that $H \approx 1$ ($H \neq \alpha$), from the scaling plot. For large driving forces ($p \gg p_c$) a value of 0.25 ± 0.01 was measured for the exponent β, in accordance with the expected EW behavior. Exponents of $\alpha \approx 0.74$, $\beta \approx 0.475$ and $z \approx 1.42$ were obtained from a $2+1$-dimensional version of the model, at the pinning threshold. Leschhorn and Tang [285] later argued that $H = 1$ for the $1+1$-dimensional model, and this is supported by their height–height correlation functions. Dong *et al.* [1076] have studied the closely related problem of the dynamics of an elastic string in a random potential. They measured a Hurst exponent of $H = 0.97 \pm 0.05$ from the string displacement height-difference correlation function, but did not report a value for the roughness exponent α. This appears to be a case where $\alpha > H$, and a failure to distinguish between these two exponents has led to some confusion.

Makse and Amaral [1077] have studied another $1 + 1$-dimensional lattice model in the QEW universality class. Their model is based on the Hamiltonian

$$\mathcal{H} = \sum_{i=1}^{i=L}[h(i+1) - h(i)]^2 - \mathcal{F}h(i) + \eta_q(i, h(i)). \qquad (5.128)$$

Columns are selected and their heights are increased by one lattice unit if the height increase would decrease the Hamiltonian. Exponents of $H \approx 0.92$, $\beta \approx 0.85$ and $\alpha \approx 1.23$ were obtained in both the pinned, $\mathcal{F} < \mathcal{F}_c$, and moving, $\mathcal{F} > \mathcal{F}_c$, regimes. The surface width $w(\ell)$ measured over a distance ℓ for moving interfaces with $\mathcal{F} > \mathcal{F}_c$ was found to scale as $w(\ell) \sim l^H(\mathcal{F} - \mathcal{F}_c)^{-\phi_w}$. The exponent ϕ_w was found to have the values $\phi_w = \phi_{ws} = 0.44 \pm 0.05$ for $\ell \ll \xi$ and $\phi_w = \phi_{wL} = 0.95 \pm 0.05$ for $\ell \gg \xi$, where ξ is the correlation length. Simulations were also carried out by adding non-linear terms to simulate processes described by the QKPZ equation. In this case, Hurst exponents of $H = H_p = 0.63 \pm 0.03$ and $H = H_m = 0.75 \pm 0.04$ were obtained in the pinned and moving, near critical regimes, respectively. In both regimes $\beta \approx H$. Values of -0.12 ± 0.06 and 0.34 ± 0.06 were estimated for the exponents ϕ_{ws} and ϕ_{wL}. For fast motion, with $\mathcal{F} \gg \mathcal{F}_c$, ordinary KPZ or EW exponents were measured, with and without non-linear terms in the Hamiltonian, respectively.

Ji and Robbins [1078, 1079] studied the dynamics of an interface with quenched disorder using a "spin-flip" model with the Hamiltonian

$$\mathcal{H} = -\sum_{i<j,j} J_{ij}s_i s_j - \sum_i (M + m_i)s_i, \qquad (5.129)$$

where $s_i = \pm 1$. Here, m_i represents the local field at the ith site, J_{ij} is the coupling constant between sites i and j and M is the external field. Quenched disorder can be introduced via the local fields m_i or the coupling constants J_{ij}. In the simulations, the field M is increased until flipping one of the spins from -1 to $+1$ would reduce $\mathcal{H}$ (until one of the spins becomes unstable). The unstable spin is then flipped, and if any of the adjacent spins become unstable, as a result, then they are also flipped. This process continues until no more unstable spins can be found. The value of M is then increased until another "down" spin on the perimeter becomes unstable, and the procedure is repeated. The entire process of increasing M and flipping spins is repeated many times in a simulation. This procedure can be regarded as a zero temperature Monte Carlo simulation with unconserved spin-flip dynamics. The model can be used to study surface growth by starting with an initially flat interface separating regions with spins of $+1$ and -1 that spans a lattice of size L. Spin flipping is then restricted to the down spins ($s = -1$) that lie on the unoccupied perimeter of the "cluster" of up spins ($s = +1$).

As figure 5.13 illustrates, the patterns generated by the random field Ising model depend strongly on the disorder. In the strong disorder limit, the patterns have the self-similar fractal geometry of a percolation cluster. In the weak disorder limit, the pattern is a compact, faceted cluster or smooth interface, depending on the initial conditions and boundary conditions. If all the "coupling constants" have the same value J, and if the local fields m_i are selected from a uniform distribution over the range $-\Delta m \leq m \leq \Delta m$, then the patterns can be characterized by a correlation length ξ that diverges at a critical value $\Delta m_c = 1$. For $\Delta m < \Delta m_c$, the pattern is faceted (figure 5.13(c)) and for $\Delta m > \Delta m_c$ the pattern is compact on short length scales ($\lambda < \xi$) and self-similar on long length scales $\lambda > \xi$. The correlation length diverges according to

$$\xi \sim (\Delta m - 1.0)^{-\nu}, \tag{5.130}$$

as $\Delta m \to 1.0$. The exponent ν has a value of 1.25 ± 0.05. For a Gaussian distribution of local fields, the behavior is quite different. For any value of $W(m)$, the rms value of the local field, there are always some sites with large enough values of the local field to pin the interface. Consequently, the growing interface is always rough on long length scales for $W(m) > 0$. For small values of $W(m)$, a crossover is found from a faceted surface on short length scales to a rough interface on long length scales. However, more complex scenarios are possible. These models provide simple examples of morphological transitions that are analogous to phase transitions and equilibrium roughening transitions.

In the $2 + 1$-dimensional case, with a uniform distribution of thresholds over the range $-\Delta m \leq m \leq \Delta m$, Ji and Robbins [1079] found a more complex morphology diagram. For $\Delta < \Delta_{c1} \approx 2.42$, a faceted surface is formed. For $\Delta_{c1} < \Delta < \Delta_{c2} \approx 3.41$, the interface is self-affine, and for $\Delta > \Delta_{c2}$ the flipped-spin pattern is a self-similar fractal with the scaling properties of 3-dimensional percolation on scales $> \xi$, where $\xi = (\Delta - \Delta_{c2})^{-\nu}$. On short length scales, $\ell < \xi$, the structure is compact. Here, ν is a non-universal exponent that has a value of about 3.0 for a uniform distribution of thresholds. In the self-affine regime, a Hurst exponent of $H \approx 2/3$ was measured, which is in agreement with the theoretical value [1070, 1071] obtained for the kinetic Ising model described by the QEW equation. The results from some of these simulations are illustrated in figure 5.13.

Nolle *et al.* [1080] used the spin-flip dynamics model, based on equation 5.129, to study the effect of the driving force (M or $\mathscr{F}$) on the surface, near the critical force $\mathscr{F}_c$, at which pinning takes place. In these simulations, the coupling constants were all the same, $J_{ij} = J$, and the quenched disorder, represented by $\{m_i\}$, was distributed uniformly over the range $-\Delta < m_i < \Delta$. In the $1 + 1$-dimensional square lattice model simulations, the initial configuration consists

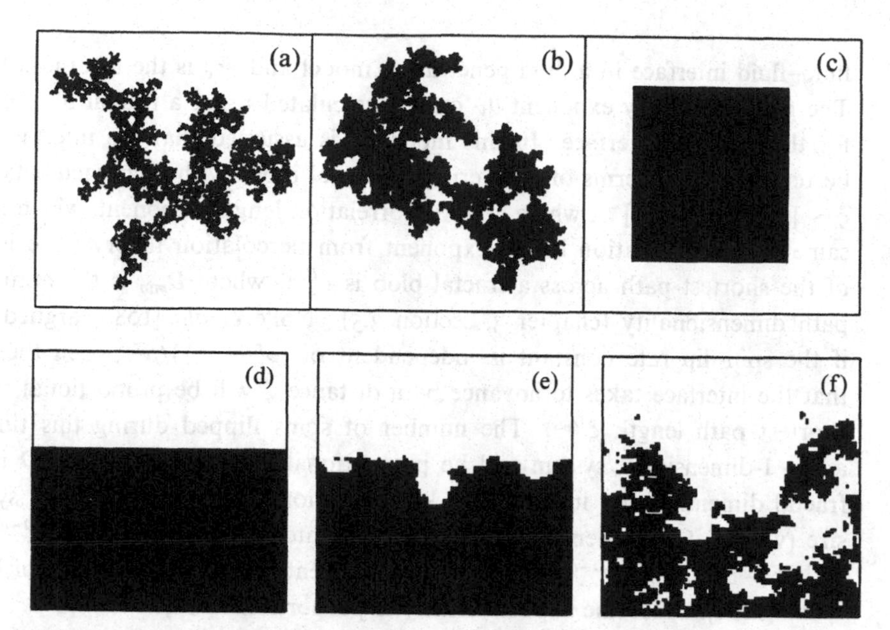

Figure 5.13 The upper row shows patterns generated by a random field Ising model with quenched disorder. Parts (a) and (b) show patterns generated on a square lattice using Gaussian distributions with rms widths of $W(m) = 15$ and $W(m) = 2.5$, respectively. Part (c) shows a flipped-spin pattern obtained with random local fields m distributed uniformly in the range $-0.7 < m < 0.7$. The lower row shows cuts through the spin-flipped regions from similar 3-dimensional, cubic lattice model simulations. The local fields were distributed uniformly in the range $-\Delta m < m < \Delta m$, with $\Delta m = 2.0$, 3.0 and 4.4 in parts (d), (e) and (f), respectively. This figure was provided by M. O. Robbins.

of two domains of "up" and "down" spins separated by a straight domain wall aligned in the "x" direction. As the simulation proceeds, only the spins on the domain wall are allowed to flip. The simulations were carried out using the zero temperature limit of the standard Metropolis Monte Carlo [188] model (chapter 1, section 1.9). At large driving forces, $\mathscr{F} \gg \mathscr{F}_c$, the interface is relatively flat, but as $\mathscr{F} \to \mathscr{F}_c$ the effects of the quenched disorder become more important and the interface becomes rougher. At $\mathscr{F} = \mathscr{F}_c$, the interface becomes self-similar with a fractal dimensionality of 4/3. In general, for $\mathscr{F} > \mathscr{F}_c$ the interface is self-similar on length scales in the range $\epsilon < \ell < \xi$ and self-affine, with a Hurst exponent of 1/2, on longer length scales. The simulation results were consistent with the idea that $\xi \sim [(\mathscr{F} - \mathscr{F}_c)/\mathscr{F}_c]^{-\nu}$, where $\nu = 4/3$ is the percolation correlation length exponent.

Nolle *et al.* [1080] have studied several versions of this model with different spin-flip algorithms. They found that such details do not change the scaling of the interface morphology but can alter the exponent θ_V that relates the interface velocity V to $(\mathscr{F} - \mathscr{F}_c)/\mathscr{F}_c$ ($V \sim [(\mathscr{F} - \mathscr{F}_c)/\mathscr{F}_c]^{\theta_V}$), where $\mathscr{F}$ is the driving force corresponding to the field M in equation 5.129 or the pressure across a fluid–fluid interface in a fluid penetration model and $\mathscr{F}_c$ is the threshold force. The growth velocity exponent θ_V can be calculated using a fractal blob model for the growing interface. In this model, it is assumed that the interface can be represented in terms of a covering of fractal blobs with a characteristic size $\xi \sim [(\mathscr{F} - \mathscr{F}_c)/\mathscr{F}_c]^{-\nu}$, where ν is the correlation length exponent, which is the same as the correlation length exponent from percolation theory. The length of the shortest path across a fractal blob is $\xi^{D_{min}}$, where D_{min} is the minimum path dimensionality (chapter 3, section 3.5). Nolle *et al.* [1080] argued that if the spin-flip rate constant is independent of $(\mathscr{F} - \mathscr{F}_c)/\mathscr{F}_c$, then the time that the interface takes to advance by a distance ξ will be proportional to the shortest path length $\xi^{D_{min}}$. The number of spins flipped during this time in a $d+1$-dimensional system will be proportional to $(L/\xi)^d\xi^D$, where D is the fractal dimensionality in the full $d+1$-dimensional space and L is the system size (width). Consequently, the spin-flipping rate is proportional to $\xi^{(D-d-D_{min})}$ or $[(\mathscr{F} - \mathscr{F}_c)/\mathscr{F}_c]^{-\nu(D-d-D_{min})}$, in good agreement with the simulations of Nolle *et al.* [1080]. Since the deposit density is proportional to $\xi^{D-(d+1)}$ (ξ^{D-2}, for a $1+1$-dimensional interface) the velocity is given by

$$V \sim [(\mathscr{F} - \mathscr{F}_c)/\mathscr{F}_c]^{-\nu(1-D_{min})}, \tag{5.131}$$

a result that can be obtained directly from the ratio between the minimum path length across a blob of size ξ and the blob diameter. Equation 5.131 was obtained using the assumption that the spin-flip rate constant is independent of $(\mathscr{F} - \mathscr{F}_c)/\mathscr{F}_c$. Consequently, equation 5.131 may not be valid for all spin-flip rules. Nolle *et al.* also studied the noise associated with the advancing interface. The interface advances at a mean velocity V given by equation 5.131. During this process, regions of size s sites fill in times of $s^{D_{min}/D}$. As a result, the power spectrum of the growth velocity has the form $S(\omega) \sim \omega^{-a}$, with $a \approx 1.55$ at high frequencies and a white noise spectrum at frequencies for which $\omega t_\xi \ll 1$, where $t_\xi \sim \xi^{D_{min}}$ is the time that it takes the interface to advance by a distance ξ. The exponent a has not been related to other characteristic exponents. A good data collapse was found using the scaling form

$$S(\omega) = Lv^2 f(\omega t_\xi, L/\xi), \tag{5.132}$$

where $v = \xi^{D-2}V$. For models such as Eden growth on a percolating lattice, the velocity near to the percolation threshold is given by $V \sim \xi^{1-D_{min}} \sim [(p - p_c)]^{-\nu(1-D_{min})}$ (compare with equation 5.131), where p is the site occupation probability. For an interface propagating along the diagonal direction on a square lattice, the distance from any point on the interface at time $t_\circ$ to the nearest point on the interface at an earlier time $t_\circ - \delta t$, measured on a path connecting nearest-neighbor occupied sites or nodes, will be equal to δt if $p > p_d$, where p_d is the directed percolation threshold. For $p > p_d$, the growth velocity is constant ($V = V_m = 2^{-1/2}$) for a process in which each unoccupied and unblocked perimeter site is occupied at each step during which δt is incremented by 1. For $p < p_d$, near to the directed percolation threshold, $V_m - V \sim (p_d - p)^{\nu_\perp}$ [1081].

A model, in which the dynamics of the penetration of a fluid–fluid interface through a 2-dimensional random array of circular obstacles was simulated [1082, 1083, 1084], gave quite different results [1085] than those obtained from the kinetic Ising model. In the most extensively investigated versions of this model, circular obstacles with radii r uniformly distributed in the range $r_1 < r < r_2$ were placed on every node of a regular triangular lattice, with a bond length of 1.0 (models based on square lattices and models in which the discs were randomly displaced from the nodes of a regular lattice were also explored). The maximum disc radius r_2 was set to a value less than 0.5, to avoid overlaps and contacts between the discs. Most of the simulations were carried out with $r_1 = 0.05$ and $r_2 = 0.49$. This disc radius distribution seems to give representative results.[4] The interface in this model consists of circular arcs connecting nearby discs. This interface advances by means of three basic instabilities called "bursts", "touches" and "overlaps". A burst takes place when there is no arc between the nearby particles that satisfies the contact angle conditions and has a radius of curvature $\mathscr{K}$ which satisfies the stability condition $\mathscr{K} = P/\Gamma$, where P (the driving force $\mathscr{F}$ in this model) is the pressure across the interface and Γ is the surface tension. A touch instability takes place when an arc between two discs contacts a third disc, and an overlap instability takes place when two arcs contact each other. The effects of viscosity are not included in the model, since it is intended to simulate low velocity invasion processes in which capillary effects are dominant. All of the unstable arcs are identified, and the geometry is sequentially updated along the interface, at all of these unstable arcs, at each stage in the simulation. The model can also be used to study invasion percolation-like processes by either adjusting the pressure continuously so that only one arc can advance at a time or by increasing the pressure monotonically, from a low initial value, so that the interface evolves via jumps consisting of the advance of single arcs and of the other arcs that become unstable as a result. A detailed description of this class of models has been provided by Cieplak and Robbins [1086].

The patterns generated by the model depend on both the contact angle θ' and the pressure $\mathscr{F}$ across the interface. There is a critical pressure $\mathscr{F}_c$ for each contact angle θ', and $\mathscr{F}_c$ also depends on the details of the model geometry. If $\mathscr{F} < \mathscr{F}_c$, the interface becomes pinned, and only a finite region is invaded. At large contact angles ($\theta' > \theta'_c$) the model generates invasion percolation-like patterns. Simulations with a contact angle of 25°, corresponding to wetting invasion ($\theta' < \theta'_c$, with $\mathscr{F} \geq \mathscr{F}_c$, or $P \geq P_c$) [1085] showed a crossover from a self-affine surface with $H \approx 0.8$ on short length scales to another self-affine structure,

4. However, a direct transition between faceted and percolation-like interface morphologies appears to be possible if the distribution is very narrow. This behavior is not characteristic of more disordered disc arrays.

with $H \approx 1/2$, on long length scales. The crossover length scale ξ increases as $\mathcal{F} \rightarrow \mathcal{F}_c$ and the simulation results indicate that $\xi \sim [(\mathcal{F} - \mathcal{F}_c)/\mathcal{F}_c]^{-\nu}$, with $\nu \approx 1.3$, consistent with the percolation value of $\nu = 4/3$. A similar crossover has been seen in some wetting invasion experiments [1049]. This behavior should be contrasted with that found for the Ising model with small values of $\delta\mathcal{F} = \mathcal{F} - \mathcal{F}_c$, described in the preceding paragraph, where a crossover from a self-similar structure on short length scales to self-affine structure on long length scales was found. Nolle *et al.* concluded that interface growth processes that can be represented by these two models can easily be distinguished, and that the exponents of $H \approx 0.75$, found in experiments, are not a consequence of a crossover from percolation to self-affine growth. In the fluid–fluid interface model, the self-affine geometry appears without any artificial restriction on overhangs in the interface. Overhangs are most prominent if $\mathcal{F} - \mathcal{F}_c$ is small, but non-zero; they become minor features in the interface if $\mathcal{F} = \mathcal{F}_c$ and if $\mathcal{F} - \mathcal{F}_c$ is large. This is the most realistic model that has been used to study the scaling behavior associated with the propagation of fluid–fluid interfaces through porous media. An attractive alternative would be to use two-phase lattice gas simulations [1087], but it would be difficult to carry out simulations on large enough time and length scales to obtain accurate estimates of the scaling exponents.

The distribution of the noise, or growth velocity fluctuations $\eta(\mathbf{x}, t)$ (for $\eta(\mathbf{x}, t) > 0$) could be represented by the power law given in equation 5.111, with an exponent of $\mu = 0.7 \pm 0.1$ for the random field Ising model, based on the Hamiltonian given in equation 5.129, and $\mu = 2.7 \pm 0.2$ for the random disc array wetting invasion model. The results obtained for the distribution of noise (local interface velocities calculated using equation 5.110) associated with this model is in excellent agreement with the experimental measurements of Horváth *et al.* [1050] ($\mu = 2.67 \pm 0.19$). This suggests a strong connection between these quenched disorder processes and the motion of an interface driven by non-Gaussian noise. However, these simulation and experimental results suggest that the non-Guassian, power law noise is a consequence of the roughening dynamics associated with the penetration of the interface through the disordered medium, rather than the cause of the surface roughness.

Robbins *et al.* [1082, 1083] have investigated the "invasion percolation" version of this model in which the driving force $\mathcal{F}$ is slowly increased from a small value. This procedure differs from standard invasion percolation models and experiments, in which the volume of an incompressible fluid is slowly increased by invading the pore space that can be reached by overcoming the smallest threshold or widest pore throat. However, the process simulated by increasing the pressure in small steps is closely related to invasion percolation. As the contact angle, measured in the invading fluid, was increased, a transition was found from a compact pattern with a self-affine surface for $\theta' < \theta'_c$ to a self-similar pattern with a self-similar surface for $\theta' > \theta'_c$ (non-wetting invasion).

At large contact angles, the interface migrates by means of bursts, and contacts between the arcs connecting two adjacent discs and a third disc. At smaller contact angles arc–arc overlaps become more important. If $\theta' < \theta'_c$, the arc–arc overlap mechanism becomes dominant, and the propagation of the interface becomes a cooperative, non-local process. Under these conditions, there appears to be an effective surface tension that controls the curvature of the interface on length scales larger than $\epsilon = 1$ (the distance between adjacent disc centers). Large negative curvatures (concave to the outside) are suppressed by the dynamics of the model because they lead to more arc–arc contacts. The pressure required to move the interface has the form $\mathscr{F} = \mathscr{F}_\circ + \Gamma'\mathscr{K}'$, where Γ' is an effective macroscopic surface tension, $\mathscr{K}'$ is the "local" curvature measured on scales larger than ϵ, and $\mathscr{F}_\circ$ is the pressure required to move a "flat" interface. For this reason, if an invasion simulation is carried out using a slowly increasing pressure, starting with a small bubble, the interface will move continuously once the excess pressure associated with the curvature of the surface of the initial bubble has been overcome.

For $\theta' > \theta'_c$, the structure was quite uniform on short length scales $\ell < \xi$, with a self-similar fractal structure on long length scales. At very large contact angles ($\theta' \approx 180°$), the correlation length was approximately equal to the distance between the centers of adjacent discs. The simulation results were consistent with the idea that $\xi \sim (\theta' - \theta'_c)^{-\nu}$, with a correlation length exponent of $\nu \approx 2.0$ [1086], and a critical contact angle that depended on model details. On long length scales, or for $\theta' \gg \theta'_c$, the displacement pattern has an effective fractal dimensionality of 1.89 ± 0.02 [1086]. This value for D is in good agreement with that associated with ordinary percolation and invasion percolation without trapping ($D = 91/48 \approx 1.8958\ldots$), but it is substantially larger than the value of $D \approx 1.82$ found in invasion percolation model simulations with trapping (both site and bond invasion percolation on a square lattice). Cieplak and Robbins [1086] also measured a fractal dimensionality of $D \approx 1.89$ for a closely related invasion percolation model, with trapping, in which the fluid–fluid interface always penetrated the widest throat on the interface at each step in the simulation. This result was surprising since "trapping" of enclosed defender fluid regions was included in both of the models studied by Cieplak and Robbins. Experiments carried out using quasi 2-dimensional arrays of spheres in a Hele-Shaw cell [1088] are not accurate enough to distinguish between the two fractal dimensionalities, but they favor the larger value. The effects of trapping are subtle; there is a need for more work on the effects of trapping on lattices other than the square lattice and on other simple invasion percolation models.[5]

Other 1+1-dimensional "pinning" models have been investigated by Sneppen

5. On a 2-dimensional triangular lattice, both the external perimeter and the hull have a fractal dimensionality of 7/4. This may lead to weaker trapping on a triangular lattice than on a square lattice and to a higher fractal dimensionality.

[1089]. As in the model of Jensen and Procaccia [1066] described above, these are solid-on-solid models with quenched disorder in which a maximum step size of 1 is allowed in the advancing interface. At the start of a simulation, "pinning forces" $\eta_q(x,h)$ are randomly selected from a Gaussian distribution and assigned to the lattice sites. In model I, the site with the smallest pinning force that can grow without violating the step height restrictions is found and filled at each stage in the simulation. This simple model generates a propagating interface with a roughness exponent of $H = 1.00 \pm 0.01$ and the exponent β has a value of 0.95 ± 0.05. This result ($H = 1$ and $\beta = 1$) can be anticipated, since the advance of the interface is controlled by the largest pinning forces or thresholds.

Sneppen [1089] also explored a closely related but more interesting model, model II (now known as the "Sneppen model"), in which the site with the smallest pinning force is selected and filled, irrespective of the heights in neighboring columns, during each stage in the simulation. After a site has been filled, the heights of neighboring columns are adjusted upwards, working out independently from both sides of the newly filled site that had the smallest pinning force at lateral position x ($h(x-1) \rightarrow h(x-1)+1, h(x-2) \rightarrow h(x-2)+1, \ldots, h(x-n) \rightarrow h(x-n)+1, \ldots$, and $h(x+1) \rightarrow h(x+1)+1, h(x+2) \rightarrow h(x+2)+1, \ldots, h(x+n) \rightarrow h(x+n)+1, \ldots$) until the restricted step height conditions are satisfied along the entire interface. This prevents the overhangs, characteristic of the invasion percolation model discussed in chapter 3, section 3.5.2, from developing, and keeps the interface self-affine. This model gives a roughness exponent of $\alpha = 0.63 \pm 0.02$ and a surface width growth exponent of $\beta = 0.9 \pm 0.01$. Using the arguments presented by Buldyrev *et al.* [1020], the roughness exponent is associated with the ratio of the directed percolation correlation length exponents ($\alpha = \nu_\perp/\nu_\parallel \approx 0.63$). Although Sneppen argued that the exponent β is different for these two models, it is not evident that the simulations alone are sufficient to distinguish between these two models, and the strong possibility that they belong to the same universality class remains. This issue has been discussed by Tang and Leschhorn [1090] and Sneppen and Jensen [1091].

The evolution of the surface fluctuations in the saturated regime ($\xi_\parallel = L$) can be characterized by the correlation function

$$C(t) = < [h(\mathbf{x}, t_\circ + t) - h(\mathbf{x}, t_\circ)] - < [h(\mathbf{x}, t_\circ + t) - h(\mathbf{x}, t_\circ)] >>, \quad (5.133)$$

which scales as $C(t) \sim t^{\beta_c}$, with $\beta_c = 0.69 \pm 0.05$ [1092]. This implies that the exponent ($\beta \approx 0.9$) that describes how the surface width grows from an initially flat line is different from the exponent β_c that describes how the surface fluctuations evolve with respect to a rough surface, in the asymptotic regime. In addition, the exponent that characterizes the power law growth of

$$C_\infty(t) =< max_{\mathbf{x}}[h(\mathbf{x}, t_\circ + t) - h(\mathbf{x}, t_\circ)] - < [h(\mathbf{x}, t_\circ + t) - h(\mathbf{x}, t_\circ)] >>, (5.134)$$

or

$$C'_{\infty}(t) = < max_{\mathbf{x}}[h(\mathbf{x}, t_{\circ}+t) - h(\mathbf{x}, t_{\circ})] - min_{\mathbf{x}}[h(\mathbf{x}, t_{\circ}+t) - h(\mathbf{x}, t_{\circ})] >, \quad (5.135)$$

where $< \ldots >$ implies averaging over all $t_{\circ}$, has yet another value of $\beta_c^{\infty} = 0.40 \pm 0.05$ [1092]. Several other exponents characterizing the interface dynamics were measured [1092]. The probability $P_s(\delta t)$ that a site at some lateral position x will be selected at time $t_{\circ} + \delta t$ after a site was selected at the same lateral position at time $t_{\circ}$ is given by

$$P_s(\delta t) \sim \delta t^{-\delta}, \quad (5.136)$$

with $\delta = 0.62 \pm 0.03$. The probability $P_r(\delta t)$ that the first return to a previously selected lateral position x will take place after a time δt is given by

$$P_r(\delta t) \sim \delta t^{-\gamma}, \quad (5.137)$$

with $\gamma = 1.20 \pm 0.1$. The distribution of lateral distances δx, separating succeeding growth events, is given by

$$P_x(\delta x) \sim \delta x^{-\phi_x}, \quad (5.138)$$

with $\phi_x = 2.25 \pm 0.05$. Finally, Olami *et al.* [1093] found that the probability $P(\eta_q)$ that the selected growth site will have a quenched noise (pinning strength) of η_q depended algebraically on $\eta_{qc} - \eta_q$:

$$P_\eta(\eta_q) \sim (\eta_{qc} - \eta_q)^{\mu}, \quad (5.139)$$

with $\mu = 0.9 \pm 0.1$ and $\eta_{qc} \approx 0.4615$. The size distribution of the avalanches (the number of sites that have to be filled after each growth event to maintain the step height restrictions) was found to have an exponential form with a mean size of only ≈ 4 [1092, 1093]. It should be no surprise that the exponents associated with the Sneppen model are not all independent exponents. They are related by scaling relationships, and it is believed [1093, 1094] that they can all be calculated from the directed percolated exponents or directed surface model exponents for $d \geq 2$. However, the details of exactly how this should be done have not yet been completely worked out. The mapping between directed percolation and models such as the Sneppen model is not rigorous.

5.4.2 Universality Classes

Several attempts have been made to define universality classes for the quenched disorder models. Many of them appear to belong to previously known universality classes such as percolation, invasion percolation and directed percolation, at least as far as the scaling exponents are concerned. In some cases, the behavior is very similar to that associated with the noise driven models described above.

The simulations do not always lead to unambiguous results, and this adds to the difficulty of classifying some of these models.

Amaral *et al.* [1069, 1095] have studied several $1+1$-dimensional, quenched disorder, interface growth models, by measuring the growth velocity as a function of the angle of inclination or tilt of the surface $< \nabla h >$ and the driving force $\mathscr{F}$. Based on theoretical arguments and computer simulation results, they proposed that the dependence of the growth velocity $V(\mathscr{F}, < \nabla h >)$ on $\mathscr{F}$ and $< \nabla h >$ for processes described by the QKPZ equation could be represented by the scaling form

$$V(\mathscr{F}, < \nabla h >) = (\mathscr{F} - \mathscr{F}_c)^{\theta_V} f[< \nabla h >^2 /(\mathscr{F} - \mathscr{F}_c)^{(\theta_V + \phi_V)}]. \qquad (5.140)$$

A good data collapse was obtained for $V(\mathscr{F}, < \nabla h >)$ for several $1+1$-dimensional, quenched disorder, interface growth models. The scaling function $f(x)$ in equation 5.140 has the form $f(x) = const.$ for $x \ll 1$ and $f(x) \sim x^{\theta_V/(\theta_V+\phi_V)}$ for $x \gg 1$. Equation 5.140, and the form of the scaling function $f(x)$, were motivated by the observation that the growth velocity of horizontal surfaces scales as $(\mathscr{F} - \mathscr{F}_c)^{\theta_V}$ and that the coefficient b of the non-linear term $b(\nabla h(\mathbf{x}, t))^2$ in the QKPZ equation scales as $b \sim (\mathscr{F} - \mathscr{F}_c)^{-\phi_V}$ in the $\mathscr{F} \to \mathscr{F}_c$ limit. Equation 5.140 describes the crossover from $V(\mathscr{F}, < \nabla h >) \sim (\mathscr{F} - \mathscr{F}_c)^{\theta_V}$ in the small slope, $x \ll 1$, limit to $V(\mathscr{F}, < \nabla h >) \sim (\mathscr{F} - \mathscr{F}_c)^{-\phi_V} < \nabla h >^2$ at larger slopes, where $x \gg 1$, and it is consistent with the idea [1069] that $V(\mathscr{F}, < \nabla h >) \sim (\mathscr{F} - \mathscr{F}_c)^{\theta_V} + c(\mathscr{F} - \mathscr{F}_c)^{-\phi_V} < \nabla h >^2$, where c is a model dependent constant. Amaral *et al.* showed that, for the $1+1$-dimensional, quenched disorder models that they studied, the coefficient b of the non-linear term in the QKPZ equation (equation 5.121) either approaches zero, and the equation of motion simplifies to the QEW equation,[6] or approaches infinity as the pinning transition is approached from above ($\mathscr{F} > \mathscr{F}_c$). This behavior identifies two universality classes associated with the QEW and QKPZ equations. If $b \sim [(\mathscr{F} - \mathscr{F}_c)/\mathscr{F}_c]^{-\phi_V}$ and ϕ_V is negative, then $b \to 0$ as $\mathscr{F} \to \mathscr{F}_c$, and the process belongs to the "random field Ising model universality class" [1085] with a Hurst exponent of 1 for $d = 1$ [1070]. These class I models can be associated with the QEW equation. If $b \sim [(\mathscr{F} - \mathscr{F}_c)/\mathscr{F}_c]^{-\phi_V}$, with positive ϕ_V, then the Hurst exponent is $H \approx 0.63$ for $d = 1$, and the process belongs to the "directed percolation", or directed surfaces, universality class [1020, 1021], corresponding to the QKPZ equation.

Amaral *et al.* [1095] found that, for processes described by the QEW equation, the roughness exponent α and the surface width growth exponent β have the same values in the pinned, $\mathscr{F} > \mathscr{F}_c$, and moving, $\mathscr{F} < \mathscr{F}_c$, regimes, near to the pinning transition at $\mathscr{F} = \mathscr{F}_c$. For the QKPZ processes, values of 0.63 ± 0.03 and 0.75 ± 0.04 were estimated for α in the pinned and moving, near-critical

6. Models in which $b(\mathscr{F}) = 0$ for all $\mathscr{F}$ also belong to the QEW universality class.

regimes, respectively. Similarly, values of 0.67 ± 0.05 and 0.74 ± 0.06 were measured for β if $\mathscr{F} < \mathscr{F}_c$ or $\mathscr{F} > \mathscr{F}_c$, respectively. This generalizes the results of Makse and Amaral [1077].

Tang *et al.* [1096] have proposed a different classification procedure based on the slope dependence of the critical force $\mathscr{F}_c$ instead of the slope dependence of the growth velocity. The non-linear term in the QKPZ equation can have similar origins to the corresponding term in the KPZ equation. In this event, the coefficient b is proportional to the front velocity and vanishes at the pinning threshold. In addition, Tang *et al.* argued that anisotropy in the porous medium is a possible source of non-linearity ($b \neq 0$). Based on these ideas, they identified three universality classes. In the isotropic case, class I, the non-linearity vanishes as $\mathscr{F} \to \mathscr{F}_c$ or $\delta\mathscr{F} = \mathscr{F} - \mathscr{F}_c \to 0$, and the interface dynamics can be described by the QEW equation with $H = (4-d)/3$. The other two classes correspond to anisotropic systems and depend on the anisotropy of the pinning forces $\eta_q(x, [h(x,t)])$. Class IIa corresponds to processes in which the "hard" direction, along which blocking sites form a directed percolation path, is parallel to the advancing interface. In this case larger thresholds are required to block the advance of an inclined interface and the motion of the interface can be described by the QKPZ equation. Tang *et al.* [1096] suggested that these processes belong to the "directed percolation universality class", characterized by the exponents $H = \alpha \approx 0.63$ and $\beta \approx 0.63$ in the $1+1$-dimensional case. They proposed the scaling form

$$V(\mathscr{F}, <\nabla h>) = (\mathscr{F} - \mathscr{F}_c(0))^{\theta_V} g[<\nabla h> /(\mathscr{F} - \mathscr{F}_c(0))^{\nu_\parallel(1-\alpha)}], \quad (5.141)$$

similar to that given in equation 5.140, where $V(\mathscr{F}, <\nabla h>)$ is the velocity, $<\nabla h>$ is the average slope of the interface and $\mathscr{F}_c(0) = \mathscr{F}_c(\nabla h = 0)$ is the critical vertical driving force for a horizontal interface. A comparison between equations 5.140 and 5.141 implies that $\theta_V + \phi = 2\nu_\parallel(1-\alpha)$. Equation 5.141 is based on the idea that there is a characteristic slope $\nabla h^*(\mathscr{F} - \mathscr{F}_c(0)) = \xi_\perp/\xi_\parallel \sim (\mathscr{F} - \mathscr{F}_c(0))^{\nu_\parallel(1-H)}$, and that equation 5.141 can be written as $V(\mathscr{F}, <\nabla h>) = (\mathscr{F} - \mathscr{F}_c(0))^{\theta_V} g'[<\nabla h> /\nabla h^*(\mathscr{F} - \mathscr{F}_c(0))]$. The exponent θ_V is given by the exponent scaling relationship $\theta_V = \nu_\parallel(z - H)$,[7] and the coefficient of the non-linear term grows as $b \sim (\delta\mathscr{F})^{\theta_V - 2\nu_\parallel(1-H)} \sim (\delta\mathscr{F})^{\nu_\parallel(z-H-2)}$, for small $<\nabla h>$. For $\mathscr{F} \to \mathscr{F}_c(0)$, the front velocity $V(\mathscr{F}, <\nabla h>)$ is independent of $\mathscr{F}$, and it follows that the scaling function $g(x)$ in equation 5.141 has the form $g(x) \sim x^{\theta_V/[\nu_\parallel(1-H)]}$

7. For $\mathscr{F} > \mathscr{F}_c$, the interface can be represented by a self-affine fractal blob model, in which an array of blobs of size $\xi_\parallel^d \times \xi_\perp$ lie on an essentially flat d-dimensional surface. The interface can be considered to grow by filling new blobs. The time required to fill a layer of blobs on the interface is given by $\delta t_b \sim \xi_\parallel^z$, and during this time the interface advances by an amount δh_b, where $\delta h_b \sim \xi_\parallel^\alpha$. Consequently, the front velocity V is given by $V \approx \delta h_b/\delta t_b \sim \xi_\parallel^{\alpha-z} \sim (\mathscr{F} - \mathscr{F}_c)^{-\nu_\parallel(\alpha-z)}$. Since the front velocity also scales as $V \sim (\mathscr{F} - \mathscr{F}_c)^{\theta_V}$, it follows that $\theta_V = \nu_\parallel(z-\alpha)$, or $\theta_V = \nu_\parallel(z-H)$, in those cases where $\alpha = H$.

for $x \gg 1$ (or $\nabla h \gg \nabla h^*$) so that $V \sim <\nabla h>^{\theta_V/[\nu_\parallel(1-H)]}$, and, for $d = 1$, where $z = 1$, $V \sim <\nabla h>$. In an anisotropic system, a higher density of blocking sites is needed to arrest the interface in a direction aligned with respect to the hard, $<\nabla h = 0>$, direction of propagation than in other "inclined" directions. Consequently, the critical driving force $\mathcal{F}_c$ is expected to decrease with increasing slope $<\nabla h>$. Since the velocity $V(\mathcal{F}, <\nabla h>)$ is zero when $\mathcal{F} = \mathcal{F}_c(\nabla h)$, equation 5.141 indicates that $\mathcal{F}_c(\nabla h) - \mathcal{F}_c(\nabla h = 0) = -|\nabla h|^{1/[\nu_\parallel(1-H)]}$. For small values of x (or $\nabla h \ll \nabla h^*$), the scaling function $g(x)$ in equation 5.141 is expected to have the form $g(x) \sim x^2$ so that $V \sim (\nabla h)^2$. This implies that the coefficient b of the non-linear term in the QKPZ equation scales as $b \sim \delta\mathcal{F}^{\theta_V - 2\nu_\parallel(1-H)}$, or $b \sim \delta\mathcal{F}^{-\phi_V}$.

If the advancing interface is not aligned along the hard direction, the rotational symmetry in the $\mathbf{x}$ space is lost, allowing new terms to appear in the surface growth equation. This leads to a new universality, class (IIb), with different exponents. Tang *et al.* concluded that the Hurst exponent for this class is given by $H_\parallel = H_{KPZ}(d-1)/z_{KPZ}(d-1)$, where $H_\parallel$ is the Hurst exponent in the tilt direction and $H_\perp = H_\parallel/\eta$, where $H_\perp$ is the Hurst exponent in the other $d-1$ directions and $\eta = 1/z_{KPZ}(d-1)$. Here, $H_{KPZ}(d-1)$ and $z_{KPZ}(d-1)$ are the Hurst exponent and dynamic exponent for the $(d-1)+1$-dimensional KPZ process. These relationships were based on the idea that a perturbation to the interface at the critical driving force will move downhill with a constant velocity ($z_\parallel = 1$), and thus the dynamics of the corresponding avalanches, which propagate along the d-dimensional interface with a constant velocity, can be described by the $(d-1)+1$-dimensional KPZ equation. The perturbation does not move far in the uphill direction because the perturbation reduces the local slope in the uphill direction and this increases the critical pinning force $\mathcal{F}_c(\nabla h)$ ($\mathcal{F}_c(\nabla h) - \mathcal{F}_c(0) \sim -|\nabla h|^{1/[\nu(1-H)]}$, where ν is the correlation length exponent). These ideas were found to be consistent with simulations carried out using the quenched disorder models of Tang and Leschhorn [1021], Sneppen [1089] and Buldyrev *et al.* [1097], described earlier in this section, with *inclined* initial surfaces. For the $1+1$-dimensional simulation, a Hurst exponent of $H = 1/2$ was obtained, instead of the value of $\alpha = H \approx 0.63$ measured for a horizontal initial surface. For $d = 2$, the simulation results indicated that $H_\parallel = \alpha_\parallel \approx 1/3$, $H_\perp = \alpha_\perp \approx 1/2$ and $\beta \approx 1/3$, so that $z_\parallel = 1$, in accordance with the theoretical predictions.

Makse [1098] has carried out simulations using a $1+1$-dimensional solid-on-solid growth model, which explicitly includes fluctuations in the lateral x as well as the longitudinal y directions. The net forces that act on the ith column, resulting from both the driving force and the pinning interactions, are

$$\mathcal{F}'_y(i) = k_y[h(i+1) + h(i-1) - 2h(i)] + \eta_q(i, h(i)) + \mathcal{F} \tag{5.142}$$

and

$$\mathscr{F}'_x(i) = -k_x + \eta_q(i, h(i)), \tag{5.143}$$

where the noise is distributed uniformly over the range $-\Delta < \eta_q < \Delta$. The interface moves forwards, in the y direction if $\mathscr{F}'_y(i) > 0$ or to the side if $\mathscr{F}'_x(i) > 0$. The motion to the side takes place only if the column in the direction of motion is shorter than the ith column. If $k_x > \Delta$, the model belongs to the QEW universality class, and if $k_x < \Delta$ the model belongs to the QKPZ, directed percolation universality class. In the $\mathscr{F} \to \mathscr{F}_c$ limit, the ratio n_x/n_y between the number of lateral fluctuations n_x and the number of forward fluctuations n_y diverges as $(\mathscr{F} - \mathscr{F}_c)^a$, where $a \approx -0.18$. This implies that, for driving forces near to the critical value $\mathscr{F}_c$, the interface advances primarily by lateral advances that engulf regions surrounded by sites with large pinning forces. Makse carried out simulations with "helical" boundary conditions to study the behavior of the model with inclined surfaces. As $\mathscr{F}$ approached $\mathscr{F}_c$ from the $\mathscr{F} > \mathscr{F}_c$ side, the growth became dominated by localized "avalanches". The mean avalanche size $< s >$ could be represented quite well by $< s > = s_\circ + \lambda(\delta\mathscr{F}) < \nabla h >^2$, and $\lambda(\delta\mathscr{F})$ was found to scale as $\lambda(\delta\mathscr{F}) \sim \delta\mathscr{F}^{-\phi_s}$. Since the interface velocity V is proportional to $< s >$, the exponent ϕ_s corresponds to the exponent ϕ_V, which describes the dependence of the coefficient b of the non-linear term in the QKPZ equation on $< \nabla h >$ ($b \sim < \nabla h >^{-\phi_V}$). This idea was confirmed by the numerical values obtained for ϕ_s and ϕ_V ($\phi_s \approx \phi_V \approx 0.64$). The behavior exhibited by this model appears to be generic for models in the QKPZ/directed percolation university class. At larger driving forces, growth via lateral fluctuations becomes less important, and the characteristic, near-critical behavior gives way to KPZ behavior.

5.4.3 Exponent Scaling Relationships

There is a close but, as yet, imprecise relationship between several of the growth models outlined above and the directed percolation model or a directed surface model that is closely related to directed percolation for $d > 2$.[8] Much of the effort to understand this class of processes has focused on the Sneppen model. The relationship between this class of models and the directed percolation model can be seen by examining the advance of the interface in the Sneppen model, described earlier in this section. If the interface is advanced according to the rules of this model, but pinning forces larger than η_m cannot be overcome, the interface will stop advancing if $\eta_m < \eta_c$, where η_c is the critical pinning force. In the Sneppen model the advancing interface is always a directed path on the square lattice, because of the step height restrictions. Every threshold

8. The structures formed by $d > 2$-dimensional directed percolation models do not form continuous $d-1$-dimensional sheets.

or pinning force on the line or surface at which the advance of the growth process is arrested will be greater than η_m and will lie on the backbone of a directed percolation cluster [1099], above the directed percolation threshold. If the arrested interface is then forced to advance by overcoming the smallest pinning force with $\eta > \eta_m$ or by randomly removing one of the pinning sites on the arrest surface, the interface will continue to advance until it is again stopped in another configuration, in which $\eta > \eta_m$ along the entire advancing interface. This process can be repeated many times to generate a sequence of arrest fronts for any selected value or values of η_m. The arrest fronts for any particular η_m form a network that separates the regions that are invaded between successive arrest surfaces. The union of all of these arrest surfaces for any particular driving force corresponding to η_m will then lie on a subset of the directed percolation backbone. In practice, the interface is never arrested in the Sneppen model. However, the advancing surface will, from time to time, pass through a configuration in which $\eta > \eta_m$ along the entire surface and each of these configurations will be a "η_m arrest surface" which lies on a directed percolation cluster backbone with $(p_c - p) \sim (\eta_c - \eta_m) = \delta_\eta$, where p is the percolation probability. In the large system size, $L \to \infty$, limit only a vanishingly small fraction of all of the pinning forces along the interface will lie below η_c in the long time, stastically stationary regime. Under these conditions, the Hurst exponent and roughness exponent will be the same as that for a path on the backbone of a critical directed percolation with $H = \alpha = \nu_\perp/\nu_\parallel \approx 0.63$, where $\nu_\perp$ and $\nu_\parallel$ are the directed percolation correlation length exponents. For $d > 1$ the arrest surfaces lie on a network of self-affine directed surfaces with roughness exponents of $\alpha = 0.48 \pm 0.03$, 0.38 ± 0.04 and 0.27 ± 0.05 for $d = 2$, 3 and 4, respectively [1097].

The spreading into the spaces between arrest surfaces proceeds via lateral fluctuations. The simulations of Makse [1098] imply that these lateral fluctuations and the associated directed percolation growth process becomes dominant only near the pinning threshold.

In higher dimensions, the regions between the arrest surfaces may have a complex, fractal structure in the lateral directions parallel to the interface (perpendicular to the direction of propagation). After the interface has broken through the arrest surface it will spread to fill the region between the breached arrest surface and the adjacent arrest surface. Havlin *et al.* [1100] have studied this process and have suggested that the exponent z associated with this lateral spreading is the same as the minimum path dimensionality $D_{min}(d)$ on a d-dimensional percolation cluster, chapter 3, section 3.5. This is based on the idea that the region between two adjacent arrest surfaces is effectively d-dimensional, because the ratio $\xi_\perp/\xi_\parallel$ between its height and width approaches zero as $\mathscr{F} \to \mathscr{F}_c$ or $p \to p_c$, in the corresponding "directed percolation" process. In order to spread over the region between two adjacent arrest surfaces, the growth process must find paths that avoid blocked regions. This conjecture was supported by numerical studies that indicated that $z \approx D_{min}(d)$ for $1 \leq d \leq 6$. For $d > 6$, $D_{min} = 2$.

Several different scaling theories [1093, 1094, 1101, 1102, 1103, 1104, 1105] have been developed to relate the exponents that characterize the dynamics of the Sneppen model to the directed percolation model. These scaling theories have focused attention on the gaps (called avalanches [1101] or associated processes [1105]) between successive arrest surfaces. For example, Leschhorn and Tang [1101] derived the scaling relationship $\beta_c^{\infty} = \alpha/(1+\alpha)$, where β_c^{∞} is the exponent describing the growth of the function $C'_{\infty}(t)$, defined in equation 5.135. Their scaling argument was based on the idea that most of the growth during a time t is confined to a region of size $\xi_{\perp}\xi_{\|} = \xi_{\perp}^{(1+\alpha)/\alpha} \approx t$ so that the maximum of $h(x,t_\circ) - h(x,t_\circ+t)$ is proportional to $t^{\alpha/(1+\alpha)}$. This is in good agreement with the value of 0.40 ± 0.05 obtained by Sneppen and Jensen [1092]. Leschhorn and Tang [1101] also derived the relationship $\phi_x = 1 + (2/\nu_{\|})$ between the exponent ϕ_x defined in equation 5.138 and the directed percolation exponents. This relationship leads to the estimate $\phi_x \approx 2.15$, which is in satisfactory agreement with the value of 2.25 ± 0.05 reported by Sneppen and Jensen [1092].

The distribution of the sizes of the gaps between successive arrest surfaces (avalanches or associated processes) plays an important role in the scaling theories. It is assumed that the gap sizes have a power law distribution of the form [1093]

$$N_s(f_m) = s^{-\tau} g(s/S), \tag{5.144}$$

where $N_s\delta s$ is the number of gaps with sizes in the range $s - \delta s/2$ to $s + \delta s/2$ in the limit $\delta s \to 0$. The characteristic gap size is given by

$$S \sim (\eta_c - \eta_m)^{-\nu}, \tag{5.145}$$

where the exponent ν is given by $\nu = \nu_{\perp} + d\nu_{\|}$, because the largest gaps are expected to have a thickness of $\xi_{\perp} \sim (\eta_c - \eta_m)^{-\nu_{\perp}}$ and a width of $\nu_{\|} \sim (\eta_c - \eta_m)^{-\nu_{\|}}$. For $d > 1$, $\nu \leq \nu_{\perp} + d\nu_{\|}$, because the gaps may have a fractal shape in the lateral directions,[9] for sufficiently large values of d. Here, attention will be confined to the $1 + 1$-dimensional case and it will be assumed that $\alpha = H$. The approaches used to develop scaling theories under these restrictions can easily be generalized to $d > 1$ and to processes for which $\alpha \neq H$. The scaling function $g(x)$ in equation 5.144 is constant for $x \ll 1$ and decreases faster than any power of x for $x \gg 1$. It follows directly from equations 5.144 and 5.145 that the mean gap size is given by

$$< s > \sim (\eta_c - \eta_m)^{-\nu(2-\tau)} \sim \delta\eta^{-\zeta}. \tag{5.146}$$

9. It appears that $D = d$ for $d = 2$ and that the gaps have self-similar fractal boundaries [1100, 1097, 1103].

Olami *et al.* [1093] calculated the exponent τ in equation 5.144 based on the idea that the distribution of lengths N_ℓ for the gaps between successive arrest surfaces is the same as the distribution of gaps in a horizontal cut through a self-affine fractal with a Hurst exponent of $H = \nu_\perp/\nu_\parallel$ ($N_\ell \sim \ell^{-(2-H)}$, from equation 2.189). Since $s \sim \ell^{1+H}$, the exponent τ in equation 5.144 is given by

$$\tau = 2/(1 + H), \tag{5.147}$$

according to this argument. This value for τ was then used to relate the exponent ν to the directed percolation exponents, leading to the result

$$\nu = (1 + H)/H, \tag{5.148}$$

which is not consistent with the idea that $\nu = \nu_\perp + \nu_\parallel$.

Maslov and Paczuski [1094] have proposed a different scaling theory for processes in the Sneppen model universality class. By focusing on the growth of the maximum threshold[10] $\eta_M = max\{\eta_m\}$ that has been overcome as the interface evolves from a flat surface, they obtained two expressions for the time (number of invaded sites, growth events, or small alavanches) t^* that it takes for the correlation length $\xi_\parallel$ to reach the system size L. One of these expressions, $t^* \sim L^{d+\nu_\perp/\nu_\parallel}$, comes from purely geometrical considerations. The other comes from the equation

$$\partial\eta_M/\partial t \approx (1 - \eta_M)/(L^d < s >) \approx [(1 - \eta_M)/(L^d)](\eta_c - \eta_M)^\zeta \tag{5.149}$$

for the growth of η_M, where the exponent ζ is defined by equation 5.146. This equation is based on the idea that the interface advances from an arrest surface in which the minimum threshold η_M on the interface is $\eta_M(n)$ to the next higher value ($\eta_M = \eta_M(n + 1)$) by filling a gap with an average size $< s >$, given by equation 5.146. As η_M increases, and approaches η_c, the average number of sites $< s >$ filled between successive increases in η_M also increases. The number of avalanches, or growth events, required to fill a gap scales linearly with the mean gap size $< s >$, and the average increase in η_M when a new gap is invaded is proportional to $(1 - \eta_M)/L^d$. The factor of $1 - \eta_M$ comes about because the thresholds along the arrest surface (along the interface, after the gap between successive arrest surfaces with $\eta_M = \eta_M(n)$ and $\eta_M = \eta_M(n + 1)$ has been filled) are randomly distributed over the range $\eta_M \leq \eta \leq 1$. Integration of equation 5.149 leads to the result $t \sim L^d \delta\eta^{1-\zeta}$ at late times, at which $(1-\eta_M)$ is essentially

10. Here, η_M is the *largest* of the minimum thresholds that have been penetrated during the advance of the interface. In the earliest stages η_M is small, since the pinning forces along the interface are randomly distributed in the range $0 < \eta < 1$. At this stage only small changes in the interface are required to increase η_M. As η_M grows, the gaps between the interfaces at which successive increases in η_M take place increase in size. At the instant at which η_M increases, the pinning forces along the entire interface are larger than η_M. In an infinite system, η_M approaches, but does not exceed, the critical threshold value η_c. The successive values of η_M, $\eta_M(n)$ form a sequence of increasing numbers ($\{\eta_M(n)\} = \eta_M(1), \eta_M(2), \eta_M(3), \ldots$, where $\eta_M(1) = 0$ and $\eta_M(\infty) = \eta_c$ in an infinite, $L \to \infty$, system).

constant and the effects of corrections to scaling are small. Here, $\delta\eta = \eta_c - \eta_M(t)$, and $\eta_M(t)$ is the value of η_M at time t. Since the time t^* corresponds to the time at which $\delta\eta$ decreases to $L^{-1/\nu_\parallel}$, it follows that $t^* \sim L^d L^{(\zeta-1)/\nu_\parallel}$. A comparison of the two expressions for t^* leads to the exponent relationship $(\zeta - 1) = \nu_\perp$, or $\zeta = 1 + \nu_\perp$. Combining this result with $\zeta = \nu(2-\tau)$ from equation 5.146 and $\nu = d\nu_\parallel + \nu_\perp$ leads to the prediction that

$$\tau = 1 + \frac{1-(1/\nu_\parallel)}{1+\nu_\perp\nu_\parallel} = 1 + [(\nu_\parallel - 1)/(\nu_\parallel + \nu_\perp)]. \qquad (5.150)$$

Maslov and Paczuski [1094] also derived the relationship

$$\phi_x = 1 + (1+\nu_\perp)/\nu_\parallel \qquad (5.151)$$

between the exponent ϕ_x defined in equation 5.138 and the directed percolation exponents. This is different from the relationship $\phi_x = 1 + (2/\nu_\parallel)$ obtained by Leschhorn and Tang [1101]. Unfortunately, simulation results are not accurate enough to distinguish unambiguously between these scaling model predictions.

The values obtained for the exponent τ by Maslov and Paczuski, using equation 5.150, and the relationship $\nu = \nu_\perp + \nu_\parallel$ used by Maslov and Paczuski, and most others, are different from those predicted by Olami *et al.* [1093]. Computer simulation results appear to favor the theoretical results of Maslov and Paczuski. However, the results obtained from the two different scaling theories are not different enough for computer simulations to be used to distinguish unambiguously between them, if all sources of uncertainty are taken into account. The results obtained from these two scaling models are a consequence of the fundamentally different assumptions outlined above. This discrepancy points to a general weakness of scaling theories. However, this weakness should not be overemphasized. Scaling theories are usually robust and reliable, but they do involve assumptions that are not always well founded or easy to test.

The scaling argument used by Huber *et al.* [1104] to calculate the sizes of the gaps between the arrest lines in the Sneppen model begins with the distribution of the holes in the skeleton of a directed percolation cluster, above the percolation threshold. This distribution is obtained using a scaling approach similar to that described in chapter 3, to calculate the expression given in equation 3.90 for the hole size distribution in a percolation cluster or other self-similar fractals. In the case of the skeleton of a directed percolation cluster, the total area occupied by a cluster with a length L scales as L^{1+H}, where $H = \nu_\perp/\nu_\parallel$ is the Hurst exponent. The area occupied by a hole of length ℓ scales as ℓ^{1+H}, and the total number of holes scales as $L^{D_\parallel+H}$, where $D_\parallel$ is the fractal dimensionality of a horizontal cut, parallel to the direction of the directed percolation cluster. Under these conditions, equation 3.88 must be replaced by

$$L^{1+H} = L^{D_\parallel+H} \int_0^\infty \ell^{1+H} \ell^{-[(\tau_{hs}-1)(1+H)+1]} f(\ell/L) d\ell, \qquad (5.152)$$

where τ_{hs} is the exponent describing the power law distribution of hole sizes. Making the substitution $x = \ell/L$ in equation 5.152 leads to the relationship

$$L^{1+H} = L^{D_{\parallel}+H} L^{1+H} L^{(1-\tau_{sh})(H+1)} L \int_0^{\infty} x^{(2-\tau_{hs})(H+1)+1} f(x) dx. \qquad (5.153)$$

Consequently, the size distribution exponent for the holes is given by

$$\tau_{hs} = 1 + [(D_{\parallel} + H)/(1 + H)] \qquad (5.154)$$

if the integral in equation 5.153 is finite (which can be justified after τ_{hs} has been calculated).

The fractal dimensionality of the horizontal cut is given by $D_{\parallel} = 1 - (\beta/\nu_{\parallel})$, where the exponent β indicates how the density of the cluster ρ grows with $p - p_c$ ($\rho \sim (p - p_c)^{\beta}$, with $\beta \approx 0.277$ for directed percolation – compare with equation 3.69, chapter 3).

However, the propagation of the interface depends on the distribution of holes in contact with the interface, not the distribution of holes in the entire network. The exponent describing the distribution of contacting holes τ_{cs} can be obtained in the same manner as the exponent τ_{hs}, by replacing the factor of $L^{D_{\parallel}+H}$ in equation 5.153 by a factor of $L = L^1$, because the number of holes of size 1 is proportional to the length of the interface. This leads to the result $1 + (H + 1)(1 - \tau_{cs}) = 0$ or

$$\tau_{cs} = 1 + [1/(1 + H)]. \qquad (5.155)$$

Huber *et al.* argued that the probability of invading a gap with a length $\ell_{\parallel}$ in contact with the arrest surface, is proportional to the number of singly connected red bonds (chapter 3, section 3.5) on a path on the directed cluster backbone, connecting the two points at the end of the gap of length $\ell_{\parallel}$. This is the number of points of contact between the gap and the arrest surface and is given by $n_R \sim \ell_{\parallel}^{1/\nu_{\parallel}}$ (compare with equation 3.73). Since the probability of invading a gap of length $\ell_{\parallel}$ is proportional to $\ell_{\parallel}^{1/\nu_{\parallel}}$, the probability of invading a gap of size s is proportional to $s^{1/((1+H)\nu_{\parallel})}$. Consequently, the exponent describing the distribution of gap sizes is given by $\tau_{gs} = \tau_{cs} - \{[1/(1 + H)]/\nu_{\parallel}\}$ or

$$\tau_{gs} = 1 + [(\nu_{\parallel} - 1)/(\nu_{\parallel} + \nu_{\perp})], \qquad (5.156)$$

which is the result obtained by Maslov and Paczuski [1094].

At the time of writing, the situation is quite fluid. The scaling theories that have been proposed for this problem [1093, 1094, 1101, 1105] are not all in agreement for some or all of the exponents. In addition, there are subtleties that diminish confidence in some of the scaling arguments. For example, the distribution of gaps between the arrest surfaces is not the same as the distribution of holes in the directed cluster backbone. The arrest surfaces are a subset of the directed cluster backbone and a gap may contain many holes.

At this stage, computer simulation results are not accurate enough either to distinguish between the results of the different scaling theories or to test directly their assumptions. It is not yet clear if there is a single, robust universality class for "directed percolation surface growth processes" or if the scaling exponents are sensitive to model details. Olami *et al.* have found evidence for the idea that the Hurst exponent for a typical advancing interface may differ from that of a critical interface, for models such as those of Sneppen [1089] and Buldyrev *et al.*, in a manner that depends on model details. In addition, Huber *et al.* [1104] have measured a value of 1.05 ± 0.02 for the exponent $\nu_{\perp}$ that is different from the directed percolation value of 1.097 ± 0.001. However, the total uncertainties associated with the computer simulations are too large to rule out universality.

5.5 Experiments

The growth of rough surfaces in thin film technology is often an undesirable consequence of attempts to prepare dense films with smooth surfaces. In other areas such as adhesion, catalysis and some aesthetical applications, controlled surface roughness may be desirable or essential. The study of the scaling behavior associated with the growth of rough surfaces has been primarily a theoretical and numerical modeling enterprise. Fortunately, the scope of experimental work in this area has been growing rapidly, though there is still an urgent need for more definitive experiments. Most experimental studies, in which attempts have been made to quantitatively characterize the scaling properties of rough surfaces, have been carried out using surfaces generated by fracture processes. Fracture is a complex, non-local, non-linear process with a rich phenomenology. The development of a satisfactory, general theoretical understanding of fracture phenomena is still far from complete. Even some of the most simple fracture processes appear to have at least the complexity of diffusion-limited aggregation (DLA), discussed in chapter 3. The characterization of fracture surfaces using fractal geometry will be discussed elsewhere [57].

As is the case in the study of self-similar fractals, scattering is one of the most important approaches to the study of rough surfaces on microscopic length scales, and rough surfaces can also be studied on longer length scales using correspondingly longer wavelengths, such as radar [1106]. In addition, the newer scanning tunneling microscopy (STM) and atomic force microscopy (AFM) technologies have been used to obtain 3-dimensional images of microscopic rough surfaces. These newer methods are ideally suited for the study of surfaces, but they are of relatively little value in the study of self-similar fractals or very rough surfaces for which overhangs are important. Stylus profilometry provides information that is similar to that obtained from STM and AFM, but on longer length scales. Similar information may be obtained using confocal microscopy, with less effort required for sample preparation [1107]. At present, confocal microscopy does not appear to be capable of providing a representation of the geometry of rough surfaces that is as accurate as those obtained from stylus profilometry or the slit island method, described in chapter 2, section 2.7.3. However, it has considerable potential for further development and is much more convenient. The slit island method combined with high resolution optical imaging, is also a robust and reliable method.

Nitrogen adsorbtion isotherms have also been used to investigate surface roughness on short length scales [1108, 1109, 1110, 1111]. While this can be a valuable adjunct to other roughness measurements, the interpretation of N_2 adsorbtion experiments depends on adsorbtion models, and the results are model dependent.

Most experimental methods are useful over only a limited range of length scales, and it can be dangerous to try to piece together data acquired from different sources. However, experiments have the potential of being able to explore a much wider range of length scales than practical $2+1$-dimensional simulations. This potential has not yet been fully realized. With only one or two possible exceptions, there is no single experiment in which linear behavior in the $\log-\log$ plots used to determine the Hurst exponent has been seen over a range corresponding to two decades of length scales in the vertical, height, direction. Indeed, there are few cases in which truly linear $\log-\log$ behavior, with good statistics, covering one decade in the height has been found. Consequently, it may be some time before unambiguous comparisons between experiments and the results from theory and simulations can be made. At this stage, the idea that some rough interfaces are self-affine is motivated more by theoretical considerations than by compelling experimental results.

5.5.1 Fluid–Fluid Displacement Experiments

Several fluid–fluid displacement experiments give results that can be described in terms of simple, self-affine scaling scenarios. However, they are less well understood than the invasion percolation process, described in chapter 3, section 3.5.2, that leads to self-similar patterns. Some of these experiments are described in this section. Many fluid–fluid displacement experiments were motivated by the growing theoretical interest in the dynamics of rough interfaces. In many cases, the experiments are relatively simple to perform, but the processes that take place may be complex and poorly understood. In addition, experiments suffer from the same difficulties as do simulations, including crossovers, finite-size effects and corrections to scaling. It is also much less easy to vary and control important parameters.

One of the first experiments to draw attention to the fact that more than just the KPZ equation was needed to explain the growth of self-affine interface roughness was the work of Rubio *et al.* [1048] on the displacement of air by dyed water in a pseudo 2-dimensional porous medium consisting of glass beads, with nominal diameters of 100, 200 or 350 μm, packed into a thin horizontal cell, constructed from glass sheets separated by a Teflon[11] gasket, with interior dimensions of 55cm $\times$ 17cm $\times$ 0.15cm deep. In these experiments, the water wets the porous medium more strongly than air. The water was injected into the cell via one of its 17cm long edges, and the central portion of the fluid–fluid interface was digitized and analyzed. The Hurst exponent H was obtained by measuring the dependence of $w(\ell)$ on ℓ, where $w(\ell)$ is the interface width, measured over a distance ℓ (equations 2.179 and 2.180). A value of 0.73 ± 0.03 was obtained for H, independent of the bead diameter ϵ and capillary number Ca_p over the ranges $100\ \mu\text{m} \leq \epsilon \leq 350\ \mu\text{m}$, and $\approx 1.75 \times 10^{-3} \leq Ca_p \leq \approx 4 \times 10^{-2}$, studied in these experiments. Here, Ca_p is the pore level capillary number defined in equation 4.15. However, the amplitude A of ℓ^H in equation 2.180 ($w(\ell) = A\ell^H$) was found to depend algebraically on the capillary number ($A \sim Ca_p^{-0.47\pm0.06}$) over the one-and-a-half decade range of capillary numbers that were studied in this work. A decrease in the surface width $\xi_\perp$ with increasing capillary number is expected, since the interface is stabilized by viscous pressure drops and destabilized by the randomness of the porous medium.

Horváth *et al.* [1049, 1112] reanalyzed the interfaces of Rubio, and obtained a higher value of $H = 0.91 \pm 0.08$ for the Hurst exponent, using the same general approach. The significance of this discrepancy is not clear, but it does appear to illustrate the sensitivity of the effective Hurst exponent to details in the data analysis. A value of 0.88 ± 0.08 was also obtained from experiments similar to those of Rubio *et al.* In subsequent experiments [1049], air was displaced by 4% H_2O/96% glycerin (η = 180 cP, $\Gamma \approx 65$ dyne cm^{-1}) in a 24 cm $\times$ 100 cm Hele-Shaw cell containing a monolayer of 220 μm diameter glass spheres glued to one plate. The glycerin was injected via one of the shorter edges of the cell. The exponent β was obtained from the time dependence of the $x = 0$ component of the height difference correlation function

$$\begin{aligned}[C_X(x,t)]^2 &= <\{[h(x_\circ,t_\circ)-<h>(t_\circ)] \\ &- [h(x_\circ+x,t_\circ+t)-<h>(t_\circ+t)]\}^2>. \end{aligned} \tag{5.157}$$

In constant injection rate experiments it was found that $C_X(0,t) \sim <h>^\beta \sim t^\beta$, with $\beta \approx 0.65$. The correlation function $C_X(x,0)$ was found to grow as x^H, with $H \approx 0.81$ for small x and $H \approx 0.49$ for large x. The short length scale exponents

11. Teflon is a DuPont trademark for polytetrafluoroethylene.

$H = 0.81$ and $\beta = 0.65$ satisfy the scaling exponent relationship $H + (H/\beta) = 2$ quite well ($H + (H/\beta) \approx 2.06$). However, the values obtained for H and β were both much higher than the values predicted by the $1+1$-dimensional KPZ equation ($H = 1/2$ and $\beta = 1/3$).

Similar experiments, in which aqueous fluids such as coffee, ink and food colorings were absorbed into various kinds of paper until the fluid stopped rising and the interface did not change, were carried out by Buldyrev *et al.* [1020, 1052]. The wetted sheets were then dried, and the interface between the wetted and non-wetted regions, indicated by the coloring substances, was digitized and analyzed. A roughness exponent of 0.63 ± 0.04 was obtained, under all conditions of temperature, humidity and fluid explored. More recently, $2+1$-dimensional experiments were carried out by Buldyrev *et al.* [1113] using paper rolls that could be unrolled after the experiment to "see" and analyze the interface between the wetted and unwetted regions, and the 3-dimensional water absorbing porous materials. The water fronts in the 3-dimensional porous materials were analyzed by slicing the material in a direction parallel to the direction in which the front advances. The Hurst exponent was measured from the height difference correlation function $C_2(x)$ (chapter 2, sections 2.7.2 and 2.7.3) and the dependence of the variance in the surface height on the lateral distance ℓ over which the variance was measured.

A Hurst exponent of 0.5 ± 0.05 was obtained from both $2+1$-dimensional experiments, using both analysis methods. This is larger than the value of about 0.4 expected for a $2+1$-dimensional KPZ process. A $2+1$-dimensional version of the model of Buldyrev *et al.* [1052] was also studied and was found to give results in good agreement with the experiments ($H = 0.52 \pm 0.03$).

In the $1+1$-dimensional experiments of Buldyrev *et al.* [1052], evaporation of the imbibed liquid is an important process. Eventually, the fluid invasion front stopped rising at a height h_c, when the forces driving the front could no longer overcome the quenched disorder pinning forces associated with the random medium (paper). In general, h_c increased as the evaporation rate decreased. Amaral *et al.* [1114] interpreted the behavior of the fluid invasion process with evaporation in terms of a directed invasion percolation model. They assumed that the invasion pressure decreases exponentially with increasing height and that, if the height is much smaller than the characteristic decay length, there will be a constant invasion pressure gradient ∇P. Based on a directed percolation model [542, 540, 1115], they concluded that the width of the interface is given by

$$\xi_\perp \sim \nabla P_i^{-\gamma}, \tag{5.158}$$

where ∇P_i is the pressure gradient across the interface, $\gamma = \nu_\perp/(\nu_\perp + 1) \approx 0.523$ and $\nu_\perp \approx 1.097$ is the perpendicular correlation length exponent for directed percolation. The scaling argument leading to equation 5.158 is essentially the same as that used to obtain the relationship $\xi \sim \nabla p^{-(\nu/(\nu+1))}$ in chapter 3, section 3.5.3, where ∇p is the occupation probability gradient and ν is the standard percolation correlation length exponent. The experimental results and results from simulations using the model of Buldyrev *et al.* [1020], with a gradient in the density of pinning sites, indicated that the width of the pinned wetting front, at the final stage at which the front can no longer migrate, can be represented by the scaling form

$$w(\ell, \nabla P) = \ell^H f(\ell/[(\nabla P)^{-\nu/H}]), \tag{5.159}$$

with $f(x) = const.$ for $x \ll 1$ and $f(x) \sim x^{-H}$ for $x \gg 1$. In the analysis of the experiments, it was assumed that the gradient ∇P was proportional to the evaporation rate. Despite the apparent success of the gradient-directed percolation model, it is not clear how well this model captures the physical processes taking place in this complex experimental system.

Family *et al.* [1116] have studied the imbibition of water into several different types of paper. In these experiments, the paper sheets were suspended horizontally over a bath of water, to reduce the effects of evaporation. The growth of the wetting front width w was found to depend algebraically on both the time t and the mean distance $< h >$ traveled by the wetting front ($\xi_\perp \sim < h >^\beta$ with $\beta \approx 0.4$ to 0.5 and $\xi_\perp \sim t^{\beta'}$ with $\beta' \approx 0.3$ to 0.4). The mean distance $< h >$ traveled by the wetting front was observed to grow as $< h > \sim t^{\approx 0.7}$; this is consistent with the ratio β/β'. Hurst exponents in the range $0.62 \leq H \leq 0.78$ were also measured from the dependence of the front width $w(\ell)$ on the length scale ℓ over which it was measured. However, the range of length scales over which $w(\ell) \sim \ell^H$ was small, and quite different behavior, which appears to be non-scaling, was found on longer length scales. This long length scale behavior may be a result of a crossover, but this crossover was not investigated in detail.

Horváth and Stanley [1117] have carried out experiments in which evaporation was controlled by confining a vertical sheet of filter paper between two parallel, 40 cm × 30 cm, transparent polymer sheets separated by a gap of 3 mm and closed with side walls. The lower end of the cell was immersed in a large vessel of liquid, and the filter paper entered the top of the cell through a slit in a polyethylene film. The shape of the wetting front was recorded using a CCD camera, and the mean height $< h >$ was calculated in real time. In these experiments, the dry filter paper was moved downwards, under computer control, using a stepping motor to maintain the mean height of the interface at a fixed height $< h >$ above the surface of the bulk liquid. A set of rollers was used to keep the paper flat. The downward velocity V needed to maintain a constant value for $< h >$ was found to scale as $V \sim < h >^{-a}$, with $a = 1.59 \pm 0.01$. Under these conditions, the surface width $\xi_\perp$, measured from the height difference correlation function $C_2(x)$, was found to grow as $\xi_\perp \sim t^\beta$, with $\beta = 0.56 \pm 0.03$. This value for β is somewhat smaller than the values of about 0.6–0.7 obtained from

$1+1$-dimensional QKPZ model simulations. At the time of writing, the uncertainty in the value for β remains quite large. Horváth and Stanley attributed the non-linear $b(\nabla h(x,t))^2$ term in these experiments to the effects of gravity acting on the liquid, which leads to different effective driving forces in the horizontal and vertical directions. It would be interesting to carry out similar experiments in a horizontal cell.

The displacement of air by water in a Hele-Shaw cell, filled with a porous medium, has been studied as a function of the strength of viscous forces relative to the surface forces (the pore level capillary number Ca_p) by He *et al.* [1118]. A growing $1+1$-dimensional self-affine surface can be described in terms of a line of self-affine "blobs" with a size of $\xi_\parallel \times \xi_\perp$. He *et al.* argued that the capillary force f_{cap}, across a single pore, has a characteristic magnitude of $P_{cap}\epsilon^2 \approx \Gamma\epsilon$, where P_{cap} is the characteristic capillary pressure ($P_{cap} \approx \Gamma/\epsilon$) and ϵ is the pore size. In a region of length $\xi_\parallel$, there will be $N = \xi_\parallel b/\epsilon^2$ such pores, where b is the interior thickness of the cell, so that the fluctuation in capillary forces over regions of size $\xi_\parallel$ will be given by

$$\delta F_{cap} \approx f_{cap} N^{1/2} \sim \Gamma\epsilon(\xi_\parallel b/\epsilon^2)^{1/2} \approx P_{cap}\epsilon(b\xi_\parallel)^{1/2}. \tag{5.160}$$

These fluctuations in the capillary forces can be considered to be responsible for the destabilization of the advancing front, and are opposed by stabilizing forces due to viscous flow in the confined porous medium. The viscous forces accelerate lagging regions in the interface and retard the leading parts of the interface, relative to the overall motion. The magnitude of the difference between the viscous forces acting on lagging and leading regions of size $\xi_\parallel$ is given by

$$\delta F_{vis} \approx \xi_\parallel b \delta P \approx \xi_\parallel b \xi_\perp \nabla P_{vis}, \tag{5.161}$$

where ∇P_{vis} is the pressure gradient in the viscous fluid. The pore level capillary number can be defined as $Ca_p = \epsilon \nabla P_{vis}/P_{cap}$. For small values of the front width $\xi_\perp$, the fluctuations in the capillary pressure over regions of size $\xi_\parallel$ will exceed the stabilizing viscous force differences over regions of the same size, and, consequently, the interface roughness will grow. This growth will continue until $\delta F_{cap} \approx \delta F_{vis}$. It follows, from equations 5.160 and 5.161, that this will happen when

$$P_{cap}\epsilon(b\xi_\parallel)^{1/2} \approx \xi_\parallel b \xi_\perp \nabla P_{vis}. \tag{5.162}$$

Assuming, on physical grounds, that the roughness exponent α and the Hurst exponent H are equal, the correlation lengths $\xi_\parallel$ and $\xi_\perp$ are related by $\xi_\parallel = \epsilon'(\xi_\perp/\epsilon')^{1/H}$, where ϵ' is the short length cut-off for the fractal scaling (approximately the bead size). This relationship between $\xi_\parallel$ and $\xi_\perp$, with the definition of the capillary number and equation 5.162, gives

$$\epsilon\epsilon'^{(1-H)/2} b^{1/2} \xi_\perp^{1/2H} \approx Ca_p \epsilon'^{-1/H} b \xi_\perp^{1+(1/H)} \tag{5.163}$$

or

$$\xi_\perp^{1+(1/2H)} = Ca_p^{-1} b^{-1/2} \epsilon^2 \epsilon'^{(H-1)/2H}, \tag{5.164}$$

so that

$$(\xi_\perp \epsilon)^{(1+2H)/2H} \approx Ca_p^{-1} b^{-1/2} \epsilon^{1/2} \epsilon^{-(1+2H)/2H} \epsilon^{3H/2H} \epsilon'^{(H-1)/2H} \tag{5.165}$$

or

$$(\xi_\perp/\epsilon)^{(1+2H)/2H} \approx Ca_p^{-1} b^{-1/2} \epsilon^{1/2} \epsilon^{-(H-1)/2H} \epsilon'^{(H-1)/2H}, \tag{5.166}$$

and [1118]

$$(\xi_\perp/\epsilon) \sim (Ca_p b^{1/2}/\epsilon^{1/2})^{-2H/(1+2H)} (\epsilon'/\epsilon)^{(1-H)/(1+2H)}. \tag{5.167}$$

It is reasonable to expect that $\epsilon' \approx \epsilon$, so that [1118]

$$(\xi_{\perp s}/\epsilon) \sim (Ca_p b^{1/2}/\epsilon^{1/2})^{-2H/(1+2H)}, \tag{5.168}$$

where $\xi_{\perp s}$ is the asymptotic, saturated, surface width.

In the propagation of a 2 + 1-dimensional interface into a 3-dimensional porous medium, the fluctuation of the capillary forces over an area of $\xi_\parallel^2$ is given by $\delta F_{cap} \approx f_{cap}(\xi_\parallel/\epsilon) \sim \Gamma \xi_\parallel \approx P_{cap} \epsilon \xi_\parallel$, and the viscous force acting on a blob of size $\xi_\parallel \times \xi_\parallel \times \xi_\perp$ is simply $\xi_\parallel^2 \delta P = \xi_\parallel^2 \times \xi_\perp \nabla P_{vis}$. Following the lines used to obtain equation 5.167, it can quite easily be shown that [1118]

$$(\xi_{\perp s}/\epsilon) \sim (Ca_p)^{-H/(1+H)} (\epsilon'/\epsilon)^{(1-H)/(1+H)} = c(Ca_p)^{-H/(1+H)}. \tag{5.169}$$

He *et al.* [1118] used a horizontal cell with a size of 26 cm × 116 cm and a gap of 0.9 mm. The cell was filled with 0.36 ± 0.06 mm diameter glass beads, and the water entered the cell via a 26 cm long groove cut into the lower (acrylic) wall of the cell. The upper wall of the cell was made out of glass. Deionized water was first injected into the cell, via the 26 cm long groove, at a high rate of 20 ml min^{-1}, until the interface had advanced 20 cm into the cell. This set up a smooth "initial" interface in accordance with equation 5.168. The injection was then continued, at constant rates in the range 0.01 to 10 ml min^{-1}. The corresponding interface velocities V were in the range 1.58 μm s^{-1} $< V <$ 1580 μm s^{-1}, and individual experiments required 6 minutes to 100 hours. He *et al.* found that the width of the interface, measured during the second half of each experiment, increased strongly with decreasing capillary number, and could be represented by the power law

$$\xi_{\perp s} \sim Ca_p^{-q}, \tag{5.170}$$

where the exponent q lay within the range $1/2 \leq q \leq 2/3$ expected from equation 5.168, if $1/2 \leq H \leq 1$. He *et al.* found that there are quite large fluctuations in the Hurst exponent H during these experiments, but that H generally increased during an experiment from an "initial" value of about 0.65

and reached a more or less constant value of about 0.8 ± 0.05 during the last half. The Hurst exponent was measured from the dependence of the interface width $w(\ell)$ on the length scale ℓ over which it was measured. In some cases, power law behavior was found over almost two decades in ℓ. In addition, there was a tendency for the exponent H to decrease with increasing capillary number. These experiments gave a new perspective on the work of Rubio *et al.* [1048] and Horváth *et al.* [1049, 1050, 1112] outlined above. It is not clear if these experiments point to crossovers between simple limiting behaviors or to a more serious revision of the nascent universality classes outlined earlier in this chapter. Further investigation, along the direction of these experiments, is needed.

The model developed by Robbins *et al.* [1082, 1083, 1086] to simulate the penetration of a fluid–fluid interface through a random $1+1$-dimensional porous medium may provide an explanation for the experiments of Rubio *et al.* [1048, 1119] and Horváth *et al.* [1112]. In particular, the formation of a self-affine interface with a Hurst exponent of $H \approx 0.8$ (described earlier in this chapter) when the driving force $\mathcal{F}$ is larger than or equal to the critical pinning force $\mathcal{F}_c$, for a fluid with a contact angle θ' which is less than θ'_c, appears to be a very significant result. The model appears to represent quite well the invasion of a wetting fluid into a porous medium, and the Hurst exponent is similar to those measured in the wetting invasion experiments.

Very similar interfaces generated by the propagation of combustion fronts have been studied by Zhang *et al.* [1120]. In these experiments optical lens cleaning paper with a thickness of $\approx 45\ \mu$m and areal density of $9.1\ \mathrm{g\,m^{-2}}$ was impregnated with 0.87–$1.6\ \mathrm{g\,m^{-2}}$ potassium nitrate. The impregnated paper was dried, held vertically in a frame and ignited along its lower edge by an electrically heated wire. A flameless, slow burning combustion front, with velocities in the range 5.5–$8.2\ \mathrm{mm\,s^{-1}}$, propagated vertically. A Hurst exponent was measured from the dependence of the surface width $w(\ell)$ on the length scale ℓ over which it was measured. The data could be represented quite well by $w(\ell) \sim \ell^H$, with $H \approx 0.7$, over about one-and-a-half decades of ℓ. Zhang *et al.* indicated that these results might be understood in terms of a KPZ-like interface growth process with “power law noise”. This idea is supported by the direct evidence obtained by Horváth *et al.* [1050] for such a noise distribution in their wetting front experiments, discussed earlier in this chapter. However, the relationship, if any, between the dynamics of combustion fronts and wetting fronts is not well understood.

Provatas *et al.* [1121] have used β radiography to study density fluctuations in hand made paper. They found that, for thin paper with an areal density of $7.2\ \mathrm{g\ m^{-2}}$, the density fluctuations $\delta\rho(\mathbf{r}) = \rho(\mathbf{r}) - <\rho(\mathbf{r})>$ could be described by a correlation function $C_\delta(r) =< \delta\rho(\mathbf{r}_\circ)\delta\rho(\mathbf{r}_\circ + \mathbf{r} >_{|\mathbf{r}|=r}$ with the form $C_\delta(r) \sim r^{-\alpha}$, where $\alpha \approx 0.4$, on length scales extending to more than ten times the fiber length. For thicker paper, the range of length scales over which the correlation function $C_\delta(r)$ could be approximated by a power law was reduced, and for papers with areal densities greater than about 40 g m–2 the evidence for a power law regime was weak. For thick paper, with an areal density of 86.3 g m^{-2}, the correlation function decayed rapidly with increasing distance r at distances greater than one or two fiber lengths. At low areal densities, the effective power law could be the result of a crossover from the single fiber fractal dimensionality of 1 to a fractal dimensionality of 2 on long length scales, without any genuine scaling of the correlation function. On the other hand, processes such as aggregation could lead to power law correlations. In any event, density correlations extending out to more than ten fiber lengths (more than 2.7 cm in this case) could have important implications for the propagation of fluid invasion fronts and combustion fronts in paper materials. The effects of these density correlations should be taken into account in the interpretation of the experiments in terms of either theoretical models or computer simulation results, particularly in the case of experiments with low areal density paper.

5.5.2 The Growth of Cell Colonies

The Eden model [348, 349] (chapter 3, section 3.4.1) was originally proposed as a model for the growth of cell colonies, and it provides a motivation for studying the geometry of cell colonies to determine if they exhibit fractal scaling properties. Fujikawa and Matsushita [1122, 1123] have shown that several species of bacteria form cell colonies with randomly branched structures, like DLA clusters [181], when grown on thin layers of agar gel with low nutrient concentrations. Similar results were also obtained by Matsuyama *et al.* [1124]. If the nutrient concentration is increased, the bacterial cell colonies become more compact and resemble "Eden" clusters more than DLA clusters. They have a more or less compact internal structure with a rough surface.

Vicsek *et al.* [1051] have studied colonies of *Escherichia coli* and *Bacillus subtilis*. The colonies were inoculated along a straight line in a 5 mm thick layer of nutrient rich agar gel. A growth front propagated away from the inoculation line, and this front was characterized using digitized images of the cell colonies. The dependence of the interface width $w(\ell)$ on the segment length (ℓ) was measured (equation 2.179). At intermediate length scales, the dependence of $w(\ell)$ on ℓ could be represented quite well by equation 2.180, with a Hurst exponent of $H = 0.78 \pm 0.07$. This value for H is substantially larger than the value of $H = 1/2$ associated with the Eden growth model, but it is quite similar to that found in some of the fluid–fluid displacement experiments, described earlier in this section, and in models with quenched disorder or non-Gaussian noise.

5.5.3 Phase Boundaries and Grain Boundaries

The growth of the crystalline/amorphous interface in thin layers of amorphous gallium arsenide implanted with As^+ ions has been studied by Licoppe *et al.* [1125]. The ion implantation was carried out at energies of 190, 110 and 40 keV and doses of 8×10^{14}, 2×10^{14} and 1.3×10^{14} atoms cm^{-2}, respectively. This leads to a relatively flat implantation profile and an amorphous layer 900 Å thick lying on a crystalline substrate. Partial recrystallization was induced by a limited time heat treatment at 300 °C. The progress of the recrystallization front was observed using high resolution electron microscopy of cross-sections perpendicular to the plane of the amorphous layer and substrate. Time resolved reflectivity measurements were used to measure the mean position of the crystalline/amorphous interface and the magnitude of the interface roughness ($\xi_\perp$) [1125]. The interpretation of the time resolved reflectivity (TRR) measurements was based on simulations carried out using a $2+1$-dimensional independent column model [939, 996], which was used to generate model surfaces for calculation of the TRR signal. The results of these calculations were then used to establish a relationship between the TRR modulation amplitude and the roughness amplitude. In this manner, the dependence of $\xi_\perp$ on the mean distance $\overline{h}$ traveled by the interface was determined. Licoppe *et al.* [1125] found that the dependence of $\xi_\perp$ on $\overline{h}$ could be described quite well by $\xi_\perp \sim \overline{h}^\beta$, with $\beta = 0.50 \pm 0.05$. This value for β is in good agreement with the result $\beta = 1/2$ obtained from the independent column model, but is not consistent with the results obtained from most more "realistic" surface growth models.

Lereah *et al.* [1126] have studied the propagation of the crystalline/amorphous interface in 250 Å thick films of Ge/Al alloys. The preparation of the films is described in chapter 4, section 4.2.2. The growth of the interface was stopped by cooling, and the crystallization front was examined using atomic scale resolution electron microscopy. The interface was found to be rough down to atomic length scales. An effective fractal dimensionality of about 1.2 was measured over length scales in the range 10 Å to 1000 Å, using projections of the interface in the 250 Å thick films onto a plane parallel to the plane of the film. This fractal dimensionality was measured using a "coastline analysis" in which the apparent length of the coastline was measured as a function of the length scale of the measurement. Lereah *et al.* [1126] concluded that the growth of the crystalline/amorphous interface might be described in terms of the Eden model, and the growth of the entire pattern has been modeled successfully using a "DLA in an Eden box" model (chapter 4, section 4.2.2) [871]. However, a fractal dimensionality of 1.2 corresponds to a Hurst exponent of 0.83, from equation 2.218, that is far from the $1+1$-dimensional Eden model value of 0.5. This effective fractal dimensionality may be a result of the crossover from the short length scale fractal dimensionality of 1.5 to the long range fractal dimensionality of 1.0 for a $H = 1/2$, self-affine, Eden model surface. In any event, the fact that the film thickness lies in the middle of the range of length scales over which the fractal dimensionality was measured complicates any interpretation of these results.

The growth of grain boundaries between gold thin films prepared by the deposition of gold vapor generated by the flash evaporation of high purity (99.9999%) Au wire in a high vacuum ($\approx 10^{-5}$ Pa) onto a NaCl single crystal was studied by Fitzsimmons and Burkel [1127]. Two gold-coated NaCl single crystals were pressed into contact with each other in a jig designed to control their relative orientation and were then annealed in air at a temperature of 250 °C for 2 hours. This procedure generated large area (001) twist grain boundaries. The NaCl substrates were then removed by dissolution in water, and the resulting gold films were placed on polished silicon wafers. In these experiments, the bicrystal specimens were prepared from a thick ($\approx$ 390 nm) film and a thin ($\approx$ 24 nm) film and the thick film Au crystals were placed in contact with the silicon. The bicrystal films and single crystal films were studied using 1.11 Å ($\Delta\lambda/\lambda = 10^{-3}$) synchrotron radiation x-ray reflectivity.

The single crystal, unlaminated gold films were found to have a density of 86%, relative to bulk gold, and a rough surface. The diffuse scattering was analyzed using the approach of Sinha *et al.* [1128], discussed below, and was found to be consistent with scattering from a self-affine surface with a Hurst exponent of 0.87 ± 0.13 and a horizontal correlation length ($\xi_{\parallel}$) of about 1000 Å, for the thick films. The diffuse scattering from the (001) twist grain boundaries was also analyzed. A Hurst exponent of $0.55 - 0.77$ and a horizontal correlation length ($\xi_{\parallel}$) of about 3000–4000 Å were measured. There are substantial uncertainties associated with these experiments and their interpretation. The analysis of scattering from specimens containing more than one rough interface is a challenging undertaking, and, like all other scattering experiments, the interpretation depends on the selection of an appropriate model and a good understanding of all of the sources of scattering. In this case, the origin of the diffuse scattering from the gold–gold interface is not well understood, but it is most probably due to density fluctuations resulting from a high void concentration at the interface. In addition, it is not known if these artificially produced grain boundaries are characteristic of "natural" grain boundaries. At this stage, the origin of the grain boundary roughness is not known. It could be primarily a consequence of the rough surfaces of the gold films used to fabricate the bicrystal specimens with twist grain boundaries, or it could also be a consequence of the growth of the grain boundaries during the annealing process.

Honjo and Ohta [1129] have studied the growth of NH_4Cl during the evaporation of a saturated aqueous NH_4Cl solution, which was held in a reservoir at a temperature of 40°C and fed to a horizontal sapphire sheet held at 40.42±0.02°C via holes with a of 50μm (vertical) × 7mm (horizontal) rectangular cross-section. The solution evaporated on the sapphire sheet and soon became supersaturated. As a result, NH_4Cl crystals grew horizontally. The NH_4Cl solution flowed from the reservoir to the advancing crystallization front via "grain boundaries" in the otherwise continuous polycrystalline NH_4Cl film. This process differs from most other growth phenomena that have been studied in the laboratory, since the new material is transported through the growing "deposit", *behind* the advancing interface. Such processes are common in nature. Hurst exponents in the range $0.79 \leq H \leq 0.91$ were measured from the dependence of the variance of the surface "height" (the dependence of the surface width $w(\ell)$) on the length scale ℓ over which it was measured. Power law behavior was found over more than one order of magnitude in ℓ.

A general discussion of pattern formation during the advance of phase boundaries in processes such as directional solidification, viscous fingering and dendritic growth is beyond the scope of this book. Information on these topics can be found in articles by Jasnow and Viñals [167], Langer and Turski [1130], Jackson and Hunt [1131], and in references contained in these articles.

5.5.4 Deposition Experiments

There is a wealth of qualitative information concerning the structure of rough surfaces produced by vapor phase deposition (see the review by Thornton [1132], for example). However, it is only quite recently that experiments have been carried out to characterize the scaling aspects of the growth of rough surfaces, in quantitative terms. In most vapor deposition processes, a thin layer of material, typically up to several thousand angstroms thick, is deposited on a more or less smooth surface. For these systems, STM, AFM and scattering techniques have been used to characterize the surface roughness. Unlike scattering experiments, the interpretation of STM and AFM experiments is, in principle, quite direct, since a height field $h(\mathbf{x})$ is measured and it can be analyzed in many ways. A theoretical foundation for the interpretation of scattering from self-affine surfaces was established by Sinha *et al.* [1128], and a clear exposition, with further development of the theory, has been presented by Yang *et al.* [1133, 1134]. Light, x-ray and neutron scattering have become the most widely used approaches to the characterization of self-similar structures on microscopic length scales. It appears that scattering techniques, particularly x-ray scattering, will also become one of the most powerful techniques for studying microscopic self-affinity. Synchrotron radiation x-ray sources are usually required to obtain good data. Unfortunately, beam attenuation limits x-ray scattering to surfaces with a width $\xi_\perp$ smaller than ≈ 100 Å [1135]. As a result of this restriction, scaling covering more than about one decade in the surface height fluctuations cannot be explored using this approach alone.

Most of the recent work on deposition has been focused on the use of techniques such as molecular beam epitaxy to obtain dense, crystalline films with smooth surfaces. The use of computer models and theoretical approaches, similar to those described in this chapter, to obtain a better understanding of molecular beam epitaxy, including deviations from the desired film structures and ways of controlling film growth, are rapidly growing areas that have already led to important advances. However, the main focus of this work is not on fractal geometry or scaling, and, for the most part, this area is beyond the scope of this book.

Most analytical studies of scattering from self-affine surfaces have been based on the assumption that the height–height correlation function $G_2(x)$ has the form

$$G_2(x) = \xi_\perp^2 [e^{\{-(x/\xi_\parallel)^{2H}\}}], \tag{5.171}$$

corresponding to a height difference correlation function of the form

$$[C_2(x)]^2 = 2\xi_\perp^2 f(x/\xi_\parallel) = 2\xi_\perp^2 [1 - e^{\{-(x/\xi_\parallel)^{2H}\}}], \tag{5.172}$$

and that the fluctuations of the surface height about the mean surface height have a Gaussian form. The function $f(y) = f(x/\xi_\parallel) = 1 - e^{\{-(x/\xi_\parallel)^{2H}\}}$ in equation 5.172 has the form $f(y) \sim y^{2H}$ for $y \ll 1$ and $f(y) = 1$ for $y \gg 1$. There is, in general, no reason why either of these assumptions should be true for a real surface. In scattering from self-similar fractals, the form chosen for the cut-off function, for the power law two-point density–density correlation function, can influence the effective value reported for the fractal dimensionality. Similar effects can be anticipated for scattering from self-affine surfaces. The specific form for the cut-off function $f(y)$ implied by equation 5.171 is not unique. For example, Palasantzas and Krim [1136] have proposed three alternative forms and Palasantzas [1137, 1138] has proposed a different form for the height–height correlation function $G_2(x)$ (or $[C_2(x)]^2$), based on equilibrium roughening models. However, the results obtained using correlation functions of the form given in equations 5.171 or 5.172 are probably adequate for most amorphous surfaces, and the scaling relationships used in the determination of the Hurst exponent H should be generally valid. For the surfaces of crystalline solids, lattice effects have to be taken into account, at least on short length scales. In practice, overhangs and porosity in the underlying deposit may cause more serious problems with the interpretation of scattering experiments.

The theoretical work of Sinha *et al.* [1128] and Wong and Bray [1139] addressed the problem of x-ray scattering from a single-valued, self-affine rough surface of a homogeneous material using a weak scattering, Born approximation. This model is appropriate for scattering from an amorphous solid that is homogeneous on the length scales of the order of $|\mathbf{q}|^{-1}$, where $\mathbf{q}$ is the momentum transfer vector. It is convenient to discuss scattering experiments in terms of the components $q_\perp$ and $q_\parallel$ of the momentum transfer vector parallel and perpendicular to the plane of the rough surface. The wave number k is defined as $k = 2\pi/\lambda$, and $q = |\mathbf{q}| = (4\pi/\lambda)\sin(\theta_s/2)$, where $\mathbf{q} = \mathbf{k}_r - \mathbf{k}_i$ is the difference between the wave vectors of the incident and scattered or reflected x-rays. Here, θ_s is the scattering angle ($\theta_s = \pi - (\theta_r + \theta_i)$, where θ_i and θ_r are the angle of incidence and angle of reflection, respectively). In terms of the scattering geometry, $q_\perp = |\mathbf{q}|\cos(\theta_q) = (2\pi/\lambda)[\sin(\pi/2 - \theta_i) + \sin(\pi/2 - \theta_r)]$ and $q_\parallel = |\mathbf{q}|\sin(\theta_q)$, where $\theta_q = (\theta_i - \theta_r)/2$ is the angle between $\mathbf{q}$ and the normal to the surface. The components in the x and y directions are given by $q_x = |\mathbf{q}|\sin(\theta_q)\cos\phi_a$ and $q_y = |\mathbf{q}|\sin(\theta_q)\sin\phi_a$, where ϕ_a is the azimuthal angle.

Many experiments are carried out at near-grazing incidence, where θ_i, the angle of incidence, is close to $\pi/2$. Because the x-ray refractive index of solids is slightly less than unity the angle of incidence must be below the critical angle of incidence for total *external* reflection. Below the critical angle of incidence, the x-ray scattering is determined by the electron density distribution. The scattered intensity consists of an instrumentally broadened, specular ($\theta_r = \theta_i$) "δ" peak superimposed on a broad, diffuse scattering, background peak. The intensity of the specular δ peak is determined by the width $\xi_\perp$ of the surface and disappears if the surface is self-affine on sufficiently large length scales. The diffuse scattering, on the other hand, depends on the height difference correlation function. The relationship between the intensity S_{spec} of the specular δ peak and the rms roughness $\xi_\perp$ has been known for many years [1128, 1140, 1141] and can be expressed in the form

$$S_{spec} \sim e^{-[q_\perp^2 \xi_\perp^2]}. \tag{5.173}$$

The exponential function in equation 5.173 is sometimes called the static Debye–Waller factor. Sinha *et al.* also demonstrated that the Hurst exponent could be determined by measuring the diffuse scattering at several different scattering angles θ_s to control $q_\perp$ ($q_\perp \approx 2\pi\theta_s/\lambda$) by fitting the experimental scattering intensities $S(q_\parallel, q_\perp)$ with intensities calculated from the H dependent height difference correlation function model.

The structure factor for a self-affine surface, with continuous height variables, can be described in terms of the scaling form[12]

$$S(\lambda^{-1}\mathbf{q}_\parallel, \lambda^{-H} q_\perp) = \lambda^2 S(\mathbf{q}_\parallel, q_\perp). \tag{5.174}$$

12. This equation is a direct consequence of the invariance of the surface under the rescaling $x \to \lambda x$, $h \to \lambda^H h$. The scattered intensity can be estimated by covering the surface with blobs of size $q_\parallel^{-1} \times q_\parallel^{-1} \times q_\perp^{-1}$ and adding the contributions from each of the blobs. Since the scattering contribution of each blob is proportional to the square of the amount of interface in the blob ($S_b \sim \lambda^4$), and the number of blobs in the beam is proportional to λ^{-2}, it follows that the total scattered intensity is proportional to λ^2.

Consequently, the width of the diffuse scattering $S(q_{\parallel})$ peak measured at a fixed $q_{\perp}$ will be given by

$$w_{q_{\parallel}} \sim q_{\perp}^{1/H}, \tag{5.175}$$

and the Hurst exponent can be obtained by measuring the width of the diffuse scattering at a series of scattering angles. This can be accomplished by using the "rocking curve" approach in which the scattering angle is fixed and the sample is rocked to change $\theta_i - \theta_r$ or $q_{\parallel}$ at fixed values of $q_{\perp}$. It also follows [1136, 1142] from equation 5.174 that for $q_{\parallel} = 0$

$$S(q_{\parallel} = 0, q_{\perp}) \sim q_{\perp}^{-[2+(1/H)]}. \tag{5.176}$$

In practice, experiments cannot be carried out in the $q_{\parallel} = 0$ limit because of the $q_{\parallel} = 0$ delta peak. However, H can be determined by measuring the intensity of the diffuse peak at sufficiently small values of $q_{\parallel}$.

The structure factor for a growing surface can be described by the scaling form

$$S(\lambda^{-1}\mathbf{q}_{\parallel}, \lambda^{-\alpha}q_{\perp}, \lambda^{z}t) = \lambda^{2}S(\mathbf{q}_{\parallel}, q_{\perp}, t). \tag{5.177}$$

For small values of $q_{\perp}$, the line width of the broad diffuse component is inversely proportional to the lateral correlation length $\xi_{\parallel}$. Consequently, the dynamical exponent can be determined from the relationship

$$w_{q_{\parallel}} \sim t^{-1/z} \sim \xi_{\parallel}^{-1}. \tag{5.178}$$

Similarly, the width $w_{q_{\perp}}$ of the diffuse scattering peak measured in the $q_{\perp}$ direction from $S(q_{\parallel} = 0, q_{\perp})$ is given by [1141]

$$w_{q_{\perp}} \sim \xi_{\perp}^{-1}. \tag{5.179}$$

The width in the $q_{\perp}$ direction can be measured in reflection high energy electron diffraction (RHEED) experiments or by using a "vertical" position sensitive detector in x-ray scattering experiments.

Yang *et al.* [1133, 1143] showed how the surfaces of crystalline solids with stepped surfaces could be characterized using scattering methods. The scattering behavior can be described in terms of the scattering phase ϕ_p, where $\phi_p = q_{\perp}a$ and a is the vertical spacing in a crystalline solid or the height of the steps in the surface of a crystalline solid. The scattering behavior depends on $[\phi]$, where $[\phi]$ is ϕ_p modulo 2π such that $-\pi \le [\phi] \le \pi$ and $\phi_p = \pi$ corresponds to the out of phase condition. A theoretical approach based on a continuous distribution of surface heights fails except under in-phase, constructive interference scattering conditions, where $[\phi] \approx 0$ and the diffraction is not sensitive to surface steps. Consequently, equation 5.175 can be used to estimate the Hurst exponent, and can be replaced by

$$w_{q_{\parallel}} \sim |[\phi]|^{1/H} \tag{5.180}$$

if $|[\phi]| \ll \pi$. Equation 5.173 also survives except under near-to-out-of-phase ($|[\phi]| >\approx \pi/2$) scattering conditions, and can be written in the more useful form

$$R = e^{-\{[\phi]^2 \langle \xi_\perp(t) \rangle^2\}}, \tag{5.181}$$

where R is the ratio between the integrated intensity of the instrumentally broadened δ peak and the integrated intensity of the diffuse scattering profile [1133, 1143]. Measurement of the ratio R as a function of time during a surface growth process allows the growth of the surface width to be determined and provides an estimate of the surface width growth exponent β. An alternative procedure is to determine $\xi_\perp(t)$ from the dependence of R on $[\phi]$, at each stage in the growth process.

In processes such as molecular beam epitaxy discretization, representing the crystalline nature of the solid introduces a short length scale cut-off to the self-affine scaling, which has important effects under "out-of-phase" diffraction conditions where $[\phi]$ is large. Under these conditions, the scattering structure factor is given by

$$S(\mathbf{q}_\parallel, q_\perp, t) = \int e^{i\mathbf{q}_\parallel \cdot \mathbf{x}} C_1(\mathbf{x}, \phi_p, t) d\mathbf{x}, \tag{5.182}$$

where

$$C_1(\mathbf{x}, \phi_p, t) =< e^{iq_\perp a}[h(\mathbf{x}_\circ, t) - h(\mathbf{x}_\circ + \mathbf{x}, t)] > . \tag{5.183}$$

Near to the large $[\phi]$, out-of-phase conditions, where the specular, $q_\parallel = 0$, δ peak disappears, the diffuse scattering is a consequence of scattering by the steps, and the line shape can be interpreted in terms of interference between these scattering contributions and can be used to measure quantities such as terrace widths.

The approach to the interpretation of scattering experiments discussed in this section is quite general. It has been applied to helium atom scattering [1144, 1145] as well as to x-ray and electron scattering. The simple relationships given in equations 5.175, 5.176, 5.180 and 5.181 do not completely obviate the need to use surface models in the interpretation of surface scattering experiments. They are asymptotic relationships subject to corrections and crossovers. More accurate and reliable results may be obtained using scattering models that take into account cut-off effects, including the long range form of the height difference correlation function and the discretization of the surface height. Scattering models are also needed for the interpretation of scattering from two or more interfaces or systems in which scattering from both voids and other heterogeneities as well as interfaces is important.

5.5.4.1 Epitaxial Growth

Because of its relative simplicity and the precise quantitative techniques that have been developed to study it, molecular beam epitaxy provides the best opportunity for detailed comparisons between experiments, simulations and theory. However, molecular beam epitaxy explores only the early stages of surface roughening. For truly rough, self-affine surfaces, the attractive simplicity of molecular beam epitaxy is lost, and the power of some experimental methods is diminished.

A good example of the use of scattering/diffraction to study the formation of rough surfaces is provided by the work of He *et al.* [1146] on the growth of iron films on Fe(001) using high resolution, low energy electron diffraction (HRLEED). A "buffer" film of about 100 atomic layers of Fe was first deposited on a Au(001) surface. At this stage, the film surface was essentially flat and contained steps confined to the first two atomic layers. The intensity of the (00) electron diffraction beam, measured under out-of-phase diffraction conditions ($[\phi] \approx \pi$, $S(q_{\|}, \phi_p \approx 3\pi)$) as a function of $q_{\|}$, consisted of a sharp, instrumentally broadened "δ" peak superimposed on a broad, diffuse scattering background. Additional iron was then deposited at a rate of ≈ 2 monolayers min^{-1} and was characterized by HRLEED every 3 min. No significant annealing was observed for long periods after each deposition. As the deposition proceeded, the intensity of the $q_{\|} = 0$, δ peak decreased in response to the roughening of the surface. The growth of the surface width $\xi_{\perp}$ was measured from the dependence of the ratio R of the integrated delta peak intensity to the integrated intensity of the diffuse background profile using equation 5.181. In this manner, the surface width was found to grow as t^{β}, with $\beta \approx 0.22 \pm 0.02$ [1133], for times in the 10 min to 100 min range. A value of 0.79 ± 0.05 was obtained for the Hurst exponent H by comparing the measured dependence of the width of the diffuse peak $w_{q_{\|}}$ on the phase $[\phi]$ with that calculated using a discrete lattice model for the height difference correlation function. However, the width of the diffuse peak was measured over a range of only about one decade, and it appears that the total uncertainty in H is probably greater than ± 0.05.

Similar behavior was found by Yang *et al.* [1147] in their HRLEED study of the molecular beam epitaxial growth of Si on Si(111). Silicon vapor, provided by electron beam bombardment of a silicon rod, was deposited onto the Si(111) substrate with a miscut angle of $< 0.1°$ and a corresponding average terrace width of ≥ 1000 Å. The surface was cleaned in ultra-high vacuum by rapid heating to 1200 °C followed by rapid cooling. The HRLEED technique can be used to measure very small surface inclinations, and Yang *et al.* found that the average local slope grew as $(\ln t)^{1/2}$. The (00) beam was scanned along the $[\bar{1}12]$ direction under almost-out-of-phase diffraction conditions, with

$q_{\perp}a \approx 7.01\pi$, where a is the bilayer step height. The intensity was measured as a function of $q_{\parallel}$, the component of the momentum transfer vector parallel to the surface. The HRLEED line shape $S(q_{\parallel}, q_{\perp} \approx 7\pi/a)$ for the initial surface was a sharp peak, determined by the instrument resolution. Yang *et al.* found that, for a deposition rate of 7 bilayers min^{-1} and a substrate temperature of $\approx$ 275 °C, an initial transient period of layer by layer growth was followed by roughening. The line shape consisted of an instrumentally broadened delta peak superimposed on a broad, diffuse scattering peak. As the surface roughened, the intensity of the δ peak diminished in response to the roughening. The exponent β describing the growth of the surface width was determined to have a value of ≈ 0.25 from the time dependence of the ratio R between the integrated intensity of the δ peak and the diffuse scattering peak using equation 5.181. Hurst exponents in the range $0.95 \leq H \leq 1.1$ were measured from the dependence of the full width at half-maximum height $w_{q_{\parallel}}$ of the diffuse scattering peak on $[\phi]$ ($[\phi] = (8 - q_{\perp}a/\pi)$) using equation 5.180. Under out-of-phase conditions where $[\phi] \approx \pi$, the width of the diffuse scattering peak is proportional to the average local slope or the average step density. In these experiments, the surface inclination was obtained from the dependence of $w_{q_{\parallel}}$ on the deposition time, for $q_{\perp}a = 7.08\pi$ ($[\phi] = 0.92$). Unfortunately, the scaling reported in the paper by Yang *et al.* covered much less than a decade in time or length scales.

These results are consistent with a surface growth process that can be described by the MH equation (equation 5.18) with deposition noise. In this case, the Hurst exponent has a value of $H = 1$ and the exponent β describing the growth of the surface width has a value of 1/4. Based mainly on the observation that the average local slope grew as $(\ln t)^{1/2}$, Yang *et al.* concluded that the surface growth was driven by statistical fluctuations rather than by Schwoebel barriers [1148, 1149, 1150], which lead to the formation of mounds with a constant slope. The roughening behavior was not observed at a slower deposition rate of 1 bilayer min^{-1} or at a higher temperature of 350 °C. In the 350 °C experiments, the δ peak at $q_{\parallel} = 0$ showed little decay, indicating that the roughness was restricted to only one or two levels on the surface.

Schwoebel barriers play an important role in deposition on terraced surfaces with surface diffusion. Because of these barriers, the sticking probability on the advancing face of a step can be very different for an atom approaching from the upper terrace than for one approaching from the lower terrace. In most cases, there will be a "barrier" separating the upper terrace from the potential "well" at the face of the step. This barrier can be regarded as a consequence of the attractive interactions between an adsorbed atom and the surface. In order to pass from the upper terrace to the face of the step, the atom must move through a configuration in which it is near to fewer atoms in the surface than an atom sitting on a flat terrace surface. In any event, the approach from the upper terrace can be expected to be different from the approach from the lower terrace, because of the lack of symmetry.

The heteroepitaxial growth of Pb on Cu(100) at low temperatures has been studied experimentally by Zheng and Vidali [1144] using helium beam scattering. This is a system with a large lattice mismatch between the substrate and the deposited material. In these experiments, 99.999% pure Pb was deposited onto a cold, 150 K, copper substrate that had been prepared by argon ion sputtering and annealing at 850 K. The average terrace width on the substrate was 400 to 500 Å, and the Pb was deposited at rates of 0.1 or 0.05 monolayers s^{-1} in a vacuum of 10^{-10} Torr. The surface roughened after the first 16 monolayers had been deposited. Analysis of the scattering experiments indicated a Hurst exponent of $H \approx 1$ and a surface width that grew as $\xi_\perp \sim t^\beta$, with $\beta \approx 0.3$. These results seem to be consistent with other molecular beam epitaxy (MBE) experiments. Zheng and Vidali [1144], and several other authors, have suggested that the results of MBE experiments are consistent with the MH equation 5.18 with deposition noise which predicts that $H = 1$ and $\beta = 1/4$. However, as Zheng and Vidali point out, the asymptotic, long time behavior might be quite different. Indeed, the width of the interface grew only slowly after ≈ 40 monolayers had been deposited, and this is consistent with an asymptotic growth process that can be described by the EW equation.

Tong *et al.* [1151] have studied CuCl thin films grown by molecular beam epitaxy on $CaF_2(111)$ substrates, with substrate temperatures of 110 °C and 80 °C, at a low deposition rate of about 25 Å min^{-1}. Samples grown to average thicknesses of 60 Å and 120 Å were examined, using atomic force microscopy to map the height in 50 000 Å × 50 000 Å areas. At early times, the deposit consisted of "islands" on an essentially bare substrate. At a nominal thickness of 60 Å, and a substrate temperature of 110 °C, the islands had a thickness of about 300 Å and a width of about 3000 Å. The islands were approximately twice as wide and high in the 120 Å film, prepared under the same conditions. At a substrate temperature of 80 °C, the "islands" were much smaller and the island size did not grow noticeably with increasing film thickness. A small increase was found under careful examination. Tong *et al.* [1151, 1152] interpreted their results in terms of the "shadowing-diffusion" equation of Karunasiri *et al.* [948] (equation 5.233) discussed below (section 5.7.3). The reciprocal space height correlation function $C_h^{(2)}(\mathbf{k}, t)$, defined as

$$C_h^{(2)}(\mathbf{k}, t) \sim \int e^{i\mathbf{k}\cdot\mathbf{x}} < [h(\mathbf{x}_\circ) - h(\mathbf{x}_\circ + \mathbf{x})]^2 > d\mathbf{x} \tag{5.184}$$

or

$$C_h^{(2)}(\mathbf{k}, t) \sim \int e^{i\mathbf{k}\cdot\mathbf{x}} < [h(\mathbf{x}_\circ)h(\mathbf{x}_\circ + \mathbf{x}) - < h >^2] > d\mathbf{x}, \tag{5.185}$$

was measured and found to exhibit a k^{-4} dependence at large k ($k = |\mathbf{k}|$), corresponding to a Hurst exponent of $H = 1$, in accordance with the theory. However, the range of scaling was much smaller for the 80 °C films than for the 110 °C films. It is interesting that the work of Tong *et al.* [1151] arose from failed attempts to prepare thin, uniform films of CuCl on $CaF_2(111)$ for the study of fast switching devices and quantum well studies. The construction of uniform thin layers for the purpose of developing devices or the preperation of systems for experimental studies is often frustrated by the development of surface roughness and/or non-uniform coverage.

In later experiments [1152], Tong *et al.* measured Hurst exponents in the range $0.8 \le H \le 0.9$ from the dependence of the interface width $w(\ell)$ on the scale ℓ over which the width was calculated, for surfaces formed by the deposition of CuCl onto CaF_2 at substrate temperatures of 80 and 110 °C. The films were grown to thicknesses of 6, 12 and 40 nm, with a deposition rate of 2.5 nm minute^{-1}. The measurements were carried out for deposition at these temperatures because 110 °C is the highest temperature at which a significant amount of CuCl will stick to $CaF_2(111)$ and 80 °C is the lowest temperature at which oriented crystalline films will grow. The surface characterization was based mainly on 5000 Å × 5000 Å and 50 000 Å × 50 000 Å scans.

High resolution, low energy electron diffraction (HRLEED) has been used by Fang *et al.* [1153] to study the growth of Pb on Pb(110). During the deposition process, the Pb(110) surface was maintained at a temperature of 363 K, which is above the preroughening temperature of 353 K and below the roughening temperature of 415 K. The Pb(110) sample was chemically etched to remove an oxide layer before it was placed in the UHV chamber, sputtered with Ar^+ ions and annealed until no impurities could be detected by Auger electron spectroscopy. The lead was deposited at rates of 5 and 25 monolayers min^{-1}. The interface width $\xi_\perp$ was determined from the diffraction intensity $S(q_\parallel, q_\perp, t)$ by carrying out the scattering measurements after different deposition times. A new Pb(110) surface was prepared after each stage of deposition by heating the substrate and deposited film from earlier stages in an experiment to above the surface melting temperature and cooling slowly to anneal the film. Experiments were also carried out by depositing more material on previously grown surfaces that were not annealed (for the high deposition rate experiments). The diffraction phase ϕ_p was controlled via the electron energy. In the low deposition rate experiments, the surface width $\xi_\perp$ was measured from the phase dependence of the $q_\parallel \approx 0$ δ peak intensity using equation 5.181, at each stage in the growth process. A surface width growth exponent of $\beta = 0.77 \pm 0.05$ was obtained from the time dependence of $\xi_\perp$ for times in the range $1 \le t \le 5$ min. A value of 1.33 ± 0.05 was obtained for the Hurst exponent from the dependence of the full width at half-maximum $w_{q_\parallel}$ in the scattering peak $I(q_\parallel)$, on the phase angle $[\phi]$ measured away from the in-phase condition $[\phi] = 0$ using equation 5.180.

In the high deposition rate experiments, H decreased from a value of about 1.18 ± 0.05 at early stages (≈ 0.5 min) to 0.98 ± 0.05 at late stages (2.5–8.4 min). During the later stages of growth, diffraction from low surface energy Pb(111) facets was observed, and the formation of these facets was correlated with the decrease of H to a value of 1.0. The significance of the effective Hurst exponents, with values larger than unity, obtained from these experiments is not clear. The peak widths $w_{q_\parallel}$ extended over less than a decade, and for most data sets the "scaling regime" was only about half a decade. Consequently, the large values obtained for H most probably reflect the uncertainties inherent in the determination of the Hurst exponent. In these experiments the δ peak was quite broad due to instrumental effects. The difficulties involved in the separation of the scattering into a Gaussian δ peak and a "power law Lorentzian" diffuse peak must add to the uncertainties.

Reflection high energy electron diffraction (RHEED) is a technique that is commonly used to study surfaces [1154] and has been applied by Chevrier *et al.* [1155] to the study of iron films grown on (111) silicon surfaces at a temperature of 50 °C by molecular beam epitaxy. The width of the rough surface $\xi_\perp$ was obtained, based on the idea that the full width $w_{q_\perp}$ at half-maximum of the RHEED diffraction peaks measured in the direction perpendicular to the surface is proportional to $\xi_\perp^{-1}$. These experiments indicate that the exponent β obtained from the dependence of $\xi_\perp$ on the film thickness has a value of about 0.23 to 0.30. However, after a deposit thickness of about a few hundred angstroms had been reached, the growth of $\xi_\perp$ deviated substantially from the power law proposed by Chevrier *et al.* and was consistent with linear growth of $\xi_\perp$, with increasing film thickness. The values of 0.23 to 0.30 estimated for the exponent β are consistent with a number of the simple growth models discussed in this chapter. Unfortunately, the Hurst exponent was not measured. Similar results were obtained by MBE deposition of iron on Fe(111) at 50 °C.

Chevrier *et al.* [1156] used a similar approach to study the epitaxial growth of Si on Si(111) at temperatures in the range 250 °C to 400 °C. The surface width $\xi_\perp$ was observed to grow linearly with film thickness for deposition at about 3 Å min^{-1} up to a total thickness of about 500Å. The rate of growth of the surface width $\partial\xi_\perp/\partial h$ could be fit quite well by the Arrhenius form $\xi_\perp = hAe^{-(E_a/k_BT)}$ over this temperature range, with an activation energy of $E_a = 0.65 \pm 0.07$ eV, which is close to the activation energy for diffusion of silicon atoms on Si(111). These results, particularly the large values of $\beta \approx 1$, appear to be very different from those reported by Yang *et al.* [1133] (described above) for the MBE deposition of Si on Si(111) at ≈ 275 °C. However, the exponent of $\beta \approx 0.25$ was obtained by Yang *et al.* for film thicknesses in the $35 \leq h \leq 100$ Å range, and the surface width increased much more rapidly with increasing deposit thickness for $h > 100$ Å. The value of ≈ 1 reported by Chevrier *et al.* for β at 300 °C was obtained from data recorded at film thicknesses of $h \geq 100$ Å. It also appears, from the work of Yang *et al.* and Chevrier *et al.*, that the growth of the surface roughness is quite sensitive to both the surface temperature and the deposition rate in the ranges of these parameters used in the experiments. Consequently, there is no fundamental discrepancy between these two experimental studies, but more work is needed on the crossover between the thin and thick film regimes and the dependence of the roughening dynamics on experimental conditions. These experiments appear to be represented surprisingly well by a simple $1 + 1$-dimensional surface growth model developed by Krug and Schimschak [1157], which includes the destabilizing effects of fluctuations (deposition noise, diffusion noise and nucleation noise) on step flow.

In an earlier study, Eaglesham and Gilmer [1158] investigated the MBE growth of silicon at temperatures of 250 to 350 °C. The silicon was deposited at a rate of 1.0 Å s^{-1} with Ge marker layers, which were deposited at a rate of 0.1 Å s^{-1} and had a thickness of approximately one monolayer. The Ge marker layers made it possible to follow the surface growth using transmission electron microscopy to study vertical sections through the deposited films, after the growth process was completed. A transition from the epitaxial growth of crystalline films to the growth of amorphous material, discussed below, was observed. In a typical sample, the surface roughness, measured from the width of the Ge marker layers, remained below the resolution of $\approx$ 17 Å during the early, epitaxial growth regime. In the amorphous, upper part of the films, a Ge marker width of 45 ± 5 Å was measured, independent of the film thickness on which the Ge marker layer was deposited. Although the surface width was below the instrumental resolution during the early stage, epitaxial growth process, it could be determined from the intensity of the dark Ge lines which were expected to decrease linearly with increasing width of the Ge marker layers, under the conditions used in these experiments. In this way, the width of the surface was obtained from the intensity of the marker layers and was found to grow linearly with deposit thicknesses in the range 50 Å to 300 Å.

Epitaxial growth on "vicinal" surfaces that consist of a series of low steps with a height of one or two atoms separating large flat terraces[13] can often be described in terms of the "flow" of a series of well separated steps that separate broad terraces. However, the structure of vicinal surfaces is not always that simple [1159]. Under some circumstances, the dynamics of the step edges can be described in terms of a $1 + 1$-dimensional surface growth process. For example, Wu *et al.* [1160] have used scanning tunneling microscopy to study the kinetic roughening of steps on vicinal Si(001), miscut towards the (100) direction.

13. Vicinal surfaces can be thought of in terms of a surface that is slightly tilted or miscut with respect to a flat, high symmetry, low index crystal face forming terraces with flat, high symmetry tops separated by vertical steps. The terrace width ξ_{tw} is given by $\xi_{tw} = a/\tan\theta$, where θ is the tilt angle and a is the step height (the unit cell size in the vertical direction).

Silicon was deposited at rates in the range 0.003 to 0.020 monolayers s^{-1}, onto the vicinal surfaces. At a temperature of 350 °C, the steps developed into a fingered pattern. The large, approximately 1000 : 1, anisotropy of the diffusion coefficient on the terrace surfaces and the presence of two types of steps in the nominally (100) step edges appear to play an important role in the development of this morphology. These features are related to the formation of Si dimers that are rotated by 90° with respect to each other on alternating terraces. This results in an instability, controlled by "interactions" between adjacent steps. Measurements of the step edge surface widths $w(\ell, t)$, during the deposition process, indicated that $w(\ell) \sim \ell^H$ for $\ell \ll \xi_\parallel$ with $H = 0.45 \pm 0.1$ and $w(\ell) = \xi_\perp \sim t^\beta$ for $\ell \gg \xi_\parallel$, with $\beta = 0.29 \pm 0.08$. However, the development of $w(\ell, t)$, during the experiments, indicated that the increase of the surface width $\xi_\perp$ could be attributed mainly to the growth of the *amplitude* of $w(\ell, t)$ rather than to the growth of the correlation length $\xi_\parallel$. Consequently, it appears that the step growth dynamics is not a $1 + 1$-dimensional KPZ or EW process, despite the agreement between the measured values of the exponents H and β and the $1 + 1$-dimensional KPZ values of $1/2$ and $1/3$ or EW model values of $1/2$ and $1/4$.

Lanczycki and Das Sarma [1161] have attempted to simulate this experiment using $1 + 1$-dimensional, square lattice ballistic deposition and solid-on-solid deposition models, with site dependent, thermally activated hopping along the $1 + 1$-dimensional growth front to represent diffusion. A better agreement with the experiments was obtained from the simulations based on the ballistic deposition model than from simulations based on the solid-on-solid model, using realistic activation and interactions energies. However, the models lack the complexity of the experiments (for example, the anisotropic diffusion on the terrace surfaces is not represented), and the comparison between the simulations and the experiments is not completely convincing. The solid-on-solid model simulations allowed atoms to jump up or down "high" steps between column tops on the $1 + 1$-dimensional growth front to maintain consistency with the deposition rules. The ballistic deposition model belongs to the $1+1$-dimensional KPZ universality class, while the asymptotic behavior of the solid-on-solid model probably belongs to the EW universality class.

5.5.4.2 Other Thin Film Growth Processes

Much of the interest in the morphological aspects of surface growth was initially focused on the columnar morphologies that develop during deposition onto relatively cold surfaces and the dependence of the columnar morphology on experimental parameters such as the angle of incidence. The columnar morphology becomes more prominent as the angle of incidence increases. For example, Bellac *et al.* [1162] have studied the thin films formed by the dc magnetron sputtering of chromium, at a rate of 5 nm minute^{-1} onto glass microscope slides, at near-grazing angles of incidence, with an angular dispersion of about 10° about a nominal angle of incidence of $\theta \approx 89°$. The Cr films were examined using atomic force microscopy. The AFM studies indicated a columnar morphology, without the pronounced anisotropy seen in the grazing incidence ballistic deposition simulations, described below. The surface width was found to grow linearly with film thickness ($\beta \approx 1$) and an effective Hurst exponent of $H \approx 1$ was measured. The large effective Hurst exponent appears to be a result of the lack of small length scale roughness, which suggests that surface diffusion plays an important role in the growth of the columnar morphology in these experiments.

A transition from an ordered epitaxial growth process at short times or small deposit thicknesses to the growth of an amorphous film at larger thicknesses has been studied by Eaglesham *et al.* [1163]. This transition was found during the growth of Si films on smooth Si(100) surfaces at temperatures in the range 30 to 300 °C. They were carried out using deposition rates in the range 0.05 to $10\,\text{Å}\,\text{s}^{-1}$ at pressures of 10^{-10} to 10^{-9} Torr. The transition took place at a height h_{epi} that depended on both the temperature T and the deposition rate ϕ_d. The limiting epitaxial thickness could be represented by the empirical relationship

$$h_{epi} = A(\phi_d)e^{-E_a(\phi_d)/T}. \tag{5.186}$$

Values of 0.4 eV and 1.5 eV were obtained for the "activation energy" $E_a(\phi_d)$ for $\phi_d = 0.07\,\text{Å}\,\text{s}^{-1}$ and $\phi_d = 11\,\text{Å}\,\text{s}^{-1}$, respectively. The surface roughness was not characterized quantitatively. However, a transmission electron microscope image of a vertical cross-section of a film grown at a temperature of 300 °C (figure 2 in the paper of Eaglesham *et al.* [1163]) shows a striking saw-tooth-like interface between the crystalline epitaxial region and the amorphous region, which implies a large Hurst exponent for the *interface*. The shape of this interface was interpreted in terms of local nucleation of the amorphous phase, rather than a process in which a very rough surface becomes amorphous everywhere at more or less the same time. Eaglesham *et al.* [1164] found similar behavior in GaAs films at 200 °C. In this case, the thickness of the epitaxial layer depended strongly on the composition (Ga/As ratio). In this system, the transition height h_{epi} increases from < 200 Å at 210 °C to > 5000 Å at 220 °C at a fixed composition. Below the transition at ≈ 215 °C, the growth of h_{epi} with increasing temperature could be described by an Arrhenius form. In both cases, Eaglesham *et al.* suggested that the transition to an amorphous layer is associated with kinetic roughening of the surface. Extrapolation of the temperature dependence of h_{epi} suggests that it should be possible to grow thin epitaxial Si films down to surprisingly low temperatures.

Quite different morphologies were observed by Vatel *et al.* [1165] during their studies of the surfaces of polysilicon films deposited onto Si(100) via SiH_4 decomposition, chemical vapor deposition. The polysilicon films were deposited onto silicon wafers that had been N-doped to a conductivity of 5–8 Ω, by 40 keV phosphorus implantation followed by annealing at 1050 °C and "total deoxidation". The chemical vapor deposition was carried out at temperatures of 580 °C or 620 °C, until the deposited film had grown to a thickness of ≈ 380 nm. This required deposition times of 95 minutes at 580 °C or 43.5 minutes at 620 °C, corresponding to deposition rates of 42Å min^{-1} and 86Å min^{-1}, respectively. The films were then doped to 10^{16} atom cm^{-2} using 80 keV phosphorus implantation and annealed at 800 °C for 30 minutes. Areas ranging from 0.64 μm × 0.64 μm to 25.6 μm × 25.6 μm were scanned using atomic force microscopy. Figure 5.14 shows surfaces reconstructed from the AFM scans. The corresponding power spectra could be approximately described as a power law over about one decade in q, or length scale. Hurst exponents of $H \approx 0.77$ and $H \approx 0.84$ were reported for the 580 °C and 620 °C samples, respectively. The substrate was analyzed in a similar manner, and it was found to have a Hurst exponent of about 0.26, but the surface width $\xi_\perp$ was much smaller than that of the polysilicon films. The films deposited at 580 °C and 620 °C were also examined using transmission electron microscopy. A Hurst exponent of $H \approx 0.25$ was measured for the substrate surface, which had a much smaller width $\xi_\perp$ than the film surface.

Transmission electron micrographs of the surfaces grown at 580 °C revealed a polycrystalline material with irregular, relatively ill-defined grain boundaries. In the film deposited at 620 °C, the grain boundaries were flatter and better defined. The distribution of grain sizes was smaller in the 125 – 375 nm thick, 620 °C samples than in the 30 – 400 nm thick, 580 °C samples.

The deposition of high melting temperature metals onto cold substrates gives very different morphologies than those associated with molecular beam epitaxy. Because of the smaller surface mobilities and the presence of voids in the deposited films, KPZ-like growth might be expected in the $t \to \infty$ limit. The deposition of metal vapors onto a variety of surfaces has been explored extensively by Krim and colleagues, under various conditions, and a summary of this work can be found in a review [1135]. Some of the earlier pioneering work was carried out with commercial samples that were exposed to air and for which preparation conditions were not available [1108]. In addition, the substrate roughness played an important role in some of the early samples [1135]. However these studies prepared the way for and motivated future work.

In one scanning tunneling microscopy (STM) study by Krim *et al.* [1166], a Hurst exponent of 0.96 ± 0.03 was obtained, for a commercial gold electrode from the dependence of the surface width $w(\ell)$ on the length scale ℓ. A power law relationship ($w(\ell) \sim \ell^H$) was obtained over a three decade range of ℓ by combining data recorded using different STM scan heads to obtain data over different length scales. This appears to be one of the exceptional cases in which scaling over two decades or more of vertical length scales has been seen in a single experiment. However, it is not clear if the problem of

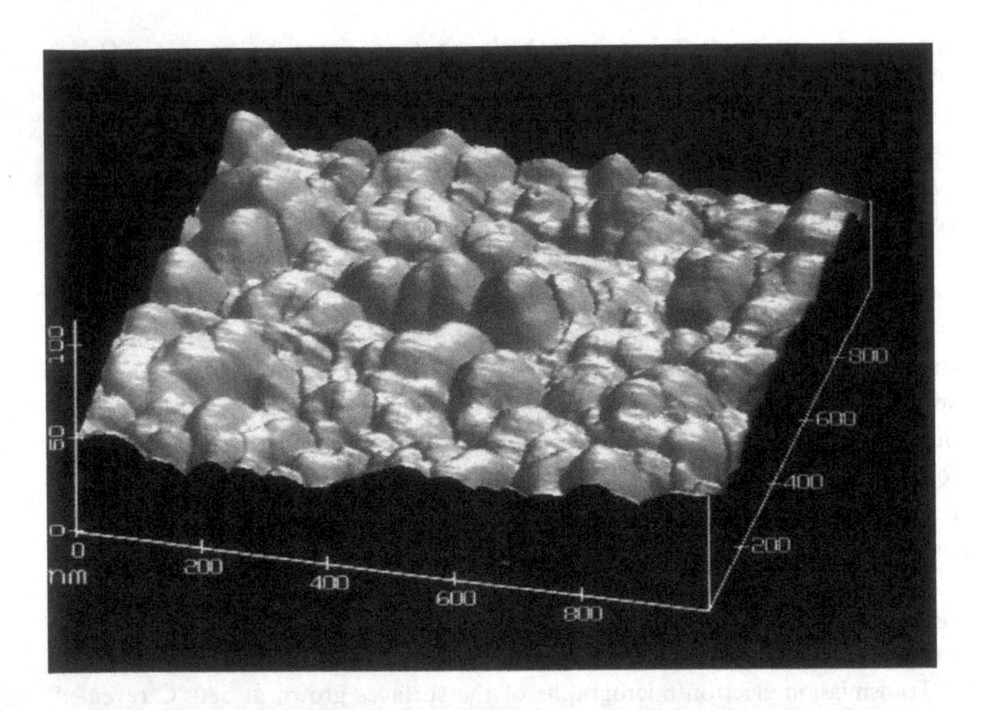

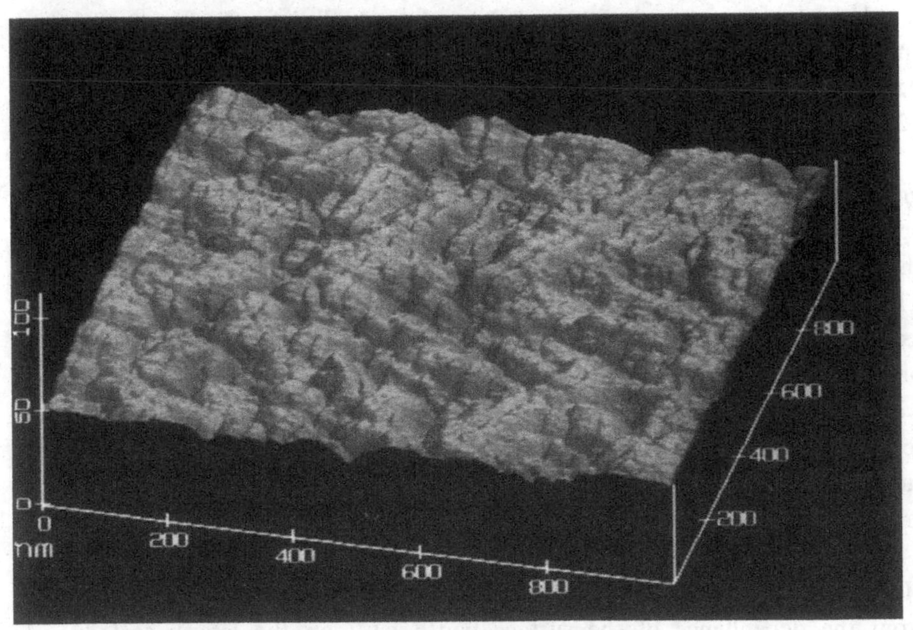

Figure 5.14 Polysilicon surfaces reconstructed from AFM scans of 1 μm $\times$ 1 μm areas. The top part shows the surface of a film generated at 620 °C and the bottom part shows the surface of a film generated at 580 °C. This figure was provided by O. Vatel.

calibrating the different scan heads has been completely overcome, and the significance of a Hurst exponent so close to unity is not certain. In addition, the lack of knowledge concerning the preparation conditions detracts from the significance of this result. Nevertheless, this experiment does represent an important technological advance, and it indicates that it may be possible to obtain high quality data covering more than three orders of magnitude in the vertical length scale for large values of H, in some systems.

In a later study [1111], 99.999% pure Ag was deposited at a rate of 0.3 Å s^{-1}, via collimated thermal evaporation in a vacuum of 10^{-7} Torr, onto optically polished quartz crystal micro-balance blanks held at a temperature of 80 K. After the films had grown to a thickness of 450–2500 Å, they were transferred *in situ*, at room temperature, without exposure to air, to a gas adsorbtion tip. The deposited films were studied via N_2 adsorbtion, using a quartz crystal micro-balance at a temperature of 77 K by measuring the N_2 pressure dependence of the resonant frequency. The adsorbtion measurements were interpreted in terms of a flat film surface for deposition at normal incidence. However, the adsorbtion data were interpreted in terms of a self-similar or self-affine fractal structure for deposition at an angle of incidence of 5° (measured from the normal). While the interpretation of the adsorbtion data is not unambiguous, the extreme sensitivity to the deposition conditions is evident. This sensitivity is surprising and exacerbates the difficulties encountered when results from different laboratories are compared. In earlier related work [1167], a Hurst exponent of ≈ 0.46 was obtained for a 1100 Å thick film deposited at 0.3 Å s^{-1} onto a 80 K substrate at an angle of incidence of 5°, using x-ray reflection measurements. Adsorbtion measurements using a film deposited at normal incidence, under otherwise identical conditions, indicated no evidence of surface roughness.

Herrasti *et al.* [1168] have investigated the deposition of gold vapor, generated in an evaporation chamber, onto a cold, 298 K, glass substrate at near-normal angles of incidence θ in the range $2° \leq \theta \leq 25°$. Films with a mean thickness $\overline{h}$ of 30 nm to 1000 nm were prepared under these conditions, at a deposition rate of 30 nm s^{-1}, which is high compared with those used in MBE experiments. The resulting deposit surfaces were studied by scanning tunneling microscopy (STM). The glass substrates were cleaned sequentially in ultrasonic baths with water and organic solvents and were examined with atomic force microscopy. They were found to have surface corrugations $\xi_\perp$ of less than 0.5 nm. The STM studies of the deposited films revealed a columnar microstructure like that illustrated in figure 5.15, with a branched pattern of voids between the "columns". The surface width $\xi_\perp$ was obtained from the STM images. This correlation length increased with increasing height $\overline{h}$ for small heights, $\overline{h} \leq \overline{h}_c$, but saturated for larger values of $\overline{h}$. This saturation was attributed to growth of the horizontal correlation length $\xi_\parallel$ to lengths greater than the finite STM window size ℓ. The Hurst exponent H was estimated to have a value of about 0.34 ± 0.035 from the power law dependence of the saturation value of the surface width $w(\ell, \overline{h} > \overline{h}_c)$ on ℓ ($w(\ell) \sim \ell^H$, over about a decade in ℓ) for length scales greater than ξ_c, where ξ_c is the column width. The Hurst exponent was also estimated by measuring the self-similar fractal dimensionality of surface contours using the area–perimeter relationship given in equation 2.190, and a similar result, $H \approx 1/3$, was obtained using equation 2.188. If equation 2.192 is used instead of equation 2.188 to relate the Hurst exponent to the single contour fractal dimensionality, a value of less than zero is obtained for H. Herrasti *et al.* also measured the distribution of height fluctuations $\delta h = h(\mathbf{x}) - h_{min}$. They found that these fluctuations could be described in terms of the exponential form $N(\delta h) = Ae^{-k\delta h}$ for $\delta h > 0.3(h_{max} - h_{min})$, where h_{min} and h_{max} are the maximum and minimum heights.

It was not possible to obtain a reliable estimate for the exponent β describing the growth of $\xi_\perp$ for $\overline{h} \ll \overline{h}_c$. The STM images indicated that surface diffusion and/or other surface reconstruction processes played an important role in the surface growth process. Although the deposits were porous, they were composed of rounded grains with a characteristic size (length scale, ξ_c) of about 50 nm. This analysis of the STM surfaces is consistent with a KPZ growth process. In a continuation of this work, Salvarezza *et al.* [1169] found a crossover from $H \approx 0.9$ for $\ell < \xi_c$ to $H \approx 0.35$ for $\ell > \xi_c$, with $\xi_c \approx 40$ nm. They also measured the dependence of the correlation length $\xi_\parallel$ on the deposit thickness $\overline{h}$. The parallel correlation length was obtained from the dependence of the surface width $w(\ell)$ on the sweep width ℓ ($w(\ell)$ saturates when $\ell \geq \xi_\parallel$). They found that the correlation length $\xi_\parallel$ stopped growing when the average deposit thickness reached a value of about 500 nm. This implies that the surface evolved towards a steady state in which both $\xi_\perp$ and $\xi_\parallel$ reached stationary values of $\xi_\perp(t \to \infty) \approx 2$ nm and $\xi_\parallel(t \to \infty) \approx 500$ nm.

Mitchell and Bonnell [1170] have analyzed STM images of gold sputter deposited onto sodium borosilicate glass, at an Ar pressure of 150 μm Hg, to a thickness of about 50 nm and obtained a Hurst exponent of about 0.75. The surface was reconstructed from 1000 STM profiles taken from an area of 3000 Å × 3000 Å and analyzed by calculating the power spectrum via Fourier analysis. The power spectrum showed good linearity on a log–log plot over about one decade in wave number, corresponding to one decade of length scales. The main emphasis of this work was on the fractal analysis of STM images. The authors provided little information on the sample preparation conditions.

A fractal dimension of 2.3 was measured for silver electrodes plated onto quartz crystals by Pfeifer *et al.* [1108], using a multilayer N_2 adsorption technique at a temperature of 77 K. The adsorption process was analyzed using a self-similar fractal model, so it is not clear if this result should be interpreted

Figure 5.15
3-Dimensional images of vapor deposited gold films obtained by scanning tunneling microscopy. Parts (a), (b) and (c) show areas of 640 nm × 640 nm, 640 nm × 640 nm and 510 nm × 510 nm, respectively. The corresponding average thicknesses $\bar{h}$ were 30 nm, 160 nm and 850 nm, respectively. The vertical bar, in part (c), indicates 18 nm in the direction normal to the substrate. This figure was provided by P. Herrasti.

as a Hurst exponent of about 0.7. A reinterpretation of these experiments in terms of self-affine fractal geometry by Kardar and Indekeu [1109] gave a Hurst exponent of about 0.37, but this does not agree with the STM results of Pfeifer *et al.* [1108], who used a similar sample, from which a Hurst exponent of 0.70 ± 0.10 was obtained for lateral length scales in the 5 to 50 Å range.

You *et al.* [1171] have studied the rough surfaces of sputter-deposited gold films using *in situ* x-ray reflectivity and STM. The *in situ* x-ray reflectivity studies were carried out using a silicon (111) substrate with an rms surface roughness of 3 Å at temperatures of 220, 300 and 350 K and Ar pressures of 1 and 10 mTorr, with a deposition rate of ≈ 0.5 Å s^{-1}. A height difference correlation function $C_2(x)$ was calculated from the STM scans of a 5000 Å × 5000 Å area of a 3500 Å thick sample (figure 5.16) deposited at 10 mTorr and 220 K. This correlation function, shown in figure 5.16(b), was consistent with self-affine roughness on horizontal length scales from a few tens of angstroms to a few thousand angstroms and vertical scales from a few angstroms to a few tens of angstroms. A Hurst exponent of about 0.4 was obtained, in good agreement with the values obtained from simulations carried out using 2 + 1-dimensional computer models in the KPZ universality class. The x-ray scattering measurements were in good agreement with near-specular scattering functions $S(q_{\|}, t)$ calculated using a model that included both the gold–vacuum and gold–substrate interfaces using the Hurst exponent of 0.4 obtained from the STM analysis. The *in situ* x-ray reflectivity measurements indicated approximately algebraic growth of the interface width, over one to one-and-a-half decades of time, but with an exponent (β) of about 0.4, rather than the value of about 0.25 expected for a 2 + 1-dimensional KPZ growth process. The surface width was obtained by fitting the specular reflectivity data, using a model that included both the gold–vacuum and gold–substrate interfaces. Although the Hurst exponents obtained for gold films by Herrasti *et al.* [1168] and You *et al.* [1171] are similar, the surfaces shown in figures 5.15 and 5.16, respectively, appear to be very different, reflecting the differences in deposition rate, substrate temperature and other experimental details.

Thompson *et al.* [1142] have used x-ray reflectivity to study the growth of surface roughness during the high vacuum vapor deposition of Ag at a rate of ≈ 0.03 nm s^{-1}, onto a room temperature Si substrate. A Hurst exponent of $H = 0.63 \pm 0.05$ and a surface width growth exponent of $\beta = 0.26 \pm 0.05$ were measured from the specular and diffuse scattering, respectively. The measurement of H was based on the dependence of the near-specular, diffuse scattering intensity $I(q_{\|} = 0, q_{\perp})$ on the perpendicular component of the scattering vector $q_{\perp}$. The Hurst exponent was determined using equation 5.176, after correcting for the dependence of the illuminated area on $q_{\perp}$. Film thicknesses in the range $10 \leq h \leq 200$ nm were characterized. A Hurst exponent of $H = 0.78 \pm 0.014$ was reported as a result of STM studies of Ag films prepared under the same

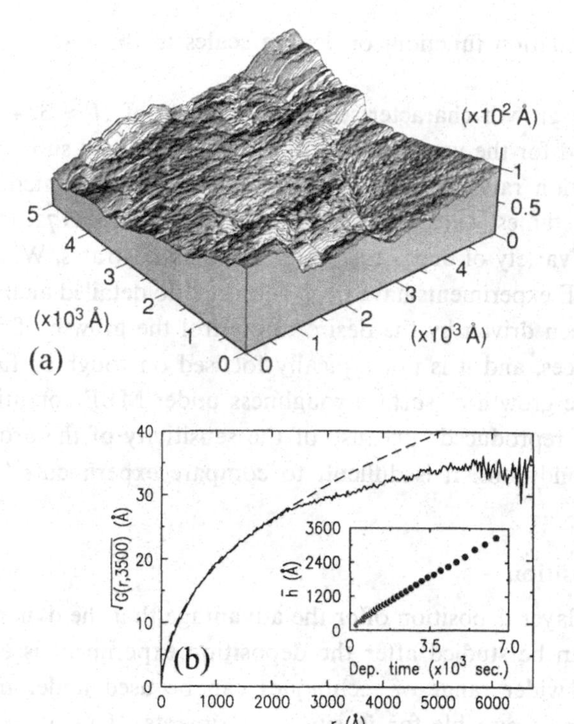

Figure 5.16 A surface reconstructed from STM measurements on a 3500 Å thick gold film sputter-deposited in 10 mTorr argon at 220 K is shown in part (a). Part (b) shows the height difference correlation function $C_2(x)$ calculated from the STM surface. The dashed curve is a power law fit with $H = 0.43$. This figure was provided by H. You.

conditions. Palasantzas [1137] obtained a Hurst exponent of $H = 0.68 \pm 0.02$ for an 80 nm thick film obtained under what appear to be identical conditions, using a different cut-off function for the height–height correlation function used in the analysis of STM data. In a related study, Palasantzas and Krim [1172] measured exponents of $H = 0.82 \pm 0.05$ and $\beta = 0.29 \pm 0.06$ for the deposition of silver onto polished quartz under very similar conditions. In this work, STM images of the surfaces of films with thicknesses up to 700 nm were analyzed. Scaling was found over about one decade of length scales and about one-and-a-half decades of time.

Very similar results ($H \approx 1.0$ and $\beta \approx 0.26$) were obtained by Ernst *et al.* [1145], by using He beam scattering to study the deposition of Cu at a dose rate of 0.008 monolayers s^{-1} onto Cu(100) substrates at 106 K. At a somewhat higher temperature of 200 K, the growth exponent β increased to about 0.56, while H remained at about 1.0.

Palasantzas [1137, 1138] also measured the roughness of Ag films deposited by thermal evaporation of Ag, with a background pressure of 5×10^{-7} Torr, onto a polished quartz crystal at a rate of 0.03 nm s^{-1}. The quartz substrate was held at a temperature of 106 K, during the deposition process, and was examined after the film had reached a thickness of 100 nm, using scanning tunneling microscopy under dry N_2. The surface was reconstructed using 500 nm scans with 400 points per scan. A Hurst exponent of 0.72 ± 0.05 was measured from the height difference correlation function, on length scales in the $\epsilon < x \ll \xi_{\|}$ regime.

It appears that surface growth characterized by exponents of $H \approx 3/4$ and $\beta \approx 1/4$ is well established for the vapor deposition of Ag onto cold substrates at relatively slow deposition rates. Palasantzas and Krim [1172] pointed out that several experimental studies [1142, 1145, 1146, 1152, 1155, 1169, 1172, 1173] gave similar results for a variety of deposited materials and substrates. While a very large number of MBE experiments have been subjected to detailed analysis, most of this work has been driven by the desire to control the growth of very smooth, defect free surfaces, and it is not typically focused on rough surfaces. Few, if any, studies of the growth of surface roughness under MBE conditions have been independently reproduced. Because of the sensitivity of the growth process to the growth conditions, it is difficult to compare experiments from different sources.

5.5.4.3 Multilayer Deposition

Films generated by multilayer deposition offer the advantage that the dynamics of the growth process can be studied after the deposition experiment is completed. Consequently, a wider range of techniques can be used under more favorable conditions than is possible for *in situ* experiments. This approach was used by Eaglesham and Gilmer [1158] in their study of Si epitaxial growth using thin Ge marker layers, which was described earlier in this section. However, this method is probability of limited general value because of the need to be confident that the marker layers do not seriously perturb the growth process.

Another example is provided by the experiments of Miller *et al.* [1174] in which sputtered films of NbN were labeled by depositing thin layers of AlN at regular intervals. The NbN and AlN were deposited by sputtering Nb or Al targets in Ar/N_2 onto polished sapphire substrates, at rates of about 2 nm s^{-1}, with a substrate temperature of 300 °C. Under these conditions, the NbN layers are crystalline but have a very high defect density, indicating a low surface mobility. The films were then sectioned and milled with 5 keV Ar^+ ions at a beam current of 0.8 mA to obtain thin sections for transmission electron microscopy (TEM). Using this technique, a permanent record of the progression of the surface was obtained. Figure 5.17 shows a TEM micrograph of a section that was thinned to about 40 nm. The conical, columnar internal structure revealed in this picture is quite striking. The column tops have a parabolic shape. Miller *et al.* found that the surface width grew with increasing time, with an exponent β of about 0.27, in good agreement with $2 + 1$-dimensional models in the KPZ universality class. However, it is not clear if the morphology shown in figure 5.17 can be generated by a model in the KPZ universality class.

Figure 5.17 A TEM micrograph of a 40 nm thick vertical slice through a film containing 101 2 nm thick layers of AlN and 100 layers of 3 nm thick NbN. This figure was provided by D. J. Miller.

A similar approach was taken by Fullerton *et al.* [1175], in their study of Nb/Si multilayer deposits. In this work, the Nb/Si multilayers were prepared by magnetron sputter deposition onto room temperature sapphire substrates. The systems studied included [≈ 35 Å Nb/≈ 30 Å Si]$_{10}$, [≈ 35 Å Nb/≈ 30 Å Si]$_{40}$ and [≈ 70 Å Nb/≈ 60 Å Si]$_{20}$, where the subscript n ($[\ldots]_n$) indicates the number of binary layers in the multilayer system. The sputter deposition was carried out in argon at pressures of 3 mTorr to 15 mTorr. At low pressures in this range, essentially flat continuous layers were obtained. At pressures > 9 mTorr, patterns similar to those reported by and Miller *et al.* were obtained. The multilayer deposits were examined using small- and large-angle x-ray diffraction as well as transmission electron microscopy. The low-angle x-ray diffraction spectra were analyzed using a model in which the interface width increased with increasing number of layers. A good fit to the experimental results was obtained with a model in which the interface width increased as t^{β} with $\beta \approx 0.5$.

The use of wide-angle diffuse x-ray scattering with grazing incidence and exit angles to study the surface geometry of W/Si multilayer films has been described by Salditt *et al.* [1176]. The multilayer deposit was grown by magnetron sputtering onto a polished silicon wafer and consisted of 150 bilayers with thicknesses of 22.8 Å (10.8 Å Si and 12.0 Å W). The scattering from these multilayers can be understood in terms of the height–height intra-layer correlation function $G_2(r)$, which was *not* assumed to have the form given in equation 5.171, and the height–height cross-correlation function or inter-layer correlation function $G_2^X(r,\delta)$. The height–height cross-correlation function $G_2^X(r,\delta)$ is defined as

$$G_2^X(\mathbf{r},i,j) = < h_i(\mathbf{r}_\circ)h_j(\mathbf{r}_\circ + \mathbf{r}) >, \tag{5.187}$$

where $h_i(\mathbf{r})$ is the height field, measured with respect to the mean height, for the ith interface. Here, $G_2^X(r,\delta) = G_2^X(\mathbf{r},i,j)$, where $\delta = | < h_i > - < h_j > |$ and $< h_i >$ is the mean height of the ith interface.

To measure the self-correlation function, a position sensitive detector was placed so that $q_\perp = 2\pi/a$, $(|[\phi]| \approx 0)$, where a is the multilayer periodicity, and the dependence of the scattered intensity on $q_\| = (q_x^2 + q_y^2)^{1/2}$ was measured at constant $q_\perp$ by rotating a vertically oriented position sensitive detector about the normal to the sample surface, keeping the angle of incidence and angle of reflection constant. Salditt *et al.* showed that their experimental data could be interpreted in terms of a correlation function of the form $G_2(\mathbf{r}) = A - B\ln r$ for scattering vectors that were dominated by scattering from the interfaces. This form would be expected for scattering from a $2+1$-dimensional EW interface. Salditt *et al.* were also able to measure the cross-correlations. Analysis of the decay of the cross-correlation with increasing separation δ indicated that the W/Si and Si/W interfaces were not equivalent. One component smoothed out roughness and the other did not. Based on these scattering results, it was not possible to tell which was which.

Swaddling *et al.* [1177] have used diffuse x-ray scattering to study rare-earth superlattices grown by molecular beam epitaxy. The rare-earth metals were deposited onto a Nb layer grown on a sapphire substrate. The systems studied can be described by using the notation $[\mathrm{Ho}_i\mathrm{M}_j]_n$, where i is the number of atomic planes of holmium and j is the number of atomic planes of M = Y or Lu, in each bilayer. The entire deposit consisted of n such bilayers. A variety of systems with $32 \leq n \leq 50$ and 23–60 atomic planes in each bilayer, grown on substrates at 300 °C, were examined. The x-ray scattering was interpreted in terms of conformally rough interfaces (interfaces that are related to each other by translation in the growth direction) with Hurst exponents in the range $0.78 \leq H \leq 0.95$, for ten superlattice systems. Exponents of $H = 0.93$, 0.87 and 0.83 were measured, for samples with a nominal composition $[Ho_{40}Y_{15}]_{15}$, grown on substrates at 200, 400 and 600 °C, respectively. Although this suggests that H decreases with increasing temperature, the differences between the Hurst exponents are too small to be confident in this trend.

5.5.4.4 Colloidal Deposition

Although the development of ballistic deposition models was initially stimulated by an interest in sediments formed by the deposition of small particles from colloidal dispersions [414, 417, 418, 987], experiments of this type have been neglected, compared with the enormous effort devoted to work on the deposition of atoms and molecules. However, Salverezza *et al.* [1179] have studied the surfaces formed by the deposition of 440 nm diameter silica particles onto a polished polymethylmethacrylate substrate, using an atomic force microscope. The amorphous silica particles were prepared with a small polydispersity, and were deposited from an aqueous dispersion at a pH of 6 to 7. After deposition for periods of up to 20 days, the remaining dispersion was removed and the deposit was dried before the surface structure was characterized. The 25×25 μm^2 images indicated that the surface consisted mainly of ordered (111) domains with a characteristic size of ξ_d. For the thicker samples the domain size ξ_d reached a value of 15 to 20 μm, and some defects could be clearly seen. Measurements of the surface width $w(h,\ell)$ as a function of the deposit thickness h and the scan length ℓ were consistent with the idea that the process could be described by the $2+1$-dimensional EW equation for length scales larger than the sphere size ϵ and smaller than the domain size ($w(h,\ell \gg \xi_{\|}) \sim [\log(h)]^{1/2}$ for $5 \times 10^2 \leq h \leq 5 \times 10^4$ μm and $w(h,\ell) \sim [\log \ell]^{1/2}$ for $\epsilon < \ell < \xi_d$).

Similarly, Lei *et al.* [1178] have obtained evidence for EW growth in a $1+1$-dimensional experiment in which ≈ 0.8 μm diameter polystyrene spheres, with a polydispersity of less than 2% were attracted (for unknown reasons) from a stable dispersion into the wedge formed between the upper surface of a droplet of the dispersion and a wetted glass substrate. In these experiments, the dispersions were prepared with a particle volume fraction of $\phi = 0.05$, and the deposition process was recorded using an optical microscope with a video recorder. The width of the surface was found to grow as t^β with $\beta = 0.26 \pm 0.03$. After $t \approx 100$ s, the dependence of the width $w(t,\ell)$ was observed to scale as $w(t,\ell) \sim \ell^H$, with $H \approx 0.51$. However, after $t \approx 300$ s, the dependence of $w(t,\ell)$ on the length ℓ over which it was measured was found to scale with an effective exponent of $H \approx 0.71$.

These experiments cannot be well represented by the off-lattice ballistic deposition models with steepest descent restructuring to a local minimum, described earlier in this chapter (sections 5.2.2 and 5.3.1) and in chapter 3 (section 3.6). It is evident from the $2+1$-dimensional experiments of Salverezza *et al.* [1179] that the deposit has a much more ordered structure than that generated by the simulations. Similarly, Lei *et al.* [1178] reported that the deposits formed in their experiments had a 2-dimensional close-packed crystal structure. Consequently, the dependence of the deposit density on the angle of inclination should be much smaller in the experiments than in the simulations, and the possible crossover from EW to KPZ behavior can be expected to occur at very late stages (beyond the limits of practical experiments) if at all.

5.5.5 Erosion Experiments

Although the "growth" of rough surfaces, during processes such as thin film deposition, has been the main focus of attention, processes such as corrosion, erosion and wear are equally important, and are more familiar. These processes are, in most respects, similar to the surface growth processes discussed above. However, there is one important difference: in deposition, the most exposed, leading parts of the interface can be expected to receive the largest flux of material from outside, and grow more quickly, while in erosion processes these exposed parts of the surface are more likely to be preferentially removed. Consequently, it can be expected that erosion processes will generally lead to a slower growth in the surface width $\xi_\perp$ than deposition processes. An extreme example of this is the very smooth surfaces produced by electropolishing, which should be contrasted with the dendritic surfaces produced by electrodeposition. Similarly, if a simple ballistic deposition model in the KPZ universality class is "converted" into a ballistic etching model in which the first filled site that is encountered by the incoming particle (the first filled that becomes a nearest neighbor to the etching particle as it "falls" along a column of the lattice) is annihilated, the resulting growth model belongs to the EW universality class. However, processes such as corrosion can generate very rough surfaces via mechanisms such as passivation/depassivation reactions and selective dissolution of grain boundaries, regions with different compositions and regions with different stresses or strain histories that do not have obvious counterparts in deposition processes. There are many ways in which rough surfaces can be generated by the removal of material from a compact body. Such processes have been neglected in comparison with "growth" processes, and they deserve more attention. The dissolution, corrosion and erosion of inhomogeneous materials and porous materials is also of great interest and practical importance.

A particularly important example of the generation of rough surfaces in which erosion plays an important role is provided by the surfaces of the Earth, the Moon and the planets. Natural surfaces such as these and models for these surfaces have been analyzed extensively using field data, computer simulation results and data from laboratory scale physical models [1180]. The scaling properties of these surfaces and the models used to study them are discussed elsewhere [57].

5.5.5.1 Ion Bombardment

Depending on the materials and conditions, ion bombardment can lead to either the roughening or smoothing of surfaces. In general, this is a complex process that does not seem to be well described by simple models. However, it is reasonable to expect that relatively simple limiting regimes may be found and that they can be understood in terms of simple models. This may lead to valuable insights into the more complex behavior that is often found in experiments directed towards practical applications.

Eklund *et al.* [1181, 1182] have used a scanning tunneling microscope (STM) to examine graphite surfaces roughened by ion bombardment with 5 keV Ar^+ ions. Typical STM scans, used to reconstruct the surface, covered an area of 2400 Å × 2400 Å. Examples are shown in figure 5.18. A total of over 1000 STM surfaces were constructed from the STM scans in this study. Highly oriented pyrolytic graphite exposed in the (0001) direction was sputter etched with a raster scanned 5 keV Ar^+ ion beam, with an angle of incidence of 60° from the normal. The STM images of the initial cleaved surfaces showed atomically flat areas over many thousands of square angstroms. This allowed the early stages of the roughening process to be studied. Ion fluxes of $\varphi_1 = 6.9 \times 10^{13}$ ion cm^{-2} s^{-1}, $\varphi_2 = 3.5 \times 10^{14}$ ion cm^{-2} s^{-1} and $\varphi_3 = 6.9 \times 10^{14}$ ion cm^{-2} s^{-1} were used, with total ion fluences of about 10^{16} ion cm^{-2}, 10^{17} ion cm^{-2} and 10^{18} ion cm^{-2}. Approximately 4, 40 and 400 monolayers were removed, using the fluxes φ_1, φ_2 and φ_3, respectively, and the corresponding fluences, at 300 K. Experiments were also carried out at 600 K and 900 K.

For low doses, Hurst exponents in the range $0.2 \geq H \geq 0.4$ were measured. Some of these surfaces are shown in figure 5.18. The surfaces were analyzed by measuring the height–height correlation function $G(x) = < [h(\mathbf{x}_\circ)h(\mathbf{x}_\circ + \mathbf{x}) - < h >^2] >$ and the Fourier space correlation function $C_h^{(2)}(\mathbf{k}, t)$, defined in equation 5.184.

The growth of the surface was found to be consistent with the scaling form

$$C_h^{(2)}(\mathbf{k}, t) = S(k, t) = k^{-(2H+2)} f(tk^z). \tag{5.188}$$

The scaling function $f(x)$ has the form $f(y) = y^{(2H+2)/z}$ for $y \ll 1$ and $f(y) = const.$ for $y \gg 1$. In the short time regime, $y = tk^z \ll 1$, the correlation function $C_h^{(2)}(\mathbf{k}, t)$ grows as $t^{(2H+2)/z}$. In most cases, the correlation function $C_h^{(2)}(\mathbf{k}, t)$ obtained from individual data sets exhibited power law behavior over about a decade in k, but for some surfaces the scaling range was shorter. However, by combining data obtained from STM scans over areas ranging from 50Å × 50Å to 2400Å × 2400Å, Eklund *et al.* were able to obtain evidence for scaling over about three decades in k. Even though all of the surface measurements were made using the same instrument, the correlation functions did not overlap exactly, and it is difficult to assess the quality of this extended power law behavior and determine an accurate exponent. The scaling form in equation 5.188 is equivalent to the scaling form $C_2(x) = x^H f(tx^{-z}) = x^H g(x/\xi_{\parallel(t)})$ for the height difference correlation function, with $f(y) = const.$ for $y \ll 1$ The values obtained for the exponent H in equation 5.188 were in the range $0.25 \leq H \leq 0.45$. The exponent z was not measured accurately, but the data were found to be consistent with the scaling relationship $z = 2 - H$ (or $z = 2 - \alpha$, equation 5.42). At the higher fluences, the surfaces were dominated by features that had a flattened dome-

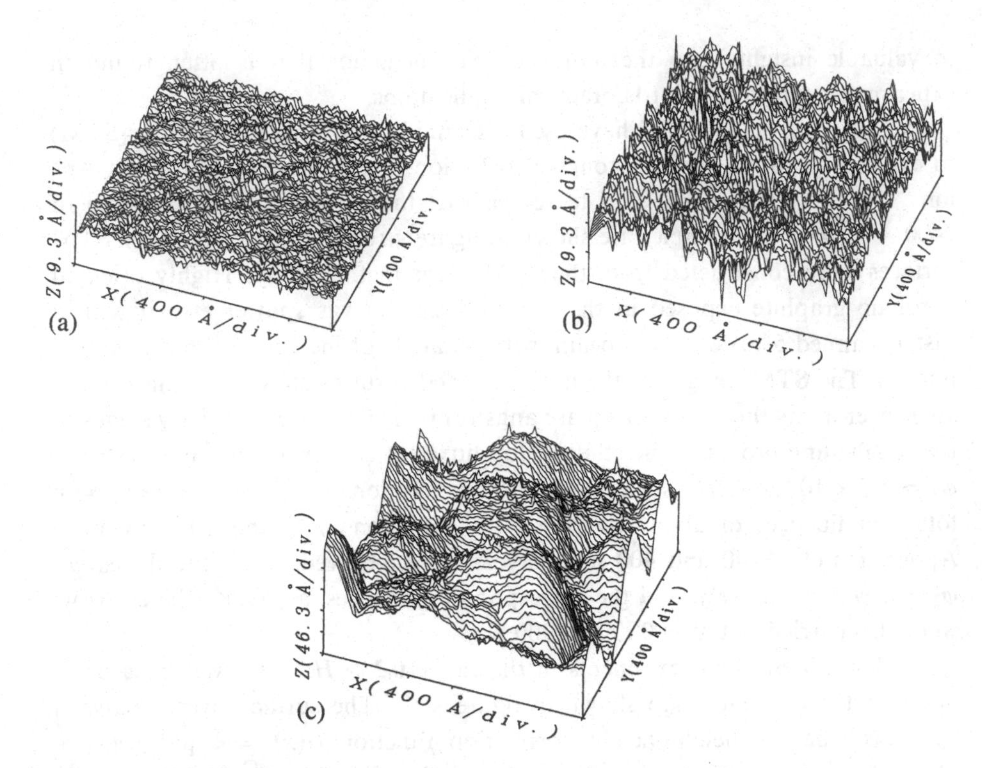

Figure 5.18 Reconstructed surfaces obtained from STM scans of graphite surfaces after sputtering with Ar^+ ions. An area of 2400 Å × 2400 Å is shown in each part of the figure. The sputtering flux was 6.9×10^{13} ion cm^{-2} s^{-1}. Parts (a), (b) and (c) show the surfaces after ion fluences of $10^{16}, 10^{17}$ and 10^{18} ion cm^{-2}, respectively. The corresponding vertical (z) scales are 18.6, 27.9 and 231.5 Å. This figure was provided by R. S. Williams.

like or hillock-like shape (figure 5.18), with a lateral correlation length $\xi_{\parallel}$ that approached the size of the STM scan. In these experiments, the width of the surface was found to increase more rapidly than the 1/2 power of the etching time and an effective value for β that is closer to 1 than 1/2 was estimated from the experiments at the two lowest fluxes (φ_1 and φ_2). The amplitude of the surface roughness decreased substantially with increasing temperature.

Eklund *et al.* pointed out that the evolution of a surface under etching can be described by the same sort of Langevin equations that are used for growth. They identified several mechanisms for material transport and the terms that they contribute to the Langevin equation. These mechanisms and their corresponding contributions are surface diffusion ($\nabla^4 h(\mathbf{x}, t)$), bulk diffusion ("$\nabla^3 h(\mathbf{x}, t)$"),[14] sputter redeposition ($\nabla^2 h(\mathbf{x}, t)$) and sputter removal ($R\Omega(\mathbf{x}, \{h\})$).

14. A term of the form $-\mathscr{D}_b \nabla^3 h(\mathbf{x}, t)$, where $\mathscr{D}_b$ is the bulk diffusion coefficient, is not allowed in the surface growth equation because it violates the $\mathbf{x} \to -\mathbf{x}$ symmetry. However, a perturbation with a wavevector $\mathbf{k}$ will decay, exponentially with a rate constant proportional to $\mathscr{D}_b k^3$, due to bulk diffusion alone.

If bulk diffusion is negligible, the Langevin equation becomes

$$\partial h(\mathbf{x},t)/\partial t = a\nabla^2 h(\mathbf{x},t) - c\nabla^4 h(\mathbf{x},t) + R\Omega(\mathbf{x},\{h\}) + \eta(\mathbf{x},t), \quad (5.189)$$

where $R\Omega(\mathbf{x},\{h\})$ describes the shadowing effect (preferential impact on the tops of the "hillocks"). Here Ω is the exposure solid angle and R is the average rate of material removal. Eklund *et al.* concluded that the large effective value measured for the exponent β that describes the growth of the surface width could not be modeled successfully with this linear equation, and they suggested that additional terms of the form $[h(\mathbf{x},t)]^n$ or $[\nabla h(\mathbf{x},t)]^n$, with $n > 1$, must also be included. Large effective exponents over the range of length scales covered in these experiments could also be a consequence of crossover effects resulting from the changing relative importance of the terms in the surface growth equation as the time scale and length scales change.

Krim *et al.* [1166] used STM to measure the structure of rough surfaces generated from initially smooth Fe(100) surfaces by ion bombardment. The smooth surfaces were generated by depositing about 2000 Å of Fe onto a MgO(001) substrate held at 150 °C. The iron film was then sputtered in an ultra-high vacuum chamber, to remove an oxide layer formed by exposure to air, and annealed at 600 °C, to generate a flat surface, which was examined using the STM. The Fe surface was then bombarded at room temperature at an angle of 25° from the normal with 5 keV Ar^+ ion fluxes of about 5×10^{14} cm^{-2} s^{-1}. The surfaces were examined using STM, for a large number of fluences in the range $10^{17} - 10^{18}$ ion cm^{-2}. A Hurst exponent of $H = 0.53 \pm 0.02$ was measured from the dependence of the surface width $w(\ell)$ on the scan length ℓ. A linear relationship between $\log w(\ell)$ and $\log \ell$ was found over about one-and-a-half orders of magnitude in ℓ.

In another study, Chason and Mayer [1183] used energy dispersive x-ray reflectivity to investigate rough surfaces obtained by the 1 keV Xe^+ ion bombardment of 35 nm thick SiO_2 films on Si, prepared by dry oxidation of Si. Under these conditions, they found that the roughness $\xi_\perp$ grew linearly, from an initial value of about 4 nm, with the amount of material sputtered ($\beta = 1$). On the other hand, rough SiO_2 surfaces produced by Xe^+ sputtering followed by thermal desorption at 500 °C became smoother when sputtered with H^+ ions with energies in the 200 to 1000 eV range. Chason *et al.* [1184] also used the energy dispersive x-ray reflectivity technique to measure a surface width growth exponent of $\beta \approx 0.5$ during the 150, 300 and 500 eV H^+ ion bombardment of Si surfaces at a temperature of 300 °C. In a more recent study, Chason and Warwick [1185] used the energy dispersive x-ray reflectivity approach to study rough surfaces formed by the 500 eV Xe^+ ion bombardment of Ge(001). The growth of the surface width $\xi_\perp$ was measured, and was found to be consistent with a power law of the form $\xi_\perp \sim t^\beta$, with $\beta \approx 1/3$.

5.5.5.2 Plasma Etching

Interaction of solid surfaces with plasmas can lead to both deposition and etching. Plasma treatment of surfaces is important in many areas of technology, including microfabrication, the promotion of adhesion and coating processes. Plasma etching is important in areas such power generation and space flight.

Pétri *et al.* [1186] have used atomic force microscopy to study the evolution of cleaned silicon (100) surfaces under SF_6 plasma etching. The surface roughness was measured on 256 × 256 grid points covering an area of 3.2 μm × 3.2 μm, corresponding to 125 Å sampling intervals. They found that the growth of the surface roughness $\xi_\perp$ could be represented by the empirical equation

$$\xi_\perp \sim E^{-1/2}(\varphi^+/\varphi^n)^\eta t^\beta, \tag{5.190}$$

where E is the ion energy, φ^+ is the ion flux and φ^n is the reactive (fluorine) neutral flux. The exponent β was found to have a value close to 1.0 and $\eta \approx 0.45$. However, the data appear to be quite scattered and the uncertainties are large. The spectral densities $S(k)$ of the rough surfaces were calculated. The power spectra indicated a crossover from a self-affine geometry on short length scales to a flat surface on longer length scales. The exponent ω describing the form of $S(k)$ at large k ($S(k) \sim k^{-\omega}$) was found to depend on conditions such as the bias voltage and etching time. In most cases, the spectral density $S(k)$ had a power law form over about a decade in k, and in some cases the scaling range was less than a decade. A broad range of values ($1.7 \leq \omega \leq 4.0$) was measured for the spectral density exponent ω. This exponent was found to increase steadily from the lower value of 1.7 to $\omega \approx 4.0$ as the magnitude of the bias voltage was increased. On the other hand, the correlation length $\xi_\parallel$ remained essentially constant. A similar trend, from $\omega \approx 2.5$ ($H \approx 0.25$) to $\omega \approx 4.0$ ($H = 1$) was observed with increasing exposure times. The range of scaling was small, about one decade or less. However, the overall trends, particularly the dependence of ω on the bias voltage, appear to be significant. For exposure times larger than 600 s, the roughness became too large for the AFM equipment.

5.5.5.3 Polishing and Wear

During the past few years, an ever increasing number of studies reporting the fractal roughness of worn surfaces have appeared. Because the range of scaling is very limited in most studies, the uncertainties are large. A variety of surfaces manufactured by processes such as machining, polishing and grinding have been analyzed using power spectra, height–height correlation functions etc., and the results have been presented in terms of self-affine scaling models. In most cases, the evidence for fractal scaling is weak. In some cases, only the estimated Hurst exponents, without supporting log – log plots or other indications of the range of scaling, have been presented. Under these circumstances, neither the quality of the data nor the quality of the analysis can be assessed.

There is a need for a more systematic and critical approach to both the experiments and data analysis used to explore the morphology of material wear processes.

Sinha *et al.* [1128] have measured a Hurst exponent H of 0.20 ± 0.02 with an rms roughness ($\xi_\perp$) of 7.6 Å and a horizontal correlation length ($\xi_\parallel$) of about 7000 Å for polished Pyrex glass, using 1.46 Å wavelength synchrotron radiation x-rays with large angles of incidence ($(\pi/2) - \theta_i$ as low as 0.12°, corresponding to near-grazing incidence). The main emphasis of this work was concerned with a theoretical analysis of x-ray and neutron scattering from self-affine rough surfaces, so the implications of these results were not discussed. Dumas *et al.* [1187] have measured a Hurst exponent of $H \approx 0.35$ for the surfaces of optical glasses using both atomic force microscopy and light scattering. Quite convincing scaling was found in the power spectrum calculated from the AFM surfaces over almost two decades in the lateral length scale (from 0.25 μm to the scan length of 25.6 μm). The light scattering data did not exhibit scaling over such a wide range of length scales, but there was excellent agreement between the Hurst exponents obtained using these two quite different methods.

The roughness of surfaces generated by larger scale processes has also been examined quantitatively. For example, Srinivasan *et al.* [1188] have analyzed electron micrographs of sapphire ($10\bar{1}0$) surfaces eroded by SiC and Al_2O_3 particles. The surfaces were air blasted with 63 μm mean diameter particles at normal incidence, with a velocity of 90 m s^{-1}. Evidence was presented to support the idea that the sapphire surfaces generated in this manner are fractal, using techniques that have also been applied to fractography [1189]. However, it is difficult to assess the significance of these results. It is apparent that additional experiments on rough surfaces generated by this type of erosion process would be valuable.

Stupak and Donavan [1190] have analyzed the wear surfaces generated by forcing a razor blade into the surface of a rotating rubber wheel. The experiments were carried out at a temperature of 25 °C, with a tangential velocity of 33 mm s^{-1}. The friction forces ranged from 650 to 2150 N per meter of wheel width. The wear rate was found to grow as W^a with increasing frictional work W, with $a \approx 1.83$, $a \approx 1.65$ and $a \approx 1.64$ for natural rubber, polybutadiene and polystyrene/polybutadiene, respectively. Profilometer traces were recorded and magnified by a factor of 10, in the vertical direction. The profilometer traces were analyzed by constructing a physical model and rolling discs with different diameters ℓ over the model. A pencil lead was inserted through a hole in each of the discs, and the length of the path $p(\ell)$ followed by the pencil lead was measured. A linear dependence of $\log p(\ell)$ on $\log \ell$ was observed over about half a decade in ℓ. Effective fractal dimensionalities ranging from $D \approx 1.13$ ($H \approx 0.87$) for the polystyrene/polybutadiene blend at the lowest value of the friction work to $D \approx 1.58$ ($H \approx 0.42$) for the polybutadiene rubber at $W = 1790$ N m were reported. For most materials and wear rates, $D \approx 1.5$ ($H \approx 0.5$).

5.5.6 Electrochemical Deposition

The now quite extensive study of disorderly dendritic patterns generated by electrochemical deposition is described in chapter 4. Kahanda *et al.* [1191] have studied the slow electrochemical deposition of copper in a quasi 2-dimensional cell. This process leads to a growth front morphology that is quite different from the dendritic morphologies formed under faster growth conditions that are described in chapter 4. Kahanda *et al.* [1191] reported that the growth fronts formed under very slow growth conditions were self-affine. Copper was deposited at front velocities, in the range 0.01 mm hour^{-1} to 0.2 mm hour^{-1}, in a thin, 25 mm × 13 mm × 25 μm cell. Copper cathodes and anodes were used with a 2 M aqueous $CuSO_4$ electrolyte. The electrodeposition was carried out with a low overpotential of $\delta V = 0.1$–0.5 V. At the highest value of δV, a branched, DLA-like pattern was formed (chapters 3 and 4). At the lower values of the overpotential, a columnar pattern grew, with narrow crevices separating neighboring columns. Fourier transformation of the curve $h_{max}(x)$, where $h_{max}(x)$ is the "highest" point on the surface at each lateral position x, gave a power law power-spectrum $S(k)$, over about two decades in k, corresponding to a self-affine surface with $0.52 \leq H \leq 0.58$, over almost two decades in the horizontal length scale ℓ. The range of length scales over which scaling was observed lay between the cell thickness and the characteristic column thickness. The value obtained for the Hurst exponent was in good agreement with the value of 1/2 expected for a 1 + 1-dimensional KPZ growth process. However, the effective value obtained for the exponent β describing the growth of the surface width w was ≥ 1, in poor agreement with 1 + 1-dimensional KPZ value of 1/3.

Keblinski [1192] has criticized this interpretation and has pointed out that a k^{-2} power spectrum, corresponding to $H = 1/2$ for a self-affine surface, can be obtained from a simple, non-fractal model for the prominent columnar structure described by Kahanda *et al.* According to this interpretation, the main contribution to the power spectrum comes from the large scale columnar structure, not the shorter scale roughness.

The growth of the columnar structure appears to be consistent with a Mullins–Sekerka instability, described in appendix A, section A.1, in which the modes of the Fourier transform of $h(x)$ ($F(k) = \mathscr{F}(h(x))$, where $\mathscr{F}$ indicates Fourier transformation) grow exponentially:

$$F(k,t) \sim e^{\upsilon(k)t}. \tag{5.191}$$

In this equation, $v(k)$ has the form

$$v(k) = v_\circ k(1 - \xi_c \xi_{\mathscr{D}} k^2), \tag{5.192}$$

where ξ_c is the capillary length and $\xi_{\mathscr{D}}$ is the "diffusion length". Kahanda *et al.* used the cell thickness of 25 μm in place of $\xi_{\mathscr{D}}$.

Iwamoto *et al.* [1193] have also studied the electrodeposition of Cu under slow growth conditions. In these experiments, Cu was electrodeposited onto copper plates from a stirred aqueous 0.3 M $CuSO_4$/1.2 M H_2SO_4 solutions, at current densities in the 1.2×10^{-2} to 4.8×10^{-2} A cm^{-2} range, corresponding to mean growth velocities of 0.25–1.0 μm min^{-1}. The rough surfaces generated by this $2+1$-dimensional growth process were studied using atomic force microscopy images obtained over areas ranging from 5 μm $\times$ 5 μm to 50 μm $\times$ 50 μm. A Hurst exponent of $H = 0.87 \pm 0.05$ and a surface width growth exponent of $\beta = 0.45 \pm 0.05$ were reported. These results were based on the dependence of the surface width $w(\ell, t)$ on the measurement length scale ℓ and the time scale, respectively. The corresponding log – log plots were linear over about one decade of w. At the higher current density of 4.8×10^{-2} A cm^{-2} the growth process became unstable and a self-similar fractal-like morphology evolved. It is possible that the apparently self-affine surfaces formed at lower current densities are characteristic of a transient regime preceding the formation of a different, perhaps DLA-like, morphology in these experiments and those of Kahanda *et al.*

Pastor and Rubio [1194] have studied the electrodeposition of Cu from 0.5 M aqueous $CuSO_4$ in a thin (10 μm) cell at current densities in the 20 mA cm^{-2} to 100 mAcm^{-2} range. The experiments were carried out at a temperature of 315 °C controlled to $\pm$4 mK. At short times, a dense Cu deposit with a self-affine surface was formed. As the deposition process continued, the effective Hurst exponent, measured from the dependence of $w(\ell)$ on ℓ, increased, as did the correlation length $\xi_\perp$. The Hurst exponent saturated at a value of $H = 0.78 \pm 0.05$. At later stages in the growth process, a fingered pattern was formed. As the deposition process proceeded, in these constant current experiments, the voltage V_x across the cell increased from an initial value of 0.1 V to 0.2 V. The fingering instability appeared when V_x reached a value of 0.3 ± 0.05 V. After the transition from self-affine growth to fingering, the voltage V_x increased rapidly and the most unstable mode in the power spectrum, calculated from the surface of the deposit, grew exponentially with increasing time.

The surface of the deposit was characterized by measuring the dependence of surface width $w(\ell, t)$ on the measurement length scale ℓ and time t. After an initial transient, $w(\ell, t)$ grew exponentially with increasing time, ($w(\ell, t) \sim e^{kt} w'(\ell)$), before the fingering instability appeared, for a quite wide range of length scales ($\ell = 0.6$ μm to 230 μm). This indicates that the *amplitude* of the surface roughness grows with increasing time and that $H \neq \alpha$. This appears to be a complex growth process that cannot be understood in terms of the simple growth models of self-affine growth scenarios described in this chapter.

Aguilar *et al.* [1195] have used scanning tunneling microscopy to study the surface of an electrodeposited CoP alloy. Areas of 250 nm × 250 nm were represented by 512 × 512 pixel height fields, and a fractal dimensionality of ≈ 1.29 was obtained for "lake" perimeters in the surface.

Like the electrochemical processes discussed in chapter 4, the formation of rough surfaces during electrochemical deposition is a complex, poorly understood phenomenon.

5.5.7 Corrosion and Oxidation

A rich phenomenology and a wide range of morphologies are associated with corrosion processes [1196, 1197], and fractal geometry can be used to characterize corrosion fronts. For example, Holten *et al.* [1198] have analyzed Al corrosion fronts generated by the electrochemical corrosion of 50 μm thick epoxy coated Al foil, sandwiched between two parallel glass sheets, with one, initially straight, edge exposed to 1 M aqueous NaCl raised to a pH of 12 by adding 0.01 M NaOH. The corrosion was carried out at potentials of 0.0 and 1.6 V, with respect to a standard calomel electrode and Hurst exponents of 0.65 ± 0.05 were measured using the dependence of the front width $w(\ell)$ on the distance ℓ over which the width was measured.

The corrosion of rolled aluminum 2024 in aerated 3% aqueous NaCl was studied by Tretheway and Roberge [1199]. The surface of an Al strip was prepared by grinding to a 600 grit finish and was immersed to within 30 mm of the top of the 100 mm long sample in the aerated NaCl solution. The surface roughness was studied using a profilometer with a 0.2 μm diamond tip. Profiles were recorded at several locations in both the submerged regions and the splash zone. Large data sets, consisting of 32 000 points covering a length of 8 mm, were obtained and the Hurst exponents were obtained from an R/S analysis (chapter 2, section 2.7.3.5). Unfortunately, the authors did not show any of the log–log plots so it is not possible to assess the reliability of the Hurst exponents. A Hurst exponent of ≈ 0.5 to ≈ 0.6 was reported for the "original" ground surface and Hurst exponents ranging from ≈ 0.6 to ≈ 0.8 were found in the corroded regions. The largest Hurst exponents were measured for profiles taken in the spray zone, above the electrolyte.

A variety of different morphologies has been observed in pitting corrosion, including "bright" pits with smooth surfaces and pits with rough surfaces. It seems likely that morphology transitions, similar to those found in electrochemical deposition, will also be found in corrosion. Because of the large economic impact of corrosion [1196, 1197], and the complex morphologies associated with corrosion, the application of fractal geometry and scaling concepts to this area deserves more attention.

Yoshinobu *et al.* [1200] have used atomic force microscopy to characterize SiO_2/Si interfaces, generated by the oxidation of silicon. In these experiments, clean Si(100) wafers were dry oxidized at a temperature of 900 °C to form a 16 μm thick SiO_2 layer. The sample was then annealed for 15 minutes under N_2 at 900 °C and a specimen cut from the wafer was etched in 5% HF until the surface became hydrophobic. Because of the very different etch rates for SiO_2 and Si in 5% HF, the surface generated by this process was believed to be a good representation of the SiO_2/Si interface. Hurst exponents in the range $0.3 \leq H \leq 0.5$ were measured for horizontal distances in the range $1 \leq \ell \leq 100$ nm, from the dependence of the surface width $w(\ell)$ on the length scale ℓ over which the width was measured. On longer scales, $\ell > 100$ nm, $w(\ell)$ is essentially constant. These results can be interpreted in terms of a self-affine surface with a long length scale cut-off at $\xi_\parallel \approx 100$ nm and $\xi_\perp \approx 0.4$ nm. A power spectrum analysis of profiles led to a power spectrum of the form $S(k) \sim k^{-\gamma}$ with $2.0 \leq \gamma \leq 2.5$ over wavelengths in the same range. This corresponds to Hurst exponents in the range $0.5 \leq H \leq 0.75$. The agreement between the two methods is not good, but the results are consistent with a value of about 0.5 for H.

5.5.8 Some General Comments

This section demonstrates that evidence for self-affine geometry and power law growth of the characteristic correlation lengths with increasing time has been found in a very wide range of experimental studies. The quality of the evidence for simple scaling behavior varies enormously from study to study, and many of the results discussed here should be regarded as "leads", that should stimulate additional study, rather than completed pieces of work. Confidence in the results obtained using indirect methods such as scattering and methods such as AFM, STM and TEM that are subject to uncertainty on very small length scales can be enhanced if more than one experimental method is used. The connections, if any, between the experiments discussed in this section and the simple models discussed in the rest of this chapter are often not apparent. This problem is exacerbated in those cases where only one of the two exponents, H and β, was determined. In some cases, the uncertainties may be compounded by the existence of "preasymptotic scaling" and complex crossover behavior, before the asymptotic regime is reached. In practice, true asymptotic behavior is not attainable in experiments; new physical processes become important over sufficiently long time scales or length scales. At this stage, it would be valuable to select a few of the most promising leads, carry out more careful and extensive experiments in an attempt to extend the range of scaling and study the growth mechanism(s) in more detail. It would be interesting to see if the approach developed by Lam and Sander [954] (described in section 5.1.3) to determine the Langevin equation from simulation results could be applied to high quality experimental data. The experimental study of self-affine rough surfaces is evolving rapidly, and it seems probable that much progress will be made during the coming years. However, it may be some time before a coherent picture emerges for many important processes.

5.6 Thin Film Growth Models

The development of most surface growth models has been motivated, at least in part, by the results of experimental studies. In recent years, the main impetus has come from thin film technology and the propagation of fluid–fluid displacement fronts through porous media. In both of these areas, there is now a healthy synergistic interaction between experimental work, theory and simulations. Experimental studies of the roughness of evolving interfaces have led to results that are inconsistent with the older surface growth models and have stimulated the development of new models that better represent the observed phenomena. The newer, more realistic, models have led to the hope that at least some surface growth experiments can be quantitatively understood in terms of relatively simple models and theoretical approaches. This in turn has stimulated additional experimental studies and has led to the development of an exciting and rapidly growing research area that is driven by the idea that we may soon have a much better understanding of important practical problems. In particular, the challenge of developing ways of controlling thin film surfaces has provided a major incentive for much of the work described in this section.

The ballistic deposition models described above provide a surprisingly realistic representation of the growth of some thin, and thick, films, but such primitive models cannot be expected to reproduce the rich phenomenology found in real systems. Many models have been developed in recent years that are intended to represent the physics of thin film growth in more detail or in more physically realistic terms. The growth of thin films, at relatively high deposition rates, has been classified by Thornton [1132] into three "zones" based on the substrate temperature T and the associated morphologies. Zone I corresponds to low temperature deposition ($T < T_1$, where $T_1 = 0.25 - 0.30\ T_m$ and T_m is the film melting temperature). The characteristics of the associated zone I morphology include porous films and rough surfaces. In zone II ($T_1 < T < T_2$, where $T_2 \approx 0.45\ T_m$), the film is quite dense and consists of a packing of conical or columnar grains separated by a network of grain boundaries. The grain boundaries appear to be associated with minima in the advancing surface and may be regions of relatively low density. In zone III, $T > T_2$, the film has a dense, poly-crystalline internal structure with a more or less flat surface. Molecular beam epitaxy experiments are carried out using low deposition rates in an attempt to generate defect free, crystalline films with smooth surfaces.

5.6.1 The Effects of Surface Diffusion

While simple ballistic deposition models appear to capture some of the essential features of processes such as vapor deposition, under Thornton zone I conditions, they do not provide an adequate representation of surface diffusion processes. In principle, this problem could be solved by resorting to molecular dynamics simulations, but although molecular dynamics simulations [1201, 1202, 1203, 1204] are making important contributions to our understanding of thin film growth, 3-dimensional simulations of this type, on a scale large enough to obtain accurate values for the exponents α, β and H are still well beyond the capabilities of modern computers. The deposition rates in practical molecular dynamics simulations are many orders of magnitude faster than those used in most experiments, and many of the molecular dynamics simulations have been carried out using $1+1$-dimensional models. A variety of models have been introduced with the objective of incorporating surface diffusion into the simple surface growth models described above. For the most part, these models use a variety of *ad hoc* rules to represent the effects of surface diffusion. It is not clear if this approach has resulted in much progress towards a realistic model for processes such as molecular beam epitaxy. However, these models have revealed a broad range of scaling behaviors and appear to be making an important contribution to the development of a better theoretical understanding of the growth of rough surfaces. Rapid progress has been made towards the goal of obtaining a better general understanding of the role played by surface diffusion in surface growth in general, and in molecular beam epitaxy in particular. A variety of models in which surface transport phenomena are included in a more realistic manner have recently been investigated, and substantial progress is being made towards the successful simulation of molecular beam epitaxy. For many of these models, only the $1+1$-dimensional versions have been studied. Although these $1+1$-dimensional models may give some idea about what will happen in the more physically relevant $2+1$-dimensional case, they cannot be regarded as serious attempts to model thin film growth. Under some circumstances, such as the relatively orderly growth of surfaces via "step flow", $1+1$-dimensional models can provide valuable insight. However, these models cannot be expected to provide a realistic representation of the ways in which the step flow regime breaks down, leading to the growth of truly rough surfaces. Some of the $2+1$-dimensional model simulations were carried out on a scale that was too small to be confident that the asymptotic regime was reached. However, experiments may not be carried out under asymptotic conditions either, and, in many cases, the early stage behavior or intermediate stage, pre-asymptotic, scaling behavior may be of much more practical importance. In many experiments and applications, only several hundred or fewer atomic layers are deposited. This is far from asymptotia. In many important processes diffusion and "annealing" within the bulk of the film are important and their neglect in computer models is not always justified. However, such processes appear to play a minor role in the growth of crystalline films by molecular beam epitaxy.

5.6.1.1 Surface Currents

Important insights into surface growth processes have been obtained by focusing attention on the slope dependence of the net surface current $\mathbf{j}_s$ [637]. Krug *et al.* [961] have studied a variety of $1+1$-dimensional and $2+1$-dimensional solid-on-solid models in which the deposited particles undergo a local restructuring to find a more "favorable" position in which they stick permanently, immediately following deposition. Depending on the microscopic rules for this short range "diffusion", non-equilibrium surface currents may be generated during deposition. The behavior of the models depends crucially on the relationship between the mass current $\mathbf{j}_s$ along the surface and the inclination of the surface. In the absence of detailed balance constraints, a net current in an uphill or downhill direction can be expected, because of the asymmetry induced by tilting the surface with respect to the incoming particle beam. For processes of this type, the growth of the interface can be described by the equation

$$\partial h(\mathbf{x},t)/\partial t = -\nabla \mathbf{j}_s(\mathbf{x},t) + \eta(\mathbf{x},t). \tag{5.193}$$

The first term on the right-hand side of equation 5.193 describes mass conserving transport on the growing surface, while the second term describes the random depositional flux and diffusion noise. Equation 5.193 is based on the idea that all processes, appart from the deposition processes represented by the "$\eta(\mathbf{x},t)$" term in equation 5.193, conserve the volume of the deposited film. This implies that the density is constant on all but short length scales that are not described by equation 5.193, and that evaporation/desorption processes do not take place. Equation 5.193 also implies that there is no net material transport within the bulk of the growing film. These restrictions are realistic for a wide range of MBE processes that are carried out at temperatures that are too low for evaporation to be important but are high enough for a crystalline, essentially void free, film to grow. Models that satisfy these conditions have been called ideal MBE models. If these restrictions are not satisfied, the interface growth equation may contain a non-linear term of the form $[\nabla h(\mathbf{x},t)]^2$, and the asymptotic behavior will be in the KPZ universality class. Equation 5.193 can be regarded as a generic molecular beam epitaxy equation, and the asymptotic behavior depends on the surface flow $\mathbf{j}_s(\mathbf{x},t)$. It is often assumed that the surface current $\mathbf{j}_s(\mathbf{x},t)$ has the form [1205]

$$\mathbf{j}_s(\mathbf{x},t) = \nabla f[h(\mathbf{x},t), \nabla^2 h(\mathbf{x},t), (\nabla h(\mathbf{x},t))^2 \ldots], \tag{5.194}$$

where $f[h(\mathbf{x},t), \nabla^2 h(\mathbf{x},t), (\nabla h(\mathbf{x},t))^2 \ldots]$ is a function of $h(\mathbf{x},t)$, $\nabla^2 h(\mathbf{x},t)$, $(\nabla h(\mathbf{x},t))^2$, etc. The EW equation and equation 5.199 can be represented by equations 5.194, but the KPZ equation cannot. However, as Krug *et al.* [637, 961] have pointed out, it is not necessary for the surface current to have the form of a gradient of a "chemical potential", and Krug [637] has shown that the non-equilibrium surface current is an odd function of $\nabla h(\mathbf{x},t)$ with the form $\mathbf{j}_s^{NE} = -(a_1 + a_3(\nabla h(\mathbf{x},t))^2 + \ldots)\nabla h(\mathbf{x},t)$.

If the surface current decreases with increasing inclination ($\partial \mathbf{j}_s(m)/\partial m < 0$), where $m = \langle \nabla h \rangle$ and the sign of $\langle \nabla h \rangle$ is important), then the surface growth is stabilized, leading to scaling in the EW universality class. Since $\mathbf{j}_s(m)$ must be zero when $m = 0$ this implies that the surface current must be a downhill current, for small slopes. If, on the other hand, the surface current increases with increasing inclination, the growth process is destabilized and may become non-linear in real systems. The measurement of surface currents is important not only because of the insight that it provides but also because surface currents that are much too small to be revealed by their influence on the scaling properties of growing surfaces can be reliably measured. In general, the slope dependence of the surface current can be quite complex. There can be one or more slopes at which the slope dependence changes sign, and the slope dependence of the surface current can depend on the orientation of the slope on the surface if the underlying material or the surface is anisotropic. The surface current must be zero on an uninclined surface, by symmetry. For any particular model, the problem of determining the slope dependence of the surface current from the rules of the model remains. It can be anticipated that slope dependent currents will be a general feature of this class of models, and it seems reasonable to suppose this this generality will extend to more complex physical processes.

Analysis of inclination dependent surface currents has proven to be invaluable in the determination of the asymptotic behavior of surface growth models and in the development of a better understanding of surface growth in general. However, this is not a panacea; the measurement of surface currents may be susceptible to both statistical uncertainties (when the surface current is very small) and finite-size effects.

Surface diffusion plays an important role in growth on slightly miscut, stepped surfaces. Schwoebel barriers that reduce the probability that a diffusing atom on a terrace will step down to a lower terrace stabilize the terraced morphology. Particles deposited on a terrace will add preferentially to the edge of the upper step. If one step begins to lag, the terrace in front of it will become broader and the growth velocity of the lagging step will increase because of the larger number of particles deposited on the broader terrace. This stabilizing mechanism results in a "step flow" process in which all of the steps migrate across the surface at a constant velocity. An important consequence of this mechanism is that a net flux of particles is generated in an "upstairs" direction. If the the Schwoebel effect is strong, this flux **j** is proportional to the terrace width ξ_t [995, 949, 637], in the step flow regime, so that $\mathbf{j}_s \sim 1/(\nabla h(\mathbf{x},t)) \sim 1/m$, where $m = \nabla h(\mathbf{x},t)$ is the slope of the surface. Because the uphill diffusional current decreases with increasing slope ($\partial \mathbf{j}_s(m)/\partial m < 0$), the surface growth in the step flow limit is stabilized by surface diffusion and can be described by the EW equation. It follows from the dependence of the surface current on the slope ($\mathbf{j} \sim 1/(\nabla h(\mathbf{x},t) \sim 1/m)$) that $\partial h(\mathbf{x},t)/\partial t \sim (1/m^2)\nabla^2 h(\mathbf{x},t)$, on length scales longer than the terrace width ξ_t.

In this step flow limit [1206], the surface roughness grows only logarithmically, starting with an idealized geometry consisting of equally spaced parallel monoatomic steps separating high symmetry faces. However, island formation will occur on a completely flat surface and will also occur if the temperature is lowered sufficiently in an experiment or if the miscut angle is too small and the terraces are too wide. Adsorbtion at the edges of the islands reduces the uphill current, and for small enough slopes the uphill surface current increases with increasing slope. The dependence of the surface current on the slope m crosses over from $\mathbf{j}_s \sim 1/m$ at large slopes to $\mathbf{j}_s \sim m$ at small slopes, with island formation. The interpolation formula

$$\mathbf{j}_s \sim m/(1+m^2) \tag{5.195}$$

has been proposed [1207].[15] If desorbtion of the deposited atoms plays an important role, the rate of desorbtion will be larger on vicinal surfaces with broad terraces and the growth rate will depend on the slope m. Under these conditions, the growth of the surface roughness can be described in terms of an anisotropic generalization of the KPZ equation, and power law growth of the surface roughness is expected [1208].

Krug and Schimschak [1157] have studied a $1+1$-dimensional step flow model and have found that the step flow behavior is always destabilized by island formation after a time $t^* \sim \xi_{tw}^{-2}(\mathscr{D}_s/\varphi)^{3/4}$, where ξ_{tw} is the step spacing (terrace width), $\mathscr{D}_s$ is the surface diffusion coefficient and φ is the deposition flux density. An early stage scaling regime in which the surface width $\xi_\perp$ grows as $t^{1/4}$ is followed by an intermediate regime in which the surface width grows linearly with increasing deposit thickness, before the asymptotic $t > t^*$ regime in which $\xi_\perp \sim t^{1/2}$ is entered. Other instabilities [1209], related to transport in the direction perpendicular to the terrace edges, may be more important in $2+1$-dimensional surface growth.

15. Johnson *et al.* [1207] proposed the more general form $\mathbf{j}_s = C_1 m/[1+(C_2 m)^2]$, where C_1 and C_2 are constants that depend on the experimental conditions and materials. Equation 5.195 has been used in simulations.

5.6.1.2 MBE Growth Equations

In general, the equation for surface growth under the "ideal MBE" conditions, discussed earlier in this section, can be obtained from equation 5.193. If it is also assumed that the surface current can be expressed in terms of the gradient of a chemical potential, the growth equation can be obtained from the dependence of the surface current on the height derivatives $\nabla(\mathbf{x}, t), \nabla^3(\mathbf{x}, t), \ldots$. Taking into account essential symmetries, the equilibrium "ideal MBE" surface growth equation containing terms up to fourth order in ∇ has the form [1210, 1211, 1212]

$$\begin{aligned} \partial h(\mathbf{x}, t)/\partial t &= a\nabla^2 h(\mathbf{x}, t) - c\nabla^4 h(\mathbf{x}, t) + e\nabla^2(\nabla h(\mathbf{x}, t))^2 \\ &+ f\nabla(\nabla h(\mathbf{x}, t))^3 + \eta_\varphi(\mathbf{x}, t) + \eta_\mathscr{D}(\mathbf{x}, t), \end{aligned} \tag{5.196}$$

where $\eta_\varphi(\mathbf{x}, t)$ is the deposition noise and $\eta_\mathscr{D}(\mathbf{x}, t)$ is the diffusion noise. The EW term "$a\nabla^2 h(\mathbf{x}, t)$" is usually attributed to evaporation of atoms deposited on the surface [945, 947], and this process can be neglected at typical MBE temperatures. However, studies of the surface current for simple models that do not contain evaporation demonstrate that the EW term can have other sources. This can be considerated to be a consequence of non-equilibrium processes for which the surface current cannot be calculated from a chemical potential [961]. The "$e\nabla^2(\nabla h(\mathbf{x}, t))^2$" term can arise from the inclination dependence of the adatom "non-equilibrium chemical potential" [637]. Villain [949] justified the inclusion of a term of this form in the growth equation, based on a slope dependence of the density of diffusing adatoms. The "$f\nabla(\nabla h(\mathbf{x}, t))^3$" term on the right-hand side of equation 5.196 appears as a correction to the EW term in an expansion of the surface current in the gradient $\nabla h(\mathbf{x}, t)$ and may be expected to appear in conjunction with the EW term. A term with the form "$-\tilde{\mathscr{D}}_b\nabla^3 h(\mathbf{x}, t)$", where $\mathscr{D}_b$ is the bulk diffusion coefficient, can appear as a result of volume diffusion [945, 947].

In general, the problem of determining the correct continuum growth equation from the discrete algorithmic rules in a simple model is far from simple. In some cases, quite subtle changes in the model may change the continuum growth equation. Very often, these changes are not apparent, and, in too many cases, the models are not described clearly or carefully enough. New relevant terms may appear with small coefficients, so that the asymptotic scaling behavior is not seen in practical scale simulations, and, in some cases, the dynamic exponent z may be too large for the asymptotic regime to be reached. It has generally been assumed that the form of the Langevin equation does not change if the dimensionality d of the lateral coordinate system is changed. While the correspondence between $1 + 1$-dimensional and $d + 1$-dimensional models with $d > 1$ may be clear in the case of the simple ballistic deposition model, this is not generally true, and a variety of $d + 1$-dimensional models may "correspond" to a particular $1 + 1$-dimensional model. If this assumption were true, it would be of considerable value, since it is much easier to determine the appropriate continuum equation of motion for $1+1$-dimensional models than for the corresponding $d \geq 2$ models. Very often, a continuum equation has been proposed for a model based solely on the agreement between the exponents calculated from the equation and the effective exponents obtained from computer simulations. Despite these difficulties, rapid progress is being made through a combination of theoretical methods and simulations that are carried out for the purpose of determining which terms exist in the continuum limit equation.

Equation 5.196 lies in the EW universality class, but on short length scales the $-c\nabla^4 h(\mathbf{x},t)$ term may be dominant and there will be a characteristic crossover length scale $\xi_{\|x}$ given by $\xi_{\|x} \sim (c/a)^{1/2}$. The characteristic EW scaling will emerge only on length scales $\ell \gg \xi_{\|x}$. This can be seen from the behavior of the terms $a\nabla^2 h(\mathbf{x},t)$ and $c\nabla^4 h(\mathbf{x},t)$ under the transformation given in equations 5.8, 5.9 and 5.10, which generates a rescaled equation containing the terms[16] $\lambda^{\alpha-2} a\nabla^2 h(\mathbf{x},t)$ and $\lambda^{\alpha-4} c\nabla^4 h(\mathbf{x},t)$. This could be important in molecular beam epitaxy since the coefficient of the $a\nabla^2 h(\mathbf{x},t)$ EW equation term could be very small.

Hunt *et al.* [1215] have used the equation of motion

$$\partial h(\mathbf{x})/\partial t = -\nabla g^{-1} \nabla h(\mathbf{x}) - \nabla^4 h(\mathbf{x}), \tag{5.197}$$

with $g = 1 + m^2$, to represent molecular beam epitaxy in the presence of Schwoebel barriers. This equation is based on the idea that the uphill slope dependent surface current can be approximated by the interpolation formula $\mathbf{j}_s \sim m/(1+m^2)$, given in equation 5.195. The second term on the right-hand side of equation 5.197 represents surface diffusion. The solutions to the $1+1$-dimensional version of equation 5.197 describe rapid growth to form a train of mounds, which then evolve more slowly by coalescence of adjacent mounds. The characteristic width ξ_m of the mounds formed in the later stages of growth increases slowly ($\xi_m \sim t^{1/4}$). This simple model captures most of the essential features associated with the growth of mounds or "slugs" in molecular beam epitaxy [1207, 1216] on essentially flat, singular surfaces. However, in the numerical solutions of equation 5.197, the average slope of the faces of the mounds increases with increasing time while it reaches a constant, low value that appears to correspond to the smallest, critical slope for step flow [1217, 1218]. An excellent example of the $t^{1/4}$ growth of mounds that can be attributed to an uphill current due to the Schwoebel barriers is provided by the work of Thürmer *et al.* [1173] on the heteroepitaxial growth of Fe(001) films on Mg(001). In this

16. It has been shown that even without the "$a\nabla^2 h(\mathbf{x},t)$" term, the solutions to the $1+1$-dimensional version of equation 5.196 exhibit EW-like scaling on long length scales [1213, 1212, 1214]. However, if the coefficient f of the $f\nabla(\nabla^3 h(\mathbf{x},t))$ term is also zero, the LDS equation (equation 5.199, discussed below) is obtained and the asymptotic scaling no longer corresponds to that associated with the EW equation.

example, the mounds have a pyramidal shape with (012) faces. The results of these experiments are in very good agreement with a model developed by Siegert and Plischke [1219], which is based on a form for the surface current that is different from that used by Hunt *et al.* [1215] and allows the effects of large slopes and lateral anisotropies to be included. According to this model $H = 1$ and $\beta = 1/4$.

Quite a large number of experiments can be characterized in terms of Hurst exponents in the range $0.7 \leq H \leq 1$ and surface width growth exponents of $\beta \approx 1/4$. It appears that this behavior can arise during the crossover from early stage growth to an asymptotic EW regime with downhill surface currents or can be a result of uphill currents at small but non-zero surface inclinations, resulting from Schwoebel barriers.

5.6.1.3 Quenched Diffusion Models

A number of surface growth models have been motivated by the desire to provide a more realistic representation of processes such as molecular beam epitaxy and, at the same time, retain enough of the speed of the simple models described earlier in this chapter, to allow large scale simulations to be carried out. The compromise between the simplicity needed to explore scaling behavior and the complexity needed to model real systems has resulted in a series of surface growth models. In the most simple of these models, a randomly deposited site, representing an atom, is allowed to explore its local environment immediately after deposition, but once it has been added to the growing deposit it is not allowed to move again. All of these simple models are solid-on-solid lattice models in which a vacant site on top of one of the columns of filled sites is occupied, at each stage in the simulation, to represent the growth process. The models differ only in the manner in which a column is selected for growth from the columns in the immediate neighborhood of a randomly selected column on which the deposited particle is initially placed. These models represent, in a crude manner, the non-equilibrium surface dynamics of a deposited atom, but there is no equilibrium surface dynamics in the absence of deposition. These simple models exhibit surprisingly complex behavior that is exquisitely sensitive to model details. Although these models are far from realistic, they do capture some of the most essential aspects of MBE growth and have contributed much to the rapidly developing understanding of surface growth under MBE conditions. The first model of this type was based on the $1+1$-dimensional independent column model. In this model, particles are initially deposited on a randomly selected column i and then transferred to the lowest column at positions $i-1$, i or $i+1$ [939]. This model belongs to the EW universality class and has been described above.

Wolf and Villain [967] developed a somewhat more realistic $1+1$-dimensional ballistic deposition model in which an attempt is made to include the effects

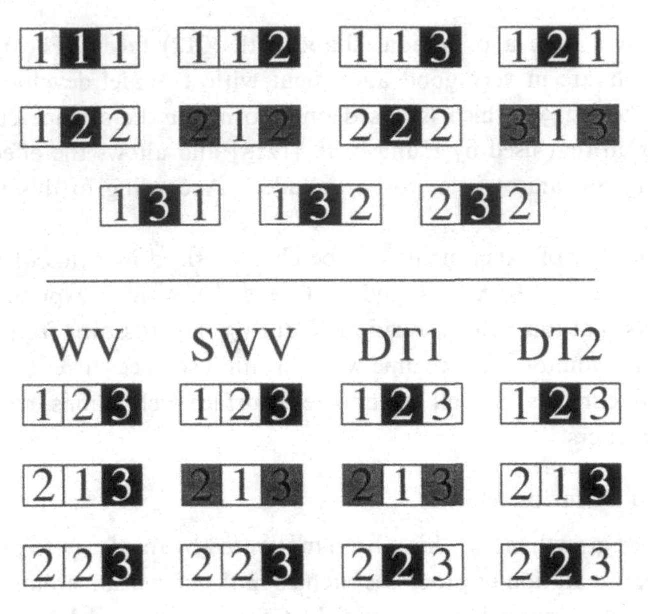

Figure 5.19 Simple 1 + 1-dimensional, quenched diffusion, solid-on-solid models. This figure shows the deposition rules in the Wolf–Villain (WV), symmetric Wolf–Villain (SWV) and two versions of the Das Sarma–Tamborenea model (DT1 and DT2). In all of these models, particles, represented by filled lattice sites, are initially deposited into the unoccupied site at the top of the randomly selected ith column of filled sites. The particle is either added to the ith column or transferred to one of the nearest-neighbor columns according to the rules illustrated in the figure, before a new column is selected and the process is repeated. Each cluster of three boxes shows the coordination numbers of the empty sites on the tops of the columns at $i-1$, i (the column in the middle) and $i+1$. If the deposited particle ends up on a specific column top and the height of that column is increased by one lattice unit, then the coordination number of that column is shown in white on a black background. If a random selection is made between two column tops, then their coordination numbers are shown in black on a gray background. The height of one of these columns will be increased by one lattice unit. The coordination numbers of the column tops that cannot be filled are indicated by black letters on a white background. The top part of the figure shows configurations that lead to the same results in all four models. The bottom part shows the coordination number configurations that distinguish the four models. Configurations that are related to those in the figure by symmetry and configurations that cannot occur are not shown. Periodic boundary conditions are used in the lateral direction.

of surface diffusion. This is a solid-on-solid model in which the "particles" are deposited at the top of the highest occupied site in a randomly chosen column, at lateral position i on a square lattice. After a particle has been deposited on the top of the column at i, it is transferred to a column at position $i+1$ or $i-1$, if the transfer would increase the coordination number of the deposited particle. If the coordinate number cannot be increased in this manner, the added particle remains at the top of column i and the height of column i is increased by one lattice unit. If the columns at both $i+1$ and $i-1$ have the same coordination number, which is larger than the coordination number at position i, then one of these two columns is selected randomly and its height is increased by one lattice unit. If one of the columns $i+1$ or $i-1$ would lead to the highest coordination number, then that column is always selected and its height is increased by one lattice unit. This model is illustrated in figure 5.19. The Wolf–Villain model can easily be extended to higher-dimensionality spaces by randomly selecting from the nearest-neighbor column tops with the highest coordination number, if this highest coordination number is higher than that associated with the initially selected column top. This model is intended to represent short range diffusive transport at an "intermediate" temperature. It is supposed that the deposited atom is "hot" because of its kinetic energy and attraction to the lattice, but that it is rapidly thermalized by interaction with the lattice. The surface growth phenomena associated with this simple model are surprisingly complex and some aspects of the growth of the model surfaces are quite realistic.

The Wolf–Villain model [967] is very similar to the independent column model [939, 996] with restructuring (transfer) to the top of the lowest column at positions $i+1$, i or $i-1$ or to a simple modification of this model in which the deposited particle is transferred to the lowest column at positions $i+1$ or $i-1$ and the height of that column is incremented if $min\{h(i+1), h(i-1)\} < h(i)$. In all of these models, the particle cannot be transferred to a site that is higher than that in which it is initially deposited,[17] in the $1+1$-dimensional case. However, there is a crucial difference. In the independent column model with restructuring, the particles move towards the nearest height minimum and tend to minimize *height differences*. This generates a term of the form $a\nabla^2 h(\mathbf{x})$ in the continuum limit surface growth equation. In the Wolf–Villain [967] model, the surface diffusion tends to minimize the *curvature*, and this generates a term of the form $-c\nabla^4 h(\mathbf{x})$. This example illustrates that quite subtle changes in the growth algorithm can lead to surface growth processes with different continuum limit equations of motion and different scaling behavior. Figure 5.20 illustrates the differences between the surfaces generated by these two models. This emphasizes the need to describe growth algorithms in clear and unambiguous terms.[18] It

17. This does not mean that there can be no uphill surface current. A local step to a lower level can take place in the uphill direction on an inclined surface.
18. I, and most probably many others, have had the experience of finding that a computer program that I have written to simulate a non-equilibrium growth process does not do *exactly* what was intended. The computer program is then corrected and the sometimes lengthy simulations are repeated, only to find that the error has little effect on the outcome and the characteristic exponents remain the same. This "error tolerance" can be attributed to the robust universality of the class of model that is being investigated. Unfortunately (?), this universality is not general and other models are extremely sensitive to model details.

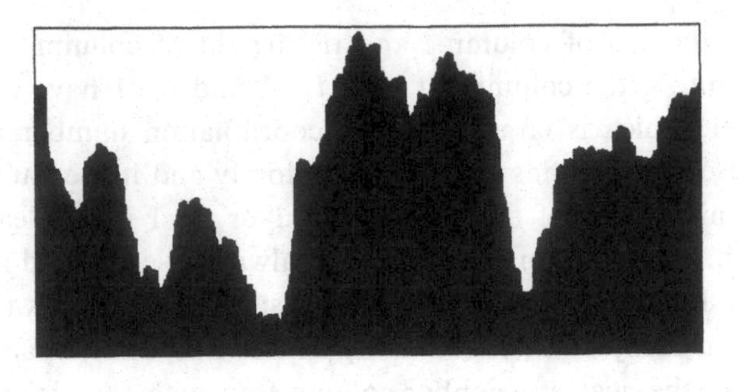

Figure 5.20
Deposits grown using $1+1$-dimensional solid-on-solid models for deposition with transfer to the more favorable nearest-neighbor columns. The top part of the figure shows a pattern obtained using the Wolf–Villain model with transfer to *nn* column tops with larger coordination numbers, and the lower part shows a pattern obtained using the Family model for transfer to *nn* column tops at lower heights. Each simulation was carried out on a square lattice with a width of $L = 512$ lattice units, and the surface is shown after a height of 5×10^5 lattice units was reached.

should be possible to repeat *exactly* the same simulations, without any additional information and without uncertainty. Unfortunately, these standards are often not met, even in papers appearing in the most prestigious journals. The confusion and uncertainty resulting from the inadequate and incomplete description of models can substantially slow progress towards a clearer understanding. The publication of results based on poorly defined models should not be tolerated.

Wolf and Villain carried out simulations using their model with finite strip widths and periodic boundary conditions. A value of 0.365 ± 0.015 was measured for the surface width growth exponent β from the dependence of $(< h^2 > - < h >^2)^{1/2}$ on the "time". A value of 1.4 ± 0.1 was obtained for the exponent α, which describes the dependence of the width of the interface $\xi_\perp$ on the lateral system size L in the long time, stationary regime limit in which $t^{1/z} \gg L$ and the width fluctuates about a constant, time independent value. The exponent z (α/β) was found to have a value of 3.8 ± 0.5. Wolf and Villain argued that this model can be described by the MH equation, which leads to the exponents $\alpha = (4-d)/2$ and $z = 4$, for $d < 4$ (equations 5.29 and 5.30). However, it now appears that, while this equation may describe the early time behavior, this model is much more complex and the effective exponents measured by Wolf and Villain are dominated by small system size and finite time effects. It turns out [637, 961] that the $1+1$-dimensional Wolf–Villain model rules generates a small, downhill surface current $\mathbf{j}_s$ that leads to a crossover to EW model behavior on long time and length scales. The asymptotic behaviors of the $1+1$-dimensional Wolf–Villain model and the $1+1$-dimensional independent column model with restructuring do, after all, belong to the same universality class.

An unrealistic feature of the Wolf–Villain model is the appearance of very large steps in the surface of the deposit. Were it not for the eventual crossover to an asymptotic EW regime, these step heights would become arbitrarily large. It appears that the early stages of growth can be described quite well by the MH equation, despite what appears to be a strong violation of the small slope assumption used in the derivation of this equation. Since $\alpha > 1$ for growing $1+1$-dimensional surfaces described by the MH equation, $\xi_\perp/\xi_\parallel$ increases with increasing time. Although the divergence of $\xi_\perp/\xi_\parallel$ does not continue indefinitely in the Wolf–Villain model, $\xi_\perp/\xi_\parallel$ does increase algebraically with increasing time for the closely related "symmetric Wolf–Villain" model, described below and illustrated in figure 5.19. In any event, it is unrealistic to suppose that atoms would jump from column top to column top, with rates that are independent of vertical distance between the column tops and that they would not stick to the sides of the columns.

It is particularly difficult to obtain reliable results from $d+1$-dimensional Wolf–Villain model simulations with $d > 2$, and only effective exponents are available [1220]. There appears to be an instability with respect to the formation of large "mounds" that leads to effective values for the exponents α, H and β that are larger than expected. This was attributed to the presence of an *uphill* surface current $\mathbf{j}_s$ that increases with increasing slope and generates a term with the form $a\nabla^2 h(\mathbf{x}, t)$ in the right-hand side of the Langevin equation, with a *negative* coefficient a. Šmilauer and Kotrla [1220] carried out simulations using inclined surfaces and measured uphill currents for small slopes and downhill currents for large slopes. They found that the slopes on the sides of the mounds that developed on the surface corresponded to the critical slope at which $\mathbf{j}_s = 0$. Much larger scale simulations may be needed to see the asymptotic behavior of these models, and such simulations would be a major computational challenge. However, the observation that the slopes on the sides of the mounds correspond to an inclination with a zero surface current indicates that the asymptotic regime may already have been reached.

Schroeder *et al.* [1221] obtained an important insight into the behavior of the Wolf–Villain model by focusing attention on the rms step height or the amplitude A of the height difference correlation function ($A = C_2(x=1)$, where C_2 is the height difference correlation function). They found that $A \sim t^\kappa$, with $\kappa = 0.19 \pm 0.01$ for times less than $t^* \approx 10^6$. At longer times, $t \gg t^*$, the amplitude A saturated to an L independent value, consistent with an asymptotic EW universality class behavior. At early times, the evolution of the correlation function could be represented by the scaling form given in equation 2.169

with the exponents $\alpha = 1.45 \pm 0.05$, $\beta = 0.365 \pm 0.005$, $H = 0.75 \pm 0.05$ and $z = 3.9 \pm 0.15$. This scaling form can also be written in the form

$$C_2(x,t) = t^{\kappa} x^{H} f_2(x/\xi_{\|}(t)), \tag{5.198}$$

where $\kappa = \zeta/z \approx 0.19 \pm 0.01$ and the exponent ζ is defined in equation 2.169. The exponent relationship $z = d + 2\alpha$, given in equation 5.50 is satisfied quite well.

Šmilauer and Kotrla [1222] have carried out very large scale simulations using the $1+1$-dimensional Wolf–Villain model and a $2+1$-dimensional version of the model. Values of ≈ 0.19 and ≈ 0.0475 were measured from the $1+1$-dimensional and $2+1$-dimensional simulations for the exponent κ describing the growth of the amplitude of the height difference correlation function at short times. An exponent as small as 0.0475 could indicate a logarithmic growth of the the amplitude of the height difference correlation function, and this would be consistent with the theoretical work of Das Sarma *et al.* [1223] described in section 5.6.3. For the $1+1$-dimensional model, the surface width was found to grow as $\xi_\perp \sim t^\beta$, with $\beta \approx 3/8$ at early times. The later time behavior was consistent with a crossover to an asymptotic EW growth behavior. For the $1+1$-dimensional simulations there was evidence of an intermediate scaling regime characterized by the exponents $H \approx 0.75$ and $\beta \approx 0.33$. Schroeder *et al.* [1221] also found evidence for a similar scaling regime, at late times, in their $1+1$-dimensional simulations. It is not clear at this time if this is a genuine intermediate scaling regime in which a term such as $e\nabla^2(\nabla h(\mathbf{x},t))^2$ or $f\nabla(\nabla h(\mathbf{x},t))^3$ plays an important role in the Langevin equation, or if it is part of the crossover from the early time, MH regime to the asymptotic EW regime.

In the case of the $2+1$-dimensional simulations, a crossover was found from a scaling regime with $H \approx 0.65$, measured from the height difference correlation function, and $\beta \approx 0.22$ to a regime in which the surface width grew more slowly. The early stage, 2+1-dimensional growth exponents are similar to those obtained from the non-linear Langevin equation $\partial h(\mathbf{x},t)\partial t = \nabla^2(\nabla h(\mathbf{x},t))^2 + \eta_D(\mathbf{x},t)$. For this equation, the exponents $\alpha = (4-d)/3$ and $\beta = (4-d)/(8+d)$ can easily be found, based on the invariance to the transformation given in equations 5.8 to 5.10. These early stage exponents are similar to those measured in a variety of deposition experiments [1172].

Park *et al.* [1224] have also carried out extensive $1+1$-dimensional Wolf–Villain model simulations. They found behavior that was consistent with crossover to a growth regime with a smaller value of β, but there was no clear indication of an asymptotic EW regime. This study also led to evidence for a critical system size L, which led Park *et al.* to suggest that the surface growth is dominated by a flux of the form $\mathbf{j}_s \sim \nabla^3 h(\mathbf{x},t)$ on short length scales ($L < L^* < \approx 120$ lattice units) and $\mathbf{j}_s \sim -\nabla h(\mathbf{x},t)$ for $L > L^*$. Krug *et al.* [961]

have predicted that $L^* \approx 2\pi(c/a)^{1/2}$, where a and $-c$ are the coefficients of the $\nabla^2 h$ and $\nabla^4 h$ terms in the Langevin equation, respectively.

The direct evidence for an asymptotic EW regime is not compelling, particularly in the $2+1$-dimensional case. However, confidence in this idea is enhanced by the fact that in both the $1+1$-dimensional and $2+1$-dimensional simulations, the purported crossover to the asymptotic EW regime takes place when the growth heights or times reach the values of 10^6 and $\approx 2 \times 10^4$, respectively, that were predicted by Krug *et al.* [961, 1225], based on measurement of the dependence of the downhill stabilizing surface current on the inclination of the surface. Much better direct evidence for asymptotic EW scaling behavior has been obtained from simulations carried out using modified $1+1$-dimensional Wolf–Villain model simulations [1226] in which deposited particles are transferred to the tops of nearest-neighboring columns with the largest "binding energy" given by $E_b = E_{nn}N_{nn} + E_{nnn}N_{nnn}$, where E_{nn} and E_{nnn} are the nearest-neighbor and next-nearest-neighbor "bond energies" and where N_{nn} and N_{nnn} are the number of occupied nearest-neighbor and next-nearest-neighbor sites, respectively. Similarly modified $2+1$-dimensional Wolf–Villain models [1227] also show a crossover to EW behavior at relatively short times and on relatively short length scales. Overall, it seems that the asymptotic behavior of the $1+1$-dimensional and $2+1$-dimensional Wolf–Villain models does belong to the EW universality class and that this can be understood in terms of the slope dependent surface diffusion current. However, the situation is not so clear for the earlier stages of growth, or for $d > 2$. It appears that the relative importance of the terms in the ideal MBE equation (equation 5.196, discussed below) change with time and length scale, in a manner that is not yet clear.

Das Sarma and Tamborenea [1228] investigated a series of $1+1$-dimensional, square lattice, solid-on-solid models in which, after a particle had been deposited, it was allowed to move, with probability P, to the nearest kink site, within a finite distance ℓ. Here, a kink site is an unoccupied perimeter site with two or more occupied nearest neighbors on the square lattice. In the most simple, $\ell = 1$, $P = 1$, version of this model, a particle, initially deposited into the unoccupied perimeter site on the top of the ith column, is transferred to one of the nearest-neighbor columns if the unoccupied perimeter site on the top of column i is not a kink site and the unoccupied perimeter site in the top of column $i-1$ or $i+1$ are kink sites. Ties between the unoccupied sites at the tops of the columns at positions $i-1$ and $i+1$ are broken by random selection. This model is defined in more detail in figure 5.19, where it is labeled "DT1". From these simulations a value of 0.375 ± 0.005 was obtained for the surface width growth exponent β, irrespective of the values used for the parameters ℓ with $\ell \leq 3$ and P in the range $0.2 \leq P \leq 1$. A value of 1.47 ± 0.1 was also measured for the exponent α. Thus, it appeared, for a while, that the models of Das Sarma and Tamborenea and the models of Wolf and Villain belong to the same universality class, described by the MH equation, with deposition noise. However, the behavior of these models has turned out to be much more subtle.

Krug [637] has studied two versions of the $\ell = 1$, $P = 1$ Das Sarma–Tamborenea model (DT1 and DT2). Model DT1 is the simplified Das Sarma–Tamborenea model described in the preceding paragraph. Model DT2 differs from model DT1 only in the case where the unoccupied perimeter site on the randomly selected ith column is a non-kink site with a coordination number of 1 and the unoccupied perimeter sites on the tops of columns $i-1$ and $i+1$ have coordination numbers of 2 and 3 or 3 and 2, respectively. In model DT1 a random selection is made betwen the columns at $i-1$ and $i+1$ for this coordination number configuration, while in model DT2, the three-coordinate kink site is always filled to represent the growth process. These two models are illustrated in figure 5.19. The Wolf–Villain model differs from the DT1 and DT2 models because a particle that is initially deposited into a kink site with a coordination number of 2 is transferred to the empty site at the top of a nearest-neighbor column if the coordination number would be increased to 3. Despite their superficial similarities, Krug [637] has shown that the Wolf–Villain model, model DT1 and model DT2 belong to three different universality classes.[19] There is a downhill surface current in the Wolf and Villain model and it belongs to the EW universality class. Because of the symmetry of the algorithm, there is no net surface current in the DT1 model and it belongs to the MH equation universality class. For model DT2 there is a very small uphill current leading to a term of the form $-a\nabla^2 h(\mathbf{x}, t)$, which will eventually result in unstable growth. However, the coefficient a is so small that this will not result in any observable effects in practical scale simulations [637]. It can be seen from figure 5.19 that, for the DT1 model, a bias in the uphill or downhill direction can arise only for deposition onto the middle column of a "112", "113", "211" or "311" coordination number configuration. Krug [637] has pointed out that there must be an equal number of "11X" and "X11" configurations in a ring labeled with 1's and X's to represent the coordination numbers of the columns in a system with helical periodic boundary conditions. Consequently, the absence of a slope dependent surface current has a simple explanation for this model. Most of the Das Sarma and Tamborenea [1228] models have not been thoroughly investigated, but it can be anticipated that they will exhibit complex behavior and sensitivity to model details like that of the simple ($P = 1$ and $\ell = 1$) members of this class.

19. Since these models differ only in subtle details, this might seem to be surprising. However, the biased Brownian process provides a simple, but not perfect, analogy. The unbiased Brownian process has a Hurst exponent of 1/2 (chapter 2, section 2.7), but if the bias is towards the origin, the asymptotic Hurst exponent is zero, and $H = 1$ if the bias is away from the origin. In this case, the origins of the different universality classes are obvious. In more complex models the source of the bias may not be as clear.

Similarly, it has been shown that there is no net downhill or uphill flux in the Wolf–Villain model, if the two-coordinate and three-coordinate column tops are treated equally [961]. Here, this model is called the "symmetric Wolf–Villain model" and it is illustrated in figure 5.19. It is only the rare configurations in which a deposited particle with a coordination number of 1 is neighbored by a column with a two-coordinate site and a column with a three-coordinate site that contribute to the breakdown of the MH equation for the ordinary Wolf–Villain model. This provides another example of the important effect that "insignificant" details can have on the behavior of deposition models.

Quite similar models have been studied by Kang and Evans [1229, 1230], using triangular lattices with and without "downward funneling" (restructuring), immediately following deposition. The model can be thought of in terms of a packing of monodisperse, hard discs on a base consisting of a horizontal row of contacting discs to form a regular structure in which the disc centers are at the nodes of a triangular lattice. Each kink site (corresponding to an unoccupied position in which a disc would contact three or more discs in the deposit) has a "catchment area" (catchment line for $d = 1$) corresponding to the region in which a disc placed on the surface would reach the kink if it followed a path of steepest descent on the surface. In the model without restructuring, the particles are initially deposited randomly at a local minimum (a position in which the deposited disc contacts two discs that are both at a lower height in the packing). If the local minimum is not also a kink site, the disc is transferred to the kink site whose catchment area extends closest to the position of initial deposition, if the distance to the nearest catchment area is within a distance ℓ, which is fixed at the start of the simulation. In the event of a tie between the catchment areas associated with nearby kink sites, the disc is transferred randomly to one or the other corresponding kink site. In the models with restructuring, the particles are initially deposited randomly on the surface, in the same way as in an off-lattice ballistic deposition model, and follow a path of steepest descent on the surface until a local minimum is found (this is the downward funneling). If this does not bring them to a kink site, they are then transferred to the kink site with the nearest catchment area, within a distance ℓ.

In these models, the effective value of the exponent β appears to increase from the expected KPZ value of about $1/3$, without restructuring, or the EW value of $1/4$, with restructuring, for $\ell = 0$ to a value of at least $3/8$ for large ℓ. For the larger values of ℓ, effective exponents greater than $3/8$ were found, but the interpretation of this result is not clear since the data did not show simple scaling in this regime. Like the other simple, quenched diffusion models, this class of models may exhibit complex behavior that is sensitive to model details. Measurement of the slope dependence of the surface currents would be worthwhile.

In another model investigated by Kang and Evans, each particle is deposited into a randomly selected local minimum and is then transferred, with probability P, to a randomly selected nearest-neighbor site on the lattice, from which it follows a path of steepest descent, if it is not initially deposited in an "adsorbing" kink site. Values of 0.33, 0.28 and 0.25 were obtained for the exponent β from simulations carried out using the $1+1$-dimensional version of this model with $P = 0$, $P = 1/2$ and $P = 1$, respectively. The value of $\beta = 0.28$ obtained for $P = 1/2$ is almost certainly an effective exponent representing a crossover between the KPZ and EW scaling regimes.

Simulations to explore $2+1$-dimensional models closely related to the $1+1$-dimensional models of Wolf and Villain and of Das and Tamborenea have been carried out by Kotrla *et al.* [1231]. The "diffusion steps" are restricted to transfer to nearest-neighbor columns. In model 1, the deposited particle is transferred to the nearest-neighbor (*nn*) column with the highest coordination number, if that coordination number is higher than that of the site in which the particle is initially deposited, and "ties" are broken by a random selection. In the $2+1$-dimensional case, the larger coordination number site may be at a higher position on the surface than the site at which the particle was initially deposited. Model 2 is like model 1 but particles are not allowed to move to new sites at higher positions. A variant of model 2 (model 2a) was studied in which the deposited particle can transfer to nearest-neighbor column top positions with the same coordination number as that of the site at the top of the initially selected column if there are no *nn* column tops with higher coordination numbers at heights that are equal or lower than that of the initially selected column top. In this event, a random selection is made from the initially selected column top position and the *nn* column top positions that have the same coordination number as the initially selected column top and are at equal or lower heights. Model 3 differs from model 1 only in the way in which ties are broken. If there are two (or more) *nn* column top positions with the same highest coordinate number, then the highest of these positions is always chosen. If both positions also have the same height, the ties can be broken randomly. Unfortunately, these models were not unambiguously defined. For model 1, the exponent β has a value of about 0.21, and for model 2 $\beta \approx 0.18$. A roughness exponent of $\alpha = 0.66 \pm 0.03$ was obtained for model 1, but the saturated, $t \rightarrow \infty$, surface width grows only logarithmically with the system size, $L_x = L_y$, for models 2 and 2a. For model 1, a value of 0.21 ± 0.02 was obtained for the exponent β, so that $z = 3.20 \pm 0.46$. These exponents are consistent with the scaling relationships $\alpha + z = 4$ and $2\alpha + d = z$ [967]. These simulations were carried out on a relatively small scale, with a maximum value of 128 for L_x ($L_x = L_y$). In light of the experience acquired by working on the related $1+1$-dimensional models, complex behavior and sensitivity to model details should be anticipated. It seems likely that much larger scale simulations with the measurement of the slope dependence of surface diffusion currents will be needed to unravel the potential complexities of these models.

Park *et al.* [1232] have investigated a version of the Wolf–Villain model with a built-in uphill bias. In this $1+1$-dimensional model, a particle deposited into the unoccupied site at the top of the ith column is transferred to the top of the neighboring column at $i-1$ or $i+1$ with the highest coordination number, if the transfer would increase the coordination number of the particle. Ties between the unoccupied sites on the tops of columns lateral positions $i-1$ or $i+1$ with the same coordinate number are broken by moving the deposited particle with a bias in favor of the tallest of the columns of filled sites at positions $i-1$ or $i+1$. If the coordination numbers of the sites at the tops of the columns at $i-1$ and $i+1$ are both equal and both higher than the coordination number at the tops of the columns at lateral position i, then the particle is deposited on top of the tallest of the two columns with a probability of $(1+P)/2$ and on the top of the shortest column with a probability of $(1-P)/2$. This model generates a grooved surface, consisting of broad "plateaus" separated by deep, narrow "canyons". The plateau tops and canyon bottoms grow with different velocities, so that $\beta = 1$. Park *et al.* studied the dependence of the morphology on the bias parameter P and system size L. They found that, for each value of P, there is a characteristic length scale associated with the separation between adjacent grooves. Similarly, for each system size L, there is a critical bias P_c, below which the surface is stable and above which it is grooved. This can be understood in terms of a competition between the downward current associated with the Wolf–Villain rules (discussed below) and the upward current induced by the bias. Park *et al.* [1232] argued that the upward current $j_\uparrow$ is proportional to $P\xi_g$, where ξ_g is the characteristic distance between grooves, because the plateaus between the grooves consist of a series of steps between levels of increasing height as the distance from a groove increases. The curvature driven downward current $j_\downarrow$ is constant over a length scale of order ξ_g since there is only approximately one position at the edge of each plateau where the coordination number at $i-1$ or $i+1$ is greater than that at i. Consequently, the transition takes place when $j_\uparrow = j_\downarrow$ or $P_c\xi_g \sim 1$. The simulation results were consistent with this idea.

Figure 5.21(b) was generated using a model in which the particle deposited in column i is transferred to the column at $i-1$, i or $i+1$ in which the unoccupied column top site has the largest coordination number. If the coordination numbers of the columns at $i-1$ and $i+1$ are both equal and larger than the coordination number for column i, then the higher of these two column tops is selected, with probability $(1+P)/2$ (if both columns have the same height one of them is selected randomly) and the column top at either $i-1$ or $i+1$ is selected randomly with probability $(1-P)/2$. If the coordination number of column i is the largest but is equal to that of column $i-1$ or $i+1$, then the higher of the two column tops with the largest coordination number is selected,

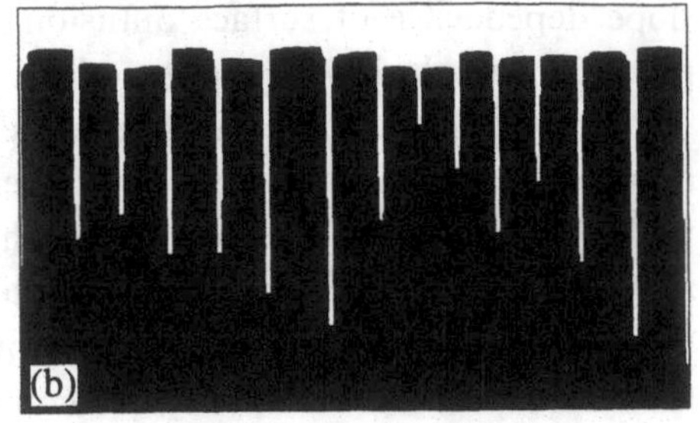

Figure 5.21 Patterns obtained from $1+1$-dimensional solid-on-solid growth models with local surface diffusion when a mean height of 5×10^5 lattice units was reached. Part (a) shows results from model DT1, defined in figure 5.19, that was designed to ensure that there is no inclination dependent surface current and part (b) shows a pattern generated using a related model with an uphill surface current. The uphill bias parameter P was given a value of 0.1. Each simulation was carried out on a square lattice, with a width of $L = 512$ lattice units. In part (b), the vertical length scale is 25× the horizontal length scale. This makes the regions between the "canyons" appear to be flat topped.

with probability $(1+P)/2$ (ties are broken by random selection) and the particle is added to the top of column i with probability $(1-P)/2$. If all the column tops have the same coordination number, then the higher of the three column tops is selected, with probability $(1+P)/2$ (ties are broken by random selection) and the particle is added to the top of column i with probability $1-P$. It is not clear if this is equivalent to the model of Park *et al.* [1232], but it is closely related and behaves in a similar manner.

Lai and Das Sarma [1210] and Das Sarma and Ghaisas [1233] have explored a variety of $1+1$-dimensional, square lattice and $2+1$-dimensional, cubic lattice, solid-on-solid models with a variety of rules for restructuring after the initial deposition event. Based on the effective values for the exponent β, they assigned these models to different universality classes (different Langevin equations). These include the universality classes described above and those defined by the evolution equations

$$\partial h(\mathbf{x},t)/\partial t = -c\nabla^4 h(\mathbf{x},t) + e\nabla^2(\nabla h(\mathbf{x},t))^2 + \eta(\mathbf{x},t) \tag{5.199}$$

($\beta = (4-d)/(8+d)$ or $\beta = 0.2$ for $d = 2$) and

$$\partial h(\mathbf{x},t)/\partial t = -c\nabla^4 h(\mathbf{x},t) + f\nabla(\nabla h(\mathbf{x},t))^3 + \eta(\mathbf{x},t) \tag{5.200}$$

($\beta = (4-d)/2(4+d)$ or $\beta = 0.1666\ldots$ for $d = 2$). Equation 5.199 is often referred to as the Lai-Das Sarma equation, or the LDS equation. Although the statistical uncertainties given by Das Sarma and Ghaisas are quite small, it is not clear if the total uncertainties in β, including statistical uncertainties, preasymptotic scaling and uncertainties resulting from finite-size effects, are small enough to justify this classification. The difficulties of establishing the relationships between simple growth models and their corresponding continuum equations of motion were not generally appreciated at the time that this work was carried out. Plischke *et al.* [1234] have found that most of these models have either downhill or uphill inclination dependent surface currents and that their asymptotic behaviors belong to either the EW universality class, or they exhibit unstable growth.

5.6.1.4 Continuous Diffusion Models

A further step towards the development of more realistic models for the growth of surfaces during deposition processes is to allow atoms or sites at the interface to change positions continuously, as new particles are being deposited. The increased realism comes at the cost of smaller scale simulations. In these models, the diffusion process is represented by transferring a particle, represented by a filled site, on the surface of the deposit to an adjacent vacant site on the unoccupied perimeter of the deposit. Almost all simulations of this type have been carried out using activated hopping models or the simple Metropolis or heat bath Monte Carlo algorithms described in chapter 1, section 1.9. In the activated hopping models, the hopping rate is assumed to have the Arrhenius form $k(\Omega_l) = Ae^{-E_a(\Omega_l)/k_BT}$, and it is assumed that the activation energy $E_a\Omega_l$ depends on only the local configuration of occupied sites Ω_l. In many cases, E_a is taken to be proportional to the number of filled nearest-neighbor sites (the number of "bonds" that must be broken to move the corresponding particle). These procedures are consistent with microscopic reversibility in the absence of deposition, provided that bias is not inadvertently introduced by the way in which the filled and empty site pairs are selected.

Pure surface diffusion models, intended to represent the behavior of equilibrium systems, should satisfy the detailed balance requirement. However, the evolution of a surface, driven by deposition, is not an equilibrium process and is not required to satisfy detailed balance. Krug *et al.* [961] emphasized that non-equilibrium conditions generate surface currents $\mathbf{j}_s(\mathbf{x}, t)$ that cannot be be calculated thermodynamically from a free energy functional. In general terms, this non-equilibrium surface flow comes about because, under non-equilibrium conditions, there is a bias in favor of flow in an "uphill" or "downhill" direction. This is a consequence of the loss of symmetry with respect to reflection about the vertical axis due to tilting of the surface with respect to the direction of incidence of the incoming particles. This effect can also be considered to be a result of the loss of detailed balance in non-equilibrium processes.

Rácz *et al.* [958, 969] have shown how subtle changes in discreet algorithms can change the continuum limit equation of motion and the asymptotic scaling behavior in simple solid-on-solid, surface diffusion models, in much the same way that subtle changes in the quenched diffusion models for MBE surface growth, described above, can have dramatic effects on long time behavior. In these models, particles are moved randomly between nearest-neighbor or next-nearest-neighbor column tops, subject to the restraint $h(i)-h(i+1) \leq \Delta h$, where Δh is a step height restriction parameter of the model. In the first model, a trial move is attempted and executed if the step height restriction is not violated. If the step height restriction is violated, then another randomly selected trial move is attempted. This model satisfies detailed balance. The second model is very similar, except that if a move from the column top at i_1 to the column top at i_2 is attempted and fails to satisfy the step height restriction, the move from i_2 to i_1 is attempted. This model no longer satisfies detailed balance. As a result, the reflection symmetry $h(i) \rightarrow -h(i)$ is broken and a term of the form $e\nabla^2(\nabla h(\mathbf{x},t))^2$, which is forbidden by symmetry in the detailed balance model, appears in the continuum limit growth equation. An even more dramatic consequence of this "minor" change in the model is the appearance of a surface current that depends on the inclination of the surface [961]. These surface currents dominate the asymptotic behavior. The nature of this slope dependent current [961] and the evolution of the surface [969] depends on the height restriction Δh.

Das Sarma and Tamborenea [1228] have studied a continuous diffusion, 1+1-dimensional, square lattice, solid-on-solid model in which all of the surface sites are allowed to hop at all times, with hopping rates given by

$$R = R_0 e^{-E_a/k_B T}, \tag{5.201}$$

where E_a is a site dependent activation energy ($E_a = E_{\circ a} + nE_{ba}$, where E_{ba} is the binding energy per bond and n is the coordination number or number of occupied nearest neighbors). In this model, the occupied site at the top of a column is allowed to move to the top of one of the nearest-neighbor columns, with rates given by equation 5.201. The hopping rates in this model do not depend on the height differences between neighboring columns. This model is equivalent to a model for surface growth studied by Gilmer and Bennema [1235]. Values of 0.3 eV and 1.0 eV were used for $E_{\circ a}$ and E_{ba}, respectively, at a temperature of 500 K. Das Sarma and Tamborenea measured the exponent β describing the growth of the surface width for different ratios between the deposition rate and the hopping rate. A value of about 0.375 was obtained when the deposition rate and the hopping rate were of comparable magnitude. A value of 1.47 ± 0.1 was obtained for the roughness exponent α that relates the surface width to the lateral system size L, in the long time limit. This value is much different from that associated with the standard ballistic deposition model and agrees well with the theoretical value of $\alpha = (4-d)/2$ for growth processes described by the MH equation.

Wilby *et al.* [1236] have carried out $1+1$-dimensional, $2+1$-dimensional and $3+1$-dimensional simulations using a continuous diffusion model [1235] that appears to correspond to equation 5.199. In this solid-on-solid model, particles are transfered between nearest-neighbor columns using an Arrhenius, activated hopping model with activation energies of $E_\circ + nE_b$, where n is the number of occupied sites at the same level ($0 \leq n \leq 2d$). The model parameters were selected to simulate deposition at a rate of 1.0 monolayer s^{-1} at an "intermediate temperature". Values of 0.333 ± 0.010, 0.200 ± 0.012 and 0.092 ± 0.020 were obtained for the exponent β, from simulations carried out with $d = 1, 2$ and 3, respectively. In addition, a crossover from $\beta \approx 3/8$, at early times, to $\beta \approx 1/3$, at late times, was observed in the $1+1$-dimensional case. This suggests that there might be a crossover from a "linear" regime described by the MH equation to a non-linear asymptotic regime described by equation 5.199. This may be a consequence of a small coefficient e for the non-linear term in equation 5.199. In light of more recent work that has revealed the complex behavior described above for simple quenched diffusion, solid-on-solid MBE models, it is not clear how these results should be interpreted. The behavior of these models is very similar to the early stages in the quenched diffusion models described earlier in this section, and the asymptotic behavior is uncertain. The observed behavior might be characteristic of an intermediate scaling regime described by equation 5.199, or it might be part of crossover to an asymptotic EW growth regime. Additional work, including the measurement of the slope dependence of surface currents, is needed. Unfortunately, neither the Hurst exponent nor the roughness exponent was measured. Wilby *et al.* [1236, 1237] also studied the behavior of the model without deposition. They found that $\beta = 0.13 \pm 0.01$ for the $1 + 1$-dimensional model and interpreted this result in terms of the linear MH equation with diffusion noise, for which $\beta = 1/8$.

Johnson *et al.* [1207] have used a lattice model with both nearest-neighbor and next-nearest-neighbor interactions to simulate thin film growth by molecular beam epitaxy. The hopping rates are proportional to e^{-E_a/k_BT}, where E_a is the activation energy. In this model, the activation energy for an occupied site is the sum of its nearest-neighbor interactions plus the difference between the next-nearest-neighbor interactions associated with the occupied site and the next-nearest-neighbor interactions in the destination site, if the next-nearest-neighbor interaction energy increases ($E_a = n_{nn}E_{nn} + max(\delta n_{nnn}, 0)E_{nnn}$, where E_{nn} and E_{nnn} are the nearest-neighbor and next-nearest-neighbor interaction energies, n_{nn} is the number of nearest neighbors before a jump and δn_{nnn} is the change in the number of next-nearest-neighbor interactions). This procedure allows the effects of Schwoebel barriers to be included and satisfies detailed balance. Simulations were carried out using model parameters that corresponded to the MBE growth of GaAs and generated a positive Schwoebel barrier. Deposition on flat surfaces resulted in the formation of large mounds with steep sides. However, if the initial surface was inclined with respect to the lattice, so that the initial surface consisted of a series of steps, stable step flow growth could be observed if $\xi_n > \xi_{tw}$, where ξ_{tw} is the width of the terraces between the steps and ξ_n is the average distance between 2-dimensional islands that grow on a singular surface (a flat, low index surface with no steps) during the formation of the first layer. The model results were found to be consistent with experiments in which surfaces generated by MBE deposition of GaAs on GaAs(001) were examined by scanning tunneling microscopy and atomic force microscopy. A transition from the formation of elongated mounds to stable step flow was observed as the miscut angle was increased. In typical experiments, a "500 bilayer" film was deposited at temperatures of 550 to 600 °C. The mounds were found to form only if the growth proceeded in a layer-by-layer fashion. Mounds were found to form on slightly miscut surfaces. If the temperature was raised, the mounds flattened and, at a sufficiently high temperature, the mounds on a miscut surface disappeared.

Johnson *et al.* [1207] measured the slope dependence of the surface current in their simulations. They found that $\mathbf{j}_s$ increased essentially linearly with increasing slope but reached a maximum when $m\xi_n \approx 1$, where m is the slope and ξ_n is the nucleation distance. This reflects the instability with respect to roughening due to island formation at very small slopes, $m < 1/\xi_n$, and the stable step-flow growth for $m > 1/\xi_n$. This led Johnson *et al.* to propose the scaling form $\mathbf{j}_s = \varphi\xi_n^2 m f(\xi_n m)$, where φ is the incident flux density, with $f(x) \sim x^{-2}$ for $x \gg 1$, so that $\mathbf{j}_s$ does not depend on the nucleation distance, and $\mathbf{j}_s \sim 1/m$. The scaling function $f(x)$ has the form $f(x) = const.$ for $x \ll 1$ so that $\mathbf{j}_s \sim m$. Equation 5.195 is a simple example of this scaling form.

The important role played by Schwoebel barriers in thin film growth has also been emphasized by Zhang *et al.* [1238]. They carried out simulations using a $1+1$-dimensional model in which the "atoms" moved with probabilities given by

$$P_i = e^{-(E_i - E_\circ)/k_B T}. \qquad (5.202)$$

The energy $E_\circ$ was chosen so that atoms on a flat region of the surface moved at every attempt. The energy E_i is proportional to the number of bonds that must be broken to separate the atom from the lattice *plus* an additional contribution of E_s, to represent the Schwoebel barrier, if the atom attempts to move from the top of an "island" to the island or terrace below. Consequently, the activation energy E_i has the form $E_i = n_i E_b + (E_s)_i$, where n_i is the coordination number of the particle before the attempted move and $(E_s)_i$ is the Schwoebel barrier for the move. A value of 0.5 eV was used for E_b in these simulations. For small values of E_s, at $T = 600$ K, the surface width $\xi_\perp$ oscillated as the deposit grew in a layer-by-layer manner. As E_s was increased, a scaling regime in which $\xi_\perp$ grew as $t^{1/2}$ was entered. The value obtained for the exponent β was not in agreement with the values expected from either the EW or KPZ model. The simulations were carried out on a quite small scale. Consequently, it is not clear how well the effective exponents reflect the asymptotic values. In one series of simulations, the barrier E_s was held fixed at 0.2 eV and simulations were carried out at several temperatures. At low temperatures a columnar or deeply grooved morphology developed. As the temperature was raised the effect of the Schwoebel barriers was reduced and the surface became flatter.

A further step towards the development of a better understanding of the effects of surface diffusion processes was taken by Siegert and Plischke [1239]. They explored a solid-on-solid model in which particles are deposited randomly and the interface is allowed to relax between deposition events. Deposition is carried out with probability P_τ, and a diffusion step is attempted with probability $1-P_\tau$ during each step in the simulation. In the relaxation step, which is completely independent of the deposition step, one of the columns of the lattice (column i) is selected randomly and a particle is moved from the top of the ith column to one of its nearest neighbors, column j, which is randomly selected, with a probability given by

$$W_{i\to j} = [e^{\Delta\mathscr{H}/k_BT} + 1]^{-1}, \quad (5.203)$$

where $\Delta\mathscr{H}$ is the change in the Hamiltonian H given by

$$\mathscr{H} = J\sum_{\{nn\}} |h_i - h_j|^n, \quad (5.204)$$

and the sum is over all nearest-neighbor column pairs. Values of 1, 2 or 4 were used for the exponent n in equation 5.204. In the absence of deposition, this model simulates thermal roughening with exponents of $H = 1/2$ and $z = 2$.

Simulations were carried out using the $1+1$-dimensional model with deposition "rates" of $P_\tau = 0.01$ and $P_\tau = 0.1$ at temperatures of $J/k_BT = 1$ and $J/k_BT = 0.01$. Values of 1.2 ± 0.2, 3.6 ± 0.4 and 0.35 ± 0.01 were obtained for the exponents α, z and β from the $1+1$-dimensional model results, with $n = 2$ [1239]. Because the exponent z is so large, saturation could not be reached for large systems, and the measurement of the roughness exponent from the dependence of the saturated surface width $\xi_\perp(L, t \to \infty)$ on the system size L was practical for only quite small values of the lateral system size L. Siegert and Plischke measured the exponents H and α from the correlation function

$$S(k, t_\circ, t) = < h_k(t_\circ + t)h_{-k}(t_\circ) >, \quad (5.205)$$

where

$$h_{-k}(t) = L^{-d/2}\sum_n [h_n(t) - < h > (t)]e^{ink}. \quad (5.206)$$

For sufficiently large values of $t_\circ$, the correlation function $S(k,t)$ has the form

$$S(k, t_\circ, t = 0) = k^{-(2H+d)} f(k^z t_\circ), \quad (5.207)$$

for a $d+1$-dimensional surface, and the exponents can be obtained via a data collapse. The scaling function $f(y)$ has the form $f(x) = const.$ for $x \gg 1$. In those cases where $\alpha > H$, the amplitude of the structure factor will grow with increasing time in the same way that the amplitude of the height difference correlation function grows. The correlation function $S(k, t_\circ, t = 0)$ can then be represented by the scaling form

$$S(k, t_\circ, t = 0) = t_\circ^{2\kappa} k^{-(2H+d)} f(k^z t_\circ). \tag{5.208}$$

In this case, both the Hurst exponent H and the roughness exponent $\alpha = H + z\kappa$ can be measured from the shape of the structure factor and the data collapse, respectively. The dynamical exponent z was calculated from the ratio $\Phi(k, t_\circ \to \infty, t) = S(k, t_\circ \to \infty, t)/S(k, t_\circ \to \infty, t = 0)$, based on the assumption that $\Phi(k, t_\circ \to \infty, t)$ is a function of $k^z t$ ($\Phi(k, t_\circ \to \infty, t) = g(k^z t)$). A quite accurate value for z can then be obtained from a data collapse of the curves $\Phi(k, t_\circ \to \infty, t_n)$ measured at a number of stages or times t_n, during a single simulation.

The results obtained for $n = 1$ were quite different. In this case, the growth of the surface appears to belong to the EW universality class for both $1+1$-dimensional and $2+1$-dimensional simulations ($H = 1/2$ and $z = 2$ for $d = 1$ and logarithmic growth of $\xi_\perp$ for $d = 2$). The role played by the exponent n in equation 5.204 can be understood in terms of the dynamics on a surface consisting of large terraces with well separated step edges. For $n = 1$, the energy does not change if the particle reaches the upper edge of a step. However, for $n > 1$ there is an energy increase and this creates a "Schwoebel barrier" [1149, 1150] that inhibits jumps between terraces. A more detailed analysis for rough surfaces, with less well defined steps [1240], indicated that there is a negative Schwoebel effect for $n = 1$. This biases the particles in a "downhill" direction and accounts for the EW behavior for $n = 1$. For $n = 2$, there is no Schwoebel barrier and, for $n = 4$, there is a positive Schwoebel effect and large slopes are formed. The results obtained from the $2+1$-dimensional simulations with $n = 4$ were consistent with a Hurst exponent of $H \approx 1$ and the surface evolved towards a single pyramidal mound in a finite system.

Siegert and Plischke [1240] also carried out simulations for the $n = 1$, $1+1$-dimensional case, using an "Arrhenius dynamics" model in which the transition probabilities are proportional to $e^{-E_a/k_B T}$. In this model, the activation energy is assumed to be Jn_b, where J is the coupling constant in the Hamiltonian and n_b is the number of bonds between the hopping particle and the adjacent columns. In the absence of deposition, this model gives a value of about 4 for the dynamic exponent z and a Hurst exponent of about $1/2$, consistent with a process described by the MH equation with conserved, diffusion noise. In the presence of deposition, large steps developed on the surface and the correlation function $S(k, t_\circ, t = 0)$ indicated that $H \approx 1$ and $\alpha > H$. For this model Siegert and Plischke measured the mean step height and found that it increased as $C_2(1, t) \sim t^\kappa$, with $\kappa \approx 0.15$. These results appear to be consistent with a process described by the MH equation in the presence of deposition noise (with all the usual caveats). Schwoebel and anti-Schwoebel effects can easily be incorporated by controlling the energy barrier from the top of a "terrace" to the edge of a step. In the presence of Schwoebel or anti-Schwoebel barriers, with deposition, the behavior was similar to that found for the Metropolis dynamics model (equations 5.203 and 5.204) with $n = 4$ and $n = 1$, respectively.

A simplified version of the model of Siegert and Plischke based on the Hamiltonian

$$\mathscr{H} = J\left[\sum_{nn} |h_i - h_j|^2 + g|h_i - h_j|^4\right], \tag{5.209}$$

has been studied by Family and Lam [1241]. In this model, only the newly deposited particles are allowed to move. The simulations were carried out at a small temperature $J/k_BT = 0.01$. They found scaling behavior that was consistent with that obtained using the original model of Siegert and Plischke in which all the surface particles can move. Values of ≈ 0.375 and ≈ 1.5 were measured for β and α, respectively, from simulations carried out with $g = 0$ $(n = 2)$. For $g = 1$, exponents of $\beta \approx 0.75$ and $\alpha = 3$ were measured, from the dependence of the surface width $w(L, t)$ on its arguments. Family and Lam carried out a renormalization group analysis based on the full surface diffusion equation (equation 5.23), with deposition noise. This simulation showed that the Langevin equation is unstable with respect to the formation of a "grooved" surface with large slopes.

The 1 + 1-dimensional model of Siegert and Plischke [1239] was one of the models used by Krug *et al.* [961] in their study of the dependence of the surface diffusion mass current on surface inclination. This work showed that slope dependent surface currents leading to either asymptotic EW universality class growth or unstable growth are a general feature of this class of models and a broad range of other MBE models as well. On the other hand, behavior characteristic of the MH equation is not generic. It can be expected only in very simple models where the surface current $\mathbf{j}_s$ may be zero, because of the "symmetry" of the growth rules. It is not unusual for the surface diffusion current to be quite small, leading to well-defined MH universality class behavior during the early stages of growth. However, this early stage behavior is unstable. A wide range of MBE models can be understood in terms of this simple, intuitive picture. This is an important advance that brings us closer to a simple physical understanding of a variety of surface growth models and the growth of real surfaces under typical MBE conditions.

5.6.2 Step Edge Dynamics

In vicinal surfaces, the individual steps are not straight and the dynamics of the step edges is of interest. Under appropriate limiting conditions, the dynamics of the step edges can be represented by $1+1$-dimensional surface growth models, similar to those described above. However, a broader range of phenomena may come into play. In general, the diffusion will be anisotropic and material may be added from both the upper and lower steps. This problem has been explored by Salditt and Spohn [1242], using a cubic lattice Monte Carlo model with the Hamiltonian

$$\mathcal{H} = \mathcal{H}_\circ + \mathcal{H}_1, \tag{5.210}$$

where

$$\mathcal{H}_\circ = -J\left[\sum_{nn} E_{ij}h_ih_j + \sum_i h_i\right]. \tag{5.211}$$

In this model, the heights are restricted to 0, 1 or 2 and the interactions E_{ij} are given by $E_{ij} = 0$, if i or j is 0, $E_{22} = 2$ and all the other interactions have a value of 1. Equation 5.211 simply counts the number of "bonds" between nearest-neighbor sites on the cubic lattice. The term $\mathcal{H}_1$ in the Hamiltonian is given by

$$\mathcal{H}_1 = -\mu\sum_i h_i, \tag{5.212}$$

where μ is a chemical potential. The dynamics of the model consists of transfer of height between adjacent columns, subject to the height restrictions. The allowed exchanges $((h_i,h_j) \to (h'_i,h'_j))$ are $(1,0) \leftrightarrow (0,1)$, $(2,0) \leftrightarrow (1,1)$, $(0,2) \leftrightarrow (1,1)$ and $(1,2) \leftrightarrow (2,1)$. The dynamics is simulated using a Metropolis Monte Carlo algorithm (chapter 1). In this model, deposition is represented by the non-conserved processes $0 \to 1$ and $1 \to 2$ with a constant rate of deposition (constant flux density φ). Evaporation ($(1 \to 0)$ and $(2 \to 1)$) is also allowed and the evaporation rates are controlled by detailed balance using the Hamiltonian and the reduced temperature (k_BT).

The system is started with two flat surfaces with heights of 0 and 1 separated by a straight step, and the chemical potential is adjusted, to a value of μ_e, so that the $h = 0$ and $h = 1$ phases coexist. The step growth process can then be driven by increasing the depositional flux beyond its equilibrium value φ_e or by increasing the chemical potential beyond μ_e. In the absence of deposition and evaporation, the edge of the step will roughen thermally. Salditt and Spohn argued that, on large length scales, the shape of the step is described by the $1+1$-dimensional KPZ equation. On shorter length scales, a variety of different regimes, with their own characteristic scaling properties, can be identified and a rich variety of patterns can be formed with simple modifications of this model.

For example, if atoms that contact the face of the step are not allowed to diffuse along the face or desorb from the face, then DLA-like patterns (chapter 3) are formed. However, the DLA-like structure does not persist to long length scales. If a high Schwoebel barrier is assumed to be present, then a simplified "one-sided" growth model can be used. This class of models should provide a promising approach to the simulations of experiments such as those of Wu *et al.* [1160] on the roughening of steps during epitaxial growth.

5.6.3 Anomalous Scaling

Simulations carried out using simple models for MBE growth attracted attention to the possibility that the Hurst exponent H and the roughness exponent α might be different. This "anomalous" scaling appears most dramatically in those cases in which the saturated surface width $\xi_\perp(L, t \to \infty)$ grows as L^α with $\alpha > 1$. These anomalous growth processes can be described in terms of a height difference correlation function $C^{(2)}(x)$ that has the scaling form given in equation 2.169 or the equivalent form

$$C^{(2)}(x) \sim x^\alpha f'(x/\xi_\parallel), \tag{5.213}$$

where the scaling function $f'(y)$ has the form $f'(y) = y^{-\zeta}$ for $y \ll 1$ and $\alpha = H + \zeta$, so that $C^{(2)}(x) \sim x^H$. For $y \gg 1$, $f'(y) = y^{-\alpha}$ so that $C^{(2)}(x)$ has a constant value for $x \gg \xi_\parallel$. Das Sarma *et al.* [1223] have shown that for processes described by the MH equation (equation 5.18) with deposition noise, the scaling function $f'(y)$ in equation 5.213 has the form

$$f'(y) = \begin{cases} y^{-1/2} \text{ for } d = 1,\ y \ll 1 \\ |\ln y|^{1/2} \text{ for } d = 2,\ y \ll 1 \\ \textit{const.} \text{ for } d > 2,\ y \ll 1 \\ y^{-\alpha} \text{ for all } d,\ y \gg 1. \end{cases} \tag{5.214}$$

These equations and the relation $\alpha = (4 - d)/2$, obtained from equation 5.18, imply that the Hurst exponent H measured from the height difference correlation function $C_2(\ell)$ will have the form $C_2(\ell) \sim \ell^H$, with $H = 1$ for $d = 1$, $H = 1$ (with logarithmic corrections) for $d = 2$ and $H = (4-d)/2$ for $d > 2$. The Hurst exponent H lies in the range $0 \leq H \leq 1$, expected for self-affine fractals, and the "super-rough" scaling of the surface width ($\xi_\perp \sim L^\alpha$ with $\alpha > 1$) does not imply a Hurst exponent greater than unity, but includes the effects of an amplitude that also depends on L.

Das Sarma *et al.* [1223] showed that the height difference correlation functions C_2 for several "MBE" models for which the surface width $\xi_\perp$ obeys the scaling form given in equation 2.177, or $\xi_\perp(L, t) = L^\alpha f(t/\xi_\parallel)$, with $\alpha > 1$, could be described accurately using this scaling model. Although attention was focused onto this "anomalous" scaling behavior because of the super-roughening characterized by a roughness exponent α with a value greater than unity, there is no reason why this behavior should be confined to processes with $\alpha > 1$. The exponents α and H should both be measured for any new model. This is most easily accomplished via the height difference correlation functions.

Krug [284] has studied a particularly simple square lattice model of this type that ensures the absence of inclination dependent surface currents (model DT1, described above). A deposit generated using this model is shown in figure 5.21(a). Krug found two distinct scaling regimes for this model. The height difference correlation functions can be represented by the scaling form

$$C_q(x) = \xi_{\parallel}^{\zeta_q} x^{H_{ql}} f(x/\xi_{\parallel}), \tag{5.215}$$

given in equations 2.169 and 2.168. Here, H_{ql} is a q dependent local Hurst exponent and the scaling function $f(y)$ has the form $f(y) = const.$ for $y \ll 1$ and $f(y) = y^{-H_{ql}}$ for $y \gg 1$. Since equation 5.215 can also be written in the form $C_q(x) = x^{H_{ql}+\zeta_q} g_q(x/\xi_{\parallel})$, it is a generalization of equation 5.213, with $\alpha = H_{ql} + \zeta_q$. The scaling function $g_q(y)$ has the form $g_q(y) \sim y^{-\zeta_q}$ for $y \ll 1$ and $g_q(y) = y^{-(H_{ql}+\zeta_q)}$ for $y \gg 1$.

The exponents ζ_q and H_q satisfy the scaling relationship $\zeta_q + H_q = \alpha = 3/2$. As with other models of this type, the diffusion length in the lateral direction is of the order of one lattice unit while the diffusion length in the vertical direction increases with increasing time. The simulations indicated that the surface widths $\xi_{\perp}^{(q)} =< |h- < h > |^q >^{1/q}$ exhibit simple scaling with the form $\xi_{\perp}^{(q)} \sim t^{\beta}$, with $\beta \approx 0.37$. The step height distribution was investigated by measuring the moments $\sigma_q =< |h_i - h_{i+1}|^q >^{1/q}$, which were found to grow as $\sigma_q \sim \xi_{\parallel}(t)^{\zeta_q} \sim t^{\zeta_q/z}$ with effective values of $\zeta_1 = 0.55 \pm 0.02$, $\zeta_2 = 0.73 \pm 0.04$, $\zeta_3 = 0.90 \pm 0.07$ and $\zeta_4 = 1.08 \pm 0.11$. The step size distribution $N_{\delta h}$, where $\delta h = |h(i) - h_{i+1}|$, was found to have a stretched exponential form $N_{\delta h} \sim e^{-(k\delta h)^{\gamma}}$, with an exponent γ that decreased with increasing system size L. The numerical data were consistent with $\gamma \sim 1/\ln L$. The rich local scaling behavior found for this model may be related to the presence of the deep narrow grooves that can be seen in figure 5.21(a). Some of the characteristic features of this model are captured by the linear, MH Langevin equation. However, this equation predicts a Gaussian rather than a "stretched exponential" size distribution of step heights. Krug also investigated a model in which the newly deposited particles attempt to find a site with the largest local curvature $\mathscr{K}_i$ defined as $\mathscr{K}_i = h(i-1) + h(i+1) - 2h(i)$. In this model, pairs of sites (in columns i and $i+1$) are selected randomly and the particle is always deposited on the column with the largest value of $\mathscr{K}$. Ties are broken by a random selection. This model can be described in terms of the MH equation, and the step height distribution is Gaussian.

5.6.4 Porous and Amorphous Films

The breakdown of the relatively orderly MBE process is heralded by the growth of an amorphous film on top of the crystalline structure deposited during MBE growth. This transition is often quite abrupt [1163] (section 5.5.4.2). The growth of amorphous films cannot be satisfactorily represented by a lattice model. However, the transition from crystalline epitaxial growth to the growth of an amorphous material is mimicked by the appearance of voids in lattice model simulations. When defects appear, or amorphous growth starts, mass conservation no longer assures volume conservation and the "KPZ" non-linear term $b(\nabla h(\mathbf{x},t))^2$ appears in the continuum limit surface growth equation.

Yan [1243] added surface diffusion to the standard $1+1$-dimensional ballistic deposition model. In this model, the deposited particle is allowed to follow a random walk along the surface of the deposit until it reaches a kink site, with two occupied nearest neighbors, or a "trapping site", with three occupied nearest neighbors. In either of these events, the deposited particle stops and is incorporated irreversibly into the deposit. If the deposited particle moves l_{rw} steps before reaching a kink or a trap, it is also stopped and incorporated into the deposit. Unfortunately, the model was not described in enough detail to allow these simulations to be repeated. A roughness exponent of $\alpha = 1/2$ was measured from the dependence of the surface width $\xi_\perp$ on the strip width L for all values of l_{rw}, but the effective value of the exponent β describing the growth of $\xi_\perp$ crosses over from a value close to, but slightly larger than 0.375, at early times, to a value of $1/3$ at late times. This work raised the possibility that the exponents seen in other, related, models may be characteristic of a "long lived" transient and that the asymptotic behavior might lie in the KPZ universality class. More recent work has confirmed this idea. This behavior is similar to the crossover to EW behavior, found in corresponding solid-on-solid models, with "diffusion". The dependence of the density ρ, in the KPZ porous growth regime, inside the deposit, far from the surface, on the number of steps l_{rw} in the surface random walk was also determined. The simulation results suggest that $1-\rho \sim l_{rw}^{-a}$, with $a \approx 1.25$, for large values of l_{rw}. Yan also examined the corresponding solid-on-solid model. For this model, exponents of $\beta \approx 1/3$ and $H = 1/2$ were obtained, for all non-zero values of l_{rw}. It appears that this model also belongs to the KPZ universality class. However, the origin of the non-linear $b(\nabla h(\mathbf{x},t))^2$ term on long length scales is not apparent, and the value of $\approx 1/3$ for the exponent β could be part of a crossover from an early stage $1+1$-dimensional MH equation exponent of $3/8$ to a late stage EW equation exponent of $\beta = 1/4$.

The unrealistic features of the models of Wolf and Villain [967], Das Sarma and Tamborenea [1228] and Kotrla *et al.* [1231] that have been described earlier in this section motivated Kessler *et al.* [1244] to study a model for molecular beam epitaxy using a growth and "surface diffusion" model in which particles are first deposited according to the standard ballistic deposition model rules. Between the deposition steps, randomly selected surface particles (sites that are not fully coordinated) are allowed to take "diffusion" steps. In each diffusion step, the particle is transferred to a position selected randomly from the most highly coordinated unoccupied sites in the $d+1$-dimensional hypercube, with sides of length $2L_{\mathscr{D}}+1$, centered on its last position, before the step is taken. The surface diffusion algorithm then depends on the length $2L_{\mathscr{D}}+1$ and the average number of diffusion steps $N_{\mathscr{D}}$ that a surface site takes before a monolayer of sites has been added. The behavior associated with the $1+1$-dimensional model is quite complex and does not correspond to that obtained by Das Sarma and Tamborenea [1228] or Wolf and Villain [967]. After a short time transient, there is a power law regime with $\beta \approx 1/4$. The size of this power law regime increases with $L_{\mathscr{D}}$. This power law growth regime is followed by a regime in which the surface width grows rapidly with increasing time, and finally, at even longer times, the surface roughness evolves according to the KPZ scenario. This crossover is associated with a crossover from a dense morphology at short times to a coarse-grained, porous structure in the KPZ regime.

A similar crossover has been found in a simple epitaxial growth model devised by Schimschak and Krug [1245]. In this model, solid-on-solid deposition rules are used in the deposition stage, but defects are allowed to form as a result of surface diffusion. The short time/short length scale behavior is like that found in the corresponding solid-on-solid model, but the asymptotic behavior is described by the KPZ equation. This $1+1$-dimensional model generates realistic looking deposit morphologies that are like a vertical cut through a thin film. The crossover from the short time to the late time scaling behavior takes place over a relatively short range of time scales but it is not accompanied by a change in morphology.

A more realistic and more complicated model developed by Das Sarma *et al.* [1246] also generates a crossover to asymptotic KPZ equation behavior. This model includes ballistic aggregation deposition rules and allows voids, vacancies and overhangs to form. The deposit is "annealed", while deposition is taking place, using a thermally activated hopping model that allows for exchanges between all sites on the unoccupied perimeter, including defect and void sites, and their neighboring (*nn* and *nnn*) occupied sites. The activation energy is assumed to depend on the nearest-neighbor coordination number of the filled site. If simulations are carried out with realistic parameters (deposition rate, activation energies and temperature), the annealing mechanism generates a dense deposit with few voids during the early stages of growth, and the growth of the surface is like that of a solid-on-solid model. Eventually, after a time t^* that depends on the model parameters, large voids develop and the asymptotic KPZ behavior emerges. On a $\log - \log$ scale, the defect density increases quite abruptly at $t = t^*$ and appears to reach a constant, asymptotic value. In the 2+1-dimensional case, the crossover time t^* increases with increasing temperature, but the asymptotic defect density of about 2/3 is insensitive to the temperature. As Das Sarma *et al.* pointed out, this unrealistically high defect density is a consequence of the lattice used in the simulations and the film would become amorphous in a real system. Under some circumstances, amorphous film can have densities as high as or higher than those of the corresponding crystalline materials. However, the densities of the crystalline and amorphous phases are generally different. Both $1+1$-dimensional and $2+1$-dimensional square lattice and cubic lattice model simulations were carried out. The crossover can be quite complicated and, particularly in the $2+1$-dimensional case, the surface width grows as $\xi_\perp \sim t^{\beta^*}$, with a large effective values for β^*, in the crossover. Another interesting result of the simulations was an Arrhenius form $t^* \sim e^{(-E_a/T)}$ for the dependence of the crossover time on the temperature. Both of these features are reminiscent of the behavior observed by Eaglesham *et al.* [1158, 1163] and others during low temperature epitaxial growth. Some of this experimental work is outlined earlier in this chapter. The activation energy E_a for surface roughening has been interpreted in terms of a defect-nucleation activation energy [1246].

The crossover to asymptotic KPZ behavior, accompanied by the formation of a porous deposit structure, seems to be a general feature of these models. The asymptotic KPZ behavior is a consequence of the growth of a porous structure and the concomitant generation of the non-linear $b[\nabla(\mathbf{x},t)]^2$ term in the surface growth equation due to a slope dependent growth velocity. For a porous film, mass conserving surface currents do not conserve the volume of the deposit and the growth velocity becomes dependent on the inclination of the surface.

5.6.5 Anisotropic Surfaces

For anisotropic surfaces, the KPZ equation may be generalized to

$$\begin{aligned}\partial h(\mathbf{x},t)/\partial t &= a_\parallel \partial^2 h(\mathbf{x},t)/\partial x_\parallel^2 + a_\perp \partial^2 h(\mathbf{x},t)/\partial x_\perp^2 + b_\parallel(\partial h(\mathbf{x},t)/\partial x_\parallel)^2 \\ &+ b_\perp(\partial h(\mathbf{x},t)/\partial x_\perp)^2 + \eta(\mathbf{x},t). \end{aligned} \quad (5.216)$$

This surface growth equation was proposed by Villain [949] to describe the dynamics of growing vicinal surfaces consisting of a series of low steps separating large flat terraces. A theoretical analysis of the anisotropic KPZ equation by Wolf [1208] was based on a one loop renormalization calculation, assuming that the surface could be rescaled by the transformation

$$x \to \lambda x \quad (5.217)$$

$$y \to \lambda \cdot \lambda^\chi y \quad (5.218)$$

$$h \to \lambda^\alpha h \quad (5.219)$$

and

$$t \rightarrow \lambda^{z_x} t, \tag{5.220}$$

where $\chi = [(z_y/z_x) - 1]$ for a $2+1$-dimensional surface with anisotropic scaling in the lateral directions.

This theory indicated the expected result that if $b_\parallel$ and $b_\perp$ have the same sign, then the surface will be self-affine and the surface width $\xi_\perp$ will grow algebraically. On the other hand, if $b_\parallel$ and $b_\perp$ have opposite signs or if one of them is zero, the theoretical work of Wolf led to the unexpected conclusion that the surface roughness and its growth will exhibit the logarithmic behavior characteristic of the EW model.

Halpin-Healey and Assdah [1247] have numerically integrated a simplified version of equation 5.216,

$$\begin{aligned} \partial h(\mathbf{x},t)/\partial t &= a\nabla^2 h(\mathbf{x},t) + b_\parallel(\partial h(\mathbf{x},t)/\partial x_\parallel)^2 + b_\perp(\partial h(\mathbf{x},t)/\partial x_\perp)^2 \\ &\quad + \eta(\mathbf{x},t), \end{aligned} \tag{5.221}$$

with non-conserved depositional noise. These simulations lend strong support to the theoretical work of Wolf. Similar, larger scale simulations, with the same result, have been carried out by Moser and Wolf [1248].

5.6.6 The Huygens Principle Model

Messier and Yehoda [924, 1249, 1250] showed that a variety of deposition processes led to the growth of a rough surface that left behind a relatively dense deposit with a microstructure consisting of "columnar" grains. For the case of SiC grown under a variety of conditions, they found that the column widths could be characterized by a correlation length $\xi_\parallel$ that grows algebraically with increasing time or film thickness, with an exponent $p = 1/z$ of about 0.73. Starting with an initially rough surface, Tang *et al.* [1251] simulated the evolution of the surface using the classical Huygens principle, well known in optics [1252]. They found power law growth of $\xi_\parallel$ ($\xi_\parallel \sim t^p$) using both $1+1$-dimensional and $2+1$-dimensional models. The value for the exponent p was found to depend on the nature of the initial roughness.

Tang *et al.* [1251] developed a theoretical model for the $1+1$-dimensional algorithm and confirmed the results obtained from their theory by computer simulations. Theoretical values of 1/2, 1/3 and 2/3 were found for the exponent p that describes the growth of the widths of the column tops, in the lateral direction, for uncorrelated initial heights selected randomly from Gaussian, uniform and power law (with $\tau = 3$) distributions, respectively. For the $2+1$-dimensional model, theoretical values of 1/2, 1/3 and 3/4 were found for p using Gaussian, uniform and power law (with $\tau = 4$) distributions, respectively. Simulations and theoretical analysis were also carried out for growth processes in which the initially rough surfaces were self-affine fractals. In the 1 + 1-dimensional model, a value of $p = 2/3$ was obtained starting with an initial surface with $H = 1/2$. For the 2 + 1-dimensional model, $p = 1/2$ for an initial Hurst exponent of 0 and $p = 1$ for an initial Hurst exponent of 1. Tang *et al.* also showed that if noise is added to the deterministic growth process, the noise induced height fluctuations dominate the surface growth if $p < 3/4$, in both the 1+1-dimensional and 2+1-dimensional cases and that the exponent p becomes equal to 3/4 in the asymptotic, noise dominated, limit. These theoretical results are consistent with the experimental observations of Messier and Yehoda for the deposition of SiC onto a cold substrate. At higher temperatures, smaller effective values were obtained for p. The van der Drift [1253] model described later in this chapter is also based on the Huygens principle construction.

Sensitivity to initial conditions is characteristic of deterministic models such as the Huygens principle models. For this reason, deterministic models are usually less universal than their stochastic counterparts.

5.7 Oblique Incidence and Shadowing Models

Many surface growth processes lead to strong instabilities and the generation of very rough surfaces. Under these conditions, the surface can no longer be represented well by a single-valued function $f(\mathbf{x})$, and the shadowing or screening of one part of the surface by another becomes an important aspect of the growth process. In this respect, these models are more like the DLA and screened growth models discussed in chapter 3 than the surface growth models discussed earlier in this chapter. These "very rough" surfaces can often be studied using models similar to those described earlier in this chapter, but the Langevin equation approach to a theoretical understanding must either be modified or abandoned in favor of other approaches in which the shadowing is included more explicitly.

5.7.1 Oblique Incidence Ballistic Deposition Models

It has been known for many years that the deposition of a metal vapor beam onto a cold surface at non-normal incidence accentuates the characteristic "columnar" morphology frequently associated with deposition processes and results in a deposit with a density that is lower than that obtained at normal incidence. As the angle of incidence θ_i approaches 90° (grazing incidence) distinct, widely separated needles are formed [1254, 1255, 1256, 1257]. The angle of growth ϕ_g which characterizes the orientation of the needles, measured from the normal to the coarse-grained surface, is quite different from the angle of incidence. Until quite recently, it was believed that the angles θ_i and ϕ_g were related by the "tangent rule"

$$\tan(\phi_g) = (1/2)\tan(\theta_i), \tag{5.222}$$

first proposed by Nieuwenhuizen and Haanstra [1258]. It is now apparent from computer simulations [288, 1259, 1260], theoretical considerations [1260] and a variety of experimental studies such as those of Hashimoto *et al.* [1261] and Fujiwara *et al.* [1262] that no such universal rule relating ϕ_g and θ_i exists. Krug [1263] (see also [1264]) has shown that, for ballistic deposition, the angles θ_i and ϕ_g are related to the dependence of the deposit density $\rho(\theta_i)$ on the angle of incidence θ_i by

$$\tan(\phi_g) = \tan(\theta_i) + [d\rho(\theta_i)/d(\theta_i)]/\rho(\theta_i). \tag{5.223}$$

Excellent agreement was obtained between equation 5.223 and the results from both 1 + 1-dimensional and 2 + 1-dimensional ballistic deposition model simulations [938, 1060]. A careful comparison between equation 5.223 and experimental results would be an important contribution.

Instead of simulating directly the deposition of an inclined collimated "beam" of particles onto a horizontal surface, it is more convenient to simulate the vertical deposition of particles onto an inclined substrate. In this way, the simplicity and efficiency of the "standard" ballistic deposition model, described above, can be retained. Figures 5.22(a) and (b) show parts of the patterns generated at an angle of inclination of 87.5° by a 1 + 1-dimensional square lattice model and a 1 + 1-dimensional off-lattice model, respectively. The patterns consist of an assembly of columns or clusters that are well separated from each other and grow at a well defined angle ϕ_g with respect to the normal to the substrate. The patterns generated by the lattice and off-lattice models are similar, but the angle of growth ϕ_g, which does not depend much on the angle of incidence, for large angles of incidence, is quite different.

These patterns can be characterized in terms of the distribution of cluster sizes

$$N_s = s^{-\tau_s}, \tag{5.224}$$

where $N_s\delta s$ is the number of clusters of size s (particles or lattice sites) in the range $s - (\delta s/2)$ to $s + (\delta s/2)$ and by the exponents $\nu_\perp$ and $\nu_\parallel$ that describe the asymptotic algebraic dependence of the cluster widths (w) and heights (h) on the cluster size (s)

$$w(s) \sim s^{\nu_\perp} \tag{5.225}$$

and

$$h(s) \sim s^{\nu_\parallel}. \tag{5.226}$$

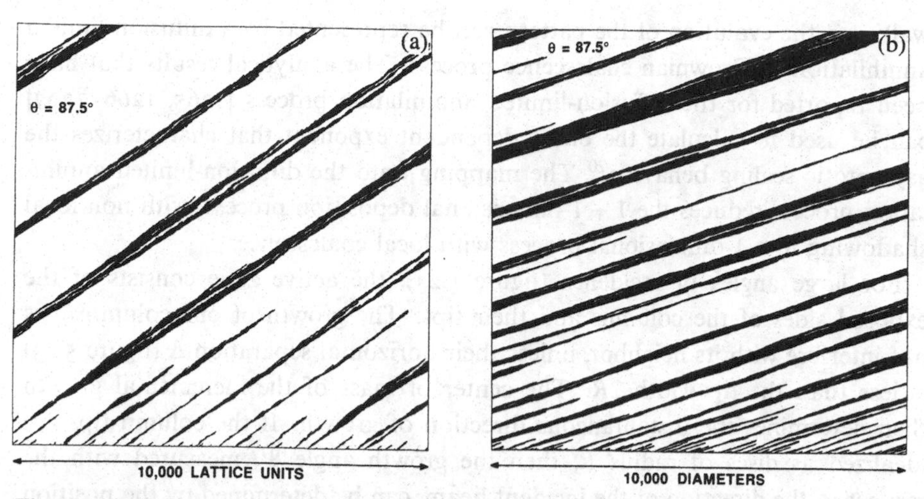

Figure 5.22 Patterns generated by $1+1$-dimensional ballistic deposition models at a near-grazing angle of incidence of $\theta_i = 87.5°$. Part (a) shows a pattern obtained from a square lattice model simulation, and part (b) shows an off-lattice deposit. The large difference between the growth angles is clear.

Although the deposits have a very low mean density as $\theta_i \to 90°$, both the individual clusters and the entire deposit are uniform on sufficiently long length scales, so that the exponents $\nu_\perp$, $\nu_\parallel$ and τ_s satisfy the scaling relationships

$$\nu_\parallel + d\nu_\perp = 1 \tag{5.227}$$

and

$$\tau_s = 2 - \nu_\parallel, \tag{5.228}$$

where d is the dimensionality of the substrate. For $d = 1$, these exponents have the exact, universal values $\nu_\parallel = 2/3$, $\nu_\perp = 1/3$ and $\tau_s = 4/3$, in the $\theta_i \to 90°$ limit.

These exponents can be calculated by mapping the column shadowing problem onto a diffusion-limited particle coalescence process, as is illustrated in figure 5.23. The growth of the columnar pattern can be understood in terms of the paths followed by the column tips. These motions can be followed via their projection onto a line perpendicular to the incoming particle flux represented by the baseline in figure 5.23. When one column shadows another, the "lower" column stops growing. In terms of the projections of the leading edges of the tips onto the baseline, this happens when the two points corresponding to the tip positions meet. The shadowing can be represented as a coalescence of the two points. The motion of the coalesced point then represents the trajectory of the leading edge of the surviving tip. The motions of the points in the baseline correspond to a random walk superimposed on a drift at constant velocity, to the left in figure 5.23. In the moving frame, the particles execute a random walk and the evolution of the pattern can be represented by a diffusion-limited annihilation or Brownian coalescence process. The analytical results that have been reported for the diffusion-limited annihilation process [1265, 1266, 1267] can be used to calculate the one independent exponent that characterizes the asymptotic scaling behavior.[20] The mapping onto the diffusion-limited annihilation process reduces the $1+1$-dimensional deposition process, with non-local shadowing to a 1-dimensional process with local coalescence.

For large angles of incidence (figure 5.23), the active zone consists of the exposed sides of the columns and their tips. The growth of one column does not interfere with its neighbor, unless their horizontal separation Δ (figure 5.23) is less than the tip width $2R$. The center of mass of the incremental growth then determines the instantaneous direction of growth. If the column tips are idealized as discs of radius R, then the growth angle $\tilde{\Psi}$, measured with the respect to the direction of the incident beam, can be determined by the position of the center of mass of a growth increment on the disc with respect to the center of the disc. The growth direction calculated in this manner is given by

$$\tilde{\Psi} = (1/\Delta) \int_{R-\Delta}^{R} \sin^{-1}(x/R)dx. \tag{5.229}$$

Were it not for this partial screening, the columns would grow in the direction of the incident flux. The average direction of growth Ψ can be obtained by integrating the instantaneous growth directions $\tilde{\Psi}$ over the distribution of horizontal separations $P(\Delta, t)$, obtained from the particle coalescence model

$$\Psi = \int_{0}^{2R} \tilde{\Psi}(\Delta)P(\Delta, t)d\Delta. \tag{5.230}$$

In this way, it can be shown [1260] that, unless the column width R is proportional to the column separations ℓ (the correlation length $\xi_{\parallel}$), then $\Psi \to 0$ for $R \ll \ell \approx \xi_{\parallel}$ or $\Psi \to \pi/2$ for $R \gg \ell$. An intermediate angle, as observed in the simulations, can only emerge if R and ℓ grow in the same way, both as $t^{1/2}$. The argument sketched here does not depend on the angle of incidence θ_i, so Ψ and ϕ_g do not depend on θ_i, for well separated columns at large angles of incidence.

Deposition patterns consisting of well separated "trees" were also generated by $2+1$-dimensional ballistic deposition models at oblique incidence or simulations of deposition onto an inclined substrate [1268, 1269], illustrated in figure 5.24. Cross-sections through a deposit, grown with an angle of incidence of 87.5°, are shown in figure 5.25. In the structures generated by this model, the "trees" have lengths l, widths w and thicknesses Δ that grow with different powers of the cluster sizes or tree sizes ($l \sim s^{\nu_{\parallel}}$, $w \sim s^{\nu_y}$ and $\Delta \sim s^{\nu_x}$). Since the individual trees are compact, $\nu_{\parallel} + \nu_y + \nu_x = 1$. The entire pattern can be compared with

20. It follows from the scaling properties of the Brownian process ($H = 1/2$, chapter 2, section 2.7) that, after a time t, each surviving column will have shadowed the dead columns in a length with a mean size proportional to $t^{1/2}$. Consequently, $\nu_{\perp}/\nu_{\parallel} = 1/2$.

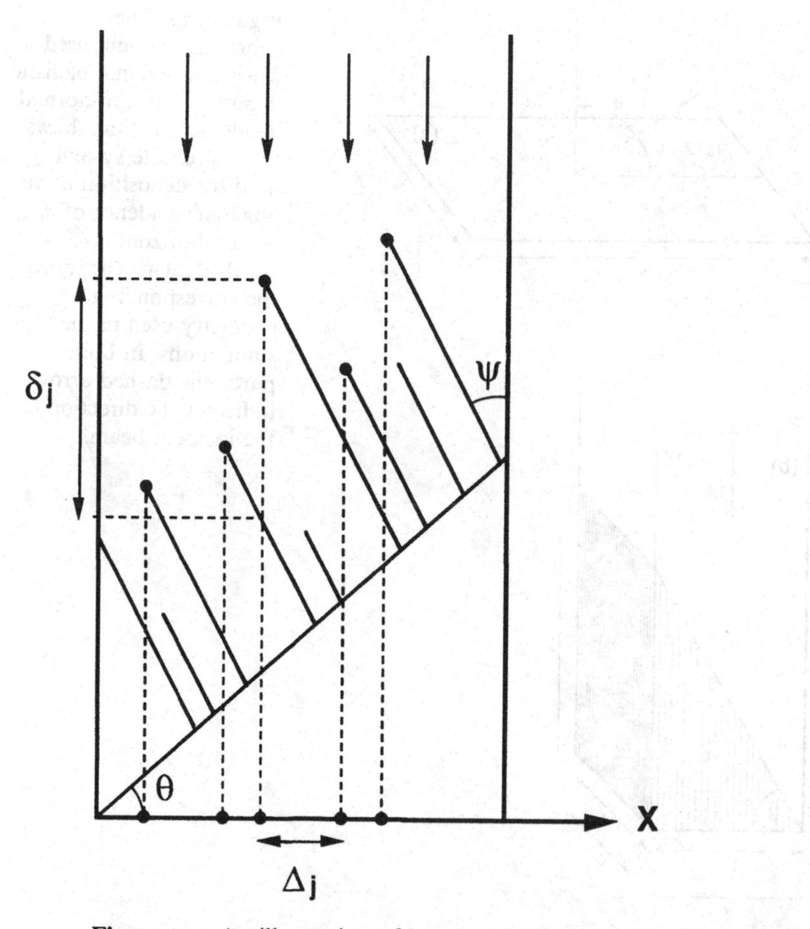

Figure 5.23 An illustration of how 1 + 1-dimensional ballistic deposition at oblique incidence can be related to the problem of 1-dimensional coalescence of Brownian particles. In this mapping, attention is focused on the tips of the growing columns. The leading edges of the tips are projected onto particles (points) lying on a horizontal line. Because of the random deposition of particles, the motion of the tips can be represented in terms of Brownian motion with respect to a constant drift velocity and the corresponding particles follow random walk paths, in a moving frame, on the horizontal line. When one column completely shadows another, the corresponding particles meet and coalesce, to represent the dynamics of the surviving column. Here, θ_i is the angle of inclination, equal to the angle of incidence, and $\theta_i - \psi$ is the angle of growth, with respect to the normal to the substrate. The arrows indicate the direction of the incident particle beam.

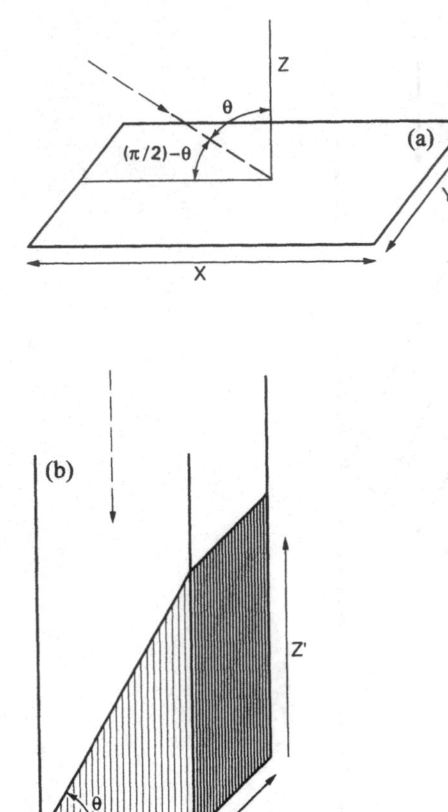

Figure 5.24 The coordinate system used in 2 + 1-dimensional ballistic deposition at non-normal incidence. Part (a) shows the coordinate system used for deposition at an angle of incidence of θ_i onto a horizontal substrate. Part (b) shows the corresponding geometry used in the simulations. In both parts, the dashed arrow indicates the direction of the incident beam.

the pattern of overlapping scales on a fish or shingles on a roof. Any two of the three exponents ($\nu_{\parallel}, \nu_x$ or ν_y) are sufficient to define the scaling structure of the deposit in the grazing incidence limit. The distribution of sizes (N_s) and heights (N_h) are given by

$$N_s \sim s^{-\tau_s} \tag{5.231}$$

$$N_h \sim h^{-\tau_h}, \tag{5.232}$$

where the exponents τ_s and τ_h are given by $\tau_s = 2 - \nu_{\parallel}$ and $\tau_h = (\tau_s - 1)/\nu_{\parallel}$.

Figure 5.26 provides the key to understanding these exponents. This figure shows the location of the large step heights in an advancing surface generated by the 2 + 1-dimensional model. These steps correspond to the advancing edges of the scales. When one edge overtakes another, the overtaken scale is "shadowed" and stops growing. Consequently, the dynamics of the leading edges of the "scales" can be described in terms of a coalescing domain boundary problem. The dynamics of the individual scale edges corresponds to a 1 + 1-

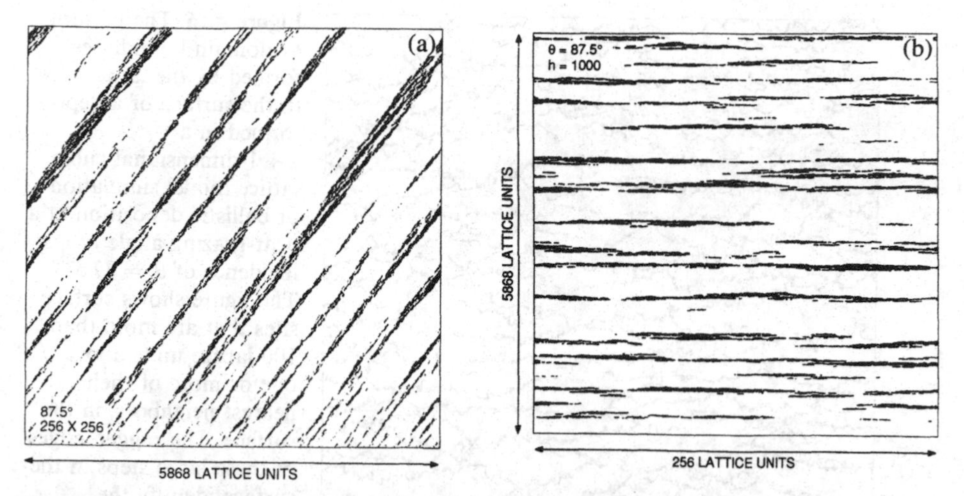

Figure 5.25 Cross-sections through deposits generated using a $2+1$-dimensional cubic lattice model for ballistic deposition onto a planar substrate at an angle of incidence of 87.5°. The coordinate system used in the simulation is shown in figure 5.24 and the lattice size L was 256 lattice units. Part (a) shows a cross-section through the (x, z) plane, and part (b) shows a cross-section parallel to the substrate, at a height of 1000 lattice units.

dimensional KPZ growth process. It then follows, from the solution of the $1+1$-dimensional KPZ problem, that $\nu_\parallel = 1/2$, $\nu_y = 1/3$ and $\nu_x = 1/6$. In both the $1+1$-dimensional and $2+1$-dimensional cases, the geometry of the grazing incidence problem allows the asymptotics of the d+1-dimensional surface growth problem to be mapped into a d-dimensional coalescence problem. This makes an analytical solution to the surface growth process possible, in the grazing incidence limit. For the $2+1$-dimensional process, the theoretical values for these exponents are in good agreement with simulation results ($\tau_s = 1.501 \pm 0.002$, $\nu_\parallel = 0.505 \pm 0.002$, $\nu_y = 0.347 \pm 0.001$, $\tau_h = 1.012 \pm 0.001$).

This model requires a generalization of the Family–Vicsek [292] scaling form. If the simulation is carried out using a cubic lattice with a cross-section of $L_x \times L_y$, then, if $L_x \approx L_y$, the surface correlation length ξ_y, which grows as $t^{2/3}$, reaches L_y before the correlation length ξ_x, which grows as $t^{1/3}$, reaches L_x. A first crossover occurs at time $t_y \sim L_y^{3/2}$. After this crossover, the surface grows in the same way as in the $1+1$-dimensional grazing incidence deposition model described above. A second crossover is found when the quasi $1+1$-dimensional competition between the clusters has eliminated all but one of these clusters. This occurs after a time $t_x \sim L_x^2 L_y^{1/2}$. In general, the interplay between the lengths L_x, L_y, ξ_x and ξ_y generates a variety of crossovers and scaling regimes [1270].

The "columns", "scales" or "clusters" are prominent features and play an important role in the development of an understanding of grazing incidence

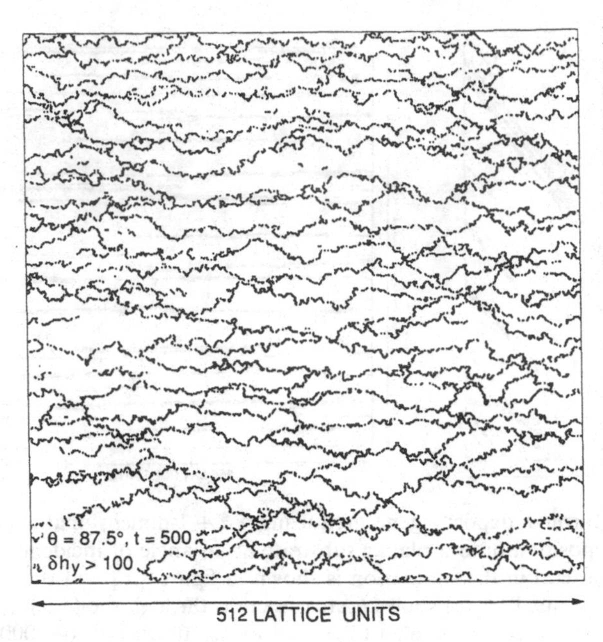

Figure 5.26 The pattern of domain boundaries formed by the large steps in the surface of a deposit formed in a $2+1$-dimensional cubic lattice model simulation of ballistic deposition at a near-grazing angle of incidence of $\theta_i = 87.5°$. The figure shows surface sites that are more than 100 lattice units above one or more of their nearest neighbors, in the surface. The ragged edges of these large steps in the surface identify the leading edges in the array of overlapping "trees" or "scales".

ballistic deposition. In off-lattice ballistic deposition, each particle is connected to only one previously deposited particle, when it is added to the deposit. Consequently, the deposit can be subdivided into non-contacting "trees" or "clusters". An example is shown in figure 5.3. However, for small angles of inclination, these clusters are not readily apparent by visual inspection. Even in lattice models, clusters can be identified if each deposited site is considered to be connected to only one of its previously deposited neighbors (chapter 2, section 2.3). Figure 5.27 shows a cross-section through one of the clusters generated in a $2+1$-dimensional, cubic lattice, normal incidence ballistic deposition simulation. The exponents $\nu_\perp$, $\nu_\parallel$ and τ_s are universal [1271], and are related to the surface growth exponents by $\nu_\parallel/\nu_\perp = z$ [1260]. This relationship and $\nu_\parallel + d\nu_\perp = 1$ provide exact values for $\nu_\parallel$ and $\nu_\perp$ ($\nu_\perp = 2/5$ and $\nu_\parallel = 3/5$) for $d = 1$ and numerical estimates for $d = 2$ that are in good agreement with the values measured directly ($\nu_\perp \approx 0.405$, $\nu_\parallel \approx 0.610$ for $d = 1$ and $\nu_\perp \approx 0.283$, $\nu_\parallel \approx 0.452$ for $d = 2$).

In many deposition processes, the incoming beam of particles is uncollimated or poorly collimated. If $1+1$-dimensional simulations are carried out in which the angles of incidence are selected randomly from the range $-\pi/2 \le \theta_i \le \pi/2$, then a strong screening effect develops and the pattern becomes dominated by fewer and fewer large "trees" or clusters [1272]. More recent work by Tang and Liang [1273] indicated that, for uniform deposition at angles of incidence in the range $-[(\pi/2) - \theta_m] \le \theta \le [(\pi/2) - \theta_m]$, there is a critical value θ_{mc}, for θ_m, above which the surface growth dynam-

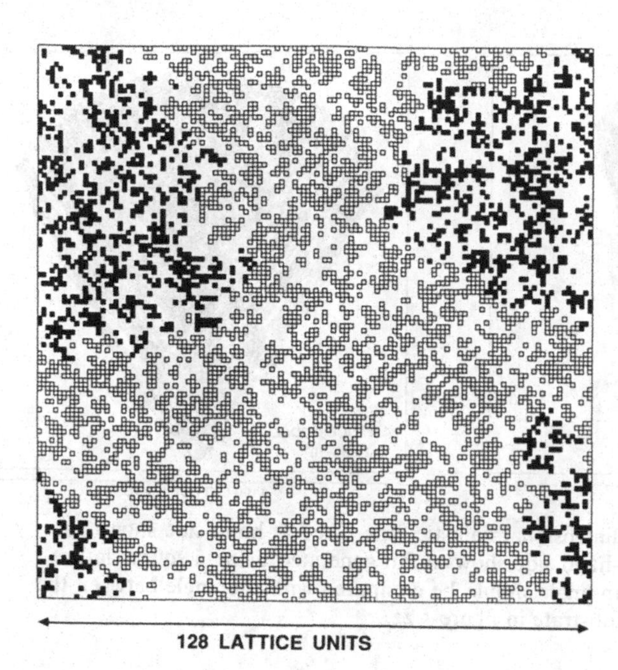

Figure 5.27 A cross-section through a deposit generated by the cubic lattice ballistic deposition model at normal incidence. The small scale simulation was carried out on a base of size 128 × 128 lattice units with periodic boundary conditions. The cut was taken at a height of 1000 lattice units, and one of the clusters of sites is indicated by filled lattice cells.

ics can be described by the KPZ equation. For $\theta_m < \theta_c \approx 10°$, the distribution of the cluster sizes (number of particles) has the power law form $N_s \sim s^{-3/2}$ and the distribution of cluster heights is given by $N_h \sim h^{-2}$. The simulations of Tang and Liang indicated that the width of the surface $\xi_\perp$ grows as t^β, with an effective value of $\beta \approx 0.7$. However, a simple scaling model indicates that $\beta = 1$ in the asymptotic limit, and this is consistent with the slow increase in the effective value of the exponent β with increasing time.

5.7.2 Ballistic Fans

Fascinating structures are generated by ballistic aggregation models, like those described in chapter 3, section 3.4.2, in which the growing aggregate starts from a single particle or lattice site and the added particles are all incident from the same direction. This model was first explored by Bensimon *et al.* [1274] and by Liang and Kadanoff [1275]. Figure 5.28 shows fans grown using 2-dimensional off-lattice and square lattice models. The overall shapes of the fans generated by these two models are quite different. The angle at the base of the fan is twice the angle $((\pi/2) - \phi_g)$ between the columns and the substrate in the corresponding grazing incidence ballistic deposition model. Figure 5.29 shows cross-sections through fans grown on a cubic lattice.

The first theory for the overall, coarse-grained envelope shape of ballistic fans was developed by Ramanlal and Sander [1276], who used the tangent rule given

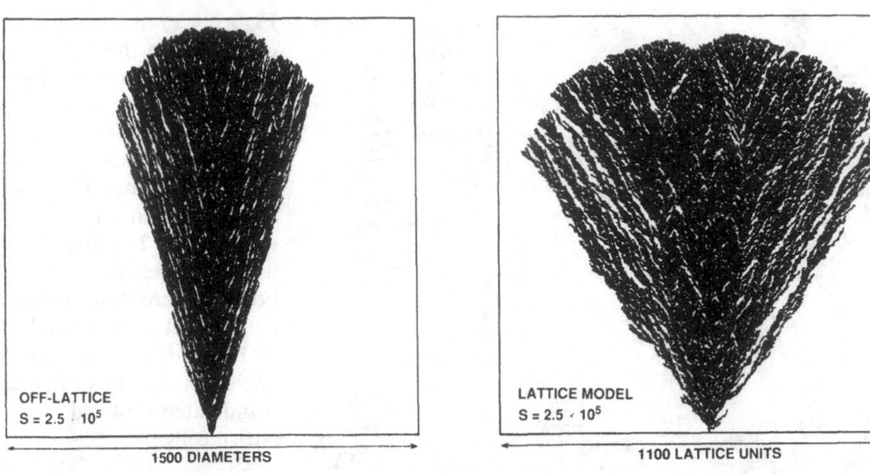

Figure 5.28 Small scale, 2-dimensional ballistic fans. The left-hand side shows an off-lattice fan and the right-hand side shows a fan generated using a square lattice model. The characteristic fan angle should be compared with the angle between the columnar clusters and the substrate in figure 5.22.

in equation 5.222 to determine the growth velocity as a function of position on the envelope. In more general terms, the shape of the fan envelope can be calculated [1277] from the dependence of the growth velocity $V(\nabla h(\mathbf{x}))$ on the slope $\nabla h(\mathbf{x})$ of the surface, using the Wulff construction [558]. In ballistic deposition the growth velocity $V(\nabla h(\mathbf{x}))$ for an inclined surface is given by $V(\nabla h(\mathbf{x})) \sim 1/\rho(\nabla h(\mathbf{x}))$, where $\rho(\nabla h(\mathbf{x}))$ describes the dependence of the deposit density on the slope of the surface ∇h. Consequently, the direction dependence of the envelope velocity can be obtained from $\rho(\nabla h(\mathbf{x}))$.

5.7.3 Shadowing Models

Several models that place primary emphasis on complete or partial shadowing have been developed. Karunasiri *et al.* [948] have studied a 1 + 1-dimensional model in which the surface growth is described by the equation

$$\partial h(\mathbf{x},t)/\partial t = -\mathscr{D}\nabla^4 h(\mathbf{x},t) + R\theta_X(\mathbf{x},\{h\}) + \eta(\mathbf{x},t). \tag{5.233}$$

The first term on the right-hand side represents the effects of the surface diffusion "current" and the second term represents the deposition intensity, where R is the deposition rate and $\theta_X(\mathbf{x},\{h\})$ is the exposure angle. This model is intended to represent processes such as sputter deposition, with a non-collimated incident flux, rather than molecular beam epitaxy with a collimated beam. In the absence of shadowing, $\theta_X(\mathbf{x},\{h\})$ is replaced by $<\theta_X(\mathbf{x},\{h\})>$, the growth of the surface width fluctuations can be described by the MH equation (equation 5.18) and $\xi_\perp \sim t^{3/8}\mathscr{D}^{-1/8} < \eta^2 >^{1/2}$ [1278], so that $\beta = 0.375$.

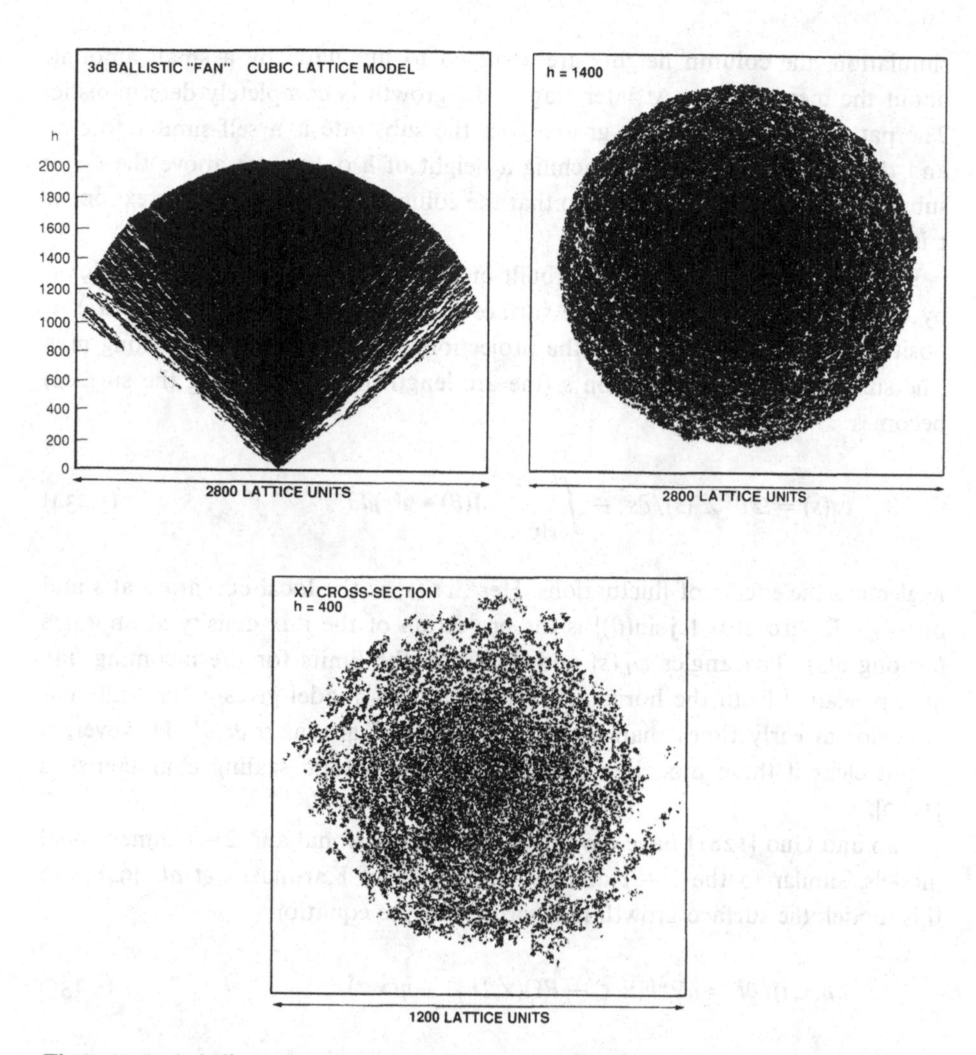

Figure 5.29 A 3-dimensional, cubic lattice model ballistic fan. The left-hand panel on the top row shows a vertical cut through the origin, and the right-hand panel shows a horizontal cut above the maximum width. A horizontal cut below the maximum width is shown in the lower panel.

When the shadowing effect is dominant, this model becomes equivalent to the "grass model" of Karunasiri *et al.* [948, 1278] in which the 1-dimensional substrate is divided into L_x equal intervals and columns are grown in these intervals, in the $+y$ direction, perpendicular to the substrate. Unlike most other lattice models, the column heights are not discretized in the grass model. The column growth velocities are assumed to be proportional to the angle to the outside that is not blocked by neighboring columns so that the growth rate for a column is proportional to the probability that its top will be contacted by a point following randomly selected ballistic trajectories. At the start of a simulation, the column heights are assumed to fluctuate, by a small amount, about the mean height. At later stages, the growth is completely deterministic. The pattern of columns that grows from the substrate is a self-similar fractal, and the number of columns reaching a height of h or greater above the $t = 0$ substrate is proportional to h^{-1}, so that the column height distribution exponent τ is 2.0.

Bales and Zangwill [1279] have built on the model of Karunasiri *et al.* [948] by calculating the velocity of the surface $v_n(s)$ along the local normal $\mathbf{n}(s)$ at position s on the surface, from the projection of the incoming flux along $\mathbf{n}(s)$. The surface velocity at position s (the arc length measured along the surface) becomes

$$v_n(s) = \mathscr{D}\partial^2\mathscr{K}(s)/\partial s^2 - \int_{\phi_R(s)}^{\pi-\phi_L(s)} \mathbf{J}(\theta) \bullet \mathbf{n}(s) d\theta, \tag{5.234}$$

neglecting the effects of fluctuations. Here, $\mathscr{K}(s)$ is the local curvature at s and $\mathbf{J}(\theta) = -R/2[\mathbf{i}\cos(\theta) + \mathbf{j}\sin(\theta)]$ is the projection of the flux density at an angle θ along $\mathbf{n}(s)$. The angles $\phi_L(s)$ and $\phi_R(s)$ are the limits for the incoming flux at s, measured from the horizontal substrate. This model gives a very different behavior at early times than does the model of Karunasiri *et al.* However, it is not clear if these models share the same asymptotic scaling characteristics [1280].

Yao and Guo [1281] have developed $1+1$-dimensional and $2+1$-dimensional models, similar to the $1 + 1$-dimensional model of Karunasiri *et al.* [948]. In this model, the surface growth is described by the equation

$$\partial h(\mathbf{x}, t)/\partial t = a\nabla^2 h(\mathbf{x}, t) + R\Omega(\mathbf{x}, \{h\}) + \eta(\mathbf{x}, t). \tag{5.235}$$

The first term on the right-hand side represents the dynamics of surface evaporation, which is assumed to be the dominant "annealing" process. In this respect, the model differs from that of Karunasiri *et al.* [948], in which surface diffusion is assumed to be dominant. The second term represents the deposition intensity, where R is the overall deposition rate and $\Omega(\mathbf{x}, \{h\})$ is the exposure angle or solid exposure angle in the $2 + 1$-dimensional model.

Most effort was concentrated on the $2 + 1$-dimensional case. Because of computational difficulties associated with the solution of equation 5.235, particularly the calculation of the solid exposure angle $\Omega(\mathbf{x}, \{h\})$ that describes the non-local screening of the incoming material flux, the $2 + 1$-dimensional simulations were carried out on a relatively small scale. After a short transient, the surface width grows linearly with increasing time ($\beta = 1$). The surfaces generated by this model consist of a disordered array of rounded peaks that grow and coarsen with increasing time. There is a characteristic wave vector k^* that decays as

$k^* \approx t^{-p}$, with $p = 0.33 \pm 0.02$ or $z \approx 3$, and the circularly averaged power spectrum $S(k,t)$ can be represented quite well by the scaling form

$$S(k,t) = t^{\delta} f(k/k^*) = t^{\delta} f(kt^p), \tag{5.236}$$

despite the relatively small scale of the simulations. The decay of the characteristic wave vector k corresponds to a $t^{1/3}$ growth of the lateral correlation length $\xi_{\|}$. This can be interpreted in terms of the growth of a characteristic column width. Since the power spectrum $S(k,t)$ is related to the surface width $\xi_{\perp}$ by $\int S(k,t)dk \sim \xi_{\perp}^2$, it follows that $\delta = 2\beta + dp$. This exponent scaling relationship was satisfied by both the 2+1-dimensional ($d = 2$) and 1+1-dimensional ($d = 1$) models of Yao and Guo [1281, 1282]. The surface width $\xi_{\perp}$ grows linearly with time, after a short transient, so that $\beta = 1$ and $\delta = 8/3$. A quite good data collapse was obtained for the surface power spectrum, using the scaling form in equation 5.236, with $p = 1/3$ and $\beta = 1$. The results of the $1+1$-dimensional simulations can also be described in terms of equation 5.236, with $p \approx 0.7$.

Yao and Guo [1282] also studied the growth of a $1+1$-dimensional interface with the evolution equation

$$\partial h(\mathbf{x},t)/\partial t = a\nabla^2 h(\mathbf{x},t) + b(\nabla h(\mathbf{x},t))^2 + R\Omega(\mathbf{x},\{h\}) + \eta(\mathbf{x},t). \tag{5.237}$$

The numerical solutions of this equation were columnar structures. A quite good data collapse for the structure factor $S(k,t)$ was obtained using the scaling form given in equation 5.236, with $p = 1$ and $\delta = 3$.

Meakin and Krug [1283] have developed a hierarchical algorithm, for the $1+1$-dimensional problem, that allows all the exposure angles to be calculated in a time of order $L\log(L)$, where L is the number of columns or needles. This model was used to carry out simulations with column growth velocities v_i given by

$$dh_i/dt = v_i = f(\theta_{Xi}), \tag{5.238}$$

where θ_{Xi} is the exposure angle of the ith column. If $f(\theta_X)$ is a monotonically decreasing function of θ, the distribution of exposure angles $P(\theta_X)$ becomes essentially bimodal as the system evolves. The peak at $\theta_X \approx 0$, corresponding to needles that are almost completely screened from further growth, increases at the expense of the peak of "active" needles at $\theta_X \approx \pi$ corresponding to a decreasing number of essentially unscreened needles. There are very few "partly screened" needles. The theoretical analysis of this class of models [1283] focuses on the transfer of population from the active to the inactive peak, in the exposure angle distribution. The general theoretical approach is similar that used by Rossi [628, 629], and outlined in chapter 3, section 3.7.5, to analyze the competitive growth of needle arrays in a Laplacian field.

Karunasiri *et al.* [948] generalized the grass model, without noise, to include the effects of surface diffusion. The equation of motion for a column is then given by

$$dh/dt = |\theta_X| - (\partial^2 \mathscr{K}/\partial s^2), \tag{5.239}$$

where $\theta_X = \theta_2 - \theta_1$ is the "exposure" angle, $\mathscr{K}$ is the "curvature" and s is the arc length.

Lichter and Chen [1284] emphasized the role played by mobility of the deposited particles in the formation of columnar microstructure with an angle of growth (ϕ_g) that is different from the angle of incidence (θ_i). However, the ballistic deposition model simulations described in this chapter (section 5.2) indicate that mobility after deposition may not play an essential role in the generation of columnar microstructure. Lichter and Chen based their analysis on the equation

$$\partial c(\mathbf{x}, t)/\partial t = J\rho \cos(\theta_i) - J\rho \nabla h(x, t) \sin(\theta_i) + \mathscr{D}\partial^2 c/\partial x^2 - Ac, \tag{5.240}$$

for the rate of change of adatom concentration $c(\mathbf{x}, t)$ at position $(\mathbf{x}, h(t))$ on the growing surface. Here, ρ is the density of the deposited material, $\mathscr{D}$ is the surface diffusion coefficient and J is the incident flux density. The parameter A sets the time scale for the immobilization of adsorbed particles and their incorporation into the immobile deposit, so that

$$\rho \nabla h(x, t) = Ac. \tag{5.241}$$

From equations 5.240 and 5.241 it follows that

$$\partial^2 h/\partial t^2 + A\partial h/\partial t = JA \cos\theta_i - JA\nabla h(x, t) \sin\theta_i + \mathscr{D}(\partial^3 h/\partial x^2 \partial t). \tag{5.242}$$

Using the initial condition

$$h(x, 0) = h_0 + h_1 \sin(kx), \tag{5.243}$$

Lichter and Chen looked for solutions of equation 5.242 with the form

$$h(x, t) = Vt + h_1 e^{\sigma t} \sin[k(x - Ut)]. \tag{5.244}$$

The wavelength corresponding to the maximum growth rate was obtained, in the $\rho \to 0$ limit, by substituting equation 5.244 in equation 5.242. Following this procedure, the angle of growth (ϕ_g) for the columnar microstructure can be obtained from

$$\tan(\phi_g) = U/(V + h_1\sigma), \tag{5.245}$$

where σ is the growth constant in equation 5.243 and $h_1\sigma$ is the peak velocity V_{max} relative to the mean velocity V ($V_{max}(t = 0) = V + h_1\sigma$) at time $t = 0$. Equation 5.245 leads to the approximate result

$$\tan(\phi_g) = (3/2) \tan\theta_i/(1 + \Phi \sin\theta_i \tan\theta_i), \tag{5.246}$$

where

$$\Phi = (4/27)h_1 J/\mathscr{D}, \tag{5.247}$$

which is accurate over a broad range of parameters ($\mathscr{D}, J$ and h_1). The results of Lichter and Chen are approximately consistent with the tangent rule ($\tan(\phi_g) = (1/2)\tan(\theta_i)$) [1258], for small values of Φ. Since equation 5.242 can describe the evolution of the surface only at early times, at which the values of the slope $\nabla h(x,t)$ are small, the relevance of these results to large angles of incidence and late times conditions, where the columnar morphology becomes most distinctive, must be questioned. In particular, "shadowing" effects that are not included in equation 5.242 appear to play an important role in the columnar morphology regime.

Nagatani [1285] has introduced an idealized 1 + 1-dimensional model for columnar growth processes in which columns grow in an independent manner, in a direction perpendicular to the substrate, except that if one column shadows another, the shaded column stops growing. Shadowing occurs when

$$h(i+j) > (j-i)\Delta h + h(i), \tag{5.248}$$

where $h(i)$ is the height of a "needle" growing at the lateral position $x = i$. This model can be interpreted in terms of growth via a collimated beam, incident from the right-hand side, with an angle of incidence θ_i given by $\cot\theta_i = \Delta h$. Nagatani studied a model of this type in which the growth velocities of the unshaded columns are given by

$$dh(i)/dt = \eta(i,t), \tag{5.249}$$

where $\eta(i,t)$ is an uncorrelated random process. He carried out simulations, using this model, and measured the distribution of column heights N_h and the total "mass" of the deposit within a distance h measured from the substrate, $M(h)$. Both of these quantities were found to depend algebraically on h,

$$N_h \sim h^{-\tau}, \tag{5.250}$$

$$M(h) \sim h^{\upsilon}, \tag{5.251}$$

and the exponents τ and υ satisfied the simple scaling law $\tau + \upsilon = 2$. Nagatani also found that the exponents τ and β depended weakly on the parameter Δh and supported this result with a position-space Monte Carlo renormalization group analysis.

Several closely related models with the same shadowing rule were investigated by Meakin and Krug [1270]. In these models, the growth velocities can be described either by the stochastic equation

$$dh(i)/dt = h(i)^a \eta(i,t) \tag{5.252}$$

or by the deterministic growth equation

$$dh(i)/dt = h(i)^a, \tag{5.253}$$

where $\eta(i,t)$ is an uncorrelated random process. In "model I" simulations, unshaded columns are picked randomly with probabilities given by $P_i \sim h(i)^a$ and the height of the selected column is increased by one unit. In "model II" simulations, unshaded columns are picked with equal probabilities and the heights of the selected columns are increased by an amount $\delta h(i)$ given by $\delta h(i) \sim h(i)^a$. In a modified version of this model (model IIa), the column heights are incremented by $\delta h(i) \sim rh(i)^a$, where r is a random number uniformly distributed over the range $0 < r < 1$. In model III, the growth is deterministic. However, the initial heights $\{h_i(0)\}$ are selected randomly from some distribution. In this model, the columns grow for a short time δt with constant velocities, calculated from equation 5.253, before shaded columns are removed from the list of growing columns.

For simulations carried out with a growth exponent (a in equation 5.252) of zero, the simulation results indicated, contrary to the results of Nagatani [1285], that $\tau = 3/2$ and $\upsilon = 1/2$, for all values of Δh. The exact values for these exponents, $\tau = 3/2$ and $\upsilon = 1/2$, can be obtained theoretically by mapping the shadowing process onto Brownian coalescence on a line [1260] in the manner that was described earlier in this chapter, for the $1 + 1$-dimensional grazing incidence ballistic deposition model. Consequently, the findings of Nagatani must be attributed to finite-size effects. This also indicates potential problems with application of the Monte Carlo renormalization group approach to non-equilibrium growth processes such as the much more challenging DLA process (chapter 3).

For model II, the theoretical analysis [1270] indicated that the size distribution exponent τ should have a value given by

$$\tau = (3 + a)/2, \tag{5.254}$$

where a is the growth velocity exponent in equation 5.252. The computer simulations gave results that were in good agreement with this exponent relationship. The behavior is more complex for model I. The simulations indicated that $\tau = 3/2$ for $a \leq 1/2$ and all values of Δh. This result can be understood since the noise alone generates height fluctuations of the order of $h^{1/2}$, so that the height dependence of the growth rate is irrelevant for $a \leq 1/2$. For values of a greater than $1/2$, it was difficult to approach the asymptotic regime in the simulations, and the theoretical analysis involved a subtle interplay between the noise induced and deterministic growth mechanisms. At this stage, an understanding of this regime remains to be developed.

A deterministic model was also explored [1270] in which the unshadowed growth velocities are given by equation 5.253, $dh(i)/dt = h(i)^a$. As for the other models, shaded columns stop growing. For a uniform distribution of initial heights $h(i,0)$ over the unit interval, the exponent τ is given by $\tau = 1 + (a/2)$. Quite different behavior was found for other initial distributions [1270].

For the deterministic model (model III), the asymptotic behavior is determined solely by the initial distribution $P(h(0))$ and the growth rules. The asymptotic behavior depends on the initial distribution of the *largest* initial heights. If this part of the distribution is uniform, the asymptotic height distribution has the form $N_h \sim h^{-\tau}$ with

$$\tau = 1 + (a/2). \tag{5.255}$$

If the distribution has the form $P(h) = (1+\nu)(1-h)^{\nu}$, then

$$\tau = 1 + a\left[\frac{1+\nu}{2+\nu}\right]. \tag{5.256}$$

A quite different behavior was found for initial distributions of the form $P(h) = \mu h^{-(\mu+1)}$, for $h \geq 1$. In this case,

$$\tau = 1 + \frac{a\mu}{a+\mu-1}. \tag{5.257}$$

For exponential and Gaussian distributions, corresponding to a power law in the $\mu \to \infty$ limit, $\tau = 1 + a$, within logarithmic corrections. A growth exponent of $a = 1$ corresponds to exponential growth. In this case, the exponent for an initial power law distribution is given by

$$\tau = 1 + \frac{\mu}{\mu - 1}. \tag{5.258}$$

This does not agree with equation 5.257, which indicates that the order in which the limits $a \to 1$ and $h \to \infty$ are taken is important. On the other hand, for growth with $a = 1$ and an initial distribution given by $P(h) = (1+\nu)(1-h)^{\nu}$ both simulations and theoretical arguments give a final power law distribution with an exponent τ, given by

$$\tau = 1 + \frac{1+\nu}{2+\nu}, \tag{5.259}$$

which is the $a \to 1$ limit of equation 5.256. A more complete review of competitive needle growth models, which emphasises the scaling theory approach has been provided by Krug [637].

5.8 Cluster Shapes and Faceted Growth

Krug *et al.* [1286] have analyzed the universal growth shapes for growth models that exhibit a morphological transition related to directed percolation. For $p > p_c$, there is directed percolation (and a smooth surface) within a "cone" with an opening half angle of θ, so that, at "time" t the facet length is $2bt$, where $b = \tan\theta$. At the threshold ($p = p_c$) anomalous scaling is found ($H + z < 2$) [1287]. For $p > p_c$ the facet advances with a velocity V^* ($V^* = 1/\sqrt{2}$ for the (11) facet in square lattice directed percolation) and the growth in other directions is faster. Krug *et al.* [1287] have used a scaling theory to analyze the anomalous roughening and cluster shape near the $p = p_c$, $m = 0$ critical point, where m is the slope with respect to the plane of the $p > p_c$ facet. They found that the growth velocity is given by $V_m \approx V^* + c|m|$ for $p > p_c$, if m is small, corresponding to the appearance of the facet. At the $p = p_c$ directed percolation threshold, the growth velocity is given by $V^* - V(m) \sim |m|^{\nu_\parallel/\nu_\perp}$, where $\nu_\parallel$ and $\nu_\perp$ are the directed percolation exponents (chapter 3, section 3.5.4). This leads to a cluster shape that is flatter than a circle of radius V^*t near the $m = 0$ tangent. For the case of growth on a 3-dimensional lattice, the facet boundaries can also be faceted, and this faceting of the facets is initiated at a 2-dimensional directed percolation transition.

A transition between rough and faceted growth has also been found [1277] for a modified model for $1+1$-dimensional ballistic deposition onto inclined substrates. In this model, a fraction p of the sites in the active zone are filled at each stage in the growth process. The evolution of the deposit is given by

$$h(x, t+1) = \begin{array}{ll} max[h(x+1, t), h(x, t) + 1, h(x-1, t)], & \text{with } \mathscr{P} = p \\ h(x, t), & \text{with } \mathscr{P} = 1 - p, \end{array} \tag{5.260}$$

where $\mathscr{P}$ is the probability. For $\theta_i > \theta_{ic}$ and $p > p_c$, the deposit consists of distinct columns with faceted tips. Throughout the entire range $\theta_i > \theta_{ic}$ and $p > p_c$, the exponents $\nu_\perp$ and $\nu_\parallel$ (equation 5.225 and equation 5.226) have the values $\nu_\perp = 1/3$ and $\nu_\parallel = 2/3$, characteristic of $1+1$-dimensional ballistic deposition in the grazing incidence limit. The critical values θ_{ic} and p_c have values of $\theta_{ic} = 45°$ and $p_c \approx 0.705489$, respectively. This suggests that the apparent dependence of these exponents on θ_i in the standard ballistic deposition model is a consequence of the fluctuations at the tips of the growing columns that are absent in faceted columns. In the model, the rough/faceted transition is also related to directed percolation on a square lattice [540, 542]. Some of the deposition patterns obtained from simulations carried out using this model are shown in figure 5.30.

Kertész and Wolf [1287] have proposed that, near to such kinetic roughening transitions, the surface width can be described in terms of the scaling form

$$\xi_\perp(\epsilon, L, t) = \xi^{\alpha'} f(L/\xi, t/\xi^{z'}), \tag{5.261}$$

where α' and z' are characteristic exponents for the roughening dynamics at the transition, and ξ is a characteristic length that depends algebraically on

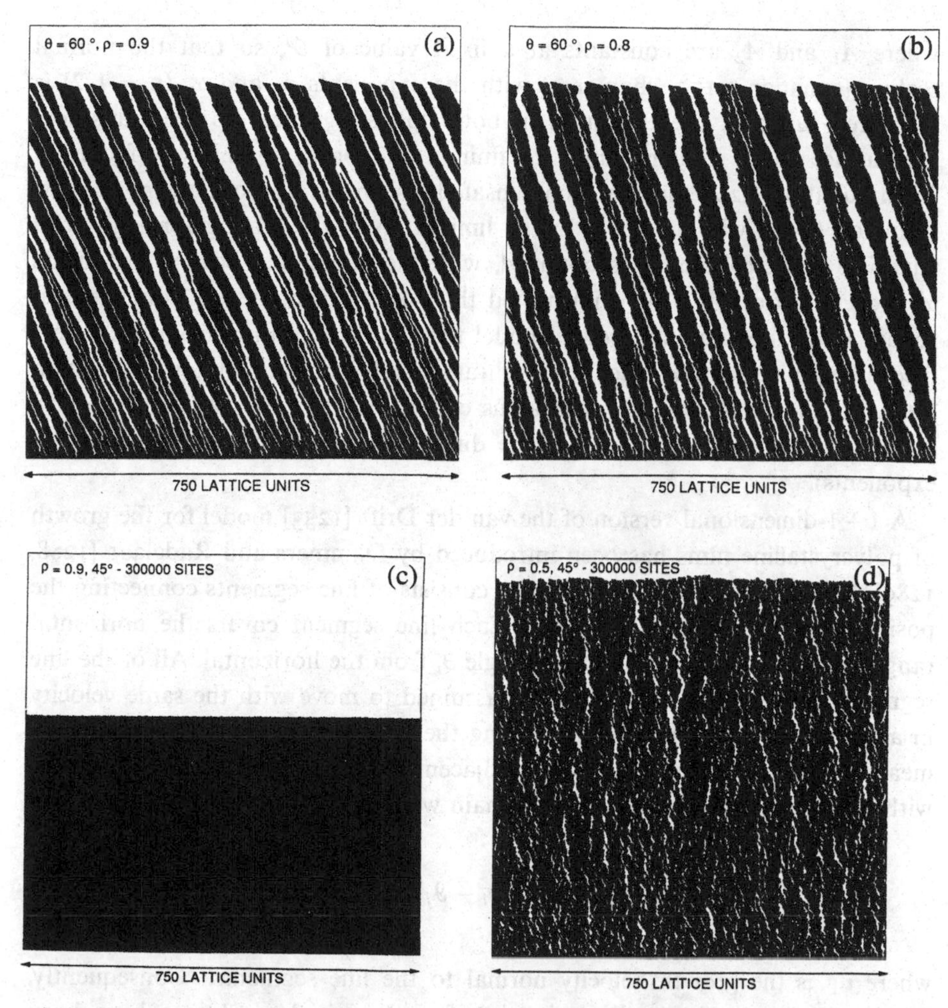

Figure 5.30 Patterns generated by a 1 + 1-dimensional, square lattice ballistic deposition model. In this model the substrate is inclined at an angle of θ_i (the angle of incidence) and a fraction ρ of the active zone sites are filled simultaneously at each stage in the simulation. If $\theta_i > 45°$ and $\rho > \rho_c$, where ρ_c is the directed percolation threshold, the surface is faceted as is shown in parts (a)–(c). Part (d) shows a deposit generated in the rough surface regime.

$\epsilon = |\mathcal{P} - \mathcal{P}_c|$ $(\xi \sim (|\mathcal{P} - \mathcal{P}_c|)^{\nu'})$. Here, $\mathcal{P}$ is a parameter that controls the transition, and $\mathcal{P} = \mathcal{P}_c$ at the transition. The scaling function $f(x, y)$ has the form

$$f_n(x, y) = x^{\alpha_n} g(y/x^{z_n}), \tag{5.262}$$

for $x \gg 1$ and $y \gg 1$. Consequently, the surface width can be described by the scaling form

$$\xi_\perp(\epsilon, L, t) = \xi^{\alpha' - \alpha_n} L^{\alpha_n} g(\xi^{z_n - z'} t/L^{z_n}) = A_1 L^{\alpha_n} g(\xi^{A_2} t/L^{z_n}), \tag{5.263}$$

where A_1 and A_2 are constants at a fixed value of $\mathscr{P}$, so that the normal scaling (equation 2.177, chapter 2) with the exponents z_n and α_n $(n = 1,2)$ is recovered far from the transition in both "phases". Near to the transition, the surface width remains finite for finite t and/or L, so that the factor of ξ^{α} in equation 5.263 must be compensated for by an appropriate power law form for $f(x,y)$ with $y \sim x^{z'}$. In the limit $L \to 0$, the scaling form given in equation 2.177, chapter 2, is obtained, with the anomalous scaling exponents α' and z'. Kertész and Wolf illustrated this scaling using a $1+1$-dimensional square lattice polynuclear growth model that has a transition from a flat surface at high nucleation rates to a self-affine KPZ surface at low nucleation rates. At the transition, the anomalous exponents are $\beta = 0.28$ and $z = 1.58$ ($z = \nu_{\parallel}/\nu_{\perp}$, where $\nu_{\parallel}$ and $\nu_{\perp}$ are the directed percolation correlation length exponents).

A $1+1$-dimensional version of the van der Drift [1253] model for the growth of polycrystalline films has been introduced by Dammers and Radelaar [1288, 1289]. In this model, the initial surface consists of line segments connecting the positions $(x_1,h_1),(x_2,h_2),\ldots,(x_n,h_n)$. Each line segment covers the horizontal range $x_i - x_{i+1}$ and is inclined at an angle ϑ_i from the horizontal. All of the line segments representing the surface are assumed to move with the same velocity in an outwards pointing direction, along the local normal to the surface. This means that the point separating the adjacent ith and jth line segments moves with a lateral velocity component (domain wall velocity) of

$$v_{ij} = v_{\perp} \cos((\vartheta_i + \vartheta_j)/2)/\cos((\vartheta_i - \vartheta_j)/2), \tag{5.264}$$

where $v_{\perp}$ is the growth velocity normal to the line segments. Consequently, this model is completely deterministic, after the initial conditions have been specified. Starting from the edges of a large number of equally spaced, randomly oriented squares, the mean separation $< \delta x >$ between adjacent grain boundaries at the surface grows as $t^{1/2}$. For some starting configurations (a narrow distribution of inclination angles near 45°), a quite long intermediate scaling regime was found with an effective exponent (for the growth of $< \delta x >$) that is substantially larger than 0.5 (close to 1.0). Thijssen *et al.* [1290, 1291] have theoretically explored a corresponding $2+1$-dimensional model. They showed that the exponent p describing the growth of the characteristic length scale $\xi_{\parallel}$, associated with the surface grains, has an asymptotic value of $p = 1/z = 0.4$ ($\xi_{\parallel} \sim t^{0.4}$). Simulations [1291] are in agreement with this theoretical prediction for the asymptotic growth of ξ for both cubic and cone-shaped "crystals".

A crossover from $t^{1.0}$ to $t^{1/2}$ behavior, consistent with the simulation results of Dammers *et al.*, was found for the $1+1$-dimensional case.

5.9 Additional Information

Interest in the growth of self-affine rough surfaces is quite recent, and there are relatively few books and reviews in this area. Family and Vicsek [1292] have published a collection of influential original papers with an introduction and comments that direct the reader to the main scientific issues and results in the individual papers. Several reviews have been published including those by Krug and Spohn [555], Meakin [1293] and Halpin-Healey and Zhang [934]. The review of Krug and Spohn is focused mainly on the theoretical aspects of surface growth, while that of Meakin is concerned more with simulations and experiments. The review of Zhang and Halpin-Healey is concerned primarily with theoretical issues. It provides the most up to date and detailed review of the theory of surface growth, with an emphasis on directed polymers. Despite its informal style, much of this review is quite specialized and assumes that the reader has a strong theoretical background. Krug [637] has provided a lucid exposition of selected aspects of surface growth in which the application of scaling theory is emphasized. A review of the experimental aspects of self-affine surface growth, by Krim and Palasantzas [1135], has recently appeared. The relative brevity of this review reflects the dearth of experimental data, compared with the large volume of theoretical and simulation results. Barabási and Stanley [933] have recently written the only monograph on surface growth in which fractal geometry and scaling are emphasized. This book, entitled *Fractal concepts in surface growth*, explains the powerful renormalization group methods that are beyond the scope of this chapter. A book on diffraction from rough surfaces by Yang *et al.* [1134], which emphasizes applications to self-affine surfaces, has also appeared. The growth of fractal surfaces is discussed in another recent book, *Fractal surfaces* by Russ [1294], but much more emphasis is placed on the analysis of natural surfaces and surfaces obtained from laboratory experiments and manufacturing processes.

Appendix A

Instabilities

A determination of the stability of simple solutions to moving-boundary equations with respect to shape perturbations is an important step in the investigation of a wide range of pattern-formation processes. The pioneering work of Mullins and Sekerka on the stability of the growth of solidification fronts and of Saffman and Taylor on moving fluid–fluid interfaces were major advances. The basic approach is to analyze the initial, short time, growth and/or decay of an infinitesimally small perturbation as a function of the characteristic length scale or wavelength of the perturbation. Although the linear stability approach, exemplified by this work, is not always sufficient, it is a basic tool in theoretical morphogenesis.

A.1 The Mullins–Sekerka Instability

Mullins and Sekerka showed that all of the parabolic solutions obtained by Ivantsov [152, 153], for the solidification problem, are linearly unstable. They considered the stability of spherical [159] and planar [149] surfaces growing in supersaturated environments, under diffusion-limited growth conditions and solidification processes controlled by thermal diffusion. In the mass diffusion controlled case, they employed a one-sided growth model in which the velocity of the interface was assumed to be given by equation 1.55. It was also assumed that the dimensionless supersaturation $\Delta = (C_\infty - C_{eq})/\Delta C$ is small, $\Delta \ll 1$, so that the Peclet number is small and the diffusion equation can be replaced by the Laplace equation, $\nabla^2\Phi = 0$, in the vicinity of the growing front. This means that the problem can be expressed in terms of surface growth velocities, given by equation 1.98, with fixed, local equilibrium, concentration boundary conditions on the advancing interface and $\Phi = \Phi_\infty = 0$ at infinity (Φ is the dimensionless field U, described in chapter 1, section 1.4).

It is intuitively obvious that, if a growing spherical surface is perturbed by placing a "bump" on it, the gradient $\nabla_n\Phi$ near the "top" of the bump will be larger than elsewhere on the surface, so that the bump will grow at a faster rate. This means that, other things being equal, a spherical shape is unstable with respect to shape perturbations under these growth conditions. However, the equilibrium concentration $C(\mathbf{x}_s)$ of the solute in the solvent at position $\mathbf{x}_s$, adjacent to the interface, will be larger near to a more strongly curved surface, as described by equation 1.58, which can be written as

$$\Phi(\mathbf{x}_s) = \Delta - \xi_{cC}\mathcal{K}(\mathbf{x}_s), \tag{A.1}$$

where ξ_{cC} is the capillary length. The larger equilibrium concentration associated with the curved top of the bump will reduce the concentration gradient and stabilize the interface. The eventual fate of the bump or perturbation will depend on the relative strengths of these two opposing effects.

By expressing the shape of a growing particle in terms of a spherical harmonic perturbation on a spherical surface,

$$r(\theta,\phi) = r_\circ + \delta Y_{lm}(\theta,\phi), \tag{A.2}$$

where $Y_{lm}(\theta,\phi)$ is the mth spherical harmonic function of order l, Mullins and Sekerka were able to show that, to first order in δ, the scalar field Φ satisfying the Laplace equation, the boundary conditions corresponding to equations 1.58 and equation A.2 has the form

$$\Phi(r,\theta,\phi) = (A/r) + (B(l)\delta Y_{lm}/r^{l+1}) + \Phi_\infty, \tag{A.3}$$

where A and $B(l)$ depend on the capillary length, the supersaturation and the order ℓ. From this expression, the normal growth velocity can be calculated via equation 1.98. If the concentration at equilibrium with a sphere of radius $r_\circ$ is much smaller than the concentration in the growing solid, the solutions have the form

$$\begin{aligned} d\delta(l)/dt &\sim \mathcal{D}(l-1)\Delta C\{[1/(\xi_{\mathcal{D}}r_\circ)] - (l+1)(l+2)\xi_{cC}/(r_\circ)^3\}\delta(l) \\ &= v(l,r_\circ)\delta(l), \end{aligned} \tag{A.4}$$

where $\xi_{\mathcal{D}}$ is the diffusion length. The first term on the right-hand side of equation A.4 expresses the destabilizing effect of the preferential growth at the more "exposed" positions on those parts of the interface that have already grown the farthest, and the slower growth of those parts that have grown the least and are screened by the more exposed parts of the growing interface. The second term describes the surface energy effects that tend to retard the growth of those parts of the surface with positive (convex) curvature and enhance the growth of regions of negative curvature.

Because of the linear nature of the equations involved, the growth of an arbitrary perturbation,

$$r(\theta, \phi) = r_\circ + \sum_{lm} \delta_{lm} Y_{lm}(\theta, \phi), \tag{A.5}$$

can be obtained by linear superposition of the growth of the individual spherical harmonic pertutbations.

If the interface is not driven by the growth process it will become smoother, via transport of material from the exposed parts of the surface into the depressions. This process is similar to Ostwald ripening, described elsewhere [58]. In the absence of the stabilizing effects of the interfacial energy (if the surface free energy Γ is zero) all perturbations will be unstable. Equation A.4 indicates that, for each value of l, there is a critical value $r_\circ^*(l)$ for $r_\circ$. In general, small spheres will be stable with respect to shape perturbations ($r_\circ < r_\circ^*(l)$) and large spheres will be unstable ($r_\circ > r_\circ^*(l)$). This is in accordance with the intuitive idea that surface effects are more important for small particles than for large ones.

It follows from equation A.4 that the growth process will become marginally unstable when $\{[1/(\xi_{\mathscr{D}} r_\circ)] - (l+1)(l+2)\}\xi_{cC}/(r_\circ)^3 = 0$ or $\xi_{\mathscr{D}}\xi_{cC} \approx r_\circ^2 \approx \mathscr{K}^{-2}$, where $\mathscr{K}$ is the curvature. The stability with respect to perturbations can be expressed in terms of the stability parameter

$$\sigma = \xi_{\mathscr{D}}\xi_{cC}\mathscr{K}^2. \tag{A.6}$$

The growth becomes unstable if $\sigma < \sigma^*$ and is stable for $\sigma > \sigma^*$. It has been postulated [1295, 1296, 1297] that for the case of dendritic growth, a dendrite tip grows in such a manner that its tip is always marginally stable against deformation ($\sigma = \sigma^*$), so that the radius of curvature of the dendrite tip would be given by

$$r_\circ \sim (\xi_{\mathscr{D}}\xi_{cC})^{1/2} \sim \xi_{MS}. \tag{A.7}$$

The characteristic length scale $\xi_{MS} = (\xi_{\mathscr{D}}\xi_{cC})^{1/2}$ is often called the Mullins–Sekerka length. Although the marginal stability principle, for pattern selection, has now been abandoned, equation A.7 remains a useful guide.

In the growth of a solid pattern from a melt, the stability of the interface is determined by the competing stabilizing effects of surface tension and the destabilizing effect of the temperature gradient near to the surface. For this process the linear stability analysis of Mullins and Sekerka shows that the appropriate stability parameter σ_T is given by

$$\sigma_T = \xi_{\mathscr{D}T}\xi_{cT}\mathscr{K}^2, \tag{A.8}$$

where $\xi_{\mathscr{D}T}$ is the thermal diffusion length and ξ_{cT} is the thermal capillary length.

An additional complication comes from the fact that heat can flow in both the solid and liquid phases. This is not a serious problem, and the solution has the same form as equation A.4, but the rate constant $v(l, r_\circ)$ does depend on the thermal conductivity ratio $k(s)/k(l)$ ($k = \mathscr{D}_T \rho C_h$, where $\mathscr{D}_T$ is the thermal diffusion coefficient, ρ is the density and C_h is the heat capacity). If the transport of heat in the solid is ignored, the one-sided approximation can be used and the two models become equivalent.

Huang and Glicksman [148] have reviewed the results of several theoretical studies of this process, including those based on the marginal stability idea. They all lead to the conclusion that the interface is unstable to perturbations on length scales larger than λ, where

$$\lambda^2 = A\xi_{cT}\xi_{\mathscr{D}_T}. \tag{A.9}$$

Despite significant differences in these theories, including the basic geometry of the unperturbed interface, the values obtained for the parameter A in equation A.9 were remarkably similar.

The same approach can be used in the case of an advancing planar interface [149]. In this case, a sinusoidal perturbation to the flat interface is used. The linear nature of the problem allows the growth of different Fourier components to be decoupled and the solution to the general problem can be obtained by superposition. The growth of the amplitude $\delta A(k)$ of a perturbation with a wave vector k ($k = 2\pi/\lambda$, where λ is the wavelength) is given by

$$d\delta A(k)/dt \sim v_\circ k[1 - B\xi_c\xi_{\mathscr{D}}k^2]d\delta A(k), \tag{A.10}$$

where B is a constant of order 1. Consequently, the amplitude of the component with a wave vector k grows as $\delta A(k) = \delta A_\circ e^{(v_\circ W(k)t)} = \delta A_\circ e^{v(k)t}$, provided that $\delta A(k)$ remains small, where $W(k) = k(1 - B\xi_c\xi_{\mathscr{D}}k^2)$ and

$$v(k) = v_\circ k(1 - B\xi_c\xi_{\mathscr{D}}k^2). \tag{A.11}$$

The dispersion function $W(k)$ is shown in figure A.1. This figure shows that long wavelength perturbations with $k < k^*$, where $k^* = 1/(B\xi_c\xi_{\mathscr{D}})^{1/2}$ are unstable and that the most unstable perturbation is found at $k_{max} = 1/(3B\xi_c\xi_{\mathscr{D}})^{1/2} \approx 1/\xi_{MS}$. For values of k greater than k^*, the effects of surface energy are dominant and these short wavelength perturbations decay. The initial stages of unstable growth can be described in terms of a morphology length $\xi_m \approx \xi_{MS} \approx 1/k_{max}$. Under the low Peclet number conditions assumed in the derivation of equation A.11, $\xi_c \ll \xi_m \ll \xi_{\mathscr{D}}$. For many growth processes ξ_m is also a characteristic length scale for the late stages of growth. The rate constant $v(k)$ for the growth of perturbations on a planar surface is equivalent to the rate constant $v(l, r_\circ)$ defined in equation A.4 for perturbations on a growing spherical particle for large $r_\circ$ and l ($v(k) = v(l, r_\circ)$ for $k = l/r_\circ$ in the $l \to \infty$, $r_\circ \to \infty$ limit).

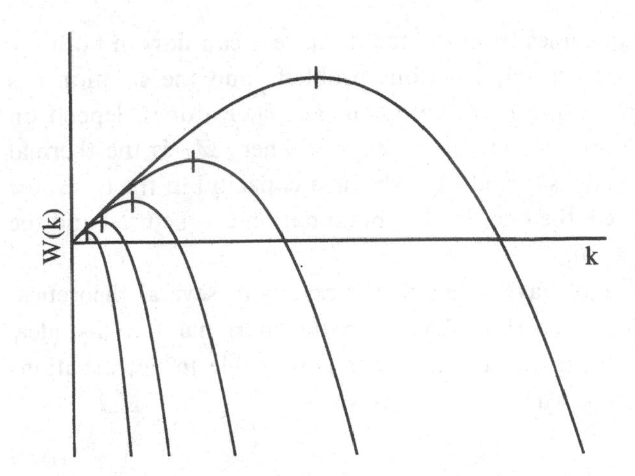

Figure A.1 The dispersion function $W(k) = k(1 - B\xi_c\xi_{\mathscr{D}}k^2)$ for several values of $B\xi_c\xi_{\mathscr{D}}$. For each curve, the maximum in $W(k)$ is found at $k = 1/(3B\xi_c\xi_{\mathscr{D}})^{1/2}$, and this value of k is indicated by a vertical line at the maximum of each curve.

In their analysis of the stability of planar surfaces, Mullins and Sekerka included both concentration and temperature fields. The analysis was based on the assumption of isotropic bulk and surface properties and local equilibrium at the interface (as in the perturbed spherical growth problem [159]). It was assumed that the interface moves with a constant velocity V and that the concentration (or temperature) field attains a stationary form in a coordinate system moving with the interface (at velocity V). The temperature $T(\mathbf{x}_s)$ adjacent to the interface is given by equation 1.77, and the growth velocity of the interface is given by equation 1.56.

For a flat interface, with a sinusoidal perturbation of the form $h(x) = \langle h \rangle + \delta \sin(kx)$, the curvature $\mathscr{K}(x)$ is given by

$$\mathscr{K}(x) = k^2\delta \sin(kx). \tag{A.12}$$

Mullins and Sekerka assumed that the temperature $T(\mathbf{x})_s$ and concentration $C(\mathbf{x})_s$ at position $(x, h(x))$ on the external surface of the advancing front had the form

$$T(\mathbf{x})_s = T_\circ + a\delta \sin(kx) \tag{A.13}$$

and

$$C(\mathbf{x})_s = C_\circ + b\delta \sin(kx), \tag{A.14}$$

where $T_\circ$ and $C_\circ$ are the values for a flat interface. These forms for $T(\mathbf{x})_s$ and $C(\mathbf{x})_s$ were justified by showing that results obtained for the concentration field $C(\mathbf{x})$ and temperature field $T(\mathbf{x})$ were consistent with the basic transport equations and boundary conditions.

At short times, a perturbation with a wave number of k will either grow or decay exponentially ($\delta(k,t) = \delta_\circ e^{v(k)t}$), with a growth constant of $v(k)$. The results obtained by Mullins and Sekerka, for the general case, with both thermal and material transport, were quite complex. On an intuitive level, the evolution of the interface can be considered to be a consequence of three competing effects. The surface energy effects always stabilize the interface and ensure that small wavelength perturbations decay. The temperature gradient could either stabilize or destabilize the interface, but Mullins and Sekerka considered the case of a positive, stabilizing temperature gradient in the direction of the propagation of the interface so that the temperature increases with increasing distance from the interface in the liquid phase. The third effect is the destabilizing effect of the distribution of solute at the interface. This is a destabilizing effect that becomes small at long wavelengths, because of the long time scales associated with chemical diffusion over long distances. At the longest length scales, the effect of the stabilizing temperature gradient dominates, because thermal diffusion coefficients are much larger than chemical diffusion coefficients. Consequently, the dispersion function $W(k)$ is different from that shown in figure A.1, because $W(k)$ becomes negative for small k. This indicates that, under these conditions, instability (growth of the perturbation) can occur over only a limited range of wavelengths or wave numbers, or the growth may be stable. Under unstable growth conditions, there will, in general, be a fastest growing wave number k_{max}. Because of the exponential nature of the growth process, this wave number will dominate the *early stages* of pattern formation.

The results obtained by Mullins and Sekerka indicate that the stability criterion for the growth of perturbations can be expressed in terms of a stability function $\mathcal{S}(k, \xi_c, \xi_T, \xi_{\mathcal{D}c})$. Here ξ_T is a thermal length related to the temperature gradient $\nabla(T)$ (for example, $\xi_T = \Delta T/\nabla(T)$, where ΔT is the freezing temperature range, for the solidification of an alloy). The interface will be flat for

$$\mathcal{S}(k, \xi_c, \xi_T, \xi_{\mathcal{D}c}) < 0, \tag{A.15}$$

and unstable for $\mathcal{S}(k, \xi_c, \xi_T, \xi_{\mathcal{D}c}) > 0$. In the simple case illustrated in figure A.1 the stability function $\mathcal{S}(k, \xi_c, \xi_{\mathcal{D}})$ can be written as $\mathcal{S}(k, \xi_c, \xi_{\mathcal{D}} = W(k) = k(1 - B\xi_c\xi_{\mathcal{D}}k^2))$. The thermal capillary length $\xi_{\mathcal{D}_T}$ does not appear in the stability function $\mathcal{S}(k)$, because Mullins and Sekerka considered the case where $V/k\mathcal{D}_T \ll 1$.

A qualitative understanding of non-equilibrium pattern formation during processes such as solidification can be expressed in terms of characteristic length scales such as the chemical and thermal diffusion lengths, the capillary lengths and other characteristic lengths such as ξ_T. Trivedi and Kurz [1298] have discussed the application of this approach to the solidification processes under low velocity and high velocity growth conditions. In this case, a series of morphology transitions takes place during directional solidification, as the diffusion length is decreased relative to the capillary and thermal length by changing the growth velocity. At very low velocities, $\xi_{\mathcal{D}} > \xi_T \gg \xi_c$, and under these conditions the interface is stabilized by the temperature gradient and the interface is flat. At high velocities $\xi_T \gg \xi_c > \xi_{\mathscr{D}}$, and the interface is again stable. At intermediate velocities, cellular and dendritic patterns are formed. The transitions between these morphologies depend on the equilibrium solute distribution coefficient as well as on the characteristic lengths.

At high growth velocities, non-equilibrium effects at the growing interface may become important. The partition of materials between the growing pattern and the medium into which it is growing may be perturbed, and the kinetics of molecular attachment may start to modify the temperature and/or composition at the interface. The overall effect is to increase the capillary length ξ_c and decrease the thermal length ξ_T [1298].

A.2 The Saffman–Taylor Problem

Saffman and Taylor [160] analyzed the displacement of a viscous fluid by a less viscous fluid in a confined space enclosed by two closely spaced, vertical, parallel walls and compared the results of their analysis with experiments. The experiments were carried out in a vertical channel or Hele-Shaw cell [1299] of dimensions 0.09 cm × 12 cm × 38 cm. The analysis was based on the idea that fluid flow in a Hele-Shaw cell can be described by Darcy's law (equation 1.79).

Saffman and Taylor addressed the problem of the stability of a horizontal, planar interface moving with velocity V in the vertical y direction, by analyzing the evolution of perturbations that were assumed to have the form

$$y = \delta e^{ikx+v(k)t}, \tag{A.16}$$

where x is the horizontal position on the interface and k is the wave number of the perturbation. Their analysis was based on the assumptions that the incompressible fluids are separated by a sharp interface and that the component of the fluid velocity normal to the fluid–fluid interface is continuous. From this it follows that

$$\partial\Phi(1)/\partial y = \partial\Phi(2)/\partial y = V + \delta v(k)e^{inx+v(k)t}, \tag{A.17}$$

where

$$\Phi(i) = (k(i)/\mu(i))(p + \rho(i)gh) = (k(i)/\mu(i))\Phi_h(i). \tag{A.18}$$

In equation A.18 $k(i)$ is the permeability of the porous medium or Hele-Shaw cell to the ith fluid, $\mu(i)$ is the viscosity of the ith fluid and $\Phi_h(i)$ is the hydraulic potential. The hydraulic potential is given by $\Phi_h = p + \rho(i)gy$, where p is the pressure in the fluid, $\rho(i)$ is the density of the ith fluid and g is the acceleration due to gravity. The quantity $\Phi(i)$ is the "velocity potential" in fluid i. The solutions to the Laplace equation, $\nabla^2\Phi(i) = 0$, that satisfy equation A.17 and for which the effects of the perturbations vanish at infinity are

$$\Phi(1) = Vy - (\delta v(k)/k)e^{-ikx-ky+v(k)t} \tag{A.19}$$

and

$$\Phi(2) = Vy + (\delta v(k)/k)e^{ikx+ky+v(k)t}, \tag{A.20}$$

where "(1)" indicates the upper fluid and "(2)" indicates the lower fluid. By requiring that the pressure be the same on both sides of the fluid–fluid interface, Saffman and Taylor showed that

$$\begin{aligned}(v(k)/k)[\mu(1)/k(1) + \mu(2)/k(2)] = \\ (\rho(1) - \rho(2))g + [\mu(1)/k(1) + \mu(2)/k(2)]V,\end{aligned} \tag{A.21}$$

to first order in δ. If the right-hand side of equation A.21 is positive, then $v(k)$ is positive ($[\mu(1)/k(1) + \mu(2)/k(2)]$ is always positive) and the perturbation grows. Similarly, if the right-hand side of equation A.21 is negative the interface is stable.

The effect of an interfacial energy or surface tension Γ is to stabilize the interface. For a zero contact angle, in a Hele-Shaw cell, the pressure difference across the interface is given by

$$\Delta P = \Gamma((2/b) + \mathcal{K}), \tag{A.22}$$

where b is the size of the gap between the two walls of the cell and $\mathcal{K} = d^2y/dx^2$ is the curvature in the plane of the cell. Saffman and Taylor argued that since b is constant everywhere only the second term on the right-hand side of equation A.22 need be retained and equation A.21 can be replaced with

$$\begin{aligned}(v(k)/k)\{[\mu(1)/k(1)] + [\mu(2)/k(2)]\} = \\ (\rho(1) - \rho(2))g + \{[\mu(1)/k(1)] + [\mu(2)/k(2)]\}V - \Gamma\mathcal{K}.\end{aligned} \tag{A.23}$$

However, the curvature perpendicular to the plane of the cell is much larger than the curvature in the plane of the cell (except at very large capillary numbers) and perturbations such as grooves in the cell walls or threads inside the cell, which act via local variations in the perpendicular curvature can have an important effect [1300]. Putting the Hele-Shaw cell permeability $(b^2/12)$ in equation A.23, the equation

$$\begin{aligned}(12/b^2)(v(k))[\mu(1) + \mu(2)] = \\ k\{(\rho(1) - \rho(2))g + 12/b^2[\mu(1) + \mu(2)]V\} - k^3\Gamma\end{aligned} \tag{A.24}$$

can be obtained. This equation, which has the same form as equation A.11, indicates that the interface will be stable on short length scales or wavelengths,

$k > k_c$, and unstable on long length scales $k < k_c$, where the critical wave number k_c is given by

$$k_c = \{[12V(\mu(1) - \mu(2)) + b^2 g(\rho(1) - \rho(2))]/(\Gamma b^2)\}^{1/2}. \tag{A.25}$$

The wave number for the most unstable perturbation is given by $k_{max} = k_c/3^{1/2}$.

An important special case is that of displacement of a viscous fluid by an essentially non-viscous fluid in a horizontal Hele-Shaw cell ($g = 0$, $\mu(2) = 0$, $\mu(1) = \mu$). In this case, equation A.24 becomes

$$(12/b^2)(v(k))\mu = k\left[12/b^2\mu V\right] - \Gamma k^3 \tag{A.26}$$

or

$$v(k) = kV\left[1 - \frac{b^2\Gamma k^2}{12\mu V}\right], \tag{A.27}$$

and equation A.25 becomes

$$k_c = [12\mu V/(\Gamma b^2)]^{1/2}, \tag{A.28}$$

and $k_{max} = 2[\mu V/(\Gamma b^2)]^{1/2}$, which is consistent with the rough estimate for the finger width, given in equation 1.83. This equation can be interpreted in terms of a "competition" between the stabilizing effects of surface tension and finite cell wall separation b and the destabilizing effect of the pressure distribution in the viscous fluid, generated by the displacement velocity and fluid viscosity.

Saffman and Taylor also analyzed the problem of the penetration of a "bubble" or finger of fluid 2 (of density $\rho(2)$ and viscosity $\mu(2)$) into a surrounding fluid (fluid 1) in a horizontal Hele-Shaw cell. Without loss of generality, a channel width of 2 was used ($-1 \leq x \leq 1$). They assumed that the finger had an asymptotic width $2w$, far behind the advancing tip, and had a symmetric shape with respect to inversion about a plane at $x = 0$. The velocity of the bubble is U and the velocity of fluid 1, well ahead of the finger, is V. Conservation of the incompressible fluids requires that $U = wV$. Saffman and Taylor assumed that the pressure change across the interface is constant. This is equivalent to assuming that $2/b \gg \mathcal{K}$, and that the surface tension and contact angles are the same for moving and stationary interfaces. Edge effects, associated with the solid sides of the Hele-Shaw cell, were also neglected. Saffman and Taylor looked for stationary (shape preserving) solutions for the finger surface. The analysis was carried out in terms of a stream function Ψ defined by

$$v_x = \partial\Phi/\partial x = \partial\Psi/\partial y \tag{A.29}$$

and

$$v_y = \partial\Phi/\partial y = -\partial\Psi/\partial x, \tag{A.30}$$

where v_x and v_y are the components of the fluid velocity. The boundary conditions $\Psi(1) = -V$ at $x = -1$ and $\Psi(1) = V$ at $x = 1$ were used at the cell edges and $\Psi(1) = \Psi(2) = Uy$ at the fluid–fluid interface. The boundary conditions were completed by taking $\Phi(1) = Uy$ at $y = -\infty$ and $\Phi(2) = Vy$ at $y = \infty$.

For the case in which the finger (fluid 2) is of zero viscosity and density, Saffman and Taylor found a family of solutions represented by the parametric equation

$$y = ((1 - w)/\pi)\ln[(1/2)(1 + \cos(\pi x/w))] \tag{A.31}$$

In the horizontal Hele-Shaw cell experiments of Saffman and Taylor, oil with a viscosity of 2.75 P was injected into colored glycerin with a viscosity of 8 P, using a cell with internal dimensions of 2.24cm ×91cm ×0.08cm. A single finger, with a width equal to about half that of the cell ($w = 1/2$), was formed and the shape of the advancing end of the finger was found to be in excellent agreement with equation A.31. At lower injection rates or smaller capillary numbers Ca ($Ca = \mu U/\Gamma$), a wider finger was formed and the shape was not found to be in such good agreement with equation A.31. Experiments carried out using several fluid pairs indicated that the finger width parameter w approached a limiting value of slightly larger than 1/2 at large injection velocities and that the width was a universal function of $\mu U/\Gamma$. On a more detailed level, the problem of viscous fingering presents a range of challenges and controversies [1301].

The 2-dimensional approximation implied by equation A.22 is not always justified. Under most circumstances, the curvature perpendicular to the plane of the cell ($\mathcal{K}_\perp \sim 1/b$) makes the largest contribution to the pressure jump ΔP and can only be neglected if this contribution is constant. In general, $\mathcal{K}_\perp$ can be expected to be a locally varying quantity that depends on the local velocity. Park and Homsy [151] found that

$$\Delta P = \left[\frac{2\Gamma}{b}(1 + 3.80Ca^{2/3})\right] + \pi\Gamma\mathcal{K}_\parallel/4, \tag{A.32}$$

assuming that the displaced fluid wets the cell walls. At low capillary numbers (small interface velocities), the capillary number dependent term in equation A.32 will be negligible. In the low capillary number limit, equation A.28 implies that the radius of curvature r_c for a viscous tip will scale as $r_c/b \sim Ca^{-1/2}$, for sufficiently small capillary numbers, $Ca \ll Ca_c$. For large values of the capillary number, the reduced radius of curvature r_c/b approaches a constant value. This is confirmed by experiments [1302], which indicate that

$$r_c/b = f(Ca), \tag{A.33}$$

with $f(Ca) \sim Ca^{-1/2}$ for $Ca/Ca_c \ll 1$, and $f(Ca) = const.$ for $Ca/Ca_c \gg 1$.

The strong effect of perturbations to the cell depth b is illustrated by the experiments of Yokoyama *et al.* [1302], in which the viscous finger was stabilized by a “V”-shaped groove, with a width of about 3 mm and a depth of 1 mm, running along the mid-line of the Hele-Shaw cell. This introduces anisotropy into the process and stabilizes the parabolic tip of the otherwise dendritic finger. If the groove ended part way down the cell, the finger tip was found to flatten and split, soon after the end of the groove had been reached.

Appendix B

Multifractals

The concept of multifractal scaling can be used to characterize a wide variety of systems in which a measure is distributed over a fractal or Euclidean set in a very non-uniform manner. In practical applications, the measure may describe the distribution of electric field, electric current, stress, chemical concentration and other intensive quantities. In the context of non-equilibrium growth, the distribution of growth probabilities or growth velocities is an important quantity that can often be described in terms of a multifractal scaling model, at least in the case of simple, fractal growth models. The basic nature of multifractals was discussed in chapter 2, section 2.8. In general terms, a multifractal consists of an infinite number of interpenetrating fractal subsets that fill a part of the entire system, called the support of the measure. Each of the subsets can be identified by a scaling index α, which is the Holder exponent (chapter 2, section 2.8) for the subset, and the size of the region occupied by each of the subsets scales with the system size or resolution as if it had a fractal dimensionality $D_f(\alpha)$ or $f(\alpha)$. The function $D_f(\alpha)$ describes how the multifractal changes as it is examined at different resolutions $\lambda = L/\epsilon$. Since the measure in a region of size ϵ, with a scaling index of α, is given by $\mu(\epsilon,\alpha) \sim \epsilon^{\alpha}$, this measure can be thought of as arising from a singularity of strength α lying in that region [343]. However, the value of this notion is diminished because singularities of all strengths will be found in the region of size ϵ on further subdivision.

Multifractal analysis has been applied to a very wide range of phenomena, including the distribution of dissipation in turbulent flows [340], molecular spectroscopy [1303], electronic wavefunctions in disordered systems [1304], the distribution of voltage drops across the elements of random resistor networks

[1305] and the distribution of rapidity among the particles resulting from high energy nuclear collisions [26, 1306].

B.1 Generation of Simple Multifractal Sets

The basic nature of multifractal sets can be illustrated using the simple random multiplicative or curdling [2] model shown in figure B.1. Initially, the measure $\mu(x)$ is distributed uniformly over the unit interval so that $\mu(x,\epsilon) = \epsilon$ for all x in the range $0 \leq x \leq 1$, where $\mu(x,\epsilon)$ is the measure in an interval of size ϵ at position x. In the first stage in the construction, the measure is redistributed so that the density of the measure in the interval $[0 - 1/2]$ is r times as large as the density of the measure in the interval $[1/2 - 1]$. The density of the measure in the interval $[0 - 1/2]$ or $[1/2 - 1]$ is $\rho_1 = 2/[1 + (1/r)]$ and the density of the measure in the other interval, of length 1/2, is $\rho_2 = 2/(1 + r)$, so that the measure on the entire unit interval is conserved. In the second stage of the construction, the measure in the intervals $[0 - 1/2]$ and $[1/2 - 1]$ is redistributed in a similar manner to generate four regions of length 1/4 in which the density of the measure is ρ_1^2, $\rho_1\rho_2$, $\rho_1\rho_2$ and ρ_2^2, respectively. After n generations, the premultifractal measure consists of 2^n regions of size $\epsilon_n = 2^{-n}$, in which the measure has the form $\mu(\epsilon_n) = 2^{-n}\rho_1^j\rho_2^k$, with $j + k = n$, and the density or intensity of the measure is $\rho_1^j\rho_2^k$. The generation process can be carried out in either a systematic or random manner, with respect to the way in which the measure is distributed between the left- and right-hand sides.

The index j (or k) can be used to identify the subsets in the nth generation premultifractal. The size or "volume" of the subset with index j, consisting of that part of the premultifractal in which the density of the measure is $\rho_1^j\rho_2^k$ or $\rho_1^j\rho_2^{n-j}$, is given by

$$V_{j,n} = 2^{-n}n!/(j!(n-j)!), \tag{B.1}$$

and the volume of this subset scales with the cut-off length scale ϵ like a self-similar fractal with a fractal dimensionality given by

$$D_j = d - (\lim_{n\to\infty} \log V_{j,n}/\log\epsilon_n) = d - [\lim_{n\to\infty} \log V_{j,n}/\log(2^{-n})]. \tag{B.2}$$

Using the asymptotic Stirling approximation, $\ln n! = n\ln(n)$, it can quite easily be shown from equations B.1 and B.2 that

$$D_j = D(\zeta) = -[\zeta\ln\zeta + (1-\zeta)\ln(1-\zeta)]/\ln 2, \tag{B.3}$$

where $\zeta = j/n$. In this case, the union of all of the fractal subsets completely covers the unit interval. The fractal dimensionality of the support is 1 and the fractal dimensionality of the largest subset, corresponding to $\zeta = 1/2$, is also 1. Although a random multiplicative process generates a multifractal with easily

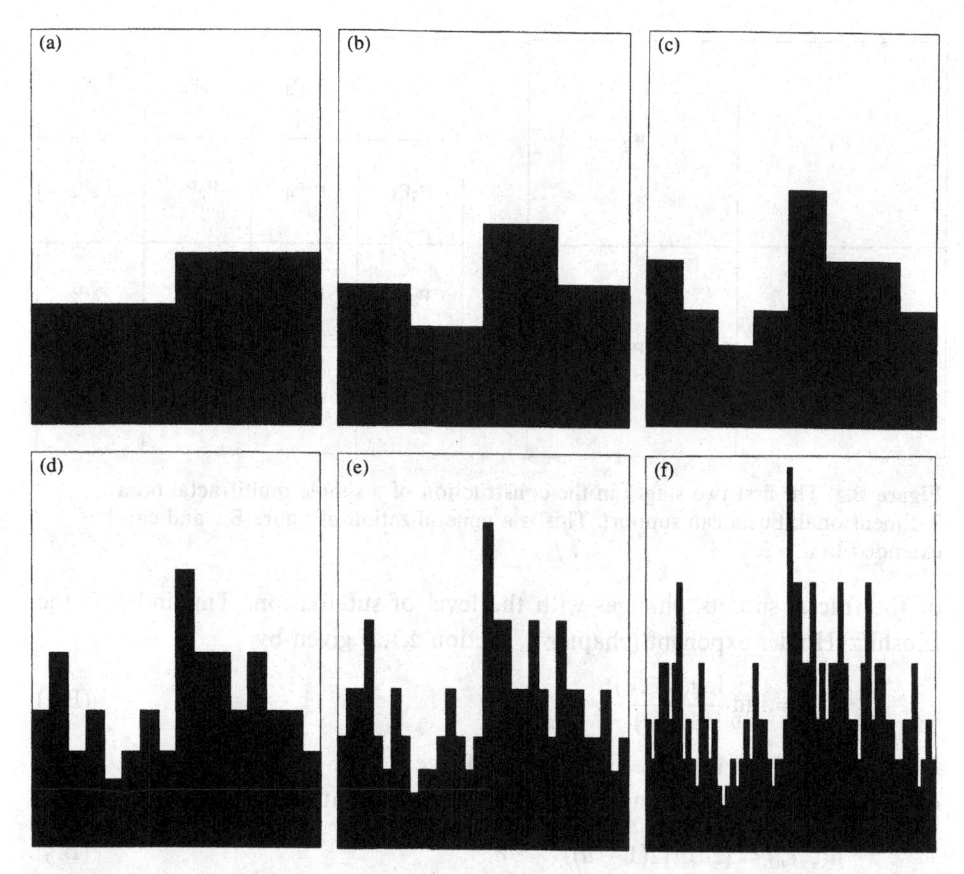

Figure B.1 Six stages in the construction of a simple multifractal on a 1-dimensional, Euclidean support. The density ratio r is 0.7.

identifiable prefractal subsets, at any level of subdivision, the measure could be redistributed to give a distribution that would not contain fractal subsets. For example, the regions of size ϵ_n could be scrambled randomly or placed in order of decreasing $\mu(\epsilon_n)$. Equation B.2 could still be used to describe how the volume of each of the subsets scales with ϵ, but the exponent D_j could not be interpreted as the fractal dimensionality of the subset.

Although the index j identifies the subsets at any particular level of subdivision, it is more useful to characterize the asymptotic $n \to \infty$ multifractal measure in terms of $\zeta = j/n$, via the fractal dimensionalities of the subsets as a function of ζ. The fractal dimensionality of these subsets is given by equation B.3. This allows the fractal measure to be characterized in terms of a single function $D(\zeta)$, at any level of subdivision (the index ζ and $D(\zeta)$ do not depend on the level of subdivision in the $n \to \infty$ limit).

The characterization of the multifractal is completed by the scaling exponent $\alpha(\zeta)$ that describes how the measure associated with regions of size ϵ_n, in each

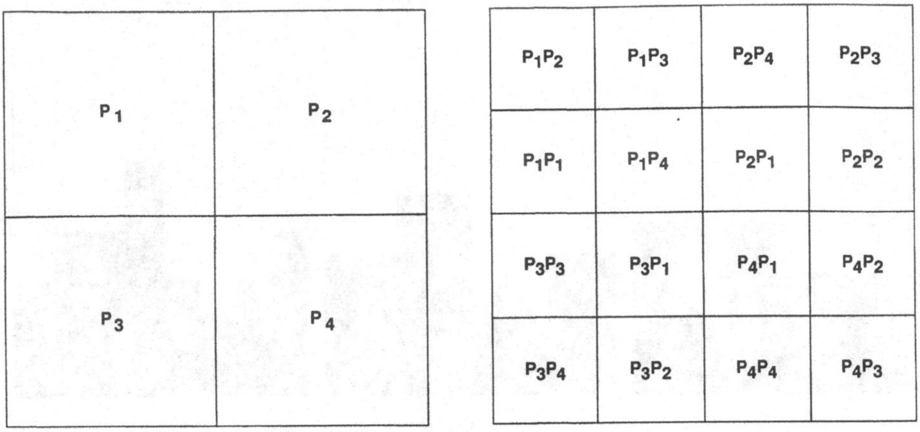

Figure B.2 The first two stages in the construction of a simple multifractal on a 2-dimensional, Euclidean support. This is a generalization of figure B.1 and can be extended to $d > 2$.

of the fractal subsets, changes with the level of subdivision. This index is the Lipshitz–Holder exponent (chapter 2, section 2.1.4) given by

$$\alpha(\zeta) = \lim_{\epsilon \to 0} \frac{\log(\mu(\zeta, \epsilon))}{\log(\epsilon)}. \tag{B.4}$$

Here, $\mu(\zeta, \epsilon)$ is the measure in a region of size ϵ, that has an index of ζ. In the case of the example illustrated in figure B.1, the quantity $\mu(\zeta, \epsilon)$ is given by

$$\mu(\zeta, \epsilon_n) = (2a)^{n\zeta}(2(1-a))^{n(1-\zeta)}n^{-2}, \tag{B.5}$$

where $a = r/(1+r)$. Since $\epsilon_n = 2^{-n}$, it follows from equations B.4 and B.5 that

$$\alpha(\zeta) = -\frac{\zeta \log(a) + (1-\zeta)\log(1-a)}{\log 2}. \tag{B.6}$$

This means that the scaling properties of each subset can be described by $D(\zeta)$ and $\alpha(\zeta)$ or by the function $D(\alpha)$, $D_f(\alpha)$ or $f(\alpha)$, since there is a unique relationship between α and ζ.

This procedure for generating multifractals can easily be extended to 2-dimensional and higher-dimensionality spaces. Figure B.2 shows the first two stages in the generation of a premultifractal on a 2-dimensional support. In the first stage of the generation process, a unit square is divided into four equal quadrants and the measure is distributed among these quadrants with relative densities of P_1, P_2, P_3 and P_4, in random order. In the second stage, each of the quadrants is subdivided in the same manner. After n stages, the square has been subdivided into 2^{2n} regions of size $\epsilon = 2^{-n}$, and the density of the measure in each region of size ϵ has the form $P_1^i P_2^j P_3^k P_4^l$, with $i + j + k + l = n$.

At any stage in the process, the system can be divided into subsets in which the density of the measure is equal. Figure B.3 shows some of these subsets generated with $P_2 = P_3 = cP_1$ and $P_4 = c^2P_1$, after ten stages. In this case, the density of the measure in each region of size $\epsilon = 2^{-10}$ is given by $\rho_i \sim c^m$, where m is an integer in the range $0 \leq m \leq 20$. This figure shows how some of the subsets are distributed throughout the system and how they interpenetrate each other. As the subdivision process continues, all values of the index α will appear within each of these subsets, and the values of α that characterize these subsets at this level of subdivision will appear within all of the other subsets. In general, there is no need for these subsets to have a simple self-similar fractal structure in position space for the distribution of the measure in probability space to be multifractal. It is only necessary for the subsets to scale like a self-similar fractal in the sense that the amount of the (normalized) measure associated with a particular value of α scales as $\lambda^{-D_f(\alpha)}$.

In the example shown in figure B.3, the structure of the tenth-generation subsets were analyzed by measuring their two-point density-correlation functions, after mapping onto a $(2^{10} \times 2^{10})$-site square lattice, so that each region of size ϵ occupies a single lattice site [1307]. Figure B.4(a) shows the two-point density–density correlation functions for the $m = 2, 4, 6, 8$ and 10 subsets. All of these correlation functions have a power law form on length scales larger than a few lattice units (a few times ϵ). This is consistent with the idea that each subset does have a self-similar structure, in this case. However, it is also apparent that as the subdivision process continues beyond the nth stage, regions with all values of α will appear within all of the regions of size ϵ_n that are generated at the nth stage. Consequently, the regions that carry the "singularities" of strength α differ from typical self-similar fractals because there are no completely empty regions on all scales. The distribution of each singularity of strength α over the support with dimensionality D_X will be extremely inhomogeneous. This inhomogeneity may lead to the power law form for the correlation functions without completely empty regions on all scales. Since the distribution of $\{p_i\}$ is log-binomial[1] for this model, it can quite easily be shown, using the approach that was used to calculate the fractal dimensionalities for the 1-dimensional example discussed above, that the dependence of the fraction of the area covered by the measure in the subsets on the resolution is given by $V_m \sim \lambda^{D_m - d}$, where D_m, the "fractal dimensionality" of the mth subset is given by

$$D_m = h(x) = -2[x \ln x + (1 - x) \ln(1 - x)] / \ln 2, \tag{B.7}$$

where $x = m/2n$. Figure B.4(b) compares the fractal dimensionalities obtained from the correlation functions shown in figure B.4(a) with the function $h(x)$ in equation B.7. In this case, it appears that the subsets are indeed self-similar fractals with fractal dimensionalities given by $h(x)$ or $D_f(\alpha)$. In the asymptotic, $n \to \infty$ limit, the multifractal measure can be characterized by the area covered

1. Because the large moments of the distribution are sensitive to the extreme values, the log-binomial distribution cannot be replaced by a log-normal distribution [1308].

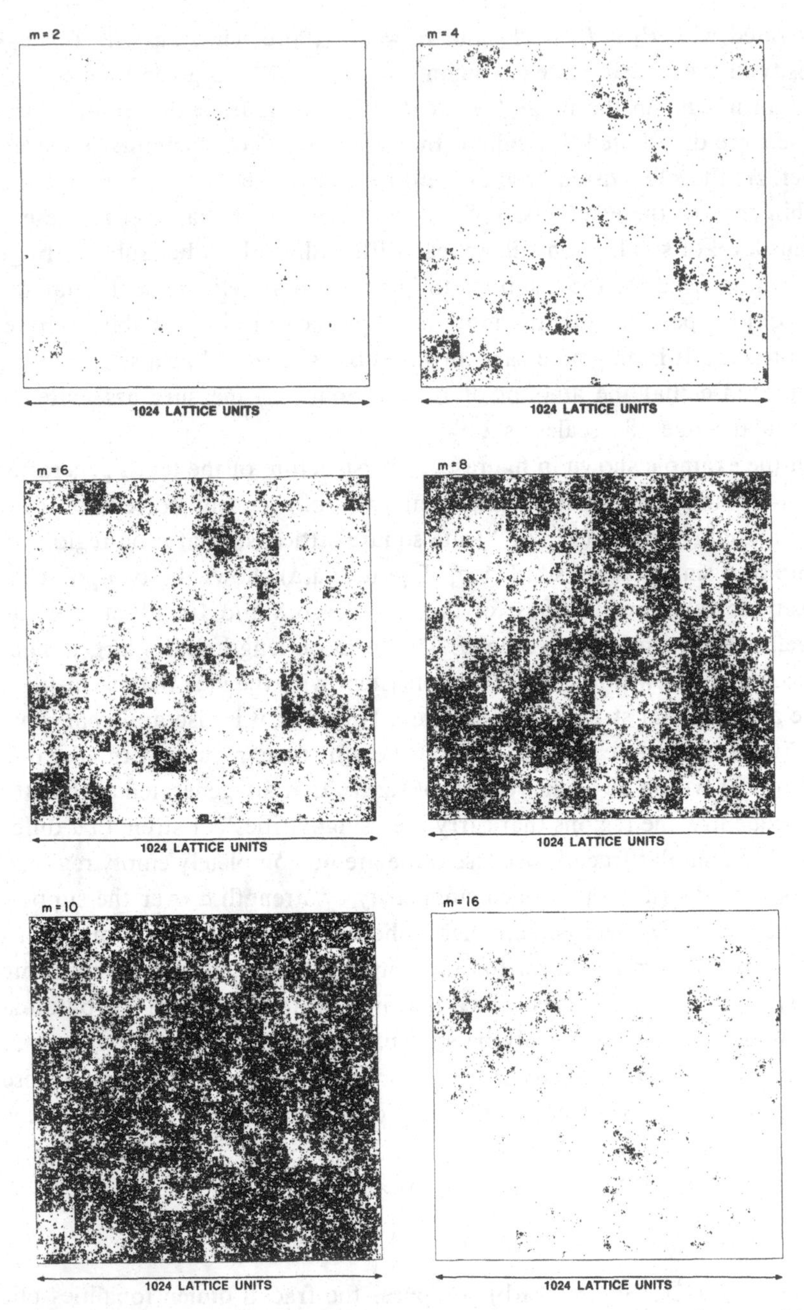

Figure B.3 Some of the subsets generated using the procedure illustrated in figure B.1 with $P_2 = P_3 = cP_1$ and $P_4 = c^2P_1$. This figure shows the prefractal subsets with densities of $(P_1)^{10}c^m$, with $m = 2$, 4, 6, 8, 10 and 16, after ten generations. At this stage, the premultifractal can be mapped onto a (1024×1024)-site square lattice of filled and empty sites.

by that part of the measure that lies in a narrow range of x (x to $x + \delta x$, with $\delta x \to 0$). The number of subsets that lie in this range is proportional to n, and it follows that the scaling of this part of the measure with the resolution is given by

$$f(x) = -\{2[x \ln x + (1-x)\ln(1-x)]/\ln 2\} + [\ln(n)/(n \ln 2)]. \qquad \text{(B.8)}$$

In the limit $n \to \infty$, $f(x) \to h(x)$. Since the index α is linearly related to x, $D_f(\alpha)$ has the same asymptotic form.

Simple multifractal models, like those described in this section, have been used to describe turbulence and other dynamical systems [340, 1309, 1310, 1311]. Similar models have also been proposed for porous media [398, 659].

B.2 Characterization of Multifractal Sets

Multifractal distributions can be analyzed in a variety of ways. The most popular, but not necessarily the best, approach[2] is to measure the moments of the measure $p_i(\epsilon)$ or $\mu_i(\epsilon, L)$ as a function of the resolution L/ϵ, where $p_i(\epsilon)$ is the measure in the ith region of size ϵ, normalized so that $\sum_i p_i(\epsilon) = 1$. This approach has been discussed in chapter 2.

An alternative approach to the characterization of fractal measures is based on the histogram of the normalized measure $\{p_i(\epsilon, L)\}$ determined at different resolutions, $\lambda = L/\epsilon$ [1312]. The histogram for the partition of the measure distributed over a system of size L into regions of size ϵ can be represented by the distribution $N(\log p, \epsilon, L)$, where $N(\log p, \epsilon, L)\delta \log p$ is the number of regions of size ϵ, in a system with an upper cut-off length scale of L, for which $\log p_i$ lies in the range $\log p$ to $\log p + \delta \log p$. For a multifractal measure, the index α is given by

$$\alpha = -\log p/\log \lambda \qquad \text{(B.9)}$$

and the fractal dimensionality of the subset associated with this index is given by

$$D_f(\alpha) = \log N(\alpha)/\log \lambda, \qquad \text{(B.10)}$$

where $N(\alpha)\delta\alpha$ is the number of regions of size ϵ for which the Holder exponent lies in the range α to $\alpha + \delta\alpha$.[3]

2. The most serious disadvantage of using the moments $Z_q(\lambda)$ to measure the $D_f(\alpha)$ is that the Legendre transform method constructs a convex envelope for the singularities. If, for example, the measure consists of two subsets with fractal dimensionalities of D_1 and D_2, and Holder exponents of α_1 and α_2, respectively, then the Legendre transform will generate a curve between the points (α_1, D_1) and (α_2, D_2), despite the fact that there are no subsets with $\alpha_1 < \alpha < \alpha_2$. The curve is generated because the function $\tau(q)$ consists of two linear segments with slopes of α_1 and α_2. In practice, the region where these two segments join becomes rounded for finite resolution λ, and all values of $\alpha = d\tau(q)/dq$ in the range $\alpha_1 < \alpha < \alpha_2$ are generated.

3. Multifractal measures can also be characterized by the function $\mathscr{C}(\alpha)$ [1313] defined as $\mathscr{C}(\alpha) = (\log P(\alpha)/\log \lambda)$, where $P(\alpha)$ is the probability that a region of size ϵ will have a Holder exponent of α. The functions $\mathscr{C}(\alpha)$ and $D_f(\alpha)$ are related by $\mathscr{C}(\alpha) = D_f(\alpha) - d$, where d is the Euclidean dimensionality of the embedding space.

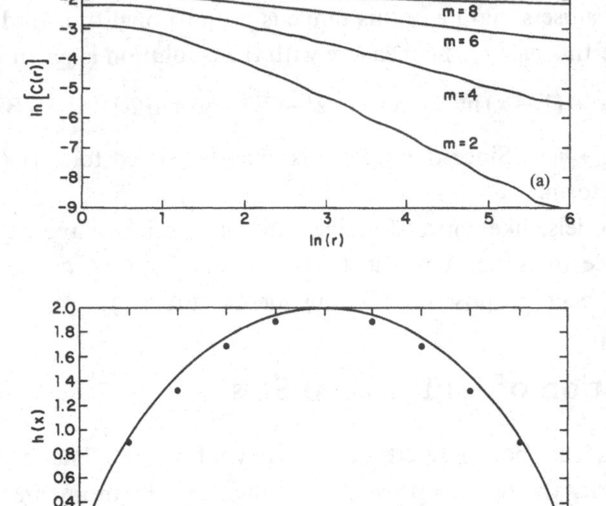

Figure B.4 Part (a) shows the two-point density-density correlation functions for tenth-generation premultifractal subsets like those shown in figure B.3. The correlation functions for the $m = 2, 4, 6, 8$ and 10 subsets shown in this figure were obtained from a large number of realizations of the random multiplicative process. Part (b) compares the fractal dimensionalities obtained from these correlation functions with equation B.7.

The distribution $N(\alpha)$ can be obtained from $N(\log p)$ using the relationship $N(\alpha) = N(-\log p)d(-\log p)/d\alpha$. It follows from equation B.9 that $d(\log p)/d\alpha = \log\lambda$ so that $N(\alpha) = N(-\log p)\log\lambda$. Combining this relationship with equation B.10 it can be seen that $\log[N(-\log p)\log\lambda]/\log\lambda = N(\alpha)/\log\lambda = D_f(\alpha)$. Consequently, the distribution $N(\log p, \epsilon, L)$ can be described in terms of the *multifractal* scaling form

$$\log[N(-\log p)\log\lambda]/\log\lambda = N(\alpha)/\log\lambda) = f'(-\log p/\log\lambda), \qquad \text{(B.11)}$$

or

$$\log[N(\log p)\log\lambda]/\log\lambda = f(\log p/\log\lambda), \qquad \text{(B.12)}$$

where the scaling function $f(x)$ in equation B.12 is the function $D_f(-\alpha)$. Consequently, an inhomogeneous measure can be characterized by plotting $\log N(\log p)/\log\lambda$ as a function of $\log p/\log\lambda$, for different values of ϵ, L and $\lambda = L/\epsilon$. If a good data collapse is obtained, there is *prima facie* evidence that the distribution of the measure is multifractal and the function $D_f(\alpha)$ is obtained directly as the scaling function $f(x)$, $(D_f(\alpha) = f(-x))$. Equation B.12 implies that the number of regions of size ϵ with normalized measures in the range λ^{α} to $\lambda^{\alpha+\delta\alpha}$ is $\lambda^{f(\alpha)}\delta\alpha$, to within logarithmic corrections. If the logarithmic corrections are ignored then

$$f(-\log p/\log\lambda) = f(\alpha) = D_f(\alpha) = \log[N(-\log p)]/\log\lambda. \qquad \text{(B.13)}$$

This approach is illustrated in figure B.5, for the distribution of the growth probability measure on clusters generated using the screened growth model (chapter 3). In these simulations [462, 544], the growth probabilities in all of the unoccupied perimeter sites were recorded for 32 clusters grown with a fractal dimensionality of 1.75 at 15 sizes in the range $20 \leq M \leq 15\,000$ sites. Figure B.5(a) shows the distribution of the growth probabilities $N(\ln p)$, where $N(\ln p)\delta \ln p$ is the number of sites for which the log of the growth probability lies in the range $\ln p$ to $\ln p + \delta \ln p$. Figure B.5(b) shows how the distributions $N(\ln p, \epsilon, L)$ were scaled. In this and other growth models, it is convenient to vary the resolution λ by keeping ϵ fixed at one lattice unit (the growth probabilities are measured in each of the unoccupied perimeter sites) and changing L by examining clusters of different sizes M, as they grow. In this case $N(\ln p)$ was scaled using the scaling form

$$\log[< N(\log p) > \log M] = \log M f'(\log p / \log M), \tag{B.14}$$

where $< \ldots >$ indicates averaging over all of the clusters. The scaling forms given in equations B.12 and B.14 are very closely related, but they become equivalent only in the $M \to \infty$ ($\lambda \to \infty$) limit. Since $M \sim L^D$, the scaling forms $D_f(\alpha)$ and $f'(x)$ are related by

$$D_f(\alpha) = Df'(-Dx). \tag{B.15}$$

In practice, there may be important corrections to scaling, which are related to the lacunarities of the fractal subsets and finite-size effects. At the time of writing, these corrections are not well understood, and an assessment of the reliability of the multifractal scaling function $f'(x)$ or $D_f(\alpha)$ is based solely on the quality of the data collapse. Since the minimum growth probability decreases with increasing resolution according to the algebraic relationship

$$p_{min} \sim M^{-\gamma_{-\infty}} \sim \lambda^{-D\gamma_{-\infty}}, \tag{B.16}$$

the support for the measure D_X should have the same fractal dimensionality as that of the unoccupied perimeter of the cluster, which is also the fractal dimensionality of the cluster, in this case. Consequently, the maximum in $D_f(\alpha)$ is expected at $D_f(\alpha)_{max} = D$ and the maximum in the scaling function f' in equation B.14 is expected at $f'_{max} = 1$. Figure B.5(b) shows the data collapse obtained for clusters of eight sizes in the range $1000 \leq M \leq 15\,000$ sites.

In practice, it is often necessary to work at finite resolutions λ. In this case, the *statistical* uncertainties can be reduced by analyzing the distribution of the measure over a large ensemble of samples with the same system size L and level of subdivision. The distribution $N(\log p)$ can then be averaged over all of the samples and $N(\log p)$ can be extended to extreme values of p that are not present in a typical member of the ensemble. This is illustrated in figure B.6, which shows the distribution of normalized interaction forces $N(\ln F^n_{\|})$ between the particles in random fractal aggregates and a quiescent fluid through which the aggregates are moving [1314, 1315]. In this example, the "parallel" components $F_{\|}$ of the

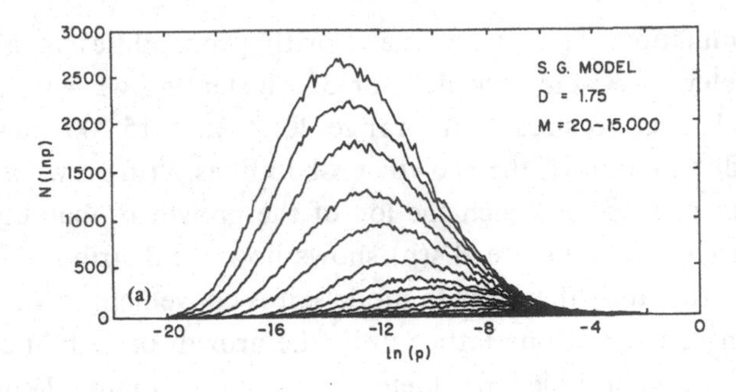

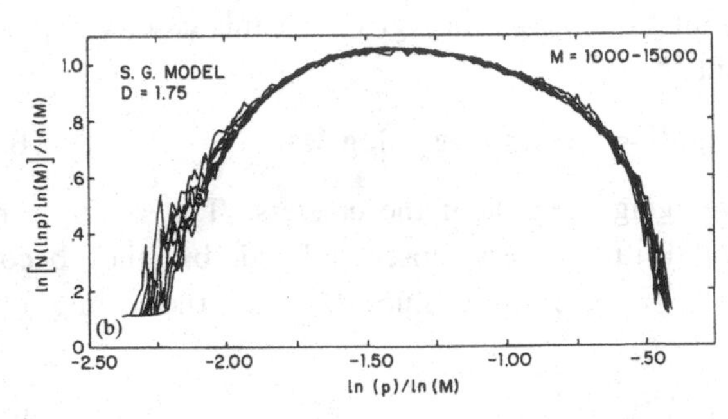

Figure B.5 Multifractal scaling of the distribution of growth probabilities on the external perimeter of screened growth clusters with a fractal dimensionality of $D = 1.75$. Part (a) shows the distribution $N(\ln p)$ averaged over 32 clusters, after they have grown to 15 sizes in the range $20 \leq M \leq 15\,000$ sites, and part (b) shows the multifractal scaling for the eight largest cluster sizes with $1000 \leq M \leq 15\,000$, using the scaling form given in equation B.14.

forces in the direction of the velocity of the aggregate, relative to the fluid at infinity, were used to define a measure on the fractal. The aggregates, with a dimensionality of $D \simeq 1.78$, were generated using a 3-dimensional, off-lattice, diffusion-limited, cluster–cluster aggregation model [58]. Figure B.6(a) shows the distribution of the normalized parallel force components $N(\ln F_{\parallel}^{n})$, averaged over a large number of clusters, for four cluster sizes. Figure B.6(b) shows how these force component distributions can be scaled using the scaling form

$$\ln[N(\ln p)\ln M] = \ln M f'(\ln p/\ln M), \tag{B.17}$$

where M is the cluster size and $p = F_{\parallel}^{n}$. For sufficiently small forces, the multifractal scaling function $D_f(\alpha) = Df'(-\alpha/D)$ is negative. These negative fractal dimensionalities indicate how rare the smallest forces, corresponding to the largest values of α, become as the cluster size or resolution increases.

The negative part of the multifractal distribution $D_f(\alpha)$ reflects the statistics of rare events. The rare events may correspond to samples in which the concentration of the measure becomes very large or very small, so that $D_f(\alpha)$ may become negative at small or large values of α. The appearance of negative parts of $D_f(\alpha)$ seems quite natural if the histogram approach is used and $N(\ln p)$ is averaged over many samples. Similarly, the negative parts of $D_f(\alpha)$ can be expected to appear if a fractal measure is subdivided into many regions and the *normalized* histogram is calculated for each region and then averaged [1316]. Negative parts do not appear in the $D_f(\alpha)$ curve obtained from the growth probabilities for screened growth model clusters because there are no perimeter sites with anomalously low or high growth probabilities.

In those cases where $D_f(\alpha)$ is not a continuous, convex function, these features will be seen in the histograms, and a spurious convex envelope will not be generated. The scaling of the histograms $N(\log p)$ is a more conservative procedure for the characterization of multifractals than the calculation of $D_f(\alpha)$ from the partition function using the Legendre transformation approach. This histogram scaling approach requires more effort, but the results are more reliable and informative.

A "direct" method for determining $D_f(\alpha)$ has been proposed and tested by Chhabra and Jensen [1317]. This method is based on the family of normalized measures $\mu_i(q,\epsilon)$ defined as

$$\mu_i(q,\epsilon) = [p_i(\epsilon)]^q / \sum_j [p_j(\epsilon)]^q. \tag{B.18}$$

The fractal dimensionality of the support of $\mu(q)$ is given by

$$D_f(q) = f(q) = \lim_{\epsilon\to 0}\left[\sum_i \mu_i(q,\epsilon)\log(\mu_i(q,\epsilon))\right] / \log\epsilon, \tag{B.19}$$

and the average value of the strength of the corresponding singularity is given by

$$\alpha(q) = \lim_{\epsilon\to 0}\left[\sum_i \mu_i(q,\epsilon)\log(p_i(\epsilon))\right] / \log\epsilon. \tag{B.20}$$

This allows the multifractal function $D_f(\alpha)$ to be determined directly from the data, without the Legendre transformation of $\tau(q)$. However, this approach does not avoid the generation of a convex envelope in those cases where the $D_f(\alpha)$ "curve" consists of discrete points corresponding to measures consisting of subsets with discrete values of α.

Multifractals can also be characterized by the correlation functions C_{pq} [1318] defined as

$$C_{pq}(r) =< (\rho_\mu(\mathbf{r}_\circ))^p(\rho_\mu(\mathbf{r}_\circ+\mathbf{r}))^q >_{|\mathbf{r}|=r}, \tag{B.21}$$

where $(\rho_\mu(\mathbf{r}_\circ))$ is the density of the measure at position $\mathbf{r}$. These correlation functions have the form

$$C_{pq}(r) \sim r^{-\alpha_{pq}}. \tag{B.22}$$

There is no general relationship between the scaling properties of the probability distributions associated with multifractals expressed by the exponents D_q, $\tau(q)$ or $D_f(\alpha)$ and the spatial distribution of the fractal subsets or the correlations between them. Consequently, there is no general relationship between α_{pq} and the multifractal scaling exponents.

Cates and Deutsch [1319] studied the related correlation functions

$$\overline{C}_{pq}(r) = [< (\rho_\mu(\mathbf{r}_\circ))^p(\rho_\mu(\mathbf{r}_\circ+\mathbf{r}))^q >_X]_{|\mathbf{r}|=r}, \tag{B.23}$$

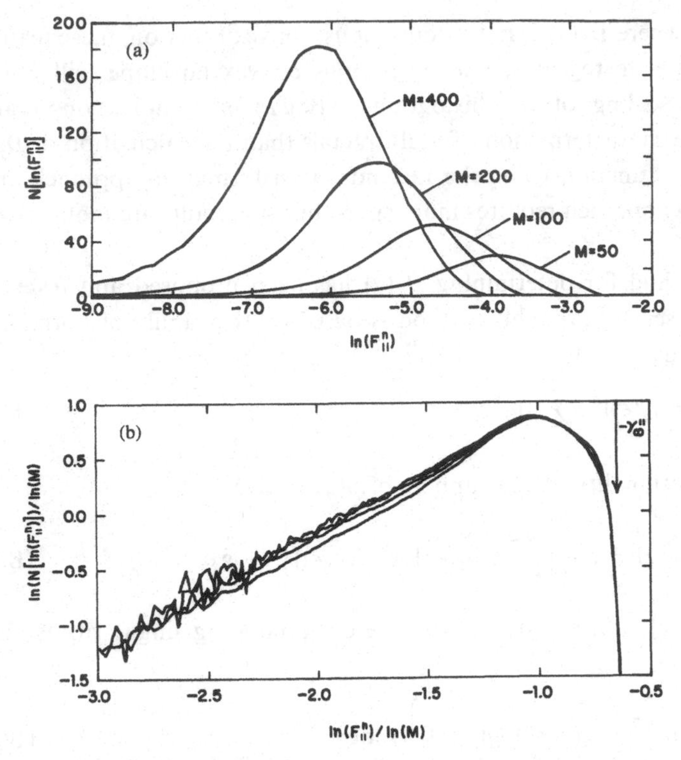

Figure B.6 The distribution of forces on fractal aggregates with a fractal dimensionality of $D \simeq 1.78$ moving through a quiescent fluid. Part (a) shows the distribution of force components parallel to the direction of motion of the cluster with respect to the fluid, averaged over a large number of clusters with sizes M of 50, 100, 200 and 400 particles. The forces have been normalized so that $\sum_{i=1}^{i=M} F_i^n = 1$ for each cluster. Part (b) shows the multifractal scaling of these force distributions, and illustrates the extension of the scaling function $f'(x)$ to negative values resulting from the averaging of very small forces over the ensemble of clusters.

where $< \ldots >_X$ implies averaging over only those regions that are part of the support of the measure, with a fractal dimensionality of $D_\circ = D_X$. The correlation functions $C_{pq}(r)$ and $\overline{C}_{pq}(r)$ are related by

$$\overline{C}_{pq}(r) = C_{pq}(r)[(\epsilon/L)^{(D_X-d)}(\epsilon/r)^{(D_X-d)}], \tag{B.24}$$

where d is the fractal dimensionality of the embedding space. Consequently, the correlation functions $\overline{C}_{pq}(r)$ have the form

$$\overline{C}_{pq}(r) \sim r^{-\overline{\alpha}_{pq}}, \tag{B.25}$$

where $\overline{\alpha}_{pq} = \alpha_{pq} + D_X - d$. Based on a random multiplicative model, Cates and Deutsch [1319] suggested that

$$\overline{\alpha}_{pq} = D_X + \tau(q) + \tau(p) - \tau(p+q) \tag{B.26}$$

and that the amplitude of the correlation function $\overline{c}_{pq}$ is given by $\overline{A}_{pq} \simeq \lambda^{-(D_x+\tau(pq))}$. For uniform measures $\tau_q = (q-1)D_x$ and $\overline{\alpha}_{pq} = 0$. However, Meneveau and Chhabra [1320] and Lee and Halsey [1321] have shown that this relationship is not generally correct.

B.3 Applications to Non-Equilibrium Growth

The nature of the active zone, the region in which growth is concentrated, is the focus of a broad range of research in the area of fractal growth. It is widely believed, with good reason, that a better understanding of the active zone will lead to a better understanding of the growth dynamics and the nature of the structure left behind when growth has ceased or the growth front has passed. The active zone, in many fractal growth processes, can be described in terms of a multifractal model. This is illustrated by analysis of the growth probability distribution for the screened growth model that was discussed earlier in this appendix. One of the issues that arises immediately, in the context of the interaction of fractal structures with their surroundings and pattern formation via surface growth or the propagation of an active zone, is the effective size of the surfaces. For example, in the growth of DLA clusters, the growth probabilities at the tips of the most exposed arms of the clusters is very much larger than the growth probabilities elsewhere in the cluster, and most of the perimeter of the cluster plays essentially no role in the growth process. Thus it is apparent that the size of the surface that is involved in the growth of the cluster is much smaller than that of the entire surface.

Although little is known about the harmonic measure for DLA, there are two useful checks on simulation results: the Makarov [663] result $D_1 = 1$ for 2-dimensional DLA and the relationship $D = 2D_{(q=3)} + 2 - d$ found by Halsey [664]. Equation 2.234, with $D_1 = 1$, indicates that a straight line passing through the origin should be tangent to $D_f(\alpha)$ at the point $(1,1)$.

If the surface of the cluster is divided into regions of size ϵ and the growth probabilities $\{p_i\}$ associated with these regions are determined, then the size of the surface can be described in terms of these probabilities. For a lattice model, it is convenient to use the unoccupied perimeter sites as the regions of size ϵ, and in off-lattice models the probability of attachment to the ith particle in the cluster can be used as $p_i(\epsilon)$. If the growth probability is distributed uniformly, $p_i = 1/s_p$, where s_p is the size of the entire perimeter or the number of regions of size ϵ. Under these conditions,

$$s_p = \sum_{i=1}^{i=s_p} p_i^j / \sum_{i=1}^{i=s_p} p_i^{j+1}, \qquad \text{(B.27)}$$

and

$$s_p = \left[1/\sum_{i=1}^{i=s_p} p_i^{j+1}\right]^{1/j} . \quad \text{(B.28)}$$

For non-uniform distributions, equation B.28 leads to an infinite family of surface sizes[4] m_j given by

$$m_j = \left[1/\sum_{i=1}^{i=M} p_i^{j+1}\right]^{1/j} = (Z_{q=j+1})^{-1/j} \sim (L/\epsilon)^{D_{j+1}}, \quad \text{(B.29)}$$

where Z is the partition sum, defined in equation 2.232 in chapter 2. Here, the surface areas m_j are measured in terms of the area $a(\epsilon)$.

4. The "definition" of this family of surfaces can be motivated by considering processes in which a fixed amount Q of some field or substance is distributed over a complex surface. If the rate of a process, such as a chemical reaction, in a region of size ϵ is proportional to a power q of the *local* surface concentration or density, then the total rate of the process is given by

$$\begin{aligned} R_q &= k_q Q^q a(\epsilon)^{-(q-1)} \sum_i p_i(\epsilon)^q \\ &= k_q Q^q a(\epsilon)^{-(q-1)} Z_q \sim a(\epsilon)^{-(q-1)} (L/\epsilon)^{-\tau(q)}, \end{aligned} \quad \text{(B.30)}$$

where k_q is a rate constant and $a(\epsilon)$ is the amount of surface in a region of size ϵ. For a uniform (hyperspherical) surface

$$R_q \sim k_q Q^q A^{-(q-1)}, \quad \text{(B.31)}$$

where A is the surface area. Equations B.30 and B.31 indicate that the effective area A_q for the qth order process is given by $Z_q \sim A_q^{-(q-1)}$, or

$$A_q \sim a(\epsilon) Z_q^{(-1/(q-1))} \sim a(\epsilon)(L/\epsilon)^{D_q} \sim a(\epsilon)\left[\sum_i p_i(\epsilon)^q\right]^{-1/(q-1)} . \quad \text{(B.32)}$$

For example, if most of the surface sites on the support with a fractal dimensionality of D_X carry more or less the same measure, but a subset with a fractal dimensionality of D_s carries a measure which scales as $(L/\epsilon)^{-\alpha}$, then

$$\sum_i p_i^q = a(\epsilon)(L/\epsilon)^{D_X(1-q)} + b(L/\epsilon)^{D_s - q\alpha}. \quad \text{(B.33)}$$

In the limit $L/\epsilon \to \infty$ and $\alpha < D$, the term on the right-hand side of equation B.33 dominates for large q and

$$D_q = (q\alpha - D_s)/(q-1). \quad \text{(B.34)}$$

This implies that $\tau_q = q\alpha - D_s$, and there is a constant gap between the values of τ_q for successive integer values of q. For smaller values of q, where $D_X(1-q)$ is larger than $D_s - q\alpha$,

$$D_q = D_X \quad \text{(B.35)}$$

and $\tau_q = (q-1)D_X$. Consequently, τ_q consists of two lines meeting at the point where $D_X(1-q) = D_s - q\alpha$ ($q = (D_X - D_s)/(D_X - \alpha)$) and $\tau_q = D_X(\alpha - D_s)/(D_X - \alpha)$). It follows from equation B.32 that for small q ($D_X(1-q) > D_s - q\alpha$) the support dominates and the effective area is given by $A \sim a(\epsilon)L^{D_X}$, while for large q $A \sim a(\epsilon)(L/\epsilon)^{(D_s - q\alpha)/(1-q)}$ and, for $q \to \infty$, $A \sim a(\epsilon)(L/\epsilon)^{\alpha}$.

In general, a set of probabilities $\{p_i\}$ can be defined for any distribution on the surface and a family of effective surface sizes $m_j(x)$ can be defined for each process or distribution. The growth probability distribution for screened growth model clusters, DLA clusters and other growth models, as well as the distribution of the random walk hitting probabilities or harmonic measure for self-similar fractals, can be described in terms of the dependence of the moments M_j on the resolution λ. In particular, for growing clusters of size M, the moments $m_j(\epsilon, M)$ grow algebraically with the cluster sizes M, so that

$$m_j(\epsilon, M) \sim M^{\gamma_j}, \qquad \text{(B.36)}$$

if the size of the subdivision ϵ is fixed at one lattice unit or particle diameter. It follows immediately from the definition of the moments m_j that

$$\gamma_j = D_{q=j+1}/D_X \qquad \text{(B.37)}$$

and that the exponents D_q and τ_q (and hence $D_f(\alpha)$) can be calculated from the scaling of the moments m_j with the cluster size. Figure B.7(a) shows the algebraic dependence of the moments m_j of the growth probability distribution on the cluster size M for screened growth model clusters with $D = 1.25$. In the limits $j \to \infty$ and $j \to -\infty$, the moments m_j are completely dominated by the largest and smallest growth probabilities (p_{max} and p_{min}), respectively. In these limits,

$$p_{max} \sim M^{-\gamma_\infty} \qquad \text{(B.38)}$$

and

$$p_{min} \sim M^{-\gamma_{-\infty}}. \qquad \text{(B.39)}$$

For the screened growth model, the dependence of the maximum and minimum growth probabilities can be measured directly. The values obtained in this manner are consistent with the idea that $\alpha_{q\to\infty} = \alpha_{min} = \gamma_\infty D_x$ and $\alpha_{q\to-\infty} = \alpha_{max} = \gamma_{-\infty} D_x$. For the DLA model, where the most exposed parts of the cluster have the highest growth probabilities and growth at the most exposed site increases the cluster radius by an amount comparable to the particle size, $p_{max} \sim M^{(1/D)-1}$ [662] or

$$\gamma_\infty = 1 - (1/D). \qquad \text{(B.40)}$$

Direct measurements of the maximum growth probability as a function of the cluster size indicate that equation B.40 does not hold for the screened growth model [462].

The situation for DLA is very different. While that part of the measure corresponding to the larger growth probabilities (the small α part of the measure for a multifractal measure) appears to be well described by multifractal scaling, the smaller growth probabilities decrease faster than any power of the cluster

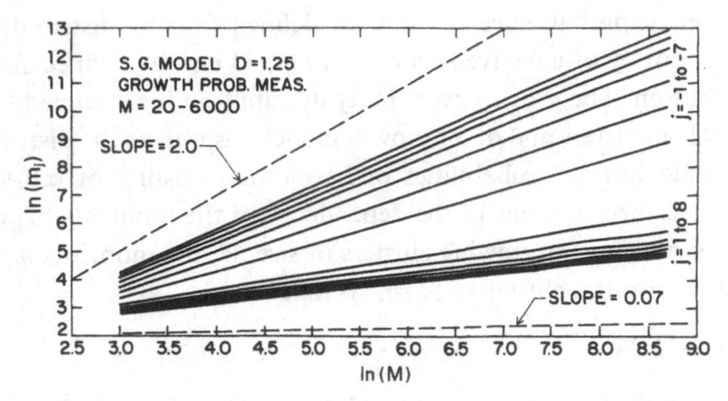

Figure B.7 The dependence of the moments of the growth probability distribution m_j on the cluster size (M) for screened growth model clusters with a fractal dimensionality of 1.25. The lower dashed line has the slope expected for the $j = \infty$ moment. A direct measurement of the dependence of the maximum growth probability on the cluster size indicated that $\gamma_\infty = 0.073 \pm 0.006$.

size, with increasing cluster size. The growth process is dominated by those regions that have the largest growth rates or probabilities. The scaling properties of that part of the growth probability measure corresponding to large growth probabilities (small values of α or the "left side" of $D_f(\alpha)$) can be measured by "probing" the surface of a DLA cluster of size M with a large number of random walk trajectories, starting from outside the region occupied by the cluster. The random walks are terminated when they first contact the cluster, and a record is kept of the number of times each unoccupied perimeter site or particle in the cluster is reached. This provides a statistical sampling of the growth probability measure. The efficiency of this method can be improved using the techniques described in chapter 3, section 3.4.3, for the generation of very large DLA clusters. This method allows the growth probability distribution to be explored for quite large DLA clusters (much larger than those for which the growth probability distribution can be measured by numerically solving the Laplace equation). The harmonic measure for any geometry can be estimated in the same way [1318, 1322].

Using this approach, Halsey *et al.* [1323] measured the generalized dimensionalities D_q with $2 \leq q \leq 8$ for the harmonic measure on 50000 particle, off-lattice DLA clusters. The results of these measurements were used to support the idea [1323] that the harmonic measure or growth probability measure for DLA clusters could be described by a multifractal scaling model. Halsey *et al.* showed that the measured values of the D_q were consistent with a very simple model for the growth probability distribution consisting of singularities of "strength" $\alpha \simeq 0.705$ ($\alpha = D - 1$) lying on a set with a fractal dimensionality of $f \simeq 0.42$

that sits on a non-singular background with a fractal dimensionality of 1. This model predicts that

$$D_q = min[1, (q\alpha - f)/(q - 1)] \tag{B.41}$$

(equations B.34 and B.35). A hierarchical wedge model for the growth of 2-dimensional DLA clusters was proposed and a relationship between the fractal dimensionality f of the wedge tips and the strength of the singularity at the wedge tips was obtained. The point $\alpha = 0.704, f = 0.42$ lies close to the locus of the points (f, α) obtained from this simple model. The value of 0.704 obtained for α implies that the maximum growth probability scales as $P_{max} \sim L^{-\alpha}$ and that the size of the cluster grows with increasing number of sites or number of particles, like a self-similar fractal with a fractal dimensionality of $D = 1 + \alpha = 1.704$, in good agreement with the value of $D \simeq 1.715$ obtained from 2-dimensional DLA simulations.

The random walk probe method was also used to measure $D_f(\alpha)$ using the contact probability histogram [1322]. A $D_f(\alpha)$ curve with a minimum α of $\alpha_{min} \simeq 0.4D$ ($\alpha_{min} \simeq D[1 - (1/D)] \simeq D - 1$) and a maximum value of $D_X \simeq D$ was obtained [1324] for 2-dimensional off-lattice DLA clusters containing up to 50000 particles. However, the "right side" of the $D_f(\alpha)$ curve cannot be explored at all using this method, even for small clusters. For this purpose the growth probabilities must be calculated using relaxation [1325], exact enumeration of random walks [396], Green's function [1326] or other accurate methods. Even with these much more accurate methods, the shapes reported for the right, low growth probability part of $D_f(\alpha)$ vary substantially [1325, 1326, 1327, 1328]. This may be a result of large finite-size corrections, numerical inaccuracies for the smallest growth probabilities, the fact that the smallest growth probabilities do not scale as a power of the cluster size and, in some cases, lattice anisotropy effects.

The screened parts of the cluster become even more strongly screened as the cluster grows. Consequently, the distribution of the smallest growth probabilities is characteristic of the structure of the cluster, but provides no indication of its future growth. Despite this, considerable effort has been devoted to developing a better description of the evolution of the smallest growth probabilities. This effort appears to be misguided from a DLA point of view. However, it may be justified in terms of developing a better understanding of physico-chemical processes taking place on the surfaces of fractals. The "high density" parts of the harmonic measure will be the most important, in most cases. However, there are some processes, such as the transport of material from a fractal aggregate into a flow field [1329], that may depend on the "low density" parts of related multifractal measures [1314]. An additional complication comes about as a result of the complex, preasymptotic scaling properties of DLA itself (chapter

3, section 3.4.5). A study of more simple, statistically self-similar fractals might be more rewarding.

Although some studies suggest that the minimum growth probability p_{min} for 2-dimensional DLA does scale as a power of the cluster size M ($p_{min} \sim M^{-\alpha_{max}/D} \sim M^{-\gamma_{-\infty}}$) [1330, 1331], it appears that p_{min} decreases faster than any power of M as M increases. For 2-dimensional DLA, the minimum growth probability will occur in a configuration that consists of a long, narrow, spiral channel. For this configuration p_{min} decays as $p_{min} \sim e^{-kM}$. However, this configuration becomes so rare in the $M \to \infty$ limit that it does not contribute, even to the large negative moments.

The non-power-law behavior is closely related to a breakdown in the multifractal scaling for $q < q_c$. For $q > q_c$, $Z_q = \sum_i p_i^q$ scales as $(L/\epsilon)^{-\tau(q)}$, but for $q < q_c$ this scaling breaks down because of the dominant effect of the smallest growth probability on Z_q. Because of the analogy between Z_q and a partition function in statistical mechanics [69, 1332, 1333], the change in behavior at $q = q_c$ is often called a "phase transition" [1334]. In this analogy, $\tau(q)$ plays the role of the free energy and q corresponds to $1/T$, where T is the temperature.

Lee and Stanley [1334] have calculated the growth probabilities for *all* of the DLA configurations in boxes of size $L \times L$ containing $L^2 - L$ bonds, with periodic boundary directions in the lateral direction. This calculation was based on a real-space renormalization model for DLA [1335, 1336] and symmetry was used to reduce the number of configurations to a manageable level (from $\simeq 3 \times 10^{17}$ to 9361, for $L = 5$). The critical value of q ($q_c \simeq -1$) was taken as the point on the $\tau(q)$ curves at which $\tau(q)$ became system size dependent. The partition function

$$Z_q(L) = \sum_j C_j \sum_i (p_{i,j})^q \tag{B.42}$$

was used to calculate the exponents τ_q. Here the index j identifies the jth configuration, C_j is its weight and $p_{i,j}$ is the growth probability in the ith perimeter site of the jth configuration. The smallest growth probability was found to decrease as $p_{min} \sim e^{-kL^2}$. For a DLA cluster, this result would be expected for a spiral configuration. For simulations carried out using different boundary conditions an equivalent configuration containing a single long channel will give a minimum growth probability of the same form [1337]. However, as Lee *et al.* [1337] emphasized, this is not a typical DLA configuration.

Havlin *et al.* [1338] obtained a value of $q_c \simeq -0.4$ for the critical value of q at which multifractal scaling breaks down. This estimate was based on analysis of a single 1000-site DLA cluster. Using clusters that were intermediate in both size and number between those of Lee *et al.* and Havlin *et al.*, Wolf [1327] obtained a value of $\simeq 0$ for q_c using a very accurate numerical method. Wolf later confirmed this result using somewhat larger clusters containing up to 220 particles [1339].

Based on simplified models for the structure of DLA clusters, an exponentially decreasing minimum growth probability of the form

$$p_{min} \sim exp(-s^a) \tag{B.43}$$

has been proposed [1340]. Schwarzer *et al.* [1341] have proposed that

$$\ln p_{min} \sim -(\ln s)^b \tag{B.44}$$

with $b \simeq 2$, based on numerical studies using triangular lattice and square lattice clusters containing up to 2500 sites. Evertsz *et al.* [1342] have shown how these and other forms for $\ln p_{min}(s)$ can be estimated for a wide range of "fjord" models using a simple geometric construction procedure that can be applied to any shape. In the work of Schwarzer *et al.* [1341], the growth probabilities were obtained using an exact enumeration method [395, 396] (chapter 3, section 3.3.4) for the random walks. Schwarzer *et al.* also proposed a non-random, hierarchical model for the structure of the trapping fjords in DLA. This model gives the form of equation B.44 for $p_{min}(s)$, with $b = 2$, or $-\ln(p_{min}(L)) \sim (\ln L)^2$, where L is a length characteristic of the overall cluster size. Simulations using off-lattice clusters with sizes ranging from 753 to 20 000 particles were consistent with equation B.44, and a value of 2.15 ± 0.22 was obtained for b [1343]. In this study, the off-lattice clusters were discretized onto a square lattice and the lattice Laplace equation was solved numerically to obtain the values of the scalar Laplacian field in all of the unoccupied perimeter sites.

Wolf [1327] also found that the dependence of the minimum growth probability on cluster size could be better represented by equation B.44, than either an exponential or power law decay, for 500 clusters grown to a size of 250 unoccupied perimeter lattice sites. The data could be fitted quite well with $b = 2.0$. Equation B.44 ($p_{min} \sim e^{-(\ln s)^b}$) suggests a multiplicative, hierarchical screening model.

Mandelbrot *et al.* [1344] have shown how a "left-sided multifractal measure" with a minimum probability that decreases as a stretched exponential form of the resolution λ and a transition at $q = 0$ can be constructed using a hierarchical (self-similar) deterministic model. In the case of a measure distributed on a unit interval, the generator for the multifractal redistributes the measure among an infinite number of sub-intervals with an infinite number of lengths reduction factors and intensity multipliers. It appears that other forms that have been proposed for the decrease in the minimum growth probability with increasing cluster size in DLA, such as equation B.44, are also consistent with a left-sided, self-similar, multifractal measure [1344].

On the other hand, Ball *et al.* have developed theoretical arguments [1345] and presented numerical data [1346] from relatively large, 10^5-particle, off-lattice DLA clusters that indicate that $D_f(\alpha)$ has a linear form for large α with a finite, but undetermined, upper bound. This implies that the minimum growth probability decreases with a power of the cluster size.

If off-lattice clusters are not discretized onto a lattice, the growth probability measure can be obtained by solving the continuum Laplace equation with the boundary condition $\Phi = 0$ on the DLA cluster that has been fattened by one particle radius and $\Phi = 1$ at infinity, where Φ is the Laplacian field. As Evertsz *et al.* have shown [1347], extremely small, non-zero, growth probabilities can be found in very small clusters. In numerical studies, the Laplace equation can be solved using a lattice with grid elements of size ϵ, smaller than the particle size. Evertsz *et al.* found that the distribution $P(\alpha)$ of Holder exponents[5] α obtained for small clusters had a power law form for large α, with an effective exponent that depended on the cut-off length ϵ. This power law distribution of Holder exponents corresponds to a $D_f(\alpha)$ curve that decreases linearly with increasing α on the right-hand, low density side. This is similar to the result found by Ball and Blumenfeld [1345] using clusters of 10^5 sites.

The distribution of growth probabilities for $(2+1)$-dimensional diffusion limited deposition [462] and 3-dimensional DLA [1348] have been estimated using the random walk probe method. More accurate studies using 3-dimensional, off-lattice DLA clusters containing up to 15 000 particles indicate that there is no "phase transition" in this case [1349, 1350]. The entire distribution of growth probabilities exhibits multifractal scaling, and the minimum growth probability decreases as a power of the cluster size. These results were obtained by discretizing 3-dimensional off-lattice DLA clusters onto a cubic lattice and solving the discretized Laplace equation with DLA boundary conditions.

When a new unoccupied site is added to the active zone of a growing DLA cluster, it will typically have a relatively large growth probability, corresponding to a point on the left-hand, high growth probability part of the $D_f(\alpha)$ curve. As the cluster grows, the point corresponding to a particular particle will jump from place to place on the $D_f(\alpha)$ curve. Most jumps will be very small and will move the point to a position further to the right on the $D_f(\alpha)$ curve. Occasionally a nearby site will be filled or a site near the mouth of a fjord that contains the particle will be filled and a large jump will take place. Ball and Blunt [1351] have proposed a typical scenario for fractal growth with multifractal growth probability distributions. According to this scenario, a new active zone site is "born" with a screening exponent α of $\alpha_\circ$. On average the value of α associated with this site remains essentially constant until on the order of $(L/\epsilon)^{\alpha_\circ}$ additional particles have been added. This is the time that is required for an additional

5. The Holder exponent was defined as $\alpha = -\ln[\mu_i(\epsilon)/\epsilon]$ (or $\alpha = -\log_M[\mu_i(\epsilon)/\epsilon]$ for the normalized Holder exponent), where $\mu_i(\epsilon)$ is the harmonic measure in the i region of size ϵ on the perimeter of a cluster of M particles.

particle to be added at a position near enough to screen the active zone site. After a number of new particles on the order of $(L/\epsilon)^{\alpha_o}$ have been added, the point on the $D_f(\alpha)$ curve will have moved to the "screened", right-hand side of the $D_f(\alpha)$ curve.

Ball and Blunt found that the most exposed positions far on the left side of the $D_f(\alpha)$ curve very often became the most strongly screened. This behavior is dramatically illustrated by the visualization on the cover of the November 8, 1990 issue of *Nature* containing an article by Mandelbrot and Evertsz [342] on the growth probability distribution in 2-dimensional DLA. The figure shows a concentration of the small growth probabilities onto regions near to the main branches of the growing cluster. These main branches were the most rapidly growing tips at earlier stages in the growth. In practice, this process may proceed via several smaller jumps, but a large jump is often dominant. At very long times, the point moves near to the peak of the $D_f(\alpha)$ curve, where almost all of the active zone sites lie. This simple model for the screening dynamics of fractal growth leads to the prediction that the number of sites from which growth will ever occur is proportional to $\lambda^{D_f(D)}$, and $D_f(D)$ $(D_f(\alpha = D))$ is determined by the $q = 1/2$ moment of the distribution of growth probabilities. Meakin and Witten [836] found that the saturated size of the "old growth/new growth interface" in small, square lattice DLA clusters grew essentially linearly with λ. For DLA, $D_f(D)$ lies near the fractal dimensionality of the support D_X, and D_X is significantly smaller than the fractal dimensionality of the cluster. This means that the idea that growth will occur at $\sim \lambda^{D_f(D)}$ sites is not completely inconsistent with the simulation results of Meakin and Witten. A modification of this scenario is needed for processes such as 2-dimensional DLA, for which the growth distribution can be described by a left-sided multifractal $D_f(\alpha)$ [1352]. In this case, strongly screened sites "disappear" from the active zone.

B.3.1 Quenched and Annealed Averages

In cases where data are collected from a large number of samples or regions of size L, this data can be averaged in several ways. One approach is to normalized the measure $\{\mu_i(\epsilon)\}$ for each sample individually, and to calculate the "quenched average" of the partition function $Z_q(j)$, defined as

$$< \log Z_q(\epsilon) >_q = (1/N) \sum_{j=1}^{j=N} \log \left[\sum_i (\mu^n_{i,j}\epsilon)^q \right] = (1/N) \sum_{j=1}^{j=N} \log Z_q(j), \quad \text{(B.45)}$$

where $\mu^n_{i,j}$ is the normalized measure in the ith region of size ϵ in the jth region of the sample and there are N samples. The normalized measure is given by

$$\mu^n_i(\epsilon) = \mu_i(\epsilon) / \sum_{i=1}^{i=M} \mu_i(\epsilon), \quad \text{(B.46)}$$

where M is the number of regions of size ϵ.

An alternative is to calculate the annealed averages

$$\log < Z_q(\epsilon) >_a = \log \left\{ (1/N) \sum_{j=1}^{j=N} \left[\sum_i \mu_{i,j}^n(\epsilon)^q \right] \right\}$$
$$= \log \left[(1/N) \sum_{j=1}^{j=N} Z_q(j) \right]. \quad \text{(B.47)}$$

The quenched and annealed averages correspond to averaging the logs of the moments or taking the log of the average. These different averaging procedures can lead to widely different results.

B.3.2 Mass Multifractals

Multifractal scaling has also been used to characterize the spatial distribution of mass in inhomogeneous fractal structures that have a uniform distribution of density or measure over the occupied regions or support in the embedding space [1353, 1354]. The distribution of mass after partitioning into regions of size ℓ can be described in terms of a partition function, like that defined in equation 2.229,

$$Z_q(\ell) = \sum_i (M_i(\ell)/M)^q, \quad \text{(B.48)}$$

where $M_i(\ell)$ is the mass in the ith region of size ℓ and $M = \sum M_i(\ell)$ is the total mass. In practice, the mass distribution or pattern is covered with boxes of size ℓ, with $\epsilon \ll \ell \ll L$, and the number of boxes $N(\alpha)\delta\alpha$, with $-\log p / \log \lambda$ lying in the range α to $\alpha + \delta\alpha$ is "counted", where $p = M_i(\ell)/M$ and $\lambda = L/\ell$. If $N(\alpha) \sim \lambda^{f(\alpha)}$ for subsets labeled by the index α, then the mass distribution is a multifractal. The multifractal "spectrum" $f(\alpha)$ can then be calculated via the partition sum, using the Legendre transform, by scaling the mass histograms obtained at different resolutions (λ), or by other approaches.

The application of this concept to random fractals generated by growth processes has been controversial. It is difficult to verify multifractal scaling numerically because of the requirement that $\epsilon \ll \ell \ll L$. Care must be taken in calculating the exponents τ_d or D_q for negative q. Regions of size ℓ that barely overlap the cluster or pattern will contribute the most to the partition function $Z_q(\ell)$; consequently $Z_q(\ell)$ is dominated by rare "events" for $q < 0$. This problem can be reduced by using boxes centered on randomly selected *occupied positions*.

A fractal with mass or geometric multifractal scaling can easily be constructed using a generator similar to that used to obtain the Cantor set described in chapter 2, section 2.1. For example, if a generator consisting of a line covering the unit interval, with the second 1/4 removed, is used in place of a generator consisting of a line covering the unit interval, with the middle third removed, then a geometric multifractal will be generated if each line segment is replaced by a scaled version of the generator, at every stage in the generation process [1353, 1354]. In practice, it can be quite difficult to distinguish this type of multifractal scaling from the density fluctuations that are associated with the lacunarity of simple statistical self-similar fractals, which can be described in terms of the simple scaling forms given in equations 2.11 and 2.12.

Muzy *et al.* [1355] have used an optical wavelet transform approach to analyze the mass distribution in 50 000-particle off-lattice DLA clusters and DLA-like electrodeposition patterns. In both cases they found that $D_q \sim 5/3$ for $-20 \leq q \leq 20$. Quite different results were obtained earlier by Vicsek *et al.* [1356] by calculating the partition functions $Z_q(\ell)$ using boxes of size ℓ centered on 50 000 randomly selected particles in the central regions (3000×3000 particle diameters) of five 10^6-particle, off-lattice DLA clusters. In this work D_q was found to rise from $D_{10} \simeq 1.70$ to $D_{-10} \simeq 1.83$, with decreasing q.

References

1. Johannes Kepler. *The six-cornered snowflake*. Oxford University Press, Oxford, 1966. Translation by Colin Hardie of the 1611 Latin text "A new year's gift or on the six-cornered snowflake".

2. Benoit B. Mandelbrot. *The fractal geometry of nature*. W. H. Freeman, New York, 1983.

3. H. Eugene Stanley. *Introduction to phase transitions and critical phenomena*. Oxford University Press, Oxford, 1971.

4. P. G. De Gennes. *Scaling concepts in polymer physics*. Cornell University Press, Ithaca, New York, 1979.

5. Lewis F. Richardson. Atmospheric diffusion shown on a distance-neighbour graph. *Proceedings of The Royal Society, London*, A110:709–737, 1926.

6. M. F. Shlesinger, B. J. West and J. Klafter. Lévy dynamics of enhanced diffusion: Applications to turbulence. *Physical Review Letters*, 58:1100–1103, 1987.

7. B. Widom. Surface tension and molecular correlations near the critical point. *Journal of Chemical Physics*, 43:3892–3897, 1965.

8. B. Widom. Equation of state in the neighborhood of the critical state. *Journal of Chemical Physics*, 43:3898–3905, 1965.

9. Benoit B. Mandelbrot. Statistique céleste. - Corrélations et texture dans un nouveau modèle d'univers hiérarchisé basé sur les ensembles trémas. *Comptes Rendus de l'Académie des Sciences Paris*, A288:81–83, 1979.

10. Andrei Linde. The self-reproducing inflationary universe. *Scientific American*, pages 32–39, November 1994.

11. Andrei D. Linde. *Inflation and quantum cosmology*. Academic Press, Boston, 1990.

12. Andrei Linde, Dmitri Linde and Arthur Mezhlumian. From the big bang theory to the theory of a stationary universe. *Physical Review*, D49:1783–1826, 1994.

13. Xiaochun Luo and David N. Scramm. Fractals and cosmological large-scale structure. *Science*, 256:513–515, 1992.

14. Philip E. Seiden and Lawrence S. Schulman. Percolation model of galactic structure. *Advances in Physics*, 39:1–54, 1993.

15. Lawrence S. Schulman and Philip E. Seiden. Percolation and galaxies. *Science*, 233:425–431, 1986.

16. Tamás Vicsek and Alexander S. Szalay. Fractal distribution of galaxies modeled by a cellular-automaton-type stochastic process. *Physical Review Letters*, 58:2818–2821, 1987.

17. Arjun Berera and Li-Zhi Fang. Stochastic fluctuations and structure formation in the

universe. *Physical Review Letters*, 72:458–461, 1994.

18. M. Batty, P. Longley and S. Fotheringham. Urban growth and form: scaling, fractal geometry and diffusion-limited aggregation. *Environment and Planning*, 21:1447–1472, 1989.

19. M. Batty and P. Longley. *Fractal cities: a geometry of form and function*. Academic Press, San Diego, 1994.

20. A. Hollingsworth. Storm hunting with fractals. *Nature*, 319:11–12, 1986.

21. S. Lovejoy, D. Shertzer and P. Ladoy. Fractal characterization of inhomogeneous geophysical measuring networks. *Nature*, 319:43–44, 1986.

22. C. Nicolis and G. Nicolis. Weather forecasting still tricky. *Nature*, 321:567, 1986.

23. D. V. Sibba Rao, C. Bapu Reddy and N. Krishna Brahman. Gravity modelling of Satpura Gondwana Basin (Central India). *Journal of the Association of Exploration Geophysicists*, 8:169–178, 1987.

24. G. Korvin, D. M. Boyd and R. O'Dowd. Fractal characterization of the South Australian gravity station network. *Geophysics Journal International*, 100:535–539, 1990.

25. C. Nicolis. Optimizing the global observational network: A dynamical approach. *Journal of Applied Meteorology*, 32:1751–1759, 1993.

26. A. Bialas and R. Peschanski. Moments of rapidity distributions as a measure of short-range fluctuations in high-energy collisions. *Nuclear Physics*, B273:703–718, 1986.

27. A. Bialas and R. Peschanski. Intermittency in multiparticle production at high energy. *Nuclear Physics*, B308:857–867, 1988.

28. D. Ghosh, A. Deb, B. Biswas, J. Roychowdhury, P. Ghosh and J. Roy. Evidence of fractality in proton emission in nucleus-nucleus interactions. *Europhysics Letters*, 23:91–97, 1993.

29. P. L. Jain, K. Sengupta and G. Singh. Multifractality of inelastic nuclear interactions at 1-6 TeV. *Physics Letters*, B241:273–277, 1990.

30. R. K. Shivpuri and Neeti Parashar. Target and energy dependence of intermittency and multifractality in interactions at cosmic ray energies. *Physical Review*, D49:219–229, 1994.

31. R. Albrecht and 35 others, WA80 Collaboration. Fluctuations and intermittency in 200 A GeV $^{16}O + (C, Au)$ reactions. *Physics Letters*, B221:427–431, 1989.

32. I. V. Ajinenko and 36 others, EHS/NA22 Collaboration. Intermittency patterns in π^+p and K^+p collisions at 250 GeV/c. *Physics Letters*, B222:306–310, 1989.

33. Ph. Brax and R. Peschanski. Multifractal analysis of intermittency and phase transitions in multiparticle dynamics. *Nuclear Physics*, B346:65–83, 1990.

34. A. Anufriev and D. Sokoloff. Fractal properties of geodynamo models. *Geophysical and Astrophysical Fluid Dynamics*, 74:207–223, 1994.

35. R. F. Smalley Jr., J.-L. Chatelain, D. L. Turcotte and R. Prévot. A fractal approach to the clustering of earthquakes: Applications to the seismicity of the New Hebrides. *Bulletin of the Seismological Society of America*, 77:1368–1381, 1987.

36. U. Fano and G. Racah. *Irreducible tensorial sets*. Academic Press, New York, 1959.

37. M. E. Rose. *Elementary theory of angular momentum*. Wiley, New York, 1957.

38. K. A. Holsapple. The scaling of impact processes in planetary sciences. *Annual Reviews of Earth and Planetary Science*, 21:333–373, 1993.

39. D. E. Gault. Experimental impact "craters" formed in water: gravity scaling realized. *Eos Transactions of the American Geophysical Union*, 59:1121, 1978.

40. K. L. R. Olevson. Energy balances for transient water craters. *United States Geological Survey*, 650-D:189–194, 1969.

41. A. M. Worthington and R. S. Cole. Impact with a liquid surface studied by the aid of instantaneous photography. *Philosphical Transactions of the Royal Society, London*, A189:137–148, 1897.

42. A. M. Worthington. A study of splashes. Longmans, Green & Co., London, 1908.

43. O. E. Engel. Collision of liquid drops with liquids. Technical Note 89, United States National Bureau of Standards, 1961.

44. O. E. Engel. Collisions of liquid drops with

liquid, II, crater depths in fluid impacts, Technical Report WADD-TR-60-475, United States National Bureau of Standards, 1962.

45. K. A. Holsapple and R. M. Schmidt. On the scaling of crater dimensions 2. Impact processes. *Journal of Geophysical Research*, 87:1849–1870, 1982.

46. Robert M. Schmidt and Kevin R. Housen. Some recent advances in the scaling of impact and explosion cratering. *International Journal of Impact Engineering*, 5:543–560, 1987.

47. John D. O'Keefe and Thomas J. Ahrens. Planetary cratering mechanisms. *Journal of Geophysical Research*, 98:17011–17028, 1993.

48. J. K. Dienes and J. M. Walsh. Theory of impact: Some general principles and the method of Eulerian codes. In R. Kinslow, editor, *High velocity impact phenomena*, pages 46–104. Academic Press, New York, 1970.

49. E. Buckingham. On physically similar systems: Illustrations of the use of dimensionless equations. *Physical Review, Second Series*, 4:345–376, 1914.

50. Henry L. Langhaar. *Dimensional analysis and theory of models*. Wiley, New York, 1951.

51. L. I. Sedov. *Similarity and dimensional methods in mechanics*. CRC, Boca Raton, tenth edition, 1993.

52. V. V. Belousov and M. V. Gzovsky. Experimental tectonics. In L. H. Ahrens, F. Press, S. K. Runcorn and H. C. Urey, editors, *Physics and chemistry of the Earth*, pages 409–499. Pergamon Press, Oxford, 1965.

53. Ruud Weijermars and Harro Schmeling. Scaling of Newtonian and non-Newtonian fluid dynamics without inertia for quantitative modelling of rock flow due to gravity (including the concept of rheological similarity). *Physics of the Earth and Planetary Interiors*, 43:316–330, 1986.

54. James Hall. On the vertical position and convolutions of certain strata and their relation with granite. *Transactions of the Royal Society of Edinburgh*, 7:79–108, 1815.

55. Stanley A. Schumm, M. Paul Mosley and William E. Weaver. *Experimental fluvial geomorphology*. Wiley, New York, 1987.

56. P. J. Ashworth, J. L. Best, J. O. Leddy and G. W. Geehan. The physical modelling of braided rivers and deposition of fine-grained sediment. In M. J. Kirkby, editor, *Process models and theoretical geomorphology*, pages 115–139. Wiley, Chichester, 1994.

57. Paul Meakin. *Applications of fractals and scaling*. Cambridge University Press, to be published.

58. Paul Meakin. *Fractal aggregates and the kinetics of aggregation*. Cambridge University Press, to be published.

59. George Kingsley Zipf. *Human behavior and the principal of least-effort. An introduction to human ecology*. Addison–Wesley, Cambridge, Massachusetts, 1949.

60. Hugh Pennington. That's the way the cookie crumbles. *New Scientist*, December 2:53–54, 1995.

61. D. J. Bird and 20 others. Evidence for correlated changes in the spectrum and composition of cosmic rays at extremely high energies. *Physical Review Letters*, 71:3401–3404, 1993.

62. J. S. Rowlinson. *J. D. van der Waals: On the continuity of the gaseous and liquid states*. North Holland, Amsterdam, 1988.

63. D'Arcy Wentworth Thompson. *On growth and form*. Cambridge University Press, Cambridge, 1942. A new edition.

64. Kevin Patrick Galvin. *Growth and coalescence in condensation*. PhD thesis, University of London, Imperial College of Science, Technology and Medicine, 1990.

65. Paul Meakin. Droplet deposition, growth and coalescence. *Reports on Progress in Physics*, 55:113–155, 1992.

66. Alan Spry. *Metamorphic textures*. Pergamon Press, Oxford, 1969.

67. Robert M. May. Simple mathematical models with very complicated dynamics. *Nature*, 261:459–467, 1976.

68. Mitchell J. Feigenbaum. Universal behavior in nonlinear systems. *Los Alamos Science*, Summer 1980:4–27, 1980.

69. Tomas Bohr and David Rand. The entropy function for characteristic exponents. *Physica*, D25:387–398, 1987.

70. B. Hallet. Spatial self-organization in geomorphology: From periodic bedforms and pattern ground to scale-invariant

topography. *Earth Science Reviews*, 29:57–75, 1990.

71. J. S. Langer. Interfacial instabilities and dendritic growth. *Supplement of the Progress of Theoretical Physics*, 64:463–476, 1978.

72. E. Ben-Jacob, P. Garik, T. Mueller and D. Grier. Characterization of morphology transitions in diffusion-controlled systems. *Physical Review*, A38:1370–1380, 1988.

73. Eshel Ben-Jacob and Peter Garik. The formation of patterns in non-equilibrium growth. *Science*, 343:523–530, 1990.

74. Ofer Shochet and Eshel Ben-Jacob. Coexistence of morphologies in diffusive patterning. *Physical Review*, E48:R4168–R4171, 1993.

75. J. S. Langer. Dendrites, viscous fingers and the theory of pattern formation. *Science*, 243:1150–1156, 1989.

76. David A. Kessler, Joel Koplik and Herbert Levine. Pattern selection in fingered growth phenomena. *Advances in Physics*, 37:255–339, 1988.

77. S.-K. Chan, H.-H. Reimer and M. Kahlweit. On the stationary growth shapes of NH_4Cl dendrites. *Journal of Crystal Growth*, 32:303–315, 1976.

78. Ph. H. Kuenen. Experimental abrasion of pebbles, 2: Rolling by current. *Journal of Geology*, 64:336–368, 1956.

79. R. C. Furneaux, W. R. Rigby and A. P. Davidson. The formation of controlled-porosity membranes from anodically oxidized aluminium. *Nature*, 337:147–149, 1989.

80. J. P. O'Sullivan and G. C. Wood. The morphology and mechanism of formation of porous anodic films on aluminium. *Proceedings of the Royal Society, London*, A317:511–543, 1970.

81. Hideki Masuda and Kenji Fukuda. Ordered metal nanohole arrays made by a two-step replication of honeycomb structures of anodic alumina. *Nature*, 268:1466–1468, 1995.

82. Yasuo Noguchi, Hiroshi Tabuchi and Hitoshi Hasegawa. Physical factors controlling the formation of patterned ground on Haleakala, Maui. *Geografiska Annaler*, A69:329–342, 1987.

83. K. A. Jackson and J. D. Hunt. Lamellar and rod eutectic growth. *Transactions of the Metallurgical Society of AIME*, 236:1129–1142, 1966.

84. W. Albers. Physical properties and design for non-structural applications of composite material. In G. Piatti, editor, *Advances in composite materials*, chapter 9, pages 185–207. Applied Science Publishers, 1978.

85. K. Kassner and C. Misbah. Similarity laws in eutectic growth. *Physical Review Letters*, 66:445–448, 1991.

86. R. Trivedi. Interdendritic spacing: Part II. A comparison between theory and experiment. *Metallurgical Transactions*, 15A:977–994, 1984.

87. John Bechhoefer and Albert Libchaber. Testing shape selection in directional solidification. *Physical Review*, B35:1393–1396, 1987.

88. P. Kurowski, C. Guthmann and S. de Cheveigné. Shapes, wavelength selection and the cellular-dendritic "transition" in directional solidification. *Physical Review*, A42:7368–7376, 1990.

89. G. Faivre, S. de Cheveigne, C. Guthmann and P. Kurowski. Solitary tilt waves in thin lamellar eutectics. *Europhysics Letters*, 9:779–784, 1989.

90. K. J. Dormer. *Fundamental tissue geometry for biologists*. Cambridge University Press, Cambridge, 1980.

91. Frederic T. Lewis. The correlation between cell division and the shapes and sizes of prismatic cells in the epedermis of cucumis. *The Anatomical Record*, 38:341–376, 1928.

92. D. A. Aboav. The arrangement of grains in a polycrystal. *Metallography*, 3:383–390, 1970.

93. Valérie de Lapparent, Margaret J. Geller and John P. Huchra. A slice of the universe. *The Astrophysical Journal*, 302:L1–L5, 1986.

94. S. Hales. *Vegetable staticks*. Innys and Woodward, London, 1712.

95. H. Möwald. Direct characterization of monolayers at the air-water interface. *Thin Solid Films*, 159:1–15, 1988.

96. Masami Hasegawa and Masaharu Tanemura. On the pattern of space division by territories. *Annals of the Institute of Statistical Mathematics*, B28:509–519, 1976.

97. Toyoichi Tanaka. Kinetics of phase transition in polymer gels. *Physica*, A140:261–268, 1986.

98. Toyoichi Tanaka, Shao-Tang Sun, Yoshitsugu Hirokawa, Seiji Katayama, John Kucera, Yoshiharu Hirose and Takayuki Amiya. Mechanical instability of gels at the phase transition. *Nature*, 325:796–798, 1987.

99. M. Georges Voronoï à Varsovie. Nouvelles applications des paramétres continus à la théorie des formes quadritiques: Deuxième mémoire. Recherches sur les parallélloèdres primitifs. *Journal für die Reine und Angewandte Mathematik*, 134:198–287, 1908.

100. G. Lejeune Dirichlet. Über die reduction der positiven quadratischen formen mit drei unbestimmten ganzen zahlen. *Journal für die Reine und Angewandte Mathematik*, 40:209–232, 1950.

101. Y. Termonia. Monte Carlo diffusion model of polymer coagulation. *Physical Review Letters*, 72:3678–3681, 1994.

102. Gilles Widawski, Michel Rawiso and Bernard François. Self-organized honeycomb morphology of star-polymer polystyrene films. *Nature*, 369:387–389, 1994.

103. George H. Markstein. Experimental and theoretical studies of flame-front stability. *Journal of the Aeronautical Sciences*, 18:199–209, 1951.

104. David A. Noever. Statistics of premixed flame cells. *Physical Review*, A44:968–974, 1991.

105. Eli Raz, S. G. Lipson and Eshel Ben-Jacob. New periodic morphologies observed during dendritic growth of ammonium chloride in thin layers. *Journal of Crystal Growth*, 8:637–646, 1991.

106. Eberhard Bodenschatz, John R. de Bruyn, Guenter Ahlers and David S. Cannell. Transitions between patterns in thermal convection. *Physical Review Letters*, 67:3078–3081, 1991.

107. E. P. Wigner and F. Seitz. On the constitution of metallic sodium. *Physical Review*, 43:804–810, 1933.

108. Cyril Stanley Smith. Grain shape and other metallurgical applications of topology. In *Metal interfaces*, pages 65–108. American Society for Metals, Cleveland, Ohio, 1952. A Seminar on Metal Interfaces held during the thirty-third National Metal Congress and Exhibition, Detroit, October 13 to 19, 1951.

109. N. Rivier. Statistical crystallography Structure of random cellular networks. *Philosophical Magazine*, 52:795–819, 1985.

110. D. Weire and N. Rivier. Soap, cells and statics - Random patterns in two dimensions. *Contemporary Physics*, 25:59–99, 1984.

111. N. Rivier. Order and disorder in packings and froths. In D. Bideau and A. Hansen, editors, *Disorder in granular media*, pages 55–102. Elsevier, Amsterdam, 1993.

112. D. Weire. Some remarks on the arrangement of grains in a polycrystal. *Metallography*, 7:157–160, 1974.

113. H. S. M. Coxeter. *Regular polytopes.* Macmillan, New York, 1958.

114. Joel Stavans and James A. Glazier. Soap froth revisited: Dynamic scaling in the two-dimensional froth. *Physical Review Letters*, 62:1318–1321, 1989.

115. N. Rivier and A. Lissowski. On the correlation between sizes and shapes of cells in epithelial mosaics. *Journal of Physics*, A15:L143–L148, 1982.

116. D. ter Haar. *Elements of statistical mechanics.* Holt Rinehart and Winston, New York, 1961.

117. M. C. Cross and P. C. Hohenbergh. Pattern formation outside of equilibrium. *Reviews of Modern Physics*, 65:851–1112, 1994.

118. Joseph E. Avron and Dov Levine. Geometry of foams: 2D dynamics and 3D statics. *Physical Review Letters*, 69:208–211, 1992.

119. D. Weaire, F. Bolton, P. Molho and J. A. Clazier. Investigation of an elementary model for magnetic froth. *Journal of Physics: Condensed Matter*, 3:2101–2114, 1991.

120. R. Wolfe and J. C. North. Planar domains in ion-implanted magnetic bubble garnets revealed by ferrofluid. *Applied Physics Letters*, 25:122–124, 1974.

121. F. C. Frank. The influence of dislocations on crystal growth. *Discussions of the Faraday Society*, 5:48–54, 1949.

122. Stéphane Douady and Yves Couder. Phyllotaxis as a self-organized growth process. In Juan Manuel Garcia Ruiz, Enrique Louis, Paul Meakin and Leonard M. Sander, editors, *Growth patterns in physical sciences and biology*, pages 185–207. Plenum, New York, 1993. NATO ASI Series B304.

123. A. T. Winfree. The prehistory of the Belousov–Zhabotinsky oscillator. *Journal of Chemical Education*, 61:661–663, 1984.

124. R. J. Field and M. Burgers. *Oscillations and traveling waves in chemical systems.* Wiley, New York, 1985.

125. A. M. Zhabotinski. The early period of systematic studies of oscillation and waves in chemical systems. In Richard J. Field and Mária Burgere, editors, *Oscillations and traveling waves in chemical systems*, pages 1–6. Wiley, New York, 1985.

126. B. P. Belousov. A periodic reaction and its mechanism. In Richard J. Field and Mária Burgere, editors, *Oscillations and traveling waves in chemical systems*, pages 605–613. Wiley, New York, 1985.

127. Arun V. Holden, Mario Markus and Hans. G. Othmer, editors. *Nonlinear wave processes in excitable media*, New York, 1991. Plenum. NATO ASI Series B244.

128. P. Pieranski, L. Beliard, J.-Ph. Tournellec, X. Leoncini, C. Furtlehner, H. Dumoulin, E. Riou, B. Jouvin, J.-P. Fénerol, Ph. Palaric, J. Heuving, B. Cartier and I. Kraus. Physics of smectic membranes. *Physica*, A194:364–389, 1993.

129. M. Seul, L. R. Monar, L. O'Gorman and R. Wolfe. Morphology and local structure in labyrinthrine stripe domain phase. *Science*, 254:1616–1618, 1991.

130. Helmuth Möwald. Surfactant layers at water surfaces. *Reports on Progress in Physics*, 56:653–685, 1993.

131. Chris S. Henkee, Edwin L. Thomas and Lewis J. Fetters. The effects of surface constraints on the ordering of block copolymer domains. *Journal of Materials Science*, 23:1685–1694, 1988.

132. Henri Bénard. Les tourbillons cellulaires dans une nappe liquide transportant de la chaleur par convection en régime permanent. *Annales de Chimie et de Physique*, 7:62–144, 1901.

133. Lord Rayleigh. On convection currents in a horizontal layer of fluid, when the higher temperature is on the underside. *Philosophical Magazine*, 32:529–546, 1916.

134. R. E. Rosensweig. *Ferrohydrodynamics.* Cambridge University Press, Cambridge, 1985.

135. A. Tsebers and E. Blūms. Long-range magnetic forces in two-dimensional hydrodynamics of magnetic fluid pattern formation. *Chemical Engineering Communications*, 67:69–88, 1988.

136. J.-C. Bacri, A. Cebers and R. Perzynski. Behavior of a magnetic fluid microdrop in a rotating field. *Physical Review Letters*, 72:2705–2708, 1994.

137. G. K. Batchelor. *An introduction to fluid dynamics.* Cambridge University Press, Cambridge, 1967.

138. J. B. Swift and P. C. Hohenbergh. Hydrodynamic fluctuations at the convective instability. *Physical Review*, A15:319–328, 1977.

139. Y. Kuramoto. *Chemical oscillations, waves and turbulence.* Springer, Berlin, 1984.

140. G. I. Sivashinsky. Nonlinear analysis of hydrodynamic instability in laminar flames - I. Derivation of the basic equations. *Acta Astronautica*, 4:1177–1206, 1977.

141. J. R. A. Pearson. On convection cells induced by surface tension. *Journal of Fluid Mechanics*, 4:489–500, 1958.

142. M. E. Glicksman, E. Winsa, R. C. Hahn, T. A. Lograsso, S. H. Tirmizi and M. E. Selleck. Isothermal dendritic growth - a proposed microgravity experiment. *Metallurgical Transactions*, 19A:1945–1953, 1988.

143. M. E. Gliksman, M. B. Koss, L. T. Bushnell, J. C. Lacombe and E. A. Winsa. Dendritic growth in terrestrial and microgravity conditions. In Fereydoon Family, Paul Meakin, Bernard Sapoval and Richard Wool, editors, *Fractal aspects of materials*, pages 13–22. Materials Research Society, Pittsburg, 1995. Materials Research Society Symposium Proceedings Volume 367.

144. J. J. Favier, J. P. Garandet, A. Rouzaud and D. Camel. Mass transport phenomena during solidification in microgravity; preliminary results from the first Mephisto flight experiment. *Journal of Crystal Growth*, 140:237–243, 1994.

145. D. T. J. Hurle. Hydrodynamics in crystal growth. In E. Kaldis and H. J. Scheel, editors, *Crystal growth and materials*, pages 549–569. North Holland, Amsterdam, 1977.

146. J. S. Langer. Instabilities and pattern formation in crystal growth. *Reviews of Modern Physics*, 52:1–28, 1980.

147. M. E. Glicksman, R. J. Schaefer and J. D. Ayers. Dendrite growth - a test of theory. *Metallurgical Transactions*, A7:1747–1759, 1976.

148. S. C. Huang and M. E. Glicksman. Fundamentals of dendritic solidification-I Steady state tip growth. *Acta Metallurgica*, 29:701–715, 1981.

149. W. W. Mullins and R. F. Sekerka. Stability of a planar interface during solidification of a dilute binary alloy. *Journal of Applied Physics*, 35:444–451, 1964.

150. E. Hürlimann, R. Trittibach, U. Bisang and J. H. Bilgram. Integral parameters of xenon dendrites. *Physical Review*, A46:6579–6595, 1992.

151. C.-W. Park and G. M. Homsy. Two-phase displacement in Hele-Shaw cells: Theory. *Journal of Fluid Mechanics*, 139:291–308, 1984.

152. G. P. Ivantsov. Temperature field around a spheroidal, cylindrical and acicular crystal growing in a supercooled melt. *Doklady Akademii Nauk SSSR*, 58:567, 1947. Soviet Physics Doklady.

153. G. P. Ivantsov. Temperature field around a spheroidal cylindrical and acicular crystal growing in a supercooled melt. In Pierre Pelcé, editor, *Dynamics of curved fronts*, pages 243–245. Academic Press, Boston, 1988. Translation from the Russian by Protrans.

154. G. Horvay and J. W. Cahn. Dendritic and spheroid growth. *Acta Metallurgica*, 9:695–705, 1961.

155. S.-C. Huang and M. E. Glicksman. Fundamentals of dendritic solidification-II Development of sidebranch structure. *Acta Metallurgica*, 29:717–734, 1981.

156. P. Pelce and Y. Pomeau. Dendrites in the small undercooling limit. *Studies in Applied Mathematics*, 74:245–258, 1986.

157. E. Brener, H. Müller-Krumbhaar and D. Tempkin. Kinetic phase diagram and scaling relations for stationary diffusional growth. *Europhysics Letters*, 17:535–540, 1992.

158. Angelo Barbieri, Daniel. C. Hong and J. S. Langer. Velocity selection in the symmetric model of dendritic growth. *Physical Review*, A35:1802–1808, 1987.

159. W. W. Mullins and R. F. Sekerka. Morphological stability of a particle growing by diffusion or heat flow. *Journal of Applied Physics*, 34:323–329, 1963.

160. P. G. Saffman and G. I. Taylor. The penetration of a fluid into porous medium or Hele-Shaw cell containing a more viscous fluid. *Proceeding of the Royal Society, London*, A245:312–329, 1958.

161. H. Guo, D. C. Hong and D. A. Kurtze. Surface-tension-driven nonlinear instability in viscous fingers. *Physical Review Letters*, 69:1520–1523, 1992.

162. Ofer Shochet, Klaus Kassner, Eshel Ben-Jacob, S. G. Lipson and Heiner Müller-Krumbhaar. Morphology transitions during non-equilibrium growth 1. Study of equilibrium shapes and properties. *Physica*, A181:136–155, 1992.

163. Ofer Shochet, Klaus Kassner, Eshel Ben-Jacob, S. G. Lipson and Heiner Müller-Krumbhaar. Morphology transitions during non-equilibrium growth 2. Morphology diagram and characterization of the transition. *Physica*, A187:87–111, 1992.

164. T. Ihle and H. Müller-Krumbhaar. Fractal and compact growth morphologies in phase transitions with diffusion transport. *Physical Review*, E49:2972–2991, 1994.

165. David A. Kessler, Joel Koplik and Herbert Levine. Numerical simulation of two-dimensional snowflake growth. *Physical Review*, A30:2820–2823, 1984.

166. Y. Saito, G. Goldbeck-Wood and H. Müller-Krumbhaar. Numerical simulation of dendritic growth. *Physical Review*, A38:2148–2157, 1988.

167. David Jasnow and Jorge Viñals. Dynamical scaling during interfacial growth in the one-sided model. *Physical Review*, A41:6910–6921, 1990.

168. H. Guo, D. C. Hong and D. A. Kurtze. Dynamics of pattern forming systems. *Physical Review*, A46:1867–1874, 1992.

169. Hong Guo and David Jasnow. Evidence for scaling in an interfacial growth instability: The solid case. *Physical Review*, A34:5027–5034, 1986.

170. Tamás Vicsek. Pattern formation in diffusion-limited aggregation. *Physical Review Letters*, 53:2281–2284, 1984.

171. Leo P. Kadanoff. Simulating

hydrodynamics: a pedestrian model. *Journal of Statistical Physics*, 39:267–283, 1985.

172. Richard C. Brower, David A. Kessler, Joel Koplik and Herbert Levine. Geometric models of interface evolution. *Physical Review*, A29:1335–1342, 1984.

173. David A. Kessler, Joel Koplik and Herbert Levine. Geometric model of interface evolution. II. Numerical simulations. *Physical Review*, A30:3161–3174, 1984.

174. David A. Kessler, Joel Koplik and Herbert Levine. Geometric models of interface growth: III. Theory of dendritic growth. *Physical Review*, A31:1712–1717, 1985.

175. E. Ben-Jacob, N. Goldenfield, J. S. Langer and G. Schöhn. Dynamics of interfacial pattern growth. *Physical Review Letters*, 51:1930–1932, 1983.

176. E. Ben-Jacob, Nigel Goldenfield, J. S. Langer and Gerd Schön. Boundary-layer model of pattern formation in solidification. *Physical Review*, A29:330–340, 1984.

177. M. Ben Amar and Y. Pomeau. Theory of dendritic growth in weakly undercooled melt. *Europhysics Letters*, 2:307–314, 1986.

178. John W. Cartmill and Ming Yang Su. Bubble size distribution under saltwater and freshwater breaking waves. *Dynamics of Atmospheres and Oceans*, 20:25–31, 1993.

179. J. Harvey, H. I. Mathews and H. Hilman. Crystal structure and growth of metallic or metallic-oxide smoke particles produced by electric arcs. *Discussions of the Faraday Society*, 30:113–123, 1960.

180. S. R. Forrest and T. A. Witten. Long-range correlations in smoke-particle aggregates. *Journal of Physics*, A12:L109–L117, 1979.

181. T.A. Witten and L.M. Sander. Diffusion-limited aggregation, a kinetic critical pheomenon. *Physical Review Letters*, 47:1400–1403, 1981.

182. Paul Meakin. Formation of fractal clusters and networks by irreversible diffusion-limited aggregation. *Physical Review Letters*, 51:1119–1122, 1983.

183. M. Kolb, R. Botet and R. Jullien. Scaling of kinetically growing clusters. *Physical Review Letters*, 51:1123–1126, 1983.

184. M. E. Cates. Brownian dynamics of self-similar macromolecules. *Journal de Physique*, 46:1059–1077, 1985.

185. Dietrich Stauffer, Antonio Coniglio and Mirielle Adam. Gelation and critical phenomena. *Advances in Polymer Science*, 44:103–158, 1982.

186. D. S. McKenzie. Polymers and scaling. *Physics Reports*, C27:35–88, 1976.

187. I. Vattulainen, T. Ala-Nissila and K. Kamkaala. Physical tests for random numbers in simulations. *Physical Review Letters*, 73:2513–2516, 1994.

188. Nicholas Metropolis, Arianna W. Rosenbluth, Marshall N. Rosenbluth, Augusta H. Teller and Edward Teller. Equation of state calculations by fast computing machines. *Journal of Chemical Physics*, 21:1087–1092, 1953.

189. Peter Stevens. *Patterns in nature*. Little Brown, Boston, 1974.

190. Pierre Pelcé. *Dynamics of curved fronts*. Academic Press, San Diego, 1988.

191. Predag Cvitanović. *Universality in chaos*. Adam Hilger, Bristol, 1984.

192. Hao Bai-Lin. *Chaos*. World Scientific, Singapore, 1984.

193. Predag Cvitanović. *Universality in chaos*. Adam Hilger, Bristol, 1989. Second edition.

194. R. R. Nigmatullin. The generalized fractals and statistical properties of the pore space of the sedimentary rocks. *Physica Status Solidi*, B153:49–57, 1989.

195. Yuval Gefen, Benoit B. Mandelbrot and Amnon Aharony. Critical phenomena on fractal lattices. *Physical Review Letters*, 45:855–858, 1980.

196. Richard F. Voss. Random fractal forgeries. In R. A. Earnshaw, editor, *Fundamental algorithms in computer graphics*, pages 805–835. Springer, Berlin, 1985. NATO ASI series F17.

197. C. Allain and M. Cloitre. Characterizing the lacunarity of random and deterministic fractal sets. *Physical Review*, A44:3552–3558, 1991.

198. Paul Meakin and Shlomo Havlin. Fluctuations and distributions in random systems. *Physical Review*, A36:4428–4433, 1987.

199. Raphael Blumenfeld and Robin C. Ball. Probe for morphology and hierarchical correlations in scale invariant structures. *Physical Review*, E47:2298–2302, 1993.

200. François Normant and Claude Tricot.

Method for evaluating the fractal dimension of curves using convex hulls. *Physical Review*, A43:6518–6525, 1991.

201. Paul W. Schmidt. Interpretation of small-angle scattering curves proportional to a negative power of the scattering vector. *Journal of Applied Crystallography*, 15:567–569, 1982.

202. A. H. Thompson, A. J. Katz and C. E. Krohn. The microgeometry and transport properties of sedimentary rocks. *Advances in Physics*, 36:625–694, 1987.

203. James Theiler. Efficient algorithm for estimating the correlation dimension from a set of discrete points. *Physical Review*, A36:4456–4462, 1987.

204. David R. Nelson, Michael Rubinstein and Frans Spaepen. Order in two-dimensional binary random arrays. *Philosophical Magazine*, A46:105–126, 1982.

205. A. N. Kolmogorov and V. M. Tikhomirov. ϵ-entropy and ϵ-capacity of sets in a function space. *American Mathematical Society Translations Series 2*, 17:277–367, 1961. Uspechi Matematiceskich Nauk. Russian Mathematical Surveys.

206. Celso Grebogi, Steven W. McDonald, Edward Ott and James A. Yorke. Exterior dimensions of fat fractals. *Physics Letters*, A110:1–4, 1985.

207. Dina Farin, Shmuel Peleg, David Yavin and David Avnir. Applications and limitations of boundary-line fractal analysis of irregular surfaces: proteins, aggregates and porous materials. *Langmuir*, 1:399–407, 1985.

208. Lajos Nyikos, László Balázs and Robert Schiller. Fractal analysis of artistic images: from cubism to fractalism. *Fractals*, 2:143–152, 1994.

209. Peter Pfeifer and Martin Obert. Fractals: basic concepts and terminology. In David Avnir, editor, *The fractal approach to heterogeneous chemistry: Surfaces, colloids, polymers*, pages 11–43. John Wiley, New York, 1989.

210. B. B. Mandelbrot. Negative fractal dimensions and multifractals. *Physica*, A163:306–315, 1990.

211. Z. Néda and Á. Mocsy. Structures obtained by mechanical fragmentation of glass plates. Preprint, 1992.

212. A. Birovljev, L. Furuberg, J. Feder, T. Jøssang, K. J. Måløy and A. Aharony. Gravity invasion percolation in two dimensions: Experiments and simulations. *Physical Review Letters*, 67:584–587, 1991.

213. R. Lenormand and S. Bories. Description d'un méchanisme de connexion de liason destiné a l'etude du drainage avec piéage en mileaux poreux. *Comptes Rendus de l'Académie des Sciences Paris*, Série II 291:279–282, 1980.

214. D. Wilkinson and J. F. Willemsen. Invasion percolation: A new form of percolation theory. *Journal of Physics*, 16:3365–3376, 1983.

215. R. Chandler, J. Koplik, K. Lerman and J. F. Willemsen. Capillary displacement and percolation in porous media. *Journal of Fluid Mechanics*, 119:249–267, 1982.

216. H. Saleur and B. Duplantier. Exact determination of the percolation hull exponent in two dimensions. *Physical Review Letters*, 58:2325–2328, 1987.

217. Aleksandar Birovljev. *Experiments on low-rate, gravity stabilized fluid–fluid displacements in 2d porous media: Percolation approach.* Master's thesis, Department of Physics, University of Oslo, 1990.

218. Aleksandar Birovljev. *Experiments on low-rate, gravity stabilized fluid-fluid displacements in 2d porous media: percolation approach.* PhD thesis, University of Oslo, 1994.

219. Madalena M. Dias and David Wilkinson. Percolation with trapping. *Journal of Physics*, A19:3131–3146, 1986.

220. Dietrich Stauffer and Amnon Aharony. *Introduction to percolation theory.* Taylor and Francis, London, 1992.

221. Vidar Frette, Jens Feder, Torstein Jøssang and Paul Meakin. Buoyancy-driven fluid migration in porous media. *Physical Review Letters*, 68:3164–3167, 1992.

222. Paul Meakin, Jens Feder, Vidar Frette and Torstein Jøssang. Invasion percolation in a destabilizing gradient. *Physical Review*, A46:3357–3368, 1992.

223. P. Pincus. Excluded volume effects and stretched polymer chains. *Macromolecules*, 9:386–388, 1976.

224. M. Daoud and P. G. De Gennes. Statics of macromolecular solutions trapped in small

pores. *Le Journal de Physique*, 38:85–93, 1977.

225. P. J. E. Peebles. The fractal galaxy distribution. *Physica*, D38:273–278, 1989.

226. L. Pietronero. The fractal structure of the universe: Correlations of galaxies and clusters and the average mass density. *Physica*, A144:257–284, 1987.

227. Paul H. Coleman and Luciano Pietronero. The fractal structure of the universe. *Physics Reports*, 213:311–389, 1992.

228. Stefano Borgani. Scaling in the universe. *Physics Reports*, 251:1–152, 1995.

229. Paul Meakin. An Eden model for randomly branched structures. *Physica Scripta*, 45:69–74, 1992.

230. K. Sandau and H. Kurz. Modelling of vascular growth processes: a stochastic biophysical approach. *Journal of Microscopy*, 175:205–213, 1994.

231. J. E. Martin. Slow aggregation of colloidal silica. *Physical Review*, A36:3415–3426, 1987.

232. P. Dimon, S. K. Sinha, D. A. Weitz, C. R. Safinya, G. S. Smith, W. A. Varady, and H. M. Lindsay. Structure of aggregated gold colloids. *Physical Review Letters*, 57:595–598, 1986.

233. Geoffrey C. Ansell and Eric Dickinson. Short-range structure of simulated colloidal aggregates. *Physical Review*, A35:2349–2352, 1986.

234. Eric Dickinson. Short-range structure in aggregates, gels and sediments. *Journal of Colloid and Interface Science*, 118:286–289, 1987.

235. A. Hasmy, M. Foret, J. Pelous and R. Jullien. Short distance correlations in fractal aggregates: Numerical simulations and SANS experiments on silica aerogels. *Journal de Physique*, IV:365–368, 1993. Colloque C8, Supplement to Volume 3, 1993.

236. Anwar Hasmy, René Vacher and Rémi Jullien. Small angle scattering by fractal aggregates: a numerical investigation of the crossover between the fractal regime and the Porod regime. *Physical Review*, B50:1305–1309, 1994.

237. Geoffrey C. Ansell and Eric Dickinson. Brownian-dynamics simulation of the formation of colloidal aggregate and sediment structure. *Faraday Discussion of the Chemical Society*, 83:167–177, 1987.

238. M. E. Fisher. The theory of critical point singularities. In M. S. Green, editor, *Critical phenomena*, pages 1–99. Academic Press, New York, 1971. Proceedings of the 51st Enrico Fermi Summer School, Varenna.

239. Michael N. Barber. Finite-size scaling. In C. Domb and J. L. Lebowitz, editors, *Phase transitions and critical phenomena*, volume 8, pages 145–266. Academic Press, London, 1983.

240. V. Privman, editor. *Finite size scaling and numerical simulation of statistical systems.* World Scientific, Singapore, 1990.

241. John L. Cardy, editor. *Finite-size scaling.* North-Holland, Amsterdam, 1988.

242. Kenneth G. Wilson and J. Kogut. The renormalization group and the ϵ expansion. *Physics Reports*, 12:75–200, 1974.

243. R. J. Creswick, H. A. Farach and C. P. Poole. *Introduction to renormalization group methods in physics.* Wiley, New York, 1992.

244. Leo P. Kadanoff, Wolfgang Götze, David Hamblen, Robert Hecht, E. A. S. Lewis, V.V. Palciauskas, Martin Rayl, J. Swift, David Aspnes and Joseph Kane. Static phenomena near critical points: Theory and experiment. *Reviews of Modern Physics*, 39:395–431, 1967.

245. J. M. Ziman. *Models of disorder: the theoretical physics of homogeneously disordered systems.* Cambridge University Press, Cambridge, 1979.

246. R. A. Ferrell, N. Menyhárd, H. Schmidt, F. Schwabl and P Szépfalusy. Dispersion in second sound and anomalous heat conduction at the lambda point of liquid helium. *Physical Review Letters*, 18:891–893, 1967.

247. B. I. Halperin and P. C. Hohenberg. Scaling laws for dynamical phenomena near a critical point. *Journal of the Physical Society of Japan*, 26:131–135, 1968. Supplement, Proceedings of the International Conference on Statistical Mechanics.

248. B. I. Halperin and P. C. Hohenberg. Scaling laws for dynamical critical phenomena. *Physical Review*, 177:952–971, 1969.

249. Dieter Forster. *Hydrodynamic fluctuations, broken symmetry and correlation functions.* Benjamin, Reading, Massachusetts, 1975.

250. A. C. Berg. *Random walks in biology.* Princeton University Press, Princeton, 1983.

251. Ivar Nordlund. Eine neue bestimmung der Avogadroschen konstante aus der Brownschen bewegung kleiner, in wasser suspendierten quecksilberkügelchen. *Zeitschrift für Physikalische Chemie*, 87:40–62, 1914.

252. J.-P. Bouchaud, A. Georges, J. Koplik, A. Provata and S. Redner. Superdiffusion in random velocity fields. *Physical Review Letters*, 4:2503–2506, 1990.

253. A. R. Osbourne and R. Caponio. Fractal trajectories and anomalous diffusion for chaotic particle motions in 2D turbulence. *Physical Review Letters*, 64:1733–1736, 1990.

254. Franz J. Wegner. Corrections to scaling laws. *Physical Review*, B5:4529–4536, 1972.

255. Amnon Aharony, Yuval Gefen, Aharon Kapitulnik and Michael Murat. Fractal eigendimensionalities for percolation clusters. *Physical Review*, B31:4721–4724, 1985.

256. Benoit B. Mandelbrot, Yuval Gefen, Amnon Aharony and Jacques Peyriére. Fractals, their transfer matrices and their eigen-dimensional sequences. *Journal of Physics*, A18:335–354, 1985.

257. Antonio Coniglio and Marco Zannetti. Novel dynamic scaling in kinetic growth phenomena. *Physica*, A163:325–333, 1989.

258. A. Coniglio and M. Zannetti. Multiscaling in growth kinetics. *Europhysics Letters*, 10:575–580, 1989.

259. C. Amitrano, A. Coniglio, P. Meakin and M. Zannetti. Multiscaling in diffusion-limited aggregation. *Physical Review*, B44:4974–4977, 1991.

260. Peter Ossadnik. Multiscaling analysis and width of the active zone of large off-lattice DLA. *Physica*, A195:319–323, 1993.

261. R. Hilfer. Multiscaling and the classification of continuous phase transitions. *Physical Review Letters*, 68:190–192, 1992.

262. Ronald L. Shreve. Statistical law of stream numbers. *Journal of Geology*, 74:17–37, 1966.

263. R. E. Horton. Erosional development of streams and their drainage basins, hydrophysical approach to quantitative morphology. *Bulletin of the American Geological Society*, 56:275–370, 1945.

264. A. N. Strahler. Hypsometric (area-altitude) analysis of erosional topography. *Geological Society of America Bulletin*, 63:1117–1142, 1952.

265. Ewald R. Weibel. *Morphometry of the human lung*. Springer-Verlag, Berlin, 1963.

266. Einar L. Hinrichsen, Knut Jørgen Måløy, Jens Feder and Torstein Jøssang. Self-similarity and structure of DLA and viscous fingering clusters. *Journal of Physics*, A22:L271–L277, 1989.

267. Einar L. Hinrichsen. *Structure and geometry of two dimensional random systems.* PhD thesis, University of Oslo, 1989.

268. Jens Feder, Einar L. Hinrichsen, Knut Jørgen Måløy and Torstein Jøssang. Geometrical crossover and self-similarity of DLA and viscous fingering clusters. *Physica*, D38:104–111, 1989.

269. Chi-Hang Lam, H. Kaufman and B. B. Mandelbrot. Orientation of particle attachment and local isotropy in diffusion limited aggregation (DLA). *Journal of Physics*, A28:L213–L217, 1995.

270. H. H. Hardy and Richard Beier. *Fractals in reservoir engineering.* World Scientific, Singapore, 1994.

271. J. Drahoš, F. Bradka and M. Punčochář. Fractal behavior of pressure fluctuations in a bubble column. *Chemical Engineering Science*, 47:4069–4075, 1992.

272. M. A. Latifi, A. Naderifar, N. Midoux and A. L. Méhauté. Fractal behavior of local liquid–solid mass transfer fluctuations at the wall of a trickle bed reactor. *Chemical Engineering Science*, 22:3823–3829, 1994.

273. Benoit B. Mandelbrot and John W. Van Ness. Fractional Brownian motions, fractional noise and applications. *Society for Industrial and Applied Mathematics Review*, 10:422–437, 1968.

274. Benoit B. Mandelbrot and James R. Wallis. Computer experiments with fractional Gaussian noises. Part I, Averages and variances. *Water Resources Research*, 5:228–241 and 267, 1969.

275. Benoit B. Mandelbrot. A fast fractional Gaussian noise generator. *Water Resources Research*, 7:543–553, 1971.

276. M. V. Berry and Z. V. Lewis. On the Weierstrass–Mandelbrot fractal function.

Proceedings of the Royal Society, London, A370:459–484, 1980.

277. Albert Lásló Barabasi and Tamás Vicsek. Tracing a diffusion-limited aggregate: Self-affine versus self-similar scaling. *Physical Review*, A41:6881–6883, 1990.
278. Scott Painter and Lincoln Paterson. Fractional Lévy motion as a model for spatial variability in sedimentary rocks. *Geophysical Research Letters*, 21:2857–2860, 1994.
279. Albert-László Barabási and Tamás Vicsek. Multifractality of self-affine surfaces. *Physical Review*, A44:2730–2733, 1990.
280. Albert-László Barabási, Roch Bourbonnais, Mogens Jensen, János Kertész, Tamás Vicsek and Yi-Cheng Zhang. Multifractality of growing surfaces. *Physical Review*, A45:R6951–R6954, 1992.
281. Albert László Barabási, Péter Szépfalusy and Tamás Vicsek. Multifractal spectra of multi-affine functions. *Physica*, A178:17–28, 1991.
282. Michael E. Fisher. Interface wandering in adsorbed and bulk phases, pure and impure. *Journal of the Chemical Society, Faraday Transactions*, 82:1569–1603, 1986. Faraday Symposium 20.
283. Reinhard Lipowsky and Michael E. Fisher. Wetting in random systems. *Physical Review Letters*, 56:472–475, 1986.
284. Joachim Krug. Turbulent interfaces. *Physical Review Letters*, 72:2907–2910, 1994.
285. Heiko Leschhorn and Lei-Han Tang. Comment on "Elastic string in a potential". *Physical Review Letters*, 70:2973, 1993.
286. Yan-Chr Tsai and Yonathan Shapir. Kinetic roughening in surfaces of crystals growing on disordered substrates. *Physical Review Letters*, 69:1773–1776, 1992.
287. Mehran Kardar, Giorgio Parisi and Yi-Cheng Zhang. Dynamical scaling of growing interfaces. *Physical Review Letters*, 56:889–892, 1986.
288. Paul Meakin, P. Ramanlal, L. M. Sander and R. C. Ball. Ballistic deposition on surfaces. *Physical Review*, A34:5091–5103, 1986.
289. David Y. K. Ko and Flavio Seno. Simulations of deposition growth models in various dimensions: the possible importance of overhangs. *Physical Review*, E50:R1741–R1744, 1994.
290. Joachim Krug, Paul Meakin and Timothy Halpin-Healey. Amplitude universality for driven interfaces and directed polymers in random media. *Physical Review*, A45:638–653, 1992.
291. Jaques G. Amar and Fereydoon Family. Universality in surface growth: Scaling functions and amplitude ratios. *Physical Review*, A45:5378–5393, 1992.
292. F. Family and T. Vicsek. Scaling of the active zone in the Eden process on percolation networks and the ballistic deposition model. *Journal of Physics*, A18:L75–L81, 1985.
293. R. Jullien and R. Botet. Scaling properties of the surface of Eden model in $d = 2, 3, 4$. *Journal of Physics*, A18:2279–2287, 1985.
294. A. Arneodo, E. Bacry, P. V. Graves and J. F. Muzy. Characterizing long-range correlations in DNA sequences from wavelet analysis. *Physical Review Letters*, 74:3293–3296, 1995.
295. János Kertesz and Dietrich E. Wolf. Noise reduction in Eden models: II. Surface structure and intrinsic width. *Journal of Physics*, A21:747–761, 1988.
296. D. E Wolf and J. Kertész. Surface width exponents for three- and four-dimensional Eden growth. *Europhysics Letters*, 4:651–656, 1987.
297. Jin Min Kim and J. M. Kosterlitz. Growth in a restricted solid-on-solid model. *Physical Review Letters*, 62:2289–2292, 1989.
298. J. Széep, J. Cserti and J. Kertész. Monte Carlo approach to dendritic growth. *Journal of Physics*, A18:L413–L418, 1985.
299. Chao Tang. Diffusion-limited aggregation and the Saffman–Taylor problem. *Physical Review*, A31:1977–1979, 1985.
300. Dietrich E. Wolf and János Kertész. Noise reduction in Eden models: I. *Journal of Physics*, A20:L257–L261, 1987.
301. Joachim Krug and Paul Meakin. Universal finite-size effects in the rate of growth processes. *Journal of Physics*, A23:L987–L994, 1990.
302. Dieter Forster, David R. Nelson and Michael J. Stephen. Large-distance and long-time properties of a randomly stirred fluid. *Physical Review*, A16:732–749, 1977.

303. Joachim Krug. Scaling relation for a growing interface. *Physical Review*, A34:5465–5466, 1987.

304. Michael Plischke and Zoltán Rácz. Active zone of growing clusters: Diffusion-limited aggregation and the Eden model. *Physical Review Letters*, 53:415–418, 1984.

305. Jacques G. Amar and Fereydoon Family. Scaling of surface fluctuations and dynamics of surface growth models with power-law noise. *Journal of Physics*, A24:L79–L86, 1991.

306. Benoit B. Mandelbrot, Dann E. Passoja and Alvin J. Paullay. Fractal character of fracture surfaces of metals. *Nature*, 308:721–722, 1984.

307. Knut Jørgen Måløy, Alex Hansen, Einar L. Hinrichsen and Stéphane Roux. Experimental measurement of the roughness of brittle cracks. *Physical Review Letters*, 68:213–215, 1992.

308. Thor Engøy, Knut Jørgen Måløy, Alex Hansen and Stéphane Roux. Roughness of two-dimensional cracks in wood. *Physical Review Letters*, 73:834–837, 1994.

309. Alex Hansen, Thor Engøy and Knut Jøgen Måløy. Measuring Hurst exponents with the first return method. *Fractals*, 2:527–533, 1994.

310. Mitsugu Matsushita, Shunji Ouchi and Katsuya Honda. On the fractal structure of contour lines on a self-affine surface. *Journal of the Physical Society of Japan*, 60:2109–2112, 1991.

311. Sadao Isogami and Mitsugu Matsushita. Structural and statistical properties of self-avoiding fractional Brownian motion. *Journal of the Physical Society of Japan*, 61:1445–1448, 1992.

312. M. B. Isichenko and J. Kalda. Statistical topography. I. Fractal dimension of coastlines and number-area rule for islands. *Journal of Non Linear Science*, 1:255–277, 1991.

313. M. B. Isichenko and J. Kalda. Statistical topography. II. Two-dimensional transport of a passive scalar. *Journal of Non Linear Science*, 1:375–396, 1991.

314. S. E. Hough. On the use of spectral methods for the determination of fractal dimension. *Geophysical Research Letters*, 16:673–676, 1989.

315. Masahiro Nakagawa. A critical exponent method to evaluate fractal dimensions of self-affine data. *Journal of the Physical Society of Japan*, 62:4233–4239, 1993.

316. R. S. Sayles and T. R. Thomas. Surface topography as a nonstationary random process. *Nature*, 271:431–434, 1978.

317. Oleg I. Yordanov and Nicolay I. Nickolaev. Self-affinity of time series with finite domain power-law spectra. *Physical Review*, E49:R2517–R2520, 1994.

318. R. F. Voss. Random fractals: characterization and measurement. In Roger Pynn and Arne Skjeltorp, editors, *Scaling phenomena in disordered systems*, pages 1–11. Plenum, New York, 1985. NATO ASI B133.

319. B. Dubuc, J. F. Quiniou, C. Roques-Carmes, C. Tricot and S. W. Zucker. Evaluating the fractal dimension of profiles. *Physical Review*, A39:1500–1512, 1989.

320. S. Miller and R. J. Reifenberger. Improved method for fractal analysis using scanning microscopy. *Journal of Vacuum Science and Technology*, B10:1203–1207, 1992.

321. H. E. Hurst, R. P. Black and Y. M. Simaika. *Long-term storage: An experimental study*. Constable, London, 1965.

322. H. E. Hurst. A suggested statistical model of some time series which occur in nature. *Nature*, 180:494, 1957.

323. H. E. Hurst. Long term storage capacity of reservoirs. *Transactions of the American Society of Civil Engineers*, 116:770–808, 1951.

324. Benoit B. Mandelbrot and James R. Wallis. Noah, Joseph and operational hydrology. *Water Resources Research*, 4:909–918, 1968.

325. S. Alexander. Fractal surfaces. In J. R. Klafter, J. Rubin and M. F. Shlesinger, editors, *Transport and relaxation in random media*, pages 59–71. World Scientific, Singapore, 1986.

326. Benoit B. Mandelbrot. Self-affine fractals and fractal dimension. *Physica Scripta*, 32:257–260, 1985.

327. J. Feder. *Fractals*. Plenum, New York, 1988.

328. Lewis F. Richardson. The problem of contiguity: An appendix to statistics of deadly quarrels. In Anatol Rapoport, Ludwig von Bertalanfly and Richard L. Meier, editors, *General systems: yearbook of*

the Society for General Systems, volume 6, pages 139–187. The Society for General Systems, Ann Arbor, Michigan, 1961.

329. B. B. Mandelbrot. How long is the coast of Britain? Statistical self-similarity and fractional dimension. *Science*, 156:636–638, 1967.

330. F. Boger. *Rough surfaces, synthesis, analysis, visualization and applications*. PhD thesis, Department of Physics, University of Oslo, 1993.

331. E. Anguiano, M. Pancorbo and M. Aguilar. Fractal characterization by frequency analysis. I. Surfaces. *Journal of Microscopy*, 172:223–232, 1993.

332. M. Aguilar, E. Anguiano and M. Pancorbo. Fractal characterization by frequency analysis. II. A new method. *Journal of Microscopy*, 172:233–238, 1993.

333. M. Pancorbo, E. Anguiano and M. Aguilar. Fractal characterization by frequency analysis. III. Effect of noise. *Journal of Microscopy*, 176:54–62, 1994.

334. Jens Feder. Fractal time series and fractal Brownian motion. In T. Riste and D. Sherrington, editors, *Spontaneous formation of space-time structures and criticality*, pages 113–135. Kluwer, Dordrecht, 1991.

335. M. V. F. Pereira, G. C. Oliveira, C. C. G. Costa and J. Kelman. Stochastic streamflow models for hydroelectric systems. *Water Resources Research*, 20:379–390, 1984.

336. V. Klemes, R. Srikanthan and T. A. McMahon. Long-memory flow models in reservoir analysis: What is their practical value? *Water Resources Research*, 17:737–751, 1981.

337. Oscar J. Mesa and German Poveda. The Hurst effect: The scale of fluctuation approach. *Water Resources Research*, 29:3995–4002, 1993.

338. B. N. Bhattacharya, Vijay K. Gupta and Ed Waymire. The Hurst effect under trends. *Journal of Applied Probability*, 20:649–662, 1983.

339. Benoit B. Mandelbrot and James R. Wallis. Some long-run properties of geophysical records. *Water Resources Research*, 5:321–340, 1969.

340. Benoit B. Mandelbrot. Intermittent turbulence in self-similar cascades: divergence of high moments of the carrier. *Journal of Fluid Mechanics*, 62:331–358, 1974.

341. C. Meneveau and K. R. Sreenivasan. Simple multifractal cascade model for fully developed turbulence. *Physical Review Letters*, 59:1424–1247, 1987.

342. Benoit B. Mandelbrot and Carl J. G. Evertsz. The potential distribution around growing fractal clusters. *Nature*, 348:143–145, 1990.

343. Thomas C. Halsey, Mogens H. Jensen, Leo P. Kadanoff, Itamar Procaccia and Boris I. Shraiman. Fractal measures and their singularities: the characterization of strange sets. *Physical Review*, A33:1141–1151, 1986.

344. H. G. E. Hentschel and I. Procaccia. The infinite number of generalized dimensions of fractals and strange attractors. *Physica*, D8:435–444, 1983.

345. P. Grassberger. Generalizations of the Hausdorff dimension of fractal measures. *Physics Letters*, 107A:101–105, 1985.

346. B. I. Halperin, Shechao Feng and P. N. Sen. Differences between lattice and continuum percolation transport exponents. *Physical Review Letters*, 54:2391–2394, 1985.

347. B. B. Mandelbrot. Chronicle of fractal geometry (expanded and corrected). *Fractals*, 2:169–182, 1994.

348. Murray Eden. A two-dimensional growth process. In J. Neyman, editor, *4th. Berkeley symposium on mathematics, statistics and probability*, pages 223–239. University of California Press, Berkeley, 1961. Volume IV: Biology and the problems of health.

349. Murray Eden. A probabilistic model for morphogenisis. In Hupert P. Yockey, Robert L. Platzman and Henry Quastler, editors, *Symposium on information theory in biology*, pages 359–370. Pergamon, New York, 1958. Gatlinburg, Tennessee, October 29–31, 1956.

350. H. L. Frisch and J. M. Hammersley. Percolation processes and related topics. *Journal of the Society of Industrial and Applied Mathematics*, 11:894–918, 1963.

351. Scott Kirkpatrick. Percolation and conduction. *Reviews of Modern Physics*, 45:574–588, 1973.

352. J. W. Essam. Percolation theory. *Reports on*

Progress in Physics, 43:833–912, 1980.

353. D. Stauffer. Scaling theory of percolation clusters. *Physics Reports*, 54:1–74, 1979.

354. H. Kesten. *Percolation theory for mathematicians*. Birkhäuser, Boston, 1982.

355. Chao Tang. Diffusion-limited aggregation and the Saffman-Taylor problem. *Physical Review*, A31:1977–1979, 1985.

356. János Kertész and Tamás Vicsek. Diffusion-limited aggregation and regular patterns: fluctuations versus anisotropy. *Journal of Physics*, A19:L257–L262, 1986.

357. Johann Nittmann and H. Eugene Stanley. Tip splitting without interfacial tension and dendritic growth patterns arising from molecular anisotropy. *Nature*, 321:663–668, 1986.

358. M. Daoud, F. Family and G. Jannink. Dilution and polydispersity in branched polymers. *Journal de Physique Lettres*, 45:L199–L203, 1984.

359. G. Parisi and Nicolas Sourlas. Critical behavior of branched polymers in the Lee-Yang edge singularity. *Physical Review Letters*, 46:871–874, 1981.

360. Deepak Dahr. Exact solution of a directed-site animals-enumeration problem in three dimensions. *Physical Review Letters*, 51:853–856, 1983.

361. D. Stauffer. Monte Carlo study of density profile, radius and perimeter for percolation clusters and lattice animals. *Physical Review Letters*, 41:1333–1336, 1978.

362. P. L. Leath. Cluster size and boundary distribution near percolation threshold. *Physical Review*, B14:5046–5055, 1976.

363. Z. Alexandrowicz. Critically branched chains and percolation clusters. *Physics Letters*, A80:284–286, 1980.

364. Zorica V. Djordjevic, Shlomo Havlin, H. Eugene Stanley and George H. Weiss. New method for growing branched polymers and large percolation clusters below p_c. *Physical Review*, B30:478–481, 1984.

365. H. P. Peters, D. Stauffer, H. P. Hölters and K. Loewenich. Radius, perimeter and density profile for percolation clusters and lattice animals. *Zeitschrift für Physik*, B34:399–408, 1979.

366. S. Havlin, Zorica V. Djordjevic, Imtiaz Majid, H. E. Stanley and G. H. Weiss. Relation between dynamic transport properties and static topological structure for the lattice-animal model of branched polymers. *Physical Review Letters*, 53:178–181, 1984.

367. Paul J. Flory. *Statistical Mechanics of Chain Molecules*. Wiley, New York, 1969.

368. A. Peterlin. Conformation of polymer molecules. In Arthur V. Tobolsky and Herman F. Mark, editors, *Polymer science and materials*, pages 41–66. Wiley, New York, 1971.

369. George H. Weiss and Robert J. Rubin. Random walks: Theory and selected applications. *Advances in Chemical Physics*, 52:363–505, 1983.

370. Michael N. Barber and B. W. Ninham. *Random and restricted walks*. Gordon and Breach, New York, 1970.

371. Georg Pólya. Über eine aufgabe der wahrscheinlichkeitsrechnung betreffend die irrfaht im straßennetz. *Mathematische Annalen*, 84:149–160, 1921.

372. Paul Meakin. Cluster-particle aggregation with fractal (Levy flight) particle trajectories. *Physical Review*, B29:3722–3725, 1984.

373. Daniel J. Amit, G. Parisi and L. Peliti. Asymptotic behavior of the "true" self-avoiding walk. *Physical Review*, B27:1635–1645, 1983.

374. Fereydoon Family and M. Daoud. Experimental realization of true self-avoiding walks. *Physical Review*, B29:1506–1507, 1994.

375. P. J. Flory. *Principles of polymer chemistry*. Cornell University, Ithaca, New York, 1971.

376. J. C. Le Guillou and J. Zinn-Justin. Critical exponents for the n-vector model in three dimensions from field theory. *Physical Review Letters*, 39:95–98, 1977.

377. Jean Dayantis and Jean-François Palierne. Scaling exponents of the self-avoiding-walk-problem in three dimensions. *Physical Review*, B49:3217–3225, 1994.

378. F. T. Wall and J. J. Erpenbeck. New method for the statistical computation of polymer dimensions. *Journal of Chemical Physics*, 30:634–637, 1959.

379. Gerhard Zifferer. Monte Carlo simulation of tetrahedral chains, 4 size and shape of

linear and star-branched polymers. *Die Makromolekulare Chemie*, 192:1555–1566, 1991.

380. Z. Alexandrowicz. Monte Carlo of chains with excluded volume: a way to evade sample attrition. *Journal of Chemical Physics*, 51:561–565, 1969.

381. Hagai Meirovitch. A new method for simulation of real chains: scanning future steps. *Journal of Physics*, A15:L735–L741, 1982.

382. Neal Madras and Alan D. Sokal. The pivot algorithm: a highly efficient Monte Carlo method for the self-avoiding walk. *Journal of Statistical Physics*, 50:109–186, 1988.

383. K. Kremer and J. W. Lyklema. Indefinitely growing self-avoiding walk. *Physical Review Letters*, 54:267–279, 1985.

384. Abel Weinrib and S. A. Trugman. A new kinetic walk and percolation perimeters. *Physical Review*, B31:2993–2997, 1985.

385. Jean-Marc Debierre and Loïc Turban. Monte Carlo study of linear diffusion-limited aggregation. *Journal of Physics*, A19:L131–L135, 1986.

386. R. M. Bradley and D. Kung. Diffusion-limited growth of polymer chains. *Physical Review*, A34:723–725, 1986.

387. J. W. Lyklema and C. Evertsz. Reflecting and absorbing boundary conditions on the tail of the Laplacian walk. *Journal of Physics*, A19:L895–L900, 1986.

388. J. W. Lyklema, C. Evertsz and L. Pietronero. The Laplacian random walk. *Europhysics Letters*, 2:77–82, 1986.

389. Paul Meakin. Diffusion-limited polymerization and surface growth. *Physical Review*, A37:2644–2659, 1988.

390. Gregory F. Lawler. A self-avoiding random walk. *Duke Mathematical Journal*, 47:655–693, 1980.

391. Gregory F. Lawler. Random walks, harmonic measure and Laplacian growth models. In Geoffrey Grimmett, editor, *Probability and phase transition*, pages 191–208. Kluwer, Dordrecht, 1994. NATO ASI C42.

392. S. N. Majumdar. Exact fractal dimension of the loop-erased self-avoiding walk in two dimensions. *Physical Review Letters*, 68:2329–2331, 1992.

393. R. Mark Bradley, David Kung, Jean-Marc Debierre and Loïc Turban. Comment on 'Reflecting and absorbing boundary conditions on the tail of the Laplacian walk'. *Journal of Physics*, A20:3547–3550, 1987.

394. Sasuke Miyazima, Yutaka Hasegawa, Armin Bunde and H. Eugene Stanley. A generalized diffusion-limited aggregation where aggregate sites have a finite radical time. *Journal of the Physical Society of Japan*, 57:3376–3380, 1988.

395. Imtiaz Majid, Daniel Ben-Avraham, Shlomo Havlin and H. Eugene Stanley. Exact-enumeration approach to random walks on percolation clusters in two dimensions. *Physical Review*, B30:1626–1628, 1984.

396. Shlomo Havlin and Benes L. Trus. Exact enumeration method for diffusion-limited aggregation. *Journal of Physics*, A21:L731–L735, 1988.

397. Paul Meakin. Random walks on multifractal lattices. *Journal of Physics*, A20:L771–L777, 1987.

398. Paul Meakin. Fractal aggregates in geophysics. *Reviews of Geophysics*, 29:317–354, 1991.

399. Frank Schweitzer and Lutz Schimansky-Geier. Clustering of "active" walkers in a two-component system. *Physica*, A206:359–379, 1994.

400. Harald Freund and Peter Grassberger. The Red Queen's walk. *Physica*, A190:218–237, 1992.

401. R. Jullien and R. Botet. Surface thickness in the Eden model. *Physical Review Letters*, 54:2055, 1985.

402. Michael Plischke and Zoltán Rácz. Active zone of growing clusters: Diffusion-limited aggregation and the Eden model. *Physical Review Letters*, 53:415–418, 1984.

403. Zoltán Rácz and Michael Plischke. Active zone of growing clusters: Diffusion-limited aggregation and the Eden model in two and three dimensions. *Physical Review*, A31:985–994, 1985.

404. Trevor Williams and Rolf Bjerknes. Stochastic model for abnormal clone spread through epithelial basal layer. *Nature*, 236:19–21, 1972.

405. Felix Hausdorff. Dimension und äußeres maß. *Mathematische Annalen*, 79:157–179,

1918.

406. P. Meakin. The Williams-Bjerknes model: Simulation results. Unpublished, 1984.

407. Denis Mollison. Conjecture on the spread of infection in two dimensions disproved. *Nature*, 240:467–468, 1972.

408. Paul Meakin. Cluster-growth processes on a two-dimensional lattice. *Physical Review*, B28:6718–6732, 1983.

409. Paul Meakin. Invasion percolation and Eden growth on multifractal lattices. *Journal of Physics*, A21:3501–3522, 1988.

410. Paul Meakin. Eden growth on multifractal lattices. *Journal of Physics*, A20:L779–L784, 1987.

411. R. Jullien and R. Botet. *Aggregation and fractal aggregates*. World Scientific, Singapore, 1987.

412. C. Y. Wang, P. L. Lin and J. B. Bassingthwaighte. The off-lattice Eden-c model. Preprint, 1994.

413. James B. Bassingthwaighte, Larry S. Liebowitch and Bruce J. West. *Fractal physiology*. Oxford University Press, New York, 1994. The American Physiological Society Methods in Physiology Series.

414. Marjorie J. Vold. Computer simulation of floc formation in a colloidal suspension. *Journal of Colloid Science*, 18:684–695, 1963.

415. D. N. Sutherland. Comments on Vold's simulation of floc formation. *Journal of Colloid and Interface Science*, 22:300–302, 1966.

416. H. P. Hutchison and D. N. Sutherland. An open-structured random solid. *Nature*, 206:1036–1037, 1965.

417. Marjorie J. Vold. The sediment volume in dilute dispersions of spherical particles. *Journal of Physical Chemistry*, 64:1616–1619, 1960.

418. Marjorie J. Vold. Sediment volume and structure in dispersions of anisometric particles. *Journal of Physical Chemistry*, 63:1608–1612, 1960.

419. T. A. Witten and L. M. Sander. Diffusion-limited aggregation. *Physical Review*, B27:5686–5697, 1983.

420. M. Batty. Generating urban forms from diffusive growth. *Environment and Planning*, 23:511–544, 1991.

421. A. Haji-Sheikh and E. M. Sparrow. The solution of heat conduction problems by probability methods. *Journal of Heat Transfer*, 89:121–131, May 1967.

422. R. C. Ball and R. M. Brady. Large scale lattice effect in diffusion-limited aggregation. *Journal of Physics*, A18:L809–L813, 1985.

423. Paul Meakin. The structure of two-dimensional Witten-Sander aggregates. *Journal of Physics*, A18:L661–L666, 1985.

424. Susan Tolman and Paul Meakin. Off-lattice and hypercubic-lattice models for diffusion-limited aggregation in dimensionalities $2-8$. *Physical Review*, A40:428–437, 1988.

425. Paul Meakin. The growth of fractal aggregates and their fractal measures. In C. Domb and J. L. Lebowitz, editors, *Phase transitions and critical phenomena*, volume 12, pages 336–489. Academic Press, London, 1988.

426. Henry Kaufman, Alessandro Vespignani, Benoit B. Mandelbrot and Lionel Woog. Parallel diffusion-limited aggregation. *Physical Review*, E52:5602–5609, 1995.

427. Paul Meakin. Particle-cluster aggregation with fractal particle trajectories and on fractal substrates. In F. Family and D. P. Landau, editors, *Kinetics of aggregation and gelation*, pages 91–99. North-Holland, Amsterdam, 1984.

428. H. E. Stanley and Paul Meakin. Multifractal phenomena in physics and chemistry. *Nature*, 335:405–409, 1988.

429. L. Niemeyer, L. Pietronero and H. J. Wiesmann. Fractal dimension of dielectric breakdown. *Physical Review Letters*, 47:1033–1036, 1984.

430. de G. Allen. *Relaxation methods*. McGraw-Hill, New York, 1954.

431. William H. Press, Brian P. Flannery, Saul A. Teukolsky and T. Vetterling William. *Numerical recipes*. Cambridge University Press, Cambridge, 1989.

432. Mitsugu Matsushita, Katsuya Honda, Hiroyasu Toyoki, Yoshinori Hayakawa and Hiroshi Kondo. Generalization and the fractal dimensionality of diffusion-limited aggregation. *Journal of The Physical Society of Japan*, 55:2618–2626, 1986.

433. Yoshinori Hayakawa, Hiroshi Konda and Mitsugu Matsushita. Monte Carlo simulations of the generalized

diffusion-limited aggregation. *Journal of The Physical Society of Japan*, 55:2479–2482, 1986.

434. Michael Murat and Amnon Aharony. Viscous fingers and diffusion-limited aggregation near percolation. *Physical Review Letters*, 57:1875–1878, 1986.

435. L. Pietronero and H. J. Wiesmann. Stochastic model for dielectric breakdown. *Journal of Statistical Physics*, 36:909–916, 1984.

436. B. Derrida and V. Hakim. Needle models of Laplacian growth. *Physical Review*, A45:8759–8765, 1992.

437. A. Sánchez, F. Guinea, L. M. Sander, V. Hakim and E. Louis. Growth and forms of Laplacian aggregates. *Physical Review*, E48:1296–1304, 1993.

438. Fereydoon Family, Daniel E. Platt and Tamás Vicsek. Deterministic growth model of pattern formation in dendritic solidification. *Journal of Physics*, A20:L1177–L1183, 1987.

439. Fereydoon Family, Tamás Vicsek and Becky Taggett. Lattice-induced anisotropy in a diffusion-limited growth model. *Journal of Physics*, A19:L727–L732, 1986.

440. Leonid A. Turkevich and Harvey Sher. Occupancy-probability scaling in diffusion-limited aggregation. *Physical Review Letters*, 55:1026–1029, 1985.

441. Wei Wang and E. Canessa. Biharmonic pattern selection. *Physical Review*, E47:1243–1248, 1993.

442. E. Canessa and Wei Wang. Effects of long range coupling on aggregation. *Zeitschrift für Naturforsch*, 48a:945–946, 1993.

443. T. C. Halsey and Paul Meakin. Axial inhomogeneity in diffusion-limited aggregation. *Physical Review*, A32:2546–2549, 1986.

444. Hans Meinhardt. *The algorithmic beauty of sea shells*. Springer-Verlag, New York, 1995.

445. P. Meakin and L. M. Sander. Comment on "Active zone of growing clusters: Diffusion-limited aggregation and the Eden model". *Physical Review Letters*, 54:2053, 1985.

446. Paul Meakin and Tamás Vicsek. Internal structure of diffusion-limited aggregates. *Physical Review*, A32:685–688, 1985.

447. M. Kolb. Effects of the growth mechanism on the structure of aggregation clusters. *Journal de Physique*, 46:L631–L637, 1985.

448. A. Arneodo, F. Argoul, E. Bacry, J. F. Muzy and M. Tabard. Golden mean arithmetic in the fractal branching of diffusion-limited aggregates. *Physical Review Letters*, 68:3456–3459, 1992.

449. B. B. Mandelbrot, A. Vespignani and H. Kaufman. Crosscut analysis of large radial DLA: departures from self-similarity and lacunarity effects. *Europhysics Letters*, 32:199–204, 1995.

450. F. Argoul, A. Arneodo, G. Grasseau and Harry L. Swinney. Self-similarity of diffusion-limited aggregates and electrodeposition clusters. *Physical Review Letters*, 61:2558–2561, 1988.

451. F. Argoul, A. Arneodo, J. Elezgaray, G. Grasseau and R. Murenzi. Wavelet transform of fractal aggregates. *Physics Letters*, A135:327–336, 1989.

452. F. Argoul, A. Arneodo, J. Elezgaray, G. Grasseau and R. Murenzi. Wavelet analysis of the self-similarity of diffusion-limited aggregates and electrodeposition clusters. *Physical Review*, A41:5537–5560, 1990.

453. Thomas Rage and Paul Meakin. Large scale simulations of $1+1$-dimensional, off-lattice diffusion-limited deposition. Unpublished, 1995.

454. E. Erzan, L. Pietronero and A. Vespignani. The fixed scale transformation approach to fractal growth. *Reviews of Modern Physics*, 67:545–604, 1995.

455. Peter Ossadnik. Multiscaling analysis of large-scale off-lattice DLA. *Physica*, A176:454–462, 1991.

456. Paul Meakin and Shlomo Havlin. Fluctuations and distributions in random aggregates. *Physical Review*, A36:4428–4433, 1987.

457. Avidan U. Neumann and Shlomo Havlin. Distributions and moments of structural properties for percolation clusters. *Journal of Statistical Physics*, 52:203–236, 1988.

458. Paul Meakin and Susan Tolman. Diffusion-limited aggregation: Recent developments. In L. Pietronero, editor, *Fractals' physical origin and properties*, pages 137–168. Plenum, New York, 1987.

459. Benoit B. Mandelbrot. Plane DLA is not

self-similar; is it a fractal that becomes increasingly compact as it grows? *Physica*, A191:95–107, 1992.

460. D. Stauffer, A. Aharony and B. B. Mandelbrot. Self-similarity and covered neighborhoods of fractals: A random walk test. *Physica*, A196:1–5, 1993.

461. C. Amitrano, A. Coniglio, P. Meakin and M. Zannetti. Evidence for multiscaling in large DLA clusters. *Fractals*, 1:840–845, 1993.

462. Paul Meakin. Scaling properties for the growth probability measure and harmonic measure of fractal aggregates. *Physical Review*, A35:2234–2245, 1987.

463. Peter Ossadnik and Jysoo Lee. Power law tail in the radial growth probability distribution for DLA. *Journal of Physics*, A26:6789–6796, 1993.

464. Rainer Hegger and Peter Grassberger. Is diffusion-limited aggregation locally isotropic or self-affine? *Physical Review Letters*, 73:1672–1674, 1994.

465. Paul Meakin. Diffusion-controlled deposition on fibers and surfaces. *Physical Review*, A27:2616–2623, 1983.

466. P. Meakin, I. Majid, S. Havlin and H. Eugene Stanley. Topological properties of diffusion limited aggregation and cluster-cluster aggregation. *Journal of Physics*, A17:L975–L981, 1984.

467. J. Vannimenus, J. P. Nadal and H. Martin. On the spreading dimension of percolation and directed percolation. *Journal of Physics*, A17:L351–L356, 1984.

468. S. Havlin and R. Nossal. Topological properties of percolation clusters. *Journal of Physics*, A17:L427–L432, 1984.

469. A. Arneodo, F. Argoul, Y. Couder and M. Rabaud. Anisotropic Laplacian growths: From diffusion-limited aggregates to dendritic fractals. *Physical Review Letters*, 66:2332–2335, 1991.

470. P. G. Saffman and G. I. Taylor. The penetration of a fluid into a porous medium or Hele-Shaw cell containing a more viscous fluid. *Proceedings of the Royal Society, London*, 1958:312–329, 1958.

471. Martine Ben Amar. Exact self-similar shapes in viscous fingering. *Physical Review*, A43:5724–5727, 1991.

472. Paul Meakin and J. M. Deutch. The formation of surfaces by diffusion limited annihilation. *The Journal of Chemical Physics*, 85:2320–2325, 1986.

473. W. J. McG. Tegart. *The electropolishing and chemical polishing of metals in research and industry*. Pergamon, London, 1959.

474. J. Edwards. The mechanism of electropolishing of copper in phosphoric acid solutions I. Processes preceding the establishment of polishing conditions. *Journal of the Electrochemical Society*, 100:189C–194C, 1953.

475. J. Edwards. The mechanism of electropolishing of copper in phosphoric acid solutions II. The mechanism of smoothing. *Journal of the Electrochemical Society*, 100:223C–230C, 1953.

476. Carl Wagner. Contribution to the theory of electropolishing. *Journal of the Electrochemical Society*, 101:225–228, 1954.

477. Roland Lenormand, Eric Touboul and Cesar Zarcone. Numerical models and experiments on immiscible displacements in porous media. *Journal of Fluid Mechanics*, 189:165–187, 1988.

478. Joachim Krug and Paul Meakin. Kinetic roughening of Laplacian fronts. *Physical Review Letters*, 66:703–706, 1991.

479. David Bensimon, Leo P. Kadanoff, Shoudan Liang, Boris I. Shraiman and Chao Tang. Viscous flows in two dimensions. *Reviews of Modern Physics*, 58:977–999, 1986.

480. S. R. Broadbent and J. M. Hammersley. Percolation processes. Crystals and mazes. *Proceedings of the Cambridge Philosophical Society*, 53:629–641, 1957.

481. P. W. Kasteleyn and C. M. Fortuin. Phase transitions in lattice systems with random local properties. *Journal of the Physical Society of Japan*, 26 Supplement:11–14, 1969.

482. Antonio Coniglio. Fractal structure of Ising and Potts clusters: Exact results. *Physical Review Letters*, 62:3054–3057, 1989.

483. R. P. Langlands, C. Pichet, Ph. Pouliot and T. Saint-Aubin. On the universality of crossing probabilities in two-dimensional percolation. *Journal of Statistical Physics*, 67:553–574, 1992.

484. P. Grassberger. Spreading and backbone dimensions of 2*D* percolation. *Journal of Physics*, A25:5475–5484, 1992.

485. Robert M. Ziff. Spanning probability in 2D percolation. *Physical Review Letters*, 69:2670–2673, 1992.

486. John L. Cardy. Critical percolation in finite geometries. *Journal of Physics*, A25:L201–L206, 1992.

487. Dietrich Stauffer, Joan Adler and Amnon Aharony. Universality at the three-dimensional percolation threshold. *Journal of Physics*, A27:L475–L480, 1994.

488. M. P. M. Den Nijs. A relation between the temperature exponents of the eight-vertex and q-state Potts model. *Journal of Physics*, A12:1857–1868, 1979.

489. Robert M. Ziff and G. Stell. Critical behavior in three-dimensional percolation: is the percolation threshold a Lifshitz point? Technical Report 88-4, University of Michigan Laboratory for Scientific Computing, 1988.

490. Tsuneyoshi Nakayama, Kousuke Yakubo and Raymond L. Orbach. Dynamical properties of fractal networks: Scaling, numerical simulations and physical realizations. *Reviews of Modern Physics*, 66:381–443, 1994.

491. V. Privman, P. C. Hohenberg and A. Aharony. Universal critical-point amplitude relations. In C. Domb and J. L. Lebowitz, editors, *Phase transitions and critical phenomena*, volume 14, pages 1–134. Academic Press, London, 1991.

492. R. B. Pearson. Conjecture for the extended Potts model magnetic eigenvalue. *Physical Review*, B22:2579–2580, 1980.

493. P. N. Strensky, R. Mark Bradley and Jean-Mark Debierre. Scaling behavior of percolation surfaces in three dimensions. *Physical Review Letters*, 66:1330–1333, 1991.

494. P. Grassberger. Numerical studies of critical percolation in three dimensions. *Journal of Physics*, A25:5867–5888, 1992.

495. Robert Pike and H. Eugene Stanley. Order propagation near the percolation threshold. *Journal of Physics*, A14:L169–L177, 1981.

496. G. Shlifer, W. Klein, P. J. Reynolds and H. E. Stanley. Large-cell renormalization group for the backbone problem in percolation. *Journal of Physics*, A12:L169–L174, 1979.

497. Hans J. Herrmann and H. E. Stanley. Building blocks of percolation: Volatile fractals. *Physical Review Letters*, 53:1121–1124, 1984.

498. H. J. Herrmann, D. C. Hong and H. E. Stanley. Backbone and elastic backbone of percolation clusters obtained by the new method of 'burning'. *Journal of Physics*, A17:L261–L266, 1984.

499. H. Eugene Stanley. Cluster shapes at the percolation threshold: an effective cluster dimensionality and its connection with critical-point exponents. *Journal of Physics*, A10:L211–L220, 1977.

500. M. Rosso B. Sapoval and J. F. Gouyet. The fractal nature of a diffusion front and the relation to percolation. *Journal de Physique Lettres*, 46:L149–L156, 1985.

501. Antonio Coniglio, Naeem Jan, Imtiaz Majid and H. Eugene Stanley. Conformation of a polymer chain at the θ' point: connection to the external perimeter of a percolation cluster. *Physical Review*, B35:3617–3620, 1987.

502. Tal Grossman and Amnon Aharony. Structure and perimeters of percolation clusters. *Journal of Physics*, A19:L745–L751, 1986.

503. Tal Grossman and Amnon Aharony. Accessible external perimeters of percolation clusters. *Journal of Physics*, A20:L1193–1201, 1987.

504. Hans J. Herrmann and H. Eugene Stanley. The fractal dimension of the minimum path in two- and three-dimensional percolation. *Journal of Physics*, A21:L829–L833, 1988.

505. J.-P. Hovi and Amnon Aharony. Distributions of structural properties for percolation clusters. *Fractals*, 3:453–463, 1995. Mandelbrot festschrift, Curaçao.

506. Shlomo Havlin, Benes Trus, George H. Weiss and Daniel Ben-Avraham. The chemical distance distribution in percolation clusters. *Journal of Physics*, A18:L247–L249, 1985.

507. S. Havlin, R. Nossal, B. Trus and G. H. Weiss. Universal substructures of percolation clusters: The skeleton. *Journal of Physics*, A17:L957–L960, 1984.

508. P. Grassberger. On the spreading of two-dimensional percolation. *Journal of Physics*, A18:L215–L219, 1985.

509. Thomas A. Larsson. Possibly exact fractal

dimensions from conformal invariance. *Journal of Physics*, A20:L291–L297, 1987.

510. Richard Zallen and Harvey Scher. Percolation on a continuum and the localization-delocalization transition in amorphous semiconductors. *Physical Review*, B4:4471–4479, 1971.

511. M. B. Isichenko. Percolation, statistical topography and transport in random media. *Reviews of Modern Physics*, 64:961–1043, 1992.

512. M. Rosso. Concentration gradient approach to continuum percolation in two dimensions. *Journal of Physics*, A22:L131–L136, 1989.

513. M. Kolb. Crossover from standard to reduced hull for random percolation. *Physical Review*, A41:5725–5727, 1990.

514. M. Rosso, J. F. Gouyet and B. Sapoval. Determination of percolation probability from the use of a concentration gradient. *Physical Review*, B32:6053–6054, 1985.

515. Paul Meakin. Invasion percolation on substrates with correlated disorder. *Physica*, A173:305–324, 1991.

516. Liv Furuberg. *Computer simulation studies of slow drainage in porous media.* PhD thesis, University of Oslo, 1994.

517. M. Kolb and M. Rosso. Loop structure of percolation hulls. *Physical Review*, E47:3081–3086, 1993.

518. A. Brooks Harris. Effect of random defects in the critical behavior of Ising models. *Journal of Physics*, C7:1671–1692, 1974.

519. J. T. Chayes, L. Chayes, Daniel Fisher and T. Spencer. Finite-size scaling and correlation lengths for disordered systems. *Physical Review Letters*, 57:2999–3003, 1986.

520. Abel Weinrib and B. I. Halperin. Critical phenomena in systems with long-range-correlated quenched disorder. *Physical Review*, B27:413–427, 1983.

521. Robert M. Ziff, Peter. T. Cummings and G. Stell. Generation of percolation cluster perimeters by a random walk. *Journal of Physics*, A17:3009–3017, 1984.

522. Robert M. Ziff and B. Sapoval. The efficient determination of the percolation threshold by a frontier-generating walk in a gradient. *Journal of Physics*, A19:L1169–L1172, 1986.

523. M. F. Sykes and J. W. Essam. Some exact critical percolation probabilities for bond and site problems in two dimensions. *Physical Review Letters*, 10:3–4, 1963.

524. K. D. Keefer. Growth and structure of fractally rough colloids. In C. J. Brinker, D. E. Clark and D. R. Ulrich, editors, *Better ceramics through chemistry II*, pages 295–304. Materials Research Society, Pittsburgh, Pennsylvania, 1986. Materials Research Society Symposium Proceedings Volume 73.

525. K. D. Keefer and D. W. Schaefer. Growth of fractally rough colloids. *Physical Review Letters*, 56:2376–2379, 1986.

526. Thomàs Sintes, Raúl Toral and Amitabha Chakrabarti. Fractal structure of silica colloids revisited. *Journal of Physics*, A29:533–540, 1996.

527. T. M. Shaw. Drying as an immiscible displacement process with fluid counterflow. *Physical Review Letters*, 59:1671–1674, 1987.

528. Lee M. Hirsch and Arthur Thompson. Size-dependent scaling of capillary invasion including buoyancy and pore size distribution effects. *Physical Review*, E50:2069–2086, 1994.

529. R. Lenormand, C. Zarcone and A. Sarr. Mechanisms of the displacement of one fluid by another in a network of capillary ducts. *Journal of Fluid Mechanics*, 135:337–353, 1983.

530. Roland Lenormand and Cesar Zarcone. Invasion percolation in an etched network: Measurement of a fractal dimension. *Physical Review Letters*, 54:2226–2229, 1985.

531. Qi zhong Cao and Po zen Wong. Fractal interfaces in heterogeneous Eden-like growth. *Physical Review Letters*, 67:77–80, 1991.

532. Mark A. Knackstedt and Stephen F. Cox. Percolation and the pore geometry of crustal rocks. *Physical Review*, E51:R5181–R5184, 1995.

533. Ricardo Paredes V and Miguel Octavio. Invasion percolation into a percolating cluster. *Physical Review*, A46:994–1001, 1992.

534. M. Rosso B. Sapoval and J. F. Gouyet. Fractal interfaces in diffusion, invasion and corrosion. In D. Avnir, editor, *The fractal approach to heterogeneous chemistry:*

surfaces, colloids, polymers, pages 227–244. Wiley, Chichester, 1989.

535. Bernard Sapoval, Michel Rosso, Jean-François Gouyet and Jean-François Colonna. Dynamics of the creation of fractal objects by diffusion and $1/f$ noise. *Solid State Ionics*, 18&19:21–30, 1986.

536. M. Rosso, J. F. Gouyet and B. Sapoval. Gradient percolation in three dimensions and relation to diffusion fronts. *Physical Review Letters*, 57:3195–3198, 1986.

537. J.-F. Gouyet, M. Rosso and B. Sapoval. Fractal structure of diffusion and invasion fronts in three-dimensional lattices through the gradient percolation approach. *Physical Review*, B37:1832–1838, 1988.

538. J. F. Gouyet and Y. Boughaleb. Structure of noise generated on diffusion fronts. *Physical Review*, B40:4760–4768, 1981.

539. M. Kolb, J. F. Gouyet and B. Sapoval. Diffusion of interacting particles in a concentration gradient: scaling, critical slowing down and phase separation. *Europhysics Letters*, 3:33–38, 1987.

540. J. W. Essam, A. J. Guttmann and K. De'Bell. On two-dimensional directed percolation. *Journal of Physics*, A21:3815–3832, 1988.

541. Iwan Jensen. Low-density series expansions for directed percolation on square and triangular lattices. *Journal of Physics*, A29:7013-7040, 1996.

542. Wolfgang Kinzel. Directed percolation. In G. Deutscher, R. Zallen and Joan Adler, editors, *Percolation structures and processes*, chapter 18, pages 425–445. Hilger/The Israel Physical Society, Bristol/Jerusalem, 1983. Annals of the Israel Physical Society, Volume 5.

543. Per Arne Rikvold. Simulations of a stochastic model for cluster growth on a square lattice. *Physical Review*, A26:647–650, 1982.

544. Paul Meakin. Scaling of the growth probability measure for fractal structures. *Physical Review*, A34:710–713, 1986.

545. Xian-zhi Wang and Yun Huang. Formula of the generalized dimensions for the screened-growth model. *Physical Review*, A46:1035–1037, 1992.

546. Paul Meakin, Francois Leyvraz and H. Eugene Stanley. New class of screened growth aggregates with a continuously tunable fractal dimension. *Physical Review*, A31:1195–1198, 1985.

547. L. M. Sander. Theory of fractal growth processes. In F. Family and D. P. Landau, editors, *Kinetics of aggregation and gelation*, pages 13–17. North-Holland, Amsterdam, 1984.

548. Deepak Dahr. Asymptotic shape of Eden clusters. In H. Eugene Stanley and Nicole Ostrowsky, editors, *On growth and form: fractal and non-fractal patterns in physics*, pages 288–292. Martinus Nijhoff, Dordrecht, 1986. NATO ASI Series E100.

549. P. Freche, D. Stauffer and H. E. Stanley. Surface structure and anisotropy of Eden clusters. *Journal of Physics*, A18:L1163–L1168, 1985.

550. R. Hirsch and D. E. Wolf. Anisotropy and scaling of Eden clusters in two and three dimensions. *Journal of Physics*, A19:L251–L256, 1986.

551. P. Meakin, R. Jullien and R. Botet. Large scale numerical investigation of the surface of Eden clusters. *Europhysics Letters*, 1:609–615, 1986.

552. P. Devillard and H. Spohn. Kinetic shape of Ising clusters. *Europhysics Letters*, 17:113–118, 1991.

553. Daniel Richardson. Random growth in a tesselation. *Proceedings of the Cambridge Philosophical Society*, 74:515–528, 1973.

554. Richard Durrett and Thomas M. Liggett. The shape of the limit set in Richardson's growth model. *Annals of Probability*, 9:186–193, 1981.

555. J. Krug and H. Spohn. Kinetic roughening of growing surfaces. In C. Godrèche, editor, *Solids far from equilibrium: Growth morphology and defects*, pages 479–582. Cambridge University Press, Cambridge, 1991.

556. Paul Meakin. Noise-reduced and anisotropy enhanced Eden and screened-growth models. *Physical Review*, A38:418–426, 1988.

557. M. T. Batchelor and B. L. Henry. Limits to Eden growth in two and three dimensions. *Physics Letters*, A157:229–236, 1991.

558. G. Wulff. Zur frage der geschwindigkeit des wachsthums und der auflösung der krystallflächen. *Zeitschrift für*

Krystallographie und Mineralogie, 34:449–530, 1901.

559. D. E. Wolf. Wulff construction and anisotropic surface properties of two-dimensional Eden clusters. *Journal of Physics*, A20:1251–1258, 1987.

560. Robert Savit and Robert M. Ziff. Morphology of a class of kinetic growth models. *Physical Review Letters*, 55:2515–2518, 1985.

561. George Y. Onoda and John Toner. Deterministic, fractal defect structures in close packings of hard discs. *Physical Review Letters*, 57:1340–1343, 1986.

562. Paul Meakin and Rémi Jullien. Two-dimensional defect-free random packing. *Europhysics Letters*, 14:667–672, 1991.

563. E. M. Tory, C. B. Yhap and D. K. Pickard. Periodicity and fine structure in two-dimensional gravitational packing. *Particulate Science and Technology*, 3:89–99, 1985.

564. Paul Meakin and Rémi Jullien. Periodic disc packing generated by random deposition in narrow channels. *Europhysics Letters*, 15:851–856, 1991.

565. Joshua E. S. Socolar. Theory of packings of identical hard discs generated by ballistic deposition. *Europhysics Letters*, 18:39–44, 1992.

566. Johann Nittmann and H. Eugene Stanley. Role of fluctuations in viscous fingering and dendritic crystal growth: a noise-driven model with non-periodic sidebranching and no threshold for onset. *Journal of Physics*, A20:L981–L986, 1987.

567. A. Dougherty, P. D. Kaplan and J. P. Gollub. Development of side branching in dendritic crystal growth. *Physical Review Letters*, 58:1652–1655, 1987.

568. A. Dougherty J. P. Gollub. Steady-state growth of NH_4Br from solution. *Physical Review*, A38:3043–3053, 1987.

569. Fong Liu and Nigel Goldenfield. Generic features of late stage crystal growth. *Physical Review*, A42:895–903, 1990.

570. A. P. Roberts and M. A. Knackstedt. Growth in non-Laplacian fields. *Physical Review*, E47:2724–2728, 1993.

571. Paul Meakin, Jens Feder and Torstein Jøssang. Radially biased diffusion-limited aggregation. *Physical Review*, 37:1952–1964, 1991.

572. Peter Garik, Roy Richter, J. Hautman and P. Ramamlal. Deterministic solutions of fractal growth. *Physical Review*, A32:3156–3159, 1987.

573. Paul Meakin and Fereydoon Family. Diverging length scales in diffusion-limited aggregation. *Physical Review*, A34:2558–2560, 1986.

574. C. Evertsz. Self-affine nature of dielectric-breakdown model clusters in a cylinder. *Physical Review*, A41:1830–1842, 1990.

575. Paul Meakin. Effects of particle drift on diffusion-limited aggregation. *Physical Review*, B28:5221–5224, 1983.

576. J. P. Nadal, B. Derrida and J. Vannimenus. Directed diffusion-controlled aggregation versus directed animals. *Physical Review*, B30:376–383, 1984.

577. T. Nagatani. Fractal structure of drift-diffusion-limited aggregation: Renormalization-group approach. *Physical Review*, A37:3514–3519, 1988.

578. Takashi Nagatani. Laplacian growth phenomena with the third boundary condition: Crossover from dense structure to diffusion-limited aggregation fractal. *Physical Review*, A40:7286–7291, 1989.

579. Paul Meakin. Diffusion-limited aggregation growth of a cluster rotating with a constant angular velocity. Unpublished, 1984.

580. Takashi Nagatani. Growth model with phase transition: Drift-diffusion-limited aggregation. *Physical Review*, A39:438–441, 1985.

581. H. G. E. Hentschel, J. M. Deutch and P. Meakin. The effect of rotational Brownian motion on Witten-Sander aggregates. *Journal of Chemical Physics*, 85:2147–2153, 1986.

582. N. Lemke, M. G. Malcum, R. M. C. de Almeida, P. M. Mors and J. R. Iglesias. Growth and form of two-dimensional rotating aggregates. *Physical Review*, E47:3218–3224, 1993.

583. Takashi Nagatani and Francesc Saugués. Morphological changes in convection-diffusion-limited deposition. *Physical Review*, A43:2970–2976, 1991.

584. P. B. Warren, R. C. Ball and A. Boelle.

Convection-limited aggregation. *Europhysics Letters*, 29:339–344, 1995.

585. Richard F. Voss. Multiparticle diffusive fractal aggregation. *Physical Review*, B30:334–337, 1984.

586. Paul Meakin and J. M. Deutch. Monte Carlo simulation of diffusion controlled colloid growth rates in two and three dimensions. *Journal of Chemical Physics*, 80:2115–2122, 1984.

587. Herbert B. Rosenstock and Charles L. Marquardt. Cluster formation in two-dimensional random walks: Application to photolysis of silver halides. *Physical Review*, B22:5797–5809, 1980.

588. R. L. Smith and S. D. Collins. Generalized model for the diffusion-limited aggregation and Eden models of cluster growth. *Physical Review*, A39:5409–5413, 1989.

589. R. W. Bower and S. D. Collins. Fractal transitions in diffusion-limited cluster formation. *Physical Review*, 43:3165–3167, 1991.

590. E. Louis, F Guinea, O. Pla and L. M. Sander. Pattern formation in screened electrostatic fields. *Physical Review Letters*, 68:209–212, 1992.

591. Wei Wang. Effect of a critical field on screened dielectric breakdown growth. *Physical Review*, E48:476–479, 1992.

592. J. Castellá, E. Louis, F. Guinea, O. Pla and L. M. Sander. Pattern formation in screened electrostatic fields: Growth in a channel and in two dimensions. *Physical Review*, E47:2729–2735, 1993.

593. Paul Meakin. Diffusion-controlled cluster formation in 2 – 6-dimensional space. *Physical Review*, A27:1495–1507, 1983.

594. Paul Meakin. Diffusion-controlled cluster formation in two, three and four dimensions. *Physical Review*, A27:604–607, 1983.

595. Paul Meakin. Models for colloidal aggregation. *Annual Reviews of Physical Chemistry*, 39:237–267, 1988.

596. Thomas C. Halsey and Michael Leibig. Theory of branched growth. *Physical Review*, A46:7793–7809, 1992.

597. Paul Meakin. Computer simulation of growth and aggregation processes. In H. Eugene Stanley and Nicole Ostrowsky, editors, *On growth and form: fractal and non-fractal patterns in physics*, pages 111–133. Martinus Nijhoff, Dordrecht, 1986. NATO ASI Series E100.

598. Thomas C. Halsey and Michael Leibig. Electrodeposition and diffusion-limited aggregation. *Journal of Chemical Physics*, 92:3756–3767, 1989.

599. Tamás Vicsek. Pattern formation in diffusion-limited aggregation. *Physical Review Letters*, 53:2281–2284, 1984.

600. Tadeusz Hepel. Effect of surface diffusion in electrodeposition of fractal structures. *Journal of the Electrochemical Society*, 134:2685–2690, 1987.

601. Jayanth R. Banavar, Mahito Kohmoto and James Roberts. Aggregate models of pattern formation. *Physical Review*, A33:2065–2067, 1986.

602. Paul Meakin. Fractal aggregates in chemistry. *Trends in Chemical Physics*, 1:303–347, 1991.

603. M. T. Batchelor, C. R. Dun and B. L. Henry. Growth and form in the zero-noise limit of discrete Laplacian growth processes with inherent surface tension: II. The triangular lattice. *Physica*, A193:553–574, 1993.

604. K. Kassner. Sidebranching in noiseless diffusion-limited aggregation. *Fractals*, 1:205–228, 1993.

605. M. T. Batchelor and B. I. Henry. Branching in the zero-noise limit of discrete Laplacian growth processes. *Physical Review*, A45:4180–4183, 1992.

606. M. T. Batchelor and B. I. Henry. Growth and form in the zero-noise limit of discrete Laplacian growth with inherent surface tension. *Physica*, A187:551–574, 1992.

607. Shonosuke Ohta and Haruo Honjo. Homogeneous and self-similar diffusion-limited aggregation including surface-diffusion processes. *Physical Review*, A44:8425–8428, 1991.

608. Zhenyu Zhang, Xun Chen and Max G. Lagally. Bonding-geometry dependence of fractal growth on metal surfaces. *Physical Review Letters*, 73:1829–1832, 1994.

609. Paul Meakin. Diffusion-limited aggregation with a multifractal distribution of sticking probabilities. Unpublished, 1987.

610. Lincoln Paterson. Diffusion-limited aggregation and two-fluid displacements in porous media. *Physical Review Letters*,

52:1621–1624, 1984.

611. A. P. Ershov, A. L. Kupershtokh and A. Ya. Dammer. Fingering in the fast flow through porous media. *Journal de Physique II France*, 3:955–959, 1993.

612. Shoudan Liang. Random-walk simulation of flow in Hele-Shaw cells. *Physical Review*, A33:2663–2674, 1986.

613. A. Block, W. von Bloh and H. J. Schellnhuber. Aggregation by attractive particle-cluster interaction. *Journal of Physics*, A24:L1037–L1045, 1991.

614. W. von Bloh, A. Block and H. J. Schellnhuber. Growth-zone scaling properties and fjord structure of aggregates grown by particle-cluster interaction. *Physica*, A191:108–112, 1992.

615. Alan J. Hurd. Two-dimensional diffusion-limited aggregation of interacting particles. In Robert B. Laibowitz, Benoit B. Mandelbrot and Dann E. Passoja, editors, *Extended abstracts, fractal aspects of materials*, pages 51–53. Materials Research Society, Pittsburgh, Pennsylvania, 1985. Proceedings of Symposium T, Materials Research Meeting, Boston Dec. 2–Dec. 4, 1985.

616. Alan J. Hurd. Diffusion-limited aggregation of silica microspheres in two dimensions. Preprint.

617. Peter Garik. Anisotropic growth of diffusion-limited aggregates. *Physical Review*, A32:1275–1278, 1985.

618. Paul Meakin. Diffusion-controlled flocculation: The effects of attractive and repulsive interactions. *Journal of Chemical Physics*, 79:2426–2429, 1983.

619. R. Pastor-Satorras and J. M. Rubi. Particle-cluster aggregation with dipolar interactions. *Physical Review*, E51:5994–6003, 1995.

620. Zoltán Rácz and Támas Vicsek. Diffusion-controlled deposition: Cluster statistics and scaling. *Physical Review Letters*, 51:2382–2385, 1983.

621. Paul Meakin. Diffusion-controlled deposition on surfaces: cluster-size distribution, interface exponents and other properties. *Physical Review*, B30:4207–4214, 1984.

622. Paul Meakin, János Kertész and Tamás Vicsek. Noise-reduced diffusion-limited deposition. *Journal of Physics*, A21:1271–1281, 1988.

623. Mitsugu Matsushita and Paul Meakin. Cluster-size distribution of self-affine fractals. *Physical Review*, A37:3645–3648, 1988.

624. Sadao Isogami, Makoto Katori and Mitsugu Matsushita. Structural and statistical properties of competing directed percolation. *Journal of the Physical Society of Japan*, 63:2919–2929, 1994.

625. Sadao Isogami and Mitsugu Matsushita. Structural and statistical properties of directed percolation. *Journal of the Physical Society of Japan*, 62:2200–2203, 1993.

626. Robin C. Ball, Robert M. Brady, Giuseppe Rossi and Bernard R. Thompson. Anisotropy and cluster growth by diffusion-limited aggregation. *Physical Review Letters*, 55:1406–1409, 1990.

627. P. Meakin. Diffusion-limited surface deposition in the limit of large anisotropy. *Physical Review*, A33:1984–1989, 1986.

628. Giuseppe Rossi. Diffusion-limited aggregation without branching. *Physical Review*, A34:3543–3546, 1986.

629. Giuseppe Rossi. Diffusion-limited aggregation without branching: A detailed analysis. *Physical Review*, A35:2246–2253, 1987.

630. J. Krug, K. Kassner, P. Meakin and F. Family. Laplacian needle growth. *Europhysics Letters*, 24:527–532, 1993.

631. Shang keng Ma. Renormalization group by Monte Carlo methods. *Physical Review Letters*, 37:461–464, 1976.

632. Robert H. Swendsen. Monte Carlo renormalization group. *Physical Review Letters*, 42:859–861, 1979.

633. M. E. Cates. Diffusion-limited aggregation without branching in the continuum approximation. *Physical Review*, A34:5007–5009, 1986.

634. K. Kassner. Solutions to the mean-field equations of branchless diffusion-limited aggregation. *Physical Review*, A42:3637–3640, 1990.

635. David S. Graff and Leonard M. Sander. Branch-height distribution in diffusion-limited aggregation. *Physical Review*, E47:R2273–R2276, 1993.

636. Philip M. Morse and Herman Feshbach.

Methods of theoretical physics. McGraw-Hill, New York, 1953. Part I.

637. Joachim Krug. Origins of scale invariance in growth processes. Habilitation thesis, Heinrich-Heine-Universität, Düsseldorf, 1994.

638. Marileen Dogterom and Stanislas Leibler. Physical aspects of the growth and regulation of microtubule structures. *Physical Review Letters*, 70:1347–1350, 1993.

639. M. Dogterom and S. Leibler. The effect of diffusion on the collective growth of linear fibers. *Europhysics Letters*, 24:245–251, 1993.

640. Pierre Devillard and H. Eugene Stanley. First order branching in diffusion-limited aggregation. *Physical Review*, A36:5359–5364, 1987.

641. Joachim Krug. Statistical physics of growth processes. In M. Droze, A. J. McKane, J. Vannimenus and D. E. Wolf, editors, *Scale invariance and non-equilibrium dynamics*, page xx. Plenum, New York, 1995.

642. F. Family and H. G. E. Hentcschel. Asymptotic structure of diffusion-limited aggregation clusters in two dimensions. *Faraday Discussions of the Chemical Society*, 83:139–144, 1987.

643. K. Kassner and F. Family. Scaling behavior of generalized diffusion-limited aggregation: The correct form of the *m*-spoke model. *Physical Review*, A39:4797–4800, 1989.

644. R. C. Ball. Diffusion limited aggregation and its response to anisotropy. *Physica*, A140:62–69, 1986.

645. Paul Meakin. Simplified diffusion-limited aggregation models. *Physica*, A187:1–17, 1992.

646. Mark A. Peterson and James Ferry. Spontaneous symmetry breaking in needle crystal growth. *Physical Review*, A39:2740–2741, 1991.

647. Douglas A. Kurtze. Bifurcation in singular needle growth. *Physical Review*, A43:7066–7067, 1991.

648. Paul Meakin. Noise-reduced diffusion-limited aggregation. *Physical Review*, A36:332–339, 1987.

649. Susan Tolman and Paul Meakin. Two, three and four-dimensional diffusion-limited aggregation models. *Physica*, A158:801–816, 1989.

650. P. Meakin and Susan Tolman. Diffusion-limited aggregation. In M. Fleischmann, D. J. Tildesley and R. C. Ball, editors, *Fractals in the natural sciences*, pages 133–148. Princeton University Press, Princeton, New Jersey, 1990. Royal Society, London. Discussion held 19 and 20 October 1988.

651. Paul Meakin. Determinist screened growth models. *Physica*, A155:37–51, 1989.

652. Peter Garik, Kieran Mullen and Roy Richter. Models of controlled aggregation. *Physical Review*, A35:3046–3055, 1987.

653. Paul Meakin. Universality, nonuniversality and the effects of anisotropy on diffusion-limited aggregation. *Physical Review*, A33:3371–3382, 1986.

654. B. R. Thompson. Unpublished, 1987.

655. C. Moukarzel. Laplacian growth on a random lattice. *Physica*, A190:13–23, 1992.

656. Mitsugu Matsushita and Hiroshi Kondo. Diffusion-limited aggregation with tunable anisotropy. *Journal of the Physical Society of Japan*, 55:2483–2486, 1986.

657. F. Family, Y. C. Zhang and T. Vicsek. Invasion percolation in an external field: dielectric breakdown in random media. *Journal of Physics*, A19:L733–L737, 1986.

658. A. Hansen, E. L. Hinrichsen, S. Roux, H. J. Herrmann and L. De Arcangelis. Deterministic growth of diffusion-limited aggregation with quenched disorder. *Europhysics Letters*, 13:341–347, 1990.

659. Paul Meakin. Diffusion-limited aggregation on multifractal lattices: A model for fluid-fluid displacement in porous media. *Physical Review*, A36:2833–2837, 1987.

660. Paul Meakin, Michael Murat, Amnon Aharony, Jens Feder and Torstein Jøssang. Diffusion-limited aggregation near the percolation threshold. *Physica*, A155:1–20, 1989.

661. R. C. Ball and T. A. Witten. Causality bounds on the density of aggregates. *Physical Review*, A29:2966–2967, 1984.

662. François Leyvraz. The 'active perimeter' in cluster growth models: a rigorous bound. *Journal of Physics*, A18:L941–L945, 1985.

663. N. G. Makarov. On the distortion of boundary sets under conformal mappings. *Proceedings of the London Mathematical*

Society, 51:369–384, 1985.

664. Thomas C. Halsey. Some consequences of an equation of motion for diffusive growth. *Physical Review Letters*, 59:2067–2070, 1987.

665. M. Nauenberg. Critical growth velocity in diffusion-controlled aggregation. *Physical Review*, B28:449–451, 1983.

666. M. Nauenberg and L. Sander. Instabilities in continuum equations for aggregation by diffusion. *Physica*, A123:360–368, 1984.

667. R. Ball, M. Nauenberg and T. A. Witten. Diffusion-controlled aggregation in the continuum approximation. *Physical Review*, A29:2017–2020, 1984.

668. Efim Brener, Herbert Levine and Yuhai Tu. Mean-theory for diffusion-limited aggregation in low dimensions. *Physical Review Letters*, 66:1978–1981, 1991.

669. Yuhai Tu, Herbert Levine and Douglas Ridgway. Morphology transitions in a mean-theory of diffusion-limited growth. *Physical Review Letters*, 71:3838–3841, 1993.

670. Yuhai Tu and Herbert Levine. Mean-theory of the morphology transition in stochastic diffusion-limited growth. *Physical Review*, E52:5134–5141, 1995.

671. M. Muthukumar. Mean-field theory for diffusion-limited aggregation. *Physical Review Letters*, 50:839–842, 1983.

672. M. Tokuyama and K. Kawasaki. Fractal dimensions for diffusion-limited aggregation. *Physics Letters*, A100:337–340, 1984.

673. H. G. E. Hentschel. Fractal dimension of generalized diffusion-limited aggregates. *Physical Review Letters*, 52:212–215, 1984.

674. Katsuya Honda, Hiroyasu Toyoki and Mitsugu Matsushita. A theory of fractal dimensionality for generalized diffusion-limited aggregation. *Journal of the Physical Society of Japan*, 55:707–710, 1986.

675. R. C. Ball, P. W. Barker and R. Blumenfeld. Sidebranch selection in fractal growth. *Europhysics Letters*, 16:47–52, 1991.

676. Leonid A. Turkevich and Harvey Scher. Probability scaling for diffusion-limited aggregation in higher dimensions. *Physical Review*, A33:786–788, 1986.

677. C. Evertsz, H. Eskes and L. Pietronero. Intrinsic test for the cone angle ansatz in the dielectric breakdown model. *Europhysics Letters*, 10:607–613, 1989.

678. J. Szép and E. Lugosi. Needle crystal formation in two dimensions. *Journal of Physics*, A19:L1109–L1115, 1986.

679. M. T. Batchelor and B. I. Henry. Fractal dimension of zero-noise diffusion-limited aggregation. Preprint.

680. Jean-Pierre Eckmann, Paul Meakin, Itamar Procaccia and Reuven Zeitak. Growth and form of noise-reduced diffusion-limited aggregation. *Physical Review*, A39:3185–3195, 1989.

681. Jean-Pierre Eckmann, Paul Meakin, Itamar Procaccia and Reuven Zeitak. Asymptotic shape of diffusion-limited aggregates with anisotropy. *Physical Review Letters*, 65:52–55, 1990.

682. Robert Almgren, Wei-Shen Dai and Vincent Hakim. Scaling behavior in anisotropic Hele-Shaw flow. *Physical Review Letters*, 71:3461–3464, 1993.

683. A. Dougherty and R. Chen. Coarsening and the mean shape of three-dimensional dendritic crystals. *Physical Review*, A46:R4508–R4511, 1992.

684. L. M. Sander, P. Ramanlal and E. Ben-Jacob. Diffusion-limited aggregation as a deterministic growth process. *Physical Review*, A32:3160–3163, 1985.

685. Harvey Gould, Fereydoon Family and H. Eugene Stanley. Kinetics of formation of randomly branched aggregates: A renormalization-group approach. *Physical Review Letters*, 50:686–689, 1983.

686. X. R. Wang, Y. Shapir and M. Rubinstein. Analysis of multiscaling structure in diffusion-limited aggregation: A kinetic renormalization-group approach. *Physical Review*, A39:5974–5984, 1989.

687. X. R. Wang, Y. Shapir and M. Rubinstein. Kinetic renormalization group approach to diffusion limited aggregation. *Physics Letters*, A138:274–278, 1989.

688. Xian-zhi Wang and Yun Huang. Calculation of the fractal dimension of diffusion-limited aggregation by the renormalization-group approach in an arbitrary Euclidean dimension d. *Physical Review*, A46:5038–5042, 1992.

689. Takashi Nagatani. Evolution criterion and renormalization group for DLA. *Journal of*

Physics, A20:6603–6609, 1987.

690. L. Pietronero, A. Erzan and C. Evertz. Theory of fractal growth. *Physical Review Letters*, 61:861–864, 1988.

691. R. Cafiero, L. Pietronero and A. Vespignani. Persistence of screening and self-criticality in the scale invariant dynamics of diffusion limited aggregation. *Physical Review Letters*, 70:3939–3942, 1993.

692. T. C. Halsey. Diffusion-limited aggregation as branched growth. *Physical Review Letters*, 72:1228–1231, 1994.

693. H. G. E. Hentschel. Stochastic multifractility and universal scaling distributions. *Physical Review*, A50:243–261, 1994.

694. B. Shraiman and D. Bensimon. Singularities in nonlocal interface dynamics. *Physical Review*, E48:2840–2842, 1984.

695. Raphael Blumenfeld. Formulating a first-principles statistical theory of growing surfaces in two-dimensional Laplacian fields. *Physical Review*, E50:2952–2962, 1994.

696. Raphael Blumenfeld. Two-dimensional Laplacian growth can be mapped onto Hamiltonian dynamics. *Physics Letters*, A186:317–322, 1994.

697. Wei shen Dai, Leo P. Kadanoff and Su min Zhou. Interface dynamics and the motion of complex singularities. *Physical Review*, A43:6672–6682, 1991.

698. Raphael Blumenfeld and Robin C. Ball. Two-dimensional Laplacian growth as a system of creating and annihilating particles. *Physical Review*, E51:3434–3443, 1995.

699. L. Peliti. Path integral approach to birth-death processes on a lattice. *Journal de Physique*, 46:1469–1483, 1985.

700. Giorgio Parisi and Zhang Yi-Cheng. Field theories and growth models. *Journal of Statistical Physics*, 41:1, 1985.

701. David Elderfield. Field theories for kinetic growth models. *Journal of Physics*, A18:L773–L780, 1985.

702. Alexander Kuhn. *Structuration interfaciale dans des système électrochimiques*. PhD thesis, L'Université Bordeaux I, 1994.

703. Charles F. Chui. *An introduction to wavelets*. Academic Press, Boston, 1992. Wavelet analysis and its applications, Volume I.

704. A. Arneodo, G. Grasseau and M. Holschneider. Wavelet transform of multifractals. *Physical Review Letters*, 61:2281–2284, 1988.

705. Gang Li, L. M. Sander and Paul Meakin. Comment on "Self-similarity of diffusion-limited aggregates and electrodeposition clusters". *Physical Review Letters*, 63:1322, 1989.

706. Tamás Vicsek. *Fractal growth phenomena*. World Scientific, Singapore, 1989.

707. Armin Bunde and Shlomo Havlin, editors. *Fractals and disordered systems*. Springer Verlag, Berlin, 1991.

708. R. Giovanoli. Natural and synthetic manganese nodules. In I. M. Varentsov and Gy. Grasselley, editors, *Geology and geochemistry of manganese*, chapter 1 : General problems, mineralogy, geochemistry, methods, pages 195–196. E. Schweizbartsche Verlagsbuck-handlung, Stuttgart, 1980.

709. H. Van Damme. Flow and interfacial instabilities in Newtonian and colloidal fluids (or the birth, life and death of a fractal). In David Avnir, editor, *The fractal approach to heterogeneous chemistry*, pages 199–226. Wiley, Chichester, 1989.

710. Juan M. Garcia-Ruiz, Fermin Otálora, Antonio Sanchez-Navas and Francisco J. Higes-Rolando. The formation of manganese dendrites as the mineral record of flow structures. In J. H. Kruhl, editor, *Fractals and dynamic systems in geoscience*, pages 307–318. Springer-Verlag, Berlin, 1994.

711. B. Chopard, H. J. Herrmann and T. Vicsek. Structure and growth mechanism of mineral dendrites. *Nature*, 253:409–412, 1991.

712. Alan J. Charig, Frank Greenaway, Angela C. Milner, Cyril A. Walker and Peter J. Whybrow. *Archaeopteryx* is not a forgery. *Science*, 232:622–626, 1986.

713. Walter K. Zwicker and Stewart K. Kurtz. The growth of silver and copper single crystals on silicon and the selective removal of silicon by electrochemical displacement. *Acta Electronica*, 16:331–338, 1973.

714. J. H. Kaufman, A. I. Nazzal, O. R. Melroy and A. Kapitulnik. Onset of fractal growth: Statics and dynamics of diffusion-controlled polymerization. *Physical Review*, B35:1881–1890, 1987.

715. I. Mogi, S. Okubo and Y. Nakagawa. Effect of high magnetic fields on fractal growth of lead metal-leaves. *Journal of Crystal Growth*, 128:258–261, 1993.

716. Attila Imre, Zofia Vértesy, Tamás Pajkossy and Lajos Nyikos. Morphology of cobalt electrodeposits. *Fractals*, 1:59–66, 1993.

717. R. M. Brady and R. C. Ball. Fractal growth of copper electro-deposits. *Nature*, 309:225–229, 1984.

718. M. Matsushita, M. Sano, Y. Hayakawa, H. Honjo and Y. Sawada. Fractal structures of zinc metal leaves grown by electrodeposition. *Physical Review Letters*, 53:286–289, 1984.

719. G. P. Luo, Z. M. Ai, Z. H. Lu and Y. Wei. Electrodeposition of a two-dimensional silver dendritic crystal under Langmuir monolayers. *Physical Review*, E50:409–413, 1994.

720. Leila Zeiri, Shlomo Efrima and Moshe Deutsch. Interfacial electrodeposition of silver: the role of wetting. *Langmuir*, 12:5180–5187, 1996.

721. D. Grier, E. Ben-Jacob, Roy Clarke and L. M. Sander. Morphology and microstructure in electrochemical deposition of zinc. *Physical Review Letters*, 56:1264–1267, 1986.

722. Yasuji Sawada, A. Dougherty and J. P. Gollub. Dendritic and fractal patterns in electrolytic metal deposits. *Physical Review Letters*, 56:1260–1263, 1986.

723. J. H. Kaufman, C. K. Baker, A. I. Nazzal, M. Flickner, O. R. Melroy and A. Kapitulnik. Statics and dynamics of the diffusion-limited polymerization of the conducting polymer polypyrrole. *Physical Review Letters*, 56:1932–1935, 1986.

724. T. A. Witten and Yacov Kantor. Space-filling constraint on transport in random aggregates. *Physical Review*, B30:4093–4095, 1984.

725. H. Eugene Stanley and Antonio Coniglio. Flow in porous media: The "backbone" fractal at the percolation threshold. *Physical Review*, B29:522–524, 1984.

726. S. Alexander and R. O. Orbach. Density of states on fractals: "fractons". *Journal de Physique Lettres*, 43:L625–L631, 1982.

727. R. Rammal and G. Toulouse. Spectrum of the Schrödinger equation for a self-similar structure. *Physical Review Letters*, 49:1194–1197, 1982.

728. Paul Meakin and H. Eugene Stanley. Spectral dimension for the diffusion-limited aggregation model of colloid growth. *Physical Review Letters*, 51:1457–1460, 1983.

729. Masaharu Fujii, Kiyomitsu Arii and Katsumi Yoshino. The growth of dendrites of fractal patterns on a conducting polymer. *Journal of Physics: Condensed Matter*, 3:7207–7211, 1991.

730. Masaharu Fujii, Kiyomitsu Arii and Katsumi Yoshino. Branching patterns of a conducting polymer polymerized electrochemically with a constant-current source. *Journal of the Electrochemical Society*, 140:1838–1842, 1993.

731. J.-N. Chazalviel. Electrochemical aspects of the generation of ramified metallic electrodeposits. *Physical Review*, A42:7355–7367, 1990.

732. John M. Huth, Harry L. Swinney, William D. McCormick, Alexander Kuhn and Françoise Argoul. Role of convection in thin-layer electrodeposition. *Physical Review*, E51:3444–3458, 1995.

733. V. Fleury, J.-N. Chazalviel and M. Rosso. Theory and experimental evidence of electroconvection around electrochemical deposits. *Physical Review Letters*, 68:2492–2495, 1992.

734. V. Fleury, J.-N. Chazalviel and M. Rosso. Coupling of drift, diffusion and electroconvection, in the vicinity of growing electrodeposits. *Physical Review*, E48:1279–1295, 1993.

735. V. Fleury, J. H. Kaufman and D. B. Hibbert. Mechanism of a morphology transition in ramified electrochemical growth. *Nature*, 367:435–438, 1994.

736. Carol Livermore and Po-zen Wong. Convection and turbulence in strongly driven electrochemical deposition. *Physical Review Letters*, 72:3847–3850, 1994.

737. Mu Wang, Willem J. P. van Enckevort, Nai-ben Ming and Plet Bennema. Formation of a mesh-like electrodeposit induced by electroconvection. *Nature*, 367:438–441, 1994.

738. D. B. Hibbert and R. Melrose. Copper electrodeposits in paper support. *Physical Review*, A38:1036–1048, 1988.

739. J. R. Melrose and D. B. Hibbert. Electrical resistance of planar copper electrodeposits. *Physical Review*, A40:1727–1730, 1989.

740. J. R. Melrose. Pattern formation in electrochemical deposition. *Chemometrics and Intelligent Laboratory Systems*, 15:231–240, 1992.

741. Shaun N. Atchison, Robert P. Burford and D. Brynn Hibbert. Chemical effects on the morphology of supported electrodeposited metals. *Journal of Electroanalytical Chemistry*, 371:137–148, 1996.

742. S. Roy and D. Chakravorty. Silver electrodeposits in ion-exchanged oxide glasses. *Physical Review*, B47:3089–3096, 1993.

743. P. Carro, S. L. Marchiano, A. Hernándes Creus, S. González, R. C. Salvarezza and A. J. Arvia. Growth of three-dimensional silver fractal electrodeposits under damped free convection. *Physical Review*, E48:R2374–R2377, 1993.

744. A. Hernández Creus, P. Carro, S. González, R. C. Salvarezza and A. J. Arvia. A new electrochemical method for determining the fractal dimension of the surface of rough metal electrodeposits. *Journal of the Electrochemical Society*, 139:1064–1070, 1992.

745. D. J. Astley, J. A. Harrison and H. R. Thirsk. The formation of metal layers. *Journal of Electroanalytical Chemistry*, 19:325–334, 1968.

746. A. Bewick and B. Thomas. Optical and electrochemical studies of the underpotential deposition of metals. *Journal of Electroanalytical Chemistry*, 65:911–931, 1975.

747. A. Hernándes Creus, A. E. Bolzan, P. Carro, S. González, R. C. Salvarezza, S. L. Marchiano and A. J. Arvia. Mass-transport-induced kinetic transformation during the electrochemical formation of three-dimensional dendritic silver under ohmic control. *Journal of Electroanalytical Chemistry*, 336:85–97, 1992.

748. A. Bewick and B. Thomas. Optical and electrochemical studies of the underpotential deposition of metals part III. Lead deposition on silver single crystals. *Journal of Electroanalytical Chemistry*, 84:127–140, 1977.

749. M. Matsushita, Y. Hayakawa and Y. Sawada. Fractal structure and cluster statistics of zinc-metal trees deposited on a line electrode. *Physical Review*, A32:3814–3816, 1985.

750. Pedro Pablo Trigueros, Josep Claret, Francesc Mas and Francesc Sagués. Pattern morphologies in zinc electrodeposition. *Journal of Electroanalytical Chemistry*, 312:219–235, 1991.

751. Iwao Mogi, Susumu Okubo and Yasuaki Nakagawa. Dense radial growth of silver metal leaves in a high magnetic field. *Journal of the Physical Society of Japan*, 60:3200–3202, 1991.

752. A. G. Bedekar, S. F. Patil, R. C. Patil and Chitra Agashe. Influence of monomer concentration and electrode conductivity on morphology of poly(o-anisidine) films. *Journal of Physics*, D27:1727–1731, 1994.

753. A. Kuhn, F. Argoul, J. F. Muzy and A. Arneodo. Structural analysis of electroless deposits in the diffusion-limited regime. *Physical Review Letters*, 73:2998–3001, 1994.

754. F. Argoul, E. Freysz, A. Kuhn, C. Léger and L. Potin. Interferometric characterization of growth dynamics during dendritic electrodeposition of zinc. *Physical Review*, E53:1777–1788, 1996.

755. M. Ben Amar, R. Combescot and Y. Couder. Viscous fingering with adverse anisotropy: A new Saffman-Taylor finger. *Physical Review Letters*, 70:3047–3050, 1993.

756. Knut Jørgen Måløy, Jens Feder and Torstein Jøssang. Viscous fingering fractals in porous media. *Physical Review Letters*, 55:2688–2691, 1985.

757. J. P. Stokes, D. A. Weitz, J. P. Gollub, A. Dougherty, M. O. Robbins, P. M. Chaikin and H. M. Lindsay. Interfacial stability of immiscible displacement in a porous medium. *Physical Review Letters*, 57:1718–1721, 1986.

758. Gérard Daccord, Johann Nittmann and H. Eugene Stanley. Radial viscous fingering and diffusion-limited aggregation: fractal dimension and growth sites. *Physical Review Letters*, 56:336–339, 1986.

759. Johann Nittmann, Gérard Daccord and H. Eugene Stanley. Fractal growth of viscous fingers: quantitative

characterization of a fluid instability phenomenon. *Nature*, 314:141–144, 1985.

760. H. Thomé, M. Rabaud, V. Hakim and Y. Couder. The Saffman-Taylor instability: From the linear to the circular geometry. *Physics of Fluids*, A1:224–240, 1989.

761. Yves Couder. Growth patterns: From stable curved fronts to fractal structures. In Roberto Artuso, Predag Cvitanović and Giulio Casati, editors, *Chaos, order and patterns*, pages 203–227. Plenum, New York, 1991. NATO ASI Series B280.

762. E. Ben-Jacob, R. Godbey, Nigel D. Goldenfeld, J. Koplik, H. Levine, T. Mueller, and L. M. Sander. Experimental demonstration of the role of anisotropy in interfacial pattern formation. *Physical Review Letters*, 55:1315–1318, 1985.

763. Jing-Den Chen and David Wilkinson. Pore-scale viscous fingering in porous media. *Physical Review Letters*, 55:1892–1895, 1985.

764. J-D. Chen. Radial viscous fingering patterns in Hele-Shaw cells. *Experiments in Fluids*, 5:363–371, 1987.

765. A. A. Sonin and R. Bartolino. Air viscous fingers in isotropic fluid and liquid crystals obtained in lifting Hele-Shaw cell geometry. *Il Nuovo Cimento*, 15D:1–8, 1993.

766. Yoshikazu Hidaka, Minoru Suzuki, Toshiaki Murakami and Takahiro Inamura. Effects of a lead oxide annealing atmosphere on the superconducting properties of $BaPb_{0.7}Bi_{0.3}O_3$ sputtered films. *Thin Solid Films*, 106:311–319, 1983.

767. Gy. Radnoczi, T. Vicsek, L. M. Sander and D. Grier. Growth of fractal crystals in amorphous $GeSe_2$ films. *Physical Review*, A35:4012–4015, 1987.

768. W. T. Elam, S. A. Wolf, J. Sprague, D. U. Gubser, D. Van Vechten, G. L. Barz Jr. and Paul Meakin. Fractal aggregates in sputter-deposited $NbGe_2$ films. *Physical Review Letters*, 54:701–703, 1985.

769. Paul Meakin. Fractal scaling in thin film condensation and material surfaces. *CRC Critical Reviews in Solid State and Materials Science*, 13:143–189, 1987.

770. Renji Zhang, Li Li and Ziqin Wu. The fractal character of anneal-induced aggregation in bilayer films of the *Ge-Au* and *Ge-Ag* systems. *Thin Solid Films*, 208:295–303, 1992.

771. Thomas Michely, Michael Hohage, Michael Bott and George Comsa. Inversion of growth speed anisotropy in two dimensions. *Physical Review Letters*, 70:3943–3946, 1993.

772. L. J. Huang, B. X. Liu, J. R. Ding and H.-D. Li. Formation of fractal patterns in nickel based alloy films by ion beam method. In Alan J. Hurd, David A. Weitz and Benoit B. Mandelbrot, editors, *Extended abstracts, fractal aspects of materials: disordered systems*, pages 188–190. Materials Research Society, Pittsburgh, Pennsylvania, 1987. Proceedings of Symposium S, Materials Research Society Meeting, Boston Dec. 1–Dec. 4, 1987.

773. R. Q. Hwang, C. Günther, J. Schröder, S. Günther, E. Kopatzki and R. J. Behm. Nucleation and growth of thin metal films on clean and modified metal substrates studied by scanning tunneling microscopy. *Journal of Vacuum Science and Technology*, A10:1970–1980, 1992.

774. L. J. Huang and W. M. Lau. Multifractal analysis of two-dimensional carbon clusters. *Journal of Physics: Condensed Matter*, 5:7087–7094, 1993.

775. K. D. Herdt and J. H. Kallweit. Modellrechnungen zum dendriten-wachstum in partiell-kristallinen hochpolymeren. *Colloid and Polymer Science*, 260:413–424, 1982.

776. Hou Jian-Guo, Wu Zi-Qin and Bi Ling-Song. Effects of ion implantation on the annealing behavior of amorphous germanium and gold bilayers. *Thin Solid Films*, 173:77–82, 1989.

777. G. B. Kruaval and J. D. Parsons. Morphological structure of silicon carbide grown by chemical vapor deposition on titanium carbide using silane and ethylene. *Journal of the Electrochemical Society*, 141:765–771, 1994.

778. D. T. Smith, J. M. Valles, Jr. and R. B. Hallock. Comment on fractal aggregates in sputter-deposited films. *Physical Review Letters*, 54:2646, 1985.

779. Duan Jian-Zhong and Wu Zi-Qin. Fractal agglomeration during crystallization of $Pd-Si$ alloy films. *Journal of Microscopy*, 148:149–155, 1987.

780. V. M. Kosevich and L. S. Palatnik. Growth mechanism and crystal structure of vacuum

condensates. *Physics of Metals and Metallography*, 26:53–67, 1968.

781. Harald Brune, Christoph Romainczyk, Holger Röder and Klaus Kern. Mechanism of the transition from fractal to dendritic aggregates. *Nature*, 369:469–471, 1994.

782. R. Q. Hwang, J. Schröeder, C. Günther and R. J. Behm. Fractal growth of two-dimensional islands: *Au* on *Ru*(0001). *Physical Review Letters*, 67:3279–3282, 1991.

783. T. A. Witten and Paul Meakin. Diffusion-limited aggregation at multiple growth sites. *Physical Review*, B28:5632–5642, 1983.

784. Holger Röder, Elmar Hahn, Harald Brune, Jean-Pierre Bucher and Klaus Kern. Building one- and two-dimensional nanostructures by diffusion-controlled aggregation at surfaces. *Nature*, 366:141–143, 1993.

785. Pablo Jensen, Albert-Lásló Barabási, Hernán Larralde, Shlomo Havlin and H. Eugene Stanley. Controlling nanostructures. *Nature*, 367:22, 1993.

786. Michael Hohage, Michael Bott, Markus Morgenstern, Zhenyu Zhang, Thomas Michely, and George Comsa. Atomic processes in low temperature Pt-dendrite growth on Pt(111). *Physical Review Letters*, 76:2366–2369, 1996.

787. Holger Röder, Karsten Bromann, Harald Brune and Klaus Kern. Diffusion-limited aggregation with active edge diffusion. *Physical Review Letters*, 74:3217–3220, 1995.

788. Jacques Amar, Fereydoon Family and Pui-Man Lam. Dynamic scaling of the island-size distribution and percolation in a model of submonolayer molecular-beam epitaxy. *Physical Review*, B50:8782–8797, 1994.

789. C. Ratsch, A. Zangwill, P. Šmilauer and D. D. Vvedensky. Saturation and scaling in epitaxial island densities. *Physical Review Letters*, 72:3194–3197, 1994.

790. Lei-Han Tang. Island formation in submonolayer epitaxy. *Journal de Physique*, I-3:935–950, 1993.

791. G. S. Bales and D. C. Chrzan. Dynamics of irreversible island growth during submomolayer epitaxy. *Physical Review*, B50:6057–6067, 1994.

792. Yan Cheng. The fractal structure in dislocation-free bicrystal silicon. *Journal of Physics*, D27:1938–1945, 1994.

793. Laszlo Balazs, Vincent Fluery, Franck Duclos and A. Van Herpen. Fractal growth of silicon-rich domains during annealing of aluminum thin films deposited on silica. *Physical Review*, E54:599–604, 1996.

794. Masae Yasui and Mitsugu Matsushita. Morphological changes in dendritic crystal growth of ammonium chloride on agar plates. *Journal of the Physical Society of Japan*, 61:2327–2332, 1992.

795. Amita Chanda and Suresh Chanda. Experimental observation of large-size fractals in ion-conducting polymer electrolyte films. *Physical Review*, B49:633–636, 1994.

796. A. Miller, W. Knoll and H. Möwald. Fractal growth of crystalline phospholipid domains in monomolecular layers. *Physical Review Letters*, 56:2633–2636, 1986.

797. A. Miller and H. Möwald. Diffusion limited growth of crystalline domains in phospholipid monolayers. *Journal of Chemical Physics*, 86:4258–4265, 1987.

798. C. M. Knobler, K. Stine and B. G. Moore. Experimental studies of phase transitions and pattern formation in two dimensions. In A. Onuki and K. Kawasaki, editors, *Dynamics and patterns in complex fluids: New aspects of the physics-chemistry interface*, pages 130–140. Springer-Verlag, Berlin, 1990. Proceedings of the 4th Nishinomiya–Yukawa Memorial Symposium, Nishinomiya City, Japan, October 26–27 1988. Springer Proceedings in Physics 52.

799. Mu Wang, G. Wildburg, J. H. van Esch, P. Bennema, R. J. M. Nolte and H. Ringsdorf. Surface-tension-gradient-induced pattern formation in monolayers. *Physical Review Letters*, 71:4003–4006, 1993.

800. Mu Wang, Cheng Sun, Willem J. P. van Enckevort, Jan van Esch, Gerald Wildburg, Ru-Wen Peng, Nai-Ben Ming, Piet Bennema, Helmut Ringsdorf and Roeland J. M. Nolte. Pattern formation in lipid monolayers under illumination. *Physical Review*, E53:2580–2587, 1996.

801. A. Dietrich, H. Möwald, W. Rettig and G. Brezesinski. Polymorphism of a

triple-chain lecithin in two- and three-dimensional systems. *Langmuir*, 7:539–546, 1991.

802. H. E. Gaub, V. T. Moy and H. M. McConnell. Reversible formation of plastic two-dimensional lipid crystals. *Journal of Physical Chemistry*, 90:1721–1725, 1986.

803. Robert M. Weis and Harden M. McConnell. Two-dimensional chiral crystals of phospholipid. *Nature*, 310:47–49, 1984.

804. Gérard Daccord. Chemical dissolution of a porous medium by a reactive fluid. *Physical Review Letters*, 58:479–482, 1987.

805. Gérard Daccord and Roland Lenormand. Fractal patterns from chemical dissolution. *Nature*, 325:41–43, 1987.

806. Roland Lenormand, Arthur Soucémarianadin, Eric Touboul and Gérard Daccord. Three-dimensional fractals: Experimental measurements using capillary effects. *Physical Review*, A36:1855–1858, 1987.

807. Charles A. Knight. Slush on lakes. In David E. Loper, editor, *Structure and dynamics of partially solidified systems*, pages 455–465. Martinus Nijhoff, Dordrecht, 1987. NATO ASI Series E125.

808. A. J. Gow and J. W. Govoni. Ice growth on Post Pond, 1973-1982. Technical report, United States Army Cold Regions Research and Engineering Laboratory, Hanover New Hampshire, 1983. CRREL Report 83-4, 25 pp.

809. A. Tokairin. *Lake ice*. Kodansha, Tokyo, 1977. In Japanese.

810. C. G. Higgins. Drainage systems developed by sapping on Earth and Mars. *Geology*, 10:147–152, 1982.

811. C. G. Higgins. Piping and sapping: development of land forms by groundwater outflow. In R. C. La Fleur, editor, *Groundwater as a geomorphic agent*, pages 18–58. Allen and Unwin, Boston, 1984.

812. T. Dunne. Formation and controls of channel networks. *Progress in Physical Geography*, 4:211–239, 1980.

813. J. A. A. Jones. The nature of soil piping: A review of research. *British Geomorphological Research Group, Research Monograph*, 3:602–610, 1980.

814. K. Gilman and M. D. Newson. Soil pipes and pipe flow: A hydrological study in upland Wales. Technical report, British Geomorphological Research Group, Norwich, England, 1980. Research Monograph Series, 110 pp.

815. P. Cerasi, P. Mills and S. Fautrat. Erosion instability in a non consolidated porous medium. *Europhysics Letters*, 29:215–220, 1995.

816. Peter B. Kelemen, J. A. Whitehead, Einat Aharonov and Kelsey A. Jordahl. Experiments on flow focusing in soluble porous media, with applications to melt extraction from the mantle. *Journal of Geophysical Research*, 100:475–496, 1995.

817. Peter B. Kelemen, Nobumichi Shimizu and Vincent J. M. Salters. Extraction of mid-ocean-ridge basalt from the upwelling mantle by focused flow of melt in dunite channels. *Nature*, 375:747–753, 1995.

818. Stanley R. Hart. Equilibration during mantle melting: A fractal tree model. *Proceedings of the National Academy of Science USA*, 90:11914–11918, 1993.

819. Haruo Honjo, Shonosuke Ohta and Mitsugu Matsushita. Irregular fractal-like crystal growth of ammonium chloride. *Journal of the Physical Society of Japan*, 55:2487–2490, 1986.

820. Shonosuke Ohta and Haruo Honjo. Growth probability distribution in irregular fractal-like crystal growth of ammonium chloride. *Physical Review Letters*, 60:611–614, 1988.

821. David G. Grier, Kimberly Allen, Rachel S. Goldman, Leonard. M. Sander and Roy Clarke. Superlattices and long-range order in electrodeposited dendrites. *Physical Review Letters*, 64:2152–2155, 1990.

822. Haruo Honjo, Shonosuke Ohta and Mitsugu Matsushita. Phase diagram of growing succinonitrile crystal in supercooling-anisotropy phase space. *Physical Review*, A36:4555–4558, 1987.

823. Anthony D. Fowler, H. Eugene Stanley and Gérard Daccord. Disequilibrium silicate mineral textures: fractal and non-fractal features. *Nature*, 341:134–138, 1989.

824. Y. Couder, F. Argoul, A. Arnéodo, J. Maurer and M. Rabaud. Statistical properties of fractal dendrites and anisotropic diffusion-limited aggregates. *Physical Review*, A42:3499–3503, 1990.

825. W. G. Chaband. Electrical breakdown -

from liquid to amorphous solid. *Journal of Physics*, D24:56–64, 1991.

826. A. Matthews and A. R. Lefkow. Problems in the physical vapour deposition of titanium nitride. *Thin Solid Films*, 126:283–291, 1985.

827. M. T. Shaw and S. H. Shaw. Water treeing in solid dielectrics. *IEEE Transactions on Electrical Insulation*, EI-19:419–452, 1984.

828. Zensuke Iwata, Teruo Fukuda and Koji Kikuchi. Deterioration of cross-linked polyethylene due to water treeing. In *Annual report, conference on insulation and dielectric phenomena*, pages 200–210. 1972.

829. William R. Newcott. Lightning: Nature's high-voltage spectacle. *National Geographic*, 184:80–103, July 1993.

830. Victor M. Cabrera. Photographic investigations of electrical discharges in sandy media. *Journal of Electrostatics*, 30:47–56, 1993.

831. L. Niemeyer and F. Pinnekamp. Surface discharge in SF_6. In Loucas G. Christophorou, editor, *Gaseous dielectrics III*. Pergamon, New York, 1982.

832. N. Femia, L. Niemeyer and V. Tucci. Fractal characteristics of electrical discharges: experiments and simulations. *Journal of Physics D: Applied Physics*, 26:619–627, 1993.

833. Knut Jørgen Måløy, Finn Boger, Jens Feder and Torstein Jøssang. Dynamics and structure of viscous fingers in porous media. In Roger Pynn and Tormod Riste, editors, *Time dependent effects in disordered materials*, pages 111–143. Plenum Press, New York, 1987. NATO ASI Series B167.

834. J. Nittman, H. E. Stanley, E. Touboul and G. Daccord. Experimental evidence for multifractality. *Physical Review Letters*, 58:619, 1987.

835. F. Mas and F. Sagués. Scaling properties of the growth probability distribution in electrochemical deposition. *Europhysics Letters*, 17:541–546, 1992.

836. Paul Meakin and T. A. Witten. Growing interface in diffusion-limited aggregation. *Physical Review*, A28:2985–2989, 1983.

837. J. Mach, F. Mas and F. Sagués. Laplacian multifractality of the growth probability distribution in electrodeposition. *Europhysics Letters*, 25:271–276, 1994.

838. E. Ben-Jacob, G. Deutscher, P. Garik, Nigel D. Goldenfeld and Y. Lereah. Formation of a dense branching morphology in interfacial growth. *Physical Review Letters*, 57:1903–1906, 1986.

839. P. D. Haaland, A. Garscadden, B. Ganguly, S. Ibrani and J. Williams. On form and flow in dusty plasmas. *Plasma Sources Science and Technology*, 3:381–387, 1994.

840. A. Garscadden, B. Ganguly, P. D. Haaland and J. Williams. Overview of growth and behavior of clusters and particles in plasmas. *Plasma Sources Science and Technology*, 3:239–245, 1994.

841. A. Garscadden, B. Ganguly, J. Williams and P. D. Haaland. Growth and morphology of carbon grains. *Journal of Vacuum Science and Technology*, A11:1119–1125, 1993.

842. Florian Banhart. Fractal carbon filaments grown on insulators under irradiation in an electron microscope. *Philosophical Magazine Letters*, 69:45–51, 1994.

843. F. Banhart. Laplacian growth of amorphous carbon filaments in a non-diffusion-limited experiment. *Physical Review*, E52:5156–5160, 1995.

844. A. Sharma. Dendritic aggregation of metal hydrides and oxides in laser produced aerosol medium. *Physical Review*, A45:4148–4187, 1992.

845. R. Whytlaw-Gray and H. S. Patterson. *Smoke: A study of aerial disperse systems.* Arnold, London, 1932.

846. D. Barkey, P. Garik, E. Ben-Jacob, B. Miller and B. Orr. Growth velocity, the limiting current and morphology selection in electrodeposition of branched aggregates. *Journal of the Electrochemical Society*, 139:1044–1050, 1992.

847. David. G. Grier, David A. Kessler and L. M. Sander. Stability of the dense radial morphology in diffusive pattern formation. *Physical Review Letters*, 59:2315–2318, 1987.

848. David G. Grier and Daniel Mueth. Dissipation geometry and the stability of the dense radial morphology. *Physical Review*, E48:3841–3848, 1993.

849. Juan K. Lin and David G. Grier. Stability of densely branched growth in dissipative diffusion controlled systems. Preprint, 1995.

850. D. Barkey. Morphology selection and the

concentration boundary layer in electrochemical deposition. *Journal of the Electrochemical Society*, 138:2912–2917, 1991.

851. G. Deutscher and Y. Lereah. Phase separation by coupled single-crystal growth and polycrystalline fingering in $Al-Ge$: Experiment. *Physical Review Letters*, 60:1510–1513, 1988.

852. J. Erlebacher, P. C. Searson and K. Sieradzki. Computer simulation of dense-branching patterns. *Physical Review Letters*, 71:3311–3314, 1993.

853. Peter Garik, J. Hetrick, B. Orr, D. Barkey and E. Ben-Jacob. Interfacial cellular mixing and a conjecture on global deposit morphology. *Physical Review Letters*, 66:1606–1609, 1991.

854. J. D. van der Waals. Thermodynamische theorie der kapillarität unter voraussetzung stetiger dichteänderung. *Zeitschrift für Physikalische Chemie*, 13:657–725, 1894.

855. John W. Cahn and John E. Hilliard. Free energy of a nonuniform system. I. Interfacial free energy. *Journal of Chemical Physics*, 28:258–267, 1958.

856. P. G. Smith, T. G. M. van der Ven and S. G. Mason. Transient interfacial tension between two miscible liquids. *Journal of Colloid and Interface Science*, 80:302–303, 1981.

857. D. Grier, N. Hecker, E. Ben-Jacob, Roy Clarke, L. Sander and R. Wilkins. Dense radial aggregation in electrochemical deposition. *Bulletin of the American Physical Society*, 32:632, 1987.

858. Lui Lam. Waves in anisotropic media. In L. Lam and H. C. Morris, editors, *Wave phenomena*. Springer, Berlin, 1988.

859. L. Lam, R. D. Pochy and V. M. Castillo. Pattern formation in electrodeposits. In Lui Lam and Hedley C. Morris, editors, *Nonlinear structures in physical systems*, pages 11–31. Springer, New York, 1990. Proceedings of the Second Woodward Conference, San Jose State University, November 17–18, 1989.

860. P. Garik, D. Barkey, E. Ben-Jacob, E. Bochner, N. Broxholm, B. Miller, B. Orr, and R. Zamir. Laplace- and diffusion-field-controlled growth in electrochemical deposition. *Physical Review Letters*, 62:2703–2706, 1989.

861. J. R. Melrose, D. B. Hibbert and R. C. Ball. Interfacial velocity in electrochemical deposition and the Hecker transition. *Physical Review Letters*, 65:3009–3012, 1990.

862. V. Fleury, M. Rosso and J.-N. Chazalviel. Geometrical aspects of electrodeposition: The Hecker effect. *Physical Review*, A43:6908–6916, 1991.

863. Alexander Kuhn and Françoise Argoul. Determination of ionic mobilities by thin layer deposition. *Journal of Chemical Education*, 71:A273–A276, 1994.

864. A. Kuhn and A. Argoul. Influence of chemical perturbations on the surface roughness of thin layer electrodeposits. Preprint, 1993.

865. A. Kuhn and A. Argoul. Revisited experimental analysis of morphological changes in thin layer electrodeposits. Preprint, 1993.

866. V. Fleury, M. Rosso, J.-N. Chazalviel and B. Sapoval. Experimental aspects of dense morphology in copper electrodeposition. *Physical Review*, A44:6693–6705, 1991.

867. O. Zik and E. Moses. Electrodeposition: The role of concentration in the phase diagram and the Hecker effect. *Physical Review*, E53:1760–1764, 1996.

868. Y. Lereah, G. Deutscher and E. Grünbaum. Formation of dense branching morphology in the crystallization of $Al-Ge$ amorphous thin films. *Physical Review*, A44:8316–8322, 1991.

869. G. Deutscher and Y. Lereah. Crystallization of amorphous metal-insulator mixtures. *Physica*, A140:191–197, 1986.

870. S. Alexander, R. Bruinsma, R. Hilfer, G. Deutscher and Y. Lereah. Phase separation by coupled single-crystal growth and polycrystalline fingering in $Al-Ge$: Theory. *Physical Review Letters*, 60:1514–1517, 1988.

871. Yossi Lereah, Irena Zarudi, Enrique Grünbaum, Guy Deutscher, Surgey V. Buldyrev and H. Eugene Stanley. Morphology of $Ge:Al$ thin films: Experiments and model. *Physical Review*, E49:649–656, 1994.

872. Tamás Vicsek. Formation of solidification patterns in aggregation models. *Physical Review*, A32:3084–3089, 1985.

873. Hajime Tanaka and Toshio Nishi. Local phase separation at the growth front of a polymer spherulite during crystallization and nonlinear spherulitic growth in a polymer mixture with a phase diagram. *Physical Review*, A39:783–794, 1989.

874. S. N. Rauseo, P. D. Barnes and J. V. Maher. Development of radial viscous fingering patterns. *Physical Review*, A35:1245–1251, 1987.

875. Lincoln Paterson. Radial fingering in a Hele-Shaw cell. *Journal of Fluid Mechanics*, 113:513–529, 1981.

876. Yves Couder. Viscous fingering in a circular geometry. In H. Eugene Stanley and Nicole Ostrowsky, editors, *Random fluctuations and pattern growth*, pages 75–81. Kluwer, Dordrecht, 1988. NATO ASI Series E157.

877. S. E. May and J. V. Maher. Fractal dimension of radial fingering patterns. *Physical Review*, A40:1723–1726, 1989.

878. David Jasnow and Chuck Yeung. Asymptotic behavior of the viscous-fingering patterns in circular geometry. *Physical Review*, E47:1087–1093, 1993.

879. Robert H. Fariss. Fifty years of safer windshields. *Chemtech*, pages 38–43, September 1993.

880. Juan Manuel Garcia Ruiz. Natural viscous fingering. In Juan Manuel Garcia Ruiz, Enrique Louis, Paul Meakin and Leonard M. Sander, editors, *Growth patterns in physical sciences and biology*, pages 183–189. Plenum, New York, 1993. NATO ASI Series B304.

881. Yahai Tu, Herbert Levine and Douglas Ridgeway. Morphology transitions in a mean-field model of diffusion-limited growth. *Physical Review Letters*, 71:3838–3841, 1993.

882. Nigel Goldenfeld. Theory of spherulitic crystallization. *Journal of Crystal Growth*, 84:601–608, 1987.

883. H. D. Keith and F. J. Padden. A phenomenological theory of spherulitic crystallization. *Journal of Applied Physics*, 34:2409–2421, 1963.

884. R. F. Voss, R. B. Laibowitz and E. I. Allessandrini. Fractal (scaling) clusters in thin gold films near the percolation threshold. *Physical Review Letters*, 44:1441–1444, 1982.

885. Aharon Kapitulnik and Guy Deutscher. Percolation scale effects in metal-insulator thin films. *Journal of Statistical Physics*, 36:815–830, 1984.

886. Pablo Jensen, Patrice Melinon, Michel Treilleux, Jian Xiong Hu, Jean Dumas, Alain Hoareau and Bernard Cabaud. Direct observation of the infinite percolation cluster in thin films: Evidence for a double percolation process. *Physical Review*, B47:5008–5012, 1993.

887. D. A. Wollman, M. A. Dubson and Qifu Zhu. Annealed percolation: Determination of exponents in a correlated-percolation problem. *Physical Review*, B48:3713–3720, 1993.

888. A. Levy, S. Reich and P. Meakin. The shape of clusters on rectangular 2d lattices in a simple "phase separation" computer experiment. *Physics Letters*, A87:248–252, 1982.

889. Kyozi Kawasaki. Kinetics of Ising models. In C. Domb and M. S. Green, editors, *Phase transitions and critical phenomena*, volume 2, pages 443–501. Academic Press, London, 1972.

890. Kyozi Kawasaki. Diffusion constants near the critical point for time-dependent Ising models I. *Physical Review*, 145:224–230, 1966.

891. Kyozi Kawasaki. Diffusion constants near the critical point for time-dependent Ising models II. *Physical Review*, 148:375–381, 1966.

892. Kyozi Kawasaki. Diffusion constants near the critical point for time-dependent Ising models III. Self-diffusion constant. *Physical Review*, 150:285–290, 1966.

893. Roland Lenormand and Cesar Zarcone. Capillary fingering: percolation and fractal dimension. *Transport in Porous Media*, 4:599–612, 1989.

894. J. Bonnet and R. Lenormand. Realisation de micromodeles pour l'etude des ecoulements polyphasique en milieu poruex. *Revue Institut Français du Pétrole*, 42:477–480, 1977.

895. E. Clément, C. Baudet and J. P. Hulin. Multiple scale structure of non wetting fluid invasion fronts in 3D model porous media. *Journal de Physique Lettres*, 46:L1163–L1171, 1985.

896. E. Clément, C. Baudet, E. Guyon and J. P. Hulin. Invasion front structure in a 3d model porous medium under a hydrostatic pressure gradient. *Journal of Physics*, D20:608–615, 1987.

897. Ch. G. Jacquin and P. M. Adler. The fractal dimension of a gas-liquid interface in a porous medium. *Journal of Colloid and Interface Science*, 107:405–417, 1985.

898. William B. Haines. Studies in the physical properties of soil. V. The hysteresis effect in capillary properties and the modes of moisture distribution associated therewith. *Journal of Agricultural Science*, 20:97–116, 1930.

899. Jing-Den Chen, Madelena M. Dias, Samuel Patz and Lawrence M. Schwartz. Magnetic resonance imaging of immiscible-fluid displacement in porous media. *Physical Review Letters*, 61:1489–1492, 1988.

900. László Balázs. Corrosion front roughening in two-dimensional pitting of aluminum thin layers. *Physical Review*, E54:1183–1189, 1996.

901. H. Zhao and J. V. Maher. Viscoelastic effects in patterns between miscible fluids. *Physical Review*, A45:R8328–R8331, 1992.

902. Henri Van Damme, Éric Alsac and Claude Laroche. L'émergence du régime fractal en digitation viscoélastique. *Comptes Rendus de l'Académie des Sciences Paris*, Série II 309:11–18, 1989.

903. H. Van Damme, E. Alsac, C. Laroche and L. Gatineau. On the respective roles of low surface tension and non-Newtonian rheological properties in fractal fingering. *Europhysics Letters*, 5:25–30, 1988.

904. Robert C. Armstrong, R. Byron Bird and Ole Hassager. *Dynamics of polymer liquids*, volume 1. Wiley, New York, 1987. Second edition.

905. P. Coussot. Structural similarity and transition from Newtonian to non-Newtonian behavior for clay-water suspensions. *Physical Review Letters*, 74:3971–3974, 1995.

906. H. Van Damme, C. Laroche and L. Gatineau. Radial fingering in viscoelastic media, an experimental study. *Revue de Physique Appliquée*, 22:241–252, 1987.

907. E. Alsac, C. Laroche, E. Lemaire and H. Van Damme. Viscochemical fingering in a colloidal fluid. *Chemical Physics Letters*, 165:277–282, 1990.

908. H. F. Van Damme, C. Laroche, L. Gatineau and P. Levitz. Viscoelastic effects in fingering between miscible fluids. *Journal de Physique*, 48:1121–1133, 1987.

909. P. G. De Gennes. Time effects in viscoelastic fingering. *Europhysics Letters*, 3:195–197, 1987.

910. S. S. Park and D. J. Durian. Viscous and elastic fingering instabilities in foam. *Physical Review Letters*, 72:3347–3350, 1994.

911. J. D. Durian. Relaxation in aqueous foams. *Materials Research Society Bulletin*, pages 20–23, April 1994.

912. Unni Oxaal, Michael Murat, Finn Boger, Amnon Aharony, Jens Feder and Torstein Jøssang. Viscous fingering on percolation clusters. *Nature*, 329:32–37, 1987.

913. U. Oxaal. Fractal viscous fingering in inhomogeneous porous models. *Physical Review*, A44:5038–5051, 1991.

914. Unni Oxaal, Finn Boger, Jens Feder, Torstein Jøssang, Paul Meakin and Amnon Aharony. Viscous fingering in square-lattice models with two types of bonds. *Physical Review*, A44:6546–6575, 1991.

915. Marit Døvle. Dispersion in 2-d homogeneous and inhomogeneous porous media. Master's thesis, Department of Physics, University of Oslo, 1993.

916. D. K. Schwartz, S. Steinberg, J. Israelachvili and J. A. N. Zasadzinski. Growth of self-assembled monolayer by fractal aggregation. *Physical Review Letters*, 69:3354–3357, 1992.

917. F. Z. Cui, A. M. Vredenberg, F. Vitalis and F. W. Saris. Fractal patterns of phase decomposition in *Bi-Fe* thin films. Preprint, 1992.

918. T. Carrière, C. Ortiz and G. Fuchs. Fractal-like aggregation of *Au* islands induced by laser irradiation. *Journal of Applied Physics*, 70:5063–5067, 1991.

919. David Avnir, editor. *The fractal approach to heterogeneous chemistry: Surfaces, colloids, polymers.* John Wiley, New York, 1989.

920. Donald L. Turcotte. *Fractals and chaos in geology and geophysics.* Cambridge University Press, Cambridge, 1992.

921. G. Korvin, editor. *Fractal models in the earth sciences.* Elsevier, Amsterdam, 1992.

922. C. C. Barton and P. R. LaPointe, editors. *Fractal geometry and its uses in the earth sciences.* Plenum, New York, 1994.

923. S. Motojima, C. Itoh and H. Iwanaga. Crystal morphology of ternary compound $(Cr, Fe)_5Si_3$ obtained by *in situ* chemical vapour deposition. *Journal of Materials Science*, 26:1467–1472, 1991.

924. Russell Messier and Joseph E. Yehoda. Morphology of ballistically aggregated surface deposits. In Rober B. Laibowitz, Benoit B. Mandelbrot and Dann E. Passoja, editors, *Fractal aspects of materials*, pages 123–125. Materials Research Society, Pittsburgh, Pennsylvania, 1985. Extended Abstracts, Proceedings of Symposium N, Materials Research Society Fall Meeting, Boston, 1985.

925. John C. Russ and J. Christian Russ. Feature-specific measurement of surface roughness in SEM images. *Particle Characterization*, 4:22–25, 1987.

926. Deniz Ertaş and Mehran Kardar. Anisotropic scaling in depinning of a flux line. *Physical Review Letters*, 73:1703–1706, 1994.

927. Onuttom Narayan and Daniel S. Fisher. Dynamics of sliding charge-density waves in $4 - \epsilon$ dimensions. *Physical Review Letters*, 68:3615–3618, 1992.

928. Deniz Ertaş and Mehran Kardar. Dynamic roughening of directed lines. *Physical Review Letters*, 69:929–932, 1992.

929. M. Burgers. *The non-linear diffusion equation.* Riedel, Dordrecht, 1974.

930. Mehran Kardar, Giorgio Parisi and Yi-Cheng Zhang. Dynamical scaling of growing interfaces. *Physical Review Letters*, 56:889–892, 1986.

931. Bogdan Nowicki. Multiparameter representation of surface roughness. *Wear*, 102:161–176, 1985.

932. Brian Klinkenberg. Fractals and morphometric measures: is there a relationship? *Geomorphology*, 5:5–20, 1992.

933. Albert-Lászlό Barabási and H. Eugene Stanley. *Fractal concepts in surface growth.* Cambridge University Press, Cambridge, 1995.

934. Timothy Halpin-Healy and Yi-Cheng Zhang. Kinetic roughening, stochastic growth, directed percolation and all that. *Physics Reports*, 254:215–414, 1995.

935. S. F. Edwards and D. R. Wilkinson. The surface statistics of a granular aggregate. *Proceedings of the Royal Society (London)*, A381:17–31, 1982.

936. W. M. Visscher and M. Bolsterli. Random packing of equal and unequal spheres in two and three dimensions. *Nature*, 239:504–507, 1972.

937. E. M. Tory, B. H. Church, M. K. Tam and M. J. Ratner. Simulated random packing of equal spheres. *The Canadian Journal of Chemical Engineering*, 51:484–493, 1973.

938. R. Jullien and P. Meakin. Simple three-dimensional models for ballistic deposition with restructuring. *Europhysics Letters*, 4:1385–1390, 1987.

939. Fereydoon Family. Scaling of rough surfaces: effects of surface diffusion. *Journal of Physics*, A19:L441–L446, 1986.

940. Moshe Schwartz and S. F. Edwards. Nonlinear deposition: a new approach. *Europhysics Letters*, 20:301–305, 1992.

941. J. P. Bouchaud and M. E. Cates. Self-consistent approach to the Kardar-Parisi-Zhang equation. *Physical Review*, E47:R1455–R1458, 1993.

942. Yuhai Tu. Absence of finite upper critical dimension in the spherical Kardar-Parisi-Zhang model. *Physical Review Letters*, 73:3109–3112, 1994.

943. Michael Lässig and Harald Kinzelbach. Upper critical dimension of the Kardar-Parisi-Zhang equation. *Physical Review Letters*, 78:903–906, 1997.

944. Joachim Krug. Classification of some deposition and growth processes. *Journal of Physics*, A22:L769–L773, 1989.

945. Conyers Herring. Effect of change of scale on sintering phenomena. *Journal of Applied Physics*, 21:301–303, 1950.

946. W. W. Mullins. Theory of thermal grooving. *Journal of Applied Physics*, 28:333–339, 1957.

947. William W. Mullins. Flattening of a nearly plane solid surface due to capillarity. *Journal of Applied Physics*, 30:77–83, 1959.

948. R. P. U. Karunasiri, R. Bruinsma and Joseph Rudnick. Thin-film growth and the shadow instability. *Physical Review Letters*, 62:788–791, 1989.

949. J. Villain. Continuum models of crystal growth from atomic beams with and without desorbtion. *Journal de Physique*, I1:19–42, 1991.

950. Leonardo Golubović and Robijn Bruinsma. Surface diffusion and fluctuations of growing interfaces. *Physical Review Letters*, 66:321–324, 1991.

951. Henry Margeneau and George Moseley Murphy. *The mathematics of physics and chemistry*. Van Nostrand, Princeton, New Jersey, 1956. Second edition.

952. Eirik G. Flekkøy and Daniel H. Rothman. Fluctuating fluid interfaces. *Physical Review Letters*, 75:260–263, 1995.

953. Joachim Krug and Herbert Spohn. Mechanism for rough-to-rough transitions in surface growth. *Physical Review Letters*, 64:2332, 1990.

954. Chi-Hang Lam and Leonard M. Sander. Inverse method for interface problems. *Physical Review Letters*, 71:561–564, 1993.

955. Ji Li and Leonard M. Sander. Scaling properties of the Kuromoto-Sivashinsky equation. *Fractals*, 3:507–514, 1995. Proceedings of the Mandelbrot festschrift, Curaçao.

956. H. P. Bonzel. Mass transport by surface self-diffusion. In V. T. Binh, editor, *Surface mobilities on solid materials*, pages 195–241. Plenum, New York, 1983. NATO ASI.

957. David Kessler and B. G. Orr. Diffusion in MBE growth. Unpublished, 1994.

958. Zoltan Rácz, Martin Siegert and Michael Plischke. Surface-diffusion induced instabilities. In R. Jullien, J. Kertész, P. Meakin and D. Wolf, editors, *Surface disordering: Growth, roughening and phase transitions*, pages 39–44. Nova Science, Commack, New York, 1993.

959. D. D. Vvedensky, A. Zangwill, C. N. Luse and M. R. Wilby. Stochastic equations of motion for epitaxial growth. *Physical Review*, E48:852–862, 1993.

960. J. Krug, H. T. Dobbs and S. Majaniemi. Adatom mobility for the solid-on-solid model. *Zeitschrift für Physik*, B97:281–291, 1995.

961. Joachim Krug, Michael Plischke and Martin Siegert. Surface diffusion currents and the universality classes of growth. *Physical Review Letters*, 70:3271–3274, 1993.

962. David A. Huse, Christopher L. Henley and Daniel S. Fisher. Huse, Henley and Fisher respond. *Physical Review Letters*, 55:2924, 1985.

963. Terence Hwa and Erwin Frey. Exact scaling function of interface growth dynamics. *Physical Review*, A44:R7873–R7876, 1991.

964. Kyozi Kawasaki and Jim Gunton. Renormalization-group and mode-coupling theories of critical dynamics. *Physical Review*, 13:4658–4671, 1976.

965. Jacques G. Amar and Fereydoon Family. Deterministic and stochastic surface growth with a generalized nonlinearity. *Physical Review*, E47:1595–1603, 1993.

966. D. Forster, D. Nelson and M. J. Stephen. Large-distance and long-time properties of a randomly stirred fluid. *Physical Review*, A16:732–749, 1977.

967. D. E. Wolf and J. Villain. Growth with surface diffusion. *Europhysics Letters*, 13:389–394, 1990.

968. Tao Sun, Hong Guo and Martin Grant. Dynamics of driven interfaces with a conservation law. *Physical Review*, A40:6763–6766, 1989.

969. Z. Rácz, M. Siegert, D. Liu and M. Plischke. Scaling properties of driven interfaces: symmetries, conservation laws and the role of constraint. *Physical Review*, A43:5275–5283, 1991.

970. Ernesto Medina, Terence Hwa, Mehran Kardar and Yi-Cheng Zhang. Burgers equation with correlated noise: Renormalization-group analysis and application to directed polymers. *Physical Review*, A39:3053–3075, 1989.

971. Y.-C. Zhang. Non-universal roughening of kinetic self-affine interfaces. *Journal de Physique, France*, 51:2129–2134, 1990.

972. G. I. Sivashinsky. Instabilities, pattern formation and turbulence in flames. *Annual Reviews of Fluid Mechanics*, 15:179–199, 1983.

973. V. Yakhot. Large-scale properties of unstable systems governed by the Kuramoto-Sivashinski equation. *Physical Review*, A24:642–644, 1981.

974. Stéphane Zaleski. A stochastic model for the large scale dynamics of some fluctuating interfaces. *Physica*, D34:427–438, 1989.

975. K. Sneppen, J. Krug, M. H. Jensen,

C. Jayaprakash and T. Bohr. Dynamic scaling and crossover analysis for the Kuromoto-Sivashinsky equation. *Physical Review*, A46:R7351–R7354, 1992.

976. F. Hayot, C. Jayaprakash and Ch. Josserand. Long-wavelength properties of the Kuramoto-Sivashinsky equation. *Physical Review*, E47:911–915, 1993.

977. Victor S. L'vov and Itamar Procaccia. Comparison of the scale invariant solutions of the Kuramoto-Sivashinsky and Kardar-Parisi-Zhang equations in d dimensions. *Physical Review Letters*, 69:3543–3546, 1992.

978. Itamar Procaccia, Mogens H. Jensen, Victor S. L'vov, Kim Sneppen and Reuven Zeitak. Surface roughening and the long-wavelength properties of the Kuramoto-Sivashinsky equation. *Physical Review*, A46:3220–3224, 1992.

979. Victor S. L'vov, Vladimir V. Lebedev, Miriam Paton and Itamar Procaccia. Proof of scale invarient solutions in the Kardar-Parisi-Zhang and Kuramoto-Sivashinsky equations in $1+1$ dimensions: analytical and numerical results. *Nonlinearity*, 6:25–47, 1993.

980. V. S. L'vov and V. V. Lebedev. Interaction locality and scaling solution in $d+1$ KPZ and KS models. *Europhysics Letters*, 22:419–423, 1993.

981. Victor L'vov and Itamar Procaccia. Comment on "Universal properties of the two-dimensional Kuramoto-Sivashinsky equation". *Physical Review Letters*, 72:307, 1994.

982. C. Jayaprakash, F. Hayot and Rahul Pandit. Jayaprakash, Hayot and Pandit reply. *Physical Review Letters*, 72:308, 1994.

983. C. Jayaprakash, F. Hayot and Rahul Pandit. Universal properties of the two-dimensional Kuramoto-Sivashinsky equation. *Physical Review Letters*, 71:12–15, 1994.

984. Rodolfo Cuerno and Albert Lászlo Barabási. Dynamic scaling on ion-sputtered surfaces. *Physical Review Letters*, 74:4746–4749, 1995.

985. Martin Rost and Joachim Krug. Anisotropic Kuramoto–Sivashinsky equation for surface growth and erosion. *Physical Review Letters*, 75:3894–3897, 1995.

986. P. Devillard and H. E. Stanley. Scaling properties of Eden clusters in three and four dimensions. *Physica*, A160:298–309, 1989.

987. Marjorie J. Vold. A numerical approach to the problem of sediment volume. *Journal of Colloid Science*, 14:168–174, 1959.

988. B. D. Lubachevsky, V. Privman and S. C. Roy. Morphology of amorphous layers ballistically deposited on a planar substrate. *Physical Review*, E47:48–53, 1993.

989. D. Henderson, M. H. Brodsky and P. Chaudhari. Simulation of structural anisotropy and void formation in amorphous thin films. *Applied Physics Letters*, 25:641–643, 1974.

990. S. Kim, D. Henderson and P. Chaudhari. Computer simulation of amorphous thin films of hard spheres. *Thin Solid Films*, 47:155–158, 1977.

991. A. G. Dirks and H. J. Leamy. Columnar microstructure in vapor-deposited thin films. *Thin Solid Films*, 47:219–233, 1977.

992. H. J. Leamy and A. G. Dirks. The microstructure of vapor deposited thin films. In E. Kaldis, editor, *Current topics in material science*, chapter 4, pages 309–344. North Holland, Amsterdam, 1980.

993. P. S. Joag. A new algorithm for computer simulation of off-lattice ballistic aggregates. *Journal of Physics*, A21:739–746, 1988.

994. Hyunggyu Park, Meesoon Ha and In mook Kim. Exact solutions of a restricted ballistic deposition model on a one dimensional staircase. *Physical Review*, E51:1047–1054, 1995.

995. W. K. Burton, N. Cabrera and F. C. Frank. The growth of crystals and the equilibrium structure of their surfaces. *Philosophical Transactions of the Royal Society*, 243A:299–358, 1951.

996. J. D. Weeks, G. H. Gilmer and K. A. Jackson. Analytical theory of crystal growth. *Journal of Chemical Physics*, 65:712–720, 1976.

997. S. K. Chan and N. Y. Liang. Scaling of growing self-organized surfaces. *Physical Review Letters*, 67:1122–1125, 1991.

998. J. Krug and J. E. S. Socolar. Comment on "Scaling of growing self-organized surfaces". *Physical Review Letters*, 68:722, 1992.

999. B. M. Forrest and L.-H. Tang. Hypercubic stacking: a Potts-spin model for surface growth. *Journal of Statistical Physics*,

60:181–202, 1990.

1000. Lei-Han Tang, Bruce M. Forrest and Dietrich E. Wolf. Kinetic surface roughening. II. Hypercubic-stacking models. *Physical Review*, A45:7162–7179, 1992.

1001. M. Plischke, Z. Rácz and D. Liu. Time-reversal invariance and universality of two-dimensional growth models. *Physical Review*, B35:1517–1520, 1987.

1002. G. Poupart and G. Zumofen. Dissolution in $1+1$ dimensions: a numerical study. *Journal of Physics*, A25:L1173–L1179, 1992.

1003. M. G. Fernandes, R. M. Latanison and P. C. Searson. Morphological aspects of anodic dissolution. *Physical Review*, B47:11749–11756, 1993.

1004. F. C. Frank. Nucleation-controlled growth on a one-dimensional growth of finite length. *Journal of Crystal Growth*, 22:233–236, 1974.

1005. D. J. Kashchiev. Growth kinetics of dislocation-free interfaces and growth mode of thin films. *Journal of Crystal Growth*, 40:29–46, 1977.

1006. George H. Gilmer. Transients in the rate of crystal growth. *Journal of Crystal Growth*, 49:465–474, 1980.

1007. N. Goldenfeld. Kinetics of a model for nucleation-controlled polymer crystal growth. *Journal of Physics*, A17:2807–2821, 1984.

1008. Wim van Saarloos and George H. Gilmer. Dynamical properties of long-wavelength interface fluctuations during nucleation-dominated crystal growth. *Physical Review*, B33:4927–4935, 1986.

1009. J. Krug and H. Spohn. Anomalous fluctuations in the driven and damped sine-Gordon chain. *Europhysics Letters*, 8:219–224, 1988.

1010. M. C. Bartelt and J. W. Evans. Initial evolution of Kashchiev models of thin-film growth. *Journal of Physics*, A26:2743–2754, 1993.

1011. M. Kardar and Y.-C. Zhang. Scaling of directed polymers in random media. *Physical Review Letters*, 58:2087–2090, 1987.

1012. Ehud Perlsman and Moshe Schwartz. Ultrametric tree structure in the directed polymer problem. *Europhysics Letters*, 17:11–16, 1992.

1013. J. M. Kim, M. A. Moore and A. J. Bray. Zero-temperature directed polymers in a random potential. *Physical Review*, A44:2345–2351, 1991.

1014. J. M. Kim, A. J. Bray and M. A. Moore. Domain growth, directed polymers and self-organized criticality. *Physical Review*, A45:8546–8550, 1992.

1015. J. S. Kim, S. Kim and F. J. Ma. Topographic effect of surface roughness on thin-film flow. *Journal of Applied Physics*, 73:422–428, 1993.

1016. Stéphane Roux, Alex Hansen and Einar L. Hinrichsen. A direct mapping between Eden growth model and directed polymers in random media. *Journal of Physics*, A24:L295–L300, 1991.

1017. Joachim Krug and Lei-Han Tang. Disorder induced unbinding in confined geometries. *Physical Review*, E50:104–115, 1994.

1018. Jacques G. Amar and Fereydoon Family. Numerical solution of a continuum equation for interface growth in $2+1$ dimensions. *Physical Review*, A41:3399–3402, 1990.

1019. Katsuya Honda and Hirotaka Toyoshima. Morphological change of self-affine surfaces to paraboloids to cusps during growth processes. *Physical Review*, A46:4582–4584, 1992.

1020. S. V. Buldyrev, A.-L. Barabási, F. Caserta, S. Havlin, H. E. Stanley and T. Vicsek. Anomalous interface roughening on porous media: Experiments and model. *Physical Review*, A45:R8313–R8316, 1992.

1021. Lei-Han Tang and Heiko Leschhorn. Pinning by directed percolation. *Physical Review*, A45:R8309–R8312, 1992.

1022. D. J. Gates and M. Westcott. Kinetics of polymer crystallization I. Discrete and continuum models. *Proceedings of The Royal Society, London*, A416:443–461, 1988.

1023. Hong Yan, David Kessler and L. M. Sander. Roughening phase transition in surface growth. *Physical Review Letters*, 64:926–929, 1990.

1024. Y. P. Pellegrini and R. Jullien. Roughening transition and percolation in random ballistic deposition. *Physical Review Letters*, 64:1745–1748, 1990.

1025. Jacques G. Amar and Fereydoon Family. Phase transition in a restricted solid-on-solid surface-growth model in $2+1$

dimensions. *Physical Review Letters*, 64:543–546, 1990.

1026. J. M. Kim, Tapio Ala-Nissila and J. M. Kosterlitz. Comment on "Phase transition in a restricted solid-on-solid surface-growth model in 2 + 1 dimensions". *Physical Review Letters*, 64:2333, 1990.

1027. D. A. Huse, J. G. Amar and F. Family. Relationship between a generalized restricted solid-on-solid growth model and a continuum equation for interface growth. *Physical Review*, A41:7075–7077, 1990.

1028. Jin Min Kim. Phase transition on a modified ballistic deposition model in 2 + 1 dimensions. *Physical Review Letters*, 68:1248, 1992.

1029. H. Yan, D. Kessler and L. M. Sander. Yan et al. reply. *Physical Review Letters*, 68:1249, 1992.

1030. Rémi Jullien, Yves-Patrick Pellegrini and Paul Meakin. Restricted step height ballistic deposition with sticky and non-sticky particles. In R. Jullien, J. Kertész, P. Meakin and D. Wolf, editors, *Surface disordering: Growth, roughening and phase transitions*, pages 53–59. Nova Science, Commack, New York, 1993.

1031. Jacques G. Amar and Fereydoon Family. Numerical solution of a continuum equation equation for interface growth in 2+1 dimensions. *Physical Review*, A41:3399–3401, 1990.

1032. J. Cook and B. Derrida. Polymers on disordered hierarchical lattices: A nonlinear combination of random variables. *Journal of Statistical Physics*, 57:89–139, 1989.

1033. J. M. Kim, A. J. Bray and M. A. Moore. Finite-temperature directed polymers in a random medium. *Physical Review*, A44:R4782–R4785, 1991.

1034. Thomas Nattermann and Lei-Han Tang. Kinetic surface roughening. I. The Kardar-Parisi-Zhang equation in the weak-coupling regime. *Physical Review*, A45:7156–7161, 1992.

1035. Lei-Han Tang, Thomas Nattermann and Bruce M. Forrest. Multicritical and crossover phenomena in surface growth. *Physical Review Letters*, 65:2422–2425, 1990.

1036. C. A. Doty and M. Kosterlitz. Exact dynamical exponent at the Kardar-Parisi-Zhang roughening transition. *Physical Review Letters*, 69:1979–1981, 1992.

1037. P. Meakin and R. Jullien. Spatially correlated ballistic deposition. *Europhysics Letters*, 9:71–76, 1989.

1038. Paul Meakin and Remi Jullien. Spatially correlated ballistic deposition in one- and two-dimensional surfaces. *Physical Review*, A41:983–993, 1990.

1039. Chung-Kang Peng, Shlomo Havlin, Moshe Schwartz and H. Eugene Stanley. Directed-polymer and ballistic-deposition growth with correlated noise. *Physical Review*, A44:R2239–R2242, 1991.

1040. H. G. E. Hentschel and Fereydoon Family. Scaling of open dissipative systems. *Physical Review Letters*, 66:1982–1985, 1991.

1041. Y.-C. Zhang. Replica scaling analysis of interfaces in random media. *Physical Review*, B42:4897–4900, 1990.

1042. Minchun Wu, K. Y. R. Billah and Masonobu Shinozuka. Numerical solution of the Kardar-Parisi-Zhang equation with long-range spatially correlated disorder. *Physical Review*, E51:995–998, 1995.

1043. Ning-Ning Pang, Yi-Kuo Yu and Timothy Halpin-Healey. Interfacial kinetic roughening with correlated noise. *Physical Review*, E52:3224–3227, 1995.

1044. Chi-Hang Lam, Leonard. M. Sander and Dietrich E. Wolf. Surface growth with temporally correlated noise. *Physical Review*, A46:R6128–R6131, 1992.

1045. Jacques G. Amar and Fereydoon Family. Crossover scaling in surface growth with truncated power law noise. *Journal de Physique*, I1:175–179, 1991.

1046. Pui-Man Lam and Fereydoon Family. Surface growth in a model of molecular-beam epitaxy with correlated noise. *Physical Review*, A44:4854–4860, 1991.

1047. Y.-C. Zhang. Non-universal roughening of kinetic self-affine interfaces. *Journal de Physique*, 51:2129–2134, 1990.

1048. M. A. Rubio, C. A. Edwards, A. Dougherty and J. P. Gollub. Self-affine fractal interface from immiscible displacement in porous media. *Physical Review Letters*, 63:1685–1688, 1989.

1049. Viktor K. Horváth, Fereydoon Family and Tamás Vicsek. Dynamic scaling of the interface in two-phase viscous flows in porous media. *Journal of Physics*, A24:L25–L29, 1991.

1050. Viktor K. Horváth, Fereydoon Family and Tamás Vicsek. Anomalous noise distribution of the interface in two-phase fluid flow. *Physical Review Letters*, 67:3207–3210, 1991.

1051. Tamás Vicsek, Miklós Cserzõ and Victor K. Horváth. Self-affine growth of bacterial colonies. *Physica*, A167:315–321, 1990.

1052. Sergey V. Buldyrev, Shlomo Havlin, Janos Kertész, H. Eugene Stanley and Tamas Vicsek. Ballistic deposition with power-law noise: A variant of the Zhang model. *Physical Review*, A43:7113–7116, 1991.

1053. Shlomo Havlin, Sergey V. Buldyrev, H. Eugene Stanley and George H. Weiss. Probability distribution of the interface width in surface roughening: analogy with a Lévy flight. *Journal of Physics*, A24:L925–L931, 1991.

1054. Y.-C. Zhang. Growth anomaly and its implications. *Physica*, A170:1–13, 1990.

1055. Joachim Krug. Kinetic roughening by exceptional fluctuations. *Journal de Physique*, I1:9–12, 1991.

1056. R. Bourbonnais, H. J. Herrmann and T. Vicsek. Simulations of kinetic roughening with power-law noise on the connection machine. *International Journal of Modern Physics*, C2:719–733, 1991.

1057. Chi-Hang Lam and Leonard M. Sander. Surface growth with power-law noise. *Physical Review Letters*, 69:3338–3341, 1992.

1058. Chi-Hang Lam and Leonard M. Sander. Exact scaling in surface growth with power law noise. *Physical Review*, E48:979–987, 1993.

1059. Zoltán Csahók and Tamás Vicsek. Kinetic roughening in a model of sedimentation of granular materials. *Physical Review*, A46:4577–4581, 1992.

1060. P. Meakin and R. Jullien. Restructuring effects in the rain model for random deposition. *Journal de Physique*, 48:1651–1662, 1987.

1061. Tamás Vicsek, Ellák Somfai and Mária Vicsek. Kinetic roughening with multiplicative noise. *Journal of Physics*, A25:L763–L768, 1992.

1062. J. Krug and H. Spohn. Universality classes for deterministic surface growth. *Physical Review*, A38:4271–4283, 1988.

1063. R. Bruinsma and G. Aeppli. Interface motion and nonequilibrium properties of the random-field Ising model. *Physical Review Letters*, 52:1547–1550, 1984.

1064. G. Parisi. On surface growth in random media. *Europhysics Letters*, 17:673–678, 1992.

1065. David A. Kessler, Herbert Levine and Yuhai Tu. Interface fluctuations in random media. *Physical Review*, A43:4551–4554, 1991.

1066. Mogens H. Jensen and Itamar Procaccia. Unusual exponents in interface roughening: the effects of pinning. *Journal de Physique*, II1:1139–1146, 1991.

1067. Lei-Han Tang, János Kertész and Dietrich E. Wolf. Kinetic roughening with power-law waiting time distribution. *Journal of Physics*, A24:L1193–L1200, 1991.

1068. Joel Koplik and Herbert Levine. Interface moving through a random background. *Physical Review*, B32:280–292, 1985.

1069. Luís A. Nunes Amaral, Albert-László Barabási and H. Eugene Stanley. Universality classes for interface growth with quenched disorder. *Physical Review Letters*, 73:62–65, 1994.

1070. Thomas Nattermann, Semjon Stepanow, Lei-Han Tang and Heiko Leschhorn. Dynamics of interface depinning in a disordered system. *Journal de Physique*, II2:1483–1488, 1992.

1071. Onuttom Narayan and Daniel S. Fisher. Threshold critical dynamics of driven interfaces in random media. *Physical Review*, B48:7030–7042, 1993.

1072. Zoltán Csahók, Katsuya Honda and Tamás Vicsek. Dynamics of surface roughening in disordered media. *Journal of Physics*, A26:L171–L178, 1993.

1073. Z. Csahók, K Honda, E. Somfai, E. Vicsek and T Vicsek. Dynamics of surface roughening in disordered media. *Physica*, A200:136–154, 1993.

1074. A. S. Kaganovich. Preprint, 1994.

1075. Heiko Leschhorn. Interface depinning in a disordered medium - numerical results. *Physica*, A195:324–335, 1993.

1076. M. Dong, M. C. Marchetti, A. Alan

Middleton and V. Vinokur. Elastic string in a random potential. *Physical Review Letters*, 70:662–665, 1993.

1077. H. A. Makse and L. A. Nunes Amaral. Scaling behavior of driven interfaces above the depinning transition. *Europhysics Letters*, 31:379–384, 1995.

1078. Hong Ji and Mark O. Robbins. Transition from compact to self-similar growth in disordered systems: fluid invasion and magnetic-domain growth. *Physical Review*, A44:2538–2542, 1991.

1079. Hong Ji and Mark O. Robbins. Percolative, self-affine and faceted domain growth in random three-dimensional magnets. *Physical Review*, B46:14519–14527, 1992.

1080. C. S. Nolle, Belita Koiller, Nicos Martys and Mark O. Robbins. Effect of quenched disorder on moving interfaces in two dimensions. *Physica*, A205:342–354, 1994.

1081. M. Khantha and J. M. Yeomans. Wetting velocity near the directed percolation threshold. *Journal of Physics*, A20:L325–L329, 1987.

1082. Marek Cieplak and Mark O. Robbins. Dynamical transition in quasistatic fluid invasion in porous media. *Physical Review Letters*, 60:2042–2045, 1988.

1083. Nicos Martys, Marek Cieplak and Mark O. Robbins. Critical phenomena in fluid invasion of porous media. *Physical Review Letters*, 66:1058–1061, 1991.

1084. Nicos Martys, Mark O. Robbins and Marek Cieplak. Scaling relations for interface motion through disordered media: Application to two-dimensional fluid invasion. *Physical Review*, B44:12294–12306, 1991.

1085. C. S. Nolle, Belita Koiller, Nicos Martys and Mark O. Robbins. Morphology and dynamics of interfaces in random two-dimensional media. *Physical Review Letters*, 71:2074–2077, 1993.

1086. Marek Cieplak and Mark O. Robbins. Influence of contact angle on quasistatic fluid invasion of porous media. *Physical Review*, B41:11508–11521, 1990.

1087. Daniel H. Rothman and Stéphane Zaleski. Lattice-gas models of phase separation: interfaces, phase transitions and multiphase flow. *Reviews of Modern Physics*, 66:1417–1479, 1993.

1088. Jens Feder, Torstein Jøssang, Knut Jørgen Måløy and Unni Oxaal. Models of viscous fingering. In R. Englman and Z. Jaeger, editors, *Fragmentation, form and flow in fractured media*, pages 531–546. Adam Hilger/The Israel Physical Society, Bristol/Jerusalem, 1986. Annals of the Israel Physical Society, Volume 8.

1089. Kim Sneppen. Organized pinning and interface growth in random media. *Physical Review Letters*, 69:3539–3542, 1992.

1090. Lei-Han Tang and Heiko Leschhorn. Self-organized interface depinning. *Physical Review Letters*, 70:3832, 1993.

1091. Kim Sneppen and M. H. Jensen. Sneppen and Jensen reply. *Physical Review Letters*, 70:3833, 1993.

1092. Kim Sneppen and M. H. Jensen. Colored activity in self-organized critical interface dynamics. *Physical Review Letters*, 71:101–104, 1993.

1093. Zeev Olami, Itamar Procaccia and Reuven Zeitak. Theory of self-organized interface depinning. *Physical Review*, E49:1232–1237, 1994.

1094. Sergei Maslov and Maya Paczuski. Scaling theory of depinning in the Sneppen model. *Physical Review*, E50:R643–R646, 1994.

1095. Luís A. Nunes Amaral, Albert-László Barabábasi, Hernán A. Makse and H. Eugene Stanley. Scaling properties of driven interfaces in disordered media. *Physical Review*, E52:4087–4104, 1995.

1096. Lei-Han Tang, Mehran Kardar and Deepak Dhar. Driven pinning in anisotropic media. *Physical Review Letters*, 74:920–923, 1995.

1097. Sergey V. Buldyrev, Shlomo Havlin and H. Eugene Stanley. Anisotropic percolation and the *d*-dimensional surface roughening problem. *Physica*, A200:200–211, 1993.

1098. Hernán A. Makse. Singulaities and avalanches in interface growth with quenched disorder. *Physical Review*, E52:4080–4086, 1995.

1099. Deepak Dhar, Mustansir Barma and Mohan K. Phani. Duality transformations for two-dimensional directed percolation and resistance problems. *Physical Review Letters*, 47:1238–1241, 1981.

1100. S. Havlin, L. A. Nunes Amaral, S. V. Buldyrev, S. T. Harrington and H. E. Stanley. Dynamics of surface roughening with quenched disorder. *Physical Review Letters*, 74:4205–4208, 1995.

1101. Heiko Leschhorn and Lei-Han Tang. Avalanches and correlations in driven interface depinning. *Physical Review*, E49:1238–1245, 1994.

1102. Sergei Maslov, Maya Paczuski and Per Bak. Avalanches and $1/f$ noise in evolution and growth models. *Physical Review Letters*, 73:2162–2165, 1994.

1103. Jørgen Falk, Mogens Jensen and Kim Sneppen. Intermittent dynamics and self-organized depinning in propagating fronts. *Physical Review*, E49:2804–2808, 1994.

1104. Greg Huber, Mogens H. Jensen and Kim Sneppen. Distribution of self-interactions and voids in $(1+1)$-dimensional directed percolation. *Physical Review*, E52:R2133–R2136, 1995.

1105. Zeev Olami, Itamar Procaccia and Reuven Zeitak. Interface roughening in systems with quenched disorder. *Physical Review*, E52:3402–3414, 1995.

1106. Michael K. Shepard, Robert A. Brackett and Raymond E. Arvidson. Self affine (fractal) topography: Surface parameterization and radar scattering. *Journal of Geophysical Research*, 100:11709–11718, 1995.

1107. D. A. Lange, H. M. Jennings and S. P. Shah. Analysis of surface roughness using confocal microscopy. *Journal of Materials Science*, 28:3879–3884, 1993.

1108. P. Pfeifer, Y. J. Wu, M. W. Cole and J. Krim. Multilayer adsorbtion on a fractally rough surface. *Physical Review Letters*, 62:1997–2000, 1989.

1109. Mehran Kardar and Joseph O. Indekeu. Wetting of fractally rough surfaces. *Physical Review Letters*, 65:662, 1990.

1110. C. L. Wang, J. Krim and M. F. Toney. Roughness and porosity characterization of carbon and magnetic films through adsorbtion isotherm measurements. *Journal of Vacuum Science and Technology*, 7:2481–2485, 1989.

1111. V. Panella and J. Krim. Adsorbtion isotherm study of the fractal scaling behavior of vapor-deposited silver films. *Physical Review*, E49:4179–4184, 1994.

1112. Viktor K. Horváth, Fereydoon Family and Tamás Vicsek. Comment on "Self-affine fractal interfaces from immiscible displacement in porous media". *Physical Review Letters*, 65:1388, 1990.

1113. S. V. Buldyrev, A.-L. Barabási, S. Havlin, J. Kertész, H. E. Stanley and H. S. Xenias. Anomalous interface roughening in 3D porous media: experiment and model. *Physica*, A191:220–226, 1992.

1114. L. A. Nunes Amaral, A.-L. Barabábasi, S. V. Buldyrev, S. Havlin and H. E. Stanley. New exponent characterizing the effect of evaporation on imbibition experiments. *Physical Review Letters*, 72:641–644, 1994.

1115. B. Hede, J. Kertész and T. Vicsek. Self-affine fractal clusters: Conceptual questions and numerical results for directed percolation. *Journal of Statistical Physics*, 64:829–841, 1991.

1116. Fereydoon Family, K. C. B. Chan and Jacques Amar. Dynamics of interface roughening in imbibition. In R. Jullien, J. Kertész, P. Meakin and D. Wolf, editors, *Surface disordering: growth, roughening and phase transitions*, pages 205–211. Nova, Commack, New York, 1992.

1117. Viktor K. Horváth and H. Eugene Stanley. Temporal scaling of interfaces propagating in porous media. *Physical Review*, E52:5166–5169, 1995.

1118. Shanjin He, Galathara L. M. K. S. Kahanda and Po zen Wong. Roughness of wetting fluid invasion fronts in porous media. *Physical Review Letters*, 69:3731–3734, 1992.

1119. M. A. Rubio, C. A. Edwards, A. Dougherty and J. P. Gollub. Rubio, Dougherty and Gollub reply. *Physical Review Letters*, 65:1389, 1990. Reply to V. K. Horv'ath *et al.* Comment on "Self-affine fractal interfaces from immiscible displacement in porous media".

1120. Jun Zhang, Y.-C. Zhang, P. Alstrøm and M. T. Levinsen. Modeling forest fire by paper-burning experiment, a realization of the interface mechanism. *Physica*, A189:383–389, 1992.

1121. N. Provatas, M. J. Alava and T. Ala-Nissila. Density correlations in paper. *Physical Review*, E54:R36–R38, 1996.

1122. H. Fujikawa and M. Matsushita. Fractal growth of Bacillus subtilis on agar plates. *Journal of the Physical Society of Japan*,

58:3875–3878, 1989.

1123. Mitsugu Matsushita and Hiroshi Fujikawa. Diffusion-limited growth in bacterial colony formation. *Physica*, A168:498–506, 1990.

1124. T. Matsuyama, M. Sogawa and Y. Nakagawa. Fractal spreading of *Serratia marcescens* which produces surface active exolipids. *FEMS Microbiology Letters*, 61:243–246, 1989.

1125. C. Licoppe, Y. I. Nissim and C. d'Anterroches. Growth-front instabilities in solid-state recrystallization of amorphous *GaAs* films. *Physical Review*, B37:1287–1293, 1988.

1126. Y. Lereah, J. M. Pénisson and A. Bourret. High resolution electron microscopy of crystalline-amorphous interface: An indication of Eden aggregate. *Applied Physics Letters*, 60:1682–1684, 1992.

1127. M. R. Fitzsimmons and E. Burkel. Interfacial roughness of [011] twist grain boundaries characterized with x-ray reflection. *Physical Review*, B47:8436–8452, 1993.

1128. S. Sinha, E. B. Sirota, S. Garoff and H. B. Stanley. X-ray scattering and neutron scattering from rough surfaces. *Physical Review*, B38:2297–2311, 1993.

1129. Haruro Honjo and Shonosuke Ohta. Self-affine fractal crystals from an NH_4Cl solution. *Physical Review*, E49:R1808–R1810, 1994.

1130. J. S. Langer and L. Turski. Studies in the theory of interfacial stability -1. Stationary symmetric model. *Acta Metallurgica*, 25:1113–1119, 1977.

1131. K. A. Jackson and J. D. Hunt. Transparent compounds that freeze like metals. *Acta Metallurgica*, 13:1212–1216, 1965.

1132. J. A. Thornton. High rate thick film growth. *Annual Reviews of Material Science*, 7:239–260, 1977.

1133. H.-N. Yang, T.-M. Lu and G.-C. Wang. Diffraction from surface growth fronts. *Physical Review*, B47:3911–3922, 1993.

1134. H.-N. Yang, G.-C. Wang and T.-M. Lu. *Diffraction from rough surfaces and dynamic growth fronts.* World Scientific, Singapore, 1993.

1135. J. Krim and G. Palasantzas. Experimental observations of self-affine scaling and kinetic roughening at sub-micron lengthscales. *International Journal of Modern Physics*, B9:599–631, 1995.

1136. G. Palasantzas and J. Krim. Effects of the form of the height-height correlation function on diffuse x-ray scattering from a self-affine surface. *Physical Review*, B48:2873–2877, 1993.

1137. George Palasantzas. Roughness and surface width of self-affine fractal surfaces via the K-correlation model. *Physical Review*, B48:14427–14478, 1993.

1138. George Palasantzas. Erratum: Roughness and surface width of self-affine fractal surfaces via the K-correlation model. *Physical Review*, B49:5785, 1994.

1139. Po zen Wong and Alan J. Bray. Scattering by rough surfaces. *Physical Review*, B37:7751–7758, 1988.

1140. L. G. Parratt. Surface studies of solids by total reflection of x-rays. *Physical Review*, 95:359–369, 1954.

1141. I. K. Robinson. Crystal truncation rods and surface roughness. *Physical Review*, B33:3830–3836, 1986.

1142. C. Thompson, G. Palasantzas, Y. P. Feng, S. K. Sinha and J. Krim. X-ray-reflectivity study of the kinetics of vapor-deposited silver films. *Physical Review*, B49:4902–4907, 1994.

1143. H.-N. Yang, T.-M. Lu and G.-C. Wang. Time-invariant structure factor in an epitaxial growth front. *Physical Review Letters*, 68:2612–2615, 1992.

1144. Hong Zheng and Gianfranco Vidali. Measurement of growth kinetics in a heteroepitaxial system: Pb on $Cu(100)$. *Physical Review Letters*, 74:582–585, 1995.

1145. H.-J. Ernst, F. Fabre, R. Folkerts and J. Lapujoulade. Observation of a growth instability during low temperature molecular beam epitaxy. *Physical Review Letters*, 72:112–115, 1994.

1146. Y.-L. He, H.-N. Yang, T.-M. Lu and G.-C. Wang. Measurements of dynamic scaling from epitaxial growth fronts: Fe film on $Fe(001)$. *Physical Review Letters*, 69:3770–3773, 1992.

1147. H.-N. Yang, G.-C. Wang and T.-M. Lu. Instability in low-temperature molecular-beam epitaxy growth of $Si/Si(111)$. *Physical Review Letters*, 73:2348–2351, 1995.

1148. Gert Ehrlich and F. G. Hudda. Atomic view of surface self-diffusion: Tungsten on tungsten. *Journal of Chemical Physics*, 44:1039–1049, 1966.

1149. Richard L. Schwoebel and Edward J. Shipsey. Step motion on crystal surfaces. *Journal of Applied Physics*, 37:3682–3686, 1966.

1150. Richard L. Schwoebel. Step motion on crystal surfaces II. *Journal of Applied Physics*, 40:614–618, 1969.

1151. William M. Tong, Eric J. Snyder, R. Stanley Williams, Akihisa Yanase, Yasaburo Segawa and Mark S. Anderson. Atomic force microscope studies of $CuCl$ island formation on $CaF_2(111)$ substrates. *Surface Science Letters*, 277:L63–L69, 1992.

1152. William M. Tong, R. Stanley Williams, Akihisa Yanase, Yusaburo Segawa and Mark S. Anderson. Atomic force microscope studies of growth kinetics: Scaling in the heteroepitaxy of $CuCl$ on $CaF_2(111)$. *Physical Review Letters*, 72:3374–3377, 1994.

1153. K. Fang, T.-M. Lu and G.-C. Wang. Roughening and faceting in a Pb thin film growing on the $Pb(110)$ surface. *Physical Review*, B49:8331–8339, 1994.

1154. P. K. Larsen and P. J. Dobsen, editors. *Reflection high-energy electron diffraction and reflection electron imaging of surfaces*. Plenum, New York, 1988. NATO ASI Series B188.

1155. J. Chevrier, V. Le Than, R. Buys and J. Derrien. A RHEED study of epitaxial growth of iron on a silicon surface: Experimental evidence for kinetic roughening. *Europhysics Letters*, 16:737–742, 1991.

1156. J. Chevrier, A. Cruz, N. Pinto, I. Berbezier and J. Derrien. Influence of kinetic roughening on the epitaxial growth of silicon. *Journal de Physique*, I4:1309–1324, 1994.

1157. Joachim Krug and Martin Schimschak. Metastability of step flow growth in $1 + 1$ dimensions. *Journal de Physique*, I5:1065–1086, 1995.

1158. D. J. Eaglesham and G. H. Gilmer. Roughening during Si deposition at low temperatures. In R. Jullien, J. Kertész, P. Meakin and D. Wolf, editors, *Surface disordering: Growth, roughening and phase transitions*, pages 69–75. Nova Science, Commack, New York, 1993.

1159. Ellen D.Williams. Surface steps and surface morphology: understanding macroscopic phenomena from atomic observations. *Surface Science*, 299/300:502–524, 1994.

1160. F. Wu, S. G. Jaloviar, D. E. Savage and M. G. Lagally. Roughening of steps during homoepitaxial growth on $Si(001)$. *Physical Review Letters*, 71:4190–4193, 1993.

1161. C. J. Lanczycki and S. Das Sarma. Dynamics of step roughening on vicinal surfaces. *Physical Review*, B51:4579–4584, 1995.

1162. D. Le Bellac, G. A. Niklasson and C. G. Granqvist. Scaling of surface roughness in obliquely sputtered chromium films. *Europhysics Letters*, 32:155–159, 1995.

1163. D. J. Eaglesham, H.-J. Gossmann and M. Cerullo. Limiting thickness h_{epi} for epitaxial growth and room-temperature Si growth on $Si(100)$. *Physical Review Letters*, 65:1227–1230, 1990.

1164. D. J. Eaglesham, L. N. Pfeiffer, K. W. West and D. R. Dykaar. Limiting thickness epitaxy in *GaAs* molecular beam epitaxy at 200 °C. *Applied Physics Letters*, 58:65–67, 1991.

1165. Olivier Vatel, Philippe Dumas, Frederic Chollet, Franck Salvan and Elie André. Roughness assessment of polysilicon using power spectral density. *Japanese Journal of Applied Physics*, 32:5671–5674, 1992.

1166. J. Krim, I. Heyvaert, C. Van Haesendonck and Y. Bruynseraede. Scanning tunneling microscopy observations of self-affine fractal roughness in ion-bombarded film surfaces. *Physical Review Letters*, 70:57–60, 1993.

1167. R. P. Chiarello, V. Panella, H. K. Kim and C. Thompson. X-ray reflectivity and adsorbtion isotherm study of fractal scaling in vapor-deposited films. *Physical Review Letters*, 67:3408, 1991.

1168. P. Herrasti, P. Ocón, L. Vázquez, R. C. Salvarezza, J. M. Vara and A. Arvia. Scanning-tunneling-microscopy study on the growth mode of vapor-deposited gold films. *Physical Review*, A45:7440–7447, 1992.

1169. R. C. Salvarezza, L. Vázquez, P. Herrasti, P. Ocón, J. M. Vara and A. J. Ariva. Self affine fractal vapor deposited gold surfaces characterization by scanning tunneling microscopy. *Europhysics Letters*,

20:727–732, 1992.

1170. Morgan W. Mitchell and Dawn A. Bonnell. Quantitative topographic analysis of fractal surfaces by scanning tunneling microscopy. *Journal of Material Research*, 5:2244–2254, 1990.

1171. H. You, R. P. Chiarello, H. K. Kim and K. G. Vandervoort. X-ray reflectivity and scanning-tunneling-microscopy study of kinetic roughening of sputter-deposited gold films during growth. *Physical Review Letters*, 70:2900–2903, 1993.

1172. G. Palasantzas and J. Krim. Scanning tunneling microscopy study of the thick film limit of kinematic roughening. *Physical Review Letters*, 73:3564–3567, 1994.

1173. K. Thürmer, R. Koch, M. Weber and K. H. Rieder. Dynamic evolution of pyramid structures during growth of epitaxial *Fe*(001) films. *Physical Review Letters*, 75:1767–1770, 1995.

1174. D. J. Miller, K. E. Gray, R. T. Kampwirth and J. M. Murduck. Studies of growth instabilities and roughening in sputtered *NbN* films using a multilayer decoration technique. *Europhysics Letters*, 19:27–32, 1992.

1175. Eric E. Fullerton, J. Pearson, C. H. Sowers, S. D. Bader, X. Z. Wu and S. K. Sinha. Interfacial roughness of sputtered multilayers: Nb/Si. *Physical Review*, B48:17432–17444, 1993.

1176. T. Salditt, T. H. Metzger and J. Peisl. Kinetic roughness of amorphous multilayers studied by diffuse x-ray scattering. *Physical Review Letters*, 73:2228–2231, 1994.

1177. P. P. Swaddling, D. F. McMorrow, R. A. Cowley, R. C. C. Ward and M. R. Wells. Determination of the interfacial roughness exponent in rare-earth superlattices. *Physical Review Letters*, 73:2232–2235, 1994.

1178. Xin-Ya Lei, Peng Wan, Cai-Hua Zhou and Nai-Ben Ming. Kinetic crossover of rough surface growth in a colloidal system. *Physical Review*, E54:5298–5301, 1996.

1179. R. C. Salvarezza, L. Vázquez, H. Míguez, R. Mayoral, C. López, and F. Meseguer. Edwards-Wilkinson behavior of crystal surfaces grown by sedimentation of SiO_2 nanospheres. *Physical Review Letters*, 77:4572–4575, 1996.

1180. A. Czirók, E. Somfai and T. Vicsek. Experimental evidence for self-affine roughening in a micromodel of geomorphological evolution. *Physical Review Letters*, 71:2154–2157, 1993.

1181. Elliot A. Eklund, R. Bruinsma, J. Rudnick and R. Stanley Williams. Submicron-scale surface roughening induced by ion bombardment. *Physical Review Letters*, 67:1759–1762, 1991.

1182. Elliot A. Eklund, Eric J. Snyder and R. Stanley Williams. Correlation from randomness: quantitative analysis of ion-etched graphite surfaces using the scanning tunneling microscope. *Surface Science*, 285:157–180, 1993.

1183. E. Chason and T. M. Mayer. Low energy ion bombardment induced roughening and smoothing of SiO_2 surfaces. *Applied Physics Letters*, 62:363–365, 1993.

1184. E. Chason, T. M. Mayer, A. Payne and D. Wu. *In situ* energy dispersive x-ray reflectivity measurement of H ion bombardment on SiO_2/Si and Si surfaces. *Applied Physics Letters*, 60:2353–2355, 1992.

1185. E. Chason and Warwick. *In situ* energy dispersive x-ray reflectivity measurement of H ion bombardment on SiO_2/Si and Si surfaces. *Applied Physics Letters*, 60:2353–2355, 1995.

1186. Richard Pétri, Pascal Brault, Olivier Vatel, Daniel Henry, Elie André, Philippe Dumas and Franck Salvan. Silicon roughness induced by plasma etching. *Journal of Applied Physics*, 75:7498–7506, 1994.

1187. Ph. Dumas, B. Bouffakhreddine, C. Amra, O. Vatel, E. Andre, G. Galindo and F. Salvan. Quantitative microroughness analysis down to the nanometer scale. *Europhysics Letters*, 22:717–722, 1993.

1188. Sreeram Srinivasan, John C. Russ and Ronald O. Scattergood. Fractal analysis of erosion surfaces. *Journal of Material Research*, 5:2616–2619, 1990.

1189. John C. Russ. *Computer-assisted microscopy. The measurement and analysis of images.* Plenum, New York, 1990.

1190. P. R. Stupak and J. A. Donavan. Fractal analysis of rubber wear surfaces and debris. *Journal of Materials Science*, 23:2230–2242, 1988.

1191. Galathara L. M. K. S. Kahanda, Xiao qun Zou, Robert Farrell and Po zen Wong. Columnar growth and kinetic roughening in electrochemical deposition. *Physical Review Letters*, 68:3741–3744, 1992.

1192. Pawel Keblinski. Comment on "Columnar growth and kinetic roughening in electrochemical deposition". *Physical Review Letters*, 71:805, 1993.

1193. Atsushi Iwamoto, Tatsuo Yoshinobu and Horoshi Iwasaki. Stable growth and kinetic roughening in electrochemical deposition. *Physical Review Letters*, 72:4025–4028, 1994.

1194. J. M. Pastor and Miguel Rubio. Rough growth and morphological instability of compact electrodeposits. *Physical Review Letters*, 76:1848–1851, 1996.

1195. M. Aguilar, E. Anguiano, F. Vázquez and M. Pancorbo. Study of the fractal character of surfaces by scanning tunelling microscopy: errors and limitations. *Journal of Microscopy*, 167:197–213, 1992.

1196. Herbert H. Uhlig. *Corrosion and corrosion control.* Wiley, New York, 1971.

1197. Ulock R. Evans. *The corrosion and oxidation of metals: Scientific principles and practical applications.* Edward Arnold, London, 1967.

1198. Terje Holten, Torstein Jøssang, Paul Meakin and Jens Feder. Fractal characterization of two-dimensional aluminum corrosion fronts. *Physical Review*, E50:754–759, 1994.

1199. K. R. Tretheway and Pierre R. Roberge. Towards improved quantitative characterization of corroding surfaces using fractal models. In Kenneth R. Tretheway and Pierre R. Roberge, editors, *Modelling aqueous corrosion from individual pits to system management*, pages 443–463. Kluwer, Dordrecht, 1994. NATO ASI Series E266.

1200. Tatsuo Yoshinobu, Atsushi Iwamoto and Hiroshi Iwasaki. Scaling analysis of SiO_2/Si interface roughness by atomic force microscopy. *Japanese Journal of Applied Physics*, 33:383–387, 1994.

1201. Karl-Heinz Müller. Cluster beam deposition of thin films: a molecular dynamics simulation. *Journal of Applied Physics*, 61:2516–2521, 1987.

1202. M. Schneider, A. Rahman and Ivan K. Schuller. Vapor-phase growth of amorphous materials: a molecular-dynamics study. *Physical Review*, B34:1802–1805, 1986.

1203. Brian W. Dodson. Molecular dynamics modeling of vapor-phase and very-low-energy ion-beam crystal growth processes. *CRC Critical Reviews in Solid State and Materials Science*, 16:115–130, 1990.

1204. C. C. Fang, F. Jones, R. R. Kola and V. Prasad. Stress and microstructure of sputter-deposited films: molecular dynamics simmulations and experiments. *Journal of Vacuum Science and Technology*, 11:2947–2952, 1993.

1205. Jin Min Kim and S. Das Sarma. Discrete model for conserved surface growth. *Physical Review Letters*, 72:2903–2906, 1994.

1206. Andrea K. Myers-Beaghton and Dimitri D. Vvedensky. Generalized Burton-Cabrera-Frank theory for growth and equilibration on stepped surfaces. *Physical Review*, A44:2457–2468, 1990.

1207. M. D. Johnson, C. Orme, A. W. Hunt, D. Graff, J. Sudijono, L. M. Sander and B. G. Orr. Stable and unstable growth in molecular beam epitaxy. *Physical Review Letters*, 72:116–119, 1994.

1208. Dietrich E. Wolf. Kinetic roughening of vicinal surfaces. *Physical Review Letters*, 67:1783–1786, 1991.

1209. G. S. Bales and A. Zangwill. Morphological instability of a terrace edge during step-flow growth. *Physical Review*, B41:5500–5508, 1990.

1210. Z.-W. Lai and S. Das Sarma. Kinetic growth with surface relaxation: Continuum versus atomistic models. *Physical Review Letters*, 66:2348–2351, 1991.

1211. S. Das Sarma and P. Tamborenea. Surface driven kinetic growth on one-dimensional substrates. *Physical Review*, E48:2575–2594, 1993.

1212. S. Das Sarma and Roza Kotlyar. Dynamical renormalization group analysis of fourth-order conserved growth nonlinearities. *Physical Review*, E50:R4275–R4278, 1994.

1213. Jin Min Kim and S. Das Sarma. Dynamical universality of the nonlinear conserved current equation for growing surfaces. *Physical Review*, E51:1889–1893, 1995.

1214. Abhijit K. Kshirsagar and S. V. Ghaisas. Nonlinearities in conservative growth. *Physical Review*, E53:R1325–R1327, 1996.

1215. A. W. Hunt, C. Orme, D. R. M. Williams, B. G. Orr and L. M. Sander. Instabilities in MBE growth. *Europhysics Letters*, 27:611–616, 1994.

1216. C. Orme, M. D. Johnson, J. L. Sudijono, K. T. Leung and B. G. Orr. Large scale surface structure formed during *GaAs* homoepitaxy. *Applied Physics Letters*, 64:860–862, 1994.

1217. M. D. Johnson, J. Sudijono, A. W. Hunt and B. G. Orr. Growth mode evolution during homoepitaxy of *GaAs*(001). *Applied Physics Letters*, 64:484–486, 1994.

1218. Dimitri Vvendensky. New slant on epitaxial growth. *Physics World*, March 1994:30–31, 1994.

1219. Martin Siegert and Michael Plischke. Slope selection and coarsening in molecular beam epitaxy. *Physical Review Letters*, 73:1517–1520, 1994.

1220. P. Šmilauer and M. Kotrla. Kinetic roughening in growth models with diffusion in higher dimensions. *Europhysics Letters*, 27:261–266, 1994.

1221. M. Schroeder, M. Seigert, D. E. Wolf, J. D. Shore and M. Plischke. Scaling of growing surfaces with large local slopes. *Europhysics Letters*, 24:563–568, 1993.

1222. Pavel Šmilauer and Miroslav Kotrla. Crossover effects in the Wolf-Villain model of epitaxial growth in $1+1$ and $2+1$ dimensions. *Physical Review*, B49:5769–5772, 1994.

1223. S. Das Sarma, S. V. Ghaisas and J. M. Kim. Kinetic super-roughening and anomalous dynamic scaling in nonequilibrium growth models. *Physical Review*, E49:122–125, 1994.

1224. K. Park, B. Kahng and S. S. Kim. Surface dynamics of the Wolf-Villain model for epitaxial growth in $1+1$ dimensions. *Physica*, A210:146–154, 1994.

1225. Joachim Krug, Michael Plischke and Martin Siegert. Surface diffusion currents and the universality classes of growth. *Physical Review Letters*, 71:949, 1993.

1226. C. S. Ryu and In mook Kim. Crossover behaviors in the molecular-beam epitaxial-growth model. *Physical Review*, E51:3069–3073, 1995.

1227. C. S. Ryu and In mook Kim. Solid-on-solid model with next-nearest-neighbor interaction for epitaxial growth. *Physical Review*, E52:2424–2428, 1995.

1228. S. Das Sarma and P. Tamborenea. A new universality class for kinetic growth: One-dimensional molecular-beam epitaxy. *Physical Review Letters*, 66:325–328, 1991.

1229. H. C. Kang and J. W. Evans. Scaling analysis of surface roughness in simple models for molecular-beam epitaxy. *Surface Science*, 269/270:784–789, 1992.

1230. H. C. Kang and J. W. Evans. Scaling analysis of surface roughness and Bragg oscillation decay in models for low-temperature epitaxial growth. *Surface Science*, 271:321–330, 1992.

1231. M. Kotrla, A. C. Levi and P Šmilauer. Roughness and non-linearities in $(2+1)$-dimensional growth models with diffusion. *Europhysics Letters*, 20:25–30, 1992.

1232. H. Park, A. Provata and S. Redner. Interface growth with competing surface currents. *Journal of Physics*, A24:L1391–L1397, 1991.

1233. S. Das Sarma and S. V. Ghaisas. Solid-on-solid rules for nonequilibrium growth in $2+1$ dimensions. *Physical Review Letters*, 69:3762–3765, 1992.

1234. Michael Plischke, Joel D. Shore, Michael Schroeder, Martin Siegert and Dietrich E. Wolf. Comment on "Solid-on-solid rules and models for nonequilibrium growth in $2+1$ dimensions". *Physical Review Letters*, 71:2509, 1993.

1235. G. H. Gilmer and P. Bennema. Simulation of crystal growth with surface diffusion. *Journal of Applied Physics*, 43:1347–1360, 1972.

1236. M. R. Wilby, D. D. Vvedensky and A. Zangwill. Scaling in a solid-on-solid model of epitaxial growth. *Physical Review*, B46:12896–12898, 1992.

1237. M. R. Wilby, D. D. Vvedensky and A. Zangwill. Erratum: Scaling in a solid-on-solid model of epitaxial growth. *Physical Review*, B47:16086, 1993.

1238. Zhenyu Zhang, John Detch and Horia Metiu. Surface roughness in thin-film growth: the effect of mass transport between layers. *Physical Review*, B48:4972–4975, 1993.

1239. Martin Siegert and Michael Plischke.

Instability in surface growth with diffusion. *Physical Review Letters*, 68:2035–2038, 1992.

1240. Martin Siegert and Michael Plischke. Solid-on-solid models of molecular beam epitaxy. *Physical Review*, E50:917–931, 1994.

1241. Fereydoon Family and Pui-Man Lam. Renormalization-group analysis and simulation studies of groove instability in surface growth. *Physica*, A205:272–283, 1994.

1242. Tim Salditt and Herbert Spohn. Kinetic roughening of a terrace ledge. *Physical Review*, E47:3524–3531, 1993.

1243. Hong Yan. Kinetic growth with surface diffusion: The scaling aspect. *Physical Review Letters*, 68:3048–3051, 1992.

1244. David A. Kessler, Herbert Levine and Leonard M. Sander. Molecular-beam epitaxial growth and surface diffusion. *Physical Review Letters*, 69:100–103, 1992.

1245. Martin Schimschak and Joachim Krug. Bulk defects and surface roughening in epitaxial growth. *Physical Review*, B52:8550–8563, 1995.

1246. S. Das Sarma, C. J. Lanczycki, S. V. Ghaisas and J. M. Kim. Deffect formation and crossover behavior in the dynamic scaling properties of molecular-beam epitaxy. *Physical Review*, B49:10693–10698, 1994.

1247. Timothy Halpin-Healey and Amine Assdah. On the kinetic roughening of vicinal surfaces. *Physical Review*, A46:3527–3530, 1992.

1248. Keye Moser and Dietrich E. Wolf. Kinetic roughening of vicinal surfaces. In R. Jullien, J. Kertész, P. Meakin and D. Wolf, editors, *Surface disordering: Growth, roughening and phase transitions*, pages 21–29. Nova, Commack, New York, 1992.

1249. R. Messier and J. E. Yehoda. Geometry of thin-film morphology. *Journal of Applied Physics*, 58:3739–3746, 1985.

1250. Joseph E. Yehoda and Russell Messier. Are thin film physical structures fractal? *Applications of Surface Science*, 22/23:590–595, 1985.

1251. C. Tang, S. Alexander and R. Bruinsma. Scaling theory for the growth of amorphous films. *Physical Review Letters*, 64:772–775, 1990.

1252. Francis A. Jenkins and Harvey E. White. *Fundamentals of optics*. McGraw-Hill, New York, 1957.

1253. A. van der Drift. Evolutionary selection, a principle governing growth orientation in vapour-deposited layers. *Philips Research Reports*, 22:267–288, 1967.

1254. J. Jurusik and L. Żdanowicz. Structure and growth morphology of thin amorphous films of cadmium arsenide. *Thin Solid Films*, 144:241–254, 1986.

1255. O. Geszti, L. Gosztola and É. Seyfried. Cross-sectional transmission electron microscopy study of obliquely evaporated silicon oxide thin films. *Thin Solid Films*, 136:L35–L38, 1986.

1256. M. J. Bloemer, T. L. Ferrell, M. C. Buncick and R. J. Warmack. Optical properties of submicrometer-size silver needles. *Physical Review*, B37:8015–8021, 1988.

1257. G. Mbise, G. B. Smith, G. A. Niklasson and C. G. Granqvist. Angular-selective optical properties of *Cr* films made by oblique-angle evaporation. *Applied Physics Letters*, 54:987–989, 1989.

1258. J. M. Niewenhuizen and H. B. Haanstra. Microfractography of thin films. *Philips Technical Review*, 27:87–91, 1966.

1259. Paul Meakin. Ballistic deposition onto inclined surfaces. *Physical Review*, A38:994–1004, 1988.

1260. Joachim Krug and Paul Meakin. Microstructure and surface scaling in ballistic deposition at oblique incidence. *Physical Review*, A40:2064–2077, 1989.

1261. T. Hashimota, K. Okomoto, K. Hara, M. Kayima and H. Fujiwara. Columnar structure and texture of iron films evaporated at oblique incidence. *Thin Solid Films*, 91:145–154, 1982.

1262. H. Fujiwara, K. Hara, M. Kamiya, T. Hashimoto and K. Okamoto. Comments on the tangent rule. *Thin Solid Films*, 163:387–391, 1988.

1263. Joachim Krug. The columnar growth angle in obliquely evaporated thin films. *Materialwissenschaft und Werkstofftechnik*, 26:22–26, 1995.

1264. Andrew Zangwill. Theory of growth-induced surface roughness. In H. A. Atwater and C. V. Thompson, editors, *Microstructural evolution of thin films*.

Academic Press, New York, 1995.

1265. Maury Bramson and David Griffeath. Asymptotics for interacting particle system on $\mathscr{L}^d$. *Zeitschrift für Wahrscheinlichkeitstheorie und verwandte Gebiete*, 53:183–196, 1980.

1266. Charles R. Doering and Daniel ben Avraham. Interparticle distribution functions and rate equations for diffusion-limited reactions. *Physical Review*, A38:3035–3042, 1988.

1267. John L. Spouge. Exact solutions for a diffusion-reaction process in one dimension. *Physical Review Letters*, 60:871–874, 1988.

1268. Paul Meakin and Joachim Krug. Columnar microstructure in three-dimensional ballistic deposition. *Europhysics Letters*, 11:7–12, 1990.

1269. Paul Meakin and Joachim Krug. Three-dimensional ballistic deposition at oblique incidence. *Physical Review*, A46:3390–3399, 1992.

1270. Paul Meakin and Joachim Krug. Scaling structure in simple screening models for columnar growth. *Physical Review*, A46:4654–4660, 1992.

1271. Paul Meakin. Universal scaling properties of ballistic deposition and Eden growth on surfaces. *Journal of Physics*, A20:L1113–L1119, 1987.

1272. Paul Meakin. Ballistic deposition at random incidence. Unpublished, 1985.

1273. Chao Tang and Shoudan Liang. Patterns and scaling properties in a ballistic deposition model. *Physical Review Letters*, 71:2769–2772, 1993.

1274. D. Bensimon, B. Shraiman and S. Liang. On the ballistic model of aggregation. *Physics Letters*, 102A:238–240, 1984.

1275. Shoudan Liang and Leo P. Kadanoff. Scaling in a ballistic aggregation model. *Physical Review*, A31:2628–2630, 1984.

1276. P. Ramanlal and L. M. Sander. Theory of ballistic aggregation. *Physical Review Letters*, 544:1828–1831, 1985.

1277. Joachim Krug and Paul Meakin. Columnar growth in oblique incidence ballistic deposition: faceting, noise reduction and mean-field theory. *Physical Review*, A43:900–919, 1991.

1278. R. P. U. Karunasiri, R. Bruinsma and J. Rudnick. Thin-film growth and the shadow instability. *Physical Review Letters*, 63:2767, 1989.

1279. G. S. Bales and A. Zangwill. Growth dynamics of sputter deposition. *Physical Review Letters*, 63:692, 1989.

1280. R. P. U. Karunasiri, R. Bruinsma and J. Rudnick. Karunasiri, Bruinsma and Rudnick reply. *Physical Review Letters*, 63:693, 1989.

1281. Jian Hua Yao and Hong Guo. Shadowing instability in three dimensions. *Physical Review*, E47:1007–1011, 1993.

1282. Jian Hua Yao and Christopher Roland Hong Guo. Interfacial dynamics with long-range screening. *Physical Review*, E47:3903–3912, 1993.

1283. Paul Meakin and Joachim Krug. Scaling properties of the shadowing model for sputter deposition. *Physical Review*, E47:R17–R20, 1993.

1284. Seth Lichter and Jyhmin Chen. Model for columnar microstructure in thin solid films. *Physical Review Letters*, 56:1396–1399, 1986.

1285. Takashi Nagatani. Scaling structure in a simple growth model with screening: forest formation model. *Journal of Physics*, A24:L449–L454, 1991.

1286. J. Krug, J. Kertész and D. E. Wolf. Growth shapes and directed percolation. *Europhysics Letters*, 12:113–118, 1990.

1287. János Kertész and Dietrich E. Wolf. Anomalous roughening in growth processes. *Physical Review Letters*, 62:2571–2574, 1987.

1288. A. J. Dammers and S. Radelaar. Two-dimensional computer modelling of polycrystalline film growth. *Textures and Microstructures*, 14-18:757–762, 1991.

1289. A. J. Dammers and S. Radelaar. A grain growth model for evolution of polycrystalline surfaces. *Materials Science Forum*, 94-96:345–350, 1992.

1290. J. M. Thijssen, H. J. F. Knops and A. J. Dammers. Dynamic scaling in polycrystalline growth. *Physical Review*, B45:8650–8656, 1992.

1291. J. M. Thijssen. Simulations of polycrystalline growth in $2+1$ dimensions. Preprint, 1994.

1292. F. Family and T. Vicsek. *Dynamics of fractal surfaces*. World Scientific, Singapore, 1991.

1293. Paul Meakin. The growth of rough surfaces and interfaces. *Physics Reports*, 235:189–289, 1993.

1294. John C. Russ. *Fractal surfaces*. Plenum, New York, 1994.

1295. J. S. Langer and H. Müller-Krumbhaar. Theory of dendritic growth-I. Elements of a stability analysis. *Acta Metallurgica*, 26:1681–1687, 1978.

1296. J. S. Langer and H. Müller-Krumbhaar. Theory of dendritic growth-II. Instabilities in the limit of vanishing surface tension. *Acta Metallurgica*, 26:1689–1696, 1978.

1297. H. Müller-Krumbhaar and J. S. Langer. Theory of dendritic growth-III. Effects of surface tension. *Acta Metallurgica*, 26:1697–1708, 1978.

1298. R. Trivedi and W. Kurz. Solidification microstructures: a conceptual approach. *Acta Metallurgica et Materialia*, 42:15–23, 1994.

1299. H. S. Hele-Shaw. Investigation of the nature of surface resistance of water and of stream-line motion under certain experimental conditions. *Transactions of the Institution of Naval Architects*, 40:21–46, 1898. Plus 58 figures on 17 plates.

1300. M. Rabaud, Y. Couder and N. Gerard. Dynamics and stability of anomalous Saffman–Taylor fingers. *Physical Review*, A37:935–947, 1988.

1301. P. G. Saffman. Viscous fingering in a Hele-Shaw cell. *Journal of Fluid Mechanics*, 173:73–94, 1986.

1302. Jun Yokoyama, Yoshiyuki Kitagawa, Hideaki Yamada and Mitsugu Matsushita. Dendritic growth of viscous fingers under linear anisotropy. *Physica*, A204:789–799, 1994.

1303. E. Shalev, J. Klafter, D. F. Plusquellic and D. W. Pratt. Scaling properties of molecular spectra. *Physica*, A191:186–189, 1992.

1304. Michael Schreiber and Heiko Grussbach. Multifractal wave functions at the Anderson transition. *Physical Review Letters*, 67:607–610, 1991.

1305. L. de Arcangelis, S. Redner and A. Coniglio. Anomalous voltage distribution of random restor networks and a new model for the backbone of the percolation cluster. *Physical Review*, B31:4725–4727, 1985.

1306. R. Pechanski. Intermittency in particle collisions. *International Journal of Modern Physics*, 6:3681–3722, 1991.

1307. Paul Meakin. The growth of fractal aggregates and their fractal measures. In C. Domb and J. L. Lebowitz, editors, *Phase transitions and critical phenomena*, volume 12, pages 335–489. Academic Press, London, 1988.

1308. Benoit B. Mandelbrot. Possible refinement of the lognormal hypothesis concerning the distribution of energy dissipation in intermittent turbulence. In Murray Rosenblatt and Charles Van Atta, editors, *Statistical models and turbulence*, pages 333–351. Springer, New York, 1972. Lecture Notes in Physics 12.

1309. U. Frisch, P. Sulem and M. Nelkin. A simple dynamical model of intermittency in fully developed turbulence. *Journal of Fluid Mechanics*, 87:719–736, 1978.

1310. Roberto Benzi, Giovanni Paladin, Giorgio Parisi and Angelo Vulpiani. On the multifractal nature of fully developed turbulence. *Journal of Physics*, A17:3521–3531, 1984.

1311. D. Schertzer and S. Lovejoy. Elliptical turbulence in the atmosphere. In L. J. S. Bradbury, F. Durst, B. E. Launder, F. W. Schmidt and J. H. Whitlaw, editors, *Proceedings of the 4th symposium on turbulent shear flow*. Springer, Berlin, 1983.

1312. Antonio Coniglio. An infinite hierarchy of exponents to describe growth phenomena. In Luciano Pietronero and Erio Tosati, editors, *Fractals in physics*, pages 165–168. North Holland, Amsterdam, 1985. Proceedings of the 6th International Conference on Fractals in Physics, Trieste July 9-12, 1985.

1313. Carl G. Evertsz and Benoit B. Mandelbrot. Multifractal measures. In H.-O. Peitgen, H. Jürgens and D. Saupe, editors, *Chaos and fractals*, pages 849–881. Springer, New York, 1992. Appendix B.

1314. Paul Meakin and John M. Deutch. Properties of the fractal measure describing the hydrodynamic force distribution for fractal aggregates moving in a quiescent fluid. *Journal of Chemical Physics*, 86:4648–4656, 1987.

1315. Paul Meakin. A multifractal distribution of the hydrodynamic force distribution for reaction-limited aggregates. *Journal of*

Chemical Physics, 88:2042–2048, 1988.

1316. M. E. Cates and T. A. Witten. Diffusion near absorbing fractals: Harmonic measure exponents for polymers. *Physical Review*, A35:1809–1824, 1987.

1317. Ashvin Chhabra and Roderick V. Jensen. Direct determination of the $f(\alpha)$ singularity spectrum. *Physical Review Letters*, 62:1327–1330, 1989.

1318. Paul Meakin. Dimensionalities for the harmonic and ballistic measures of fractal aggregates. *Physical Review*, A33:1365–1371, 1986.

1319. M. Cates and J. Deutsch. Spatial correlations in multifractals. *Physical Review*, A35:4907–4910, 1987.

1320. Charles Meneveau and Ashvin B. Chhabra. Two-point statistics of multifractal measures. *Physica*, A164:564–574, 1990.

1321. Sung Jong Lee and Thomas. C. Halsey. Some results on multifractal correlations. *Physica*, A164:575–592, 1990.

1322. Paul Meakin, H. Eugene Stanley, Antonio Coniglio and Thomas A. Witten. Surfaces, interfaces and screening of fractal structures. *Physical Review*, A32:2364–2367, 1985.

1323. Thomas C. Halsey, Paul Meakin and Itamar Procaccia. Scaling structure of the surface layer of diffusion-limited aggregates. *Physical Review Letters*, 56:854–857, 1986.

1324. Paul Meakin, Antonio Coniglio, H. Eugene Stanley and Thomas A. Witten. Scaling properties for the surfaces of fractal and nonfractal objects: An infinite hierarchy of critical exponents. *Physical Review*, A34:3325–3340, 1986.

1325. Y. Hayakawa, S. Sato and M. Matsushita. Scaling structure of the growth probability distribution in diffusion-limited aggregation processes. *Physical Review*, A36:1963–1966, 1987.

1326. C. Amitrano, A. Coniglio and F. di Liberto. Growth probability distribution in kinetic aggregation processes. *Physical Review Letters*, 57:1016–1019, 1986.

1327. Marek Wolf. Hitting probabilities of diffusion-limited aggregation clusters. *Physical Review*, A43:5504–5517, 1991.

1328. P. Ramanlal and L. M. Sander. Distribution of growth probabilities in fluid flow and diffusion limited aggregation. *Journal of Physics*, A21:L995–L1001, 1988.

1329. T. A. Witten. Flow and diffusion near a fractal: Partial saturation phenomena. *Physical Review Letters*, 59:900–903, 1987.

1330. B. B. Mandelbrot and T. Vicsek. Directed recursive models for fractal growth. *Journal of Physics*, A20:L377–L383, 1989.

1331. A. B. Harris and Michael Cohen. Scaling of negative moments of the growth probability of diffusion-limited aggregates. *Physical Review*, A41:971–982, 1990.

1332. M. J. Feigenbaum. Some characterizations of strange sets. *Journal of Statistical Physics*, 46:919–924, 1987.

1333. P Szépfalusy, T. Tél, A. Csordás and Z. Kovács. Phase transitions associated with dynamical properties of chaotic systems. *Physical Review*, A36:3525–3528, 1987.

1334. Jysoo Lee and H. Eugene Stanley. Phase transition in the multifractal spectrum of diffusion-limited aggregation. *Physical Review Letters*, 61:2945–2948, 1988.

1335. Takahashi Nagatani. Renormalization-group approach to multifractal structure of growth probability distribution in diffusion-limited aggregation. *Physical Review*, A36:5812–5819, 1987.

1336. T. Nagatani. A renormalization group approach to the scaling structure of diffusion-limited aggregation. *Journal of Physics*, A20:L381–L386, 1987.

1337. Jysoo Lee, Preben Alstrøm and H. Eugene Stanley. Scaling of the minimum growth probability for the "typical" diffusion-limited aggregation configuration. *Physical Review Letters*, 62:3013, 1989.

1338. Shlomo Havlin, Benes Trus, Armin Bunde and H. Eduardo Roman. Phase transition in diffusion-limited aggregation. *Physical Review Letters*, 63:1189, 1989.

1339. Marek Wolf. Size dependence of the minimum-growth probabilities of typical diffusion-limited aggregation clusters. *Physical Review*, E47:1448–1451, 1993.

1340. Raphael Blumenfeld and Amnon Aharony. Breakdown of multifractal behavior in diffusion-limited aggregation. *Physical Review Letters*, 62:2977–2980, 1989.

1341. S. Schwarzer, J. Lee, A. Bunde, S. Havlin, H. E. Roman and H. E. Stanley. Minimum growth probability of diffusion-limited aggregates. *Physical Review Letters*,

65:603–606, 1990.

1342. Carl J. G. Evertsz, Peter W. Jones and Benoit B. Mandelbrot. Behavior of the harmonic measure at the bottom of fjords. *Journal of Physics*, A24:1889–1901, 1991.

1343. S. Schwarzer, J. Lee, S. Havlin, H. E. Stanley and Paul Meakin. Distribution of growth probabilities for off-lattice diffusion-limited aggregation. *Physical Review*, A43:1134–1137, 1991.

1344. Benoit B. Mandelbrot, Carl G. Evertsz and Yoshinori Hayakawa. Exactly self-similar left-sided multifractal measures. *Physical Review*, A42:4528–4536, 1990.

1345. Robbin C. Ball and Raphael Blumenfeld. Exact results on exponential screening in two-dimensional diffusion-limited aggregation. *Physical Review*, A44:R828–R831, 1991.

1346. Robbin C. Ball, Martin Blunt and O. Rath Spivak. Diffusion-controlled growth. *Proceedings of the Royal Society*, A423:123–132, 1989.

1347. Carl J. G. Evertsz, Benoit B. Mandelbrot and Lionel Woog. Variability of the form and of the harmonic measure for small off-off-lattice diffusion-limited aggregates. *Physical Review*, A45:5798–5804, 1992.

1348. A. Block, W. von Bloh and H. J. Schellnhuber. Efficient box-counting determination of generalized fractal dimension. *Physical Review*, A42:1869–1874, 1990.

1349. Stefan Schwarzer, Marek Wolf, Shlomo Havlin, Paul Meakin and H. Eugene Stanley. Multifractal spectrum of off-lattice three-dimensional diffusion-limited aggregation. *Physical Review*, A46:R3016–R3019, 1992.

1350. Stefan Schwarzer, Shlomo Havlin and H. Eugene Stanley. Multifractal scaling of 3D diffusion-limited aggregation. *Physica*, A191:117–122, 1992.

1351. Robbin C. Ball and Martin Blunt. Dynamics of screening in multifractal growth. *Physical Review*, A41:582–589, 1990.

1352. Benoit B. Mandelbrot. New "anomalous" multiplicative multifractals: Left sided $f(\alpha)$ and the modelling of DLA. *Physica*, A168:95–111, 1990.

1353. Tamás Tél and Tamás Vicsek. Geometrical multifractility of growing structures. *Journal of Physics*, A20:L835–L840, 1987.

1354. Tamás Tél, Ágnes Fülöp and Tamás Vicsek. Determination of the fractal dimensions for geometric multifractals. *Physica*, A159:155–166, 1989.

1355. J. F. Muzy, B. Pouligny, E. Freysz, F. Argoul and A. Arneodo. Optical-diffraction measurement of fractal dimensions and $f(\alpha)$ spectrum. *Physical Review*, A45:8961–8964, 1992.

1356. T. Vicsek, F. Family and Paul Meakin. Multifractal geometry of diffusion-limited aggregates. *Europhysics Letters*, 12:217–222, 1990.

Index